Particle Toxicology

Particle Toxicology

Edited by

Ken Donaldson
Paul Borm

CRC Press
Taylor & Francis Group
Boca Raton London New York

CRC Press is an imprint of the
Taylor & Francis Group, an **informa** business

CRC Press
Taylor & Francis Group
6000 Broken Sound Parkway NW, Suite 300
Boca Raton, FL 33487-2742

First issued in paperback 2019

ISBN-13: 978-0-4893-5092-1 (hbk)
ISBN-13: 978-0-367-38961-1 (pbk)

Library of Congress Cataloging-in-Publication Data

Particle toxicology / editors, Ken Donaldson and Paul Borm.
p. ; cm.
Includes bibliographical references and index.
ISBN-13: 978-0-8493-5092-4 (hardcover : alk. paper)
ISBN-10: 0-8493-5092-1 (hardcover : alk. paper)
1. Pulmonary toxicology. 2. Particles--Toxicology. I. Donaldson, Kenneth, DSc. II. Borm, Paul.
[DNLM: 1. Air Pollutants--toxicity. 2. Inflammation. 3. Mineral Fibers--toxicity. 4. Oxidative Stress. WA 754 P2732 2007]

RC720.P373 2007
616.2'00471--dc22 2006037017

Visit the Taylor & Francis Web site at
http://www.taylorandfrancis.com

and the CRC Press Web site at
http://www.crcpress.com

Foreword

The association between lung diseases and the inhalation of dusts has been recognized throughout history, stretching back to Agricola and Paracelsus in the fifteenth and sixteenth centuries. Needless to say the scientific endeavour associated with identifying the relationship between particle characteristics and pathological processes—the essence of modern particle toxicology—awaited the development of a contemporary understanding of both lung disease and the physicochemical nature and aerodynamic behaviour of particles. These elements finally came together in the mid-twentieth century and modern approaches to understanding harmful inhaled particles can be first traced to quartz (crystalline silica) and its fibrogenic effects in the lungs. Undeniably, in a truly applied toxicology approach to the notion that the surface reactivity of quartz was the harmful entity, a whole programme of toxicology-based therapy was undertaken, using aluminium to attempt to reduce the harmfulness of the quartz in already exposed subjects.

Meanwhile the epidemic of disease caused by asbestos, the other particle source of the twentieth century, was taking hold and by late- to mid-twentieth century, an understanding of the toxicology of asbestos began. The full understanding of the asbestos hazard was, however, only realised in the 1980s and 1990s, following the rise in use of synthetic vitreous fibres in the years following the reduction in asbestos use. In these years, ground-breaking studies demonstrated the importance of length and biopersistence, which explained differences between asbestos types and placed all respirable mineral fibres in a single toxicology paradigm that embraced both asbestos and the synthetic vitreous fibres.

In the 1990s, ambient particulate matter as a regulated air pollutant (PM_{10}[1]) became the focus of global concern. This was initiated by epidemiological studies that were now able to process huge data sets on air quality and human morbidity and mortality. Both cohort and time-series studies in many countries associated substantial premature mortality and excess morbidity in urban residents to their air pollution exposure, with particles as the most potent component of the air pollution cocktail. Although the risks are low, particulate matter affects the whole population and the effects were still preset below the air quality standards. It also became evident that certain groups, such as elderly and people with respiratory and cardiovascular diseases, were at increased risk. Since then, particle toxicologists are faced with the fact that PM_{10} is a complex mixture by itself, whereas the risks identified in the epidemiologic studies are based on total mass concentrations. A further reduction of the PM levels would be very expensive and a cost effective strategy was warranted. There was an urgent need to identify the causal relationship between PM, (personal) exposure and associated health effects. This recognition stimulated governments globally, and new funding flowed into particle toxicology research to identify the critical aspects that could be linked with the health effects observed in epidemiological studies. It soon became clear that no single, omnipresent constituent could be identified that related to the variety of health effects. It turned out to be a big challenge for many because of the variability in PM_{10} (size range, surface chemistry, agglomeration, shape, charge, chemical composition, et cetera), the focus on susceptibility factors (disease, age, and gender) and the lack of good *in vitro* and animal models to mimic these factors.

The increasing emphasis of PM toxicology on the cardiovascular system as a key target for adverse effects brought an entirely new dimension. Particle toxicologists were forced to move out of their comfort zone in the respiratory tract and try to understand how inhaled particles could also affect the cardiovascular system or other target tissues such as the brain. At the end of the twentieth century and the dawn of the twenty-first century, manufactured nanoparticles[2] have come to

---

[1] Defined as mass of particles centered around an aerodynamic diameter of 10 μm.

[2] Generally defined as particles with at least one dimension less than 100 nm.

represent the new frontier for particle toxicologists based on nanotechnology's potential to produce a wide range of new particles varying in size and chemistry. Traditionally, particle dosimetry has always been linked with particle toxicology, due to the complex relationship between exposure and target dose. Unexpected translocation of nanoparticles from the respiratory system to other organs and a recognition that manufactured nanoparticles could affect the skin and the gut—depending on the type of exposure—have extended the area of research.

Throughout the fifty or so years that have seen the full flowering of the scientific discipline of particle toxicology, particle toxicologists have looked to mainstream molecular biology for their pathobiological paradigms, with the examples intra-cellular signalling pathways, inflammation biology, immunomodulation, and genotoxicity as prime examples. They have also looked to chemistry and physics for an improved understanding of the particle characteristics that drive toxicity, including the assessment of free radical production and oxidative stress—a leading paradigm for how particles affect cells. In addition they have worked in tandem with aerosol physicists and modellers to develop the dosimetric models that are so important, including the role of aerodynamic diameter in dictating the site of the deposition of particles. Particle toxicologists have also worked with epidemiologists and most recently with cardiologists and neurologists, and the net result has been to produce a truly multidisciplinary science that uses computational modelling, *in vitro* techniques, and animal and human studies to address their hypotheses.

This volume represents the view of a number of world's leading particle toxicologists in their chosen specialties, many of whom were involved in the events described above and in raising particle toxicology to the status that it has today. Their chapters address the most important aspects of particle toxicology and confirm its status as a mature science. As such, I believe that this volume is a database that provides not only a historical view, but most of all state-of-the-science concepts in a single volume. It covers the broad spectrum of particle toxicology from particle characterization, respiratory tract dosimetry, cellular responses, inflammation, fibrogenesis, cardiovascular and neurological effects, and genotoxicity. The chapters cover all kind of particle types, unlike previous books that have focused on single particle types, such as quartz or fibres and so forms an essential reference work. Particle toxicology is different from any other toxicology. Different in the sense that it has demonstrated that "dose," as defined by Paracelsus, has more dimensions than mass per volume. The book deals with the specific nature of particle toxicology in great detail, and I truthfully believe that this volume will provide the reader with a unique and practical insight into this fascinating branch of toxicology.

On behalf of the editors, Ken Donaldson and Paul Borm, I would like to thank the authors for their generous time in writing the chapters and the staff of Taylor & Francis for their excellent support in the production of the book.

Flemming R. Cassee, Ph.D.
National Institute for Public Health and the Environment
Bilthoven, The Netherlands

Preface

The toxicology of particles is an absorbing area of research in which to work and when we conceived this book, we wanted to capture some of the fascination that we feel about our profession.

We are well-pleased with the result—everyone we invited to write a chapter agreed and almost everyone delivered a manuscript—a remarkable outcome in this time of conflicting deadlines. It is difficult to keep up with the sheer quantity of data that accumulates on particle toxicology. This has resulted in polarisation of meetings and specialists into particle types, thus there are meetings on PM or nanoparticles and there can be inadequate cross-talk. This is unfortunate because of the benefits of understanding the toxicology of one particle type for understanding other particle types. This volume deals with all particle types and offers state-of-the-science reviews that should benefit practitioners of the many disciplines who are involved in particle toxicology. Particle toxicology is a "work in progress," as witnessed by the rise of nanoparticle toxicology, and has become an important area of endeavour in toxicology, pollution science, respiratory medicine and increasingly, cardiovascular medicine. This book is, therefore, timely and apposite to meeting this need for information.

We warmly thank the authors who have been involved in writing the various chapters of this book and the staff of Taylor & Francis for their invaluable and professional assistance in its realisation.

Ken Donaldson
Paul Borm

Editors

Professor Dr. Paul J.A. Borm has been with the Centre of Expertise in Life Sciences (CEL) at Zuyd University in Heerlen, The Netherlands since 2003. Although his research work has concentrated mostly on lung diseases, his activities and coordination have always included a larger array of subjects related to (occupational) health care. He is the author of more than 160 peer reviewed papers and more than 150 oral presentations on topics in occupational and environmental toxicology. Professor Borm is a member of the German MAK-commission and the Dutch Evaluation committee on Occupational Substances (DECOS). He has been an invited member of expert groups such as IARC (1996), ILSI (1998), and ECVAM (1997), and he has been the organizer of many international meetings and workshops on occupational risk factors. He is an editorial board member for *Human Experimental Toxicology* and *Inhalation Toxicology* and a co-editor of *Particle and Fibre Toxicology*.

The combination of his know-how in pharmacology, toxicology, and management of interdisciplinary research projects and teams are among his skills. In his current function at Zuyd University, he is trying to interface fundamental and applied sciences with developments and needs in the public and private sector, such as health care, functional foods, and nanotechnologies. Dr. Borm is involved in a number of large-scale projects including education in nanotechnology, technology accelerator using nanotechnology, and cell therapy. Apart from his position at Zuyd, Borm holds management contracts with start-ups (Magnamedics GmbH) and grown-ups in Life Sciences. Drug delivery and/or toxicological testing of drug delivery tools are core businesses in these activities.

Ken Donaldson is professor of respiratory toxicology in the Medical School at the University of Edinburgh, where he is co-director of the Edinburgh Lung and the Environment Group Initiative Colt Laboratory—a collaborative research institute involving the Edinburgh University Medical School, Napier University, and the Institute of Occupational Medicine, carrying out research into disease caused by inhaled agents, predominantly particles.

He has carried out 27 years of research into the inhalation toxicology of all medically important particle types—asbestos, man-made vitreous fibres, crystalline silica, nuisance dusts, ultrafine/nanoparticles, particulate air pollution (PM10), and organic dust, as well as ozone and nitrogen dioxide. He is a co-author of over 250 peer-reviewed scientific articles, book chapters, and reviews on lung disease caused by particles and fibres. Dr. Donaldson is a member of three government committees—COMEAP (Committee on the Medical Effects of Air Pollution), which advises the government on the science of air pollution; EPAQS (Expert Panel on Air Quality Standards), which provides independent advice to the government on air quality issues (ad hoc member); and the Advisory Committee on Hazardous Substances, which provides expert advice to the government on the science behind hazardous chemicals. He has advised WHO, EU, US EPA, UK, HSE, and other international bodies on the toxicology of particles. He is a registrant of the BTS/IOB Register of Toxicologists, a Eurotox-registered toxicologist, a Fellow of the Royal College of Pathologists, a Fellow of the Society of Occupational Medicine, and he has a DSc for research in toxicology of particle-related lung disease. He is the founding editor in chief, along with Paul Borm, of the journal *Particle and Fibre Toxicology*.

Contributors

Armelle Baeza-Squiban
Laboratoire de Cytophysiologie et Toxicologie Cellulaire
Université Paris 7 – Denis Didèrot
Paris, France

Peter G. Barlow
Queen's Medical Research Institute
University of Edinburgh
Edinburgh, Scotland

Kelly BéruBé
School of Biosciences
Cardiff University
Cardiff, United Kingdom

Sonja Boland
Laboratoire de Cytophysiologie et Toxicologie Cellulaire
Université Paris 7 – Denis Didèrot
Paris, France

Paul J. A. Borm
Centre of Expertise in Life Sciences (CEL)
Hogeschool Zuyd
Heerlen, Netherlands

Arnold R. Brody
Tulane University Health Sciences Center
Tulane University
New Orleans, Louisiana

David M. Brown
School of Life Sciences
Napier University
Edinburgh, Scotland

Lilian Calderón-Garcidueñas
The Center for Structural and Functional Neurosciences
University of Montana
Missoula, Montana

Vincent Castranova
Health Effects Laboratory Division
National Institute for Occupational Safety and Health
Morgantown, West Virginia

Andrew Churg
Department of Pathology
University of British Columbia
Vancouver, British Columbia, Canada

Ken Donaldson
MRC/University of Edinburgh Centre for Inflammation Research
Queen's Medical Research Institute
Edinburgh, Scotland

Steve Faux
MRC/University of Edinburgh Centre for Inflammation Research
Queen's Medical Research Institute
Edinburgh, Scotland

Peter Gehr
Institute of Anatomy
University of Bern
Bern, Switzerland

Andrew J. Ghio
U.S. Environmental Protection Agency
Research Triangle Park, North Carolina

M. Ian Gilmour
National Health and Environmental Effects Research Laboratory
U.S. Environmental Protection Agency
Research Triangle Park, North Carolina

Tom K. Hei
Center for Radiological Research
Columbia University
New York, New York

Reuben Howden
National Institute of Environmental Health Sciences
National Institutes of Health
Research Triangle Park, North Carolina

Gary R. Hutchison
Medical Research Council
Queen's Medical Research Institute
Edinburgh, Scotland

Tim Jones
School of Earth, Ocean, and Planetary Sciences
Cardiff University
Cardiff, United Kingdom

Frank J. Kelly
Pharmaceutical Science Research Division
King's College
London, United Kingdom

Steven R. Kleeberger
National Institute of Environmental Health Sciences
National Institutes of Health
Research Triangle Park, North Carolina

Wolfgang G. Kreyling
Institute of Inhalation Biology and Focus Network Aerosols and Health
GSF–National Research Center for Environment and Health
Neuherberg, Germany

Eileen Kuempel
Risk Evaluation Branch
CDC National Institute for Occupational Safety and Health
Cincinnati, Ohio

Stephen S. Leonard
Health Effects Laboratory Division
National Institute for Occupational Safety and Health
Morgantown, West Virginia

Jamie E. Levis
University of Vermont College of Medicine
University of Vermont
Burlington, Vermont

William MacNee
MRC/University of Edinburgh Centre for Inflammation Research
Queen's Medical Research Institute
Edinburgh, Scotland

Francelyne Marano
Laboratoire de Cytophysiologie et Toxicologie Cellulaire
Université Paris 7 – Denis Didèrot
Paris, France

Nicholas L. Mills
Centre for Cardiovascular Sciences
The University of Edinburgh
Edinburgh, Scotland

Winfried Möller
Institute of Inhalation Biology and Clinical Research Group "Inflammatory Lung Diseases"
GSF–National Research Center for Environment and Health
Munich, Germany

Asklepios Hospital for Respiratory Diseases
Munich-Gauting, Germany

Brooke T. Mossman
University of Vermont College of Medicine
University of Vermont
Burlington, Vermont

Ian S. Mudway
Pharmaceutical Science Research Division
King's College
London, United Kingdom

Detlef Müller-Schulte
Magnamedics GmbH
Aachen, Germany

David E. Newby
Centre for Cardiovascular Sciences
The University of Edinburgh
Edinburgh, Scotland

Günter Oberdörster
University of Rochester Medical Center
University of Rochester
Rochester, New York

Dale W. Porter
Health Effects Laboratory Division
National Institute for Occupational Safety and Health
Morgantown, West Virginia

Kenneth L. Reed
DuPont Haskell Laboratory for Health and Environmental Sciences
Newark, Delaware

William Reed
Department of Pediatrics and Center for Environmental Medicine
University of North Carolina at Chapel Hill
Chapel Hill, North Carolina

Barbara Rothen-Rutishauser
Institute of Anatomy
University of Bern
Bern, Switzerland

James M. Samet
U.S. Environmental Protection Agency
Research Triangle Park, North Carolina

Rajiv K. Saxena
School of Life Sciences
Jawaharlal Nehru University
New Delhi, India

Christie M. Sayes
DuPont Haskell Laboratory for Health and Environmental Sciences
Newark, Delaware

Roel P. F. Schins
Institut für umweltmedizinische Forschung (IUF) an der Heinrich-Heine-Universität
Düsseldorf, Germany

Samuel Schürch
Institute of Anatomy
University of Bern
Bern, Switzerland

Department of Physiology and Biophysics
University of Calgary
Calgary, Canada

Manuela Semmler-Behnke
GSF-National Research Center for Environment and Health
Neuherberg and Munich, Germany

Tina Stevens
Curriculum in Toxicology
University of North Carolina at Chapel Hill
Chapel Hill, North Carolina

Vicki Stone
School of Life Sciences
Napier University
Edinburgh, Scotland

Deborah E. Sullivan
Tulane University Health Sciences Center
Tulane University
New Orleans, Louisiana

Lang Tran
Institute of Occupational Medicine
Edinburgh, United Kingdom

David B. Warheit
DuPont Haskell Laboratory for Health and Environmental Sciences
Newark, Delaware

Table of Contents

1 An Introduction to Particle Toxicology: From Coal Mining to Nanotechnology

Paul J. A. Borm
Centre of Expertise in Life Sciences, Hogeschool Zuyd

Ken Donaldson
MRC/University of Edinburgh Centre for Inflammation Research, Queen's Medical Research Institute

CONTENTS

1.1 HISTORICAL DEVELOPMENT OF PARTICLE TOXICOLOGY

Particle research and particle toxicology have been historically closely connected to industrial activities or materials, such as coal, asbestos, manmade mineral fibers, and more recently, ambient particulate matter (Donaldson and Borm 2000) and Nanotechnology (Donaldson 2004; Kurath 2006). The Middle ages saw the first recordings of ill health associated with mining in the writings of Agricola (1494–1555) and Paracelsus (1493–1541), who noted lung diseases in miners in Bohemia and Austria, respectively (Seaton 1995). Initial studies in the modern era concerned workers employed in the coal mining and coking industry, a widespread industry producing, transporting, or burning large amounts of coal. During these processes large quantities of particles were generated, and historically, exposures to coal and coal mine dust have been described as attaining 40 mg/m^3, whereas in current mining a standard of 2–3 mg/m^3 is well maintained (Figure 1.1).

Nowadays, research on particles largely concentrates on exposure to ambient particulate matter (PM) at concentrations between 10 and 50 $\mu g/m^3$. Among these particles, the fine and the ultrafine fraction (<100 nm) are considered to be the most harmful (Peters et al. 1997a; Donaldson et al. 2005), although consensus is not yet reached as to the relative role of the different size fractions. The term ultrafine particles has gradually become intertwined with the term nanoparticles, since they embrace the same size range as particles produced by current nanotechnology (Buxton et al. 2003; Ferrari 2005).

This book contains reviews on the mechanisms and properties of various materials that we are exposed to in particulate form. Both the order and the content will allow the reader to achieve a

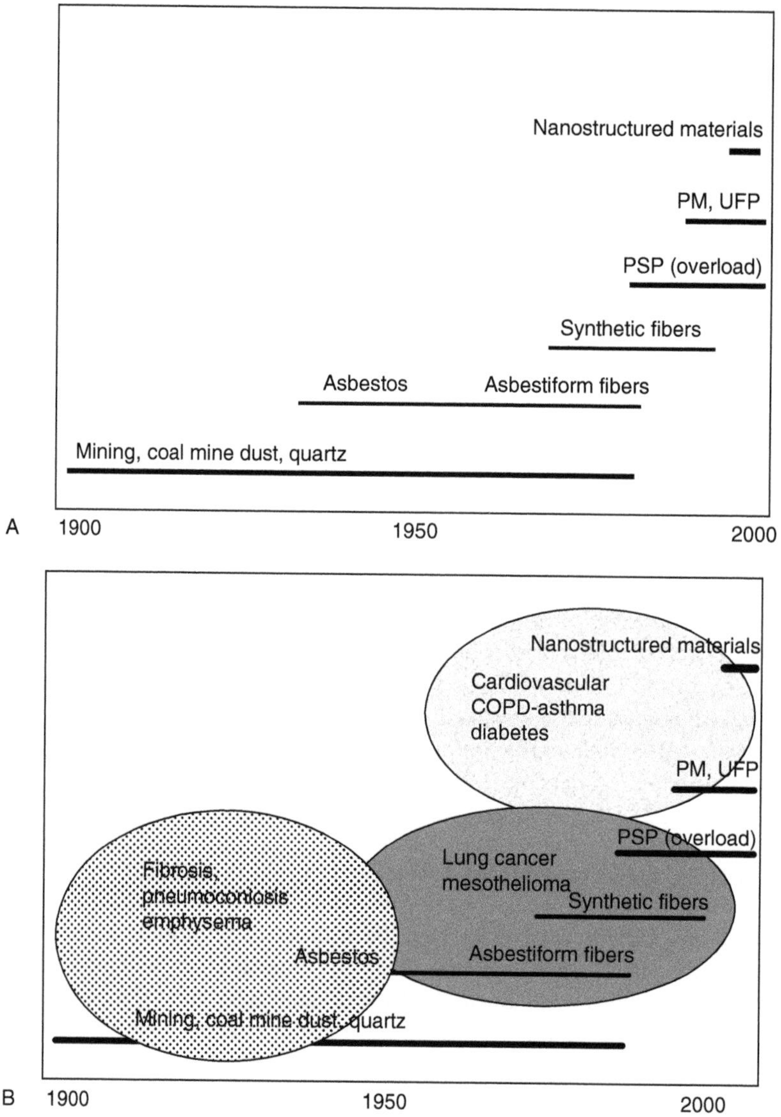

FIGURE 1.1 Historical development of particle toxicology along with technologies and (major) toxicological products emerging from these technologies. Panel A: time frame of particles driving particle toxicology, Panel B: particles along with the major outcomes that were studied.

comprehensive understanding of how particles may cause adverse health effects. These actions may be related to the material, the particulate or fibrous shape, or the specific site of deposition or translocation. This introductory chapter will try to give a brief overview of developments of particle research, from a historical perspective, from coal mining, fiber-related diseases, and ambient particulate matter to hazards imposed by nanomaterials.

1.2 THE IMPACT OF COAL AND ASBESTOS ON RESEARCH AND REGULATION

Coal mining is one of the oldest occupational activities that was, and still is, performed on a large scale. Apart from offering tremendous economic and political benefits, coal mining carried dangers from exposure to noise, heat, and airborne dusts, causing many associated diseases. Respiratory diseases caused by coal mine dust are well known from epidemiological studies in the past century and include coal workers' pneumoconiosis (review: Heppleston 1992), but also chronic bronchitis (review: Wouters et al. 1994) and emphysema (Ruckley et al. 1984; Leigh et al. 1994). The classical industrial use of coal was its heating and conversion into coke, a hard substance consisting of purely carbon. Coke ovens can be seen all over the Ruhr area and were the starting point of fertilizer production. If coke is combined with iron ore and limestone, the mixture is then heated to produce iron, which explains the combined appearance of coal mines, coke ovens, and the steel industry. Therefore, it should also be no surprise that the European Community for Steel and Coal (ECSC) was founded in 1951. Up to 1999, the ECSC was the steering agency for research on particle (i.e., coal) induced respiratory diseases in Europe. During its existence, the ECSC ran five different medical research programs with a total budget of 35.3 million euro. In its slipstream, well-known research institutes were built, of which few survive today. Their research has played a major role in producing extensive epidemiological data, the exploration of mechanisms in particle deposition (e.g., Chapter 3) and particle-induced lung diseases (e.g., Chapter 12 and Chapter 23), and by a combination of the two, performed work to explain the large inter-individual differences in disease rates between miners and coalmines. A lot of this work was based on the hypothesis that the crystalline silica (quartz) content in respirable coal mine dust was the principal agent in coal particles mediating their adverse effects. This approach also deserves some historical explanation.

The epidemic of lung disease caused by asbestos has substantially occupied particle toxicologists and continues to resonate in modern society. In the West, where asbestos is effectively regulated out of use or banned, there is a continuing rise in deaths from mesothelioma. This has been well documented all over Europe with a peak showing a 30–40 year lag pro rata with the peak of amphibole exposures (Peto et al. 1995; McElvenny et al. 2005, U.K. data). However, in the less developed world, asbestos exposure and disease burden is also increasing (Kazan-Allen 2005). Toxicologists have made a huge contribution to understanding the nature of the fiber hazard. Wagner, using erionite, showed importantly that asbestos fibers were not the only fiber capable of causing mesothelioma (Wagner et al. 1985). However, it was the rise in the use of the synthetic vitreous fibers (SVF) as replacements for asbestos that really allowed a full understanding of fiber toxicology and the unification of fiber toxicology into one understanding that embraces both asbestos and the SVF. The RCC studies on SVF were a groundbreaking set of studies that compared a number of SVF of different composition at similar length, for their pathogenicity long-term rat studies (Mast et al. 1995; Hesterberg et al. 1996; Hesterberg et al. 1998; McConnell et al. 1999). In brief, these state-of-the-science studies identified the key role of biopersistence and length in mediating adverse effects of particles and fibers. This shed light on the observation that chrysotile had been reported to be generally less harmful than the amphiboles and that in the lungs of chrysotile miners, the dominant recoverable fiber was in fact amphibole (McDonald et al. 1997). This reflected the greater solubility of chrysotile in the lung milieu—a direct consequence of its Swiss-roll structure and its acid-soluble brucite layer (Bernstein et al. 2003; Wypych et al. 2005). The importance of biopersistence in modulating the pathogenicity of long thin fibers was sealed with its enshrinement in European Legislation, which allow nonbiopersistent SVF to be exonerated as carcinogens, based on having composition that renders them soluble or following adequate testing (Council of the European Union 1997). During this time, Mossman and coworkers made considerable advances in demonstrating that asbestos could activate key oncogenes in epithelial cells, hinting at direct carcinogenic effects driven by oxidative stress (Heintz et al. 1993; Janssen et al. 1995) whilst the proinflammatory effects and their size-related effects were clearly demonstrated (Donaldson et al. 1988; Donaldson et al. 1989; Petruska et al. 1991; Ye et al. 1999) providing support for indirect carcinogenesis acting through inflammation and oxidative stress. A number of these issues are reviewed in Chapter 5 (anti-oxidant defense), Chapter 8 (oxidative stress), Chapter 12 (cell proliferation) and Chapter 6 (genotoxicity).

The fiber story continues in the form of recent biopersistence studies with pure chrysotile, showing it to indeed have a very short half-life in animal studies (Bernstein, Rogers, and Smith 2003) and in the rise in concern over carbon nanotubes, which can be very thin, very long, and very biopersistent (Donaldson et al. 2006) and which is discussed in Chapter 2 (mineralogy) and Chapter 22 (conceptual framework).

1.3 THE ROLE OF QUARTZ IN PARTICLE TOXICOLOGY

For a long time, the dust-induced disease that affected coal miners was thought to be silicosis. In the 1930s, the views of Haldane exerted great influence on this discussion. Haldane argued that silicosis, coal workers' pneumoconiosis, and bronchitis were clinically and pathologically distinct. Unfortunately, he also believed that pure coal dust was not harmful, in spite of some earlier studies on the effects of pure coal dust in coal trimmers (Collis and Gilchrist 1928), workers who are involved in the filling of bunkers and cargo holds in ships. This opinion remained in the United Kingdom and spread to the continent until the 1940s, when new reports showed that coal that was washed free of silica (Gough 1940; Hart and Aslett 1942) induced a "dust disease" that was pathologically different from silicosis, in coal trimmers and stevedores who leveled coal in the holds of ships. Interestingly, even after this epidemiological reappraisal, many epidemiological studies were conducted that concentrated on quartz content in relation to pathological category (King and Nagelschmidt 1945). Pursuing the quartz theme in coal workers' pneumoconiosis (CWP), 2% quartz mixed with anthracite failed to induce fibrosis by inhalation exposure in rats, and the same concentration of quartz alone was also without effect. In fact, after inhalation exposure in rats, clear signs of fibrosis were only seen when quartz was added to the coal dust to a level of 20% (Ross et al. 1962). Large-scale studies conducted in the course of ECSC medical programs (Table 1.1) were not able to show a consistent relation between quartz content of more than 40 respirable coal mine dusts, with *in vitro* toxicity, *in vivo* effects, or epidemiological outcome (Davis et al. 1982). Up to the 1990s, when exploration of the disease process of CWP proceeded at the molecular level, there appears to be only a quantitative difference between the response of key immunoinflammatory cells to quartz and coal mine dust (Gosset et al. 1991; Schins

TABLE 1.1
Major Differences between Exposure and Effects of Traditional (Coal) Mine Dust and Later Studies on Ambient Particulate Matter, Including Ultrafine Particles

	Mining Dusts (Coal, Asbestos)	Ambient PM/CDNP
Endpoints	Nonmalignant respiratory diseases (CWP, bronchitis, emphysema) Malignant respiratory (lung cancer, mesothelioma)	Mortality and exacerbations of existing diseases (asthma, cardiovascular, diabetes) Lung cancer
Exposure routes	Inhalation	Inhalation
Target population	Workers (mining, shipyards)	World population
Target organs	Respiratory tract	Respiratory tract Heart Circulation, liver
Particle size	Respirable fraction <5 μm	Respirable (<2.5 μm) and ultrafine (<100 nm)
Exposure levels	Currently around 2 mg/m^3 Historically up to 40 mg/m^3	Between 15 and 60 μg/m^3
Indication of excess risk	For 35 yr, 1 mg/m^3 CWP (2–3%) PMF (0.25%) Bronchitis (50%)	Increase of 10 μg/m^3, $PM_{2.5}$ Daily mortality: 0.4–1.4% Bronchitis: 5–25%

Note: CWP, coal workers' pneumoconiosis; PMF, progressive massive fibrosis; $PM_{2.5}$, Particulate matter <2.5 μm.

and Borm 1999), but many researchers still consider coal as an inert material mixed with an active principle, namely, quartz.

Notwithstanding the general reduction in mining in Western countries, research on quartz does continue. Many questions in coal-induced adverse effects still remain unanswered. Quartz has been classified as a human carcinogen (IARC 1997), but the question remains of whether mixed workplace dusts containing quartz should be considered as a carcinogen. Several research groups have addressed this question using different approaches. A series of studies has shown the variability of the natural quartz hazard with regard to inflammation and genotoxicity (Clouter et al. 2001; Bruch et al. 2004; Cakmak et al. 2004; Fubini et al. 2004). In another approach it was shown that coating the quartz surface with a small amount of aluminum, PVNO, or soluble matrix components for coal, reduced the ability of quartz to cause inflammation, DNA damage, hemolysis, and cell toxicity (Duffin et al. 2002; Schins et al. 2002; Albrecht et al. 2004).

The emphasis on quartz in the history of particle toxicology has left a legacy in that quartz remains the positive control of choice (DQ12 in Europe, Min-U-Sil® in the U.S.A.) for *in vitro* and *in vivo* studies. Quartz causes toxicity, inflammation and genotoxicity in the short- and long-term and so is used as a positive control, even in the nanoparticles era, since it can act as a check that toxicology assays are working and can detect a toxic particle.

The recognition that coalminers received very high exposures to relatively low toxicity dust raises the phenomenon of "rat lung overload." This phenomenon, seen at high exposure to low toxicity dusts in rats, is characterized by failed clearance, build up of dose, inflammation, and cancer. Although to be anticipated on the basis of Paracelsus' famous rubric, "everything is a poison—it is the dose that delineates the poison," this has been a concern for particle toxicologists using rats for hazard identification and risk assessment. Subsequently there has been much debate (Morrow 1988; Oberdörster 1992; Mauderly 1996). Originally the concept of "volumetric overload" dominated with the concept that the internal volume of macrophages occupied by particles had a inhibitory effect on phagocytosis (Morrow 1988, 1992). However, later work showed that surface area dose was the driver for onset of overload inflammation (Tran et al. 2000) and this helped to recognize how nanoparticles, with their huge surface area per unit volume, might act.

With regards to coal mine dust, in the human scenario, where there has been largest exposure to low toxicity dust, there is no evidence of lung cancer in coal miners and severe fibrosis is relatively rare. However, humans tend to interstitialize particles without much adverse effect (Nikula et al. 1997a, 1997b) which contrasts with rats, where interstitialization is linked to inflammation and adverse effects. There is general consensus that coal miners do not show the effects of rodent type overload (Kuempel et al. 2001) and that rats are unique, even amongst rodents, in showing a very extravagant pathogenic response to high lung burdens of low toxicity dust. This issue is intensively discussed in Chapter 21 on mathematical modeling.

1.4 FROM COAL MINE DUST AND ASBESTOS TO AMBIENT PARTICLES

Although some earlier well-known episodes of air pollution (Meuse Valley in 1930; London in 1952) were known to be associated with increased disease and mortality, it has taken many decades and series of epidemiological studies to convince both scientists and policymakers that, even nowadays, ambient particle exposures cause adverse effects leading to acute mortality. Dockery and coworkers in 1993 showed a relation between changes in acute mortality in the general population and variation in concentrations of PM in six different cities in the United States. This study has been reviewed and repeated, extended and updated (Samet et al. 2000), and followed by many others (Pope et al. 2004, review) but its initial findings have been confirmed in many different countries. From these studies it is estimated that per 10 $\mu g/m^3$ increase in the annual concentration of $PM_{2.5}$, mortality increases by 1.4%, while respiratory disease such as bronchitis or asthma exacerbations increase by as much as 4%. Based on the extent of these effects, particulate matter still belongs to the priority topics identified by the EU and the U.S. Health Effects Institute (HEI), the WHO program, and the U.S. National Research Council (NRC 1998). Although epidemiological evidence suggests that it is the fine ($PM_{2.5}$) and even the ultrafine ($PM_{0.1}$) fraction that contains the toxic components, there is no general agreement on this issue (Oberdörster et al. 1994; Wichmann et al. 2000). The wide number of endpoints (from attacks of asthma to death) suggests that more than one component may be driving the health effects. However, compared to traditional research in particle toxicology, the exposures are much lower and the size of the particles is different. The current challenge is, therefore, to explain why exposures to PM_{10} (mass < 10 μm) typically as low as 40 μg/m^3, compared to 5–40 mg/m^3 in the coal mines, can cause acute death in those with asthma or cardiovascular diseases.

The PM research area is expanding rapidly with many changes compared to the former coal-driven particle research (Table 1.1). It is remarkable to have witnessed the change in focus from shift-type exposure in a specific underground occupation to the entire world population on a 24-hour-per-day basis. A major difference is the change in methodological endpoints, not confined to the lung but also focused on atherosclerosis, cardiovascular abnormalities, and recently effects on the brain (see Chapter 14, Chapter 15, and Chapter 19). A second fundamental change is the particle size of interest, with its impact on ambient measurements and experimental methods. Nanoparticles (< 100 nm) are the subjects of many toxicological investigations in animals and humans (Oberdörster et al. 2005; Donaldson et al. 2006; Nel et al. 2006, reviews) and have been shown to linked to adverse effects in epidemiology (Peters et al. 1997a; Wichmann et al. 2000). Animal studies have demonstrated that inflammation at "overload" is determined by the surface area dose of inhaled particles (Tran et al. 2000), which is greatly determined by the particle size. In addition to this effect, which is mainly caused by saturation of clearance, ambient fine and ultrafine particles, in view of their origins in fuel combustion, contain a large number of soluble metals and organics that have been associated with a variety of inflammatory responses (Chapter 8 and Chapter 11). Among the suggested mechanisms of activation, the production of intracellular oxidative stress and changes in Ca^{2+} levels leading to activation of transcription factors such as NF-κB, is the best described and elucidated in Chapter 9 through Chapter 11 of this book. The activation of NF-κB is well known to lead to transcription of a number of chemokines (IL-8), cytokines (TNF-α), and other enzymes (COX-2; inducible nitric oxide synthetase, iNOS) that are able to directly or indirectly enhance oxidative stress (Donaldson et al. 1998, review).

Funding strategies and criteria have changed dramatically since the heyday of the ECSC, to a molecular approach, and the global context of coal mining has also altered. Breakthroughs typically originate from links to other disciplines, such as cardiovascular pharmacology or immunology. The translocation of ultrafine particles into the bloodstream and their direct effect on heart and vessel wall is fascinating, and exposure *in vivo* to ambient particles does affect blood vessel contraction (Brook et al. 2002; Bagate et al. 2004; Mills et al. 2005). Whether this is a direct effect of translocated particles, or an indirect effect through inflammatory mediators released from the lung, is still open for research. One of the hot issues here is whether and how ultrafine particles can pass the lung barriers, as well as their general distribution and passage of membranes in the body. This issue is being discussed in a number of chapters in this book (Chapter 3, Chapter 7, Chapter 14, and Chapter 22).

Along with the evolution in our understanding of general disease, particle toxicology has benefited from the developments in molecular medicine. As the authors have described previously (Donaldson and Borm 2000), there has been a considerable change in the experimental approach taken by particle toxicologists. At the dawn of modern particle toxicology, in the late 1970s and early 1980s, cell death and injury were measured and we attempted to relate this to disease potential, probably reflecting what we *could* measure in those early days. The rise of inflammation as a key response to particle deposition in tissue and a process in particle effects has been inexorable and inflammation now lies at the very heart of our understanding of lung and systemic disease associated with particle exposure. As assays became available, we saw an increase in genotoxicity studies and the important link between inflammation and genotoxicity was recognized and particles like asbestos and quartz were identified as both direct carcinogens and indirect carcinogens that acted through inflammation. As in all molecular medicine, there has been a huge concentration on the dysregulation of gene expression as a basis for disease, and this continues. Oxidative stress has emerged as the dominant paradigm for how particles initiate inflammation and genotoxicity, a stress further augmented by inflammatory cells releasing their arsenal of oxidants.

Together, the chapters in this book show a twenty-first century view of particle toxicology that has advanced from one where cell death and damage dominates pathological change only in the lungs, to a dynamic view where small particles, on top of effects in the lungs, may translocate to the blood and brain and reticuloendothelial system to exert effects. The gene lies firmly at the center of events and in sites where particles interact with cells, gradients of cytokines, and dynamic changes in antioxidant defense combine to modulate the expression of genes. Some of these gene products cause the antioxidant defenses to be enhanced, whilst others cause inflammatory cells to be recruited. Others stimulate mesenchymal cells to divide and lay down extracellular matrix. Inflammatory cells contribute further to the local cytokine "soup," which may travel systemically to affect the vascular bed and the vessel wall, especially atherogenesis, another inflammatory process. If particles reach the blood they may interact with platelets and the endothelium to enhance thrombosis, and even atherogenesis. Events in the vascular system may enhance atherothrombosis (Chapter 14), whilst in tissue inflammatory leukocytes may cause genetic damage to "innocent bystander" cells through release of their formidable arsenals of mediators and oxidants (Chapter 16). This proliferative and pro-mutagenic milieu is a fertile one for development of fibrotic lesions and clones of mutated cells that eventually culminate in cancer (see Chapter 4, Chapter 12, and Chapter 16). The particles themselves are phagocytosed, transported, undergo dissolution, and accumulate endogenous molecules. This dynamic view of cellular and molecular particle toxicology has been won by tremendous endeavor and insightful research across the world, which continues apace.

1.5 FROM ULTRAFINE PM TO NANOTECHNOLOGY

Translocation of nanoparticles, as mentioned earlier, is one of the tools being explored by pharmaceutical companies to deliver drugs in compartments such as the brain, which is otherwise hard to reach. Such medical applications are part of the possibilities generated by nanotechnology (Duncan 2003; Ferrari 2005). Nanotechnology is now considered to be one of the world's most promising new technologies, able to affect many industrial branches, just as coal mining did about a century ago (Borm 2002; Brune et al. 2006).

Nanotechnology is molecular manufacturing or, more simply, building things one atom or molecule at a time. Utilizing the chemical properties of atoms and molecules, nanotechnology proposes the construction of novel molecular devices possessing extraordinary properties. The trick is to manipulate atoms individually and place them exactly where needed to produce the desired structure. The major difference with current technology is that the starting points of conventional technology and nanotechnology are completely different. Instead of making things smaller by miniaturizing and using new materials, nanotechnology aims to build even smaller devices using atoms/molecules as a starting point. Technical applications now claimed feasible include catalysts, fuel cells, glues and paints, solar cells, smart drugs, and optical systems. Many researchers are tempted step beyond facts and step into visionary predictions, as did the late Dr. Richard Smalley, discoverer of buckyballs (Buckminster-fullerene, C60), and chairman of chemistry and head of the Nanotechnology Initiative at Rice University, who stated that "nanotechnology will reverse the harm done by the industrial revolution." In relation to coal dust-induced adverse health effects, it is hard to imagine how nanotechnology will be able to restore quality of life, or reverse the suffering and loss of productivity induced by respiratory diseases that afflicted thousands of coal miners.

Since only a few years toxicologists have become aware of the potential hazards of engineered nanomaterials (Borm 2002; Colvin 2003; Nel et al. 2006) especially if these will be used for medical applications (Buxton et al. 2003). Ultrafines, due to their small size, have a tremendous surface, which also makes them attractive for carrying drugs through the body. But again, the size of the surface and its chemical properties do determine their aptness for interaction with biological targets, and now many particles are produced, including dendrimers, polymers, quantum dots, magnetic carriers, and many others (Chapter 22 and Chapter 24). However, studies have demonstrated that even inert materials like gold and TiO_2 in nano size can become chemically active, due to the fact that the small size does not allow a surface of oxygen atoms alone (Jefferson and Tilley 1999). In other words, inert materials can become reactive just by making them smaller. Many nanoparticles are produced in emulsions or suspensions, and in order to maintain their single particle-based properties, are often surface-treated to prevent aggregation. As shown in earlier work with coal and quartz, these coatings may be even more important than the particle itself. However, recent work with ultrafine TiO_2, in which the surface was made hydrophobic by specific surface treatments, showed that this was less toxic and proinflammatory than its untreated ultrafine analogue (Oberdörster 2001; Höhr et al. 2002). However, every combination of surface coating and basic core material needs to be carefully evaluated in the future. The sources of evidence for the toxicity of nanoparticles (NP), originate from three classes of NP: bulk, combustion, and engineered nanoparticles (see also Chapter 22). The current discussion on engineered nanoparticles is mainly driven by data on combustion NP (diesel, ultrafines) and a small set of bulk NP (carbon blacks, TiO_2). However, that the current data set on engineered NP is growing and qualitative effects (inflammation, atherosclerosis, oxidative stress, Ca-transport, etc.) are also shown with various nano products, such as single wall nanotubes (SWNT). Some studies allow bridging of data, such as recent work (Radomksi et al. 2005) on platelet aggregation with different nanomaterials, but also including ambient PM samples. Since there is an overload of outstanding toxicology questions before we gain a conceptual understanding on nanomaterials, research, and regulation, the following issues need careful consideration:

1. Some effects of nanomaterials are probably the same qualitatively for engineered NP and others; this can be handled by checking the validity of current testing systems.
2. There is an urgent need to identify effects that are new for (engineered) NP and may occur in other populations and exposure situations than workers only.
3. Almost no data is available on ecotoxicity or ADME of NP, and this area should receive research priority.
4. When choices are to be made in testing and research, they should be driven by the application of the nanostructured materials.

1.6 IN CONCLUSION

Particle research has been historically closely connected to industrial activities or materials, such as coal, asbestos, manmade mineral fibers, and more recently, ambient particulate matter (PM) and nanostructured materials. Differences between historical and current research in particle toxicology include the exposure concentrations, particle size, target populations, endpoints, and length and extent of exposure. Inhaled particle effects are no longer confined to the lung, since particles are suggested to translocate to the blood while lung inflammation invokes systemic responses. Finally, the particle size and concentrations have both been reduced about 100-fold, from 2–5 to 20–50 mg/m^3, and from 1–2 μm to 20–100 nm (ultrafine), as domestic fuel burning has decreased and vehicle sources have increased and attention has moved from the coal mining industry to general environment. Exposures to nanoparticles in the nanotechnology sector, which continuously produces new materials in the ultrafine range, are largely unknown. Although inhalation exposure is considered to be minimal in this technology, some particles are produced for consumer applications in household products, drugs, and foodstuffs. Clearly, particle toxicology is evolving with this trend and will have to develop new concepts of understanding particle actions, measuring, and testing particles and regulation.

All these observations underscore the importance of interdisciplinary research and exchange in the area of surface properties and activity of very small particles. Both epidemiological and toxicological studies have contributed to a body of evidence suggesting that coal, asbestos, and later ultrafine particles can induce or exaggerate a number of adverse effects. The last decade of PM research has shown a close collaboration between epidemiology and toxicology to solve important questions in this area. No doubt we are facing a next decade of challenging particle research driven by nanomaterials. Now the questions imposed on us by nanomaterials demand new collaborations with disciplines like chemistry, material sciences, and engineering. The formation of so called competence centers, or networks, where nanomaterials can be made and tested, will be vital for further development of sustainable nanomaterials. Both particle toxicology and nanotechnology will benefit from better knowledge generated in the twilight zone between materials and biological molecules.

REFERENCES

Albrecht, C., Schins, R. P. F., Höhr, D., Becker, A., Shi, T., Knaapen, A., and Borm, P. J. A., Inflammatory time course following quartz instillation: role of TNFa and particle surface, *Am. J. Respir. Cell Mol. Biol.*, 31, 292–301, 2004.

Bagate, K., Meiring, J. J., Gerlofs-Nijland, M. E., Vincent, R., Cassee, F. R., and Borm, P. J. A., Vascular effect of ambient particulate matter instillation in spontaneous hypertensive rats, *Toxicol. Appl. Pharmacol.*, 197, 29–39, 2004.

Bernstein, D. M., Rogers, R., and Smith, P., The biopersistence of Canadian chrysotile asbestos following inhalation, *Inhal. Toxicol.*, 15, 1247–1274, 2003.

Borm, P. J. A., Particle toxicology: from coal mining to nanotechnology, *Inhal. Toxicol.*, 14, 311–324, 2002.

Brook, R. D., Brook, J. R., Urch, B., Vincent, R., Rajagopalan, S., and Silverman, F., Inhalation of fine particulate air pollution and ozone causes acute arterial vasoconstriction in healthy adults, *Circulation*, 105, 1534–1536, 2002.

Bruch, J., Rehn, S., Rehn, B., Borm, P. J. A., and Fubini, B., Variation of biological responses to different respirable quartz flours by a vector model, *Int. J. Hyg. Environ. Health*, 207, 1–14, 2004.

Brune, H., Ernst, H., Grunwald. A., Grunwald, W., Hofman, H., Krug, H., Janich, P., et al., *Nanotechnology-Assessment and Perspectives*, Springer-Verlag, p. 495, 2006, (ISBN 3-540-32819-X).

Buxton, D. B., Lee, S. C., Wickline, S. A., and Ferrari, M., National heart, lung, and blood institute nanotechnology working group. Recommendations of the national heart, lung, and blood institute nanotechnology working group, *Circulation*, 108 (22), 2737–2742, 2003.

Cakmak, G., Schins, R., Shi, T., Fenoglio, I., Fubini, B., and Borm, P. J. A., *In vitro* genotoxicity of commercial quartzes in A549 alveolar epithelial cells, *Int. J. Hyg. Environ. Health*, 207, 105–113, 2004.

Clouter, A., Brown, D., Borm, P., and Donaldson, K., Inflammatory effects of respirable quartz collected in workplaces versus standard DQ12 quartz: Particle surface correlates, *Toxicol. Sci.*, 63, 90–98, 2001.

Collis, E. L. and Gilchrist, J. C. Effects of dust upon coal trimmers, *J. Ind. Hyg.*, 4, 101–110, 1928.

Colvin, V. L., The potential environmental impact of engineered nanomaterials, *Nat. Biotechnol.*, 21 (10), 1166–1170, 2003.

Council of the European Union Commission Directive 97/69/EC., L343/19-L343/20. 13-12-0097. *Off. J. Eur. Commun.*, 1997.

Davis, J. M. G., Addison, J., Bruch, J., Bruyere, S., Daniel, H., and Degueldre, G., et al. Variations in cytotoxicity and mineral content between respirable mine dusts from the Belgian, British, French and German coalfields, *Ann. Occup. Hyg.*, 26, 541–549, 1982.

Dockery, D. W., Pope, C. A., Xu, X., Spengler, J. D., Ware, J. H., Fay, M. E., Ferris, B. G., and Speizer, F. E., An association between air pollution and mortality in six U.S. cities, *N. Engl. J. Med.*, 329, 1753–1759, 1993.

Donaldson, K. and Borm, P. J. A., The quartz hazard: a variable entity, *Ann. Occup. Hyg.*, 42, 287–294, 1998.

Donaldson, K. and Borm, P. J. A., Particle paradigms, *Inhal. Toxicol.*, 12 (suppl. 3), 1–6, 2000.

Donaldson, K., Bolton, R. E., Jones, A., Brown, G. M., Robertson, M. D., Slight, J., Cowie, H., and Davis, J. M., Kinetics of the bronchoalveolar leucocyte response in rats during exposure to equal airborne mass concentrations of quartz, chrysotile asbestos, or titanium dioxide 13, *Thorax*, 43, 525–533, 1988.

Donaldson, K., Brown, G. M., Brown, D. M., Bolton, R. E., and Davis, J. G., Inflammation generating potential of long and short fiber amosite asbestos samples, *Br. J. Ind. Med.*, 46, 271–276, 1989.

Donaldson, K., Stone, V., Duffin, R., Clouter, A., Schins, R. P. F., and Borm, P. J. A., The quartz hazard: effects of surface and matrix on inflammogenic activity, *J. Exp. Pathol. Toxicol. Suppl.*, 1, 109–118, 2001.

Donaldson, K., Tran, L., Jimenez, L. A., Duffin, R., Newby, D. E., Mills, N., and MacNee, W. Stone V: Combustion-derived nanoparticles: a review of their toxicology following inhalation exposure, *Part Fibre Toxicol.*, 2, 10, 2005.

Donaldson, K., Aitken, R., Tran, L., Stone, V., Duffin, R., Forrest, G., and Alexander, A., Carbon nanotubes: a review of their properties in relation to pulmonary toxicology and workplace safety, *Toxicol. Sci.*, 2006.

Dreher, K. L., Particulate matter physicochemistry and toxicology: In search of causality—a critical perspective, *Inhal. Toxicol.*, 12, 45–57, 2000.

Duffin, R., Clouter, A., Brown, D., Tran, C. L., MacNee, W., Stone, V., et al., The importance of surface area and specific reactivity in the acute pulmonary inflammatory response to particles, *Ann. Occup. Hyg.*, 46 (suppl. 1), 242–245, 2002.

Duncan R., The dawning area of polymer therapeutics. *Nature reviews- Drug discovery*, 2, 347–360, 2003.

Duncan, R. and Izzo, L., Dendrimer biocompatibility and toxicity, *Adv. Drug. Deliv. Rev.*, 57, 2215–2237, 2005.

Ferrari, M., Cancer nanotechnology: opportunities and challenges, *Nat. Rev. Cancer*, 5, 161–171, 2005.

Fubini, B., Fenoglio, I., Ceschino, R., Ghiazza, M., Martra, G., Tomatis, M., Borm, P., Schins, R. P. F., and Bruch, J., Relationships between the state of the surface if four commercial quartz flours and their biological activity *in vitro* and *in vivo*, *Int. J. Hyg. Environ. Health*, 207, 89–104, 2004.

Gosset, P., Lasalle, P., Vanhee, D., Wallaert, B., Aerts, C., Voisin, C., and Tonnel, A. B., Production of tumor necrosis factor-a and interleukin-6 by human alveolar macrophages exposed *in vitro* to coal mine dust, *Am. J. Respir. Cell Mol. Biol.*, 5, 431–436, 1991.

Gough, J., Pneumoconiosis in coal trimmers, *J. Pathol. Bacteriol.*, 51, 277–285, 1940.

Hart, P. d'. A. and Aslett, E. A., Medical Research Council. Special Report Series No. 243. Her Majesty's Stationery Office, London, 1942.

Heintz, N. H., Janssen, Y. M., and Mossman, B. T., Persistent induction of c-fos and c-jun expression by asbestos, *Proc. Natl Acad. Sci. USA*, 90, 3299–3303, 1993.

Heppleston, A. G., Coal workers' pneumoconiosis: a historical perspective on its pathogenesis, *Am. J. Ind. Med.*, 22, 905–923, 1992.

Heppleston, A. G. and Stiles, J. A., Activity of macrophage factor in collagen formation by silica, *Nature*, 214, 521–522, 1967.

Hesterberg, T. W., Miiller, W. C., Musselman, R. P., Kamstrup, O., Hamilton, R. D., and Thevenaz, P., Biopersistence of man-made vitreous fibers and crocidolite asbestos in the rat lung following inhalation, *Fundam. Appl. Toxicol.*, 29, 267–279, 1996.

Hesterberg, T. W., Hart, G. A., Chevalier, J., Miiller, W. C., Hamilton, R. D., Bauer, J., and Thevenaz, P., The importance of fiber biopersistence and lung dose in determining the chronic inhalation effects of X607, RCF1, and chrysotile asbestos in rats, *Toxicol. Appl. Pharmacol.*, 153, 68–82, 1998.

Hohr, D., Steinfartz, Y., Schins, R. P. F., Knaapen, A. M., Martra, G., Fubini, B., and Borm, P. J. A., Hydrophobic coating of ultrafine titanium dioxide reduces the acute inflammatory response after instillation in the rat, *Int. J. Hyg. Environ. Health*, 205, 239–244, 2001.

Höhr, D., Steinfartz, Y., Schins, R. P. F., Knaapen, A. M., Martra, G., Fubini, B., and Borm, P. J. A., Hydrophobic coating of ultrafine titanium dioxyde reduces the acute inflammatory response after instillation in the rat, *Int. J. Hyg. Environ. Health, 2002*, 205 (3), 239–244, 2001.

IARC, Silica, some silicates, coal dust and para-aramid fibrils, *IARC Monogr. Eval. Carcinogen. Risks Hum.*, 68, 41–242, 1997.

Janssen, Y. W., Heintz, N. H., and Mossman, B. T., Induction of c-fos and c-jun protooncogene expression by asbestos is ameliorated by *n*-acetyl-L-cysteine in mesothelial cells, *Cancer Res.*, 55, 2085–2089, 1995.

Jefferson, D. A. and Tilley, E. E. M., The structural and physical chemistry of nanoparticles, In *Particulate Matter: Properties and Effects Upon Health*, Maynard, R. L., and Howardeds, C. V., eds., Cromwell Press, Trowbridge, UK, S63–S84, 1999.

Kazan-Allen, L., Asbestos and mesothelioma: Worldwide trends, *Lung Cancer*, 49 (suppl. 1), S3–S8, 2005.

King, E. J. and Nagelschmidt, G., The mineral content of the lungs of workers from the South Wales coalfield. Chronic pulmonary disease in South Wales coalminers—III. Experimental studies, Medical Research Council, Special report series no. 250, Her Majesty's Stationery Office, London, 1–20, 1945.

Kuempel, E. D., O'Flaherty, E. J., Stayner, L. T., Smith, R. J., Green, F. H., and Vallyathan, V. A., Biomathematical model of particle clearance and retention in the lungs of coal miners, *Regul. Toxicol. Pharmacol.*, 34 (1), 69–87, 2001.

Kurath, M. and Maasen, S., Toxicology as a nanoscience?—Disciplinary identities reconsidered, *Part Fiber Toxicol.*, 3 (1), 6, 2006. Apr. 28 [Epub ahead of print].

Leigh, J., Driscoll, T. R., Cole, B. D., Beck, R. W., Hull, B. P., and Yang, J., Quantitative relation between emphysema and lung mineral content in coal workers, *Occup. Environ. Med.*, 51, 400–407, 1994.

Mast, R. W., Mcconnell, E. E., Anderson, R., Chevalier, J., Kotin, P., Bernstein, D. M., Thevenaz, P., Glass, L. R., Miiller, W. C., and Hesterberg, T. W., Studies on the chronic toxicity (Inhalation) of 4 types of refractory ceramic fiber in male fischer-344 rats, *Inhal. Toxicol.*, 7, 425–467, 1995.

Mauderly, J. L., Lung overload: the dilemma and opportunities for resolution. Special issue. Particle overload in the rat lung and lung cancer: implications for risk assessment, *Inhal. Toxicol.*, 8 Supplement, 1–28, 1996, Ref Type: Generic.

McConnell, E. E., Axten, C., Hesterberg, T. W., Chevalier, J., Miiller, W. C., Everitt, J., Oberdörster, G., Chase, G. R., Thevenaz, P., and Kotin, P., Studies on the inhalation toxicology of two fiberglasses and amosite asbestos in the syrian golden hamster. Part II. Results of chronic exposure 1, *Inhal. Toxicol.*, 11, 785–835, 1999.

McDonald, J. C. and McDonald, A. D., Chrysotile, tremolite and carcinogenicity, *Ann. Occup. Hyg.*, 41, 699–705, 1997.

McElvenny, D. M., Darnton, A. J., Price, M. J., and Hodgson, J. T., Mesothelioma mortality in Great Britain from 1968 to 2001, *Occup. Med. (London)*, 55, 79–87, 2005.

Mills, N. L., Tornqvist, H., Robinson, S. D., Gonzalez, M., Darnley, K., MacNee, W., Boon, N. A. et al., Diesel exhaust inhalation causes vascular dysfunction and impaired endogenous fibrinolysis, *Circulation*, 112 (25), 3930–3936, 2005.

Morrow, P. E., Possible mechanisms to explain dust overloading of the lungs, *Fundam. Appl. Toxicol.*, 10, 369–384, 1988.

Morrow, P. E., Dust overloading of the lungs: update and appraisal, *Toxicol. Appl. Pharmacol.*, 113, 1–12, 1992.

National Research Council. Research priorities for airborne particulate matter. I. Immediate priorities and a long-range research portfolio. Washington DC: National Academy Press, 1998.

Nel, A., Xia, T., Madler, L., and Li, N., Toxic potential of materials at the nanolevel, *Science*, 311 (5761), 622–627, 2006.

Nemmar, A., Vanbilloen, H., Hoylaerts, M. F., Hoet, P. H., Verbruggen, A., and Nemery, B., Passage of intratracheally instilled ultrafine particles into the systemic circulation of the hamster, *Am. J. Respir. Crit. Care Med.*, 164, 1665–1668, 2001.

Nikula, K. J., Avila, K. J., Griffith, W. C., and Mauderly, J. L., Lung tissue responses and sites of particle retention differ between rats and cynomolgus monkeys exposed chronically to diesel exhaust and coal dust, *Fundam. Appl. Toxicol.*, 37, 37–53, 1997a.

Nikula, K. J., Avila, K. J., Griffith, W. C., and Mauderly, J. L., Sites of particle retention and lung tissue responses to chronically inhaled diesel exhaust and coal dust in rats and cynomolgus monkeys, *Environ. Health Perspect.*, 105, 1231–1234, 1997b.

Oberdörster, G., Pulmonary effects of inhaled ultrafine particles, *Int. Arch. Occup. Environ. Health*, 74, 1–8, 2001.

Oberdörster, G., Ferin, J., and Morrow, P. E., Volumetric loading of alveolar macrophages (AM): a possible basis for diminished AM-Mediated particle clearance, *Exp. Lung Res.*, 18, 87–104, 1992.

Oberdörster, G., Galein, R. N., Ferin, J., and Weiss, B., Association of acute air pollution and acute mortality: involvement of ultrafines? *Inhal. Toxicol.*, 7, 111–124, 1994.

Oberdörster, G., Oberdörster, E., and Oberdörster, J., Nanotoxicology: an emerging discipline evolving from studies of ultrafine particles, *Environ. Health Perspect.*, 113 (7), 823–839, 2005.

Peters, A., Wichmann, H. E., Tuch, T., Heinrich, J., and Heyder, J., Respiratory effects are associated with the number of ultrafine particles, *Am. J. Respir. Crit. Care Med.*, 155, 1376–1383, 1997a.

Peto, J., Hodgson, J. T., Matthews, F. E., and Jones, J. R., Continuing increase in mesothelioma mortality in Britain 2, *Lancet*, 345, 535–539, 1995.

Petruska, J. M., Leslie, K. O., and Mossman, B. T., Enhanced lipid-peroxidation in lung lavage of rats after inhalation of asbestos, *Free Radical Biol. Med.*, 11, 425–432, 1991.

Pope, C. A. 3rd, Burnett, R. T., Thurston, G. D., Thun, M. J., Calle, E. E., Krewski, D., Godleski, J. J. Cardiovascular mortality and long-term exposure to particulate air pollution: epidemiological evidence of general pathophysiological pathways of disease, *Circulation*, 109 (1), 71–7, 2004.

Radomski, A., Jurasz, P., Alonso-Escolano, D. et al., Nanoparticle-induced platelet aggregation and vascular thrombosis, *Br. J. Pharmacol.*, 146 (6), 882–893, 2005.

Ross, H. F., King, E. J., Yognathan, M., and Nagelschmidt, G., Inhalation experiments with coal dust containing 5%, 10%, 20% and 40% quartz: tissue reactions in the lungs of rats, *Ann. Occup. Hyg.*, 5, 149–161, 1962.

Ruckley, V. A., Gauld, S. J., Chapman, J. S., Davis, J. M. G., Douglas, A. N., Fernie, J. M., Jacobsen, M., and Lamb, D., Emphysema and dust exposure in a group of coal workers, *Am. Rev. Respir. Dis.*, 129, 528–532, 1984.

Samet, J. M., Dominici, F., Curriero, F. C., Coursac, I., and Zeger, S. L., Fine particulate air pollution and mortality in 20 U.S. cities, *N. Engl. J. Med.*, 343, 1742–1749, 2000.

Schins, R. P. F. and Borm, P. J. A., Mechanisms and mediators in coal dust toxicity: a review, *Ann. Occup. Hyg.*, 43, 7–33, 1999.

Schins, R. P., Duffin, R., Hohr, D., Knaapen, A. M., Shi, T., Weishaupt, C. et al., Surface modification of quartz inhibits toxicity, particle uptake, and oxidative DNA damage in human lung epithelial cells, *Chem. Res. Toxicol.*, 15 (9), 1166–1173, 2002.

Seaton, A., Chapter 1 A short history of occupational lung disease, In *Occupational Lung Diseases*, Morgan, W. K. C. and Seaton, A., eds., Saunders, Philadelphia, PA, 1–8, 1995.

Seaton, A., MacNee, W., Donaldson, K., and Godden, D., Particulate air pollution and acute health effects, *Lancet*, 345, 176–178, 1995.

Tran, C. L., Buchanan, D., Cullen, R. T., Searl, A., Jones, A. D., and Donaldson, K., Inhalation of poorly soluble particles. II. Influence of particle surface area on inflammation and clearance, *Inhal. Toxicol.*, 12, 1113–1126, 2000.

Wagner, J. C., Skidmore, J. W., Hill, R. J., and Griffiths, D. M., Erionite exposure and mesotheliomas in rats, *Br. J. Cancer*, 51, 727–730, 1985.

Wichmann, H. E., Spix, C., Tuch, T., Wölke, T., Peters, A., Heinrich, J. et al., *Daily mortality and fine and ultrafine particles in Erfurt Germany. Part I: Role of Particle Number and Particle Mass Health Effects Institute Research Report number 98*, HEI, Cambridge, MA, 2000.

Wouters, E. F. M., Jorna, T. H. J. M., and Westenend, M., Respiratory effects of coal dust exposure: clinical effects and diagnosis, *Exp. Lung Res.*, 20, 385–394, 1994.

Wypych, F., Adad, L. B., Mattoso, N., Marangon, A. A., and Schreiner, W. H., Synthesis and characterization of disordered layered silica obtained by selective leaching of octahedral sheets from chrysotile and phlogopite structures 8, *J. Colloid Interf. Sci.*, 283, 107–112, 2005.

Ye, J., Shi, X., Jones, W., Rojanasakul, Y., Cheng, N., Schwegler-Berry, D., Baron, P., Deye, G. J., Li, C., and Castranova, V., Critical role of glass fiber length in TNF-alpha production and transcription factor activation in macrophages, *Am. J. Physiol.*, 276, L426–L434, 1999.

2 Mineralogy and Structure of Pathogenic Particles

Tim Jones
School of Earth, Ocean, and Planetary Sciences, Cardiff University

Kelly BéruBé
School of Biosciences, Cardiff University

CONTENTS

2.1 INTRODUCTION

It is estimated that the element silicon makes up 27.7% of the earth's continental crust. The majority of this silicon is in the form of crystalline silicon dioxide (SiO_2) as the polymorph quartz, a well-established respiratory hazard (Rimala, Greenbergac, and William 2005). However, silicon is also present in numerous other minerals, and therefore, "silicates and non-silicates" have been used by mineralogists as a framework, further based on structure (ortho-, ring, chain, sheet, and tecto-), to describe all minerals (Deer, Howie, and Zussman 1966). This framework has been used in this chapter, but only minerals that are known or suspected to be respiratory hazards (Guthrie and Mossman 1993) are included (Figure 2.1 and Table 2.1). The chemistry and structure of minerals is complicated by the fact that many minerals exist in solid state series and in different shape "habits." For example, the chain silicate, amphibole, cummingtonite $(Mg,Fe^{+2})_7[Si_8O_{22}](OH)_2$-grunerite $(Fe^{+2},Mg)_7[Si_8O_{22}](OH)_2$ series. Grunerite is the name for the more iron-rich end-members, and is of significance here, because in its fibrous habit it is the mineral amosite; carcinogenic asbestos (Nolan, Langer, and Wilson 1999). In addition to the chemistry and habit, the formation conditions of the minerals can have bearings on the toxicity, e.g., SiO_2. The vast majority of crystalline SiO_2 occurs as the polymorph quartz, a mineral that usually forms at relatively low temperatures and high pressures. However, if SiO_2 forms at high temperatures and low pressures, such as near the surface in a volcanic dome, it forms the SiO_2 polymorph cristobalite. Concerns over the possible toxicity of volcanic ash particles have led to research which has shown

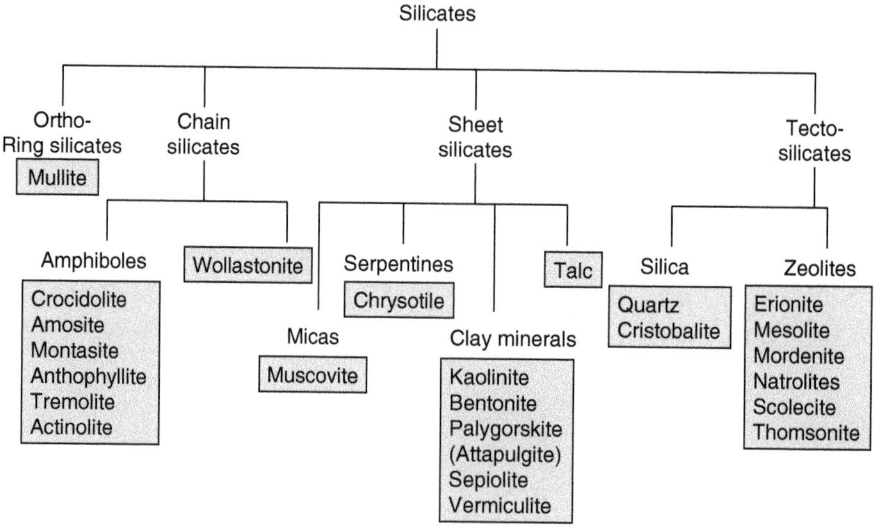

FIGURE 2.1 The distribution of some respiratory hazards within the silicates.

that respirable quartz and cristobalite have different bioreactivities in the lung (Housley et al. 2002; BéruBé et al. 2004; Forbes et al. 2004).

Minerals are the building blocks of rocks, therefore, on occasion we have to consider assemblages of minerals forming a rock type, which itself has been implicated in adverse respiratory health effects. An example of this would be "bauxite," the common name for aluminum ore derived from weathered igneous rocks. Bauxite consists of a mixture of aluminum hydroxide minerals, the most abundant of which is gibbsite, plus other major, minor, and trace minerals. There is an occupational respiratory disease unique to bauxite miners, "Shaver's Disease," but this has only been linked to the ore "bauxite," and not to any individual component mineral (Dinman 1988; Radon et al. 1999; Kraus et al. 2000). Another issue that arises when considering rock types as respiratory hazards is the matter of natural mineral contamination; vermiculite is a good example of this. Vermiculite is an important industrial and domestic material and is composed of clays that have been artificially expanded by heating. The problem is that the clay minerals, which are amenable to this process, can be naturally contaminated by potentially carcinogenic, asbestiform minerals, such as tremolite (McDonald, Harris, and Armstrong 2004). Serious precautions are therefore needed when using vermiculite, although it is probably not the vermiculite itself that is dangerous: rather, it is the trace amounts of contaminating tremolite that pose the hazard (Wright et al. 2002). In addition to recognized pathogenic minerals and rocks, this chapter includes descriptions of the chemistry and structure of pathogenic particles that, at first glance, would appear to be "non-mineral," yet upon close examination show a mineral component. An example of this would be "soot" from the combustion of fossil fuels. Soot is the most abundant component in urban $PM_{2.5}$ and has been linked to a number of respiratory issues, such as the increase in childhood asthma (Brauer et al. 2002; Nicolai et al. 2003). Examination of the carbon microspheres, that are the building blocks of soot particles, shows that they are largely composed of microcrystallites of the mineral graphite.

2.2 ORTHO- AND RING SILICATES

The orthosilicates, which include synthetic mullite, are based on SiO_4^{4-}, where the tetrahedral oxygen atoms are not shared with other tetrahedral. Also called nesosilicates, they are a diverse group of minerals. They tend to be rather hard and dense minerals with a generally poor cleavage. The ring silicates contain only three common species, and are isostructural with six-member rings

TABLE 2.1
Some Minerals and Particles and Their Known or Suspected Health Issues

Mineral	Some Known or Suspected Respiratory Health Issues
	Silicate Minerals
Synthetic mullite	Irritation of the respiratory system, silicotuberculosis
Crocidolite	Asbestosis, lung cancers, mesothelioma, pleural plaques, ferruginous bodies
Amosite	Asbestosis, lung cancers, mesothelioma, pleural plaques, ferruginous bodies
Montasite	Asbestosis, lung cancers, mesothelioma
Anthophyllite	Asbestosis, lung cancers, mesothelioma, plural calcification
Tremolite	Asbestosis, lung cancers, mesothelioma
Actinolite	Asbestosis, lung cancers, mesothelioma, non-malignant pleural lesions
Wollastonite	Lung fibrosis, respiratory morbidity, pulmonary toxicity
Chrysotile	Asbestosis, lung cancer, mesothelioma, pleural tumors
Talc	Talc pneumoconiosis (talcosis), pulmonary oedema, fibrotic pleural thickening
Muscovite	Pulmonary interstitial fibrosis, severe pneumoconosis
Kaolinite	Kaolin pneumoconiosis, simple and complicated Kaolinosis, COPD
Bentonite	Fuller's earth pneumoconiosis, "bentonite" granulomas, silicosis
Palygorskite (attapulgite)	Bronchoalveolar hyperplasia, alveolar tumors, mesothelioma
Sepiolite	Deterioration of lung function, co-carcinogen, fibrosis
Vermiculite	Natural asbestos contamination
Quartz	Fibrosis, silicosis, lung cancer
Tridymite	Fibrosis, silicosis
Cristobalite	Fibrosis, silicosis
Erionite	Lung cancer, mesothelioma, non-malignant fibrotic lung disease
Zeolites: mesolite, mordenite, natrolite, scolecite, thomsonite	Lung cancer
	Non-Silicate Minerals
Anatase	Chronic inflammation, lung tumors (rats), impaired pulmonary clearance
Apatite	Decrease in pulmonary function, hemolytic activity
Bauxite	Occupational pulmonary disability (Shaver's disease)
Fluorite	Bronchitis, silicosis, pulmonary lesions
Graphite	Graphite pneumoconiosis, focal emphysema, fibrosis and small fibrous nodules
Haematite	Siderosis
Siderite	Siderosis, "mottled" chest x-rays
	Rock Quarrying and Airborne Rock Dusts
Basalt	Pulmonary airway obstruction
Coal dust	Emphysema, coal-worker's pneumoconiosis
Dust storms	Desert lung syndrome
Pumice	Occupational silicosis, sclerosis of lymphatic glands, Liparitosis
Volcanic ash	Lymph node granuloma and delayed lung inflammation
	Urban, Rural, and Technogenic Particles
PM_{10}–$PM_{2.5}$, $PM_{2.5}$–$PM_{0.1}$	Asthma, pneumonia, bronchitis, heart failure
Nanoparticles	Asthma, pneumonia, bronchitis, heart failure
Fly ash	Asthma, pneumonia, bronchitis
Soot	Asthma, chronic bronchitis, radiological changes, skin cancer
Water-soluble component	Pulmonary inflammation, sudden cardiac death
Amorphous silicon dioxide	Inflammation, oedema, meta- and hyperplasia, fibrosis
Diatomaceous earth	Fibrosis, silicosis
Glass fiber/MMMF	Inflammation, occupational asthma, pleural plaques
Metals aerosols	Inflammation, lung cancer, metal fume fever

stacked on top of each other. Within the ortho- and ring silicates, a mineral of respiratory toxicological interest is synthetic (technogenic) mullite, a common refractory product (Carleton, Giere, and Lumpkin 2002). Particles of mullite are found in urban air, from sources such as coal-fired power stations (Giere, Carleton, and Lumpkin 2003). Mullite ($3Al_2O_3$ $2SiO_2$) occurs in most ceramic products containing alumina and silica. The mullite particles consist of needle-like interlocking crystals, growth of which is promoted by the presence of impurities. There have been concerns that workers in the ceramics industry exposed to mullite could suffer from irritation of the respiratory system (Fishman and Velichkovskii 2001; Fishman et al. 2001; Fishman 2003; Brown et al. 2005).

2.3 CHAIN SILICATES

The chain silicates are notable as they contain the carcinogenic minerals crocidolite and amosite. Chain silicates are a group of minerals with their tetrahedrons in single or multiple chains, with two oxygen atoms of each tetrahedron forming part of the adjoining tetrahedron. Amphiboles are a group of inosilicate minerals, containing hydroxyl (OH) groups, with double chains of aligned silicate tetrahedra. They exist in two different systems, orthorhombic (orthoamphibole) and monoclinic (clinoamphibole).

Crocidolite ($Na_2Fe_3^{+2}Fe_2^{+3}[Si_8O_{22}](OH)_2$) is commonly known as blue asbestos, and is the highly-fibrous form of riebeckite in the glaucophane-riebeckite group (Gibbons 2000). It has straight blue fibers that have a greater tensile strength than the most common asbestos chrysotile (sheet silicate), but less heat resistance. The fibers are prone to splintering when mechanically damaged. The size of the fibers plays an important role in asbestos-induced disease. The World Health Organization (WHO) defines fibers in the workplace as having a diameter of less than 3 μm and a length of more than 5 μm, with an aspect ratio of 1:3 or more; however, concerns have been raised about other issues such as biopersistence (Donaldson and Tran 2004). The fibers are believed to be hazardous to health since they are capable of entering and being deposited in the lungs, and have also a degree of migration capability within the lung (Stanton et al. 1981; Davis et al. 1986; Platek, Riley, and Simon 1992; Roller et al. 1996).

Amosite $(Fe^2+Mg)_7[Si_8O_{22}](OH)_2$ is the characteristic fibrous iron-rich form of cummingtonite–grunerite, and is commonly known as brown asbestos. It has straight, brittle fibers that are light gray to pale brown. It has good heat insulation properties and was commonly used in thermal systems insulation. Within the same series, the other asbestiform mineral of commercial importance is montasite, which is the softer and more magnesium-rich variety. Amosite is believed to have an equivalent health risk to crocidolite (Acheson et al. 1981; McDonald and McDonald 1996; Nolan, Langer, and Wilson 1999; Britton 2002).

Anthophyllite $Mg_7Si_8O_{22}(OH)_2$ is white to greenish-gray and is the magnesian end-member of the orthorhombic anthophyllite–gedrite series (Zeigler et al. 2002). Most anthophyllite crystals are prismatic or acicular. Fibers of anthophyllite are extremely flat and thin; the characteristic shape resembles that of a knife, and tends to be of generally uniform size. Tremolite and actinolite are asbestiform calcium amphiboles. Tremolite is a clinoamphibole and has the chemical formula, $Ca_2Mg_5Si_8O_{22}(OH)_2$. Actinolite $Ca_2(Mg,Fe)_5Si_8O_{22}(OH)_2$ has a similar composition, with iron replacing some magnesium. Wollastonite $CaSiO_3$ is a fairly common pyroxenoid inosilicate with industrial applications in ceramics and as filler. The mineral can be found as tabular crystals but is more commonly seen as lamellar radiating masses or fibrous aggregates. It is soluble in hydrochloric acid. Mechanical damage tends to result in elongate uneven splinters. All of the above fibrous minerals have recognized health issues [e.g., anthophyllite (Kiviluoto 1960; Dodson and Levin 2001; Dodson et al. 2005), tremolite (McConnell et al. 1983; Roggli et al. 2002; Luce et al. 2004), actinolite (Spurney et al. 1979; Metintas et al. 2005) and wollastonite (Huuskonen et al. 1983; Hanke et al. 1984; Wozniak et al. 1996; Tatrai et al. 2004; Maxim and McConnell 2005)].

2.4 SHEET SILICATES

The sheet silicates, or phyllosilicates, are composed of extending sheets of SiO_4 tetrahedra with the formula $(Si_2O_5)^{2-}$. Neighboring tetrahedra share the three oxygens of each tetrahedron. A second layer is formed from the apical oxygens that are attached to external ions. These external ions are in octahedral coordination. Different combinations of these two layers result in the different sheets silicates.

Chrysotile $Mg_3(Si_2O_5)(OH)_4$ is a generic term for undifferentiated, asbestiform, serpentine group species, consisting of a monoclinic mineral (clinochrysotile), and orthorhombic minerals (orthochrysotile and parachrysotile). Lizardite and antigorite are closely related serpentine group species (Kulagina and Pylev 1985). The most commonly used industrial asbestiform material, it consists of soft, silky white/green/yellow/gray bunches of flexible fibers. The fibers of chrysotile are formed by the crystalline sheet structure rolling into a scroll or coil form, giving a curly, tubular, hydrophilic fiber. This contrasts to the crystalline chain structure of the asbestiform amphiboles which exhibit a straight hydrophobic fiber. Health issues for the serpentines (chrysotile) are a well-researched issue (McDonald et al. 1993; Liddell, McDonald, and McDonald et al. 1997; Yano et al. 2001; Li et al. 2004).

Talc $Mg_3[Si_4O_{10}](OH)_2$ is a triclinic mineral composed of Si_2O_5 sheets with magnesium sandwiched between sheets in octahedral sites, and in tri-octahedral arrangement. Talc appears to be unable to form chemical-replacement series by accepting iron or aluminum into its structure. Rarely existing in an asbestiform habit, the major respiratory hazard posed by talc (Blejer and Arlon 1973; Wagner et al. 1977; Gibbs et al. 1992; Scancarello, Romeo, and Sartorelli 1996) is fine dust created by the milling to PM_{10} of soapstone and steatite. Commonly recognized as the primary ingredient in talcum powder, it is no longer recommended for dermal use (i.e., nappy powder), but it is still a significant industrial mineral. It is an important filler material for paints, rubber and insecticides (Di Lorenzo et al. 2003). In the food industry (Tomasini et al. 1988; Canessa et al. 1990), particularly in Asia, it is used as a mild abrasive in the polishing of cereal grains such as rice, potentially exposing workers handling the processed cereals to airborne talc (Breeling 1974).

Muscovite $KAl_3Si_3O_{10}(OH)_{1.8}F_{0.2}$ is the most common of the mica group minerals, and is typically found as massive crystalline "books" or in flaky grains. In the micaceous habit it has a platy texture with flexible plates. Muscovite has a sheet structure, of which basic units consist of two polymerized sheets of silica tetrahedrons positioned with the vertices of their tetrahedrons pointing toward each other and cross-linked with aluminum. Found in many different rock types, it is clear to milky-white with a pearly luster on cleavage faces, often with a sparkling appearance. Mica exposure is frequent in mines, mills, agriculture, construction, and industry and has been shown to induce pneumoconiosis (Landas and Schwartz 1991; Zinman et al. 2002). This health issue extends to people living around mica mines, as well as the workers in the associated mica-processing factories (Venter et al. 2004).

In the clay minerals group, the minerals associated with health concerns (Elmore 2003) include kaolinite, bentonite, palygorskite [attapulgite] (Huggins, Denny and Shell 1962; IARC 1980; Lipkin 1985), sepiolite (Baris, Sahin, and Erkan 1980) and vermiculite (Howard 2003). There is some debate on the actual health effects that can be attributed to the clay minerals themselves, since this group often has associated minerals with recognized adverse health effects. For example, kaolinite often has crystalline silica contamination, and vermiculite has both asbestos and crystalline silica as common contaminants.

Kaolinite $Al_2Si_2O_5(OH)_4$ is commonly known as kaolin or "China clay." The structure of kaolinite consists of tetrahedral silica sheets alternating with octahedral alumina sheets. The sheets are arranged so that the corners of the silica tetrahedrons form a common layer with the adjacent octahedral sheet. The charges within the structure are balanced, and analyses have shown that there is rarely substitution in the lattice. Visible crystals of kaolinite are extremely rare, with a typical grain size of 2–5 μm, with rare examples up to 1 mm across. When viewed under SEM, the pseudohexagonal crystals have a platy appearance: both fibers and spheres have been observed under Scanning Electron Mircroscopy (SEM). "China clay" typically contains other minerals as contaminants, in particular quartz and muscovite, the former of which has known respiratory hazards (Lapenas et al. 1984; Wagner et al. 1986; Morgan et al. 1988; Gao et al. 2000). This is a result of the typical genesis of kaolinite from the weathering of feldspar in granite, where two other common granite components are the quartz and mica. The occupational respiratory disease associated with kaolinite has been variously named as complicated pneumoconiosis, kaolin pneumoconiosis, or kaolinosis (Mossman and Craighead 1982). Anecdotal evidence suggests that kaolinite workers in the enclosed spaces of the processing and storage buildings are more prone to illness than the workers out in the quarries. The symptoms are similar to silicosis; however, postmortem shows a profusion of small opalicities, and peribronchiolar nodules transversed by fibrous bands (Oldham 1983; Lapenas et al. 1984; Sheers 1989; Rundle, Sugar, and Ogle 1993; Parsons et al. 2003).

Bentonite is the common or generic name for "Wyoming" bentonite, swelling (sodium) bentonite, and non-swelling (calcium) bentonite. Montmorillinite clay $((1/2Ca,Na)(Al,Mg,Fe)_4(Si,Al)_8O_{20}(OH)_4nH_2O)$, which is part of the smectite group of clay minerals, is the main constituent of bentonite (Gibbs and Pooley 1994). The clay is formed by the alteration of volcanic ash, and where it shows a high absorbency capacity is commonly known as "Fuller's Earth." It is usually commercially seen in a finely ground powdered form, but is occasionally available as coarse particles, and has uses as diverse as cat litter to oil-well drilling mud. The potential of bentonite to cause fibrogenicity and granulomas (Boros and Warren 1973) in the lung has been investigated. In *in vitro* studies, bentonite showed a high membrane-damaging (lysis) potential, shown as hemolytic activity in human erythrocytes (Geh et al. 2005). Human epidemiological studies reviewing workers chest x-rays showed 44% silicosis in bentonite workers (Phibbs, Sundin, and Mitchell 1971). Other studies report seven months to eight years exposures to bentonite resulting in pneumoconiosis (Rombala and Guardascione 1955).

Palygorskite $(Mg, Al)_2Si_4O_{10}(OH)4H_2O$, also known as attapulgite, occurs as a fibrous chain-structure mineral in clay deposits in hydrothermal deposits, soils, and along faults. It is of commercial importance for a range of uses, typically as an absorbent. The fiber characteristics vary with the source, but fiber lengths in commercial samples are generally less than 5 μm. It can form matted masses that resemble woven cloth. Unlike most other clay minerals, palygorskite can form large crystals. The results of studies in animals suggest that carcinogenicity is dependent on the proportion of long fibers (>5 μm) in a given dust sample (Jaurand et al. 1987; Rodelsperger et al. 1987; Meranger and Davey 1989; Renier et al. 1989). In an inhalation study in rats, in which about 20% of the fibers were longer than 6 μm, bronchoalveolar hyperplasia and benign and malignant alveolar tumors and mesotheliomas were observed. Intratracheal instillation studies with palygorskite fibers in sheep and rat lungs demonstrated significant and sustained inflammatory and fibrogenic changes (Begin et al. 1987; Lemaire et al. 1989).

Sepiolite $Mg_2H_2(SiO_3)_3XH_2O$ is a similar clay mineral to palygorskite; it occurs as a fibrous chain-structure mineral in clays, with major commercial deposits in Spain. Sepiolite fiber lengths in commercial (e.g., animal/pet litter) samples are generally less than 5 μm. Inhalation/instillation into rat lungs of short and long fibers from different geological locations (i.e., Spain [short], Finland [long], China [long]) suggested that long sepiolite fibers, with slow elimination rates, were important factors for their adverse biological (fibrosis) reaction (Bellman, Muhle, and Ernst 1997). Sepiolite appears to be strongly hemolytic in many classic assays, and may act as a cocarcinogen (Denizeau et al. 1985). Lung function has been shown to deteriorate rapidly in workers who are occupationally exposed to the commercial dust (McConnochie et al. 1993).

Vermiculite $Mg_{1.8}Fe^{2+}{}_{0.9}Al_{4.3}SiO_{10}(OH)_24(H_2O)$ is the name given to hydrated laminar magnesium–aluminum–iron silicate, a mineral that resembles mica. Vermiculite deposits contain a range of other minerals that were formed at the same time. Of particular concern are vermiculite deposits from some sources that have been found to contain amphibole asbestiform minerals (Van Gosen et al.

2002; Gunter 2004; McDonald, Harris, and Armstrong 2004; Pfau et al. 2005), such as tremolite and actinolite. When subjected to high temperatures, vermiculite has the unusual property of exfoliating or "popping" into worm-like pieces (Latin *vermiculare*: to breed worms). The process occurs as a result of the rapid conversion of contained water to steam that mechanically separates the layers. The exfoliation is the basis for the commercial and domestic use of the mineral, the increase in bulk typically grades from ×8 to ×12, but can reach as high as ×30. If the vermiculite contains asbestiform minerals, the exfoliation process releases fibers into the atmosphere. Vermiculite is used in the construction, agricultural, horticultural and industrial markets. The U.S. Environmental Protection Agency (EPA) has completed a study to evaluate the level of asbestos in domestic vermiculite attic insulation, and whether there is a risk to homeowners (USEPA 2005). There does not appear to be any indications that vermiculite itself is a respiratory hazard.

2.5 TECTO- (FRAMEWORK) SILICATES

Tectosilicates, formerly known as framework silicates, are minerals in which the silica tetrahedra share all four O^{2-} corners with adjacent tetrahedra. The result is a strong three-dimensional lattice. The large range of tectosilicates is a result of the partial substitution of Si by Al, balanced by cations such as K, Na, or Ca accommodated in the relatively open frameworks. The tectosilicates include many of the main rock-forming minerals, such as quartz and feldspars. Crystalline silica dioxide occurs in five different SiO_2 structures: quartz, cristobalite, coesite, tridymite, and stishovite. Three of the polymorphs: quartz, cristobalite, and tridymite have closely related temperature/pressure related crystallographic structures. Naturally-occurring coesite and stishovite are associated with meteorite impacts into silica dioxide-rich rocks, although coesite has also been found in kimberlite pipes in association with diamonds. From a human respiratory health perspective, only quartz, cristobalite, and tridymite are of importance (Fubini 1998; Occupational Safety & Health Administration [OSHA] 2005; Rimala, Greenbergac, and William 2005). The relative toxicities of quartz, cristobalite, and tridymite in the lung have been investigated by a number of workers, often generating different orders of toxicity. It is noteworthy that the OSHA permissible exposure limit values for cristobalite and tridymite are 0.05 mg/m^3, whereas quartz is 0.1 mg/m^3 (Castranova, Dalal, and Vallyathan 1997).

There are two dimorphs of quartz, α-quartz and β-quartz, with very similar structures. The determining factor for type of dimorph is temperature, with the boundary at 573°C. The structures are composed of networks of SiO_2 tetrahedra that are arranged in spiral chains (helices) around three- and six-fold screw axes. Alpha-quartz is trigonal and stable below 573°C. Above 573°C heat-induced agitation is sufficient to overcome a slight skewness in the structure and it converts to hexagonal β-quartz. The occupational development of acute silicosis and numerous different toxicological methods have shown that freshly-fractured silica dust is the most toxic (Castranova, Dalal, and Vallyathan 1997; Fubini 1998; Fubini and Hubbard, 2003; Rimala, Greenbergac, and William 2005). The consensus of opinion is that freshly cleaved crystal planes have surface properties that are more bioreactive with lung tissue, resulting in pulmonary disease. It is suggested that the silicon-based radicals (˙Si and Si–O˙) created on the surface are of importance (Vallyathan et al. 1995; Fubini et al. 2001). This has been supported by electron spin resonance (ESR) spectra showing high values for freshly ground silica, followed by decay in the signal with time after grinding (Fubini et al. 1990). When shattered, the fresh crystal surfaces of the SiO_2 show greater or lesser degrees, and thicknesses of structural disorder (Baumann 1979). The size of the SiO_2 particles is also critical, as smaller masses have greater surface areas. Baumann (1979) calculated that powdered crystalline SiO_2 with a surface area of 11.8 m^2/gm had a perturbed surface of approximately 9%. In addition to the surface disorder (Altree-Williams et al. 1981), there is an internal component acting as potential boundaries between crystallites, potentially providing planes of weakness when the SiO_2 is powdered.

The monoclinic polymorph α-tridymite is stable at temperatures below 870°C (Smith 1998). At temperatures between 870°C and 1470°C it is hexagonal α-tridymite. Tridymite is only metastable at normal surface temperatures tending to alter, over thousands of years, to quartz. It retains its original, overall crystal morphology; thus much "tridymite" is really quartz pseudomorphs after tridymite. Tridymite was once considered to be rare, but it is now known that different volcanic rocks (e.g., in California, Colorado, and Mexico) can contain small to microscopic crystals of tridymite (USGS 2005). Nevertheless, it is still much less common than quartz or cristobalite and occupational exposures to tridymite could be expected in workers mining or processing powdered igneous rocks. Tridymite (and cristobalite) is produced in some industrial operations when alpha quartz or amorphous silica is heated (such as foundry processes, calcining of diatomaceous earth, brick and ceramics manufacturing, and silicon carbide production (NIOSH 1974; Weill, Jones, and Parkes 1994). Burning of agricultural waste or products such as rice hulls may also cause amorphous silica to become tridymite and cristobalite (Rabovsky 1995). For example, in Asian countries, the burning of rice husk ash (RHA), as a means to cover the demands for energy and silica resource (i.e., cement industry, lightweight construction products, abrasives, and absorbents), has been shown to generate cristobalite and tridymite (Shinohara and Kohyama 2004). The silica content in airborne dust for workers in RHA production factories, in power generation plants using rice hull, and engaged in farming operations that include the burning of rice husk, are now being controlled in the workplace due to the occurrence of pneumoconiosis in workers engaged in the packing and screening of RHA products (Liu, Liu, and Li 1996).

Cristobalite is the low pressure, high temperature SiO_2 polymorph, and is relatively abundant in volcanic rocks (Baxter et al. 1999). Cristobalite is only metastable at surface temperatures; the conversion to quartz is believed to occur exceedingly slowly. It is the presence of cristobalite in volcanic rocks that has prompted many of the concerns about the possible respiratory toxicity of volcanic ash (Baxter 1999; Baxter, Bernstein, and Buist 1986; Baxter et al. 1999). This is a subject of great interest, particularly after the May 18, 1980, eruption of Mount St. Helens (Baxter et al. 1981; Beck, Brain, and Bohannon 1981). There are also concerns for workers engaged in certain industries that can produce cristobalite as a by-product, such as glass manufacture (NIOSH 2002). Most cristobalite is believed to crystallize in a high temperature phase called β-cristobalite, with an isometric symmetry. It later cools, and the crystals convert to α-cristobalite. Alpha-cristobalite has an octahedron crystal form; however, when converted to α-cristobalite, the crystals retain the outward β-cristobalite form. As with quartz, it is the freshly-fractured faces of cristobalite that are of most respiratory concern.

Erionite is a fibrous zeolite (Gottardi and Galli 1985), with the approximate formula $(K_2, Na_2, Ca)_2Al_4Si_{14}O_{36}14H_2O$. It is a hydrated potassium sodium calcium aluminum silicate. Erionite forms wool-like, fibrous masses in the hollows of rhyolitic tuffs and in basalts. Approximately 40 natural zeolites have been identified, and they are noted for their very open crystalline lattices, with large internal surface areas. They are able to lose or gain water molecules, and exchange cations without major structural changes; this ability has led to many industrial applications, including use as "molecular sieves." Erionite is the most carcinogenic mineral fiber documented in man and in rodent inhalation studies. There is scientific consensus about the adverse effects of erionite in DNA strand breaks (Eborn and Aust 1995) or mesothelioma induction (Wagner et al. 1985). The considerable toxicity of erionite may be due to its fibrous nature and size, which ensures penetration into the lungs, and its surface chemistry, which promotes the formation of hydroxyl radicals (Hansen and Mossman 1987; Mossman and Sesko 1990; Fubini and Mollo 1995; Fubini, Mollo, and Giamello 1995; Fach et al. 2002).

Epidemiological studies in populated regions with high levels of naturally occurring erionite, (i.e., the Anatolian region of Turkey), have linked environmental exposure to erionite fibers with the development of malignant mesothelioma and non-malignant fibrotic lung disease (Baris et al. 1987; Artvinli and Baris 1979). The spectrum of cancers, fibrosis, and other pulmonary abnormalities associated with exposure to erionite is markedly similar to the range of health effects described for occupational exposure to the amphibole asbestos, crocidolite, but the incidence of disease is increased dramatically. Human epidemiological studies have shown very high mortality from malignant mesothelioma in particular Turkish villages. This is an area where erionite is found in the local volcanic tuffs and the locals live in rock-built houses and caves (Lilis 1981; Artvinli and Baris 1982; Maltoni, Minardi, and Morisi 1982; Emri et al. 2002; Emri and Demir 2004). There is significant local airborne contamination from erionite, and the locals are exposed to the fibers from birth. Postmortem investigations found erionite fibers in lung tissue samples from cases of pleural mesothelioma. The inhabitants in contaminated villages had higher levels of ferruginous bodies than inhabitants of control villages (Dumortier et al. 2001). Several other natural zeolites, in addition to erionite, also can have a fibrous habit. Of particular health concern (i.e., "biologically active") are mesolite, $Na_2Ca_2Al_6Si_9O_{30}8H_2O$; mordenite, $(Ca,Na_2,K_2)Al_2Si_{10}O_{24}7(H_2O)$; natrolite, $Na_2Al_2Si_3O_{10}2H_2O$; paranatrolite, $Na_2Al_2Si_3O_{10}3H_2O$; tetranatrolite, $Na_2Al_2Si_3O_{10}2H_2O$; scolecite, $CaAl_2Si_3O_{10}3H_2O$: and thomsonite, $NaCa_2Al_5Si_5O_{20}6H_2O$ (Wright, Rom, and Moatmed 1983; Gottardi and Galli 1985; Fach et al. 2002).

2.6 NON SILICATES

The minerals anatase, rutile, and brookite are all naturally-occurring polymorphs with the same chemistry, TiO_2, but they have different structures. Anatase shares properties such as luster, hardness, and density with the other two polymorphs. It has a tetragonal symmetry, with the structure based on octahedrons of titanium oxide sharing four edges to produce a four-fold axis structure. Naturally-occurring anatase can be associated with quartz and is relatively rare in nature, occurring in cavities in schists, gneisses, granites, and other igneous rocks.

Titanium dioxide (TiO_2) is manufactured worldwide in large quantities for use in a wide range of applications. It is most widely used as a white pigment. This is due to its high refractive index and reflectance combined with its ease of dispersion in a variety of media and non-reactivity towards those media during processing and throughout product life. The two main processes for making TiO_2 pigments are the sulfate process and the chloride process. The sulfate process was the first to be developed on a commercial scale in Europe and the U.S., around 1930. It was the primary process until the early 1950s, when the chloride process was researched and developed. Currently, the chloride process accounts for ~60% of the world's TiO_2 pigment production. Pure TiO_2 is extracted from its mineral feedstock by reaction with either sulfuric acid or chlorine, and then it is milled and treated to produce a range of products that are designed for specific end uses. The majority of TiO_2 products are based on the crystal- "type rutile, with a primary particle size range of 200–300 nm. In this context it is called "pigment grade." At this particle size, TiO_2 pigments offer maximum opacity, as well as impart whiteness and brightness to the paints, coatings, papers, and plastic products in which they are used. These TiO_2 pigments are also used in many white or colored products including foods, pharmaceuticals, cosmetics, ceramics, fibers, and rubber products, to mention only a few.

One of TiO_2 properties is its efficient absorption of ultraviolet light which makes it a very effective sunscreen for use in cosmetics. Usually its opacity is not required in this application, so very low particle size material (size 10–20 nm) is used and this is commonly called "ultrafine" or UF. The final product is of a particle size that could become airborne and inhaled. Titanium dioxide is highly insoluble, non-reactive with other materials, thermally stable, and non-flammable, which has led to it being considered to pose little risk to respiratory health. This is supported by the toxicological database on TiO_2 and the fact that it has been used traditionally for many years as a "negative control" dust in many *in vitro* and *in vivo* toxicological investigations. However, this view was challenged when lung tumors were found in the lungs of rats after lifetime exposure to very high concentrations of pigment grade TiO_2 (Lee, Trochimowicz, and Reinhardt 1985) and ultrafine TiO_2 (Bermudez et al. 2004; Hext, Tomenson, and Thompson 2005). In contrast, no tumors were seen in similarly exposed mice and hamsters (Muhle et al. 1989; Warheit et al. 1997; Bermudez et al. 2002). These apparent species differences suggested that the experimentally-induced lung tumors were a rat-specific, threshold phenomenon, dependent upon lung overloading and accompanied by chronic inflammation to exert the observed tumorigenic response. The relevance of this phenomenon to human exposures remains questionable but, to date, epidemiological studies conducted do not suggest a carcinogenic effect of TiO_2 dust on the human lung (Hext, Tomenson, and Thompson 2005).

Apatite $Ca_5(PO_4)_3(F,OH,Cl)$ is a group of hexagonal minerals, usually subdivided into the three minerals; fluorapatite, chlorapatite, and hydroxylapatite. The fundamental unit of the apatites is the tetrahedral $(PO_4)^{3-}$ anionic group. The three types may partially replace each other, and it is hard to distinguish between them; therefore they are usually simply called "apatite." The most common of the three, by far, is fluorapatite. Apatite is the most common phosphate mineral, and is essential in the manufacture of phosphate-based fertilizers, and is important in the chemical and pharmaceutical industries. Concerns have been raised about the occupational respiratory hazards of phosphates (Sebastien et al. 1983) and apatite (Mikulski et al. 1994a). Epidemiological studies have been undertaken on the respiratory function of smoking and non-smoking workers exposed to apatite dust. It was concluded that occupational exposure to phosphorite and apatite dusts causes a decrease in pulmonary function in non-smoking workers (Mikulski et al. 1994b).

Aluminum ore, called bauxite, is usually formed in deeply weathered volcanic rocks, such as basalt. Bauxite is a heterogeneous material, mostly composed of several aluminum hydroxide minerals, plus varying amounts of silica, iron oxide, aluminosilicate, and other minor or trace minerals. The principal aluminum hydroxide minerals, $Al(OH)_3$ in bauxites are gibbsite and its polymorphs boehmite and diaspore. In gibbsite, the fundamental structure is a layer of aluminum ions, sandwiched between two sheets of tightly-packed hydroxy ions; only two of the three octahedrally-coordinated sites are occupied by cations. Aluminum ore dust, or bauxite dust inhalation, can result in "Shaver's Disease," corundum smelter's lung, bauxite lung, or bauxite smelters' disease (Dinman 1988; Radon et al. 1999; Kraus et al. 2000). This occupational disease in bauxite miners is a progressive form of pneumoconiosis caused by exposure to bauxite fumes which contain aluminum and silica particulates. Initially, the disease appears as alveolitis and then progresses to emphysema. Patients may develop pneumothorax (collapsed lung). It is typically seen in workers involved in the smelting of bauxite to produce corundum (crystalline form of aluminum oxide and one of the rock-forming minerals). Due to corundum's hardness, it is commonly used as an abrasive in machining, from huge machines to sandpaper. Emery is an impure and less abrasive variety.

The mineral calcium fluoride or fluorite, CaF_2, occurs as cubic crystals and cleavable masses. The calcium ions are arranged in a cubic lattice, with the fluorine ions at the center of smaller cubes derived by dividing the unit cell into eight cubes. When pure, it is colorless, however, impurities cause color, and some varieties fluoresce. It is common in limestone and dolomites, and has multiple uses in the fiberglass, ceramic, and glass industries. The respiratory hazards of fluorite are related to the fluorine and silica content. Acute inhalation has been linked to gastric, intestinal, circulatory, and nervous system problems. Chronic inhalation or ingestion can result in weight and appetite loss, anemia, and bone and teeth defects. The symptoms are therefore like those of fluorinosis, a disease associated with the ingestion of fluorine, usually in water or from contaminated land. It has been suggested that fluorite promotes silica fibrogenicity (pneumoconiosis) in the lungs (Xu et al. 1987). Fluorite miners have reported bronchitis, silicosis and pulmonary lesions (Xu et al. 1987).

The mineral graphite is a crystalline form of elemental carbon. Each carbon atom is covalently bonded, with bond length 1.42 Å, to three others in the same plane, with a bond angle of 120°. The carbon atoms thus form linked six-member rings to form flat or, occasionally, buckled planes. The sheets are usually stacked in an ABAB array, or less commonly in an ABCABC array. In comparison to the covalent bonds (1.42 Å) forming the sheets, the sheets are widely separated at a length of 3.35 Å. The space between the sheets is known as the "van der Waals gap," due to the weak van der Waals forces attracting them together. Inhalation of dust containing graphite can cause lung disease in foundry workers and workers in graphite mines or mills. Mixed dust pneumoconiosis caused by long-term occupational exposure to graphite dust is a rare disease. Only a few cases of "graphite pneumoconiosis" have been reported in literature, and these were usually diagnosed postmortem (Domej et al. 2002). The characteristic symptom of graphite pneumoconiosis is non-asbestos ferruginous bodies based on a black graphite core (Mazzucchelli, Radelfinger, and Kraft 1996).

Siderite ($FeCO_3$) and hematite (Fe_2O_3) are both commercially-important iron ores. Siderite is rarely found in the pure $FeCO_3$ form, as it commonly substitutes Fe^{2+} with elements such as Mn or Mg. Siderite is slowly soluble in dilute HCl, leaving an iron oxide. When found as a common massive ore, hematite is known as "red hematite." Hematite has a structure consisting of layers of oxygen and iron ions perpendicular to the triad axis. As with siderite, hematite (Mossman and Craighead 1982) is rarely found in a pure form, with small amounts of MnO and FeO commonly present.

Siderosis is a "silicosis-like" occupational lung disease caused by exposure to iron carbonates and iron oxides. It is prevalent in hematite and other iron-ore miners (Chen et al. 1989, 1990) and workers in the iron and steel industries. Siderosis has similar diagnostic symptoms to simple pneumoconiosis, with the exception of a striking "mottled" appearance on chest x-rays. Post-mortem examination of siderosis patients is characterized by the deep brick-red staining of the lung tissue.

2.7 ROCK QUARRYING AND AIRBORNE ROCK DUSTS

Basalt is a hard, dark igneous (volcanic) rock with less than about 52 wt.% silica (SiO_2). It is the most common rock type in the earth's crust. Huge areas of lava called "flood basalts" are found on many continents, such as the Deccan Traps in India. Given its abundance, basalt quarries are commonplace worldwide, supplying rock for road surfacing, construction, and some industries. The common minerals in basalt include olivine $(Mg, Fe^{2+})_2SiO_4$, pyroxene [Ca, Na, Fe^{2+}, Mg, and others][Ch, Al, Fe^{3+}, Mn, and others]$(Si, Al)_2O_6$, and plagioclase $(Na,Ca)Al_{1\text{-}2}Si_{3\text{-}2}O_8$, none of which is individually known to have respiratory toxicities. There are two respiratory health concerns associated with basalt. Firstly, finely-powdered quarried basalt for use in the ceramics industry, where an epidemiology study in India has shown 27% of the plant workers suffering from "basalt pneumoconiosis" (El Ghawabi et al. 1985). Secondly, a fibrous product spun out of molten basalt, basalt thin fiber (BTF) wool. Pure basalt wool is no longer crystalline basalt, but is actually non-crystalline glass with the same bulk chemistry as the original basalt. There are concerns about the similarity of the dimensions of basalt wool fibers to known carcinogens, such as asbestos, and the possibility that basalt wool could itself be a carcinogen (Adamis et al. 2001).

Coal dust consists of two main components: the organic particles and inorganic mineral grains. The coal itself is the organic component, and the different types of coal, depending upon the burial history and plant precursors, are classified as "macerals." For example, woody tissue-derived coal particles from a bituminous-rank coal would consist of the macerals "vitrinite." The rank of a coal relates to its burial history, which in turn determines the physicochemistry of the macerals. Low-rank coals are "lignites" or "brown coals"; medium-ranked coals are the "subbituminous" or "bituminous" coals; and high-ranked coals the "anthracites." During the "coalification" process during burial, as a result of temperature and pressure, the organic macerals become enriched in carbon, with the organic molecules becoming more cross-linked and refractory. The inorganic component of coal dust includes all the rock/mineral dusts that are associated with the coal and both mineral grains that were deposited at the same time as the coal, as well as minerals that formed in the coal subsequent to burial. Iron pyrite, FeS_2, is a common example of the latter type, and is the mineral that is mainly responsible for the emission of SO_2 from the burning of poor-quality coals.

The types and differing proportions of the different minerals to be found in "coal dust" thus relates to the geology of the coal mine or opencast (Figure 2.2a). Coal seams are found in sedimentary rocks, therefore are typically associated with sandstones, siltstones and shales (Figure 2.2b–Figure 2.2d). Some of the minerals found in these rock types, such as quartz and clay minerals, have recognized respiratory health effects. Therefore, when considering the possible toxic effects of "coal dust," we are in reality considering the toxicity of the mineral grains found in association with the actual coal (Jones et al. 2002). Occupational exposure to coal dust is a recognized cause of respiratory diseases such as emphysema and coal-worker's pneumoconiosis and is directly related to total exposure (Castronova and Vallyathan 2000). The highest incidence of respiratory disease is in underground coal-face miners. In general, anthracite coal mining has been associated with higher rates of pneumoconiosis than that found in bituminous miners (Ortmeyer, Baier, and Crawford 1973; Bennett et al. 1979). Anthracite coal mine dust contains more surface free radicals than

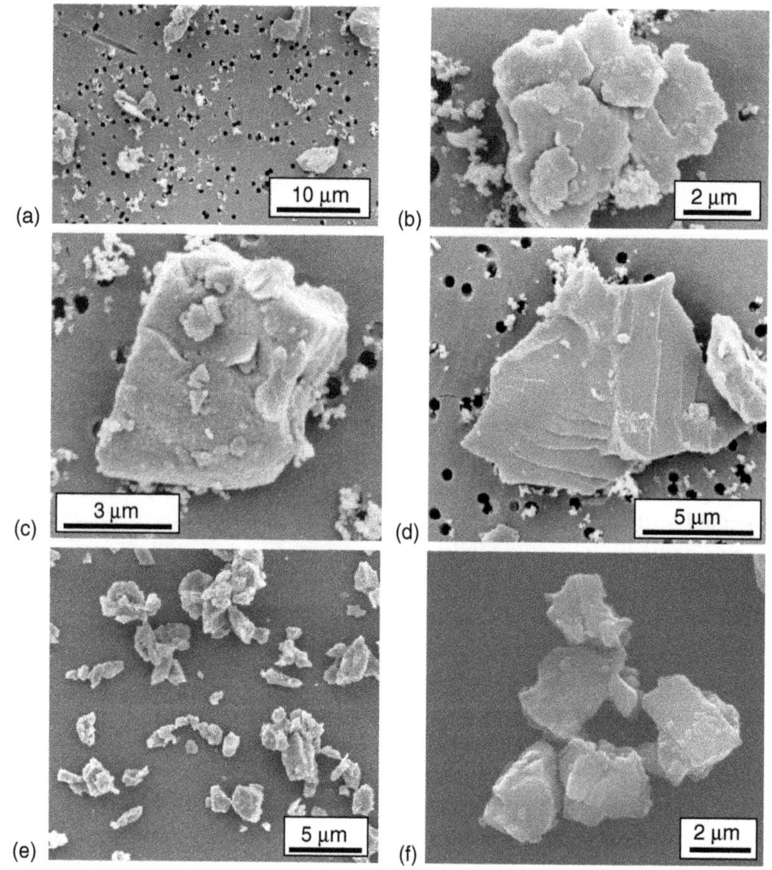

FIGURE 2.2 (a) Airborne mineral and soot particles from a coal open-cast pit collected on a polycarbonate filter. The minerals make up the bulk of the mass of the collection, however, the numbers of particles are dominated by soot particles. (b) Airborne clay mineral grain from a coal open-cast pit. (c) Airborne mineral grain from a coal open-cast coal pit. The grain shows poor cubic morphology. (d) Airborne mineral grain of microcrystalline SiO_2 from a coal open-cast coal pit. The grain shows conchoidal fracturing. (e) Airborne respirable volcanic ash particles from the Soufriere Hills volcano on the Island of Montserrat. The mineralogy cannot be determine from particle morphology, since no crystallinity can be seen. (f) Detail of respirable volcanic ash particles, the Soufriere Hills volcano, Montserrat.

bituminous coal, which may explain its higher cytotoxicity and pathogenicity (Dalal et al. 1990; Dalal et al. 1995). In addition, anthracite can have higher crystalline silica content than bituminous coal (Dalal et al. 1991). However, experimental evidence suggests that silica particles from bituminous mines may be coated with clay, rendering them less active (Wallace et al. 1994). Respirable coal mine dust has a relatively large surface area due to its small aerodynamic size and porous nature. Organic aromatic compounds present in the coal mine atmosphere, such as benzene, methylene, phenol, and phenanthrene, can be adsorbed onto the surface of coal mine dust and may affect its biologic activity.

Dust storms occur seasonally in many parts of the world, with the dust sourced from desert surfaces of loose dry sediment, that are typically receiving less than 250 mm annual rainfall and have little or no vegetation. The exposed surfaces can occur naturally, or as a result of poor agricultural practices. The "Dust Bowl" of the mid-West United States is a good example of the latter. The mineralogical makeup of dust depends on its source. Usually the source deposits consist of common minerals such as quartz, with associated clays and evaporites (salts), and the generated airborne dust reflects that composition. Silt and clay particles are commonly re-suspended from desert surfaces by wind. The larger particles (0.02 mm or larger) tend to remain suspended only minutes to hours, traveling at most a few hundred kilometers. The smaller particles can remain suspended in the atmosphere for weeks, traveling thousands of kilometers. Size fractionation of the different minerals, therefore, plays a part in the bulk mineralogy of the storms, with the smaller clay minerals an important component of the finer dusts. The respiratory health of desert inhabitants of the Saharan, Libyan, Negev, and Arabian deserts has been investigated (Nouh 1989; Hiyoshi et al. 2005). Long-term exposures can result in the development of a benign, non-progressive pneumoconiosis referred to as "Desert Lung Syndrome" (Nouh 1989). This condition is asymptomatic and does not appear to worsen with time. Researchers have suggested that the condition is benign because of the "age" of the mineral grains. Weathered dust particles have less reactive surfaces, whereas freshly fractured surfaces are more biologically reactive. A recent concern is the adverse health effects of a mixture of mineral dusts and urban PM_{10}, particularly in areas that experience both regular dust storms and high levels of man-made pollution. WHO estimates in Asia that outdoor air pollution causes more than 500,000 premature deaths a year.

Wind erosion in arid and semi-arid areas of middle and northwestern China forms the Asian Sand Dust (ASD) aerosol (Hiyoshi et al. 2005). This ASD spreads over large areas, including East China, the Korean Peninsula, and Japan. Sometimes the aerosol is transported across the Pacific Ocean to the United States (Duce et al. 1980; Husar et al. 2001; Kim, Han, and Park 2001). The sand dust aerosol originates in the sandstorms occurring in the Gobi Desert and the Ocher Plateau in spring. Both the daily observations and atmospheric concentrations of the dust aerosol have been increasing steadily in the eastern Asia region in recent years (Zhuang et al. 2001; Mori et al. 2003). ASD contains various chemical species such as sulfate or nitrate derived from alkaline soil which captures acid gases, such as sulfur oxides and nitrogen oxides (Choi et al. 2001). These gases are byproducts formed from coal and other fossil fuels combusted in industrialized eastern China. Recent epidemiologic studies have shown that ASD events are associated with an increase in daily mortality in Seoul, Korea (Kwon et al. 2002) and Taipei, Taiwan (Chen et al. 2004). ASD has also caused cardiovascular and respiratory problems in Seoul, Korea (Kwon et al. 2002).

Pumice is vesicular ejecta of feldspar-rich rhyolite lava that is poor in iron and magnesium and rich in silica. It is non-crystalline volcanic glass, with a general formula; $M_xO_y + SiO_2$, where $M = Al$, Ca, Mg and other metals with bound silica (SiO_2). Pumice has a low density and it can float on water, and is produced during eruptions of some stratovolcanoes, (e.g., Mount St. Helens and Montserrat). It occurs principally in Ethiopia, Germany, Hungary, Italy (Sicily, Lipari), Madagascar, Spain, and the United States. Some varieties, such as Lipari pumice, have a high content of total silica (71.2%–73.7%) and a fair amount of free silica (1.2%–5%).

Pumice tends to be soft, and has been used for building stone. It is also a mild abrasive, with some commercial uses and a household use in the removal of calloused skin from feet. Medical concerns are centered on the major component of amorphous silicon dioxide. When used commercially, pumice is classified as a nuisance dust, with possible aggravation of pre-existing upper respiratory problems and lung disease. Pumice stone workers or miners are liable to occupational silicosis and sclerosis of the lymphatic glands. Extreme, non-occupational, exposures can cause lungs to be vulnerable to pneumoconiosis. Apart from the characteristic signs of silicosis observed in the lungs and sclerosis of the hilar lymphatic glands, the study of some fatal cases has revealed damage to various sections of the pulmonary arterial tree. Clinical examination has revealed respiratory disorders (emphysema and sometimes pleural damage), cardiovascular disorders (cor pulmonale) and renal disorders (albuminuria, haematuria, cylindruria), as well as signs of adrenal deficiency.

The description of pneumoconiosis due to amorphous silica is rare. One of these, named "liparitosis," is related to the inhalation of pumice powder extracted in the island of Lipari (Aeolian Archipelago, Sicily). Despite its low incidence due to localized exposure, liparitosis deserves a certain interest, as it can be considered representative of pneumoconiosis derived by amorphous silica compounds, including diatomite and artificial amorphous silica, the industrial manufacturing of which is extremely widespread. Liparitosis is characterized by a chronic evolution of 20–30 years. Clinically, it is almost silent, vaguely simulating a catarrhal bronchitis. From a radiological standpoint, it is described as the progression a fine reticulation to a later stage, characterized by mass-like fibrosis in the basal lung (Castronovo 1953; Mazziotti et al. 2004).

Volcanic ash is the fine airborne solid component of volcanic eruptions (Figure 2.2e and f). It is a material that can be produced in prodigious quantities, as large volcanic events can carpet vast areas in dust many meters thick. The respiratory dangers posed by volcanic dust include both during the eruption itself, as well as the ash "clean-up," sometimes long after the actual eruption has ceased. The nature of the hazard thus changes from an environmental exposure during the eruption, to an occupational exposure for those workers engaged in such tasks as sweeping ashfall off roofs. The composition, mineralogy, and structure of volcanic ash are determined by a very large range of factors, including the tectonic setting of the volcano and the nature of the eruption. Single volcanic events are capable of producing many different types of ash over the duration of the eruption. Most ash consists of a crystalline and an amorphous component. Concerns over the possible adverse respiratory effects of volcanic ash, have now focused on the crystalline SiO_2 content, a well-established respiratory hazard. Consequently, the eruption of the Soufrière Hills volcano on Montserrat has been extensively studied (Wilson et al. 2000; Horwell et al. 2001; Housely et al. 2002; Searl, Nicholl, and Baxter 2002; Horwell et al. 2003; BéruBé et al. 2004; Forbes et al. 2004; Lee and Richards 2004), and the varying percentage of cristobalite in the different ashes produced by this volcano established. The ash that contains the highest percentages of cristobalite is the "dome-collapse pyroclastic flow" ash. A volcanic dome forms when the magma moves up the volcanic conduit to the surface, and, relatively slowly, is extruded though a vent to form a pile or "dome." This dome is still very hot, typically around 800°C, and at surface pressures. With these high temperature and low pressure conditions, the SiO_2 polymorph cristobalite forms tiny crystals in open spaces (vugs) in the rock. Eventually the dome reaches an unstable size and collapses down the side of the volcano. This fast and dangerous collapse is called a "pyroclastic flow." The violent energy of the pyroclastic flow pulverizes the dome material into small fragments, releasing much of the cristobalite as fine airborne ash.

2.8 URBAN, RURAL, AND TECHNOGENIC PARTICLES

PM_{10} is defined as airborne particulate matter with a mean aerodynamic diameter of 10 microns or less; likewise $PM_{2.5}$ has a mean aerodynamic diameter of 2.5 μ or less. The PM_{10}–$PM_{2.5}$ is known as the "coarse" fraction, and the $PM_{2.5}$–$PM_{0.1}$ as the "fine" fraction and $PM_{0.1}$ is the "ultrafine" fraction. PM_{10} is highly heterogeneous, and as such, it is futile to try and define the mineralogy or structure. However, broadly speaking, the composition of PM_{10} is controlled by factors such as weather, continental-scale influences, and regional and local influences. In urban and industrial areas, PM_{10} is dominated by road transport, industrial, and construction particles, whereas in rural areas, it is occasionally possible to recognize regional mineralogical signatures. Moreno et al. (2003) collected and analyzed crustally-derived (soil) particles from the Lizard Peninsula in Cornwall. The geology of the Lizard is unusual in the U.K. in that the serpentinite rocks contain minerals that are Mg- and Fe-rich, and these have altered in the soil to secondary minerals such as serpentine, talc, vermiculite, and tremolite. By comparison, with silicate minerals collected from London's air, a clear difference was recognized, and this enabled the "finger-printing" of local crustal PM_{10}.

Nanoparticles have been broadly defined as microscopic particles with dimensions less than 100 nm (Figure 2.2a, Figure 2.3b–Figure2.3d), however, "engineered" nanoparticles are more precisely defined as having one dimension less than 100 nm. Many nanoparticles are prone to rapid agglomeration, forming larger "particles" with dimensions much greater than 100 nm. These are termed "nanostructured particles," as long as their activity is governed by their nanoparticle components. For example, an agglomerate of TiO_2 nanoparticles forming a single "nanostructured particle" much larger than 100 nm in diameter, has significantly greater biological activity than a single crystal TiO_2 particle of the same diameter. With the development of nanotechnology, size effects of particles have gradually been considered to be important. Nanoparticles may be more toxic than larger particles of the same substance (Oberdörster et al. 2005a) because of their larger surface area, enhanced chemical reactivity, and easier penetration of cells. Nevertheless, several studies have shown that the cytotoxicity of nanosized TiO_2 was very low or negligible, as compared with other nanoparticles (Zhang et al. 1998; Peters et al. 2004; Yamamoto et al. 2004; Oberdörster, Oberdörster, Oberdörster 2005b), and the size was not the effective factor of cytotoxicity (Yamamoto et al. 2004). Probably the two nanoparticles of greatest interest to respiratory toxicologists are those composed of carbon (soot or carbon black) and TiO_2 (Humble et al. 2003). As previously described, nanoparticles (20 nm diameter) of TiO_2 will produce a persistently high inflammatory reaction in rat lungs, when compared with larger TiO_2 (250 nm diameter) particles (Oberdörster, Ferin, and Lehnert 1994; Oberdörster et al. 2005b).

Several methods, such as metallorganic and chemical vapor deposition, have been devised to produce TiO_2 nanoparticles of reproducible size. Given the bioreactivity of titanium, the expectation is that the surface of TiO_2 nanoparticles should be dominantly composed of oxygen atoms, but this is not supported by high-resolution transmission electron microscopy and the nature of the Fresnel fringe on the surface of the nanoparticles (Jefferson and Tilley 1999). The Fresnel fringe reflects the potential "drop off" at the crystal surface, and studies have indicated that nanoparticle surfaces contain titanium as well as oxygen. The implication of this is that the surface titanium has a distorted fivefold coordination, with resulting higher bioreactivity (Jefferson and Tilley 1999).

Fly ash is a generic term for particulate matter produced during combustion that primarily originates from mineral and metal contaminants in the organic fuels (Figure 2.3e–Figure2.3h). Some confusion exists between different scientific fields as to the definition of fly ash. This is a result of people describing materials as the end-product of combustion processes, rather than based on the composition of the materials themselves. For example, some archaeologists might include angular carbonaceous particles in their definition of fly ash, with this material actually representing charcoal or coke. The two main types of fly ash of interest to respiratory toxicologists are Residual Oil Fly Ash (ROFA) and solid fuel-combustion fly ash; these two materials are quite different. ROFA is a highly complex material containing transition metals, sulfates, and acids, incorporated with a resilient, particulate carbonaceous core (Ghio et al. 2002). Of particular interest and concern is the solubility, and therefore, bioavailability, of transition metals associated with these particles. The particulate carbonaceous core is effectively a soot particle, as described later in this section. Adverse respiratory health effects in humans due to occupational exposure to ROFA have been recorded (Hauser et al. 1995; Costa and Dreher 1997; Dreher et al. 1997; Kodavanti et al. 1997a; Kodavanti et al. 1997b; Kodavanti et al. 2000; Lewis et al. 2003; Antonini et al. 2004;

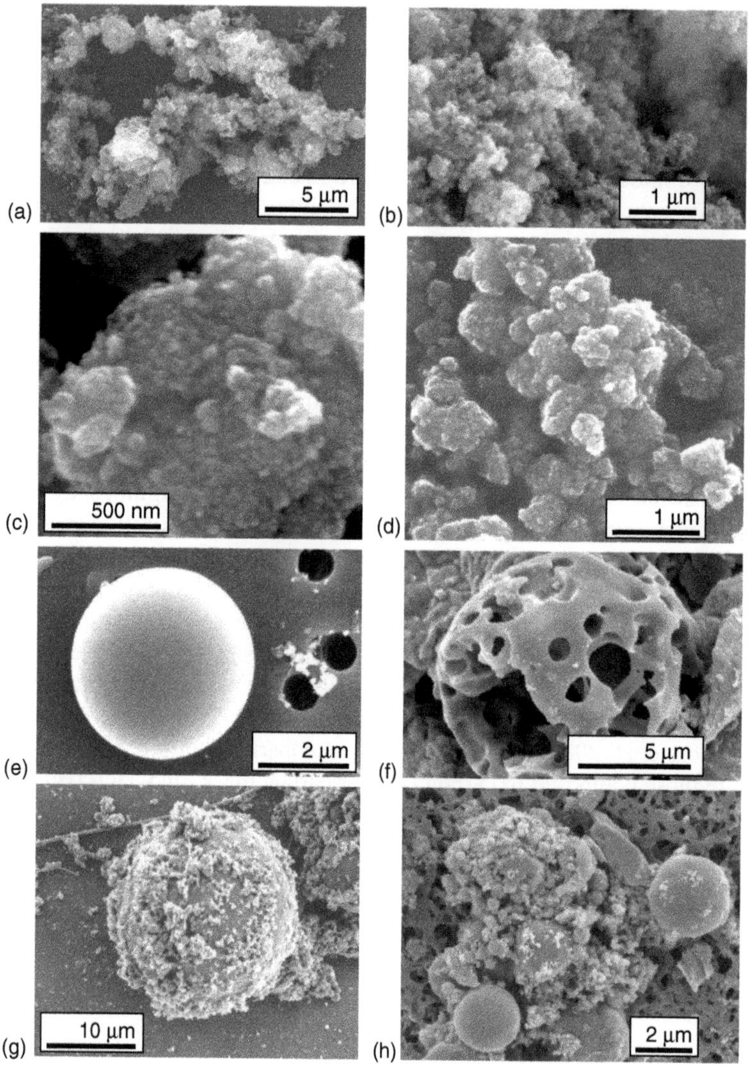

FIGURE 2.3 (a) Nickel oxide nanoparticles aggregated in nanostructured nickel oxide particles. (b) Silicon dioxide nanoparticles aggregated in nanostructured silicon dioxide particles. (c) Surface detail of nanostructured titanium dioxide particle. (d) Titanium dioxide nanoparticles aggregated in nanostructured titanium oxide particles. (e) Airborne iron microsphere collected in Port Talbot, South Wales. These are abundant in the local air, and with a mean diameter of 2 μm are almost all respirable. (f) A slightly damaged fly ash particle. This was collected in London during the 1950s, and was probably generated by the combustion of poor-quality coal for power and domestic use. (g) A sphere of fly ash that has been heavily coated in soot. (h) A composite particle consisting of metallic spheres, mineral grains, and soot.

Gardner et al. 2004; Roberts et al. 2004). ROFA is commonly used in studies evaluating the pulmonary responses to particulate matter exposure, partly due to the ease with which large homogeneous sample masses can be obtained from oil-fired power stations.

Fly ash produced by solid-fuel combustion is mainly sourced from mineral or metal contaminants in the fuel. Fly ash can thus be generated by the burning of coal, other fuel types (even municipal and commercial waste), and as a by-product of combustion-based industrial processes. Even though these combustion-using facilities usually extract the vast majority of airborne particulates in the post-combustion air-stream by bag rooms or other particle-collecting systems, fly ash is still a common airborne component, especially in industrial towns. Fly ash is composed of predominantly spherical or sub-spherical glassy particles, and thus represents solidified material that is 60%–90% amorphous glass, created from the melting of the mineral components. Most fly ash is found in the PM_{10}–$PM_{2.5}$ coarse fraction. The spheres are either solid or hollow, and often have characteristic morphologies riddled with degassing holes (Figure 2.3f). Broken fragments of the fly ash spheres are common in airborne samples. Fly ash particles generated by the metal industries often have characteristic morphologies; either granular or spherical (Figure 2.3e). It is noted that, although spherical combustion particles (i.e., smelter) generated by the iron industry have a mean diameter of around 2 μ with impaction-based collection systems, they are often found in the PM_{10}–$PM_{2.5}$ fractions, due to their high densities.

Fly ash that is produced by the burning of coal has a composition based on the physicochemistry of the coal, in particular the mineral components and the "rank," that is the temperature/pressure burial history of the coal. The amount of fly ash generated during combustion is controlled by the mineral content of the coal and the combustion conditions. Any example of extensive fly ash production into the urban atmosphere is the domestic burning of low grade, (i.e., mineral-rich), coal in pre-Clean Air Act London (Figure 2.3g), and the resulting "pea-souper smogs" (BéruBé et al. 2005). The burning of high-rank coals such as anthracites and bituminous coals will typically produce a "pozzolanic" fly ash with commercial applications in the cement industry as an additive. The glassy spheres have a high silica, alumina, and iron oxide content, and although they are mineralogically stable in the presence of water, will usually leach a proportion of any incorporated transition metals. Bulk fly ash tends to be alkaline (pH 9–12), and moving into neutral or acidic environments can significantly increase the mobility of associated metals. Fly ash produced by the burning of lower ranked coals, such as sub-bituminous coal or lignites, has pozzolanic qualities, but can also mineralogically react with water, resulting in the production of cement-type crystalline minerals. This is of concern where particle collection systems are prone to water-saturation, or biotoxicological assays use water-based delivery systems. It should also be noted that the high Ca and S content of fly ash can result in the formation of gypsum crystals under wet conditions, and it is not uncommon under electron microscopy to see small gypsum crystals on the surface of fly ash particles (Jones et al. 2001; Jones et al. in press).

The structure of soot has been extensively researched over the last 30 years, and our current understanding of the formation mechanisms means that engineers can generate soot particles with particular characteristics, such as size range, with relative ease. Soot particles are typically observed as chain-like or grape-bunch-like aggregations of primary spherical particles (Figure 2.4a and b). The aggregations can consist of just a few, through to many thousands of spheres. The mechanism of soot formation involves an initial particle nucleation from fuel pyrolysis forming polycyclic aromatic hydrocarbons, addition to the nucleus by gas molecules, coagulation by particle-particle collisions, removal of functional groups and dehydrogenation, and structural rearrangements of the condensed material. The wide spectrum of soot particle sizes is controlled by factors such as air dilution and temperature change in the combustion chamber and outlets, which are in turn controlled by factors such as engine loading, fuel type and fuel:oxygen combustion ratios. The fractal morphology of soot aggregations based on TEM and light scattering, have been investigated by many workers to obtain the fractal dimension (Lahaye and Ehrburger 1994). This was found to be in the range 1.70–1.85 and is shown in the equation:

$$N = C(R_g/r_p)^{\mathrm{Df}}$$

where N, number of primary spheres; R_g, radius of gyration; R_p, radius of the primary spheres; Df, fractal dimension.

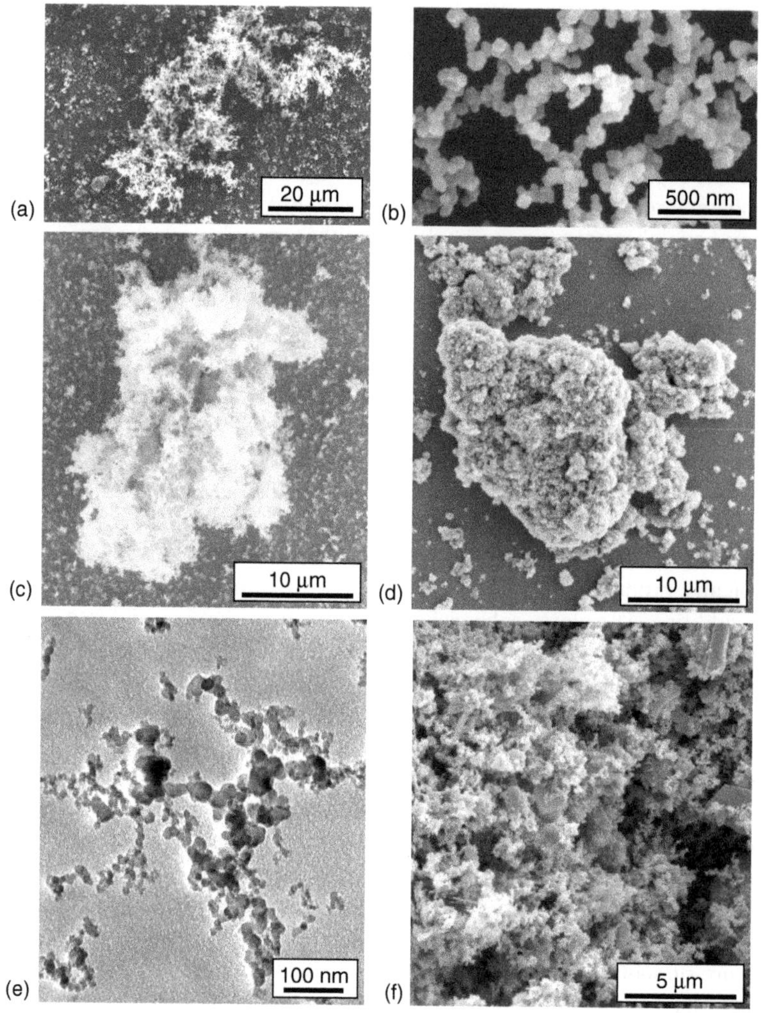

FIGURE 2.4 (a) A large soot particle that was collected in Cornwall, and was probably generated by the (illegal) burning of truck tires. (b) Detail of chains of soot microspheres that form the soot aggregates (particles). In this example the individual microspheres are exceptionally large, at up to 100 nm in diameter. (c) A large "fluffy" soot particle. This particle has not been wetted, and was collected near the exhaust pipe of an ancient and worn out diesel tractor. (d) Derived from the same source as the soot particle in Figure 2.4c, this particle has been wetted with resulting "clumped" morphology. (e) Transmission electron micrograph of a diesel exhaust particle. Capable of much higher resolution than SEM, analytical TEM is a powerful tool for the investigation of airborne particles. (f) PM_{10} collected in Birmingham, U.K., as part of the national monitoring program. The sample is dominated by vehicle exhaust (soot) particles.

The significance of this number is twofold. First, since the fractal dimension is below 2.00, most of the primary particles, up to an aggregate size threshold, should be seen in a two-dimensional projection. This is critical for two-dimensional investigation techniques such as transmission electron microscopy. Second, as the size of the soot aggregation increases, then the overall density of the aggregation decreases. This has implications for health assessments based on particulate masses, and also the ease with which primary particle surface areas can be accessed in large aggregations. A final consideration on the overall morphology of soot aggregations is the effect of wetting, this being of importance in particle collecting systems prone to occasional water saturation. When wetted, the soot aggregations change from a "flocculated" open morphology to a more compact ("clumped") appearance; a macroscopic analogy would be the effects of wetting a cotton wool ball (Figure 2.4c and Figure 2.4d). Thus, the effects of wetting will increase the density of the soot aggregations, as well as the overall decrease in size.

The structure of the individual soot spheres has been elucidated by techniques such as ultra-high resolution transmission electron microscopy. The central part of the sphere, typically with a diameter around 10 nm, is composed of concentrically-piled carbon networks in a thermodynamically unstable turbostratic state. The relatively poor organization of the individual carbon networks creates large intranetwork spaces that have the potential to hold materials such as volatile organic compounds, sulfur, and transition metals. The outer part of the sphere is better ordered, more stable, and composed of microcrystallites of orientated carbon sheets. These crystallites can be considered to be graphitic, with the planes of the sheets of graphite perpendicular to the center of the sphere. Graphitic microcrystallites are planar and around 1 nm thick and 3.5 nm wide. The region where the individual spheres collide to form the distinctive "chains," the graphitic microcrystallites in the saddle or neck of the join are perpendicular to a line drawn between the centers of the two colliding spheres. In the cooling stage of the process, vapor phase hydrocarbons condense on the sphere surface. The surface area of the soot spheres and the chemical components, either incorporated into the body of the sphere, or available on the surface, are of interest to respiratory toxicologists. For example, many toxicological studies have shown that chemically "pure" soot particles, such as carbon black (CB), have significantly less bioactivity than common urban PM_{10} soot particles (Figure 2.4e and Figure 2.4f) generated by diesel engines (Murphy 1998; BéruBé et al. 1999; Murphy, BéruBé, and Richards 1999).

Acute effects to CB exposure in humans are similar to effects caused by other insoluble particles. In general, long-term occupational exposure to CB may cause slight radiological changes; workers may develop chronic bronchitis and slight reduction in lung function. These effects can be attributed to a non-specific irritant effect of heavy dust exposure. Fibrosis has been reported in early studies under excessive dust conditions (Robertson et al. 1988; Gardiner et al. 1993; Szozda 1994; Gardiner 1995; IARC 1996; Kupper, Breitstadt, and Ulmer 1996). Epidemiology studies have not established excess cancer risk with CB exposure (IARC 1996; Parent, Siemiatycki, and Renaud 1996; Robertson 1996; Robertson and Inman 1996; Brockmann, Fischer, and Muller 1998; Straif et al. 1999; Veys 1999; USEPA 2000).

The water-soluble component of PM_{10} can be subdivided into two groups: (1) particles that are wholly liquid with ions in solution, and (2) particles with a solid core surrounded by a liquid envelope. The first group does not fall within the scope of this chapter, other than to note the "artifact" of mineral (salt) crystals that can form on particle collection systems that are prone to water saturation (Jones et al. 2001; Jones et al. in press) (Figure 2.5a–Figure2.5d). The second group is appropriate for this section, especially where the solid core is mineral (Figure 2.5e and Figure2.5f), as opposed to, for example, a soot particle. The second group includes the particles responsible for the dense urban "smogs"—a composite word from "smoke" and "fog"—which were responsible for the infamous London yellow smogs of times past. The smogs formed from solid particles generated by activities such as domestic coal burning, which acted as a nucleus for moisture. The moisture reacted with atmospheric sulfurous gases also generated by the coal burning to become acidic, thus potentially enhancing the leaching of metals from the solid nucleus. Although the London yellow smogs (Davis, Bell, and Fletcher 2002; Stone 2002; Hunt et al. 2003; Weinhold 2003; Whittaker et al. 2004; Merolla and Richards 2005) are hopefully confined to the history books, mineral grain–liquid interaction particles are still relatively commonplace, but a problem lies with recognizing them since they are usually destroyed on the collection medium. With low-energy impaction collection systems, the traces of these particles can be seen. Once the particles impact on the substrate the moisture envelope around the solid nucleus evaporates away. The particles are thus characterized by an "eroded" solid particle that was the original mineral nucleus, surrounded by usually "well-formed," water-soluble crystals. An example of the

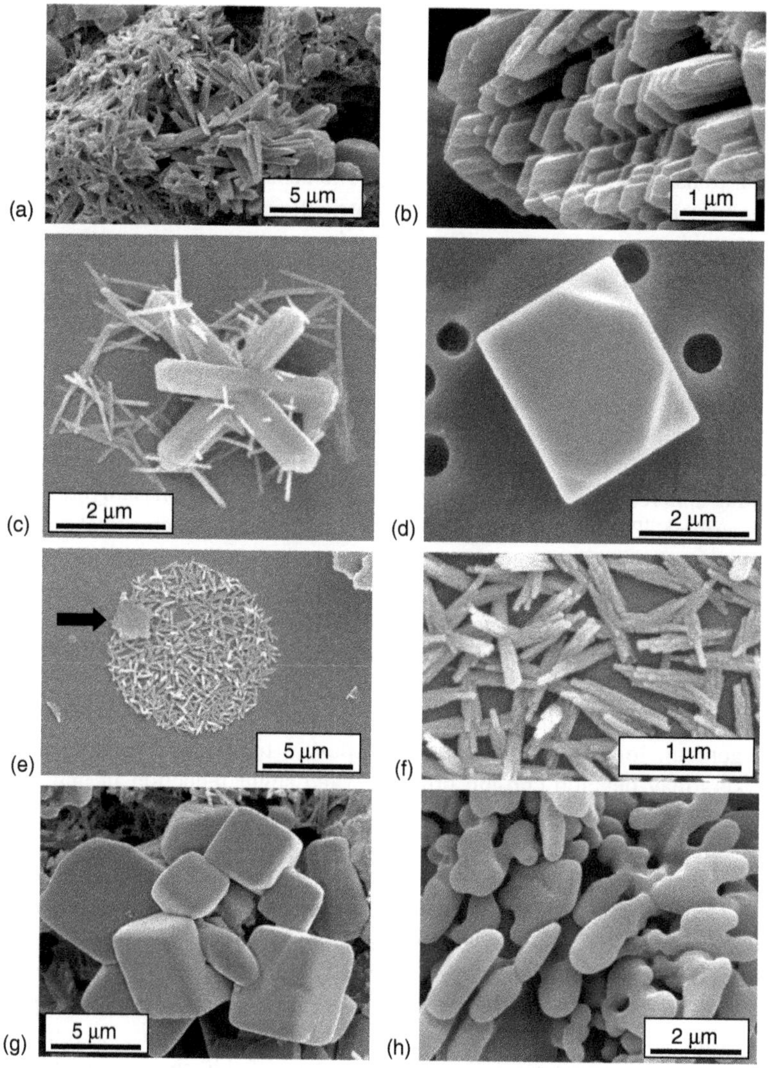

FIGURE 2.5 (a) Microcrystals of anhydrite ($CaSO_4$) on a glass fiber TEOM filter. The crystals formed *in situ* from moisture collected on the filter. (b) Microcrystals of "swallow-tail gypsum ($CaSO_4 \cdot 2H_2O$) formed *in situ* on a glass fiber TEOM filter. (c) Two phases of water-soluble salts that precipitated on a polyurethane impaction substrate from an airborne water "particle." (d) Crystal of halite (NaCl) that was collected in Cornwall, U.K. The site is directly next to the sea, and it is unknown whether the particle was collected as a solid crystal or it precipitated from a saline droplet on the impaction polycarbonate filter surface. (e) A circle of small $CaSO_4$ crystals delineate the area of a water droplet on a polyurethane impaction surface. Regularly seen on impaction surfaces the Ca^{++} is believed to be derived from the "eroded" particle of limestone (black arrow) that formed the nucleus of the particles, and the SO_4^{--} from reaction between the water envelope around the nucleus and atmospheric SO_2. (f) Detail of the $CaSO_4$ microcrystals in Figure 2.5e. (g) Halite crystals that formed *in situ* on a fiber filter. The crystals are reasonably well-formed with interlocking crystals. (h) Halite that precipitated *in situ* on a fiber filter. The rounded morphology is believed to be the result of very rapid precipitation.

typical chemistry that occurs in these particles would involve a limestone mineral nucleus. The crustally-derived mineral nucleus is covered in an envelope of water that reacts with atmospheric gases to become acidic. The acid reacts with the limestone to release CO_2 and H_2O, and release Ca^{++} and SO_4^{--} into the water envelope. The finally assemblage on the substrate is a circle of precipitated $CaSO_4$ crystals corresponding to the impact shape of the moisture droplet and an "eroded" mineral nucleus. A final note on water-soluble particles is that the speed of crystallization influences the final shape of the particles, where slowly precipitated salts tend to have good crystal forms, and rapidly precipitated particles are often unrecognizable as crystalline material (Figure 2.5g and Figure 2.5h).

A number of amorphous silicon dioxides are commercially available, such as Cabosil (Cab-*o*-Sil, Cabot Corp., U.K.), and these have been used in respiratory research to elucidate the relative importance of a material being crystalline or amorphous. Cabosil is a synthetic, amorphous, fumed silicon dioxide, with wide-ranging uses in food, cosmetics, resin thickener, anti-caking agent, etc., It is available as "treated" or "untreated," with the treatments involving the attaching of various molecules onto the particle surfaces to change important characteristics, and therefore commercial uses. Fumed silicon dioxides possess primary particle sizes of 9–30 nm, and as such, the aggregate particles can be considered to be nano-structured. Cab-*o*-Sil untreated fumed silica has two chemical groups on its surface. These are hydroxyl (silanol—the silicon analogy of an alcohol) groups and siloxane groups. Hydroxyl groups are subdivided into three subgroups: the isolated hydroxyl group, hydrogen-bonded hydroxyl groups, and geminal hydroxyl group. Approximately 40% of the surface are hydroxyl groups and are 60% siloxanes. Cab-*o*-Sil has been shown to have moderate bioreactivity in the rat lung, however, does not have the degree of toxicity of the equivalent mass of crystalline SiO_2 (Murphy et al. 1999; Murphy, BéruBé, and Richards 1999). Having stated that, it must be remembered that the crystalline SiO_2 samples will behave as normal "microscopic" samples, whereas the Cab-*o*-Sil is nano-structured, with significantly greater surface area per mass.

Diatomaceous earth is a naturally-occurring deposit derived from the siliceous skeletons (frustules) of photosynthetic aquatic microorganisms called diatoms. Diatoms are commonly between 20 and 200 μm in diameter or length, although sometimes they can be up to 2 mm long; however, the frustules are often exceedingly fragile, and can break into respirable fragments. The diatoms secrete their frustules as opal-A, a non-crystalline SiO_2. However, once deposited, this can change to opal-CT (30–50°C) over a few thousand years, and finally, over millions of years, diagenetically alters to quartz. Diatomaceous earth can be divided into freshwater and saltwater deposits. Freshwater deposits are characteristically low in crystalline silica, whereas saltwater deposits can contain high crystalline silica content. Certain industrial processes require diatomaceous earth to be "calcined" at 800–1000C to produce an end product that may contain 60% or more cristobalite. Respiratory concerns have centered on incidents of silicosis in the workers mining diatomaceous earth deposits (Harber et al. 1998; Steenland et al. 2001; Merget et al. 2002; Park et al. 2002; Honma et al. 2004). Diatomaceous earth products are also widely used in industry as filters, abrasives, and fillers, often in settings where workers could be exposed to respirable airborne particles (Checkoway et al. 1993).

Glass fiber (Stanton et al. 1977; Sincock, Delhantly, and Casey 1982) and man-made mineral fibers (MMMF) are made by spraying or extruding molten glass, furnace slag, or mineral rock, such as basalt. The material is therefore non-crystalline and has high silicon content. Respiratory health concerns have been based on the morphological similarities between MMMF and asbestos (De Vuyst et al. 1996; Brown et al. 2005). There is little clear evidence for human disease from inhalation exposure to glass fibers, although some evidence exists for certain rock wools (Merchant 1990; Hesterberg and Hart 2001; Moore et al. 2002; Baan and Grosse 2004; Wardenbach et al. 2005). The relatively large diameters of most commercial fibrous glass products of around 7.5 μm is probably significant, since the fibers cannot penetrate deep into the lung. There is also evidence that the glass fibers are not persistent in the lung, and break down rapidly. In contrast, certain slag and rockwool fibers have relatively smaller diameters and greater durability within the lungs, although this does seem to vary within different products.

Metals aerosols are produced by numerous industrial processes, and the composition of the aerosols is dictated by the processes. Of particular respiratory health interest are the metal aerosol fumes produced by welders (Newhouse, Oakes, and Wooley 1985; Honma et al. 2004;

Yu et al. 2004). Welding "smoke" is a mixture of fine particles and gases, with the components including toxic metals such as; chromium (Yu et al. 2001; Antonini et al. 2005; Sorensen et al. 2005), nickel (Antonini et al. 2005), manganese (Yu et al. 2004), beryllium (Stefaniak et al. 2004), cadmium (Binks et al. 2005; Jakubowski et al. 2005), cobalt (Dunlop et al. 2005; Krakowiak et al. 2005), copper (Santec et al. 2005), lead (Binks et al. 2005), and zinc (Karlsen et al. 1992; Fine et al. 1997; Fine et al. 2000). The metal particulates tend to be in the form of oxides, with the critical issue being their bioavailability. Particle sizes range from 5 nm to 20 μm, with 10%–30% larger than 1 μm (Zimmer, Baron, and Biswas 2002). Short-term exposure to metal aerosols can cause "metal fume fever" from 4 to 12 h after exposure. It can also cause irritation to the respiratory tract, and cause coughing, pulmonary edema, and pneumonitis (Antonini et al. 2003). Studies of long-term exposure to metal aerosols show an increased risk of lung cancer (Nemery 1990; Kelleher, Pacheco, and Newman 2000).

It is estimated that more than one million workers worldwide perform some type of welding as part of their work duties. Epidemiological studies have shown that a large number of welders experience some type of respiratory illness. Respiratory effects seen in fulltime welders have included bronchitis, siderosis, asthma, and a possible increase in the incidence of lung cancer. Pulmonary infections are increased in terms of severity, duration, and frequency among welders. Most welding materials are alloy mixtures of metals characterized by different steels that may contain iron, manganese, chromium, and nickel. Animal studies have indicated that the presence and combination of different metal constituents is an important determinant in the potential pneumotoxic responses associated with welding fumes. Animal models have demonstrated that stainless steel welding fumes, which contain significant levels of nickel and chromium, induce more lung injury and inflammation, and are retained in the lungs longer than mild steel welding fumes, which contain mostly iron. In addition, stainless steel fumes generated from welding processes, using fluxes to protect the resulting weld, contain elevated levels of soluble metals, which may affect respiratory health (Antonini et al. 2004).

REFERENCES

Acheson, E. D., Bennett, C., Gardner, M. J., and Winter, P. D., Mesothelioma in a factory using amosite and chrysotile asbestos, *Lancet*, 2, 1403–1405, 1981.

Adamis, Z., Kerenyi, T., Honma, K., Jackel, M., Tatrai, E., and Ungvary, G., Study of inflammatory responses to crocidolite and basalt wool in the rat lung, *J. Toxicol. Environ. Health A*, 62 (5), 409–415, March 9, 2001.

Altree-Williams, S., Byrnes, J. G., and Jordan, B., Amorphous surface and quantitative X-ray powder diffractometry, *Analyst*, 106, 69–75, 1981.

Antonini, J. M., Lewis, A. B., Roberts, J. R., and Whaley, D. A., Pulmonary effects of welding fumes: review of worker and experimental animal studies, *Am. J. Ind. Med.*, 43 (4), 350–360, 2003.

Antonini, J. M., Taylor, M. D., Leonard, S. S., Lawryk, N. J., Shi, X., Clarke, R. W., and Roberts, J. R., Metal composition and solubility determine lung toxicity induced by residual oil fly ash collected from different sites within a power plant, *Mol. Cell Biochem.*, 255 (1–2), 257–265, 2004.

Antonini, J. M., Leonard, S. S., Roberts, J. R., Solano-Lopez, C., Young, S. H., Shi, X., and Taylor, M. D., Effect of stainless steel manual metal arc welding fume on free radical production, DNA damage, and apoptosis induction, *Mol. Cell Biochem.*, 279 (1–2), 17–23, 2005.

Artvinli, M. and Baris, I., Malignant mesotheliomas in a small village in the Anatolian region of Turkey, *J. Natl Can. Inst.*, 63, 17–27, 1979.

Artvinli, M. and Baris, Y. I., Environmental fiber-induced pleuro-pulmonary diseases in an Anatolian village: an epidemiologic study, *Arch. Environ. Health*, 37 (3), 177–181, 1982.

Baan, R. A. and Grosse, Y., Man-made mineral (vitreous) fibers: evaluations of cancer hazards by the International Agency for Research on Cancer Monographs Programme, *Mutat. Res.*, 553 (1–2), 43–58, 2004.

Baris, Y. I., Sahin, A. A., and Erkan, M. L., Clinical and radiological study of sepiolite workers, *Arch. Environ. Health*, 35, 343–346, 1980.

Baris, Y. I., Simonato, L., Artvinli, M., Pooley, F., Saracci, R., Skidmore, J., and Wagner, C., Epidemiological and environmental evidence of the health effects of exposure to erionite fibers: a four-year study in the Cappadocian region of Turkey, *Int. J. Cancer*, 39 (1), 10–17, 1987.

Baumann, H., Characterisation of silicon dioxide surfaces by successive determinations of the solution rate, In *Health Effects of Synthetic Silica Particulates*, ASTM Special Technical Publication 731, 30–47, 1979.

Baxter, P. J., Impacts of eruptions on human health, In *Encyclopaedia of Volcanoes*, Siggurdson, H., ed., Academic Press, New York, 1035–1043, 1999.

Baxter, P. J., Ing, R., Falk, H., French, J., Stein, G. F., Bernstein, R. S., Merchant, J. A., and Allard, J., Mount St. Helens Eruptions, May 18 to June 12, 1980: an overview of the acute health impact, *J. Am. Med. Assoc.*, 246 (22), 2585–2589, 1981.

Baxter, P. J., Bernstein, R. S., and Buist, A. S., Health effects of volcanoes: an approach to evaluating the health effects of an environmental hazard, *Am. J. Pub. Health*, 76, 84–90, 1986.

Baxter, P. J., Bobadonna, C., Dupree, R., Hards, V., Kohn, S., Murphy, M., Nichols, A. et al., Cristobalite in volcanic ash of the Soufriere Hills Volcano, Montserrat, British West Indies, *Science*, 283, 1142–1145, 1999.

Beck, B. D., Brain, J. D., and Bohannon, D. E., The pulmonary toxicity of an ash sample from the Mt. St. Helens Volcano, *Exp. Lung Res.*, 2, 289–301, 1981.

Begin, R., Masse, S., Rola-Pleszczynski, M., Geoffroy, M., Martel, M., Desmarais, Y., and Sebastien, P., The lung biological activity of American attapulgite, *Environ. Res.*, 42 (2), 328–339, 1987.

Bellman, B., Muhle, H., and Ernst, H., Investigations on health-related properties of two sepiolite samples, *Environ. Health Perspect.*, 105 (5), 1049–1052, 1997.

Bennett, J. G., Dick, J. A., Kaplan, Y. S., Shand, P. A., Shennan, D. H., Thomas, D. J., and Washington, J. S., The relationship between coal rank and the prevalence of pneumoconiosis, *Br. J. Ind. Med.*, 36, 206–210, 1979.

Bermudez, E., Mangum, J. B., Asgharian, B., Wong, B. A., Reverdy, E. E., Janszen, D. B., Hext, P. M., Warheit, D. B., and Everitt, J. I., Long-term pulmonary responses of three laboratory rodent species to subchronic inhalation of pigmentary titanium dioxide particles, *Toxicol. Sci.*, 70 (1), 86–97, 2002.

Bermudez, E., Mangum, J. B., Wong, B. A., Asgharian, B., Hext, P. M., Warheit, D. B., and Everitt, J. I., Pulmonary responses of mice, rats, and hamsters to subchronic inhalation of ultrafine titanium dioxide particles, *Toxicol. Sci.*, 77 (2), 347–357, 2004.

BéruBé, K. A., Jones, T. P., Williamson, B. J., Winters, C., Morgan, A. J., Pooley, D. D., and Richards, R. J., Physicochemical characterisation of diesel exhaust particles: factors for assessing biological activity, *Atmos. Environ.*, 33 (10), 1599–1614, 1999.

BéruBé, K. A., Jones, T. P., Housley, D. G., and Richards, R. J., The respiratory toxicity of airborne volcanic ash from the Soufrière Hills volcano, Montserrat, *Min. Mag.*, 68 (1), 47–60, 2004.

BéruBé, K. A., Whittaker, A., Jones, T. P., Moreno, T., and Merolla, L., London smogs: why did they kill?, *Proc. Roy. Micro. Soc.*, 40 (3), 171–183, 2005.

Binks, K., Doll, R., Gillies, M., Holroyd, C., Jones, S. R., McGeoghegan, D., Scott, L., Wakeford, R., and Walker, P., Mortality experience of male workers at a UK tin smelter, *Occup. Med. (Lond.)*, 55 (3), 215–226, 2005.

Blejer, J. P. and Arlon, P. F., Talc: a possible occupational and environmental carcinogen, *J. Occup. Med.*, 15, 92–97, 1973.

Boros, D. L. and Warren, K. S., The bentonite granuloma. Characterization of a model system for infectious and foreign body granulomatous inflammation using soluble mycobacterial, histoplasma and schistosoma antigens, *Immunology*, 24, 511–529, 1973.

Brauer, M., Hoek, G., Van Vliet, P., Meliefste, K., Fischer, P. H., Wijga, A., Koopman, L. P. et al., Air pollution from traffic and the development of respiratory infections and asthmatic and allergic symptoms in children, *Am. J. Respir. Crit. Care Med.*, 166 (8), 1092–1098, 2002.

Breeling, J. L., Potential hazard from eating rice coated with glucose and talc, *J. Am. Med. Assoc.*, 228, 101, 1974.

Britton, M., The epidemiology of mesothelioma, *Semin. Oncol.*, 29 (1), 18–25, 2002.

Brockmann, M., Fischer, M., and Muller, K. M., Exposure to carbon black: a cancer risk? *Int. Arch. Occup. Environ. Health*, 71, 85–99, 1998.

Brown, R. C., Bellmann, B., Muhle, H., Davis, J. M. G., and Maxim, L. D., Survey of the biological effects of refractory ceramic fibres: overload and its possible consequences, *Ann. Occup. Hyg.*, 49 (4), 295–307, 2005.

Canessa, P. A., Torraca, A., Lavecchia, M. A., Patelli, M., and Poletti, V., Pneumoconiosis (silicosis) in the confectionery industry, *Sarcoidosis*, 7 (1), 75–77, 1990.

Carleton, L. E., Giere, R., and Lumpkin, G. R., Micro- and nano-chemical characterization of fly ash particles. Geological Society of America, 2002 annual meeting, *Abs. Progs. Geol. Soc. Am.*, 34 (6), 190, 2002.

Castronovo, E., Radiological eatures of pumice pneumoconiosis (liparitosis) and its pathogenic interpretation, *Riv. Inf. Mal. Prof.*, 28, 289, 1953.

Castronova, V. and Vallyathan, V., Silicosis and coal workers' pneumoconiosis, *Environ. Health Perspect.*, 108 (4), 675–684, 2000.

Castranova, V., Dalal, N. S., and Vallyathan, V., Role of surface free radicals in the pathogenicity of silicosis, In *Silica and Silica-Induced Lung Diseases*, Castranova, V., Vallyathan, V., and Wallace, W. E., eds., CRC Press, Boca Raton, FL, 1997.

Checkoway, H., Heyer, N. J., Demers, P. A., and Breslow, N. E., Mortality among workers in the diatomaceous earth industry, *Br. J. Ind. Med.*, 50 (7), 586–597, 1993.

Chen, S. Y., Hayes, R. B., Wang, J. M., Liang, S. R., and Blair, A., Nonmalignant respiratory disease among hematite mine workers in China, *Scand. J. Work. Environ. Health*, 15 (5), 319–322, 1989.

Chen, S. Y., Hayes, R. B., Liang, S. R., Li, Q. G., Stewart, P. A., and Blair, A., Mortality experience of haematite mine workers in China, *Br. J. Ind. Med.*, 47 (3), 175–181, 1990.

Chen, Y. S., Sheen, P. C., Chen, E. R., Liu, Y. K., Wu, T. N., and Yang, C. Y., Effects of Asian dust storm events on daily mortality in Taipei, Taiwan, *Environ. Res.*, 95, 151–155, 2004.

Choi, J. C., Lee, M., Chun, Y., Kin, J., and Oh, S., Chemical composition and source signature of spring aerosol in Seoul, Korea, *J. Geophys. Res.*, 106, 18067–18074, 2001.

Costa, D. L. and Dreher, K. L., Bioavailable transition metals in particulate matter mediate cardiopulmonary injury in healthy and compromised animal models, *Environ. Health Perspect.*, 105 (5), 1053–1060, 1997.

Dalal, N. S., Jafari, B., Vallyathan, V., and Green, F. H. Y., Cytotoxicity and spectroscopic investigations of organic-free radicals in fresh and stale coal dust, In *Proceedings 7th International Pneumoconiosis Conference, Part 2, 23–26 August 1988, Pittsburgh, Pennsylvania.* NIOSH Publ 90–108, National Institute for Occupational Safety and Health, Cincinnati, OH, 1470–1477, 1990.

Dalal, N. W., Jafari, B., Petersen, M., Green, F. H. Y., and Vallyathan, V., Presence of stable coal radicals in autopsied coal miners' lungs and its possible correlation to coal workers' pneumoconiosis, *Arch. Environ. Health*, 46, 366–372, 1991.

Dalal, N. S., Newman, J., Pack, D., Leonard, S., and Vallyathan, V., Hydroxyl radical generation by coal mine dust: possible implication to coal workers' pneumoconiosis (CWP), *Free Radic. Biol. Med.*, 18, 11–20, 1995.

Davis, J. M., Addison, J., Bolton, R. E., Donaldson, K., Jones, A. D., and Smith, T., The pathogenicity of long versus short fibre samples of amosite asbestos administered to rats by inhalation and intraperitoneal injection, *Br. J. Exp. Pathol.*, 67, 415–430, 1986.

Davis, D. L., Bell, M. L., and Fletcher, T., A look back at the London smog of 1952 and the half century since, *Environ. Health Perspect.*, 110 (12), A734–A735, 2002.

Deer, W. A., Howie, R. A., and Zussman, J., An introduction to the rock-forming minerals, Longman Group Limited, 1966.

Denizeau, F., Marion, M., Chevalier, G., and Cote, M. G., Absence of genotoxic effects of nonasbestos mineral fibers, *Cell Biol. Toxicol.*, 2, 23–32, 1985.

De Vuyst, P., Dumortier, P., Swaen, G. M., Pairon, J. C., and Brochard, P., Respiratory health effects of man-made vitreous (mineral) fibres, *Eur. Respir. J.*, 8 (12), 2149–2173, 1996.

Dinman, B. D., Alumina-related pulmonary disease, *J. Occup. Med.*, 30 (4), 328–335, 1988.

Di Lorenzo, L., De Tommaso, C., Lastilla, G., Massola, A., and Soleo, L., Pneumoconiosis in a female worker exposed to a primer used in the production of non-stick pans: clinical case, *Med. Lav.*, 94 (5), 459–465, 2003.

Dodson, R. F. and Levin, J. L., An unusual case of mixed-dust exposure involving a "noncommercial" asbestos, *Environ. Health Perspect.*, 109 (2), 199–203, 2001.

Dodson, R. F., Graef, R., Shepherd, S., O'Sullivan, M., and Levin, J., Asbestos burden in cases of mesothelioma from individuals from various regions of the United States, *Ultrastruct. Pathol.*, 29 (5), 415–433, 2005.

Domej, W., Foldes-Papp, Z., Schlagenhaufen, C., Wippel, R., Tilz, G. P., Krachler, M., Demel, U., Lang, J., and Urban-Woltron, H., Detection of graphite using laser microprobe mass analysis of a transbronchial biopsy from a foundry worker with mixed dust pneumoconiosis, *Wien. Klin. Wochenschr.*, 114 (5–6), 216–221, 2002.

Donaldson, K. and Tran, C. L., An introduction to the short-term toxicology of respirable industrial fibres, *Mutat. Res.*, 553 (1–2), 5–9, 2004.

Dreher, K. L., Jaskot, R. H., Lehmann, J. R., Richards, J. H., McGee, J. K., and Ghio, A. J., Soluble transition metals mediate residual oil fly ash-induced acute lung injury, *J. Toxicol. Environ. Health*, 50, 285–305, 1997.

Duce, A. R., Unni, C. K., Ray, B. J., Prospero, J. M., and Merrill, J. T., Long-range atmospheric transport of soil dust from Asia to the tropical north pacific: temporal variability, *Science*, 209, 1522–1524, 1980.

Dumortier, P., Coplu, L., Broucke, I., Emri, S., Selcuk, T., de Maertelaer, V., De Vuyst, P., and Baris, I., Erionite bodies and fibres in bronchoalveolar lavage fluid (BALF) of residents from Tuzkoy, Cappadocia, Turkey, *Occup. Environ. Med.*, 58 (4), 261–266, 2001.

Dunlop, P., Muller, N. L., Wilson, J., Flint, J., and Churg, A., Hard metal lung disease: high resolution CT and histologic correlation of the initial findings and demonstration of interval improvement, *J. Thorac. Imaging*, 20 (4), 301–304, 2005.

Eborn, S. K. and Aust, A., Effect of iron acquisition on induction of DNA single-strand breaks by erionite, a carcinogenic mineral fiber, *Arch. Biochem. Biophys.*, 316, 507–514, 1995.

El Ghawabi, S. H., Zewer, R. E., Ibrahim, S. M., and Selim, S. R., Basalt Pneumoconiosis, *J. Soc. Occup. Med.*, 35, 131–133, 1985.

Elmore, A. R., Cosmetic Ingredient Review Expert Panel. Final report on the safety assessment of aluminum silicate, calcium silicate, magnesium aluminum silicate, magnesium silicate, magnesium trisilicate, sodium magnesium silicate, zirconium silicate, attapulgite, bentonite, Fuller's earth, hectorite, kaolin, lithium magnesium silicate, lithium magnesium sodium silicate, montmorillonite, pyrophyllite, and zeolite, *Int. J. Toxicol.*, 22 (1), 37–102, 2003.

Emri, S. and Demir, A. U., Malignant pleural mesothelioma in Turkey, 2000–2002, *Lung Cancer*, 45 (1), S17–S20, 2004.

Emri, S., Demir, A., Dogan, M., Akay, H., Bozkurt, B., Carbone, M., and Baris, I., Lung diseases due to environmental exposures to erionite and asbestos in Turkey, *Toxicol. Lett.*, 127 (1–3), 251–257, 2002.

Fach, E., Waldman, W. J., Williams, M., Long, J., Meister, R. K., and Dutta, P. K., Analysis of the biological and chemical reactivity of zeolite-based aluminosilicate fibers and particulates, *Environ. Health Perspect.*, 110 (11), 1087–1096, 2002.

Fine, J. M., Gordon, T., Chen, L. C., Kinney, P., Falcone, G., and Beckett, W. S., Metal fume fever: characterization of clinical and plasma IL-6 responses in controlled human exposures to zinc oxide fume at and below the threshold limit value, *J. Occup. Environ. Med.*, 39 (8), 722–726, 1997.

Fine, J. M., Gordon, T., Chen, L. C., Kinney, P., Falcone, G., Sparer, J., and Beckett, W. S., Characterization of clinical tolerance to inhaled zinc oxide in naive subjects and sheet metal workers, *J. Occup. Environ. Med.*, 42 (11), 1085–1091, 2000.

Fishman, B. B., Features of chronic pulmonary diseases in workers engaged in the production of high aluminum mullite refractory items, *Med. Tr. Prom. Ekol.*, 7, 30–33, 2003.

Fishman, B. B. and Velichkovskii, B. T., Fibrogenicity of dust emitted by highly aluminiferous refractories, *Med. Tr. Prom. Ekol.*, 10, 13–17, 2001.

Fishman, B. B., Medik, V. A., Veber, V. R., Bastsrykina, O. V., and Prindik, A. A., Aspects of silicotuberculosis course in workers of highly aluminous mullite refractories, *Med. Tr. Prom. Ekol.*, 3, 29–34, 2001.

Forbes, L., Jarvis, D., Potts, J., and Baxter, P. J., Volcanic ash and respiratory symptoms in children on the island of Montserrat, British West Indies, *Occup. Environ. Med.*, 60 (7), 529–530, 2004.

Fubini, B., Health effects of silica, In *The Surface Properties of Silica*, Legrand, A. P., ed., Wiley, England, 1998.

Fubini, B. and Hubbard, A., Reactive oxygen species and reactive nitrogen species. Generation by silica in inflammation and fibrosis, *Free Rad. Biol. Med.*, 34 (12), 1507–1516, 2003.

Fubini, B. and Mollo, L., Role of iron in the reactivity of mineral fibers, *Toxicol. Lett.*, 82/83, 951–960, 1995.

Fubini, B., Giamello, E., Volante, M., and Bolis, V., Chemical functionalities at the silica surface determining its reactivity when inhaled. Formation and reactivity of surface radicals, *Toxicol. Ind. Health*, 6 (6), 571–598, 1990.

Fubini, B., Mollo, L., and Giamello, E., Free radical generation at the solid/liquid interface in iron containing minerals, *Free Rad. Res.*, 23, 593–614, 1995.

Fubini, B., Fenoglio, I., Elias, Z., and Poirot, O., Variability of biological responses to silicas: effect of origin, crystallinity, and state of surface on generation of reactive oxygen species and morphological transformation of mammalian cells, *J. Environ. Pathol. Toxicol. Oncol.*, 20, 95–108, 2001.

Gao, N., Keane, M. J., Ong, T., and Wallace, W. E., Effects of simulated pulmonary surfactant on the cytotoxicity and DNA-damaging activity of respirable quartz and kaolin, *J. Toxicol. Environ. Health A*, 60 (3), 153–167, 2000.

Gardiner, K., Effects on respiratory morbidity of occupational exposure to carbon black: a review, *Arch. Environ. Health*, 50 (1), 4–60, 1995.

Gardiner, K., Trethowan, N. W., Harrington, J. M., Rossiter, C. E., and Calvert, I. A., Respiratory health effects of carbon black: a survey of European carbon black workers, *Br. J. Ind. Med.*, 50, 1082–1096, 1993.

Gardner, S. Y., McGee, J. K., Kodavanti, U. P., Ledbetter, A., Everitt, J. I., Winsett, D. W., Doerfler, D. L., and Costa, D. L., Emission-particle-induced ventilatory abnormalities in a rat model of pulmonary hypertension, *Environ. Health Perspect.*, 112 (8), 872–878, 2004.

Geh, S., Yucel, R., Duffin, R., Albrecht, C., Borm, P. J., Armbruster, L., Raulf-Heimsoth, M., Bruning, T., Hoffmann, E., Rettenmeier, A. W., and Dopp, E., Cellular uptake and cytotoxic potential of respirable bentonite particles with different quartz contents and chemical modifications in human lung fibroblasts, *Arch. Toxicol.*, 1–9, 2005.

Ghio, A. J., Silbajoris, R., Carson, J. L., and Samet, J. M., Biologic effects of oil fly ash, *Environ. Health Perspect.*, 110 (1), 89–94, 2002.

Gibbons, W., Amphibole asbestos in Africa and Australia; geology, health hazard and mining legacy, *J. Geol. Soc. Lond.*, 157 (4), 851–858, 2000.

Gibbs, A. R. and Pooley, F. D., Fuller's earth (montmorillonite) pneumoconiosis, *Occup. Environ. Med.*, 51 (9), 644–646, 1994.

Gibbs, A. E., Pooley, F. D., Griffiths, D. M., Mitha, R., Craighead, J. E., and Ruttner, J. R., Talc pneumoconiosis: a pathologic and mineralogic study, *Hum. Pathol.*, 23 (12), 1344–1354, 1992.

Giere, R., Carleton, L. E., and Lumpkin, G. R., Micro- and nanochemistry of fly ash from a coal-fired power plant, *Am. Min.*, 88 (11–12(2)), 1853–1865, 2003.

Gottardi, G. and Galli, E., *Natural Zeolites*, Springer, Berlin, 1985.

Gunter, M. E., An overview of the past, present, and future concerns surrounding the asbestos content of vermiculite ore from Libby, Montana, In: Geological Society of America, 2004 Annual Meeting Abstract with Programs, Vol. 36, No. 5, p. 24, 2004.

Guthrie, G. D. Jr. and Mossman, B. T., *Health Effects of Mineral Dusts. Reviews in Mineralogy*, 28, Mineralogical Society of America, 1993.

Hanke, W., Sepulveda, M. J., Watson, A., and Jankovic, J., Respiratory morbidity in wollastonite workers, *Br. J. Ind. Med.*, 41, 474–479, 1984.

Hansen, K. and Mossman, B. T., Generation of superoxide from alveolar macrophages exposed to asbestiform and nonfibrous particles, *Can. Res.*, 47, 1681–1686, 1987.

Harber, P., Dahlgren, J., Bunn, W., Lockey, J., and Chase, G., Radiographic and spirometric findings in diatomaceous earth workers, *J. Occup. Environ. Med.*, 40 (1), 22–28, 1998.

Hauser, R., Elreedy, S., Hoppin, J. A., and Christiani, D. C., Airway obstruction in boilermakers exposed to fuel oil ash: a prospective investigation, *Am. J. Respir. Crit. Care Med.*, 152, 1478–1484, 1995.

Hesterberg, T. W. and Hart, G. A., Synthetic vitreous fibers: a review of toxicology research and its impact on hazard classification, *Crit. Rev. Toxicol.*, 31 (1), 1–53, 2001.

Hext, P. M., Tomenson, J. A., and Thompson, P., Titanium dioxide: inhalation toxicology and epidemiology, *Ann. Occup. Hyg.*, 6, 461–472, 2005.

Hiyoshi, K., Ichinose, T., Sadakane, K., Takano, H., Nishikawa, M., Mori, I., Yanagisawa, R. et al., Asian sand dust enhances ovalbumin-induced eosinophil recruitment in the alveoli and airway of mice, *Environ. Res.*, 99 (3), 361–368, 2005.

Honma, K., Abraham, J. L., Chiyotani, K., De Vuyst, P., Dumortier, P., Gibbs, A. R., Green, F. H. et al., Proposed criteria for mixed-dust pneumoconiosis: definition, descriptions, and guidelines for pathologic diagnosis and clinical correlation, *Hum. Pathol.*, 35 (12), 1515–1523, 2004.

Horwell, C. J., Brana, L. P., Sparks, R. J. S., Murphy, M. D., and Hards, V. L., A geochemical investigation of fragmentation and physical fractionation in pyroclastic flows from the Soufrière Hills volcano, Montserrat, *J. Volcan. Geothermal Res.*, 109 (4), 247–262, 2001.

Horwell, C. J., Fenoglio, I., Vala Ragnarsdottir, K., Sparks, R. S., and Fubini, B., Surface reactivity of volcanic ash from the eruption of Soufriere Hills volcano, Montserrat, West Indies with implications for health hazards, *Environ. Res.*, 93 (2), 202–215, 2003.

Housley, D. G., BéruBé, K. A., Jones, T. P., Anderson, S., Pooley, F. D., and Richards, R. J., Epithelial response to instilled Montserrat and Antiqua respirable dusts and their major mineral components, *Occup. Environ. Med.*, 59, 0–6, 2002.

Howard, T. P., Pneumoconiosis in a vermiculite end-product user, *Am. J. Ind. Med.*, 44 (2), 214–217, 2003.

Huggins, C. W., Denny, M. V., and Shell, H. R., Properties of palygorskite, an asbestiform mineral, Department of the Interior, US Bureau of Mines, Washington, DC, (Report of Investigation 6071), 1962.

Humble, S., Tucker, A. J., Boudreaux, C., King, J. A., and Snell, K., Titanium particles identified by energy-dispersive X-ray microanalysis within the lungs of a painter at autopsy, *Ultrastruct. Pathol.*, 27 (2), 127–129, 2003.

Hunt, A., Abraham, J. L., Judson, B., and Berry, C. L., Toxicologic and epidemiologic clues from the characterization of the 1952 London smog fine particulate matter in archival autopsy lung tissues, *Environ. Health Perspect.*, 111 (9), A481, 2003.

Husar, R. B., Tratt, D. B., Schichtel, B. A., Falke, S. R., Li, F., Jaffe, D., Gasso, S. et al., Asian dust events of April 1998, *J. Geophys. Res.*, 106, 18316–18330, 2001.

Huuskonen, M. S., Tossavainen, A., Koskinen, H., Zitting, A., Korhonen, O., Nickels, J., Korhonen, K., and Vaaranen, V., Wollastonite exposure and lung fibrosis, *Environ. Res.*, 30, 291–304, 1983.

International Agency for Research on Cancer, In *Biological effects of attapulgite*, Bignon , J., Sebastien, P., Gaudichet, A., and Jaurand, M. C., eds., IARC Sci. Publ. 30, 163–181, 1980.

International Agency for Research on Cancer, In *Mechanisms of Fiber Carcinogenicity*, Kane, A. B., Boffetta, P., Saracci, R., and Wilbourn, J. D., eds., Scientific Publications No. 140, 1996.

Jakubowski, M., Abramowska-Guzik, A., Szymczak, W., and Trzcinka-Ochocka, M., Influence of long-term occupational exposure to cadmium on lung function tests results, *Int. J. Occup. Med. Environ. Health*, 17 (3), 361–368, 2005.

Jaurand, M. C., Fleury, J., Monchaux, G., Nebut, M., and Bignon, J., Pleural carcinogenic potency of mineral fibers (asbestos, attapulgite) and their cytotoxicity on cultured cells, *J. Natl Cancer Inst.*, 79 (4), 797–804, 1987.

Jefferson, D. A. and Tilley, E. E. M., The structural and physical chemistry of nanoparticles, In *Particulate Matter: properties and Effects Upon Health*, Maynard, R. L. and Howard, C.V., eds., Bios Scientific Publishers, 63–84, 1999.

Jones, T. P., Williamson, B. J., BéruBé, K. A., and Richards, R. J., Microscopy and chemistry of particles collected on TEOM filters: Swansea, south Wales, 1989–1999, *Atmos. Environ.*, 35, 3573–3583, 2001.

Jones, T. P., Blackmore, P. R., Leach, M. T., BéruBé, K. A., Sexton, K., and Richards, R. J., Characterisation of airborne particulates collected within and proximal to an open-cast coal-mine: South Wales, UK, *Environ. Monitor. Assess.*, 75 (3), 293–312, 2002.

Jones, T. P., Moreno, T., BéruBé, K. A., and Richards, R. J., Physicochemical characterisation of microscopic airborne particles from south Wales: a review of the locations and methodologies, *Sci. Total Environ.*, in press.

Karlsen, J. T., Farrants, G., Torgrimsen, T., and Reith, A., Chemical composition and morphology of welding fume particles and grinding dusts, *Am. Ind. Hyg. Assoc. J.*, 53 (5), 290–297, 1992.

Kelleher, P., Pacheco, K., and Newman, L. S., Inorganic dust pneumonias: the metal-related parenchymal disorders, *Environ. Health Perspect.*, 108 (4), 685–696, 2000.

Kim, B. G., Han, J. S., and Park, S. U., Transport SO_2 and aerosol over the Yellow Sea, *Atmos. Environ.*, 35, 727–737, 2001.

Kiviluoto, R., Pleural calcification as a roentgenologic sign of non-occupational endemic anthophyllite-asbestosis, *Acta. Radiol.*, 194, 1–67, 1960.

Kodavanti, U. P., Jaskot, R. H., Costa, D. L., and Dreher, K. L., Pulmonary pro-inflammatory gene induction following acute exposure to residual oil fly ash: role of particle-associated metals, *Inhal. Toxicol.*, 9, 679–701, 1997a.

Kodavanti, U. P., Jaskot, R. H., Su, W. Y., Costa, D. L., Ghio, A. J., and Dreher, K. L., Genetic variability in combustion particle-induced chronic lung injury, *Am. J. Physiol. Lung Cell Mol. Physiol.*, 272, L521–L532, 1997b.

Kodavanti, U. P., Schladweiler, M. C., Ledbetter, A. D., Watkinson, W. P., Campen, M. J., and Winsett, D. W., The spontaneously hypertensive rat as a model of human cardiovascular disease: evidence of exacerbated cardiopulmonary injury and oxidative stress from inhaled emission particulate matter, *Toxicol. Appl. Pharmacol.*, 64, 250–263, 2000.

Krakowiak, A., Dudek, W., Tarkowski, M., Swiderska-Kielbik, S., Niescierenko, E., and Palczynski, C., Occupational asthma caused by cobalt chloride in a diamond polisher after cessation of occupational exposure: a case report, *Int. J. Occup. Med. Environ. Health*, 18 (2), 151–158, 2005.

Kraus, T., Schaller, K. H., Angerer, J., and Letzel, S., Aluminium dust-induced lung disease in the pyro-powder-producing industry: detection by high-resolution computed tomography, *Int. Arch. Occup. Environ. Health*, 73 (1), 61–64, 2000.

Kulagina, T. F. and Pylev, L. N., Morphological assessment of pleural tumors induced in rats by Dzhetygara chrysotile asbestos and [lizardite], *Gig. Tr. Prof. Zabol.*, 3, 27–29, 1985.

Kupper, H. U., Breitstadt, R., and Ulmer, W. T., Effects on the lung function of exposure to carbon black dusts, *Int. Arch. Occup. Environ. Health*, 68, 478–483, 1996.

Kwon, H. J., Cho, S. H., Chun, Y., Lagarde, F., and Pershagen, G., Effects of the Asian dust events on daily mortality in Seoul, Korea, *Environ. Res.*, 90, 1–5, 2002.

Lahaye, J. and Ehrburger-Dolle, F., Mechanisms of carbon black formation. Correlation with the morphology of aggregates, *Carbon*, 32 (7), 1319–1324, 1994.

Landas, S. K. and Schwartz, D. A., Mica-associated pulmonary interstitial fibrosis, *Am. Rev. Respir. Dis.*, 144 (3 Pt 1), 718–721, 1991.

Lapenas, D., Gale, P., Kennedy, T., Rawlings, W. Jr., and Dietrich, P., Kaolin pneumoconiosis. Radiologic, pathologic and mineralogic findings, *Am. Rev. Respir. Dis.*, 130, 282–288, 1984.

Lee, S. H. and Richards, R. J., Montserrat volcanic ash induces lymph node granuloma and delayed lung inflammation, *Toxicology*, 195 (2–3), 155–165, 2004.

Lee, K. P., Trochimowicz, H. J., and Reinhardt, C. F., Pulmonary response of rats exposed to titanium dioxide (TiO2) by inhalation for two years, *Toxicol. Appl. Pharmacol.*, 79 (2), 179–192, 1985.

Lemaire, I., Dionne, P. G., Nadeau, D., and Dunnigan, J., Rat lung reactivity to natural and man-made fibrous silicates following short-term exposure, *Environ. Res.*, 48 (2), 193–210, 1989.

Lewis, A. B., Taylor, M. D., Roberts, J. R., Leonard, S. S., Shi, X., and Antonini, J. M., Role of metal-induced reactive oxygen species generation in lung responses caused by residual oil fly ash, *J. Biosci.*, 28 (1), 13–18, 2003.

Li, L., Sun, T. D., Zhang, X., Lai, R. N., Li, X. Y., Fan, X. J., and Morinaga, K., Cohort studies on cancer mortality among workers exposed only to chrysotile asbestos: a meta-analysis, *Biomed. Environ. Sci.*, 17 (4), 459–468, 2004.

Liddell, F. D., McDonald, A. D., and McDonald, J. C., The 1891–1920 birth cohort of Quebec chrysotile miners and millers: development from 1904 and mortality to 1992, *Ann. Occup. Hyg.*, 41 (1), 13–36, 1997.

Lilis, R., Fibrous zeolites and endemic mesothelioma in Cappadocia, Turkey, . *J. Occup. Med.*, 23, 548–550, 1981.

Lipkin, L. E., Failure of attapulgite to produce tumours: prediction of this result by *in vitro* cytotoxicity test, In *In vitro Effects of Mineral Dusts*, Beck, E. G. and Bignon, J., eds., Springer, Berlin, 1985.

Liu, S., Liu, N., and Li, J., Silicosis caused by rice husk ashes, *J. Occup. Health*, 38, 57–62, 1996.

Luce, D., Billon-Galland, M. A., Bugel, I., Goldberg, P., Salomon, C., Fevotte, J., and Goldberg, M., Assessment of environmental and domestic exposure to tremolite in New Caledonia, *Arch. Environ. Health*, 59 (2), 91–100, 2004.

Maltoni, C., Minardi, F., and Morisi, L., Pleural mesotheliomas in Sprague-Dawley rats by erionite: first experimental evidence, *Environ. Res.*, 29, 238–244, 1982.

Maxim, L. D. and McConnell, E. E., A review of the toxicology and epidemiology of wollastonite, *Inhal. Toxicol.*, 17 (9), 451–466, 2005.

Mazziotti, S., Gaeta, M., Costa, C., Ascenti, G., Barbaro, M. L., Spatari, G., Settineri, N., and Barbaro, M., Computed tomography features of liparitosis: a pneumoconiosis due to amorphous silica, *Eur. Respir. J.*, 23 (2), 208–213, 2004.

Mazzucchelli, L., Radelfinger, H., and Kraft, R., Nonasbestos ferruginous bodies in sputum from a patient with graphite pneumoconiosis: a case report, *Acta Cytol.*, 40 (3), 552–554, 1996.

McConnell, E. E., Rutler, H. A., Ulland, B. M., and Moore, J. A., Chronic effects of dietary exposure to amosite asbestos and tremolite in F344 rats, *Environ. Health Perspect.*, 53, 27–44, 1983.

McConnochie, K., Bevan, C., Newcombe, R. G., Lyons, J. P., Skidmore, J. W., and Wagner, J. C., A study of Spanish sepiolite workers, *Thorax*, 48 (4), 370–374, 1993.

McDonald, J. C. and McDonald, A. D., The epidemiology of mesothelioma in historical context, *Eur. Respir. J.*, 10 (11), 2690–2691, 1996.

McDonald, J. C., Liddell, F. D., Dufresne, A., and McDonald, A. D., The 1891–1920 birth cohort of Quebec chrysotile miners and millers: Mortality 1976–1988, *Br. J. Ind. Med.*, 50 (12), 1073–1081, 1993.

McDonald, J. C., Harris, J., and Armstrong, B., Mortality in a cohort of vermiculite miners exposed to fibrous amphibole in Libby, Montana, *Occup. Environ. Med.*, 61 (4), 363–366, 2004.

Meranger, J. C. and Davey, A. B., *Non-Asbestos Fibre Content of Selected Consumer Products*, IARC Scientific Publications No. 90, 347–353, 1989.

Merchant, J. A., Human epidemiology: a review of fiber type and characteristics in the development of malignant and nonmalignant disease, *Environ. Health Perspect.*, 88, 287–293, 1990.

Merget, R., Bauer, T., Kupper, H. U., Philippou, S., Bauer, H. D., Breitstadt, R., and Bruening, T., Health hazards due to the inhalation of amorphous silica, *Arch. Toxicol.*, 75 (11–12), 625–634, 2002.

Merolla, L. and Richards, R. J., In vitro effects of water-soluble metals present in UK particulate matter, *Exp. Lung Res.*, 31 (7), 671–683, 2005.

Metintas, M., Metintas, S., Hillerdal, G., Ucgun, I., Erginel, S., Alatas, F., and Yildirim, H., Nonmalignant pleural lesions due to environmental exposure to asbestos: a field-based, cross-sectional study, *Eur. Respir. J.*, 26 (5), 875–880, 2005.

Mikulski, T., Mierzecki, A., Przepiera, A., and Swiech, Z., Hemolytic activity of phosphorite and apatite dusts *in vitro*, *Int. J. Occup. Med. Environ. Health*, 7 (1), 59–64, 1994a.

Mikulski, T., Podraza, H., Steciuk, W., and Swiech, Z., Assessment of the respiratory system in workers occupationally exposed to phosphorite and apatite dusts, *Int. J. Occup. Med. Environ. Health*, 7 (2), 119–124, 1994b.

Moore, M. A., Dumortier, P., Swaen, G. M., Pairon, J. C., and Brochard, P., Categorization and nomenclature of vitreous silicate wools, *Regul. Toxicol. Pharmacol.*, 35 (1), 1–13, 2002.

Moreno, T., Gibbons, W., Jones, T., and Richards, R., The geology of ambient aerosols: characterising urban and rural/coastal silicate $PM_{10-2.5}$ & $PM_{2.5}$ using high volume cascade collection and scanning electron microscopy, *Atmos. Environ.*, 37, 4265–4276, 2003.

Morgan, W. K., Donner, A., Higgins, I. T., Pearson, M. G., and Rawlings, W. Jr., The effects of kaolin on the lung, *Am. Rev. Respir. Dis.*, 138 (4), 813–820, 1988.

Mori, I., Nishikawa, M., Tanimura, T., and Quan, H., Change in size distribution and chemical composition of Kosa (Asian dust) aerosol during long-range transport, *Atmos. Environ.*, 37, 4253–4264, 2003.

Mossman, B. T. and Craighead, J. E., Comparative cocarcinogenic effects of crocidolite asbestos, hematite, kaolin, and carbon in implanted tracheal organ cultures, *Ann. Occup. Hyg.*, 26 (1–4), 553–567, 1982.

Mossman, B. T. and Sesko, A. M., In vitro assays to predict the pathogenicity of mineral fibers, *Toxicology*, 60, 53–61, 1990.

Muhle, H., Mermelstein, R., Dasenbrock, C., Takenaka, S., Mohr, U., Kilpper, R., MacKenzie, J., and Morrow, P., Lung response to test toner upon 2-year inhalation exposure in rats, *Exp. Pathol.*, 37 (1–4), 239–242, 1989.

Murphy, S. A., The response of lung epithelium to well characterized fine particles, *Life Sci.*, 62 (19), 1789–1799, 1998.

Murphy, S. A., BéruBé, K. A., and Richards, R. J., The bioreactivity of carbon black and diesel exhaust particles to primary epithelial type 2 and Clara cell cultures, *Occup. Environ. Med.*, 56, 813–819, 1999.

National Institute for Occupational Safety and Health, Criteria for a Recommended Standard: Occupational Exposure to Crystalline Silica, Publication No. 75–120, Department of Health, Education, and Welfare, Health Services and Mental Health Administration, Cincinnati, Ohio, 1974.

National Institute for Occupational Safety and Health, Health Effects of Occupational Exposure to Respirable Crystalline Silica, Publication No. 2002-129, Department of Health and Human Services, Centers for disease control and prevention, Cincinnati, Ohio, 2002.

Nemery, B., Metal toxicity and the respiratory tract, *Eur. Respir. J.*, 3 (2), 202–219, 1990.

Newhouse, M. L., Oakes, D., and Wooley, A. J., Mortality of welders and other craftsmen at a shipyard in northeast England, *Br. J. Ind. Med.*, 42, 406–410, 1985.

Nicolai, T., Carr, D., Weiland, S. K., Duhme, H., von Ehrenstein, O., Wagner, C., and von Mutius, E., Urban traffic and pollutant exposure related to respiratory outcomes and atropy in a large sample of children, *Eur. Respir. J.*, 21 (6), 913–915, 2003.

Nolan, R. P., Langer, A. M., and Wilson, R., A risk assessment for exposure to grunerite asbestos (amosite) in an iron ore mine, *Proc. Natl Acad. Sci.*, 96 (7), 3412–3419, 1999.

Nouh, M. S., Is the desert lung syndrome (non-occupational dust pneumoconiosis) a variant of pulmonary alveolar microlithiasis? Report of 4 cases with review of the literature, *Respiration*, 55 (2), 122–126, 1989.

Oberdörster, G., Ferin, J., and Lehnert, B. E., Correlation between particle size, *in vivo* particle persistence, and lung injury, *Environ. Health Perspect.*, 102 (5), 173–179, 1994.

Oberdörster, G., Maynard, A., Donaldson, K., Castranova, V., Fitzpatrick, J., Ausman, K., Carter, J. et al., ILSI Research Foundation/Risk Science Institute Nanomaterial Toxicity Screening Working Group, Principles for characterizing the potential human health effects from exposure to nanomaterials: elements of a screening strategy. Part. Fibre, *Toxicology*, 2, 8, 2005a.

Oberdörster, G., Oberdörster, E., and Oberdörster, J., Nanotoxicology: an emerging discipline evolving from studies of ultrafine particles, *Environ. Health Perspect.*, 113 (7), 823–839, 2005b.

Occupational Safety and Health Administration (OSHA), USA, http://www.osha.gov/SLTC/pel/, 2005.

Oldham, P. D., Pneumoconiosis in Cornish china clay workers, *Br. J. Ind. Med.*, 40, 609–619, 1983.

Ortmeyer, C. E., Baier, E. J., and Crawford, G. M. Jr., Life expectancy of Pennsylvania coal miners compensated for disability, *Arch. Environ. Health*, 27, 227–230, 1973.

Parent, M.-E., Siemiatycki, J., and Renaud, G., Case-control study of exposure to carbon black in the occupational setting and risk of lung cancer, *Am. J. Ind. Med.*, 30, 285–292, 1996.

Park, R., Rice, F., Stayner, L., Smith, R., Gilbert, S., and Checkoway, H., Exposure to crystalline silica, silicosis, and lung disease other than cancer in diatomaceous earth industry workers: a quantitative risk assessment, *Occup. Environ. Med.*, 59 (1), 36–43, 2002.

Parsons, B., Salter, L., Coe, T., Mathias, R., Richards, R. J., and Jones, T. P., Airborne particulate matter (PM10) in the china clay area, Cornwall, UK, *Min. Mag.*, 67 (2), 16–169, 2003.

Peters, K., Unger, R. E., Kirkpatrick, C. J., Gatti, A. M., and Monari, E., Effects of nano-scaled particles on endothelial cell function *in vitro*: studies on viability, proliferation and inflammation, *J. Mater. Sci. Mater. Med.*, 15 (4), 32–325, 2004.

Pfau, J. C., Sentissi, J. J., Weller, G., and Putnam, E. A., Assessment of autoimmune responses associated with asbestos exposure in Libby, Montana, USA, *Environ. Health Perspect.*, 113 (1), 25–30, 2005.

Phibbs, B. P., Sundin, R. E., and Mitchell, R. S., Silicosis in Wyoming bentonite workers, *Am. Rev. Respir. Dis.*, 103, 1–17, 1971.

Platek, S. F., Riley, R. D., and Simon, S. D., The classification of asbestos fibres by scanning electron microscopy and computer-digitizing tablet, *Ann. Occup. Hyg.*, 36 (2), 155–171, 1992.

Rabovsky, J., Biogenic amorphous silica, *Scand. J. Work Environ. Health*, 21 (2), 108–110, 1995.

Radon, K., Nowak, D., Heinrich-Ramm, R., and Szadkowski, D., Respiratory health and fluoride exposure in different parts of the modern primary aluminum industry, *Int. Arch. Occup. Environ. Health*, 72 (5), 297–303, 1999.

Renier, A., Fleury, J., Monchaux, G., Nebut, M., Bignon, J., and Jaurand, M. C., *Toxicity of an Attapulgite Sample Studied in Vivo and in Vitro*, IARC Scientific Publications 90, 180–184, 1989.

Rimala, B., Greenbergac, A. K., and William, R. N., Basic pathogenetic mechanisms in silicosis: current understanding, *Curr. Opin. Pulm. Med.*, 11, 169–173, 2005.

Roberts, J. R., Taylor, M. D., Castranova, V., Clarke, R. W., and Antonini, J. M., Soluble metals associated with residual oil fly ash increase morbidity and lung injury after bacterial infection in rats, *J. Toxicol. Environ. Health A*, 67 (3), 251–263, 2004.

Robertson, J. M., Epidemiologic studies in North American carbon black workers, *Inhal. Toxicol.*, 8, 41–50, 1996.

Robertson, J. M. and Inman, K. J., Mortality in carbon black workers in the United States, *J. Occup. Environ. Med.*, 38 (6), 569–570, 1996.

Robertson, J. M., Diaz, J. F., Fyfe, I. M., and Ingalls, T. H., A cross-sectional study of pulmonary function in carbon black workers in the United States, *Am. Ind. Hyg. Assoc. J.*, 49 (4), 161–166, 1988.

Rodelsperger, K., Bruckel, B., Manke, J., Woitowitz, H. J., and Pott, F., Potential health risks from the use of fibrous mineral absorption granulates, *Br. J. Ind. Med.*, 44 (5), 337–343, 1987.

Roggli, V. L., Vollmer, R. T., Butnor, K. J., and Sporn, T. A., Tremolite and mesothelioma, *Ann. Occup. Hyg.*, 46 (5), 447–453, 2002.

Roller, M., Pott, F., Kamino, K., Althoff, G. H., and Bellmann, B., Results of current intraperitoneal carcinogenicity studies with mineral and vitreous fibres, *Exp. Toxicol. Pathol.*, 48, 3–12, 1996.

Rombala, G. and Guardascione, V., La Silicosi da bentoni, *Med. Lav.*, 46, 480–497, 1955.

Rundle, E. M., Sugar, E. T., and Ogle, C. J., Analyses of the 1990 chest health survey of china clay workers, *Br. J. Ind. Med.*, 50, 913–919, 1993.

Santec, Z., Puvacic, S., Radovic, S., and Puvacic, Z., Vineyard pesticide induced changes in the lungs: experimental studying on rabbits, *Med. Arh.*, 59 (6), 343–345, 2005.

Scancarello, G., Romeo, R., and Sartorelli, E., Respiratory disease as a result of talc inhalation, *J. Occup. Environ. Med.*, 38 (6), 610–614, 1996.

Searl, A., Nicholl, A., and Baxter, P. J., Assessment of the exposure of islanders to ash from the Soufriere Hills volcano, Montserrat, British West Indies, *Occup. Environ. Med.*, 59 (8), 523–531, 2002.

Sebastien, P., McDonald, J. C., Cornea, G., and Gachem, A., Electron microscopical characterization of respirable airborne particles in a Tunisian phosphate mine, In *Proceedings of the 6th International Pneumoconiosis Conference, Bochum, Federal Republic of Germany, 20–23 September, 1983*, Bergban-Berufsgenossenschaft, International Labour Office, 3, 1650–1665, 1983.

Sheers, G., The china clay industry—lessons for the future of occupational health, *Respir. Med.*, 83, 173–175, 1989.

Shinohara, Y. and Kohyama, N., Quantitative Analysis of Tridymite and Cristobalite Crystallized in Rice Husk Ash by Heating, *Ind. Health*, 42, 277–285, 2004.

Sincock, A. M., Delhantly, J. D. A., and Casey, G., A comparison of the cytogenetic response to asbestos and glass fibre in Chinese hamster and human cell lines. Demonstration of growth inhibition in primary human fibroblasts, *Mutat. Res.*, 101, 257–268, 1982.

Smith, D., Opal, cristobalite and tridymite: noncrystallinity versus crystallinity, nomenclature of the silica minerals and bibliography, *Powder Diffr.*, 14, 2–19, 1998.

Sorensen, M., Schins, R. P., Hertel, O., and Loft, S., Transition metals in personal samples of PM2.5 and oxidative stress in human volunteers, *Can. Epidemiol. Biomarkers Prev.*, 14 (5), 1340–1343, 2005.

Spurney, K., Potts, F., Huth, F., Weiss, G., and Opiela, H., Identification and carcinogenic effect of fibrous actinolite of a quarry, *Staub-Reinhalt. Luft*, 39, 386–389, 1979.

Stanton, M. F., Layard, M., Tegerin, A., Miller, E., May, M., and Kent, E., Carcinogenicity of fibrous glass: pleural response in the rat in relation to fibre dimension, *J. Natl Cancer Inst.*, 58, 587–603, 1977.

Stanton, M. F., Layard, M., Tegeris, A., Miller, E., May, M., Morgan, E., and Smith, A., Relation of particle dimension to carcinogenicity in amphibole asbestoses and other fibrous minerals, *J. Natl Cancer Inst.*, 67, 965–975, 1981.

Steenland, K., Mannetje, A., Boffetta, P., Stayner, L., Attfield, M., Chen, J., Dosemeci, M. et al., Pooled exposure-response analyses and risk assessment for lung cancer in 10 cohorts of silica-exposed workers: an International Agency for Research on Cancer multi-centre study, *Cancer Causes Control*, 12 (9), 773–784, 2001.

Stefaniak, A. B., Hoover, M. D., Day, G. A., Dickerson, R. M., Peterson, E. J., Kent, M. S., Schuler, C. R., Breysse, P. N., and Scripsick, R. C., Characterization of physicochemical properties of beryllium aerosols associated with prevalence of chronic beryllium disease, *J. Environ. Monit.*, 6 (6), 523–532, 2004.

Stone, R., Air pollution, counting the cost of London's killer smog, *Science*, 298 (5601), 2106–2107, 2002. December 13.

Straif, K., Chambless, L., Weiland, S. K., Wienke, A., Bungers, M., Taeger, D., and Keil, U., Occupational risk factors for mortality from stomach and lung cancer among rubber workers: an analysis using internal controls and refined exposure assessment, *Int. J. Epidemiol.*, 28, 1037–1043, 1999.

Szozda, R., The respiratory health of carbon black workers differences between Polish, West European and American Scientific Reports, *J. UOEH*, 16 (1), 91–95, 1994.

Tatrai, E., Kovacikova, Z., Brozik, M., and Six, E., Pulmonary toxicity of wollastonite *in vivo* and *in vitro*, *J. Appl. Toxicol.*, 24 (2), 147–154, 2004.

Tomasini, M., Forni, A., Rivolta, G., Mantegazza, D., and Chiappino, G., Talcosis–asbestosis: an unusual risk in a food industry, *G. Ital. Med. Lav.*, 10 (3), 111–113, 1988.

United States Environmental Protection Agency (USEPA), Support: epidemiology Studies of Xerox Employees Exposed To Toner, with Attachments & Cover Letter Dated 11/15/00 (EPA/OTS;Doc #8EHQ-1100-0651S), 2000.

United States Environmental Protection Agency (USEPA), http://www.epa.gov/asbestos/pubs/insulation.html, 2005.

U.S. Geological Survey (USGS), http://volcanoes.usgs.gov/ash/index.html, 2005.

Vallyathan, V., Castranova, V., Pack, D., Leonard, S., Shumaker, J., Hubbs, A. F., Shoemaker, D. A., Ramsey, D. M., and Pretty, J. R., Freshly fractured quartz inhalation leads to enhanced lung injury and inflammation in rats, *Am. J. Respir. Crit. Care Med.*, 152, 1003–1009, 1995.

Van-Gosen, B. S., Lowers, H. A., Bush, H. A., Meeker, G. P., Plumlee, G. S., Brownfield, I. K., and Sutley, J. S., Reconnaissance study of the geology of U.S. vermiculite deposits; are asbestos minerals common constituents? U.S. Geological Survey Bulletin 8, 2002.

Venter, E., Nyantumbu, B., Solomon, A., and Rees, D., Radiologic abnormalities in South African mica millers: a survey of a mica milling plant in the Limpopo Province, *Int. J. Occup. Environ. Health*, 10 (3), 278–283, 2004.

Veys, C. A., A study of the mortality experience at the Stoke (UK) factory 1951–1990, *Hum. Exp. Toxicol.*, 18 (8), 513, 1999.

Wagner, J. C., Berry, G., Cooke, T. L., Hill, R. J., Pooley, F. D., and Skidmore, J. W., Animal experiments with talc, In *Inhaled particles. IV. Part 2. Proceedings of an International Symposium, Edinburgh, 22–26 September, 1975*, Walton, W. H., ed., Pergamon Press, Oxford, 647–654, 1977.

Wagner, J. C., Skifmore, J. W., Hill, R. J., and Griffiths, D. M., Erionite exposure and mesotheliomas in rats, *Br. J. Cancer*, 51, 727–730, 1985.

Wagner, J. C., Pooley, F. D., Gibbs, A., Lyons, J., Sheers, G., and Moncrieff, C. B., Inhalation of china stone and china clay dusts: relationship between the mineralogy of dust retained in the lungs and pathological changes, *Thorax*, 41 (3), 190–196, 1986.

Wallace, W. E., Harrison, J. C., Grayson, R. L., Keane, M. J., Bolsaitis, P., Kennedy, R. D., Wearden, A. Q., and Attfield, M. D., Aluminosilicate surface contamination of respirable quartz particles from coal mine dusts and from clay works dusts, *Ann. Occup. Hyg.*, 38, 439–445, 1994.

Wardenbach, P., Rodelsperger, K., Roller, M., and Muhle, H., Classification of man-made vitreous fibers: comments on the revaluation by an International Agency for Research on Cancer working group, *Regul. Toxicol. Pharmacol.*, 43 (2), 181–193, 2005.

Warheit, D. B., Hansen, J. F., Yuen, I. S., Kelly, D. P., Snajdr, S. I., and Hartsky, M. A., Inhalation of high concentrations of low toxicity dusts in rats results in impaired pulmonary clearance mechanisms and persistent inflammation, *Toxicol. Appl. Pharmacol.*, 145 (1), 10–22, 1997.

Weill, H., Jones, R. N., and Parkes, W. R., Silicosis and related diseases, In *Occupational Lung Disorders*, Parkes, W. R., ed., 3rd ed., Butterworth-Heinemann Ltd., London, 285–339, 1994.

Weinhold, B., Heavy on the metals: parsing the particulate content of the London smog, *Environ. Health Perspect.*, 111 (9), A481, 2003.

Whittaker, A., Jones, T. P., BéruBé, K. A., Maynard, R., and Richards, R. J., Killer smogs of the 1950s London: composition of the particles and their bioactivity, *Sci. Total Environ.*, 334–335, 435–445, 2004.

Wilson, M., Stone, V., Cullen, R. T., Searl, A., Maynard, R. L., and Donaldson, K., In vitro toxicology of respirable Montserrat volcanic ash, *Occup. Environ. Med.*, 57 (11), 727–733, 2000.

Wozniak, H., Wiecek, E., Tossavainen, A., Lao, I., and Kolakowski, J., Comparative studies of fibrogenic properties of wollastonite, chrysotile and crocidolite, *Med. Pr.*, 37 (5), 288–296, 1996.

Wright, W. W., Rom, W. N., and Moatmed, F., Characterization of zeolite fiber sizes using scanning electron microscopy, *Arch. Environ. Health*, 38, 99–103, 1983.

Wright, R. S., Abraham, J. L., Harber, P., Burnett, B. R., Morris, P., and West, P., Fatal asbestosis 50 years after brief high intensity exposure in a vermiculite expansion plant, *Am. J. Respir. Crit. Care Med.*, 165 (8), 1145–1149, 2002.

Xu, Y. H., Zhang, S. J., Wu, C. Q., Zhou, J., Chang, H. F., and Yu, S. Q., The pathogenicity of dust from a fluorite mine, *Environ. Res.*, 43 (2), 350–358, 1987.

Yamamoto, A., Honma, R., Sumita, M., and Hanawa, T., Cytotoxicity evaluation of ceramic particles of different sizes and shapes, *J. Biomed. Mater. Res. A*, 68 (2), 244–256, 2004.

Yano, E., Wang, Z. M., Wang, X. R., Wang, M. Z., and Lan, Y. J., Cancer mortality among workers exposed to amphibole-free chrysotile asbestos, *Am. J. Epidemiol.*, 154 (6), 538–543, 2001.

Yu, I. J., Song, K. S., Chang, H. K., Han, J. H., Kim, K. J., Chung, Y. H., Maeng, S. H. et al., Lung fibrosis in Sprague-Dawley rats, induced by exposure to manual metal arc-stainless steel welding fumes, *Toxicol. Sci.*, 63 (1), 99–106, 2001.

Yu, I. J., Song, K. S., Maeng, S. H., Kim, S. J., Sung, J. H., Han, J. H., Chung, Y. H. et al., Inflammatory and genotoxic responses during 30-day welding-fume exposure period, *Toxicol. Lett.*, 154 (1–2), 105–115, 2004.

Zeigler, T. L., Plumlee, G. S., Lamothe, P. J., Meeker, G. P., Witten, M. L., Sutley, S. J., Hinkley, T. K. et al., Mineralogical, geochemical, and toxicological variations of asbestos toxicological standards and amphibole samples from Libby, MT, In *Geological Society of America*, 2002 *Annual Meeting, Abs. w. Prog.*, 34 (6), 146, 2002.

Zhang, Q., Kusaka, Y., Sato, K., Nakakuki, K., Kohyama, N., and Donaldson, K., Differences in the extent of inflammation caused by intratracheal exposure to three ultrafine metals: role of free radicals, *J. Toxicol. Environ. Health A*, 53 (6), 423–438, 1998.

Zhuang, G. S., Guo, J. H., Yuan, H., and Zhao, C. Y., The compositions, sources, and size distribution of the dust storm from China in spring of 2000 and its impact on the global environment, *Chin. Sci. Bull.*, 46, 895–901, 2001.

Zimmer, A. T., Baron, P. A., and Biswas, P., The influence of operating parameters on number-weighted aerosol size distribution generated from a gas metal arc welding process, *J. Aerosol Sci.*, 33 (3), 519–531, 2002.

Zinman, C., Richards, G. A., Murray, J., Phillips, J. I., Rees, D. J., and Glyn-Thomas, R., Mica dust as a cause of severe pneumoconiosis, *Am. J. Ind. Med.*, 41 (2), 139–144, 2002.

3 Particle Dosimetry: Deposition and Clearance from the Respiratory Tract and Translocation Towards Extra-Pulmonary Sites

Wolfgang G. Kreyling
Institute of Inhalation Biology and Focus Network Aerosols and Health, GSF–National Research Center for Environment and Health

Winfried Möller
Institute of Inhalation Biology and Clinical Research Group "Inflammatory Lung Diseases," GSF–National Research Center for Environment and Health

Asklepios Hospital for Respiratory Diseases

Manuela Semmler-Behnke
GSF–National Research Center for Environment and Health

Günter Oberdörster
University of Rochester Medical Center, University of Rochester

CONTENTS

3.1 INTRODUCTION

Particulate matter (PM) in ambient air is a complex mixture of multiple components ranging from a few nanometers in size to tens of micrometers. Primary particles are directly emitted as liquids or solids from sources such as biomass burning, incomplete combustion of fossil fuels, volcanic eruptions, and wind-driven or traffic-related suspension of road, soil, and mineral dust, sea salt, and biological materials (plant fragments, microorganisms, pollen, etc.). Secondary particles, on the other hand, are formed by gas-to-particle conversion in the atmosphere (new particle formation by nucleation and condensation of gaseous precursors). Airborne particles undergo various physical and chemical interactions and transformations (atmospheric aging), that is, changes of particle size, structure, and composition (coagulation, restructuring, gas uptake, chemical reaction). PM develops dynamically as a reactive system in time and space, depending both on sources and weather conditions. The total particle number and mass concentrations typically vary in the range of about 10^2–10^5 cm^{-3} and 1–100 μg m^{-3}, respectively.[1] While ultrafine particles (UFPs) (<100 nm in diameter) dominate the number concentration of ambient aerosols, the mass concentration is dominated by the accumulation fraction of particles of 0.1–2.5 μm in size. In general, the predominant chemical components of air PM are sulphate, nitrate, ammonium, sea salt, mineral dust, organic compounds, and black or elemental carbon (EC), each of which typically contribute about 10%–30% of the overall mass load.

Particularly, the anthropogenic fraction of ambient particles is undergoing changes according to the changes of the anthropogenic emission sources. At most locations within industrialized countries PM mass concentration measurements showed a dramatic decline over the last five decades. Unfortunately, other metrics like number concentration are more or less not available during this time period; however, when comparative measurements are available, the decline in mass concentration was not accompanied by the decline in number concentration. The latter remained more or less constant over the decade of the 1990s in the city of Erfurt, Germany, while the PM mass concentration declined by 70%.[2] The changes observed in Erfurt were predominantly determined by the dramatic social and industrial changes in Eastern Germany as a result of the reunification of Germany. The ambient aerosol changes in Erfurt, caused by the decline of the former German Democratic Republic and German unification during the last decade, provide a unique, time-lapse motion picture for similar, but less rapid, changes in other Western urban environments during previous decades. It shows the transition from high particle mass concentrations towards low mass concentrations at stagnant total particle number concentration, but increasing number concentrations of 0.01–0.03 μm sized particles. Indeed, transitions documented by declining mass concentration parameters started in Western urban areas earlier in the 1950s and lasted several decades. Episodic measurements of urban particle number concentrations in the late 1970s showed similar bimodal distributions as the Erfurt spectra of winter 1991/1992 with prominent number concentrations in the size range of 0.1–0.5 μm.[3] The declining mass concentration resulted from the decreasing number concentration of particles larger than 0.1 μm, as confirmed by the Erfurt data indicating a spectral shift towards smaller particles.[2]

It needs to be emphasized that among the multiple components present in ambient PM, many toxic substances are found. However, none of these reach concentrations of toxicological relevance, as is known for occupational exposures.[4] Hence, adverse health effects are unlikely to be related to a single PM component, but more likely to a complex, possibly synergistic interaction of multiple components with the respiratory tract and other target organs.

In the past, the most common measurement used has been "particle mass concentration," and daily averages range nowadays from 20 to 50 $\mu g/m^3$ in European and North American cities (PM10, particle mass concentration in the size range < 10 μm aerodynamic diameter). Taking the PM size range over more than four decades into account, this mass concentration measure is determined largely by PM in the coarse and accumulation mode fraction and basically neglects UFPs < 100 nm in size. The limitation of this metric may be illustrated by the fact that the water-solubility of ambient PM may vary from 20 to 80% of PM mass, and yet the toxicity of soluble compounds is unlikely to be similar to that of the insoluble fraction.

With closer insights into particle-lung interactions, other measures, such as the concentration of particle number and surface area, need to be taken into account, depending on whether ultrafine or larger particles are to be considered. However, exposure measures may be inadequate, since it may be that the number of deposited particles per unit surface area of airways, airway bifurcations (crest at airway division), and alveoli, or the dose to a specific cell type like macrophages, determines the response for specific regions. Therefore, the use of a metric depends on the specific questions posed, requiring specifically defined measures.

On the walls (epithelium) of the respiratory tract, particles first come into contact with the mucous or serous lining fluid and its surfactant layer on top. Therefore, the fate of particle compounds that are soluble in this lining fluid needs to be distinguished from that of slower-dissolving or even insoluble compounds.

Soluble particle compounds will be dissolved, bound to proteins, often metabolized in the lining fluid and will eventually be transferred to the blood, undergoing further metabolization and possible excretion. In this way they have the potential to reach any organ[5–7] and to induce toxic effects far from their site of entry into the lungs.

3.2 RELEVANT PARTICLE PARAMETERS REGARDING PARTICLE DOSIMETRY

3.2.1 Particle Size

Figure 3.2 depicts the range of sizes of airborne ambient PM, including the nucleation-mode, Aitken-mode, accumulation-mode, and coarse-mode particles. Ambient particles < 0.1 μm, defined as UFPs in the toxicologic literature, consist of transient nuclei or Aitken nuclei. Particles with at least one dimension below 0.1 μm are commonly considered as nanoparticles (NP).

Particle transport in the gas phase is determined by its aerodynamic behavior and is quantitated by the aerodynamic diameter.[8] However, under breathing conditions in the human respiratory tract, this diameter is only defined for particles with an aerodynamic diameter larger than 0.5 μm; below this size particle transport is dominated by diffusion properties which are determined by thermodynamic properties of the gas phase; in this case particles are described according to their thermodynamic diameter as first suggested by Heyder and coworkers.[9]

Evaluation of size mode alone as a modulating factor in PM toxicity is difficult since it is not independent of chemical composition (i.e., certain size modes tend to contain certain chemical components, e.g., metals in the fine mode and crustal materials in the coarse mode). Furthermore, there are clear differences between particles in different size modes in terms of total and regional dosimetry within the respiratory tract, and subsequent pathways and rates of relocation/clearance both within and outside of the respiratory tract.

3.2.2 Particle Shape

Particle shape varies from perfect spheres, such as droplets, to fibers, and more recently, nanotubes of extremely small diameters of few nanometers and of lengths in the mm range. Non-spherical particles are described by the aspect ratio of diameter and length. Particle shape is known to affect the aerodynamic, as well as the diffusive behavior of the particles. Due to the alignment of fibers along the stream lines of the inhaled gas, particles of several 10 μm length can penetrate down to the lung periphery. Just recently it was shown that particle shape also affects phagocytosis by macrophages.[10] Alveolar macrophages (AM) with diameters in the size range of 10–15 μm are not able to completely engulf and clear such fiber particles from the alveolar epithelium. Therefore such long fibers can stay for long times in the lung and they can be the origin of cancerous transformations.[11]

3.2.3 Particle Composition, Homogeneity, Heterogeneity

The complex heterogeneity of aerosol composition does not only vary between particles, but also within individual particles. Compounds can be distinguished by their volatility or solubility in aqueous or lipid solvents, versus compounds which cannot escape the particle structure by evaporation or dissolution/desorption processes. In this case, it is helpful to distinguish the extractable shell from the stable core of these particles. Note that shells may not only exist as an outer layer around the particle core, but it can intrude into a porous core.

The core of a particle has to be distinguished from its surface. Particle matrices of solid particles may be made up of mono- and/or poly-crystalline lattice structures, versus amorphous molecular assemblies on the surface. Liquid particles may consist of different solvents (aqueous vs organic) containing numerous solutes in each of their phases. The surface may comprise of molecules of the crystalline structure extending to the particle surface, or may contain a molecular layer coating the particle core with a ligand, etc. This coating may consist of different molecular constituents than the core. Since the surface provides the interface to the outside, the surface structures and composition may undergo chemical and physical changes that will not necessarily affect the particle core.

3.2.4 Particle Density

Particle density—usually expressed in g/cm^3—refers to the mass per unit volume of the particle itself. Liquid particles, and ground or crushed particles, have a density equal to that of their parent material. Smoke and fume particles may have apparent densities significantly less than that predicted from their chemical composition. Generally, the mean particle density is the sum of the densities of all gaseous, liquid, and solid compounds multiplied by their volume fractions within the particle. Note for hollow or porous particles, the mean density may be smaller than 1 g/cm^3 because of the large volume fraction of the gas phase. Particle density is taken into account by the definition of the aerodynamic diameter of aerosol particles, which is the diameter of a unit density sphere having an equivalent settling velocity in air, to that of the particle.

Measurements of the mean apparent density of all particles of an ambient aerosol are integrating over all particle densities, and have been based on simultaneous measurements of various aerosols parameters.[12] While the particle density is an important parameter of the aerodynamic behavior during inhalation, it is of minor relevance after the particle has deposited on the epithelial wall of the respiratory tract.

3.2.5 Particle Surface Parameters

Particle surface is described by various parameters resulting in considerable confusion. The most simple surface area is calculated as the spherical surface of an equivalent diameter, which had been determined by any kind of size determining methodology. The "active surface area" is used in aerosol science and takes only the outer surface area into account, which is able to interact with other aerosol particles, for instance during coagulation.[13,14] In contrast, the Brunaver, Emmelt and Teller (BET) surface area determined by N_2 or other gas molecule absorption, describes the entire porous surface, including subnanometer pores. Considering surface area of particles that contain soluble compounds, one needs to distinguish between the original surface area and that of the insoluble core of the particles. Specific interaction with the biological environment will generally be based on the properties of the particle surface area and less likely on the core of the particle, as will be further discussed.

Surface charge is the result of the structure of polar molecular assemblies at the surface of the particle in various liquids, such that the particle appears to have a net-charge in an electrical field. Usually this phenomenon is measured as the zeta potential, which determines the particle mobility in a uniform external electrical field. Since surface molecules are much less tightened to the particle than inside molecules, the surface charge undergoes changes depending on the fluid media properties, such as ionic strength, proton concentration, and of other constituents.

3.2.6 Particle Properties and Toxicological Aspects

The toxicity of PM may be due merely to the particle's presence on biological tissues, to its chemical constituents, including any adsorbed components, or to some combination of these factors. Toxicological research has examined a number of PM physico-chemical characteristics in relation to their potential for induction of adverse biological effects. In its final report, the U.S. National Research Council's (NRC) Committee on Research Priorities for Airborne PM (National Research Council and Committee on Research Priorities for Airborne Particulate Matter 2004) provided a list of particle characteristics that may be important to health responses, including size and size distributions, mass concentration, number concentration, acidity, particle surface chemistry, particle core chemistry, metals, carbon (organic carbon (OC) and black or EC), biogenic origin, secondary inorganic aerosols, and material associated with the earth's crust. Other characteristics that have been recognized as potentially playing a role in toxicity are particle surface area, and other surface properties like charge and structure, chemical reactivity, water solubility of constituent chemicals, and geometric form of the particles. In addition, the particle number itself is likely to play an important role since the smaller particles are the more particles are dispersed at a given particle mass such that many more cells or cellular components interact with them, compared to a few large particles. This listing, of course, is based upon those characteristics that have undergone the most scrutiny by investigators in the field.

3.2.7 Exposure–Dose–Response

Of central importance in toxicology is the Exposure–Dose–Response paradigm. With respect to inhaled particulate compounds, inhaled, deposited, and retained doses need to be differentiated. The

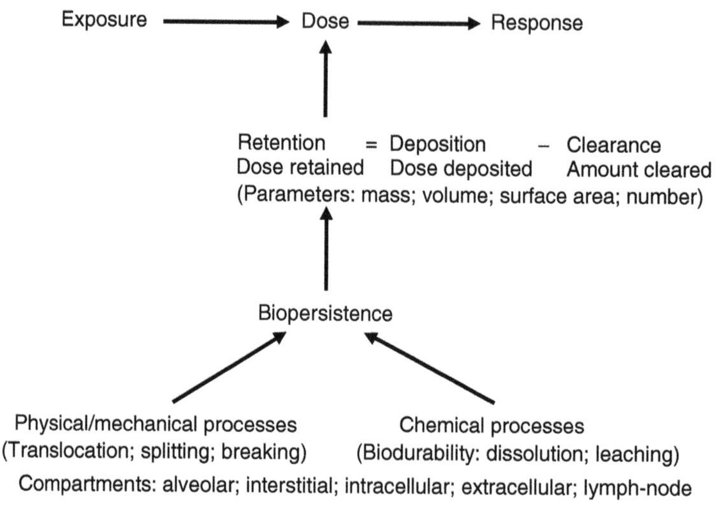

FIGURE 3.1 Biopersistence and biodurability in the lung in relation to dose parameters for exposure-dose-response relationships of inhaled nonfibrous and fibrous particles. (From Oberdörster, G., Ferin, J., and Lehnert, B. E., *Environ. Health Perspect.*, 102 (Suppl. 5), 173–179, 1994.)

initially deposited dose in a given region of the respiratory tract is subject to different clearance processes, dependent on the site of deposition and on physico-chemical characteristics of the particle in question. As a result, the dose retained at a certain time after deposition will change. Figure 3.1 depicts the different physical and chemical processes which determine the overall biopersistence of a particle, and thereby, the retained dose at a given time.[15]

The retained dose is determined by the biopersistence of a particulate fibrous or nonfibrous compound, which in turn is based on several clearance mechanisms, physical/mechanical (e.g., translocation via mucociliary escalator, alveolar macrophage phagocytosis, splitting, breaking) and chemical (leaching, dissolution). The terms retention, biopersistence, and biodurability have often been used interchangeably; however, we suggest that "biopersistence" should refer to the general *in vivo* behavior of a particle (fibrous or nonfibrous), and that retention should reflect a dosimetric term (retained dose; retention halftime). The term "biodurability" should be reserved for chemical processes occurring *in vivo*, contributing to biopersistence. The mechanisms that influence biopersistence can vary for different pulmonary compartments (e.g., alveolar vs. interstitial; intracellular vs. extracellular).

Although doses are usually expressed as particle mass, other dosemetrics such as particle volume, particle surface area, or particle number may be more appropriate, depending on the elicited response and underlying mechanisms. For example, it has been hypothesized that the retardation of alveolar macrophage-mediated clearance occurring in a condition of particle overload in the lung is due to the phagocytized particle volume in macrophages rather than particle mass.[16] Whereas this hypothesis did well explaining lung overload-induced prolonged alveolar clearance of poorly soluble low toxicity particles, it did not correlate well with the lung burden of UFPs (e.g., TiO_2) causing clearance retardation but no volumetric overload. Rather, particle surface area was the dosemetric that correlated well with altered particle retention kinetics, pulmonary inflammatory responses, and carcinogenicity induced by poorly soluble, low-toxicity particles of a range of particle sizes from fine to ultrafine.[15,17] The concept of particle surface as dosemetric has now become generally accepted to be critical for explaining the biological/toxicological/chemical activity of NPs; however, in addition to just particle surface area, it has become apparent that other particle characteristics need to be considered as well, as is discussed later in this chapter.

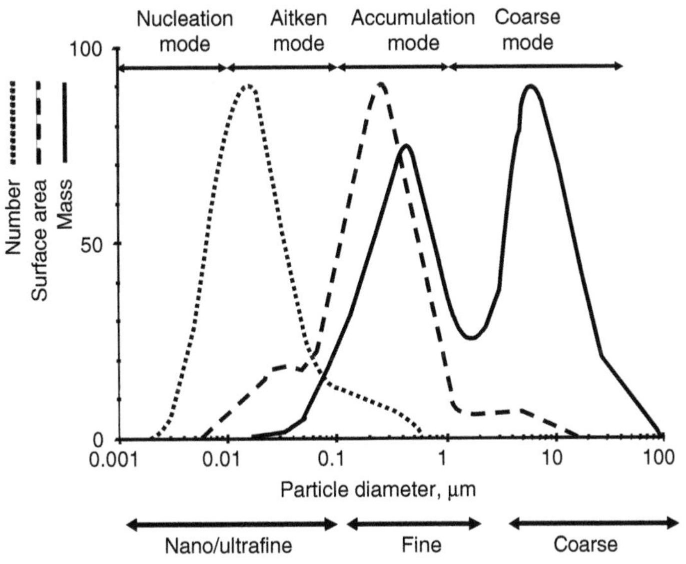

FIGURE 3.2 Particle size distribution of ambient aerosol by three characteristics: mass, surface area, and number; size ranges of different particle modes. (Based on Finlayson-Pitts and Pitts from Finlayson-Pitts, B. J. and Pitts, J. N., *Chemistry of the Upper and Lower Atmosphere: Theory, Experiments, and Applications*, Academic Press, San Diego, 2000.)

For example, different physiological clearance processes in the alveolar space vs. the tracheobronchial region (see Section 3.4) affect retention kinetics; different pH values in extracellular vs. intracellular (acidic phagolysosome) sites translate into different dissolution rates of a particle.

3.2.8 Do Man and Animals Get the Same PM Dosages and Distribution in the Lungs?

The probability of any biological effect occurring in humans or animals depends on deposition and retention of particles, as well as the metabolic fate of the particle compounds and the underlying tissue sensitivity. Extrapolation of airway and lung dosages between species must consider these differences in evaluating dose–response relationships, as was discussed recently[18–20] and which will be summarized below. Thus, extrapolation of deposition patterns from most healthy animal models can be performed at least to some extent, since the differences in anatomy and breathing conditions are widely known.

However, subsequent particle retention, redistribution within the lungs, and clearance pathways towards other organs and out of the body are based on rather complex mechanisms and differ consistently between rodent models and man. As a result, extrapolation will only be possible under limited specific conditions.

3.3 TOTAL AND REGIONAL PARTICLE DEPOSITION

3.3.1 Particle Transformation during Inhalation

While airborne the aerodynamic properties of inhaled particles, the fluid dynamics during breathing, and the anatomy of the respiratory tract will determine their deposition probability on the surfaces of the various regions and anatomical sites of the respiratory tract. While hygroscopic growth was discussed in the past to affect the size of particles, resulting in alterations of the deposition probability in the various regions of the respiratory tract,[21] chain aggregated particles may undergo structural transformation in rounding up and compacting under the surface tension of an aqueous film taken up by the aggregated surface in the high relative humidity of the respiratory tract.[22–25] Compared to the aggregated state the compacted particles change their mobility diameter, resulting in a change of the deposition probability in the respiratory tract.

3.3.2 Mechanisms of Deposition and Modeling

Particle deposition in the respiratory tract is determined predominantly by three mechanisms that move particles out of the streams of inhaled and exhaled air towards the airway walls, where they are deposited. These mechanisms are (1) sedimentation by gravitational forces acting on particles > 0.5 μm aerodynamic diameter, (2) impaction caused by their inertial mass in branching airways acting on particles > 1.5 μm aerodynamic diameter, and (3) diffusional motion of particles < 0.5 μm (thermodynamic diameter) by thermal motion of air molecules. Each mechanism becomes relevant for the given size range. These physical mechanisms of particle motion affecting particle deposition work through three important components (aerosol properties, and physiology) during breathing: (a) particle dynamics including the size and shape and its possible dynamic change during breathing, (b) geometry of the branching airways and the alveolar structures, and (c) breathing pattern determining the airflow velocity and the residence time in the respiratory tract, including breathing through the nose in comparison to breathing through the mouth.[26]

The respiratory tract can be divided into a series of three filtering units: (1) the extrathoracic region (ET) including the nose or mouth, (2) the bronchial conducting airways bronchial region (BB), and (3) the alveolar interstitial (AI) region.[21,27] Larger particles are primarily deposited in the nose and in larger airways due to impaction, because they cannot follow the air stream at bifurcations. Inhalation flow rate is a major factor controlling this mechanism. Larger particles are filtered out of the air stream and cannot penetrate down to the deep lung. Smaller particles can pass through the large airways and are deposited in the lung due to sedimentation (settlement of particles in resting air due to gravity). Long residence times in smaller structures enhance deposition due to sedimentation. Very small particles (ultrafine, smaller than 100 nm) behave like diffusing gas molecules and stochastically disperse due to Brownian motion. The smaller the particles, the more effective is Brownian motion. In addition, longer residence times enhance the deposition due to Brownian motion. In the fine particles size range of 0.3–0.5 μm there is minimal motion, either due to sedimentation or due to diffusion. This effect is seen as the lowest section in the total deposition curve in the human respiratory tract, as shown in Figure 3.3, according to the ICRP model (i.e., the Human Respiratory Tract Model of the International Commision of Radiological Protection).

The simulations in Figure 3.3 represent two different breathing scenarios: exposure during sleeping and exposure during heavy exercise. Monodisperse particles of unit density were used in the calculations. Exposure during sleeping assumes nose breathing and has a mean ventilation rate of 0.45 m^3/h. Exposure during heavy exercise represents mouth breathing and has a mean ventilation rate of 3.0 m^3/h. Total deposition is not very different between the two scenarios, but regional deposition is significantly affected. Compared to heavy exercise, sleeping has much higher extrathoracic deposition for larger particles, and in the ultrafine size range, due to nose breathing, filtering out a large fraction of the particles from the inhaled air stream. This results in low bronchial deposition of larger particles for sleeping compared to heavy exercise, which has maximal bronchial deposition at about 5 μm. For fine particles, alveolar deposition maximal at about 3 μm aerodynamic diameter with little difference between the two breathing scenarios. In the UFP size range, the two breathing scenarios have significant impact on regional deposition probability. In the alveolar region, both breathing scenarios have maximum deposition between 10 and 100 nm, where heavy exercise has higher alveolar deposition for particles smaller than 60 nm and a maximum value at about 20 nm. This is primarily caused by the lower filtering of the upstream bronchial and ETs during heavy exercise, due to shorter residence times in the particular compartments. The deposition behavior for very small particles below 10 nm gets more and more

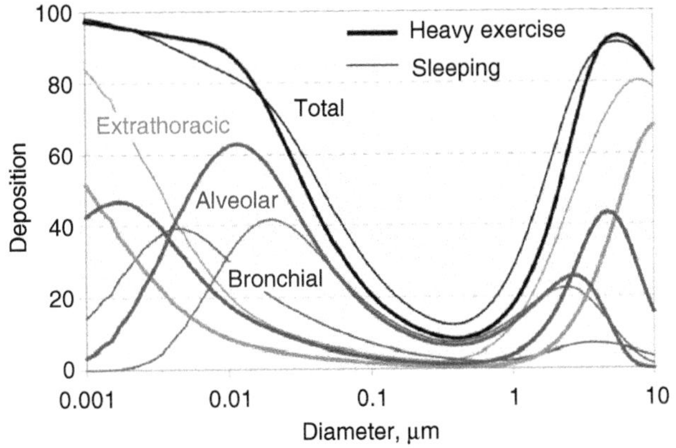

FIGURE 3.3 Total particle deposition in the human respiratory tract and corresponding fractions in the extrathoracic, bronchial, and alveolar region according to ICRP Publication 66. (From ICRP Publication 66, *Ann. ICRP*, 24 (1–3), 1–482, 1994.) Data are shown for two different breathing scenarios: exposure during sleeping, (nose breathing), and exposure during heavy exercise (mouth breathing).

significant in the bronchial and ET. Again, heavy exercise has maximum bronchial deposition at smaller particles sizes (2 nm) compared to sleeping (5 nm). The decline of bronchial and alveolar deposition for very small UFPs is due to the rising extrathoracic deposition, which prevents penetration into the lung periphery. For toxicological dose assessment, the ventilation rate has also to be taken into account, which differs between the two extreme breathing scenarios by a factor of 6.7.

The different deposition probabilities in the various regions of the respiratory tract mean that the toxicity of particles of different sizes can have different effects, according to the region of the lung being involved. Since the geometry of the conducting airways and the alveolar structures in children with developing lungs differs from the adult lungs particle, deposition probability will also change. Furthermore, lung geometry is likely to be changed in diseases like asthma, which mainly affects the larger airways, or in Chronic Obstructive Pulmonary Disease (COPD), which affects small peripheral airways and alveoli. These diseases can also cause a several-fold increase in deposition of PM in the diseased parts of the lung (see below).

Even though the fraction of fine PM is small in thoracic conducting airways, note that the PM deposition density per airway surface area may often exceed that of the gas exchange region, because of its 100-fold larger surface (adult lungs $\sim 140\ m^2$), when compared to that of airways.[28] This is demonstrated in Figure 3.4 for 20 and 250 nm particles; deposition probability is estimated in airway generations (generation 1–16) and the alveolar region (generation 16–23) based on the MPPD computer code.[27,29] For both particle sizes, deposited particle mass fractions are either plotted for each generation, or relative to the surface area of each generation. While the largest amount of both particle sizes is found in the alveolar region, the highest particle mass per epithelial surface area is found in the upper airways. In general, particles < 1 µm are well able to reach the alveolar region with its rather thin air-blood-membrane of about 2 µm thickness, and their deposition in the alveoli increases with decreasing diameter until they reach a size of 20 nm. Particularly, the very small particles are likely to be important for cardiovascular effects, as they may be able to penetrate the lung epithelial membrane and enter into the blood stream (see below).

3.3.3 Particle Deposition in Patients

Lung diseases often change the geometry of the lung, followed by changes in breathing pattern and particle deposition.[26] Airway obstructions in asthma are caused by smooth muscle contraction due

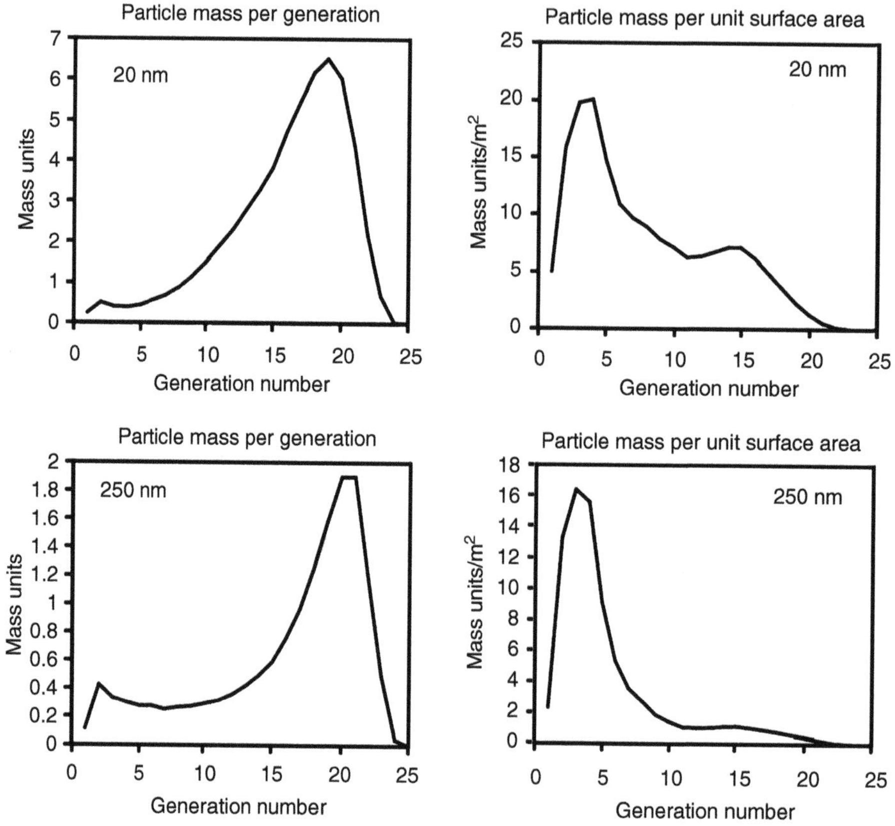

FIGURE 3.4 Deposition probability in human airways (generation 1–16) and the alveolar region (generation 16–23) based on the MPPD computer code. (From Asgharian, B., Hofmann, W., and Bergmann, R., *Aerosol Sci. Technol.* 34, 332–339, 2001; Anjilvel, S. and Asgharian, B., *Fundam. Appl. Toxicol.*, 28, 41–50, 1995.) For 250 and 20 nm particle sizes, deposited particle mass fractions are either plotted for each generation (left panels) or relative to the surface area of each generation (right panels).

to allergic responses and due to local inflammations, but the obstructions are mostly reversible. Obstructions seen in COPD patients are due to local inflammation and mucus hypersecretion, and these are mostly irreversible. Narrower airways imply higher particle deposition due to impaction of micron-sized particles at bifurcations.[30–33] On the other side, emphysema seen in many COPD patients can imply lower particle deposition in the lung periphery due to increased sizes of small airways and alveoli.[34]

Increased UFP deposition in COPD patients was also found in previous studies[35,36]; note in these studies a single breath inhalation maneuver during spontaneous breathing was used. Therefore, these studies cannot discriminate deposition between airways and lung periphery. Increased deposition of UFPs was also demonstrated in asthmatics.[33]

3.3.4 Particle Deposition in Children

There are only few studies on particle deposition in the respiratory tract of children and none in infants. The developing lungs of infants and children have smaller structures (airways) compared to adults and therefore the physical factors mentioned above cause increased particle deposition.[37–40] Higher breathing frequencies in children, leading to shorter residence times of aerosol particles in the respiratory tract, may compensate this effect in part. In the developing lung, UFP inhalation may bear an increased risk for children compared to adults, toxicological effects may impair the development and growth of certain lung structures, and may enhance the prevalence for allergic diseases.[41–43]

3.4 PARTICLE RETENTION AND CLEARANCE

3.4.1 Particle Transformation after Deposition in the Epithelial Lung Lining Fluid

Once deposited, other than aerodynamic and thermodynamic particle parameters become important.

Even if the morphology of chain-aggregated particles persists until deposition on the epithelial surface, they may either compact there or disintegrate into their primary particles, depending on their surface properties, such as surface free energy, enthalpy, etc., and on epithelial lining fluid properties, including surfactant and surface tension lowering molecules.[44]

Ambient aerosol particles, and many occupational particles, are a complex mixture of multiple chemical substances, including soluble and insoluble compounds. Because of the liquid layer of the epithelium (ELF) particle solubility and bioavailability will determine the subsequent metabolism and biokinetics of the various fractions and components of a particle. Water soluble or lipid soluble fractions and components will disperse from the particle and rapidly spread and dilute in ELF and will undergo completely different biochemical interactions with the ionic and organic constituents of the liquid epithelial layer, when compared to the remainder core of the particle. Biochemical reactions with ions, chelating molecules, and proteins, as well as with cell membrane receptors, depend on the chemical nature of such soluble compounds. These reactions may continue to change through lipid-layered cell membranes of different cell types of the respiratory epithelium and beyond this barrier into the vascular circulation. In any case, reactions occurring are not directly located next to the particle, but they are more spread. In this case, the particle had served as a vehicle to carry its soluble load to the location of particle deposition within the respiratory tract. When these water soluble or lipid soluble particle fractions and components disintegrate into ions, atoms, or molecules, it can generally be presumed that their reactivity is related to the number of those molecular units per liquid volume, (i.e., their molar concentration) or, in case of their accumulation, their concentration in distinct cell types or organ structures. This means their reactivity is determined by their mass concentration. This reactivity needs to be distinguished from catalytic reaction properties of water or lipid soluble particle fractions and components, and their metabolites, which may be less mass-concentration dependent, but which will be determined by the availability of reaction partners of the biological system.

Adverse responses of the biological environment are particularly expected with increasing mass of a soluble compound or a mixture of such soluble compounds. Since UFPs (< 100 nm) usually do not appreciably contribute to the mass of ambient aerosols and many occupational aerosols,[2] adverse reactions are only expected in case of high chronic exposure concentrations (so called "particle-overload" in the lungs) or in case of very highly reactive soluble compounds. However, since the mass of ambient and many occupational aerosols is found in the 0.1–2.5 μm particle fraction, this fraction may contain soluble compounds at a sufficient mass concentration to initiate adverse biological reactions.

Once the soluble compounds have been carried away, the insoluble core of the residual particle retains at the location of deposition. Depending on the properties of the surface, including its molecular structure and conformation, its charge, etc., reactions with the constituents of the biological fluids and cell membranes may occur. Note:

1. These reactions relate almost exclusively to the surface and not to the particle core underneath the surface.
2. These reactions are confined to the location of the particle and may cause a very strong local effect, which eventually triggers mediators released by cells such as AM, dendritic

cells (DC), polymorpho-nuclear neutrophilic granulocates (PMN), epithelial type I and type II cells.

3. The entire surface area, including the inner surface of porous particles or the total surface area of primary particles, of chain aggregates needs to be considered, since the biological fluid may penetrate into pores of nanometer sizes.

Interactions depend on the size of the solutes of ELF (e.g., a 66 kDa protein-like albumin corresponding to a mean diameter of 5–6 nm will not be able to penetrate into one-nanometer pores). However, intracellular particle dissolution in phagolysosomes of AM was proportional to the BET surface area (surface area determined by N_2 absorption) of moderately soluble, uniform, micron-sized, porous cobalt oxide particles.[45,46] The total surface area involved in biochemical reactions may be considerably larger than the outer surface area alone. We suggest calling it "bioavailable surface area" or "biochemically interactive particle surface area" to distinguish from other surfaces addressed above. The biochemically interactive surface provides the substrate for biochemical reactions with ELF, while only the part of the surface adjacent to cell membranes may react with membrane receptors. Currently this surface cannot be determined and the BET surface area seems to describe best toxicologically relevant particle effects.

3.4.2 Biochemical Reactions of Insoluble Particles with ELF

Insoluble particles or residual particle cores deposited and retained in ELF may react with ions, chelating molecules, and proteins, which may result in functional changes of the latter,as well as coating or complexing the particles by those biological constituents. Thereby, the particles may specifically bind to selected proteins. If these proteins are undergoing forces like osmotic pressure or in a concentration gradient of the complexing protein, the entire complex, including the particle, will be translocated from its original location. Due to the homeostatic balance of proteins and biomolecules originating from and departing to the adjacent microvasculature,[47–49] particularly the UFPs are finding a mechanism that may carry them from the epithelial surface into the circulation. Pathways may be paracellular across tight junctions between adjacent epithelial cells, or transcellular by transcytosis as shown by Heckel et al.[50] and are size-dependent, as discussed by Patton et al.[6] for macromolecules of different sizes. Similar mechanisms are likely to apply for different sized UFPs. A schematic in Figure 3.5 illustrates these pathways. Because of the much larger inertia of micron-sized particles, they are much less likely to participate in this translocation mechanism, see schematic Figure 3.6. Yet, 240 nm lecithin-coated polystyrene particles were able to cross the alveolo-capillary-membrane, while uncoated 240 nm polystyrene particles did not penetrate the membrane, emphasizing the importance of particle coating.[51]

Furthermore, when high doses of polystyrene particles were delivered to non-phagocytic B-16 cells in an *in vitro* test, particles as large as 500 nm were internalized by the cells via an energy-dependent process.[52] Internalization of microspheres with a diameter < 200 nm involved clathrin-coated pits. With increasing size, a shift to a mechanism that relied on caveolae-mediated internalization became apparent. Note however, that these results obtained from *in vitro* studies may differ from *in vivo* conditions.

Recently, we showed that the electrophoresis-gel patterns of proteins of blood serum or broncho-alveolar lavage fluid (BALF) binding to different UFPs appeared to differ.[53] UFPs were titanium dioxide (TiO_2, P25, Degussa), amorphous silicate (Aerosil, Degussa) or carbon particles (Printex 90, Degussa). Proteins, like albumin, bound to all three particle types. But other proteins bound only to a single type of particle. Efforts are underway to determine these proteins. Due to the wealth of biomolecules and proteins in ELF, it appears rather unlikely that particles retain in ELF without any such interactions. In addition, receptor binding at the adjacent cell membranes of phagocytic and epithelial cells will eventually result in intracellular uptake. This may provide another mechanism of removal of UFPs from the epithelial surface. In fact, it is well

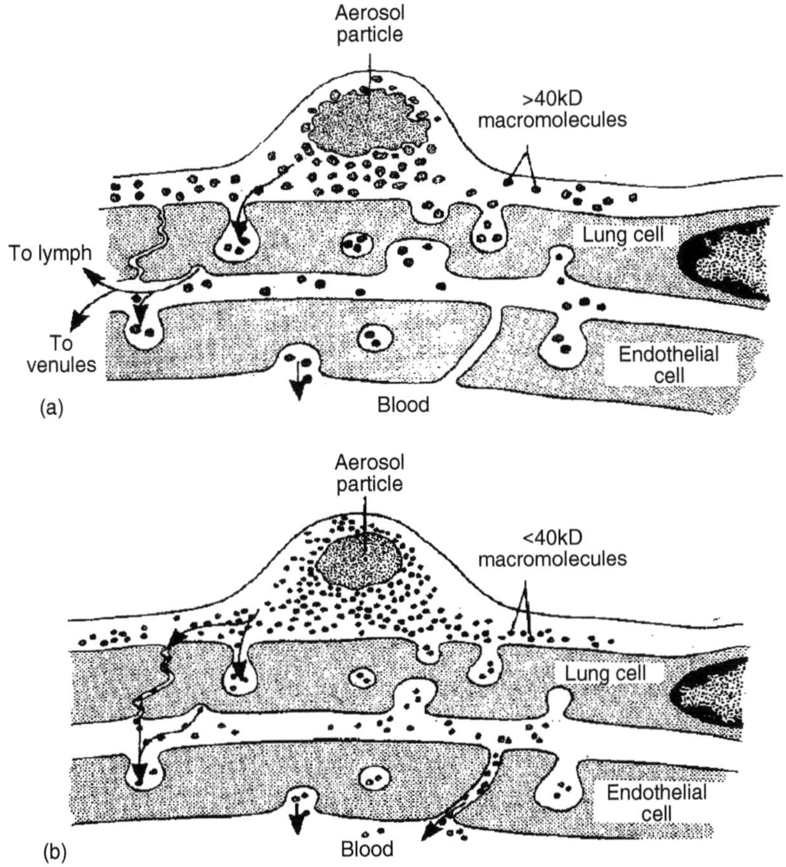

FIGURE 3.5 Suggested absorption of large (a, transcytosis) and small (b, tight junctions) macromolecules across alveolar Type I cells. (From Patton, J. S., *Adv. Drug Deliv. Rev.*, 19 (1), 3–36, 1996.)

accepted that phagocytosis of micron-sized particles is largely accomplished within 6–24 h after deposition.

3.4.3 ICRP Clearance Model

The previous model of clearance of inhaled particles published by the ICRP summarized experimental data available until 1994.[21] The main purpose of the model was to assess the dose after inhalation of radio-labeled aerosols, either during work place exposure, during experimental inhalation and clearance studies (particles, drugs), or after accidental exposure. The total clearance kinetics is described as a superposition of fractional clearance rates from different compartments of the respiratory tract, such as ET; BB, starting from the trachea (generation 0) down to generation 8; the bronchiolar region (bb); including generations 9–15; and the AI region (generations 16–23), including respiratory bronchioles and alveolar ducts. In all regions, clearance mechanisms underlie three basic routes:

- Absorption to blood (in case of soluble aerosols)
- Transport to the gastrointestinal tract (GI)
- Transport to the lymph nodes

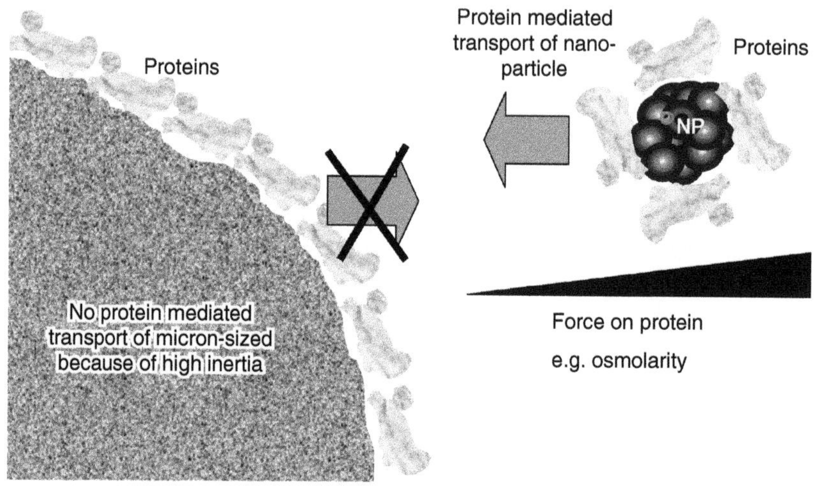

FIGURE 3.6 Hypothesis: nanoparticle-protein-complexes provide a vehicle of nanoparticle (NP) transport according to forces on the complexed protein, like in a concentration gradient of the protein. Transport of micron-sized particles is less likely because of the much higher particle inertia and less protein-binding to the smaller specific surface area compared to the properties of nanoparticles.

The rates of transport of particles depend on:

- The site of deposition in the respiratory tract
- The physicochemical composition of the particles
- The time since deposition in the respiratory tract

In addition, the model assumes that:

- Particle transport and absorption of dissolved particle material to blood are independent.
- Particle transport rates are the same for all materials.
- The rate of absorption of dissolved particle material depends on the material, but is the same in all compartments of the respiratory tract.

Figure 3.7 summarizes the basic model and gives typical data for the mechanical transport rate for insoluble particles from the different compartments of the respiratory tract.

3.4.4 Clearance from the Airways

The airways are covered by a mucus layer, which is transported by beating cilia towards the larynx, from where it is swallowed into the GI tract. Mucus transport velocities between 10–20 mm/min in the trachea, and subsequently decreasing velocities in airways of decreasing caliper confirm a fast removal of particles being deposited in the airways and on the mucus layer. Extrapolating this to the whole system of airways suggests that particles being deposited there should clear within about 24 h.

In contrast, clearance studies after shallow aerosol bolus inhalation showed that there is a fraction of long-term retained particles, which seems to depend on the geometric particle size.[54,55] Since the 1980s, there is experimental evidence that not all particles deposited on the epithelium of human airways are cleared immediately via the mucociliary transport to the larynx, but are retained for weeks and even longer time.[20,21] Different mechanisms may be responsible for slow clearance, resulting in a loss of particles from mucociliary transport, such as penetration of

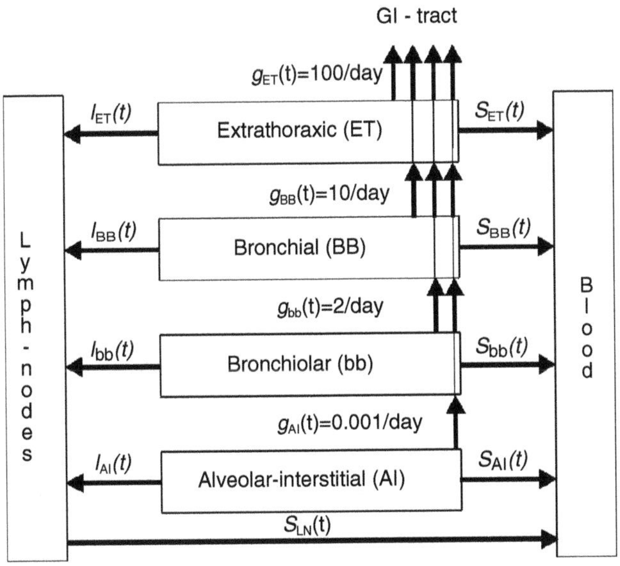

FIGURE 3.7 Schematic view of the ICRP-66 particle clearance model from the different compartments of the human lung (after from ICRP Publication 66, *Ann. ICRP*, 24 (1–3), 1–482, 1994), including particle transport to the gastrointestinal tract (indicated by transport rates g_{ET}, g_{BB}, g_{bb}, g_{AI}), and to the lymph nodes (indicated by transport rates l_{ET}, l_{BB}, l_{bb}, l_{AI}); in addition, absorption of soluble fractions to the blood circulation is indicated by absorption rates s_{ET}, s_{BB}, s_{bb}, s_{AI}.

particles through the mucus into periciliary spaces, or deposition of particles in areas being free of mucus at the actual time point of deposition. Smaller particles may have a higher probability of penetrating between cilia, where they can be engulfed by airway phagocytic cells, or where they can translocate into the epithelium. Recent studies have shown that the kinetics of long-term clearance in the airways coincides with that of alveolar clearance.[56] In a recent review, we have discussed delayed clearance from airways and its dependency on geometrical particle size in more detail.[57] The observed delayed clearance of UFPs means that there is only a negligible fast-cleared fraction of UFPs after tidal breathing; although this breathing pattern will result in both bronchiolar and alveolar deposition. Based on model predictions, bronchiolar deposition should account for about 20%–30% of total deposition.[21,58] Indeed, we have confirmed that there is negligible fast clearance within 24–72 h after inhalation in two other studies, after tidal breathing of 100 and 35 nm carbon particles; this was shown not only in healthy adult subjects, but also in asthmatic patients and in smokers.[59,60]

3.4.5 Clearance from the Lung Periphery

The absence of mucociliary action in the lung periphery results in much slower clearance kinetics. The rate-determining step for particle transport from the peripheral lungs to the larynx, and hence to the GI tract, is the time required to transport the particle from an alveolus to the most distal ciliated airways—the terminal bronchioli. Subsequent transport via mucociliary action up the ciliated airways to the larynx is considered to be comparatively rapid. Particle transport towards the ciliated airways and larynx has been determined in various animal species and man.[61] Note, the species difference on the particle clearance kinetics: transport rates (daily cleared fraction of retained particle load) from the lung periphery to the larynx of rodents vary about 0.01–0.05 d^{-1} and exceed the human transport rate by a factor of about 10.[20] A summary of studies in different animal species and man, using different particle materials, is given in Annex E of the Human Respiratory Tract Model (HRTM) of ICRP-66.[21] Efforts were made to distinguish this particle clearance pathway, particularly, from particle dissolution. As reviewed earlier, particle transport from the peripheral lung towards ciliated airways is mediated by AM, while particle transport of free (nonphagocytized) particles may play a minor role.[20] As mentioned above, the chemical stability of the inhaled particles plays a crucial role in the clearance kinetics and the overall toxicological action in the lung. This is included in the dose evaluation of the HRTM for inhaled radiolabeled aerosols, by distinguishing between insoluble, moderately soluble, and soluble materials. The transport rates of particles deposited in the different regions of the lung as shown in Figure 3.7 hold for insoluble materials.

Slowly dissolving and insoluble particles deposited in the alveolar region will be taken up by specialized defense cells, the AM, within a few hours after deposition, at least under physiological conditions in healthy lungs. These cells dispose of a number of mechanisms to digest, or dissolve or disintegrate the incorporated particles. Therefore, AM will determine the fate of these particles. However, macrophage-mediated particle removal may be impaired, especially in elderly, smokers, and subjects with lung diseases.[62] Additionally, macrophages are less able to take up UFPs ($<$ 100 nm), even in healthy lungs.[57]

In the following, we list the major pathways of particle translocation from the human alveolar epithelium, which is in part alveolar macrophage mediated. Fractions of deposited insoluble particles are indicated undergoing these pathways, as reviewed earlier.[20] Particularly, the fractions differ consistently from those which have been observed in rodents, as the most common experimental animal models. Macrophage-mediated particle pathways include:

1. Movement towards ciliated airways on the epithelium phagocytized in AM, for further removal by ciliary action; passage through the gut; and excretion (about a third of deposited insoluble particles).
2. Retention/sequestration on the epithelium phagocytized in AM, or binding to or taken up, as free particles into the epithelial lining cells (together with c, more than half of deposited insoluble particles).
3. Across the epithelial lining cells towards the spaces between underlying cells (together with b, more than half of deposited insoluble particles).
4. Across the epithelial lining cells towards the lymphatic drainage system (between 1%–5% of deposited insoluble particles).
5. Across the epithelial lining cells eventually into the blood vessels towards secondary target organs. There is evidence for UFP uptake into the blood circulation, depending on the physical structure and chemical composition of their surface. Fractions and rates of uptake are currently under debate. Identified secondary target organs are liver, spleen, kidneys, heart, and brain.

However, UFPs are less effectively taken up by macrophages, but interact to a greater extent with epithelial lining cells than larger particles. According to the larger number concentration of UFPs in the ambient aerosol—usually exceeding more than one to two orders of magnitude when compared to the concentration of fine particles—there will be many more deposited UFPs spread over the surface of an alveolus. This is likely to lead to a scattered signal of opsonization, resulting in less recognition, and subsequently, in less phagocytosis by AM, when compared to fewer micron-sized particles coated with relatively more opsonin molecules on their surfaces.

3.4.6 Clearance towards Lymph Nodes

Although it is frequently stated that particles accumulate in the tracheobronchial lymph nodes (TBLN), located mainly around the bifurcation of the trachea into the main bronchi,[21] there are no data available on the kinetics of particle transport to TBLN in man. Data mostly come from autopsies of workers. Data from dogs and monkeys suggest a rather minute accumulative transport of insoluble particles into TBLN of 1%–5% (percent of lung deposit); this transport is even lower in rodents by an order of magnitude, as was reviewed in more details previously.[20]

3.4.7 Clearance in Patients with Lung Diseases

Acute and chronic diseases can affect the clearance capacity of the respiratory tract and the residence time of foreign materials and organisms in the different regions of the respiratory tract. Acute and chronic infections can inhibit ciliary function, can cause mucus hypersecretion, and can alter the viscous properties of mucus, leading to a retardation or a complete inhibition of mucus transport.[63,64] In addition, genetic diseases can cause ciliary dysfunctions, such as primary ciliary dyskinesia (PCD), or mucus stiffening, such as cystic fibrosis (CF), followed by bacterial colonization of the airways and the formation of bronchiectasis.[65–67] In all these diseases, coughing plays a significant role in mucus clearance, particularly in the upper generations of the tracheobronchial tree.[68]

Alveolar clearance is also impaired by chronic cigarette smoke exposure and by interstitial lung diseases, such as sarcoidosis and interstitial pulmonary fibrosis.[62,69] AM are suppressed in their function of phagocytosis, signal transduction, and intracellular digestion.[70,71]

3.4.8 Clearance of Ultrafine Particles

Besides translocation, clearance of UFPs may differ from that of particles in the micrometer size range. Recent studies in healthy subjects suggest that insoluble carbon particles (which are the primary fraction of combustion particles) do not clear from the lung within days, and are therefore long-term retained.[59,60] In addition, it has been shown that after shallow bolus inhalation, only about 25% of UFPs are cleared from the conducting airways by the mucociliary escalator within three days, while the remaining fraction appears long-term retained.[72,73] This data support the hypothesis that UFPs may translocate to epithelial cells, and subsequently to other organs or the body.

However, the above mentioned human studies on the clearance of UFPs were short-term studies over a few days. No human long-term studies are currently available. In addition, animal studies on the clearance of UFPs are very limited. Surprisingly, when lung clearance of inhaled ultrafine iridium particles or micron-sized polystyrene particles was followed over six months in rats, there was no difference between the kinetics of lung retention and the major clearance pathway to the larynx and fecal excretion of UFPs and micron-sized particles.[74] This surprising result was interpreted by considerable relocation of ultrafine ^{192}Ir particles from the lung epithelium, across the epithelial membrane, and back, as suggested by Ferin and coworkers[75] when evaluating retention kinetics of inhaled ultrafine TiO_2 particles in rats, which did not happen to micron-sized particles in the rat lung, as was shown by,[76] who provided the biokinetic data obtained from micron-sized particles.

3.4.9 Non-Significant Long-Term Retention in Rat and Mouse Airways for Ultrafine Particles

However, the observed long-term UFP retention in human airways does not or only to a minor degree occur in rats. After inhalation of 15–20 and 80 nm ultrafine iridium particles by endotracheally intubated and ventilated rats, we observed that 30 and 20%, respectively, of the totally deposited particles in the thorax, were rapidly cleared from the lungs.[74,77] Note there was no extrathoracic airway deposition, but exclusive thoracic deposition because the rats were endotracheally intubated. These rapidly-cleared fractions correspond reasonably well to deposition estimates in rats based on earlier data and theoretical calculations.[29,78] Ongoing studies on adult mice demonstrate a similar pattern to that of the rats. Yet human studies were performed using ultrafine carbon particles, while iridium particles were used in rodent studies. Therefore, it cannot be excluded that the different particle materials may have caused the differences. However, when ultrafine carbon particles spiked with primary iridium particles were used in additional rat studies, a similar pattern of fast clearance from the airways was observed (Kreyling, personal communication). Hence, there seems to be a clear different clearance pattern of UFP in the airways of rats and mice when compared to those of man and dog.

3.5 SYSTEMIC TRANSLOCATION AND UPTAKE IN SECONDARY TARGET ORGANS

3.5.1 WHY IS THERE SO MUCH INTEREST IN TRANSLOCATION OF ULTRAFINE PARTICLES?

An early key study demonstrated that ultrafine TiO_2, with an average particle size of 20 nm, caused more inflammation in rat lungs than exposure to the same airborne mass concentration of fine TiO_2 with an average particle size of about 250 nm.[79,80] So far, TiO_2 had been considered as a low-toxicity dust and indeed had served as a negative control dust in many studies on the toxicology of particles. Therefore, this report was highly influential in highlighting that a material that was low in toxicity in the form of fine particles could be significantly more inflammatory in the form of UFPs. Doses ranging from 30 to 2000 μg of TiO_2 were intratracheally instilled into rats and mice; inflammation was determined by the influx of neutrophilic cells in the lavage fluid one day later. Interestingly, when the deposited TiO_2 dose was expressed as particle surface area, rather than as the traditional particle mass, there was a common dose-response relationship of the inflammatory responses of these two different sizes of TiO_2 particles in both species.[79,80] Similarly, a unique dose-response function between inflammatory neutrophilic cell influx and the surface area of administered carbon UFPs obtained from different combustion processes with different sizes and ratios of EC, vs. organic compounds, was observed in a mouse model.[81] The response curve of carbon particles showed a different slope compared to that of TiO_2 particles, indicating different surface reactivities of the two particle materials. In addition, Stöger and coworkers[81] found a lower threshold of instilled carbon particle surface area in the lungs of their mouse model of 20 cm^2, below which they did not observe an acute dose-dependent inflammatory response of the lungs. The importance of particle surface area for eliciting inflammatory responses in the lung has been confirmed by several *in vitro* studies.[82–84] This literature has been reviewed extensively during recent years.[57,85–88]

While a large surface area per se, is unlikely to cause any toxic reaction with biological systems, it provides the substrate for surface molecules and structure-initiating interactions with the biological fluids and membranes. Interactions may be negligible when the reactive surface area is small enough, but a significant response may be induced when the surface area exceeds a certain value.[81]

In addition, it appears that UFPs can be subject to transport across membrane boundaries by proteins after forming complexes, which micron-sized particles are not be able to because of their much larger inertia and size, as was discussed above. In addition, micron-sized particles are rapidly phagocytized by AM. Therefore, they are only shortly available for protein-mediated transport. Actually, modification of the surface of NPs is currently under intense investigations in the discipline of Nanomedicine—defined as medical application of nanotechnology and related research—aiming to design, test, and optimize characteristic biokinetic behaviors of medicinal NPs as diagnostic and therapeutic tools to reach high-target organ specificity[89]; for example, drug delivery to the central nervous system (CNS) via blood-borne NPs requires surface modifications in order to facilitate translocation across the tight blood-brain barrier via specific receptors (e.g., apolipoprotein coating for LDL-receptor–mediated endocytosis in

TABLE 3.1
Number and Surface Area of Different Sized Spherical Particles of Unit Density for a Given Mass of 100 ng

Particle Diameter (nm)	Particle Number	Surface Area (mm^2)
2	2.4×10^{13}	300
20	2.4×10^{10}	30
1000	1.9×10^{5}	0.006

brain capillaries).[90–92] Such highly desirable properties of NPs must be carefully weighed against potential adverse cellular responses of targeted NP drug delivery, and a rigorous toxicologic assessment is mandatory.

Besides surface area, the large number of UFPs is another issue that provides the option to disperse them into many more cells and intracellular structures than micron-sized particles of the same mass. For instance, a particle mass of 100 ng corresponds only 2.4×10^4 particles (spheres of unit density) of 2 μm diameter, but 2.4×10^8 particles of 20 nm diameter or 2.4×10^{11} particles of 2 nm diameter. Note, 20 nm particles comprise a major fraction of the number concentration of ambient aerosol particles,[2] and 2–5 nm particles are a common source of primary particles of many aggregated UFPs originating from many combustion processes.

Assuming a particle mass of 100 ng would have been accumulated in a secondary target organ as the heart or the brain under common instances, this mass would be not considered to be of any toxicological relevance for many low toxicity particles; particularly if one considers the low number of 2 μm particles in the entire organ (see Table 3.1). However, if there are about 10 times more particles in the organ than its number of cell, as it is the case for 2 nm particles (cell estimate is based on a 100 g organ and an average cellular volume of 6.5×10^{-11} cm^3 corresponding to a 5 μm diameter cell), one could hypothesize that adverse effects may be triggered. Of course, whether or not adverse effects will actually be induced will depend on many factors, including localization within subcellular structures, particle chemistry, and other surface characteristics.

3.5.2 Studies on Ultrafine Particle Translocation into Circulation and Secondary Target Organs

While a growing body of reports confirms that there is translocation of UFPs into blood circulation, and subsequent uptake in secondary target organs, the size of translocated fraction, the transport mechanisms, and the rate-determining parameters are under current debate. There is agreement that under normal conditions, excluding high-dose exposure or toxic particle exposure, micron-sized particles are not likely to be translocated towards the systemic circulation, neither through the endothelial vasculature, nor through the tracheo-bronchial lymph nodes, as was reviewed earlier by Kreyling and Scheuch.[20] More recently others and we have reviewed the translocation data of UFPs[57,88,93] and emphasized that quantitative determination of translocation requires a full account of the metabolic fate of the particles, as well as any labels that are used to indirectly recognize the particle, such as radioactive labels, fluorescent dyes, or magnetic particles. This includes the accumulation and retention in any organ or tissue of the entire body, as well as the excreted fraction. In addition, the localization of the retained UFPs on a microscopic level is of great importance for providing further insight into possible adverse reactions, like inflammatory processes.

In case of radio-labels, stable fixation of the label to the particle is required in order to avoid analysis of the metabolic fate of the label not associated with the particles. Since we have discussed this issue in detail,[57] we will particularly focus on studies that tried to use a most stable label on the particles. However, this is a challenging issue, since the immanent problem of labeling UFPs will be the growing fraction of molecules on the particle surface when the particle size decreases (e.g., at a particle size of 10 nm, already 12% of all molecules are located on the surface[88]). From the surface they are likely to come off easier compared to being inside the particles.

3.5.3 Human Data

In human studies there is the limitation that a comprehensive analysis of particle accumulation and retention in each organ and tissue is practically impossible because of the limiting resolution of existing detection systems. Therefore, much emphasis and control is required to minimize and to determine the disintegration of the particles and their labels. In addition, if UFP translocation is only a very small fraction of the delivered dose, it may not be possible to determine it because of the limits of detection; in this case only upper limits of the translocated fraction can be provided. Taking this into account, there are currently no reliable human experimental data available showing a translocated particle mass fraction beyond 1% of the delivered dose to the lungs.[59,60,73,94] However, indirect evidence of particle translocation in humans comes from recent exposure studies on healthy subjects using diluted diesel exhaust.[95] Inhalation of dilute diesel exhaust impaired two important and complementary aspects of vascular function in humans: the regulation of vascular tone, and endogenous fibrinolysis. The study does not provide answers whether the observed cardiovascular effects were initiated by translocated particles or by mediators released in the lungs and triggering the responses of the cardiovascular system. However, neither of the pathways of initiation can be excluded. This applied also similarly to the experimental thrombogenesis studies in hamsters and rats with polystyrene particles[96,97] (see below). Although these authors showed the same immediate effects after intravenous injection of the particles, making a direct particle effect more plausible.

Humans' classical pathology has reported such particle loads in secondary target organs only under heavy exposure conditions; for example, the increasingly blackening lungs with increasing age of smokers are well documented in many textbooks for more than a century. Particles in the liver and other organs of the reticuloendothelial system have been reported in coalmine and asbestos workers,[98,99] but only in those with long exposure times and high lung burdens of the materials. In these studies it was suggested that the route of translocation was via the lymphatic system into the blood circulation.

3.5.4 Animal Data

Evidence of translocated UFPs, such as gold, silver, titanium dioxide, polystyrene, and carbon, and sizes ranging from 5 to 100 nm originates from animal data, providing either evidence of particles in circulation,[100–102] or secondary target organs,[103,104] or thrombogenic effects.[96,97] There remains uncertainty about translocated fractions of UFPs beyond 5% of the delivered lung dose, because mostly of reasons mentioned above and discussed in detail earlier.[57,105] However, completely balanced studies of the biokinetics of iridium particles indicate translocated fractions of about 1% of the deposited UFP dose to the lungs of rats.[74,106] Just recently, a similar estimated fraction of 1%–2% of translocated 50 and 200 nm polystyrene particles was reported.[107] The former studies have the disadvantage of using iridium particles being extremely rare in the environment; however, the physico-chemical properties of the iridium particles labeled with ^{192}Ir gamma-emitting isotope allowed precise balanced biokinetic analyses. These test particles did not only accumulate in the liver at about 0.1%–0.5% of the lung dose, but also in spleen, kidneys, brain, and heart at similar fractions. Uptake of the 15–20 nm particles in secondary target organs was about a factor of 2–3 higher than for 80 nm particles. Interestingly, in a long-term retention study[74] ultrafine iridium particle contents in each of these secondary target organs did not increase with increasing retention time after a single 1 h exposure, but peaked after one week at about 0.5% in

TABLE 3.2
Retained Mass Fractions, as Well as the According Numbers of Insoluble Iridium Particles in the Lungs and in Secondary Target Organs, One Week and Six Months after a Single 1 h Inhalation of 15 nm Sized Iridium Particles by WKY Rats

	Retained Mass Fraction		Retained Particles Number		Particle Surface Area (cm^2)	
Organ	**One Week**	**Six Months**	**One Week**	**Six Months**	**One Week**	**Six Months**
Lungs	0.6	0.06	7.00×10^{11}	7.00×10^{10}	1.26×10^{2}	1.26×10
Liver	0.006	0.0005	7.00×10^{9}	5.83×10^{8}	1.26×10	1.05×10^{-1}
Spleen	0.004	0.0003	4.66×10^{9}	3.50×10^{8}	8.42×10^{-1}	6.32×10^{-2}
Heart	0.004	0.0005	4.66×10^{9}	5.83×10^{8}	8.42×10^{-1}	1.05×10^{-1}
Brain	0.003	0.0005	3.50×10^{9}	5.83×10^{8}	6.32×10^{-1}	1.05×10^{-1}
Kidney	0.006	0.0001	7.00×10^{9}	1.17×10^{8}	1.26×10	2.11×10^{-2}

In addition, the according surface area of the retained particles is calculated based on the BET surface area of 123 m^2/g or 1500 m^2/cm^3 (per mass or per volume of particles, respectively).

Source: From Semmler, M., Seitz, J., Erbe, F., Mayer, P., Heyder, J., Oberdörster, G., and Kreyling, W. G., *Inhal. Toxicol.*, 16 (6–7), 453–459, 2004; Kreyling, W. G., Semmler, M., Erbe, F., Mayer, P., Takenaka, S., Schulz, H., Oberdörster, G., and Ziesenis, A., *J. Toxicol. Environ. Health A.*, 65 (20), 1513–1530, 2002.

each secondary organ. Thereafter, fractions declined again and remained detectable, but below 0.1% of the initial deposit throughout the six-month-period of observation, indicating clearance mechanisms in these organs.

However, even though the mass fractions of iridium particles were rather low in secondary target organs, the number of particles is impressively high. Data at one week and six months after the single inhalation are shown for lungs and all secondary target organs—liver, spleen, heart, brain, and kidneys—in Table 3.2. More than one billion particles were found in each of the secondary target organs one week after a single 1 h exposure; and still more than 100 million particles were determined six months after the inhalation.

In addition, the surface area of the retained iridium particles was calculated based on the BET specific surface area of 123 m^2/g or 1500 m^2/cm^3 per mass or volume of particles[108] (BET surface area was determined by nitrogen absorption measurements). Since there is now evidence that the iridium particles are covered with iridium oxide,[109] the density of iridium oxide (11.7 g/cm^3) and not that of iridium (22.4 g/cm^3) was used for calculations. One week after inhalation,the total surface area of the retained iridium particles is close to 1 cm^2 and after six months data are still one–tenth of the one-week data. Compared to 1 μm-sized particles, this retained surface area of the 15 nm particles in secondary target organs is five orders of magnitude larger because of their very large specific surface area. It remains to be investigated which impact that may have, and whether the large number of retained particles and the accordingly large particle surface area, may have any adverse effects on the biological- surrounding like proteins, extracellular fluids, cells, as well as whole secondary target organs, which usually are not considered to be exposed to such foreign bodies.

It needs to be emphasized that particle uptake in the brain was not via a neuronal pathway from the olfactory mucosa in the nose, since extrathoracic airways of the rats were bypassed when ventilated through an endotracheal tube during the exposure. Therefore, principally two pathways from the lung epithelium towards the brain are possible: (1) along neuronal axons and synapses[110–112] as reviewed by Oberdörster et al.[88] or (2) via the systemic circulatory route. While the second route, via blood circulation, intuitively seems likely because of the analogy to the uptake in other secondary organs, the extremely tight blood brain barrier (BBB) represents a significant obstacle to easy translocation.[90] Only after coating with polysorbate 80 or lipo-E protein could the BBB be overcome. In contrast, the olfactory neuronal pathway circumvents the BBB permitting exposure of the olfactory bulb of the CNS and possibly deeper brain structures to inhaled UFPs. For instance, translocation of nanosized particles along olfactory nerves from the olfactory epithelium to the olfactory bulb was first reported by Howe and Bodian[113] for 0.03 μm polio virus in monkeys, and was later described for nasally deposited colloidal 0.05 μm silver-coated gold particles, moving into the olfactory bulb of squirrel monkeys.[111] Carbon UFPs were reported to translocate also along the same pathway to the CNS, based on their presence in the olfactory bulb of rats after inhalation.[112,114] And most recently, translocation of inhaled 30 nm Mn-oxide particles was compellingly demonstrated by exposing rats with the right nostril occluded, which resulted in Mn accumulation only in the left olfactory bulb of the brain (side of the open nostril). Repeat exposure to these ultrafine Mn-oxide particles resulted in significant inflammation (Elder et al. 2006). Other studies demonstrated the transport of fluorescent 40 nm polystyrene NPs from the nerve endings in the tracheal epithelium, along their neurons to their cell body in the ganglion nodosum and jugular ganglia in the neck of guinea pigs along neurons innervating the trachea.[115]

Recent studies of Nemmar and coworkers[96,116] shed light on possible prothrombotic effects in the systemic circulation as a result of activation of platelets. In their hamster model of experimentally-induced thrombus formation, they observed thrombus formation in peripheral veins and arteries after intravenous and intratracheal administration of positively charged 60 nm polystyrene NPs. They emphasized the importance of size and surface charge of test particles. In fact, the induced thrombus formation was not detectable after administration of negatively charged or neutral 60 nm, as well as 400 nm sized positively charged polystyrene test particles; astonishingly, they observed similar thrombus formation after application of diesel exhaust particles via both routes.[96] These observations were confirmed in a rat model by another group using a slightly modified approach.[97] From the fact that they were able to observe peripheral thrombus formation, both after intravenous injection and after intratracheal NP instillation into the lungs, they concluded that particle translocation from the lung epithelium to circulation was likely to be one of the underlying mechanisms triggering platelet activation. Although a direct particle effect due to alveolo-capillary translocation is a conceivable mechanism for the observed effects, the evidence is still indirect and requires confirmation in future studies. Nevertheless, the presence of these UFPs with specific properties obviously was able to trigger biological responses, which may result in adverse health effects.

3.5.5 Dose Estimate of UFP Accumulation in Secondary Target Organs during Chronic UFP Exposure

Taking this long-term UFP kinetics after a single exposure into account, we can estimate long-term accumulation in human secondary target organs during a year-long chronic exposure, based on the results obtained from the tested UFP used in our studies. Applying this approach to human exposure, however, requires great caution, as indicated by other interspecies differences observed between rodents and man, and may only be considered to represent a first rough estimate. Let us consider a daily unit dose to the lung; from this dose only 0.0005 is long-term retained as found in liver, heart, and brain after six months (see Table 3.2). During continuous exposure over a year, this daily long-term fraction will accumulate to a fraction of 0.2. The daily dose of insoluble UFP to a human lung can be estimated to be based on the following assumptions:

1. $1\times10^3\ cm^{-3}$ insoluble UFP of air (i.e., 10% of an average UFP concentration $1\times10^4\ cm^{-3}$)
2. 1×10^4 L/d daily inhaled gas volume by an adult human
3. 0.3 deposited fraction of the inhaled UFP concentration in the peripheral lungs
   a. 3×10^9 UFP/d daily dose of insoluble UFP to a human lung
   b. 3×10^{11} UFP during one year exposure (assuming about 70% clearance of these insoluble UFP from the lungs)

Based on these assumptions, 6×10^8 UFP would have accumulated in each secondary target organ during one year of continuous exposure. Although the accumulated doses in secondary target organs are three orders of magnitude less than the lung dose, it is indeed not negligible. Of course, whether or not adverse effects will actually be induced will depend on many factors, including localization within cell types, subcellular structures, particle chemistry, etc. Note, extrapolation from rodent data to man requires great caution as indicated by other interspecies differences observed between rodents and man. In addition, questions remain whether all insoluble particles translocate from the lungs to secondary organs; in particular, whether EC is translocated because it is likely to be the most prominent compound of insoluble ambient UFP. From these results, we have expected that particulate uptake should have been documented in secondary target organs of smokers or according animal models. However, we were not able to find any reference clearly distinguishing between particulate versus molecular constituents in secondary target organs. Indeed, it needs considerable efforts to identify very tiny UFP on those secondary organs for which no need existed until now. As mentioned above, particles in the liver and other organs of the reticuloendothelial system have only been reported in coalmine and asbestos workers, who had been long-term exposed to high doses.[98,99]

REFERENCES

1. Van Dingenen, R., Raes, F., Putaud, J. P., Baltensperger, U., Charron, A., Facchini, M. C., Decesari, S. et al., A European aerosol phenomenology-1: physical characteristics of particulate matter at kerbside, urban, rural and background sites in Europe, *Atmos. Environ.*, 38 (16), 2561–2577, 2004.
2. Kreyling, W. G., Tuch, T., Peters, A., Pitz, M., Heinrich, J., Stölzel, M., Cyrys, J., Heyder, J., and Wichmann, H. E., Diverging long-term trends in ambient urban particle mass and number concentrations associated with emission changes caused by the German unification, *Atmos. Environ.*, 37 (27), 3841–3848, 2003.
3. Whitby, K. T. and Sverdrup, G. M., *California Aerosols: Their Physical and Chemical Characteristics*, A Wiley-Interscience Publication, New York, Chichester, Brisbane, Toronto, 1980.
4. Valberg, P. A., Is PM more toxic than the sum of its parts? Risk-assessment toxicity factors vs. PM-mortality "effect functions," *Inhal. Toxicol.*, 16 (suppl. 1), 19–29, 2004.
5. Effros, R. M. and Mason, G. R., Measurements of pulmonary epithelial permeability *in vivo*, *Am. Rev. Respir. Dis.*, 127 (suppl. 5), S59–S65, 1983.
6. Patton, J. S., Mechanisms of macromolecule absorption by the lungs, *Adv. Drug Deliv. Rev.*, 19 (1), 3–36, 1996.
7. Patton, J. S., Fishburn, C. S., and Weers, J. G., The lungs as a portal of entry for systemic drug delivery, *Proc. Am. Thorac. Soc.*, 1 (4), 338–344, 2004.
8. Willeke, K. and Baron, P. A., *Aerosol Measurement: Principles, Techniques and Applications*, John Wiley & Sons, New York, Chichester, Weinhein, Brisbane, Singapore, Toronto, 1993.
9. Heyder, J., Gebhart, J., Rudolf, G., Schiller, C. F., and Stahlhofen, W., Deposition of particles in the human respiratory tract in the size range 0.005–15 microns, *J. Aerosol Sci.*, 17 (5), 811–825, 1986.
10. Champion, J. A. and Mitragotri, S., Role of target geometry in phagocytosis, *Proc. Natl. Acad. Sci. USA*, 103 (13), 4930–4934, 2006.
11. Warheit, D. B., Chang, L. Y., Hill, L. H., Hook, G. E., Crapo, J. D., and Brody, A. R., Pulmonary macrophage accumulation and asbestos-induced lesions at sites of fiber deposition, *Am. Rev. Respir. Dis.*, 129 (2), 301–310, 1984.
12. Pitz, M., Cyrys, J., Karg, E., Wiedensohler, A., Wichmann, H. E., and Heinrich, J., Variability of apparent particle density of an urban aerosol, *Environ. Sci. Technol.*, 37 (19), 4336–4342, 2003.
13. Gregg, S. J. and Sing, K. S. W., *Adsorption, Surface Area, and Porosity*, 2nd ed., Academic Press, London, New York, 1982.
14. Rogak, S. N., Baltensperger, U., and Flagan, R. C., Measurement of mass transfer to agglomerate aerosols, *Aerosol Sci. Technol.*, 14 (4), 447–458, 1991.
15. Oberdörster, G., Ferin, J., and Lehnert, B. E., Correlation between particle size, *in vivo* particle persistence, and lung injury, *Environ. Health Perspect.*, 102 (suppl. 5), 173–179, 1994.

16. Morrow, P. E., Possible mechanisms to explain dust overloading of the lungs, *Fundam. Appl. Toxicol.*, 10 (3), 369–384, 1988.
17. Driscoll, K. E., Role of inflammation in the development of rat lung tumors in response to chronic particle exposure, *Inhal. Toxicol.*, 8 (suppl.), 139–153, 1996.
18. Warheit, D. B., Interspecies comparisons of lung responses to inhaled particles and gases, *Crit. Rev. Toxicol.*, 20 (1), 1–29, 1989.
19. Lippmann, M. and Schlesinger, R. B., Interspecies comparisons of particle deposition and mucociliary clearance in tracheobronchial airways, *J. Toxicol. Environ. Health*, 13 (2–3), 441–469, 1984.
20. Kreyling, W. G. and Scheuch, G., Clearance of particles deposited in the lungs, In *Particle-Lung Interactions*, Gehr, P., and Heyder, J., eds., Marcel Dekker Inc., New York, Basel, 323–376, 2000.
21. ICRP Publication 66, Human respiratory tract model for radiological protection. A report of a Task Group of the International Commission on Radiological Protection, *Ann. ICRP*, 24 (1–3), 1–482, 1994.
22. Schmidt-Ott, A., Baltensperger, U., Gaeggeler, H. W., and Jost, D. T., Scaling behaviour of physical parameters describing agglomerates, *J. Aerosol Sci.*, 21 (6), 711–717, 1990.
23. Kütz, S. and Schmidt-Ott, A., Characterization of agglomerates by condensation-induced restructuring, *J. Aerosol Sci.*, 23 (suppl. 1), S357–S360, 1992.
24. Weber, A. P., Baltensperger, U., Gaggeler, H. W., and Schmidt-Ott, A., In situ characterization and structure modification of agglomerated aerosol particles, *J. Aerosol Sci.*, 27 (6), 915–929, 1996.
25. Weber, A. P. and Friedlander, S. K., In situ determination of the activation energy for restructuring of nanometer aerosol agglomerates, *J. Aerosol Sci.*, 28 (2), 179–192, 1997.
26. Schulz, H., Brand, P., and Heyder, J., Particle deposition in the respiratory tract, In *Particle-Lung Interactions*, Gehr, P., and Heyder, J., eds., Marcel Dekker Inc., New York, Basel, 229–290, 2000.
27. Asgharian, B., Hofmann, W., and Bergmann, R., Particle deposition in a multiple-path model of the human lung, *Aerosol Sci. Technol.*, 34, 332–339, 2001.
28. Winkler-Heil, R. and Hofmann, W., Deposition densities of inhalted particles in human bronchial airways, *Ann. Occup. Hyg.*, 46 (suppl. 1), S326–S328, 2002.
29. Anjilvel, S. and Asgharian, B., A multiple-path model of particle deposition in the rat lung, *Fundam. Appl. Toxicol.*, 28, 41–50, 1995.
30. Anderson, M., Svartengren, M., Bylin, G., Philipson, K., and Camner, P., Deposition in asthmatics of particles inhaled in air or in helium–oxygen, *Am. Rev. Respir. Dis.*, 147 (3), 524–528, 1993.
31. Kim, C. S. and Kang, T. C., Comparative measurement of lung deposition of inhaled fine particles in normal subjects and patients with obstructive airway disease, *Am. J. Respir. Crit. Care Med.*, 155 (3), 899–905, 1997.
32. Brown, J. S., Zeman, K. L., and Bennett, W. D., Regional deposition of coarse particles and ventilation distribution in healthy subjects and patients with cystic fibrosis, *J. Aerosol Med.*, 14 (4), 443–454, 2001.
33. Chalupa, D. C., Morrow, P. E., Oberdörster, G., Utell, M. J., and Frampton, M. W., Ultrafine particle deposition in subjects with asthma, *Environ. Health Perspect.*, 112 (8), 879–882, 2004.
34. Brand, P., Meyer, T., Sommerer, K., Weber, N., and Scheuch, G., Alveolar deposition of monodisperse aerosol particles in the lung of patients with chronic obstructive pulmonary disease, *Exp. Lung Res.*, 28 (1), 39–54, 2002.
35. Anderson, P. J., Wilson, J. D., and Hiller, F. C., Respiratory tract deposition of ultrafine particles in subjects with obstructive or restrictive lung disease, *Chest*, 97 (5), 1115–1120, 1990.
36. Brown, J. S., Zeman, K. L., and Bennett, W. D., Ultrafine particle deposition and clearance in the healthy and obstructed lung, *Am. J. Respir. Crit. Care Med.*, 166 (9), 1240–1247, 2002.
37. Schiller-Scotland, C. F., Hlawa, R., and Gebhart, J., Experimental data for total deposition in the respiratory tract of children, *Toxicol. Lett.*, 72 (1–3), 137–144, 1994.
38. Schiller-Scotland, C. F., Hlawa, R., Gebhart, J., Heyder, J., Roth, C., and Wönne, R., Particle deposition in the respiratory tract of children during spontaneous and controlled mouth breathing, *Ann. Occup. Hyg.*, 38 (suppl. 1), 117–125, 1994.
39. Bennett, W. D. and Zeman, K. L., Effect of body size on breathing pattern and fine-particle deposition in children, *J. Appl. Physiol.*, 97 (3), 821–826, 2004.

40. Asgharian, B., Menache, M. G., and Miller, F. J., Modeling age-related particle deposition in humans, *Journal of Aerosol Medicine*, 17 (3), 213–224, 2004.
41. Pope, C. A. and Dockery, D. W., Acute health effects of PM10 pollution on symptomatic and asymptomatic children, *Am. Rev. Respir. Dis.*, 145 (5), 1123–1128, 1992.
42. Bunn, H. J., Dinsdale, D., Smith, T., and Grigg, J., Ultrafine particles in alveolar macrophages from normal children, *Thorax*, 56 (12), 932–934, 2001.
43. Brauer, M., Hoek, G., Van Vliet, P., Meliefste, K., Fischer, P. H., Wijga, A., Koopman, L. P. et al., Air pollution from traffic and the development of respiratory infections and asthmatic and allergic symptoms in children, *Am. J. Respir. Crit. Care Med.*, 166 (8), 1092–1098, 2002.
44. Gehr, P., Geiser, M., Im Hof, V., and Schürch, S., Surfactant-ultrafine particle interactions: what we can learn from PM10 studies, *Philos. T. Roy. Soc. A.*, 358 (1775), 2707–2717, 2000.
45. Kreyling, W. G., Godleski, J. J., Kariya, S. T., Rose, R. M., and Brain, J. D., *In vitro* dissolution of uniform cobalt oxide particles by human and canine alveolar macrophages, *Am. J. Respir. Cell Mol. Biol.*, 2 (5), 413–422, 1990.
46. Kreyling, W. G., Intracellular particle dissolution in alveolar macrophages, *Environ. Health Perspect.*, 97, 121–126, 1992.
47. Kim, K. J., Fandy, T. E., Lee, V. H. L., Ann, D. K., Borok, Z., and Crandall, E. D., Net absorption of IgG via FcRn-mediated transcytosis across rat alveolar epithelial cell monolayers, *Am. J. Physiol. Lung Cell. Mol. Physiol.*, 287 (3), L616–L622, 2004.
48. Kim, K. J., Matsukawa, Y., Yamahara, H., Kalra, V. K., Lee, V. H. L., and Crandall, E. D., Absorption of intact albumin across rat alveolar epithelial cell monolayers, *Am. J. Physiol. Lung Cell. Mol. Physiol.*, 284 (3), L458–L465, 2003.
49. Kim, K. J. and Malik, A. B., Protein transport across the lung epithelial barrier, *Am. J. Physiol. Lung Cell. Mol. Physiol.*, 284 (2), L249–L259, 2003.
50. Heckel, K., Kiefmann, R., Dorger, M., Stoeckelhuber, M., and Goetz, A. E., Colloidal gold particles as a new *in vivo* marker of early acute lung injury, *Am. J. Physiol. Lung Cell. Mol. Physiol.*, 287 (4), L867–L878, 2004.
51. Kato, T., Yashiro, T., Murata, Y., Herbert, D. C., Oshikawa, K., Bando, M., Ohno, S., and Sugiyama, Y., Evidence that exogenous substances can be phagocytized by alveolar epithelial cells and transported into blood capillaries, *Cell Tissue Res.*, 311 (1), 47–51, 2003.
52. Rejman, J., Oberle, V., Zuhorn, I. S., and Hoekstra, D., Size-dependent internalization of particles via the pathways of clathrin- and caveolae-mediated endocytosis, *Biochem. J.*, 377 (Pt 1), 159–169, 2004.
53. Semmler, M., Regula, J., Oberdörster, G., Heyder, J., and Kreyling, W. G., Lung-lining fluid proteins bind to ultrafine insoluble particles: a potential way for particles to pass airblood barrier of the lung?, *Eur. Respir. J.*, 24 (suppl. 48), 100s, 2004.
54. Stahlhofen, W., Gebhart, J., and Heyder, J., Experimental determination of the regional deposition of aerosol particles in the human respiratory tract, *Am. Ind. Hyg. Assoc. J.*, 41 (6), 385a–398a, 1980.
55. Scheuch, G., Stahlhofen, W., and Heyder, J., An approach to deposition and clearance measurements in human airways, *J. Aerosol Med.*, 9 (1), 35–41, 1996.
56. Möller, W., Häussinger, K., Winkler-Heil, R., Stahlhofen, W., Meyer, T., Hofmann, W., and Heyder, J., Mucociliary and long-term particle clearance in the airways of healthy non-smokers, *J. Appl. Physiol.*, 97 (6), 2200–2206, 2004.
57. Kreyling, W. G., Semmler-Behnke, M., and Möller, W., Ultrafine particle-lung interactions: does size matter?, *J. Aerosol Med.*, 19 (1), 74–83, 2006.
58. Asgharian, B., Hofmann, W., and Miller, F. J., Mucociliary clearance of insoluble particles from the tracheobronchial airways of the human lung, *J. Aerosol Sci.*, 32 (6), 817–832, 2001.
59. Wiebert, P., Sanchez-Crespo, A., Seitz, J., Falk, R., Philipson, K., Kreyling, W.G., Möller, W., Sommerer, K., Larsson, S., and Svartengren, M., Negligible clearance of ultrafine particles retained in healthy and affected human lungs, *Eur. Respir. J.*, 18, 733–740, 2006.
60. Wiebert, P., Sanchez-Crespo, A., Falk, R., Philipson, K., Lundin, A., Larsson, S., Möller, W., Kreyling, W. G., and Svartengren, M., No significant translocation of inhaled 35 nm carbon particles to the circulation in humans, *Inhal. Toxicol.*, 10, 741–747, 2006.

61. Bailey, M. R., Kreyling, W. G., Andre, S., Batchelor, A., and Collier, C. G., An interspecies comparison of the lung clearance of inhaled monodisperse cobalt oxide particles, part I: objectives and summary of results, *J. Aerosol Sci.*, 20 (2), 169–188, 1989.
62. Möller, W., Barth, W., Kohlhäufl, M., Häussinger, K., Stahlhofen, W., and Heyder, J., Human alveolar long-term clearance of ferromagnetic iron-oxide microparticles in healthy and diseased subjects, *Exp. Lung Res.*, 27, 547–568, 2001.
63. Rogers, D. F., Mucus hypersecretion in chronic obstructive pulmonary disease, *Novartis Found. Symp.*, 234, 65–83, 2001.
64. Sethi, S., Bacterial infection and the pathogenesis of COPD, *Chest*, 117 (5 suppl. 1), 286S–291S, 2000.
65. Regnis, J. A., Robinson, M., Bailey, D. L., Cook, P., Hooper, P., Chan, H. K., Gonda, I., Bautovich, G., and Bye, P. T., Mucociliary clearance in patients with cystic fibrosis and in normal subjects, *Am. J. Respir. Crit. Care Med.*, 150 (1), 66–71, 1994.
66. Knowles, M. R. and Boucher, R. C., Mucus clearance as a primary innate defense mechanism for mammalian airways, *J. Clin. Invest.*, 109 (5), 571–577, 2002.
67. Möller, W., Häussinger, K., Ziegler-Heitbrock, L., and Heyder, J., Mucociliary and long-term particle clearance in airways of patients with immotile cilia, *Respir. Res.*, 7 (1), 10, 2006.
68. Foster, W. M., Mucociliary transport and cough in humans, *Pulm. Pharmacol. Ther.*, 15 (3), 277–282, 2002.
69. Cohen, D., Arai, S. F., and Brain, J. D., Smoking impairs long-term dust clearance from the lung, *Science*, 204 (4392), 514–517, 1979.
70. Kirkham, P. A., Spooner, G., Rahman, I., and Rossi, A. G., Macrophage phagocytosis of apoptotic neutrophils is compromised by matrix proteins modified by cigarette smoke and lipid peroxidation products, *Biochem. Biophys. Res. Commun.*, 318 (1), 32–37, 2004.
71. Renwick, L. C., Brown, D., Clouter, A., and Donaldson, K., Increased inflammation and altered macrophage chemotactic responses caused by two ultrafine particle types, *Occup. Environ. Med.*, 61 (5), 442–447, 2004.
72. Roth, C. and Stahlhofen, W., Clearance measurements with radioactively labeled ultrafine particles, *Ann. Occup. Hyg.*, 38 (suppl. 1), 101–106, 1994.
73. Möller, W., Felten, K., Sommerer, K., Scheuch, G., Meyer, G., Meyer, P., Häussinger, K., and Kreyling, W. G., Deposition, retention and translocation of ultrafine particles from central and peripheral lung regions, *Am. J. Respir. Crit. Care Med.*, submitted, 2006.
74. Semmler, M., Seitz, J., Erbe, F., Mayer, P., Heyder, J., Oberdörster, G., and Kreyling, W. G., Long-term clearance kinetics of inhaled ultrafine insoluble iridium particles from the rat lung, including transient translocation into secondary organs, *Inhal. Toxicol.*, 16 (6–7), 453–459, 2004.
75. Ferin, J. and Oberdörster, G., Translocation of particles from pulmonary alveoli into the interstitium, *J. Aerosol Med.*, 5 (1), 179–187, 1992.
76. Lehnert, B. E., Ortiz, J. B., Steinkamp, J. A., Tietjen, G. L., Sebring, R. J., and Oberdörster, G., Mechanisms underlying the "particle redistribution phenomen," *J. Aerosol Med.*, 5 (4), 261–277, 1992.
77. Kreyling, W. G., Semmler, M., Erbe, F., Mayer, P., Takenaka, S., Schulz, H., Oberdörster, G., and Ziesenis, A., Translocation of ultrafine insoluble iridium particles from lung epithelium to extrapulmonary organs is size dependent but very low, *J. Toxicol. Environ. Health A.*, 65 (20), 1513–1530, 2002.
78. Hofmann, W. and Asgharian, B., Comparison of mucociliary clearance velocities in human and rat lungs for extrapolating modeling, *Ann. Occup. Hyg.*, 46 (Suppl. 1), S323–S325, 2002.
79. Ferin, J., Oberdörster, G., and Penney, D. P., Pulmonary retention of ultrafine and fine particles in rats, *Am. J. Respir. Cell Mol. Biol.*, 6 (5), 535–542, 1992.
80. Oberdörster, G., Toxicology of ultrafine particles: *in vivo* studies, *Philos. T. Roy. Soc. A.*, 358 (1775), 2719–2739, 2000.
81. Stoeger, T., Reinhard, C., Takenaka, S., Schroeppel, A., Karg, E., Ritter, B., Heyder, J., and Schulz, H., Instillation of six different ultrafine carbon particles indicates a surface area threshold dose for acute lung inflammation in mice, *Environ. Health Perspect.*, 114 (3), 328–333, 2006.

82. Li, X. Y., Gilmour, P. S., Donaldson, K., and MacNee, W., Free radical activity and pro-inflammatory effects of particulate air pollution (PM10) *in vivo* and *in vitro*, *Thorax*, 51 (12), 1216–1222, 1996.
83. Faux, S. P., Tran, C. L., Miller, B. G., Jones, A. D., Monteiller, C., and Donaldson, K., Reseach Report 154 Report No. 154, 2003.
84. Beck-Speier, I., Dayal, N., Karg, E., Maier, K. L., Schumann, G., Schulz, H., Semmler, M. et al., Oxidative stress and lipid mediators induced in alveolar macrophages by ultrafine particles, *Free Radic. Biol. Med.*, 38 (8), 1080–1092, 2005.
85. Donaldson, K., Brown, D., Clouter, A., Duffin, R., MacNee, W., Renwick, L., Tran, L., and Stone, V., The pulmonary toxicology of ultrafine particles, *J. Aerosol Med.*, 15 (2), 213–220, 2002.
86. Donaldson, K., MacNee, W., Mills, N., Newby, D., and Robinson, S., Role of inflammation in cardiopulmonary health effects of PM, *Toxicol. Appl. Pharmacol.*, 207 (suppl. 2), 483–488, 2005.
87. Donaldson, K., Tran, L., Jimenez, L. A., Duffin, R., Newby, D. E., Mills, N., Macnee, W., and Stone, V., Combustion-derived nanoparticles: a review of their toxicology following inhalation exposure, *Part. Fiber Toxicol.*, 2 (10), 1–14, 2005.
88. Oberdörster, G., Oberdörster, E., and Oberdörster, J., Nanotoxicology: an emerging discipline evolving from studies of ultrafine particles, *Environ. Health Perspect.*, 113, 823–839, 2005.
89. ESF 2005, ESF forward look on Nanomedicine., European Science Foundation Policy Briefings, http://www.esf.org/newsrelease/83/SPB23Nanomedicine.pdf, 2005.
90. Kreuter, J., Nanoparticulate systems for brain delivery of drugs, *Adv. Drug Deliv. Rev.*, 47 (1), 65–81, 2001.
91. Kreuter, J., Shamenkov, D., Petrov, V., Ramge, P., Cychutek, K., Koch-Brandt, C., and Alyautdin, R., Apolipoprotein-mediated transport of nanoparticle-bound drugs across the blood-brain barrier, *J. Drug Target.*, 10 (4), 317–325, 2002.
92. Kreuter, J., Influence of the surface properties on nanoparticle-mediated transport of drugs to the brain, *J. Nanosci. Nanotech.*, 4 (5), 484–488, 2004.
93. Kreyling, W., Semmler-Behnke, M., and Möller, W., Health implications of nanoparticles, *J. Nanopart. Res.*, 8, DOI 10.1007/s11051-005-9068-z, 2006, in press.
94. Mills, N. L., Amin, N., Robinson, S. D., Anand, A., Davies, J., Patel, D., de la Fuente, J. M. et al., Do inhaled carbon nanoparticles translocate directly into the circulation in humans? *Am. J. Respir. Crit. Care Med.*, 173 (4), 426–431, 2006.
95. Mills, N. L., Tornqvist, H., Robinson, S. D., Gonzalez, M., Darnley, K., MacNee, W., Boon, N. A. et al., Diesel exhaust inhalation causes vascular dysfunction and impaired endogenous fibrinolysis, *Circulation*, 112 (25), 3930–3936, 2005.
96. Nemmar, A., Hoylaerts, M. F., Hoet, P. H., Dinsdale, D., Smith, T., Xu, H., Vermylen, J., and Nemery, B., Ultrafine particles affect experimental thrombosis in an *in vivo* hamster model, *Am. J. Respir. Crit. Care Med.*, 166 (7), 998–1004, 2002.
97. Silva, V. M., Corson, N., Elder, A., and Oberdörster, G., The rat ear vein model for investigating *in vivo* thrombogenicity of ultrafine articles (UFP), *Toxicol. Sci.*, 85, 983–989, 2005.
98. Auerbach, O., Conston, A. S., Garfinkel, L., Parks, V. R., Kaslow, H. D., and Hammond, E. C., Presence of asbestos bodies in organs other than the lung, *Chest*, 77 (2), 133–137, 1980.
99. LeFevre, M. E., Green, F. H., Joel, D. D., and Laqueur, W., Frequency of black pigment in livers and spleens of coal workers: correlation with pulmonary pathology and occupational information, *Hum. Pathol.*, 13 (12), 1121–1126, 1982.
100. Berry, J. P., Arnoux, B., and Stanislas, G., A microanalytic study of particles transport across the alveoli: role of blood platelets, *Biomedicine*, 27 (9–10), 354–357, 1977.
101. Kapp, N., Kreyling, W., Schulz, H., Im Hof, V., Gehr, P., Semmler, M., and Geiser, M., Electron energy loss spectroscopy for analysis of inhaled ultrafine particles in rat lungs, *Microsc. Res. Tech.*, 63 (5), 298–305, 2004.
102. Geiser, M., Rothen-Rutishauser, B., Kapp, N., Schurch, S., Kreyling, W., Schulz, H., Semmler, M., Hof, V. I., Heyder, J., and Gehr, P., Ultrafine particles cross cellular membranes by nonphagocytic mechanisms in lungs and in cultured cells, *Environ. Health Perspect.*, 113 (11), 1555–1560, 2005.

103. Takenaka, S., Karg, E., Roth, C., Schulz, H., Ziesenis, A., Heinzmann, U., Schramel, P., and Heyder, J., Pulmonary and systemic distribution of inhaled ultrafine silver particles in rats, *Environ. Health Perspect.*, 109 (suppl. 4), 547–551, 2001.
104. Oberdörster, G., Sharp, Z., Atudorei, V., Elder, A., Gelein, R., Lunts, A., Kreyling, W. G., and Cox, C., Extrapulmonary translocation of ultrafine carbon particles following whole-body inhalation exposure of rats, *J. Toxicol. Environ. Health A.*, 65 (20), 1531–1543, 2002.
105. Kreyling, W. G., Semmler, M., and Möller, W., Dosimetry and toxicology of ultrafine particles, *J. Aerosol Med.*, 17 (2), 140–152, 2004.
106. Kreyling, W. G., Semmler, M., Erbe, F., Mayer, P., Takenaka, S., Oberdörster, G., and Ziesenis, A., Minute translocation of inhaled ultrafine insoluble iridium particles from lung epithelium into extrapulmonary tissues, *Ann. Occup. Hyg.*, 46 (suppl. 1), S223–S226, 2002.
107. Chen, J., Tan, M., Nemmar, A., Song, W., Dong, M., Zhang, G., and Li, Y., Quantification of extrapulmonary translocation of intratracheal-instilled particles *in vivo* in rats: effect of lipopolysaccharide, *Toxicology*, 2006, doi:10.1016/j.tox.2006.02.016.
108. Roth, C., Ferron, G. A., Karg, E., Lentner, B., Schumann, G., Takenaka, S., and Heyder, J., Generation of ultrafine particles by spark discharging, *Aerosol Sci. Technol.*, 38 (3), 228–235, 2004.
109. Szymczak, W., Kreyling, W. G., Seitz, J., and Wittmaack, K., Mass spectrometric characterisation of pure and mixed ultrafine particles of iridium and carbon, *J. Aerosol Sci.*, 35 (suppl. 1), 37–38, 2004.
110. Bodian, D. and Howe, H. A., The rate of progression of poliomyelitis virus in nerves, *Bull. Johns Hopkins Hosp.*, 69, 79–85, 1941.
111. de Lorenzo, A. J. D. and Darin, J., The olfactory neuron and the blood-brain barrier, In *Taste and Smell in Vertebrates*, Wolstenholme, G. E. W., and Knight, J., eds., Churchill, London, 151–176, 1970.
112. Oberdörster, G., Sharp, Z., Atudorei, V., Elder, A., Gelein, R., Kreyling, W., and Cox, C., Translocation of inhaled ultrafine particles to the brain, *Inhal. Toxicol.*, 16 (6–7), 437–445, 2004.
113. Howe, H. A. and Bodian, D., Poliomyelitis in the chimpanzee: a clinical-pathological study, *Proc. Soc. Exp. Biol. Med.*, 43, 718–721, 1940.
114. Oberdörster, G. and Utell, M. J., Ultrafine particles in the urban air: to the respiratory tract-and beyond?, *Environ. Health Perspect.*, 110 (8), A440–A441, 2002.
115. Hunter, D. D. and Undem, B. J., Identification and substance P content of vagal afferent neurons innervating the epithelium of the guinea pig trachea, *Am. J. Respir. Crit. Care Med.*, 159 (6), 1943–1948, 1999.
116. Nemmar, A., Hoet, P. H., Dinsdale, D., Vermylen, J., Hoylaerts, M. F., and Nemery, B., Diesel exhaust particles in lung acutely enhance experimental peripheral thrombosis, *Circulation*, 107 (8), 1202–1208, 2003.
117. Finlayson-Pitts, B. J. and Pitts, J. N., *Chemistry of the Upper and Lower Atmosphere: Theory, Experiments, and Applications*, Academic Press, San Diego, 2000.

4 Particulate Air Pollutants and Small Airway Remodeling

Andrew Churg
Department of Pathology, University of British Columbia

CONTENTS

4.1 SUMMARY

Small airway remodeling is an important cause of chronic obstructive pulmonary disease (COPD) in cigarette smokers. The incidence of COPD in subjects exposed to chronically high levels of particulate air pollutants (i.e., particulate matter) appears to be increased, and morphologic studies have shown that particulate matter (PM) produces small airway remodeling that is structurally similar to that seen in cigarette smokers. Experimental studies show that PM enters airway walls and suggest that the most important component of airway remodeling, increased airway wall fibrous tissue, is driven by oxidants and iron released from PM particles. Release of the fibrogenic cytokine, $TGF\beta_1$, again through an oxidant mechanism, also appears to be involved. These processes lead to thickened, fibrotic, and distorted small airways that have increased resistance to flow, producing the clinical picture of COPD. Humans are exposed to both PM and ozone *in vivo*, and experimental data imply that this combination of exposures potentiates airway wall remodeling, but information on this question in humans is lacking.

4.2 EPIDEMIOLOGY

Although cigarette smoking is by far the most important cause of COPD, there is (somewhat limited) evidence to suggest that chronic exposure to high levels of particulate air pollutants (i.e., PM) may also cause chronic airflow obstruction (reviewed by Sunyer 2001). The most extensive studies are those of Abbey and colleagues (Abbey et al. 1991, 1995, 1998, 1999; Beeson et al. 1998) who examined a cohort of nearly 4000 nonsmoking Seventh-Day Adventists in Southern California and found that risks for the development of new cases of chronic bronchitis and COPD were directly correlated with levels of exposure to ambient PM_{10} and $PM_{2.5}$ (Abbey et al. 1991, 1995). Symptom severity also correlated with $PM_{2.5}$ and PM_{10} levels, and long-term increases in PM_{10} concentrations were associated with greater FEV_1 decrements (Abbey et al. 1998), as were increases in mortality from nonmalignant lung disease (Abbey et al. 1999). Oddly enough, these findings were largely confined to males, and minimal effects were observed in females.

The UCLA–CORD study (Detels et al. 1991; Tashkin et al. 1994), which examined lung function changes over time in three areas of Southern California, produced reasonably similar results. Declines in FEV_1 were found to be significantly greater in both smokers and nonsmokers in the Long Beach region, which had the highest levels of particulate pollutants, nitrogen oxides, and sulfates, but smoking appeared to account for the majority (about 70%) of the decline. These effects were found in men; whereas in women, effects were only seen in never-smokers. The importance of PM as opposed to ozone can be inferred from the fact that no chronic effects were seen in the Glendora region, which had high levels of ozone but not PM.

Sunyer (2001) has summarized a number of other, largely cross-sectional, studies. These have shown a reasonably consistent association of increased levels of air pollutants with increased declines in FEV_1 and/or FVC. There are, however, differences in the pollutants that seem to be most important from study to study; some report associations with PM_{10} or total suspended particulates, whereas others only find associations with levels of sulfates or acid (see Sunyer 2001, Table 3). All of these studies can be criticized on methodological grounds (reviewed in Sunyer 2001), and the variation from study to study in the specific type of air pollutant that correlates with functional changes is problematic. Nonetheless, it appears reasonable to conclude that chronically elevated levels of air pollutants, including PM, can produce COPD.

Particulate Matter emitted from cooking with biomass fuels can produce very high exposure levels (Smith 1993; Brauer et al. 1996), and it appears to produce COPD. This problem is largely confined to women in developing countries who cook with biomass fuels in enclosed spaces. Perez-Padilla et al. (1996), using a case-control approach in a series of Mexican women, reported an increased risk of chronic bronchitis and chronic airflow obstruction associated with cooking with wood. The risk of chronic bronchitis was linearly associated with hour-years of cooking with biomass fuels. Dennis et al. (1996) reached a similar conclusion for women in Colombia who cooked with biomass fuels. Because of the level of PM involved, there is probably a potential for severe disease in this setting.

4.3 ANATOMIC CHANGES IN THE AIRWAYS OF HUMANS WITH CHRONICALLY HIGH-PM EXPOSURE

There are two pathologic lesions associated with COPD in cigarette smokers: emphysema and small airway remodeling (i.e., small airways disease) (see Wright 2005 for a detailed review of the pathologic features seen in cigarette smokers). At this point, there is no convincing evidence that PM exposure produces emphysema, although high levels of exposure to a variety of occupationally encountered dusts, notably coal and silica, do (Churg and Wright 2003; Churg 2005), and, given the general similarity of mechanisms that probably apply, PM might well do the same.

Small airway remodeling is a variable combination of increased fibrous tissue, smooth muscle, and inflammatory cells in the walls of the membranous and respiratory bronchioles, such that the airway lumen becomes narrowed and/or distorted (Wiggs et al. 1992; Pare et al. 1997; Saetta, Turato, and Zuin 2000; Jeffery 2001). The consequence of this process is a marked increase in resistance to airflow in the small airways and functional evidence of chronic airflow obstruction.

A few studies have looked at pathologic changes in the small airways in individuals chronically exposed to high levels of PM. Souza et al. (1998) examined a series of autopsic lungs from forensic (violent death) cases of 34 residents of low-PM regions of Brazil and 50 residents of Sao Paolo, a region with a mean annual PM_{10} of about 80–100 μg/m^3 between 1991 and 1995. Almost all (90%) of the subjects were males, and the mean age at death was approximately 28 years. Approximately 60% were smokers, but on average, the amount of smoking (mean 7 pack-years) was low. Souza et al. used histologic sections to grade changes in the small airways and also determined the gland/wall ratio in the large airways. No differences were seen for the gland/wall ratio between high-PM and low-PM subjects regardless of smoking status, but black pigmented dust (i.e., anthracosis), inflammation, wall thickness, and mucus hypersecretion were, for the most part, greater in smokers. These features were also somewhat greater in high-PM subjects compared to low-PM subjects in both the smoking and nonsmoking groups. The authors used the pigmented dust score as a surrogate of PM exposure and found that this correlated with both airway wall inflammation and airway wall thickness.

These data imply that long-term residence in high-PM regions leads to the same type of small airway remodeling that has been shown to correlate with clinical airflow obstruction in cigarette smokers (Wright 2005). However, interpretation of this study is difficult because of confounding by smoking (more than half the subjects were smokers), which clearly produces similar changes including black pigment deposition, and a high proportion of subjects (more than 50%) with probable occupational dust exposures. Nonetheless, after adjustment for these confounders, the statistical analysis indicated an effect of PM level on pathologic changes.

Pinkerton et al. (2000) performed a somewhat similar study using examined forensic death autopsic lungs from 42 relatively young (median age = 33) Hispanic males from the Fresno County Coroner's Office. During the time the samples were collected, mean PM_{10} and $PM_{2.5}$ levels in Fresno were reasonably high (43.5 and 22 μg/m^3, respectively). On histologic examination, about half the subjects had evidence of cigarette smoke injury in the form of chronic bronchitis and more distal small airway remodeling. The distal small airways contained considerable black carbonaceous dust accompanied by birefringent particles. The highest particle loads were found in the walls of generation one respiratory bronchioles and in membranous bronchioles. The amount of retained dust, measured as visible pigment and graded on a semiquantitative scale, correlated with the degree of fibrosis in these airways.

As in the Souza study, smoking and occupational dust exposure were potential confounders, since 50% of the subjects were smokers. As well, most of them had worked in local farming/blue collar occupations, and farming in this region is notoriously dusty. Another problem is that smoking causes small airway lesions that are similar in location and appearance to those caused by dusts, and synergistic interactions between smoke and dust may amplify dust effects. These issues notwithstanding, this study lends support to the idea that chronic high-level PM can produce abnormalities in small airways.

In an attempt to overcome these problems, Churg et al. (2003) obtained autopsic lungs from another high-PM region, Mexico City (3 year mean PM_{10} = 66 μg/m^3), and compared them to lungs of subjects from Vancouver, a city with low PM (1984–1993 average = 25 μg PM_{10} and 15 μg $PM_{2.5}$ (Brook and Dann 1997)). We purposely selected only lungs from women to minimize the potential for occupational dust exposure, and further excluded subjects with occupational dust exposures by history. All of the subjects were never-smokers, and none of the Mexico City women had a history of cooking with biomass fuels. Using histologic sections, we performed a visual grading procedure for the amount of muscle and fibrous tissue in the walls of the small airways and found that the lungs from Mexico City had dramatically greater fibrosis and muscle in both the membranous and respiratory bronchioles (Figure 4.1 and Figure 4.2). The lumens of the small airways in the Mexico City lungs showed considerably greater distortion than those in the Vancouver lungs (Figure 4.1).

We also used a microdissection procedure to isolate small airway wall tissue from the lungs, and we evaluated particulate content by analytical electron microscopy. This analysis showed the presence of chained aggregates of carbonaceous spheres with the morphology of diesel exhaust particles in the airway walls.

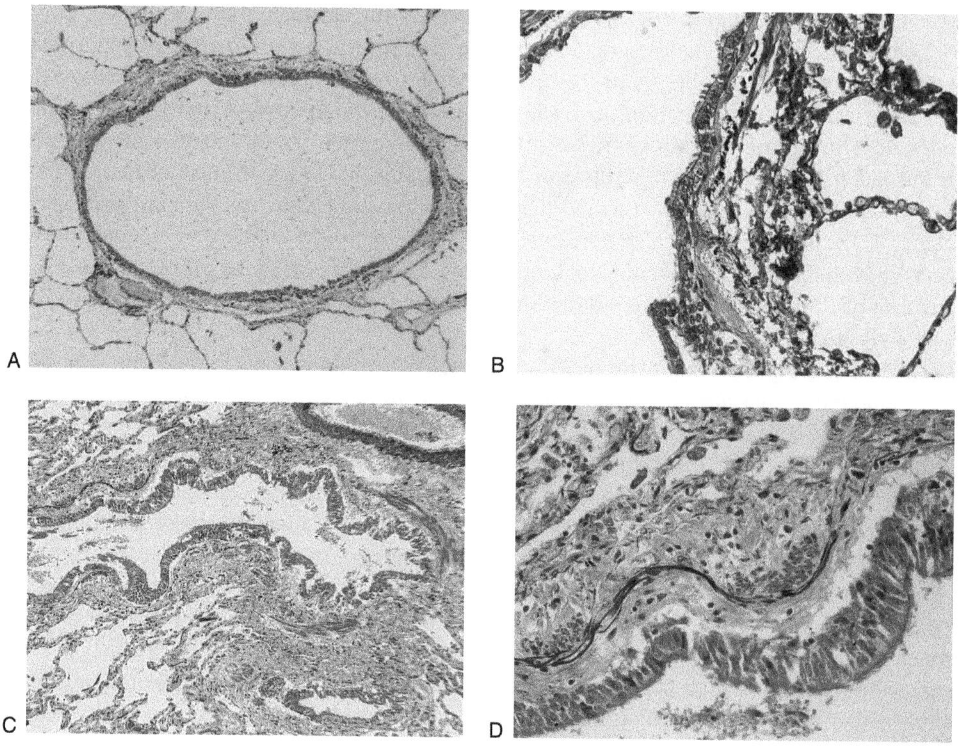

FIGURE 4.1 Representative small airways (membranous bronchioles) from Vancouver lungs (A and B) and Mexico City lungs (C and D) at low and high power. Note the much greater wall thickness and markedly increased collagen in the lungs from Mexico City. (From Churg, A. and Wright, J.L., Bronchiolitis caused by occupational and ambient atmospheric particles, *Semin. Respir. Crit. Care. Med.*, 24, 577–584, 2003. With permission.)

Our study thus suggests that, even in the absence of smoking and occupational dust exposure, chronic exposure to high levels of PM leads to small airway remodeling. As well, it is evident that PM enters and is retained in the airway walls. Whether high levels of other types of pollutants—including ozone, NO_2, SO_2, and CO, which are present in much greater concentrations in Mexico City compared to Vancouver air—played a role is uncertain, and synergistic interactions between pollutants are possible (see Interactions of PM and Ozone below). However, the lesions that we observed are essentially identical to the small airway remodeling found in workers occupationally exposed to mineral dusts in the absence of other pollutants (Churg and Wright 2003), suggesting that PM is the primary pollutant responsible for these effects.

Overall, the studies of Souza et al. (1998), Pinkerton et al. (2000), and Churg et al. (2003), although few in number, suggest that chronic exposure to high levels of PM produces distinctly visible small airway lesions, in particular, increases in fibrous tissue, muscle, and, in the respiratory bronchioles, pigmented dust, with accompanying lumenal distortion. These findings thus provide a pathologic basis for the functional abnormalities detected in persons with chronic exposures to elevated PM concentrations (see Functional Consequences and Perspective below).

4.4 EXPERIMENTAL STUDIES ON PM AND AIRWAY WALL REMODELING

Many experimental studies on the effects of PM have been published. The vast majority have dealt with acute responses, either in monolayer tissue culture systems or *in vivo*, and relatively little is known about chronic effects. However, acute effects are important to note because they form at

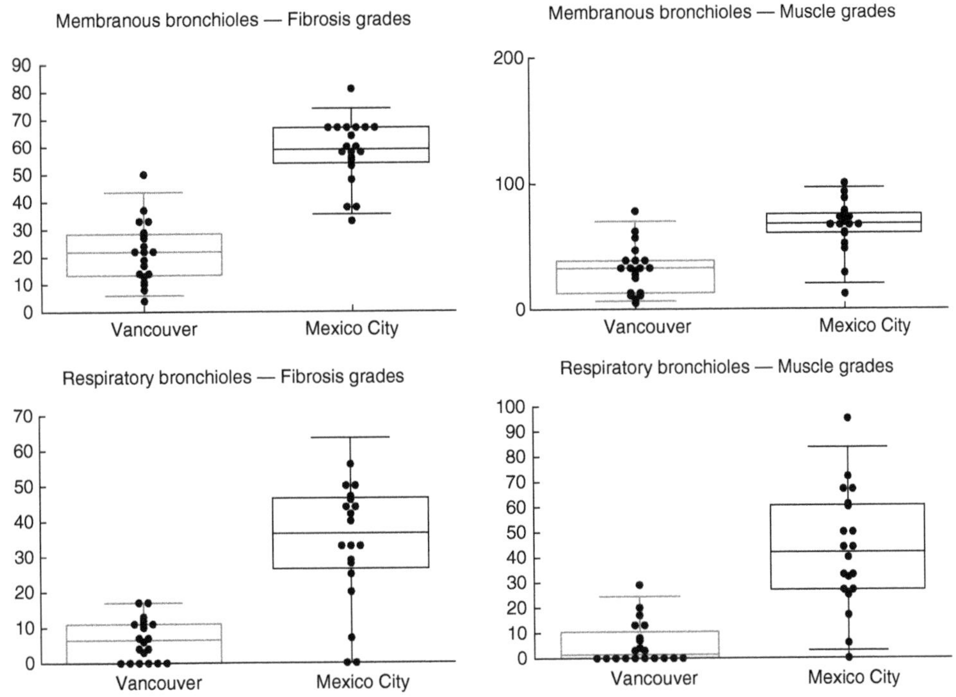

FIGURE 4.2 Results of visual grading for fibrous tissue and muscle in the walls of small airways from Vancouver and Mexico City. Data are shown as box plots: the box walls are the 25th and 75th percentiles, the horizontal line is the median value, and the whiskers show the 5th and 95th percentile. For both types of airway and both fibrous tissue and muscle, the Mexico City lungs show significantly greater abnormalities than the Vancouver lungs. (From Churg, A. and Wright, J.L., Bronchilitis caused by occupational and ambient atmospheric particles, *Semin. Respir. Crit. Care. Med.*, 24, 577–584, 2003. With premission.)

least part of the basis for the chronic changes. In a very broad sense, it is believed that most acute effects of PM relate, at some level, to oxidative stress, largely driven by transition metals, particularly iron, in the particles (Costa and Dreher 1997; Donaldson et al. 1997, 2003; Frampton et al. 1999; Jimenez et al. 2000; Molinelli et al. 2002; Tao, Gonzalez-Flecha, and Kobzik 2003; Roberts et al. 2003), although very small (ultrafine) particles probably can generate oxidants through surface effects that are independent of metal content. PM can generate oxidants in cell-free aqueous systems and also in monolayer cultures. The ability of PM to generate oxidants *in vitro* correlates with the ability to produce inflammation *in vivo* (Costa and Dreher 1997; Frampton et al. 1999; Molinelli et al. 2002; Donaldson et al. 2003; Tao, Gonzalez-Flecha, and Kobzik 2003), and the acute inflammatory response to PM reflects oxidative activation of fundamental cell signaling pathways, particularly the NF-κB and AP-1 pathways, with subsequent production of proinflammatory cytokines such as TNFα, IL-6, IL-8 (murine MIP-2), MCP-1, MIP-1α, and GM-CSF (Costa and Dreher 1997; Timblin et al. 1998; Frampton et al. 1999; Stone et al. 2000; Fujii et al. 2001; Donaldson et al. 2003; Tao, Gonzalez-Flecha, and Kobzik 2003) by macrophages and lung epithelial cells.

Examining airway fibrosis in simple model systems is problematic, since fibrosis tends to involve mediators produced by both epithelial and mesenchymal cells (and sometimes alveolar macrophages as well), along with interactions of epithelial and mesenchymal cells; merely exposing monolayer cultures of either type of cell to particulates provides little guidance about what happens *in vivo*. For these reasons, we have adopted a tracheal explant system that employs small segments of rat trachea that are exposed to dust or PM and then maintained in air organ culture for extended periods. While the trachea is obviously a model of the large airways and not the small airways that are the target of PM effects, the cells and organization of both large and small airways (ignoring the cartilage, which is irrelevant in this context) are reasonably similar, although tracheal explants lack the circular muscle layer seen in small airways.

In our tracheal explant model, explants are submerged for a brief period, usually 1 h, in a suspension of the dust of interest, and then the explants are carefully removed from the suspension so that the dust sits on the surface of the epithelial cells. Over a period of about 1 week in air organ culture, the surface particles are taken up by the epithelial cells and eventually transported to the underlying interstitial tissues. Using this model, we were able to show that amosite asbestos fibers produce airway wall remodeling in the form of increased procollagen and hydroxyproline, a measure of collagen content in the explants (Dai and Churg 2001). Gene expression of the fibrogenic growth factors PDGF-A and $TGF\beta_1$ increased over time; however, PDGF-B gene expression was not increased. Generation of active oxygen species (AOS), probably through redox cycling of surface iron on the fiber, appeared to be crucial to the development of fibrosis, since procollagen gene expression could be inhibited by deferoxamine and AOS scavengers, and, conversely, adding exogenous iron to the asbestos fiber surface increased the level of procollagen gene expression. Since asbestos fibers produce not only interstitial fibrosis (asbestosis) but also fibrosis of the small airways (small airway remodeling) *in vivo* (Churg 2005), these observations suggested that the tracheal explant model probably mimicked the fibrogenic mechanisms seen in airways *in vivo* and might be a useful way to dissect these mechanisms.

To examine the effects of PM on airway wall remodeling, we used the tracheal explant system with two different types of dust. The first was 0.12 μm TiO_2 particles, meant to be a model of relatively inert fine PM particles (Dai, Xie, and Churg 2002). By themselves, the TiO_2 particles did not increase procollagen gene expression or hydroxyproline in the explants, but if the particles were first loaded with cationic iron by incubation in an Fe(II)/Fe(III) chloride solution, then a progressive increase in procollagen gene expression (Figure 4.3) and hydroxyproline was seen over time. The hydroxyl radical scavenger tetramethylthiourea (TMTU) and deferoxamine prevented these effects, confirming that they were driven by AOS and iron. Addition of iron to the surface caused activation of NF-κB, a common effect of PM as noted above; of interest, increased levels of both IκBα phosphoserine 32/36 and IκBα phosphotyrosine were found. Iron loading also led to an NF-κB dependent increase in gene expression of prolyl-4-hydroxylase (Figure 4.4), an enzyme crucial to collagen production, thus providing one additional mechanism by which AOS may control tissue collagen levels.

As noted above, transition metals are an important component of many types of PM, and the findings just described are similar to a large number of observations that link PM toxicity to the presence of transition metals, mostly iron, but also vanadium (Costa and Dreher 1997; Quay et al. 1998; Frampton et al. 1999; Ghio et al. 1999; Jimenez et al. 2000). To determine whether iron removed from the iron-loaded TiO_2 particles might be playing a role, Dai, Xie, and Churg (2002) treated iron-loaded TiO_2 with citrate and then applied the citrate extract to tracheal explants. The extract proved to be an even stronger inducer of procollagen gene expression than the dust itself was. Thus, these results imply that leaching of redox-active iron from PM may be one mechanism behind PM-induced airway wall remodeling.

Actual PM particles are chemically much more complex than iron-loaded TiO_2. To determine whether actual PM could induce airway wall fibrogenesis, Dai et al. (2003) exposed tracheal explants to Ottawa Urban Air Particles (EHC93, closely corresponding to PM_{10}) or diesel exhaust particles (sample1650a from the National Institute of Standards and Technology). After 7 days of air organ culture, both types of PM caused increased procollagen gene expression and increased tissue hydroxyproline (Figure 4.5). AOS and NF-κB inhibitors completely blocked these effects. Both types of PM also produced increases in gene expression of $TGF\beta_1$, a powerful profibrogenic cytokine, and the TGFβ antagonist, fetuin (α2-HS-glycogprotein), prevented increases in procollagen gene expression (Figure 4.6), suggesting that induction of growth

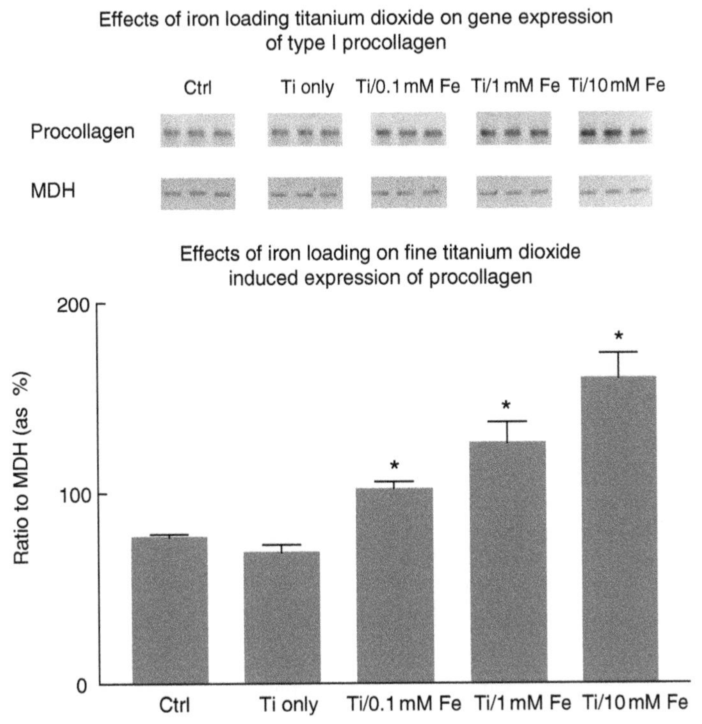

FIGURE 4.3 Effects of iron loading on fine TiO_2-induced procollagen gene expression in rat tracheal explants after 7 days in an organ culture. Adding iron to the surface changes the TiO_2 from a particle that is not fibrogenic to one that is fibrogenic, and the amount of gene expression increases in proportion to the amount of surface iron. (From Dai, J., Xie, C., and Churg, A., *Am. J. Respir. Cell Mol. Biol.*, 26, 685–693, 2002. With permission.)

factors by PM played a role in airway wall remodeling. With EHC93, TGFβ gene expression was driven by AOS and could be prevented by scavengers of AOS, but for diesel exhaust particles, TGFβ gene expression was not blocked by AOS scavengers; rather, it was blocked by inhibitors of the mitogen-activated protein kinase pathway. Thus, these findings imply that real PM particles can produce airway wall fibrosis but that different types of particles may use different pathways.

The potential role of TGFβ in PM-induced small airway remodeling is worth commenting on. TGFβ is produced with the TGFβ effector molecule complexed to a latency-associated peptide (LAP). The LAP must be dissociated from the TGFβ molecule for TGFβ ligation to its receptors and subsequent downstream TGFβ signaling to occur. Pociask, Sime, and Brody (2004) have shown that asbestos fibers can oxidize the LAP and that oxidation of the LAP releases active TGFβ; this process appears to run through iron on the surface of the asbestos fibers. Cigarette smoke, in many respects, functions as an oxidant-generating fine particle, and we have recently demonstrated that cigarette smoke also oxidizes the LAP and releases active TGFβ in a cell-free system; this process can be inhibited by AOS scavengers (Wang, Wright, and Churg 2005) (Figure 4.7). These findings raise the possibility that PM can act in a similar fashion, thus directly activating a cytokine that functions in an autocrine fashion to cause airway wall fibrosis.

4.5 INTERACTIONS OF PM AND OZONE

While it is convenient to study the effects of PM by itself in the laboratory, in real life, PM exposures are commonly accompanied by elevated levels of ozone as well as other atmospheric

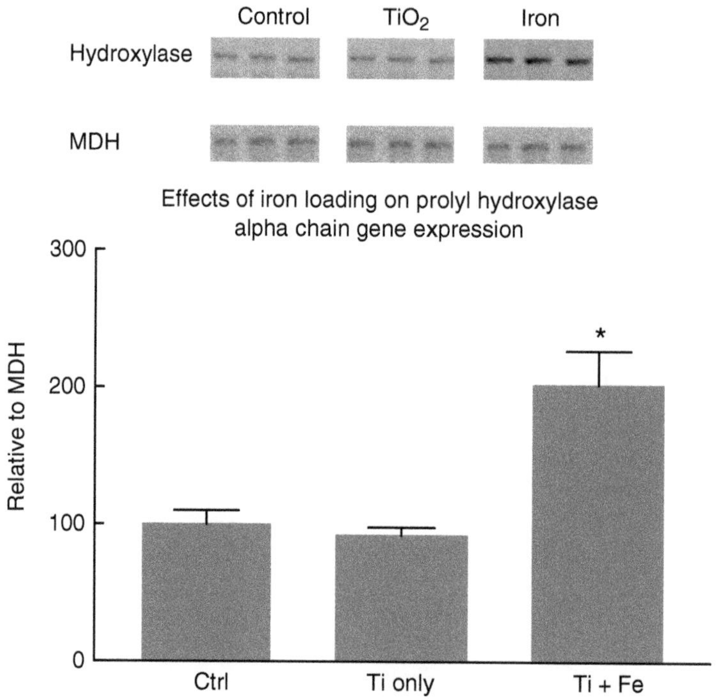

FIGURE 4.4 Effects of iron loading on fine TiO_2-induced prolyl-4-hydroxylase gene expression in rat tracheal explants after 7 days in an organ culture. Iron-loaded TiO_2 induces prolyl-4-hydroxylase, an enzyme crucial to the production of collagen. (From Dai, J., Xie, C., and Churg, A., *Am. J. Respir. Cell Mol. Biol.*, 26, 685–693, 2002. With permission.)

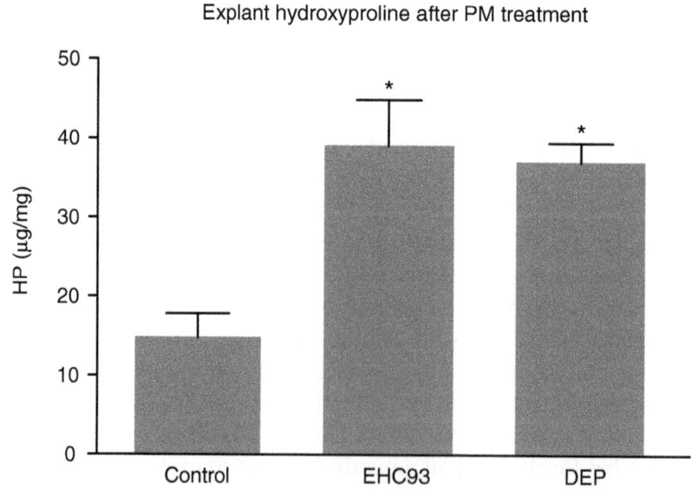

FIGURE 4.5 Effects of PM (EHC93) or diesel exhaust particles (DEP) on tracheal explant hydroxyproline, a measure of collagen content, after 7 days in an organ culture. Both dusts increase production of collagen, indicating that these particles can cause airway wall fibrosis. (From Dai, J., Xie, C., Vincent, R., and Churg, A., *Am. J. Respir. Cell Mol. Biol.*, 29, 352–358, 2003. With permission.)

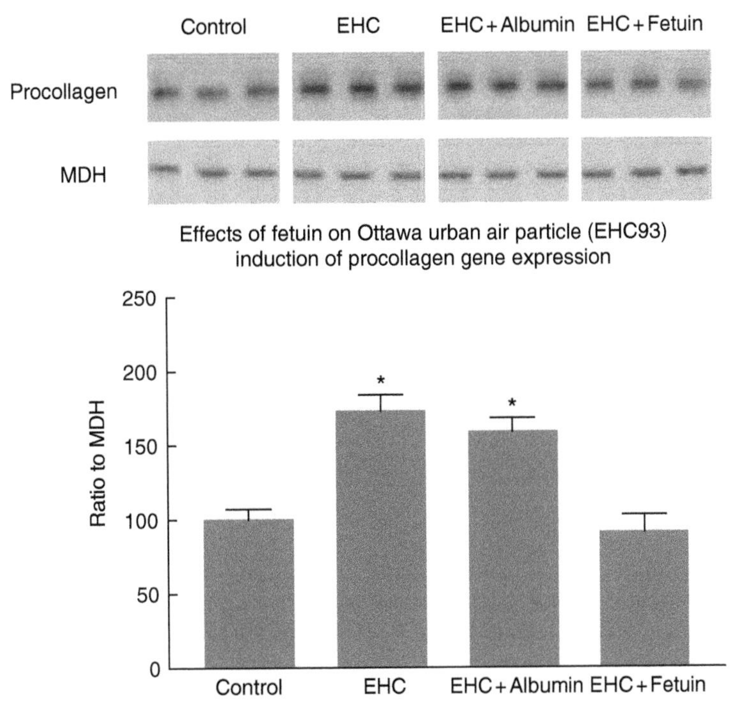

FIGURE 4.6 Effects of the TGFβ antagonist, fetuin, on PM-induced (EHC) procollagen gene expression in rat tracheal explants. Fetuin completely abolishes the increases caused by PM, indicating that TGFβ plays a role in fibrogenesis in this model. Albumin, a protein of comparable size, is included as a negative control. (From Dai, J., Xie, C., Vincent, R., and Churg, A., *Am. J. Respir. Cell Mol. Biol.*, 29, 352–358, 2003. With permission.)

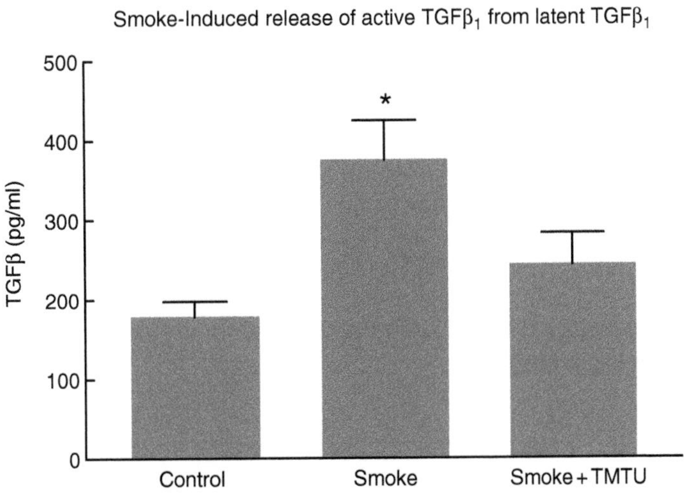

FIGURE 4.7 Effects of cigarette smoke-conditioned medium on the release of active $TGF\beta_1$ from recombinant latent $TGF\beta_1$ in a cell-free system. Exposure of latent $TGF\beta_1$ to the smoke-conditioned medium causes a release of the active TGFβ molecule, and this process is prevented by the AOS scavenger tetramethylthiourea (TMTU), indicating that it is oxidant driven. A similar process may occur with PM. (From Wang, R. D., Wright, J. L., and Churg, A., *Am. J. Respir. Cell Mol. Biol.*, 33, 387–393, 2005. With Permission.)

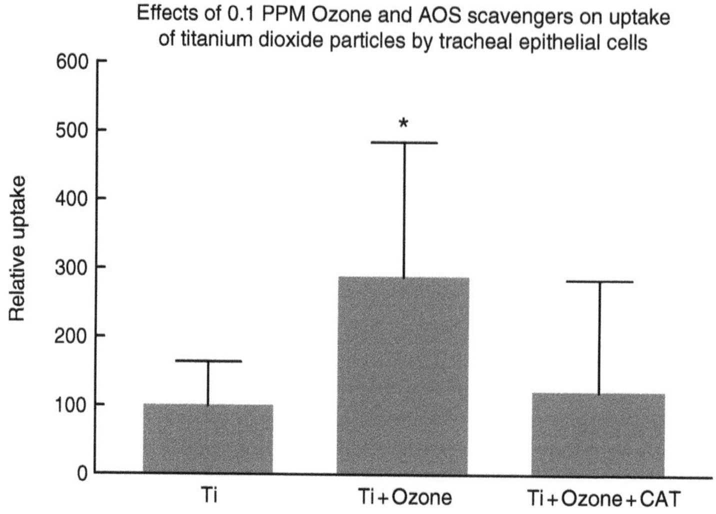

FIGURE 4.8 Effects of brief pre-exposure to ozone on subsequent uptake of fine TiO_2 particles by the epithelial cells of rat tracheal explants. Ozone increases particle uptake, and this effect is abolished by catalase, a scavenger of hydrogen peroxide. (From Churg, A., Brauer, M., and Keeling, B., *Am. J. Respir. Crit. Care Med.*, 153, 1230–1233, 1996. With permission.)

contaminants. There is evidence, albeit somewhat scanty, that PM and ozone may interact in producing damage to the lungs. Thus, Adamson, Vincent, and Bjarnason (1999) found that, in rats, exposure to PM (specifically EHC93) plus ozone resulted in additive or subadditive increases in tissue infiltration by neutrophils and macrophages. Madden et al. (2000) observed that treatment of diesel particles with 0.1 ppm ozone increased lavage total protein, LDH, and neutrophils. Jakab and Hemenway (1994) found that co-exposure to carbon black, used here as a surrogate for PM, and ozone increased ozone-induced inflammatory infiltrates, whereas carbon black alone did not cause inflammation.

In the studies of Vincent et al. (1997) and Adamson, Vincent, and Bjarnason (1999), exposure to UAP alone did not produce an increase in cell proliferation in the lung, whereas ozone alone produced a small increase, but the combination of agents induced marked synergistic increases. This process was seen in most cell types and was more marked in the bronchioles compared to the parenchyma. We found that a very brief (10 min) pre-exposure of tracheal explants to low levels of ozone increased the uptake of TiO_2 particles by the tracheal epithelial cells over the subsequent 7 days (Figure 4.8; Churg, Brauer, and Keeling 1996), an event that probably is consequential, since, as noted above, particle uptake appears to be important in the fibrogenic effects of PM in this model system.

Overall, these data suggest that PM and ozone may interact to potentiate the pathologic changes that lead to small airway remodeling.

4.6 FUNCTIONAL CONSEQUENCES AND PERSPECTIVE

The functional result of small airway remodeling is lumenal narrowing and lumenal tortuosity, leading to marked increases in airway resistance and clinical COPD. This process has been documented in detail in cigarette smokers, where correlations are found between the level of pathologic changes in the small airways and abnormalities of flow (Hogg et al. 2004; Wright 2005). There do not appear to be any similar correlative studies between pathologic abnormalities and pulmonary function in individuals exposed to high levels of PM, but the presence of airway changes similar to those seen in cigarette smokers (Souza et al. 1998; Pinkerton et al. 2000; Churg et al. 2003), and the presence of airflow obstruction in individuals who cook with biomass fuels (Dennis et al. 1996; Perez-Padilla et al. 1996) suggests that exactly the same processes are occurring.

In large populations in developed countries, cigarette smoking is clearly the dominant cause of COPD, but, as noted at the beginning of this chapter, chronic exposure to high levels of PM probably increases the overall incidence/severity of COPD. Given that cigarette smoke and PM both operate through oxidative processes, it would not be surprising if combined exposures enhanced the airway changes that lead to airflow obstruction. As described above, combined exposures to ozone and PM increase pathologic abnormalities in experimental models, but their effects *in vivo* in humans are unclear. Given the common occurrence of high exposure to both agents, further investigation of this problem is warranted.

ACKNOWLEDGMENTS

Supported by Grants 42539 and 81409 from the Canadian Institutes of Health Research.

REFERENCES

Abbey, D. E., Mills, P. K., Petersen, F. F., and Beeson, W. L., Long-term ambient concentrations of total suspended particulates and oxidants as related to incidence of chronic disease in California Seventh-Day Adventists, *Environ. Health Perspect.*, 94, 43–50, 1991.

Abbey, D. E., Ostro, B. E., Petersen, F., and Burchette, R. J., Chronic respiratory symptoms associated with estimated long-term ambient concentrations of fine particulates less than 2.5 microns in aerodynamic diameter ($PM_{2.5}$) and other air pollutants, *J. Exp. Anal. Environ. Epidemiol.*, 5, 137–159, 1995.

Abbey, D. E., Burchette, R. J., Knutsen, S. F., McDonnell, W. F., Lebowitz, M. D., and Enright, P. L., Long-term particulate and other air pollutants and lung function in nonsmokers, *Am. J. Respir. Crit. Care Med.*, 158, 289–298, 1998.

Abbey, D. E., Nishino, N., McDonnell, W. F., Burchette, R. J., Knutsen, S. F., Beeson, W., and Yang, J. X., Long-term inhalable particles and other air pollutants related to mortality in nonsmokers, *Am. J. Respir. Crit. Care Med.*, 159, 373–382, 1999.

Adamson, I. Y. R., Vincent, R., and Bjarnason, S. G., Cell injury and interstitial inflammation in rat lung after inhalation of ozone and urban particulates, *Am. J. Respir. Cell Mol. Biol.*, 20, 1067–1072, 1999.

Beeson, W. L., Abbey, D. E., and Knutsen, S. F., Long-term concentrations of ambient air pollutants and incident lung cancer in California adults: results from the AHSMOG study. Adventist health study on Smog, *Environ. Health Perspect.*, 106, 813–823, 1998.

Brauer, M., Bartlett, K., Regalado-Pineda, J., and Perez-Padilla, R., Assessment of particulate concentrations from domestic biomass combustion in rural Mexico, *Environ. Sci. Technol.*, 30, 104–109, 1996.

Brook, J. R. and Dann, T. F., The relationship among TSP, PM10, PM2.5, and inorganic constituents of atmospheric particulate matter at multiple Canadian locations, *J. Air Waste Manag. Assoc.*, 47, 2–19, 1997.

Churg, A. and Green, F. H. Y., Occupational lung disorders, In *Thurlbeck's Pathology of the Lung*, Churg, A., Myers, J., Tazelaar, H., and Wright, J. L., eds. 3rd ed., Thieme Medical Publishers, New York, 769–862, 2005.

Churg, A. and Wright, J. L., Bronchiolitis caused by occupational and ambient atmospheric particles, *Semin. Respir. Crit. Care Med.*, 24, 577–584, 2003.

Churg, A., Brauer, M., and Keeling, B., Ozone enhances the uptake of mineral particles by tracheobronchial epithelial cells in organ culture, *Am. J. Respir. Crit. Care Med.*, 153, 1230–1233, 1996.

Churg, A., Brauer, M., Carmen Avila-Casado, M., Fortoul, T. I., and Wright, J., Chronic exposure to high levels of particulate air pollution and small airways remodeling, *Environ. Health Perspect.*, 111, 714–718, 2003.

Costa, D. L. and Dreher, K. L., Bioavailable transition metals in particulate matter mediate cardiopulmonary injury in healthy and compromised animal models, *Environ. Health Perspect.*, 105 (suppl. 5), 1053–1060, 1997.

Dai, J. and Churg, A., Relationship of fiber surface iron and active oxygen species to expression of procollagen, PDGF-A, and TGFβ_1 in tracheal explants exposed to amosite asbestos, *Am. J. Respir. Cell Mol. Biol.*, 24, 427–435, 2001.

Dai, J., Xie, C., and Churg, A., Iron loading makes a non-fibrogenic model air pollutant particle fibrogenic in rat tracheal explants, *Am. J. Respir. Cell Mol. Biol.*, 26, 685–693, 2002.

Dai, J., Xie, C., Vincent, R., and Churg, A., Air pollution particles produce airway wall remodeling in rat tracheal explants, *Am. J. Respir. Cell Mol. Biol.*, 29, 352–358, 2003.

Dennis, R., Maldonado, D., Norman, S., and Baena, E., Woodsmoke exposure and risk for obstructive airways disease among women, *Chest*, 109, 115–119, 1996.

Detels, R., Tashkin, D. P., Sayre, J. W., Rokaw, S. N., Massey, F. J. Jr., Coulson, A. H., and Wegman, D. H., The UCLA population studies of CORD: X. A cohort study of changes in respiratory function associated with chronic exposure to SO_x, NO_x, and hydrocarbons, *Am. J. Public Health*, 81, 350–359, 1991.

Donaldson, K., Brown, D. M., Mitchell, C., Dineva, M., Beswick, P. H., Gilmour, P., and MacNee, W., Free radical activity of PM10: iron-mediated generation of hydroxyl radicals, *Environ. Health Perspect.*, 105, 1285–1289, 1997.

Donaldson, K., Stone, V., Borm, P. J., Jimenez, L. A., Gilmour, P. S., Schins, R. P., Knaapen, A. M. et al., Oxidative stress and calcium signaling in the adverse effects of environmental particles (PM_{10}), *Free Radic. Biol. Med.*, 34, 1369–1382, 2003.

Frampton, M. W., Ghio, A. J., Samet, J. M., Carson, J. L., Carter, J. D., and Devlin, R. B., Effects of aqueous extracts of PM_{10} filters from the Utah Valley on human airway epithelial cells, *Am. J. Physiol.*, 277, L967–L970, 1999.

Fujii, T., Hayashi, S., Hogg, J. C., Vincent, R., and Van Eeden, S. F., Particulate matter induces cytokine expression in human bronchial epithelial cells, *Am. J. Respir. Cell Mol. Biol.*, 25, 265–271, 2001.

Ghio, A. J., Steonheurner, J., Dailey, L. A., and Carter, J. D., Metals associated with both the water soluble and insoluble fractions of ambient air pollution particles catalyze an oxidative stress, *Inhal. Toxicol.*, 11, 37–49, 1999.

Hogg, J. C., Hogg, J. C., Chu, F., Utokaparch, S., Woods, R., Elliott, W. M., Buzatu, L. et al., The nature of small-airway obstruction in chronic obstructive pulmonary disease, *NEJM*, 350, 2645–2653, 2004.

Jakab, G. J. and Hemenway, D. R., Concomitant exposure to carbon black particulates enhances ozone-induced lung inflammation and suppression of alveolar macrophage phagocytosis, *J. Toxicol. Environ. Health*, 41, 221–231, 1994.

Jeffery, P. K., Remodeling in asthma and chronic obstructive lung disease, *Am. J. Respir. Crit. Care Med.*, 164, S28–S38, 2001.

Jimenez, L. A., Thompson, J., Brown, D. A., Rahman, I., Antonicelli, F., Duffin, R., Drost, E. M. et al., Activation of NF-κB by PM_{10} occurs via an iron-mediated mechanism in the absence of IκB degradation, *Toxicol. Appl. Pharmacol.*, 166, 101–110, 2000.

Madden, M. C., Richard, J. H., Dailey, L. A., Hatch, G. E., and Ghio, A. J., Effect of ozone on diesel exhaust particle toxicity in rat lung, *Toxicol. Appl. Pharmacol.*, 168, 140–148, 2000.

Molinelli, A. R., Madden, M. C., McGee, J. K., Stonehuerner, J. G., and Ghio, A. J., Effect of metal removal on the toxicity of airborne particulate matter from the Utah Valley, *Inhal. Toxicol.*, 14, 1069–1086, 2002.

Pare, P. D., Roberts, C. R., Bai, T. R., and Wiggs, B. J., The functional consequences of airway remodeling in asthma, *Monaldi Arch. Chest Dis.*, 52, 589–596, 1997.

Perez-Padilla, R., Regalado, J., Vedal, S., Pare, P., Chapela, R., Sansores, R., and Selman, M., Exposure to biomass smoke and chronic airway disease in Mexican women, *Am. J. Respir. Crit. Care Med.*, 154, 701–706, 1996.

Pinkerton, K. E., Green, F., Saki, C., Vallyathan, V., Plopper, C. G., Gopal, V., Hung, D. et al., Distribution of particulate matter and tissue remodeling in the human lung, *Environ. Health Perspect.*, 108, 1063–1069, 2000.

Pociask, D. A., Sime, P. J., and Brody, A. R., Asbestos-derived reactive oxygen species activate TGFβ_1, *Lab. Investig.*, 84, 1013–1023, 2004.

Quay, J. I., Reed, W., Samet, J., and Devlin, R. B., Air pollution particles induce IL-6 gene expression in human airway epithelial cells via NF-κB activation, *Am. J. Respir. Cell Mol. Biol.*, 19, 98–106, 1998.

Roberts, E. S., Richards, J. H., Jaskot, R., and Dreher, K. L., Oxidative stress mediates air pollution particle-induced acute lung injury and molecular pathology, *Inhal. Toxicol.*, 15, 1327–1346, 2003.

Saetta, M., Turato, G., and Zuin, R., Structural basis for airflow limitation in chronic obstructive pulmonary disease, *Sarcoidosis Vasc. Diffuse Lung Dis.*, 17, 239–245, 2000.

Smith, K., Fuel combustion, air pollution exposure and health: The situation in developing countries, *Annu. Rev. Energy Environ.*, 18, 529–566, 1993.

Souza, M. B., Saldiva, P. H. N., Pope, C. A., and Capelozzi, V. L., Respiratory changes due to long-term exposure to urban levels of air pollution, *Chest*, 113, 1312–1318, 1998.

Stone, V., Tuinman, M., Vamvakopoulos, J. E., Shaw, J., Brown, D., Petterson, S., Faux, S. P. et al., Increased calcium influx in a monocytic cell line on exposure to ultrafine carbon black, *Eur. Respir. J.*, 15, 297–303, 2000.

Sunyer, J., Urban air pollution and chronic obstructive pulmonary disease, *Eur. Respir. J.*, 17, 1024–1033, 2001.

Tao, F., Gonzalez-Flecha, B., and Kobzik, L., Reactive oxygen species in pulmonary inflammation by ambient particulates, *Free Radic. Biol. Med.*, 35, 327–340, 2003.

Tashkin, D. P., Detels, R., Simmons, M., Liu, H., Coulson, A. H., Sayre, J., and Rokaw, S., The UCLA population studies of chronic obstructive respiratory disease: XI impact of air pollution and smoking on annual change in forced expiratory volume in one second, *Am. J. Respir. Crit. Care Med.*, 149, 1209–1217, 1994.

Timblin, C., BéruBé, K., Churg, A., Driscoll, K., Gordon, T., Hemenway, D., Walsh, H. et al., Ambient particulate matter causes activation of the c-jun kinase/stress-activated protein kinase cascade and DNA synthesis in lung epithelial cells, *Cancer Res.*, 58, 4543–4547, 1998.

Vincent, R., Bjarnason, S. G., Adamson, I. Y. R., Hedgecock, C., Kumarathasan, P., Guenette, J., Potvin, M. et al., Acute pulmonary toxicity of urban particulate matter and ozone, *Am. J. Pathol.*, 151, 1563–1570, 1997.

Wang, R. D., Wright, J. L., and Churg, A., TGFβ_1 drives airway remodeling in cigarette smoke-exposed tracheal explants, *Am. J. Respir. Cell Mol. Biol.*, 33, 387–393, 2005.

Wiggs, B. R., Bosken, C., Pare, P. D., James, A., and Hogg, J. C., A model of airway narrowing in asthma and in chronic obstructive pulmonary disease, *Am. Rev. Respir. Dis.*, 145, 1251–1258, 1992.

Wright, J. L., Chronic airflow obstruction, In *Thurlbeck's Pathology of the Lung*, Churg, A., Myers, J., Tazelaar, H., and Wright, J. L., eds., 3rd ed., Thieme Medical Publishers, New York, 675–739, 2005.

5 Particle-Mediated Extracellular Oxidative Stress in the Lung

Frank J. Kelly and Ian S. Mudway
Pharmaceutical Science Research Division, King's College

CONTENTS

5.1 HEALTH EFFECTS OF PARTICULATE MATTER AND THE OXIDATIVE STRESS HYPOTHESIS

During the last decade, concerns have increased within both political and scientific communities over the impact of airborne particulate matter (PM) on public health. This concern has arisen primarily based on the findings of American epidemiological studies demonstrating an association between the mass concentration of PM (particularly particles with an aerodynamic diameter of less than 10 μm) in the air we breathe, and rates of respiratory and cardiac mortality and morbidity (Dockery et al. 1993; Pope et al. 1995; Samet et al. 2000). These associations have subsequently proven to be robust in numerous epidemiological studies worldwide (Brunekreef and Holgate 2002), with stronger associations usually associated with PM with an aerodynamic diameter of <2.5 μm. Moreover, the estimated decrease in life span associated with long-term residence in areas with high ambient PM is estimated to be between 1 and 2 years, which is large compared with other environmental risk factors. Whilst the data demonstrating PM-induced health effects is irrefutable, major questions remain concerning the mechanisms by which these compositionally heterogeneous species elicit their toxic actions. It has been argued that as exposure to a broad spectrum of particle types (vehicle emissions, cigarette and wood smoke PM, mineral dusts etc) elicits similar acute responses in humans, namely neutrophilic inflammation (Burns 1991; Salvi et al. 1999; Ghio 2000), reduced inspiratory capacity (Salvi et al. 1999; Stenfors et al. 2004) and heightened bronchial reactivity (Sherman et al. 1989; Menon et al. 1992; Nordenhall et al. 2000), they may act through a common mechanism. Currently numerous groups are investigating the hypothesis that these common mechanisms may relate to the capacity of these particles to cause damaging oxidation reactions in the lung, as well as systemically. In this review we will consider this hypothesis in detail, with specific emphasis on initial interactions between inhaled PM and components of the thin liquid layer that overlies the respiratory epithelium, the respiratory tract lining fluid (RTLF). This compartment represents the first physical interface with which PM interacts, upon deposition in the airways. A clear understanding of the initial reactions between PM constituents and components of the RTLF is thus critical in understanding any observed toxicity.

Oxidative stress is a relatively new term in biology that was first introduced by Sies (1991), and defined as "a disturbance in the pro-oxidant-antioxidant balance in favor of the former, leading to potential damage." Since then, many other definitions have been proposed, all of which attempt to explain a process which essentially involves the flow of electrons from one molecule to another within a biological setting. The importance of this process lies in the reactivity of the molecules involved. Under normal circumstances, electrons orbit around atoms in pairs, having opposite spins. When an atom has a single unpaired electron, its reactivity increases markedly, and it is referred to as a free radical. In a biological setting, free radicals are potentially very dangerous since they can react indiscriminately with neighboring molecules. This process of "electron stealing" leads to oxidation of, for example, lipids, proteins, and nucleic acids, and as a consequence, altered

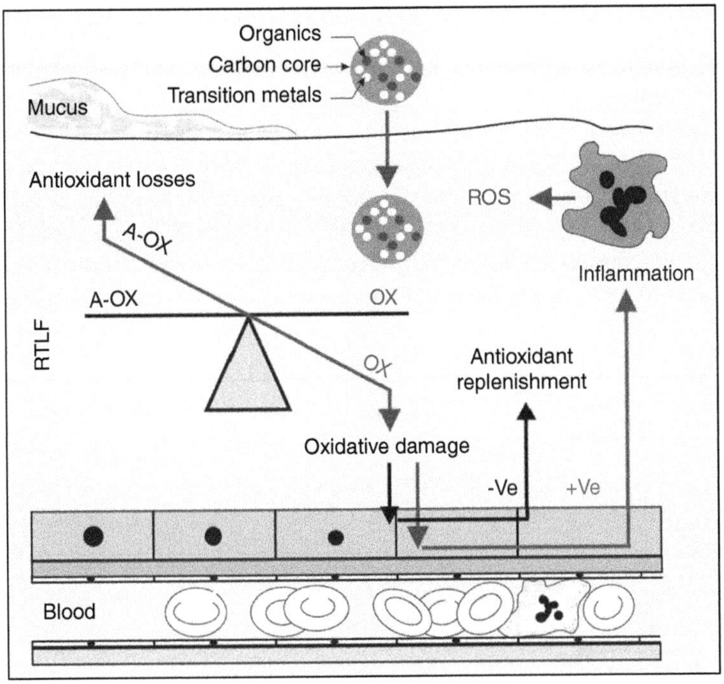

FIGURE 5.1 The imposition of oxidative stress at the air–gas interface. Under physiological conditions, the production of oxidant species (OX) in the aqueous respiratory tract lining fluid (RTLF) compartment is kept in check by the endogenous antioxidant defenses (A-OX). When the production of reactive oxygen species increases, either through the inhalation of pro-oxidants or the development of airway inflammation, the RTLF antioxidant network may become overwhelmed, permitting oxidative damage to occur to protein and lipid components of this compartment. The underlying epithelium can respond to this oxidative stress in a graded fashion, either up-regulating the expression of antioxidant defenses or mobilizing them to the RTLF, or by eliciting an inflammatory response resulting in a greater oxidative burden within the RTLF, which ultimately leads to cell death and tissue injury.

function or inactivation of these target molecules. If these reactions are numerous, they can cause extensive cellular damage, impaired cell functions, and in some cases, the influx of inflammatory cells to the sites of injury (Freeman and Crapo 1982; Halliwell and Gutteridge 1999; Droge 2002). The extent of damage is related to the availability of neutralizing antioxidant defenses, since these specialized molecules preferentially react with free radicals and give rise to products that often possess low toxicity. The damage arising from aberrant free radical activity is often loosely referred to as oxidative stress which, in its simplest form is a potentially harmful process occurring when there is an excess of free radicals, a decrease in antioxidant defenses, or a combination of these events (Figure 5.1).

There is now a strong body of evidence demonstrating that disturbances in normal cellular and extracellular redox status in the lung, in response to PM exposure, can trigger inflammation via the activation of redox-sensitive signaling pathways (Li et al. 1996; Bonvallot et al. 2001; Pourazar et al. 2005). The capacity of PM to elicit such a response has been explained by the delivery of surface adsorbed (redox active) metals (Aust et al. 2002) and organic contaminants (Squadrito et al. 2001) into the lung that drive oxidation reactions with the generation of cytotoxic reactive oxygen and nitrogen species (ROS and RNS) (Doelman and Bast 1990; Kelly 2003). The resultant airway inflammation itself then results in an increased production of oxidants by activated phagocytes, recruited to the airways, perpetuating the cycle of oxidative injury. In addition, ambient PM also contains appreciable concentrations of bacterial endotoxin that can also trigger inflammation.

5.2 ANTIOXIDANT DEFENSES AT THE AIR–LUNG INTERFACE

Owing to its function, large surface area and exposure to high partial pressures of oxygen, the lung is particularly susceptible to oxidative injury. It is therefore logical that a robust extracellular antioxidant defense system exits to protect against undue oxidation of its delicate pulmonary epithelial cells. On entering the lung, ambient pollutants do not come into direct contact with the respiratory epithelium, but rather they first interact with a liquid layer that covers the respiratory epithelium from the nasal mucosa to the alveoli, the RT4 (Figure 5.2). Human RTLF is a complex and regionally heterogeneous compartment, ranging in depth from between 1 and 10 μm in the proximal airways to 0.2–0.5 μm in the distal airways and alveoli (Cross et al. 1998). It consists of

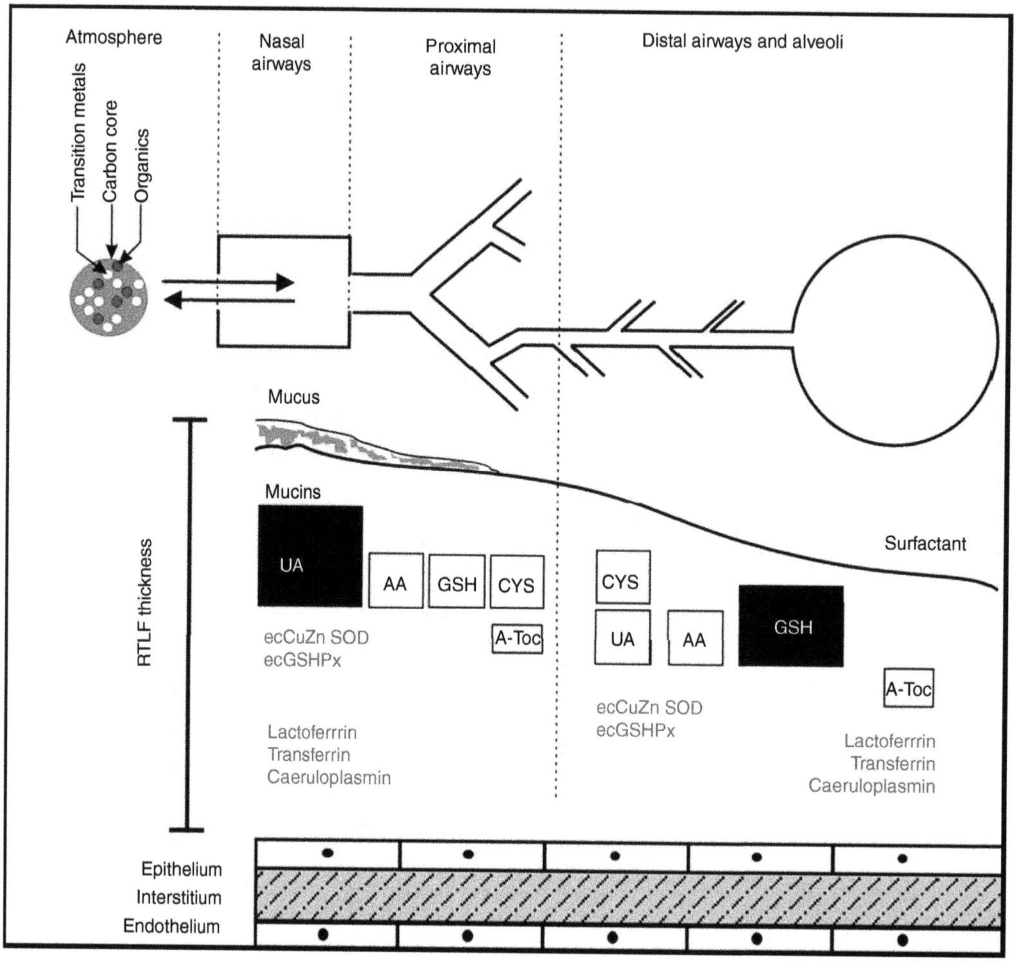

FIGURE 5.2 RTLF antioxidant network in the nasal/proximal airways and alveolar regions of the lung. Both the upper- and lower-airways RTLFs contain cysteine (CYS) and ascorbate (AA), but there are especially high concentrations of urate (UA) in the upper airways, especially the nasal mucosa. In contrast, glutathione (GSH) concentrations appear to be greatest in the terminal airways and alveoli. α-Tocopherol is present throughout the airway RTLFs, but at low concentrations. Extracellular superoxide dismutase (ecCuZnSOD) and glutathione peroxidase (ecGSH-Px) are also present in RTLFs. Catalase and glutathione reductase have also been reported in human RTLF, but it is not clear whether these are derived from cellular lysis and so are not included. Metal-ion chelator proteins within the RTLF are also illustrated, with lactoferrin emphasized in the upper airway where its concentration is highest.

secretions from underlying lung and resident immune cells, as well as plasma-derived exudates. In the nasal and proximal airways, it exists as a two-phase structure consisting of a gel and sol phase, the former consisting of thiol-rich mucopolypeptide glycoprotein (mucins). This contrasts with the distal airway and alveolar lining fluids, which are devoid of mucins, but contain surfactant lipids and proteins. Nasal, proximal, and distal lavage procedures have allowed a detailed examination of the RTLF in different lung compartments. In addition to the mucin and surfactant components, the RTLF also contains a broad spectrum of low molecular weight antioxidants, as well as small concentrations of antioxidant enzymes (Kelly et al. 1995; 1999) (Figure 5.2).

5.2.1 Small Molecular Weight Antioxidants

An array of small molecular weight antioxidants has been measured in human RTLF, including ascorbate (vitamin C) (Willis and Kratzing 1974; Skoza, Snyder, and Kikkawa 1983; van der Vliet et al. 1999), urate (Peden et al. 1990), reduced glutathione (GSH) (Cantin et al. 1987; Jenkinson, Black, and Lawrence 1988), and α-tocopherol (vitamin E) (Mudway et al. 2001). As the lavage procedure used to sample the airway introduces a large and variable dilution, the concentration of these antioxidants quoted in the literature are often difficult to interpret. Whilst many groups simply quote concentrations in the recovered lavage, others have made attempts to correct for the dilution, using a range of methods based on instilled dyes or endogenous dilution markers (Haslam and Baughma 1999). Of these, the most commonly adopted is the urea correction method that uses the ratio of urea concentrations between the lavage sample and plasma as the basis for estimating the dilution. This technique is often criticized because the instillation of a large volume of saline creates a massive concentration gradient along which urea can move from cellular and interstitial sources. One way of limiting this potential problem is to reduce the lavage dwell time and not to perform repeat sampling of the same airway segment, as is often performed in lavage procedures. The RTLF antioxidant concentration ranges quoted in this review will be based on values obtained using these methods, unless otherwise stated.

5.2.1.1 Ascorbate

As a consequence of its high solubility, ascorbate is thought to diffuse freely between aqueous extra and intracellular compartments of the body including the RTLF, where concentrations (50–150 μM) comparable to those in plasma have been measured (Kelly, Buhl, and Sandström 1999; van der Vliet 1999). In humans, this antioxidant vitamin originates solely from dietary sources in humans owing to the lack of the enzyme, gluconlactone oxidase, necessary for its biosynthesis (Halliwell and Gutteridge 1999). Differences in dietary intake may therefore explain the variability in ascorbate concentrations that have been measured in bronchoalveolar lavage (BAL) fluid from different individuals (Kelly, Buhl, and Sandström 1999). Ascorbate is an excellent reducing agent and scavenges a variety of free radicals and oxidants *in vitro*, including superoxide and peroxyl radicals, hydrogen peroxide, hypochlorous acid, and singlet oxygen. During its antioxidant activity, ascorbate readily undergoes two consecutive, but reversible, one-electron reductions. The first one-electron reduction produces the semi-dehydroascorbate radical (SDA) also known as the ascorbyl radical ($Asc^{\bullet}$) (Buettner and Jurkiewicz 1996). $Asc^{\bullet}$ can also undergo a further one-electron reduction to dehydroascorbate (DHA). The ascorbyl radical is relatively unreactive, owing to its unpaired electron being in the delocalized π-system, and in the absence of a further oxidation, two ascorbyl radicals will undergo a disproportionate reaction to regenerate one molecule of ascorbate and one molecule of DHA. The DHA, once formed, rapidly undergoes hydrolysis to 2,3-diketo-L-gulonic acid, ultimately degrading to a diverse range of products, including oxalic and L-threonic acids. Dehydroascorbate and its metabolites are potentially cytotoxic in significant concentrations, and therefore cells possess enzymes that convert either SDA or DHA back to ascorbate at the expense of GSH or NADPH. (Diliberto et al. 1982; Park and

Levine 1996). In addition, glutathione has been shown to recycle DHA nonenyzmatically (Winkler, Orselli, and Rex 1994) *in vitro*, although at a rate insufficient to prevent substantial losses of DHA. Little is currently understood about the turnover of ascorbate within the RTLF as to date, only ascorbate and dehydroascorbate have been routinely measured. At present, it is not clear whether such regenerative mechanisms exist in the RTLF, but certainly the high concentrations of glutathione in this compartment may act to limit the oxidative losses of ascorbate. Clearly in the absence of some recycling mechanism, the oxidative losses of ascorbate within the RTLF would constitute a significant drain on bodily stores of this antioxidant. Many cell types, including neutrophils, are able to take up DHA rapidly via the facilitative glucose transporters GLUT1 and GLUT3 (Rumsey et al. 1997). Once internalized, the DHA is rapidly reduced back to ascorbic acid by glutaredoxin at the expense of glutathione (Park and Levine 1996). Thus, it may be possible that DHA formed in the RTLF under normal or pathological conditions may be reduced back to ascorbate by cellular uptake by the epithelium (Nualart et al. 2003).

In addition to having a direct scavenging action, ascorbate also acts indirectly to prevent lipid peroxidation through its reaction with the membrane bound α-tocopherol radical. *In vitro* studies have demonstrated that in this way, ascorbate is able to reduce the α-tocopherol radical, thereby regenerating the vitamin E molecule (Packer, Slater, and Wilson 1979; Niki, Yamamoto, and Kamiya 1985; Doba, Burton, and Ingold 1985). This synergistic action has not yet however been reported *in vivo*. This potential interaction, together with the recycling of DHA by glutathione, illustrates that the antioxidant network within the RTLF is highly synergistic (Figure 5.3).

Whilst ascorbate represents a critical protective component of the RTLF, it also has the potential to act as a pro-oxidant. In the presence of ferric iron (Fe^{3+}) or cupric copper (Cu^{2+}) ions, ascorbate can promote reduction to ferrous (Fe^{2+}) and cuprous forms (Cu^{2+}), permitting the formation of the damaging hydroxyl radical in the presence of hydrogen peroxide.

$$Fe^{3+} + \text{ascorbate(reduced)} \rightarrow Fe^{2+} + \text{ascorbate(oxidized)}$$

$$Fe^{2+} + H_2O_2 \rightarrow Fe^{3+} + OH^{\cdot} + OH^{-}$$

This situation occurs rarely under normal physiological conditions, as both Fe and Cu are sequestered into transport and storage proteins to limit these potentially damaging reactions.

5.2.1.2 Urate

Urate is an oxidized purine base present in all RTLF compartments of the lung, but at particularly high concentrations in the nasal and proximal airways (200–500 μM) where it appears to be the predominant antioxidant defense (Cross et al. 1994; Blomburg et al. 1998). In the nasal airways it appears to be secreted from gland cells along with lactoferrin (Peden et al. 1990, 1993), but little is known about its transport into the lower airways. This antioxidant can directly scavenge hydroxyl radicals, oxyhaem oxidants formed between the reactions of hemoglobin and peroxy radicals, peroxyl radicals themselves, and singlet oxygen (Becker 1993). In these reactions it acts in a sacrificial mode, in that it is irreversibly damaged to produce a range of oxidation products, including allantoin (Ames et al. 1981). As with ascorbate, relatively little is known about the turnover of urate in the RTLF. In addition to its role as a free radical scavenger, urate has also been reported to chelate iron and protect against iron-mediated free radical damage of ascorbate and lipids (Davies et al. 1986; Ghio et al. 1994). Increased urate concentrations have been reported in rats challenged with iron salts and silica-iron in concert with increased lung xanthine oxidoreductase activities. This would imply that elevated RTLF urate might reflect a regulated adaptive response to limit Fe-induced oxidative injury (Ghio et al. 2002).

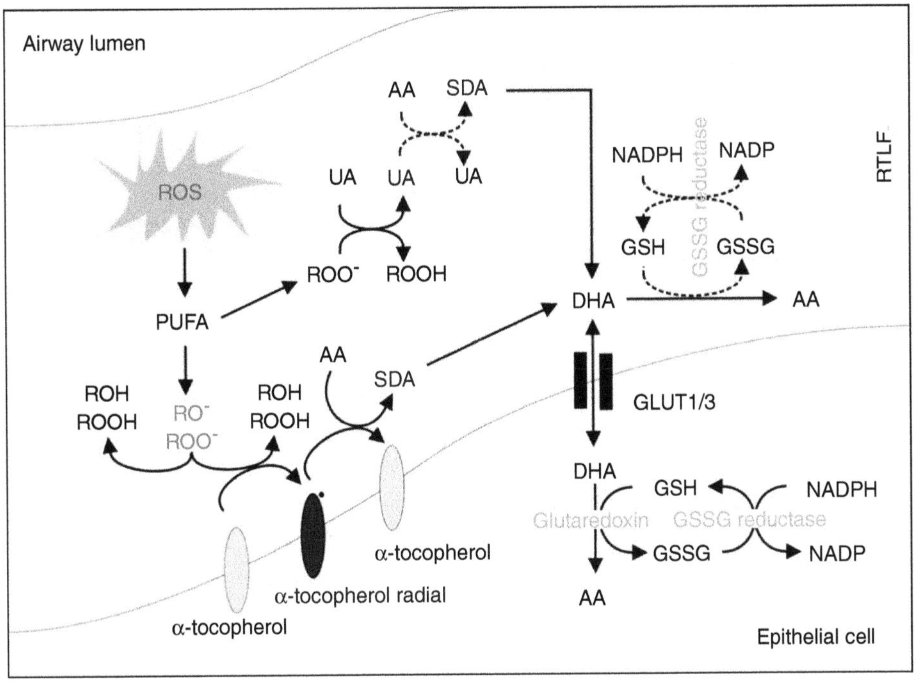

FIGURE 5.3 Cooperativity of RTLF antioxidants. Peroxyl (ROO˙) and alkoxyl (RO˙) lipid radicals formed though the oxidation of polyunsaturated fatty acids (PUFA) can be reduced to harmless alcohols (ROH, ROOH) through the scavenging action of membrane-bound α-tocopherol. The tocopherol radical can be re-reduced at the expense of ascorbate (AA), which is converted to the ascorbyl radical (SDA) and hence to dehydroascorbic acid (DHA). As DHA rapidly, it is normally rapidly reduced back to ascorbate intracellularly by glutaredoxin at the expense of the oxidation of glutathione (GSH) to glutathione disulphide. It is not known whether a similar DHA reductase activity exists within the RTLF, but GSH can recycle DHA to a limited extent in an non-enzyme catalyzed mechanism. Urate (UA) can also scavenge ROO˙ and RO˙ Radicals, and the redox potential of the urate radical suggests that UA may also be recycled by ascorbate.

5.2.1.3 Reduced Glutathione

High levels of glutathione (100–200 μM) have been reported in proximal airway and alveolar RTLFs from healthy individuals, with some reports suggesting concentrations as high as 400 μM (Cantin et al. 1987). This is around 100 times higher than normal plasma concentrations (Kelly and Richards 1999) with over 90% present in its reduced form (GSH; L-glutamyl-Lcysteinylglycine). The source of GSH has not been established, although in the light of low concentrations in plasma and poor re-adsorption from the respiratory tract, it is likely that this antioxidant is produced via a secretary mechanism from the cells in the lung (Smith, Anderson, and Shamsuddin 1992). As well as acting as a substrate for the glutathione redox cycle, glutathione is known to react with a wide range of compounds *in vitro* including hydroxyl radicals, hypochlorous acid, peroxynitrite, peroxyl radicals, carbon centred radicals, and singlet oxygen. In its reactions with free radicals, thiyl radicals are produced, which can subsequently be converted to oxidized glutathione through a radical transfer process (Halliwell and Gutteridge 1999). Glutathione is particularly good at defending against oxidants such as hypochlorous acid and hypobromous acid (Winterbourn 1985), which are released from neutrophils and eosinophils, respectively. The presence of this antioxidant in lung lining fluid may therefore be particularly important in defending the extra-cellular surface of the lung against activated inflammatory cells.

5.2.1.4 α-Tocopherol

RTLF α-Tocopherol is believed to be derived from type II cells (Rustow et al. 1993) and is present at relatively low concentrations within lung lining fluid. It is, however, a powerful antioxidant, both in terms of scavenging free radicals such as singlet oxygen, alkoxy radicals, peroxynitrite, nitrogen dioxide, ozone, and superoxide (Wang and Quinn 1999) and its ability to terminate lipid peroxidation (Witting 1980; Burton and Ingold 1981; Niki, Takahashi, and Komuro 1986; Nakamura et al. 1987). It is thought that the reactivity of α-tocopherol with organic peroxyl radicals accounts for the majority of its biological activity (Witting 1980; Burton and Ingold 1981), a reaction that yields a relatively stable lipid hydroperoxide and a vitamin E radical, thereby effectively interrupting the lipid peroxidation chain reaction (McCay 1985).

5.2.2 Mucins

In addition to providing a physical barrier and clearance mechanism to remove inhaled toxins, mucin components of the gel phase of the upper airway RTLF also have significant antioxidant properties, by virtue of their cysteine-rich domains and carbohydrate moieties (Cross, Halliwell, and Allen 1984; Hiraishi et al. 1993). Mucins have been shown to have metal binding properties (Cooper, Creeth, and Donald 1985), as well as the capacity to scavenge hydroxyl radicals (Cross, Halliwell, and Allen 1984). In addition, the thiol moieties imply that mucins should also provide considerable protections against hypochlorous acid ($HOCl^-$).

5.2.3 Enzymatic Antioxidant Defenses

RTLF also contains antioxidant enzymes, whose role it is to scavenge oxidants or to repair damage caused by ROS. These enzymatic antioxidants are a complex set of proteins and include extracellular CuZn superoxide dismutase (EC-SOD), catalase, and glutathione peroxidase (GPx).

5.2.3.1 Glutathione Peroxidase

Glutathione peroxidase (GPx) is a selenoprotein that catalyzes the reduction of hydrogen peroxide and lipid peroxides, at the expense of glutathione (Brigelius-Flohé and Traber 1999). Of the four types of GPx, namely cellular or cytosolic or classical GPx, gastrointestinal GPx, extracellular or plasma GPx, and phospholipid hydroperoxide GPx (Takebe et al. 2002), Avissar et al. (1996) has shown that RTLFs contains selenium-dependent cellular GPx and extracellular GPX, each contributing approximately 50% to the total GPx activity.

5.2.3.2 EC-SOD

EC-SOD is expressed in especially high levels in mammalian lungs by the alveolar type II pneumocytes (Oury et al. 1996), and can be found in the RTLF (Mudway and Kelly 2000) and airway epithelial cell junctions. EC-SOD catalyzes the dismutation of the superoxide free radical to hydrogen peroxide thus in concert with glutathione peroxidase, protects lung interstitium against free radical generated during inflammation (Oury et al. 1996).

5.2.3.3 Catalase

Catalase is a heme-containing tetramerous protein, primarily present in peroxisomes, but also found cytoplasm, mitochondria, and BALF (Fridovich 1998; Halliwell and Gutteridge 1999; Vallyathan et al. 2000). Catalase is important in the dismutation of hydrogen peroxide to water and oxygen (Quinlan, Evans, and Gutteridge 2002). As with many of the enzymatic antioxidants reported in RTLF, there is some uncertainty regarding whether they are authentic extracellular proteins, or are derived from cellular disruption during the lavage procedure.

5.2.4 Metal Chelation Proteins

Metal-chelating proteins present in the RTLF also perform an important antioxidant function by binding free transition metals, thus preventing them from participating in potentially damaging redox reactions (Halliwell and Gutteridge 1999).

5.2.4.1 Transferrin

Transferrin controls the transport of iron in the body. It is the predominant metal-chelating protein in the RTLF and has been shown to be a potent inhibitor of lipid peroxidation *in vivo* (Pacht and Davis 1988). Of interest, the affinity of iron for transferrin is pH-dependent, such that in plasma (pH 7.4), binding is very strong, whereas virtually no binding occurs at $pH<4.5$. This is pertinent to particle-association metal chemistry in the RTLF, in that the pH of RTLF from healthy individuals has been reported to be between 7 and 7.5, but in certain disease states, such as asthma, the pH is approximately pH 5 (Hunt et al. 2000). One might speculate, therefore the chelation of iron by transferrin to be sub-optimal in asthmatics, this would increase the presence of unbound iron and may, in turn, contribute to the low levels of RTLF antioxidants seen in asthmatics (Kelly, Buhl, and Sandström 1999; Mudway and Kelly 2000).

5.2.4.2 Lacoferrin

Lacoferrin is able to bind and transport iron and release it again at specific receptor cells in the human intestine. High concentrations have been reported in the upper airways, whilst studies have shown that concentrations are increased in the lower respiratory tract of chronic bronchitics (Thompson et al. 1990). Like transferrin, lactoferrin is a potent inhibitor of lipid peroxidation *in vivo* (Pacht and Davis 1988). Lactoferrin can bind free iron with high affinity and thus function as a local antioxidant, protecting the immune cells against the free radicals produced by them (Britigan, Serody, and Cohen 1994). Furthermore, reports suggest that lactoferrin can bind free iron released from dying cells, thereby protecting the phagocytic cells, as well as adjacent tissues, from ROS produced from the Haber–Weiss reaction (Britigan, Serody, and Cohen 1994).

5.2.4.3 Ferritin

Ferritin controls the storage of iron within the body, and thus can also provide protection against iron-generated ROS. Concentrations of ferritin in alveolar cells and on the alveolar surface are increased in patients with a variety of respiratory disorders (Stites et al. 1999) and it is likely that the presence of ferritin in the RTLF reflects the release of cell contents from dying cells. A number of inflammatory mediators can also induce intracellular up-regulation, as well as the cleavage of iron from extracellular ferritin, thereby allowing the free iron to be redox active (Reif 1992). Indeed, Stites et al. (1999) reported that oxidant injury could be promoted in lungs of patients with cystic fibrosis as a consequence of mobilizing iron.

5.2.4.4 Caeruloplasmin

Caeruloplasmin is a glycoprotein involved in serum copper transport. It represents the main antioxidant defense against copper in human plasma, but also promotes the incorporation of iron into transferrin without the formation of toxic iron products (Koc et al. 2003). It is primarily synthesized in the liver and secreted to the blood; however, recent studies have identified the lung as another major site of synthesis. In particular, Yang et al. (1996) have suggested that the airway epithelial cells are the major source of ceruloplasmin identified in RTLF.

5.3 INDUCTION OF OXIDATIVE STRESS BY INHALED PARTICLES

Particulate matter is a complex mixture of chemical components in terms of their chemical composition, dependant on the emission source and in combustion scenarios, the type of fuel being burnt. For example, ultrafine particles from sources of combustion generally comprise a carbonaceous core with absorbed substances condensed onto the surface during combustion and atmospheric processes. These substances may include organic and elemental carbon; polycyclic aromatic hydrocarbons (PAHs); metals (both redox active and non-redox active); and biological compounds such as bacterial endotoxin, as well as sulphate, nitrate, chloride, and ammonium. The composition dictates the surface reactivity of the particles, an important factor in determining particle toxicity (Fubini 1997) and its adverse health effects. Whether these PM components ultimately result in substantial oxidative damage, inflammation, and injury ultimately depends on their initial interactions with the antioxidant defenses within the RTLF. These potential interactions are illustrated in Figure 5.4. It is useful to grade the response of the lung to particle-induced oxidative stress, with low-level oxidative stress resulting in an up-regulation of endogenous extra- and intracellular responses, prior to the induction of substantial toxicity (Li et al. 2002). The various pathways by which inhaled PM can elicit injury to the airway are illustrated in Figure 5.5.

5.3.1 THE ROLE OF REDOX ACTIVE METALS

Redox-active metals are defined as those containing unpaired electrons in their *d*-orbital, and are capable of generating free radical species via redox cycling mechanisms with biological reductants. Using this definition, the entire first row elements in the d-block of the periodic table, with the exception of zinc, qualify: Sc, Ti, V, Cr, Mn, Fe, Co, Ni, and Cu. Of these, the most common components of PM are listed below.

5.3.1.1 Iron (Fe)

As outlined earlier, under normal conditions iron is sequestered in the chemically less active Fe^{3+} form in the storage protein ferritin, or is associated with transport or receptor proteins, thereby facilitating the removal of reactive iron from areas including the respiratory tract. In the presence of excess iron (for example, when associated with PM), these proteins become overwhelmed, and since the body has no way of actively excreting iron, it is possible for a free pool to accumulate in the body. Such a pool is an excellent target for Fenton chemistry (Halliwell and Gutteridge 1999), and the subsequent production of superoxide, hydrogen peroxide, and hydroxyl radicals, all of which have been associated with oxidative stress in the human body (see below). The Fenton reaction, in particular the Haber–Weiss (superoxide-driven) form, has been implicated in the pulmonary toxicity of iron (Winterbourn 1995; Turi et al. 2003).

$$Fe^{2} + O_2 \rightarrow Fe^{3} + \mathbf{O}_{2-}^{\cdot}$$

$$2O_{2-}^{\cdot} + 2H \rightarrow \mathbf{H_2O_2} + O_2$$

$$H_2O_2 + Fe^{2+} \rightarrow Fe^{3} + \mathbf{HO}^{\cdot} + OH$$

In vivo, the rate of Fe-catalyzed radical production is critically dependent on the concentration of endogenous chealators, reducing agents, and O_2 concentration. Thus, in the lung, it is likely that some transitional metal-dependent radical production will occur prior to the sequestering of the exogenous Fe in transferrin and lactoferrin. It should be noted, however, that in the presence of biological reductants such as ascorbate, these reactions are cyclic with the ferric iron being reduced to its ferrous form, promoting further ROS production.

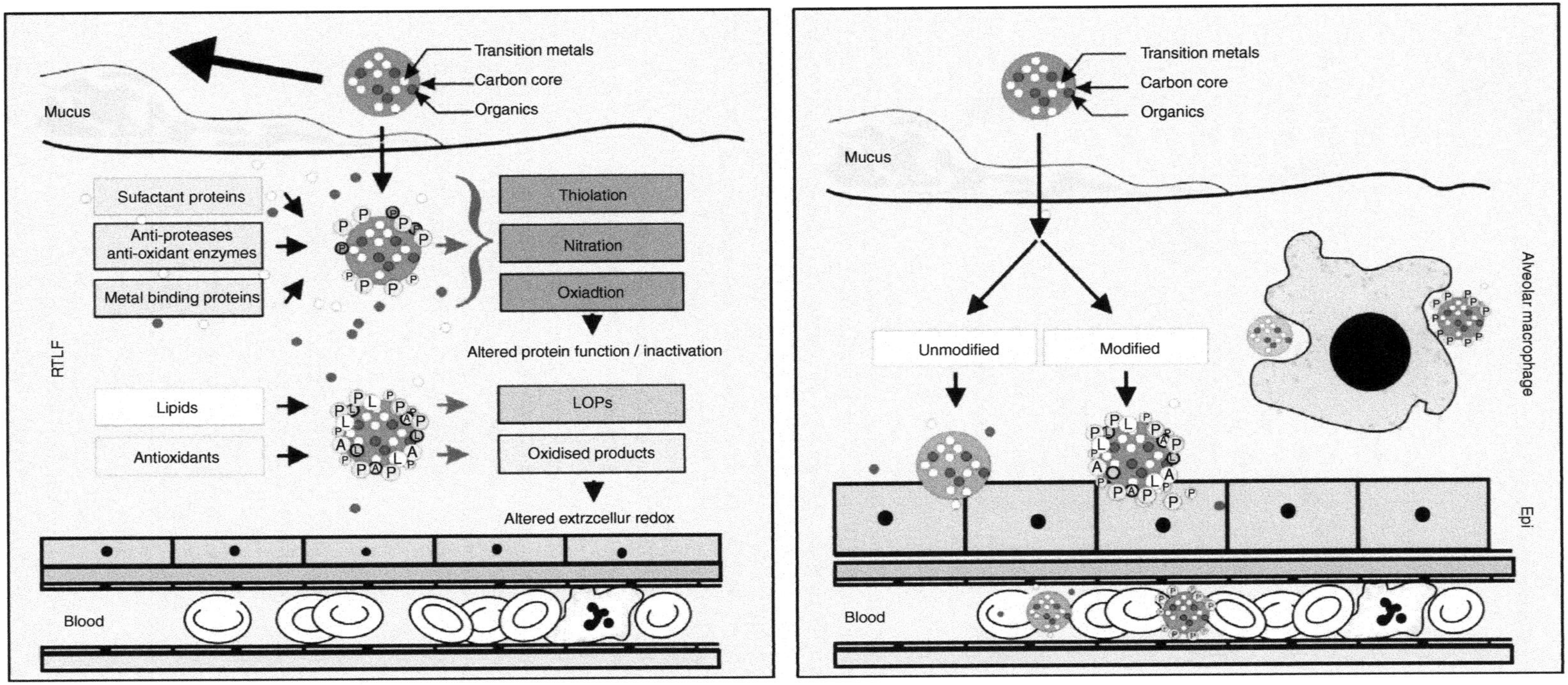

FIGURE 5.4 (**See color insert**) Particle-RTLF interactions. Inhaled particles depositing in the airways may become trapped by the layer of mucus and transported from the airways by the mucocillary elevator. Those particles that are retained in the airways can either leach soluble components into the RTLF that oxidize antioxidants, lipid, and protein components of the RTLF, or absorb the oxidized or native RTLF components onto their surface. These interactions therefore alter both the redox state of the RTLF, which will impact upon the underlying cells, as well as modifying the particle surface that ultimately reaches the underlying epithelium, or that is intercepted by alveolar macrophages within the extracellular compartment.

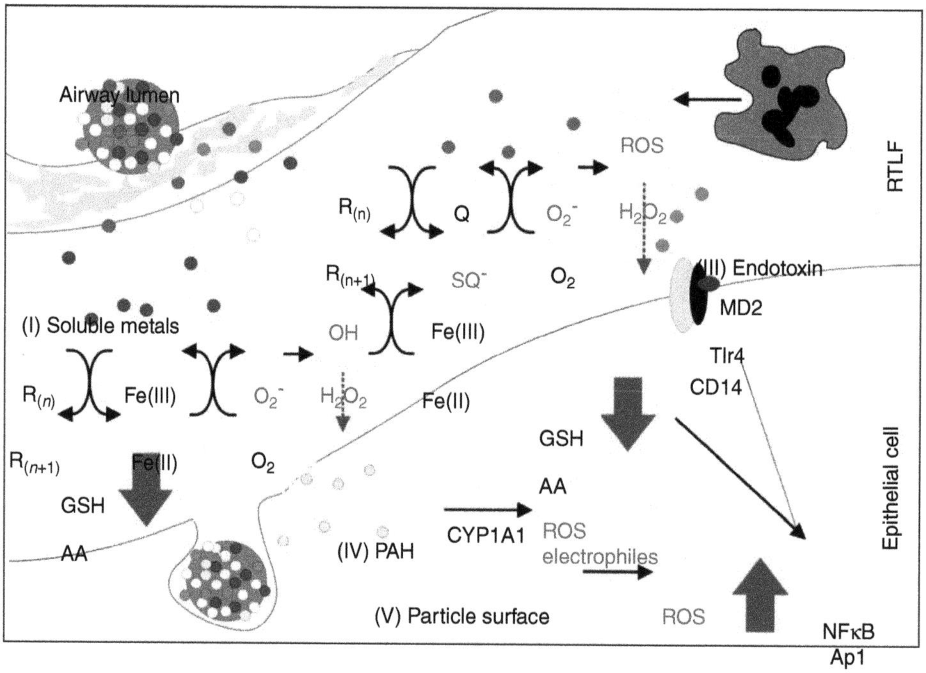

FIGURE 5.5 Pathways of particle induced toxicity at the air–lung interface. Particles elicit oxidative stress through five inter-related mechanisms: (I) Through the introduction of redox active metals such as iron, into the lung, which redox cycle in the presence of biological reductants ($R_{(n\,=\,\text{number of electrons})}$) and oxygen to yield the superoxide radical ($O_2^{\cdot-}$), hydrogen peroxide (H_2O_2), and finally the damaging hydroxyl radical ($^{\cdot}OH$). (II) Quinones on the particle surface can also redox cycle in the presence of biological reductants to form the semi-quinone radical ($SQ^{\cdot-}$) that will also yield superoxide and hydrogen peroxide. Both of these pathways will result in the loss of extracellular antioxidants and hence, an altered, at least transiently, extracellular redox. (III) Bacterial endotoxin associated with the particle surface has been shown to trigger inflammation through its interaction with the TLr4/CD14/MD2 receptor, which further adds to the oxidative burden in the airways. (IV) Polyaromatic hydrocarbons (PAHs) have no intrinsic oxidative activity, but they can undergo biotransformations intracellularly through the action of P450cyp1A1 to form reactive electrophiles and reactive oxygen species (ROS). (V) The particle surface itself has been shown to cause oxidative stress *in vivo*, though the mechanism by which this occurs is not well defined. Overall then exposure to inhaled particles results in losses of extra- and intra-cellular antioxidants, especially ascorbate and glutathione. This altered redox state results in the upregulation of redox sensitive signalling partway and transcription factors (NΦκB and AP-1) leading to the increased production of cytokines and the development of airway inflammation.

5.3.1.2 Copper (Cu)

Copper is another redox active metal commonly associated with ambient PM and, like iron, it can act as a catalyst in the formation of ROS and catalyze the peroxidation of lipids:

$$\mathrm{Cu(II)} + O_2^{\cdot-} \rightarrow \mathrm{Cu(I)} + O_2$$

$$\mathrm{Cu(I)} + H_2O_2 \rightarrow \mathrm{Cu(II)} + \cdot OH + H^-$$

$$2O_2^{\cdot-} + 2H^+ \rightarrow H_2O_2 + O_2$$

Whilst ceruloplasmin has been reported in the RTLF, its Cu binding sites are not available to bind exogenous Cu, thus it is likely that exogenous Cu will participate in free radical production *in vivo*. Although the Cu content of ambient PM is only a fraction of the Fe content, it is therefore much more likely to drive damaging oxidations reactions within the RTLF. Furthermore, Cu has been shown to be a much more effective catalyst of radical production than any of the other major PM-associated metals in reductive environment. It has been observed that Cu^{2+} and Cu^{+} can participate in a redox cycling reaction that produces hydrogen peroxide (and benzoquinone) via oxidation of hydroquinone. Reactive oxygen species produced in this manner have been linked to plasma DNA damage and cleavage (Li et al. 1995).

5.3.1.3 Manganese (Mn)

Ferroalloy production, iron and steel foundaries, and combustion from powerplants all make significant contributions to the ambient PM Mn concentrations. In addition to these sources, Mn is also present in gasoline-powered vehicle emission as a consequence of the use of methylcyclo-pentadienyl manganese tricarbonyl as an octane-increasing petrol addative. Mn has an valancy of Mn(II) in aqueous solution, althougth it can also exist in a range of other oxidation states, including Mn(III), Mn(IV), and Mn(VII). Whilst Mn has been shown to act as a catalyst for the dismutation of superoxide to hydrogen peroxide, unlike Fe and Cu, it does not promote the formation of hydroxyl radicals in the presence of H_2O_2 to any great extent (Halliwell and Gutteridge 1992).

5.3.1.4 Vanadium (V)

Vanadium represents one of the key metal components of PM, and is derived from the burning of fuel oil, along with Fe and Ni. PM pollution generated from oil powered power stations has been a major environmental issue in the United States, and as such, a considerable body of work has been performed addressing the toxicity of residual oil fly ash (ROFA) and its V component. V has several oxidation states, V(V), V(IV), V(III), and V(II), of which the 5+ form is the most common. The production of hydroxyl radicals by vanadate has been demonstrated by ESR during co-incubations with liver microsomes (Shi and Dalal 1992) which appear to occur through Fenton-like chemistry rather than the Haber–Weiss reaction (Shi and Dalal 1993).

$$\mathrm{V(V)} + \mathrm{O_2^{-}\cdot} \rightarrow \mathrm{V(IV)} + \mathrm{O_2}$$
$$\mathrm{V(IV)} + \mathrm{H_2O_2} \rightarrow \mathrm{V(V)} + \mathrm{OH^{-}} + \mathrm{OH}$$

V(IV) has also been shown to promote the decomposition of lipid peroxides. The production of ROS by V(V) can also be promoted by its reduction to V(IV) by cellular reductants, such as ascorbate and glutathione, thus raising the possibility that the reaction between PM V and anti-oxidants within the RTLF may actually sponsor, rather than inhibit, radical production in this compartment.

5.3.1.5 Nickel (Ni)

Excess nickel is a well-documented carcinogen in humans, especially when inhaled as insoluble particulates such as nickel subsulphide (Kasprzak 1991). Despite evidence that nickel compounds cause the oxidation of lipids (Athar, Hasan, and Srivastava 1987; Stinson et al. 1992), proteins (Zhuang, Huang, and Costa 1994) and nucleic acids (Stinson et al. 1992; Lynn et al. 1997), intracellular radical production (Huang, Klein, and Costa 1994), and intracellular glutathione depletion (Rodriguez et al. 1991; Herrero et al. 1993; Li, Zhao, and Chou 1993), there is little evidence that Ni undergoes redox cycling reactions. Hydrated Ni^{2+} ions do not react readily with H_2O_2 to form hydroxyl radicals and are not efficient catalysts of peroxide decomposition. However, chelation of Ni^{2+} to biological ligands alters the reduction potential, allowing it to be oxidized to Ni^{3+} by strong oxidants, such as hydrogen peroxide and organic hydroperoxides (Datta et al. 1992).

This ligand-dependent Ni^{2+}/Ni^{3+} redox cycling of nickel results in the formation of oxygen radicals via Fenton-type reactions (Klein, Frenkel, and Costa 1991; Lin, Zhuang, and Costa 1992).

$$\underset{\text{(ROOH)}}{Ni^{2+} - \text{chelate}} + H_2O_2 \rightarrow Ni(III)\text{chelate} + \underset{\text{(RO}^{\cdot}\text{)}}{OH^{\cdot}} + OH^-$$

Further evidence suggests that insoluble particulate nickel compounds, Ni_3S_2 and NiS, can also induce the release of reactive oxygen species from phagocytic cells (Costa and Mollenhauer 1980). These insoluble forms of Ni have been shown to be more carcinogenic than soluble nickel and *in vitro* studies have demonstrated that they induce greater intracellular free radical production than soluble nickel salts (Huang et al. 1994a, 1994b).

5.3.1.6 Chromium (Cr)

Chromium is a carcinogen, and has also been shown to react with H_2O_2 in Fenton type reactions to yield the hydroxyl radical (Galaris and Evangelou 2002).

$$Cr^{6+} + O_2^{\cdot -} \rightarrow Cr^{5+} + O_2$$

$$Cr^{5+} + H_2O_2 \rightarrow Cr^{6+} + OH^{\cdot} + OH^-$$

Recent studies have shown that hydroxyl radicals, generated from a Cr^{5+} intermediate, are responsible for eliciting DNA strand breaks (Stohs and Bagchi 1995).

5.3.2 The Role of Nonredox Active Metals

In addition to transition metals, particulate air pollution can contain nonredox active metals, which can influence the toxic effects of transition metals, either exacerbating or lessening the production of free radicals.

5.3.2.1 Zinc (Zn)

Although the literature regarding the oxidative potential of zinc is rather ambiguous, the general consensus from *in vitro* studies is that zinc acts in an antioxidant capacity. For example, zinc inhibits Fe^{2+} and Cu^{2+} induced formation of 2-thiobarbituric-reactive substances—a key lipid oxidation product (Zago and Oteiza 2001)—whilst Chevion et al. 1990, showed that excess zinc inhibits the toxicity of paraquat through a mechanism involving the displacement of copper from its binding sites, inhibiting free radical production at that site. Zinc has also been seen to interact favorably with negatively charged phospholipids, occupying potential binding sites, preventing the binding of redox-active metals, and thereby blocking the initiation or propagation of lipid oxidation. Evidence does exist, however, that zinc can catalyze iron- and copper-induced oxidation (Zago and Oteiza 2001), presumably by displacing iron and copper atoms from chelators and making them more bioavailable to catalyze free radical generation.

5.3.2.2 Aluminium (Al)

Al has been found to aggravate lipid peroxidation by ferrous iron, through a subtle membrane rearrangement action (Verstraeten et al. 1997; Zatta et al. 2002). It can also promote iron ROS generation, possibly by binding directly to a superoxide radical anion to form $AlO_2^{\cdot 2+}$ (Exley 2003). Furthermore, Al^{3+} has also been noted to compete with Fe^{3+} for iron ligands, altering the availability and concentration of iron, and thus potential free radical formation (Zatta et al. 2002).

5.3.2.3 Lead (Pb)

The mechanism(s) behind lead-induced oxidative stress remain elusive; however, its ability to catalyze peroxidative reactions is believed to be a major contributor. It has been observed that dose- and time-dependant increases in peroxides in microsomal membranes occur in response to lead (Lawton and Donaldson 1991), while enhanced malondialdehyde concentrations following incubation of linoic, lonolenic, and arachadonic acid with lead has been reported (Yiin and Lin 1995). In addition, lead induces ROS production through oxyhemoglobin interaction (Ribarov, Benov, and Bechev 1981), has high affinity for protein sulfhydryl groups, and can inhibit the antioxidant enzymes GPx and SOD (Gurer and Ercal 2000).

5.3.3 Particle-Associated Quinone Toxicity

In addition to metals, evidence is also accumulating to suggest that organic components carried on the particle surface play an important role in mediating a toxic effect. Ambient PM has been shown to contain stable organic radicals that have been tentatively identified as quinones (Squadrito et al. 2001), which again are highly redox active molecules that can generate ROS (Bolton et al. 2000; Kumagai et al. 2002). For example, and shown below, a semiquinone radical, $QH^{\cdot}$, can reduce oxygen to form superoxide; this will undergo dismutation to hydrogen peroxide and finally form the hydroxyl radical in the presence of "free" iron. Biological reductants, such as ascorbate, NAD(P)H, and glutathione, are then able to reduce the oxidized quinoid back to the reduced state ($QH^{\cdot}$ and the hydroquinone) enabling the reaction to cycle again, illustrating once more that the interaction between PM constituents and RTLF antioxidants are not always necessarily protective.

$$QH^{\cdot} + O_2 \rightarrow Q + O_2^{\cdot -} + H$$

$$2O_2^{\cdot -} + 2H^+ \rightarrow H_2O_2 + O_2$$

$$H_2O_2 + Fe^{2+} \rightarrow HO^{\cdot} + HO^- + Fe^{3+}$$

$$Q + e^- + H^+ \rightarrow QH^{\cdot}$$

$$QH^{\cdot} + e^- + H^+ \rightarrow QH_2$$

5.3.4 Polycyclic Aromatic Hydrocarbons (PAHs) Induced Oxidative Stress

The organic fraction of PM, particularly the fine fraction, can also include the semi-volatile PAHs. These compounds can induce oxidative stress indirectly through biotransformation by cytochrome P450 and dihydrodiol dehyrrogenase to generate the redox active quinones. PAHs and their derivatives are formed during the incomplete combustion and pyrolysis of organic material, and thus can be released into the atmosphere from natural sources (e.g., volcanic eruptions and forest fires) and anthropogenic emissions (coal, oil, and gas-burning facilities, road traffic, waste incineration and industrial activities, cigarette smoke, asphalt road, and roofing operations) (Manoli et al. 2004). A study using cultured human bronchial epithelial cells to compare the effects of native diesel exhaust particles (nDEP), organic extracts of DEP (OE-DEP), and carbonaceous particles, represented by stripped DEP (sDEP) and carbon black particles (CB), suggests that the DEP-induced inflammatory response in airway epithelial cells mainly involves organic compounds such as PAHs, which in turn induce CYP1A1 gene expression (Bonvallot et al. 2001).

5.3.5 Lipopolysacharide (LPS) Induced Oxidative Stress

Biological compounds associated with airborne particulates are derived from bacteria, viruses, fungal spores, pollen, dust mites, animal dander, and molds. The outer membrane of Gram-negative bacteria is covered with LPS molecules (endotoxin) (Park et al. 2001; Van strien et al. 2004) and these have been shown to be present on the majority of ambient fine and coarse particles, albeit with considerable variations in concentration between PM from different locations (Mueller-Anneling et al. 2004). The potential of particle-associated LPS to elicit toxicity has been observed in cultured macrophages which, when incubated with PM samples rich in LPS, underwent TNFα and oxygen radical release, as well as cell death (Monn et al. 2002).

5.3.6 Generation of ROS by Inflammatory Cells

The primary oxidant challenge induced by ambient particles in the lung initiates the influx, and subsequent activation, of inflammatory cells, such as neutrophils, to the lung. This highly orchestrated series of events can lead to a second wave of oxidative stress and added cell injury, since activated inflammatory cells also generate and release large quantities of free radicals. The events that lead to the influx of inflammatory cells probably involve a signal (possibly activated lipids in the case of ozone) that activates transcription factors, such as nuclear factor κB (NFκB), and increases the expression of a range of proinflammatory chemokines and cytokines. These signaling molecules lead to the up-regulation of a range of adhesion molecules, both on the endothelial cell surface and by the inflammatory cells, and direct movement of the immune cells from the vasculature into the lung tissue.

Immune cells, such as neutrophils, eosinophils, and macrophages, possess a membrane-bound flavoprotein, cytochrome b245 NADPH oxidase, which is induced during cell activation to produces superoxide anions. These reactive oxygen species are removed by superoxide dismutase, which in turn results in the generation of hydrogen peroxide. The latter can easily pass across cell membranes, where it may activate intracellular signaling pathways, or lead to the generation of other reactive oxygen species; for example, the hydroxyl radical or hypochlorous acid in the presence of transition metals and myeloperoxidase respectively.

5.4 The Role of Particle Size and Surface Area versus Composition as Determinants of PM Toxicity

The wide array of oxidizing species described above, which can localize on the surface of PM, undoubtedly act as catalysts in the oxidative deterioration of biological macromolecules, within the reducing environment of the lung. However, as outlined for the pathological effects of respirable fibers, surface reactivity is not the only aspect of particle toxicity. Evidence from *in vivo* exposures of model particles has indicated that particle size is also an important determinant of reactivity (Ferin, Oberdörster, and Penney 1992; Brown et al. 2001; Zhang et al. 2003).

On a mass basis, ultrafine or nanoparticles have a much greater surface area, and therefore capacity to carry toxic components into the deep lung. For example, to obtain 10 $\mu g/m^3$ of 2 μm diameter particles requires 1.2 particles per mL of air and the total surface area of particles is 24 $\mu m^2/mL$; the same airborne mass concentration of 20 nm particles requires 2.4 million particles with a surface area of 3,016 $\mu m^2/ml$. The lung is likely to respond quite differently to 2.4 million particles with such a huge surface area (and a unit size comparable to cellular macromolecules) than to a relatively small number of larger particles (Bogunia-Kubik, Sugisaka, and Murray 2002). Indeed, consistent experimental findings has led to the hypothesis that a better metric for the dose was surface area rather than mass (Oberdörster et al. 1992; Oberdörster, Ferin, and Lehnert 1994; Driscoll 1996; Tran et al. 2000; Brown et al. 2001; Oberdörster, Oberdörster, and Oberdörster 2005). Thus, if particle-induced pulmonary inflammatory processes are a function of the numbers of particles interacting with both inflammatory and epithelial cells, or a function of the particle surface area, ultrafine particles may be highly effective for inducing adverse effects, irrespective perhaps of surface oxidants. Indeed, Brown and colleagues (2000) have shown that ultrafine particles without transition metals on their surface induce marked inflammation in rat lungs.

The mechanism by which a large number of particles or a large surface area leads to increased biological activity is not known, but surface area does appear to be the metric that drives the inflammation *in vivo* caused by low toxicity particles (Duffin et al. 2002). It may be that the particle surface is a source of ROS, as demonstrated by Wilson et al. (2002), who incubated fine or ultrafine carbon black particles with a compound that undergoes activation to a fluorescent state when it is oxidized. This effect is not mediated by transition metals, or any other soluble agent (Brown et al. 2000), but is some consequence of the high surface area interacting with the biological system.

5.5 PARTICLE-INDUCED TOXICITY: LESSONS FROM CIGARETTE SMOKE AND FIBER RESEARCH

One way to improve our understanding of the mechanisms and impact of particle-induced toxicity on the lung, with the ultimate aim of decreasing individual sensitivity to air pollution, is to look at observations from other inflammatory lung conditions, induced by oxidative stress, where more extensive research has been carried out. Obviously examples include cigarette smoking, COPD, and fiber-induced toxicity.

One puff of cigarette smoke contains at least a trillion ($>10^{12}$) free radicals (Church and Pryor 1985), and it is this rich source of oxidants, together with disease-related inflammation and infection, that are believed to contribute to the well-documented evidence of oxidative stress in patients with COPD (Rahman and MacNee 1996; Repine et al. 1997). Markers of oxidative stress, found in both the lung and systemic circulation of cigarette smokers and patients with COPD, include increased 4-hydroxy-2-nonenal-modified protein (a specific peroxidation product of linoleic acid) levels in airway and alveolar epithelial cells, endothelial cells, and neutrophils (Rahman et al. 2002); greater activation of alveolar macrophages in the lungs of smokers (Peto et al. 1992); and increased levels of aldehydes in exhaled breath condensate of adults with COPD (Corradi et al. 2003). Numerous observations have also been made on the antioxidant status in cigarette smoke-induced lung damage. In 1986, Taylor et al. reported a relationship between a deficiency in plasma antioxidant activity and an abnormal FEV_1/FVC ratio in patients with COPD. Subsequently, a number of antioxidant disturbances have been observed in smokers and patients with COPD, but notable inconsistencies exist. Elevated levels of GSH have been reported in BALF in certain chronic smokers (Cantin et al. 1987; Linden et al. 1989). A marginal increase in vitamin C in bronchoalveolar lavage fluid of smokers has also been observed (Bui et al. 1992), however Pacht and Davis (1988) have demonstrated reduced levels of vitamin E. Plasma antioxidant capacity has also been shown to be decreased in smokers (Rahman 1996a, 1996b) and in association with exacerbations of COPD (Rahman, Skwarska, and MacNee 1997; Calikoglu et al. 2002); however, a number of studies have revealed increased circulating antioxidant concentrations in cigarette smokers (Bui et al. 1992). Furthermore, red blood cells from cigarette smokers has been reported to contain increased levels of SOD and catalase, and are better able to protect endothelial cells from the effects of hydrogen peroxide (Toth et al. 1986).

These inconsistencies have led to suggestions that a protective response mechanism may exist (Repine et al. 1997), whereby low-grade oxidative stress can bring about a subsequent adaptive resistance to oxidative stress, by increasing antioxidant defenses. This may explain why, although about 90% of all COPD patients are or have been smokers (Peto et al. 1992), only about 20% of cigarette smokers develop the condition. The mechanisms for the induction of antioxidant enzymes by cigarette smoke exposure are currently unknown. However, the increase in extracellular GSH is related, in part, to an elevated activity of the GSH synthetic enzyme γ-glutamylcysteine synthase.

Oxidative stress is also a central hypothetical mechanism in the proinflammatory (Donaldson and Tran 2002), fibrotic and carcinogenic (Kane 1996) effects of common respirable industrial fibers (e.g., asbestos, synthetic vitreous fibers, and refractory ceramic fibers) for which considerable, and possible transferable, knowledge has accumulated. The considerable heterogeneity in the pathogenic potential of different respirable fibers to cause fibrosis, lung cancer, and mesothelioma has now been explained by the numbers of long fibers modified by their biopersistence (Donaldson and Tran 2004). Short fibers are easily phagocytosed and "mechanically" cleared by macrophages, thus preventing the stimulation of inflammatory mediators (Donaldson et al. 1992; Ye et al. 1999). In contrast, long fibers present a potential problem since they cannot be adequately phagocytosed and result in frustrated phagocytosis or multiple phagocytosis by several macrophages. This causes failed clearance, cell stimulation, and stress that can have inflammatory (Ye et al. 1999), fibrotic and carcinogenic effects. However, a major modifying factor in the pathogenicity of long fibers is their ability to persist in the lung milieu (i.e., the biopersistence of the fiber) and this, in turn, is a function of its biosolubility, breakage into sort fibers, phagocytosis, and clearance. Furthermore, and in parallel with PM, evidence exists that some fibers have specially enhanced pathogenicity and this may be a consequence of their surface reactivity. For example, a significant component of several types of asbestos is iron, and as such, this fiber can deliver oxidative stress to cells via iron-mediated redox chemistry (Gilmour et al. 1995).

5.6 EVIDENCE OF PARTICLE-INDUCED OXIDATIVE STRESS FROM ANIMAL STUDIES

Convincing evidence to support a role for PM in eliciting adverse health effects comes from animal toxicology studies. *In vivo* animal exposures to diesel exhaust particles (DEPs) (Sagai et al. 1993; Ichinose et al. 1995) and a range of surrogate PM, such as carbon black (Li et al. 1996a; Brown et al. 2001; Gallagher et al. 2003; Gilmour et al. 2004; Renwick et al. 2004), TiO_2 (Lee, Trochimowicz, and Reinhardt 1985; Ferin, Oberdörster, and Penney 1992; Heinrich et al. 1995; Borm et al. 2000; and Renwick et al. 2004), metallic cobalt, and polystyrene (Brown et al. 2001) can induce pulmonary inflammation, oxidative stress, and ultimately fibrosis, and lung injury. Whilst an inflammatory response has been demonstrated by increased numbers of polymorphonuclear leucocytes (PMN), oxidative stress has been detected by the ability of particles to generate superoxide and hydroxyl radicals (Sagai et al. 1993; Han, Takeshita, and Utsumi 2001) and deplete antioxidants both in cells and in BAL fluid (Shvedova et al. 2005). Furthermore, in the murine lung, EC-SOD has been shown to reduce alveolar inflammation and injury induced by PM instillations (Ghio et al. 2002a; Ahmed et al. 2003).

The interaction of particles with metals to potentiate pulmonary toxicity has also been studied in animal toxicology. For example, in the rat lung ultrafine carbon black induced a significant neutrophil influx and this inflammatory effect was positively enhanced by the addition of iron salts (Wilson et al. 2002). However, much of the experimental work examining the contribution of metals to PM toxicity has made use of ROFA, which contains about 10% by weight of water-soluble Fe, Ni, and V. Instillation of high concentrations of ROFA into rat airways has been shown to induce neutrophilia, oedema, bronchial hyper-reactivity, and increased susceptibility to infection in these animals (Gavett et al. 1997; Kadiiska et al. 1997; Kodavanti et al. 2001; Antonini 2002). Equivalent responses have been observed in these studies using only the water-soluble components of these samples (Dreher et al. 1997). An increase in pulmonary antioxidants have also been reported in laboratory animals following ROFA challenge. Instillation of ROFA into mice lungs significantly increases BAL fluid total glutathione concentration, whilst this response is attenuated in transgenic mice over-expressing EC-SOD (Ghio et al. 2002a). Furthermore, previous work has shown that elevated lung extracellular GSH concentrations in PM-challenged is due to an increased activity of c-glutamylcysteine synthase (Rahman 1996a, 1996b). Urate concentrations are also increased in the RTLF of rats following instillation of metal rich PM, in association with an increased expression of xanthine oxidase, suggesting that the production of uric acid is directly regulated by iron, via regulation of the expression or activity of its enzymatic source (Ghio et al. 2002b).

5.7 CONCENTRATED AMBIENT AND DIESEL EXHAUST PARTICLE EXPOSURE IN HUMAN SUBJECTS

The use of diesel engines was promoted, especially in Europe, during the 1980s and early 1990s, to improve fuel efficiency and reduce emissions of CO, CO_2, and hydrocarbons (Nauss and the Diesel Working Group 1995). Whilst this initiative was effective in reducing these emissions, diesel engines generated up to 100 times more particles than comparable gasoline engines (Vistal 1980; Department of the Environment 1996). Moreover, electron microscope studies indicated that 80% of such particles had an aerodynamic diameter of size <0.1 μm and were, therefore, readily inhalable, with approximately 10% estimated deposited in the alveolar region of the lung (Scheepers and Bos 1992). Occupational exposure studies demonstrated that diesel exhaust induced decrements in FVC and FEV_1 (Ulfvarson et al. 1987), as well as increased eye and airways irritation, cough, and headache (Gamble, Jones, and Minshall 1987). Furthermore, epidemiological studies demonstrated impaired lung function in children attending schools near busy roads with a high density of diesel-powered vehicles (Brunekreef et al. 1997), as well as an increased prevalence of respiratory symptoms such as cough, wheeze, and physician-diagnosed asthma (van Vliet et al. 1997).

Consistent with the hypothesis that these responses could be mediated through oxidant pathways, diesel exhaust particles have been shown to generate free radicals in bronchial and nasal epithelial cells exposed to both diesel particles, as well as their organic extracts (Sagai et al. 1993; Baulig et al. 2003). Further evidence for a key role of oxidative stress in the up-regulation of pro-inflammatory cytokines has been demonstrated by the capacity of antioxidants to reduce both NFκB activation and cytokine release from cells challenged *in vitro* with DE particles, or their extracts (Abe et al. 2000; Hashimoto et al. 2000; Kawasaki et al. 2001).

Human studies have also been conducted to further our understanding of the toxic pathways mediated by diesel particulates, and as well the protective mechanisms present in the lung to combat oxidant stress. Exposure of healthy subjects to diesel exhaust (300 $\mu g/m^3$ PM_{10}, 1.6 ppm NO_2, and 4.3 ppm total hydrocarbons) under controlled chamber conditions has been shown to induce both lower airway and systemic inflammation (Salvi et al. 1999; Mudway et al. 1999), and increase numbers of neutrophils in induced sputum (Nordenhall et al. 2000). The ability of DE to modify the antioxidant defense network within the respiratory tract lining fluids has also been demonstrated by a 10-fold increase in nasal lavage ascorbate during a controlled exposure to 300 $\mu g/m^3$ PM_{10} (Blomberg et al. 1998). The magnitude of this response may suggest considerable trauma to the nasal epithelium, with ascorbate either flooding into the RTLF from the plasma pool or from necrotic/apoptopic nasal cell populations. In the absence of any other significant changes in antioxidant concentrations or oxidative stress, these findings imply that the physiological response of normal individuals to this particular DE exposure, namely increased ascorbic acid in the nasal cavity, appears to be sufficient to prevent further oxidant stress in the respiratory tract. It does not, however, appear to prevent the development of airway neutrophilia or the up-regulation of redox sensitive signaling pathways in the bronchial mucosa of these subjects (Pourazar et al. 2005).

The effects, again in healthy individuals, of more environmentally relevant concentrations of DE (100 $\mu g/m^3$ PM_{10}) have also been examined (Mudway et al. 2004). Under these conditions, where the subject responses were considerably attenuated relative to the 300 $\mu g/m^3$ challenge, an increased flux of GSH, but not urate or ascorbate, into the bronchial and nasal airways was reported, together with an increase in subjective symptoms and a mild bronchoconstriction, but no signs of airway inflammation, antioxidant depletion at any level of the respiratory tract, or lipid oxidation. These data again suggest that the healthy lung is protected against DE, and at these concentrations, the adaptive movement of thiol antioxidants onto the surface of the lung contributes to the protection. The rapid increase in RTLF GSH concentrations reported in this study prompted speculation of an unregulated release of intracellular GSH stores through cellular necrosis, though no evidence of cell loss or lysis was found post-DE exposure. Alternatively, the increase may reflect enhanced normal homeostatic mechanisms to maintain the reducing character of the extracellular compartment, or regulated GSH export, as a pre-commitment step for cellular apoptosis (van den Dobbelsteen et al. 1996).

Although diesel PM may contribute significantly to the PM airshed in many European cities, experimental exposures to diesel are not wholly reflective in ambient conditions where the PM mixture is more heterogeneous. Exposure to ambient PM at elevated concentrations can be achieved through the use of ambient fine particle concentrators that are capable of enriching the mass of ambient PM. A number of human studies have been performed, examining the airway and systemic responses of healthy and asthmatic individuals to concentrated ambient particles. These exposures have generally demonstrated relatively mild or no pulmonary responses in healthy subjects and asthmatics (Ghio, Kim, and Devlin 2000; Gong, Sioutas, and Linn 2003) with modest increases in systemic inflammation and coagulation markers. Whilst none of the published human exposure studies have examined RTLF antioxidant responses, the lack of a pronounced response may imply that the basal RTLF defenses are sufficient to prevent oxidative injury. Notably, rats exposed to 300 $\mu g/m^3$ of CAPs for 5 h, display a 2-fold increase in cardiac and pulmonary chemiluminescence that has been attributed to increased radical production, parallel to increased lung SOD and CAT25 concentrations, consistent with the view PM-induced radical production can trigger the up-regulation of antioxidant defenses (Gurgueira et al. 2002).

Whilst the findings from controlled human exposure studies indicate that the antioxidant network at the air–lung interface is capable of dealing with the oxidative challenge posed by DE at ambient concentrations, this may not be the case in the diseased lung. Further research is, however, needed to ascertain whether, indeed, basal antioxidant status may be compromised or adaptive responses attenuated in subjects with pulmonary diseases, thereby contributing an increased susceptibility to the adverse effects of air pollution.

Another subpopulation sensitive to the effects of particulate pollution may be those with a genetic basis underlying individual susceptibility to oxidative challenge. This has been postulated partly in relation to wide variations in glutathione *S*-transferase (GST) activities (Gilliand et al. 1999). Common allelic variants exist for four GST genes, GSTM1, GSTM3, GSTT1, and GSTP1 (Hirvonen 1999), and GST polymorphisms are associated with several inflammatory diseases, including asthma (Ollier et al. 1996; Strange and Fryer 1999; Fryer et al. 2000). Moreover, (Gilliland et al. 2004) have used a human nasal provocation model to examine responses to allergen, or allergen plus diesel exhaust particles, and to assess whether functional variants in the GST superfamily could account for the variation in interindividual responsiveness to diesel-exhaust particles (Gilliland et al. 2004). Individuals with GSTM1-null or GSTP1 Ile105 wildtype genotypes showed enhanced nasal allergic responses to diesel exhaust particles, whilst GSTM1-null individuals showed a significantly larger increase in IgE and histamine in nasal lavage fluid after challenge with diesel-exhaust particles or allergen, than children with a functional GSTM1 genotype. Data such as this help to further our understanding of the effects of air pollution and its influence on allergic response at the cellular and molecular level. Moreover, the findings provide further support for the concept that antioxidant defenses do have a key role in controlling the response to DE particles by scavenging reactive oxygen species.

5.8 MODELING THE INTERACTION OF AMBIENT PM WITH RTLF ANTIOXIDANTS

To address the mechanisms by which ambient PM impact upon the RTLF antioxidant network, and to identify the components driving the observed activity, researchers have utilized a range of *in vitro* modeling systems. One such approach by Mudway and coworkers has compared the oxidative activity of DEP with the transition metal rich residual oil fly ash (ROFA) particle by incubating both samples in a synthetic lung lining fluid model (Mudway et al. 2004). DE was found to have less activity than ROFA, although both significantly depleted lung lining fluid ascorbate and GSH. Furthermore, substantial protection was observed against both ascorbate and glutathione depletion, following co-incubation with the free radical scavengers, superoxide dismutase (SOD) and catalase, and with diethylenetriamine-pentaacetate (DTPA), a highly-effective chelator of redox active transition metals, including iron and copper (Tang, Yang, and Shen 1997). These samples were reported to contain no ferric or ferrous metal ions, and other metal chelators with a high affinity for iron and copper had no effect. Together these findings suggest that the DE extract contained a transition metal, excluding iron and copper, which was capable of being reduced by both ascorbate and reduced glutathione with the generation of superoxide, which then further depleted antioxidants within the synthetic RTLF. Of further interest, whereas the greatest protection against ascorbate depletion was observed with DTPA, the SOD/catalase co-incubation was more effective for GSH, suggesting that metal reduction was predominately driving ascorbate loss, whilst GSH depletion was more dependent on secondary superoxide production. Subsequent studies using this model have shown ambient coarse (2.5–10 μm) and fine (0.1–2.5 μm) PM to have high levels of oxidative activity (Duggan et al. 2002), largely related to their content of redox active transition metals (Duggan et al. 2003). Placing these results in the context of the findings from the human challenge studies with both CAPs and diesel, would suggest that the RTLF provides an effective barrier against transition metal-driven oxidative injury.

5.9 PARTICLE-INDUCED OXIDATION REACTIONS AT THE AIR–LUNG INTERFACE AS A PREDICTOR OF OBSERVED HEALTH EFFECTS

A crucial question that still remains unanswered is whether the capacity of particles to induce oxidation reactions within the RTLF predicts the magnitude of the observed health effects. The rules that govern the balance between the number of beneficial and detrimental interactions between lung lining fluid antioxidant defenses and PM oxidative components are not well established, but these may contribute, in part, to the sensitivity of individuals to air pollution (Kelly et al. 1995). The most simplistic viewpoint would be that the greater the range and concentration of antioxidant defenses on the lung surface, the greater the number of successful interactions with harmful PM, and the better the level of protection from oxidant air pollutants. If this were the case however, subgroups of the population recognized to be susceptible to air pollution should have decreased lung lining fluid antioxidant defenses. This has indeed been found in asthmatics, a susceptible subgroup in that they have markedly decreased concentrations of ascorbic acid (vitamin C) in lung lining fluid compared to healthy control subjects (Kelly, Buhl, and Sandström 1999). The same was not found to be true, however, of healthy subjects displaying the largest symptomatic responses (airway neutrophil and lung function decrements) to ozone (Mudway et al. 2001). One of the key findings to emerge from the human CAPs and diesel exposure studies is that at ambient PM concentrations, acute symptomatic responses were relatively mild, with a predominance of protective adaptive responses, including augmentation of RTLF and cellular antioxidant defenses (Mudway et al. 2004). Thus, it's also likely that individual differences in the capacity to mount these protective adaptations may also contribute to the differential sensitivity observed in the healthy populations and in patient groups. Further work is needed to validate the hypothesis that lung lining fluid antioxidants perform a critical role in reducing the toxic consequences of oxidant air pollutants.

5.10 CONCLUSIONS

The surface of the lung is bathed in an antioxidant-rich lining layer that has evolved to protect the respiratory epithelium against inhaled xenobiotics. Upon inhalation, this is the first physical interface encountered by PM and, as such, it plays a critical role in modifying the potential toxicity of these particles. The interaction between the RTLF and PM constituents are, however, far from simple. For example, there is evidence that the interaction between catalytic metals and quinones, with reductants within this compartment, may actually promote radical production. Whether this represents an effective signaling pathway, permitting the underlying epithelium to mount adaptive response, or illustrates that the RTLF defenses are not designed for many of the toxins present in "contemporary" air, remains open to question. It is clear however, that a simple concept of RTLF antioxidant being protective is too simplistic.

As with any toxic challenge, the obvious solution is to remove, or at least decrease to an acceptable level, the source of trouble. In many countries, air pollution levels have fallen in recent years, while additional measures are in place in several more regions to decrease concentrations further. It is unlikely, however, that these practical measures will completely eliminate the problem, even in the medium term. As a consequence, considerable research is now underway to improve our understanding of the impact of PM on biological systems. A better appreciation of the toxic pathways, as well as the components and characteristics of PM that mediate their effects, would allow a more targeted approach to remove the most toxic elements of air pollution, and could possibly provide a means to decrease individual sensitivity.

ACKNOWLEDGMENTS

We are grateful to Wendy Rowlands for her help in preparing this review. The authors also acknowledge the Medical Research Council, The Wellcome Trust, The European Union, and Health Effects Institute for supporting their program of work in PM toxicology.

REFERENCES

Abe, S. et al., Diesel exhaust (DE)-induced cytokine expression in human bronchial epithelial cells: a study with a new cell exposure system to freshly generated DE *in vitro*, *Am. J. Respir. Cell Mol. Biol.*, 22, 296, 2000.

Ahmed, M. N. et al., Extracellular superoxide dismutase protects lung development in hyperoxia-exposed newborn mice, *Am. J. Respir. Crit. Care Med.*, 167, 400, 2003.

Ames, B. N. et al., Uric acid provides an antioxidant defense in humans against oxidant and radical caused aging and cancer: a hypothesis, *Proc. Natl. Acad. Sci. U.S.A.*, 78, 6858, 1981.

Antonini, J. M. et al., Residual oil fly ash increases the susceptibility to infection and severely damages the lungs after pulmonary challenge with a bacterial pathogen, *Toxicol. Sci.*, 70, 110, 2002.

Athar, M., Hasan, S. K., and Srivastava, R. C., Evidence for the involvement of hydroxyl radicals in nickel mediated enhancement of lipid peroxidation: implications for nickel carcinogenesis, *Biochem. Biophys. Res. Commun.*, 147, 1276, 1987.

Aust, A. E. et al., Particle characteristics responsible for effects on human lung epithelial cells, *Res. Rep. Health Eff. Inst.*, 110, 1, 2002.

Avissar, N. et al., Extracellular glutathione peroxidase in human lung epithelial lining fluid and in lung cells, *Am. J. Physiol.*, 270, L173, 1996.

Baulig, A. et al., Involvement of reactive oxygen species in the metabolic pathways triggered by diesel exhaust particles in human airway epithelial cells, *Am. J. Physiol. Lung Cell. Mol. Physiol.*, 285, L671, 2003.

Becker, R. F., Towards the physiological function of uric acid, *Free Radic. Biol. Med.*, 14, 615, 1993.

Blomberg, A. et al., Nasal cavity lining fluid ascorbic acid concentration increases in healthy human volunteers following short term exposure to diesel exhaust, *Free Radic. Res.*, 28, 59, 1998.

Bogunia-Kubik, K., Sugisaka, M., and Murray, J., From molecular biology to nanotecnology and nanomedicine, *Biosystems*, 65, 123, 2002.

Bolton, J. L. et al., Role of quinones in toxicology, *Chem. Res. Toxicol.*, 13, 135, 2000.

Bonvallot, V. et al., Organic compounds from diesel exhaust particles elicit a proinflammatory response in human airway epithelial cells and induce cytochrome p450 1A1 expression, *Am. J. Respir. Cell Mol. Biol.*, 25, 515, 2001.

Borm, P. J. et al., Chronic inflammation and tumour formation in rats after intratrachial instillation of high doses of coal dusts, titanium dioxide and quartz, *Inhal. Toxicol.*, 12, 225, 2000.

Brigelius-Flohe, R. and Traber, M. G., Vitamin E: function and metabolism, *Faseb. J.*, 13, 1145, 1999.

Britigan, B. E., Serody, J. S., and Cohen, M. S., The role of lactoferrin as an anti-inflammatory molecule, *Adv. Exp. Med. Biol.*, 357, 143, 1994.

Brown, D. M. et al., Increased inflammation and intracellular calcium caused by ultrafine carbon black is independent of transition metals or other soluble components, *Occup. Environ. Med.*, 57, 685, 2000.

Brown, D. M. et al., Size-dependant pro-inflammatory effects of ultrafine polysteryne particles: a role for surface area and oxidative stress in the enhanced activity of ultrafines, *Toxicol. App. Pharmacol.*, 175, 191, 2001.

Brunekreef, B. and Holgate, S. T., Air pollution and health, *Lancet*, 360, 1233–1242, 2002.

Brunekreef, B. et al., Air pollution from truck traffic and lung function in children living near motorways, *Epidemiology*, 8, 298, 1997.

Buettner, G. R. and Jurkiewicz, B. A., Chemistry and biochemistry of ascorbic acid, In *Handbook of Antioxidants*, Cadenas, E., and Packer, L., eds., Marcel Dekker Inc., New York, 91–115, 1996.

Bui, M. H. et al., Dietary vitamin C intake and concentrations in the body fluids and cells of male smokers and non-smokers, *J. Nutr.*, 122, 312, 1992.

Burney, P., The origins of obstructive airways disease A role for diet? *Am. J. Respir. Crit. Care Med.*, 151, 1292, 1995.

Burns, D. M., Cigarettes and cigarette smoking, *Clin. Chest Med.*, 12, 631, 1991.

Burton, G. W. and Ingold, K. U., Auto-oxidation of biological molecules, The antioxidant activity of vitamin E and related chain breaking phenolic antioxidants *in vitro*, *J. Am. Chem. Soc.*, 103, 6472, 1981.

Calikoglu, M. et al., The levels of serum vitamin C, malonyldialdehyde and erythrocyte reduced glutathione in chronic obstructive pulmonary disease and in healthy smokers, *Clin. Chem. Lab. Med.*, 40, 1028, 2002.

Cantin, A. M. et al., Normal alveolar epithelial lining fluid contains high levels of glutathione, *J. Appl. Physiol.*, 63, 152, 1987.

Chen, R., Tunstall-Pedoe, H., and Tavendale, R., Environmental tobacco smoke and lung function in employees who never smoked: the scottish MONICA study, *Occup. Environ. Med.*, 58, 563, 2001.

Chevion, M. et al., Zinc—a redox inactive-metal provides a novel approach for protection against metal-mediated free radical induced injury: study of paraquat toxicity in *E.coli*, *Adv. Exp. Med. Biol.*, 264, 217, 1990.

Church, D. F., Pryor, W. A., Free-radical chemistry of cigarette smoke and its toxicological implications. *Environ. Health Perepect.*, 64, 111–126, 1985 Dec.

Cooper, B., Creeth, J. M., and Donald, A. S., Studies of the limited degradation of mucus glycoproteins. The mechanism of the peroxide reaction, *Biochem. J.*, 228, 615, 1985.

Corradi, M. et al., Aldehydes in exhaled breath condensate of patients with chronic obstructive pulmonary disease, *Am. J. Respir. Crit. Care Med.*, 167, 1380, 2003.

Costa, M. and Mollenhauer, H. H., Phagocytosis of nickel subsulfide particles during the early stages of neoplastic transformation in tissue culture, *Cancer Res.*, 40, 2688, 1980.

Cross, C. E., Halliwell, B., and Allen, A., Antioxidant protection: a function of tracheobronchial and gastrointestinal mucus, *Lancet*, 8390, 1328, 1984.

Cross, C. E. et al., Oxidants, antioxidants, and respiratory tract lining fluids, *Environ. Health Perspect.*, 102, 185, 1994.

Cross, C. E. et al., Oxidative stress and antioxidants at biosurfaces: plants, skin, and respiratory tract surfaces, *Environ. Health Perspect.*, 106, 1241, 1998.

Datta, A. K. et al., Enhancement by nickel(II) and -histidine of 2′-deoxyguanine oxidation with hydrogen peroxide, *Carcinogenesis*, 13, 283, 1992.

Davies, K. J. et al., Uric acid-iron ion complexes. A new aspect of the antioxidant functions of uric acid, *Biochem. J.*, 235, 747, 1986.

Department of the Environment, London, Quality of urban air review group, *The Third Report of the Quality of Urban Air Review Group*, 37, 1996.

Diliberto, E. J. Jr., et al., Tissue, subcellular, and submitochondrial distributions of semidehydroascorbate reductase: possible role of semidehydroascorbate reductase in cofactor regeneration, *J. Neurochem.*, 39, 563, 1982.

Doba, T., Burton, G. W., and Ingold, K. U., The effect of vitamin C, either alone or in the presence of vitamin E or a water soluble vitamin C analogue, upon the peroxidation of aqueous multilamellar phospholipids liposomes, *Biochem. Biophys. Acta*, 835, 298, 1985.

Dockery, D. W. et al., An association between air pollution and mortality in six U.S. cities, *N. Engl. J. Med.*, 329, 1753, 1993.

Doelman, C. J. and Bast, A., Pro- and anti-oxidant factors in rat lung cytosol, *Adv. Exp. Med. Biol.*, 264, 455, 1990.

Donaldson, K. and Tran, C. L., Inflammation caused by particles and fibers, *Inhal. Toxicol.*, 14, 5, 2002.

Donaldson, K. and Tran, C. L., An introduction to the short-term toxicology of respirable industrial fibers, *Mutat. Res.*, 553, 5, 2004.

Donaldson, K. et al., Asbestos-stimulated tumour necrosis factor release from alveolar macrophages depends on fiber length and opsonization, *J. Pathol.*, 168, 243, 1992.

Dreher, K. L. et al., Soluble transition metals mediate residual oil fly ash induced acute lung injury, *J. Toxicol. Environ. Health*, 50, 285, 1997.

Driscoll, K. E., Role of inflammation in the development of rat tumors in response to chronic particle exposure, *Inhal. Toxicol.*, 8, 139, 1996.

Droge, W., Free radicals in the physiological control of cell function, *Physiol. Rev.*, 82, 47, 2002.

Duffin, R. et al., The importance of surface area and specific reactivity in the acute pulmonary inflammatory response to particles, *Ann. Occup. Hyg.*, 46, 242, 2002.

Duggan, S. et al., Particulate matter in London air is highly oxidising, implying a significant potential risk to respiratory health, *Eur. Respir. J.*, 20, 260, 2002.

Duggan, S. et al., Transition metal rich particulate matter can subvert the endogenous defenses within the respiratory tract lining fluid, *Am. J. Respir. Crit. Care Med.*, 167, A973, 2003.

Elsayed, N. M., Antioxidant mobilization in response to oxidative stress: a dynamic environmental-nutritional interaction, *Nutrition*, 10, 828, 2001.

Exley, C., The pro-oxidant activity of aluminium, *Free Radic. Biol. Med.*, 36, 380, 2003.

Ferin, J., Oberdörster, G., and Penney, D. P., Pulmonary retention of ultrafine and fine particles in rats, *Am. J. Respir. Cell Mol. Biol.*, 6, 535, 1992.

Freeman, B. A. and Crapo, J. D., Biology of disease: free radicals and tissue injury, *Lab. Invest.*, 47, 412, 1982.

Fridovich, I., Oxygen toxicity: a radical explanation, *J. Exp. Biol.*, 201, 1203, 1998.

Fryer, A. A. et al., Polymorphism at the glutathione *S*-transferase *GSTP1* locus: a new marker for bronchial hyperresponsiveness and asthma, *Am. J. Respir. Crit. Care Med.*, 161, 1437, 2000.

Fubini, B., Surface reactivity in the pathogenic response to particulates, *Environ. Health Perspect.*, 105, 1013, 1997.

Galaris, D. and Evangelou, A., The role of oxidative stress in mechanisms of metal-induced carcinogenesis, *Crit. Rev. Oncol. Hematol.*, 42, 93, 2002.

Gallagher, J. et al., Formation of 8-oxo-7,8-dihydro-2′-deoxyguanosine in rat lung DNA following subchronic inhalation of carbon black, *Toxicol. Appl. Pharmacol.*, 190, 224, 2003.

Gamble, J., Jones, W., and Minshall, S., Epidemiological-environmental study of diesel bus garage workers: chronic effects of diesel exhaust on the respiratory system, *Environ. Res.*, 44, 6, 1987.

Gavett, S. H. et al., Metal and sulfate composition of residual oil fly ash determines airway hyperreactivity and lung injury in rats, *Environ. Res.*, 72, 162, 1997.

Ghio, A. J., Kim, C., and Devlin, R. B., Concentrated ambient air particles induce mild pulmonary inflammation in healthy human volunteers, *Am. J. Respir. Crit. Care Med.*, 162, 981, 2000.

Ghio, A. J. et al., Complexation of iron cation by sodium urate crystals and gouty inflammation, *Arch. Biochem. Biophys.*, 313, 215, 1994.

Ghio, A. J. et al., Iron regulates xanthine oxidase activity in the lung, *Am. J. Physiol. Lung Cell. Mol. Physiol.*, 283, L563, 2002.

Ghio, A. J. et al., Overexpression of extracellular superoxide dismutase decreases lung injury after exposure to oil fly ash, *Am. J. Physiol. Lung Cell. Mol. Physiol.*, 283, L211, 2002a.

Gilliland, F. D. et al., A theoretical basis for investigating ambient air pollution and children's respiratory health, *Environ. Health Perspect.*, 107, 403, 1999.

Gilliland, F. D. et al., Effect of glutathione-*S*-transferase M1 and P1 genotypes on xenobiotic enhancement of allergic responses: randomised, placebo-controlled crossover study, *Lancet*, 363, 119, 2004.

Gilmour, P. S. et al., Detection of surface free-radical activity of respirable industrial fibers using supercoiled phi-x174 rf1 plasmid DNA, *Carcinogenesis*, 16, 2973, 1995.

Gilmour, P. S. et al., Pulmonary and systemic effects of short-term inhalation exposure to ultrafine carbon black particles, *Toxicol. Appl. Pharmacol.*, 195, 35, 2004.

Gong, H., Sioutas, C., and Linn, W. S., Controlled exposures of healthy and asthmatic volunteers to concentrated ambient particles in metropolitan Los Angeles, *Res. Rep. Health Eff. Inst.*, 1, 37, 2003.

Gurer, H. and Ercal, N., Can antioxidants be beneficial in the treatment of lead poisoning? *Free Rad. Biol. Med.*, 10, 927, 2000.

Gurgueira, S. A. et al., Rapid increases in the steady-state concentration of reactive oxygen species in the lungs and heart after particulate air pollution inhalation, *Environ. Health Perspect.*, 110, 749, 2002.

Halliwell, B. and Gutteridge, J. M., Biologically relevant metal ion-dependent hydroxyl radical generation. An update, *FEBS Lett.*, 307, 108, 1992.

Halliwell, B. and Gutteridge, J. M. C., *Free Radic. Biol. Med.*, Clarendon Press, Oxford, 1999.

Han, J. Y., Takeshita, K., and Utsumi, H., Noninvasive detection of hydroxyl radical generation in lung by diesel exhaust particles, *Free Radic. Biol. Med.*, 30, 516, 2001.

Hashimoto, S. et al., Diesel exhaust particles activate p38 MAP kinase to produce interleukin 8 and RANTES by human bronchial epithelial cells and *N*-acetylcysteine attenuates p38 MAP kinase activation, *Am. J. Respir. Crit Care Med.*, 161, 280, 2000.

Haslam, P. L. and Baughman, R. P., Report of ERS task force: guidelines for measurement of acellular components and standardization of BAL, *Eur. Respir. J.*, 14, 245, 1999.

Heinrich, U. et al., Chronic inhalation of wistar rats and two different strains of mice to diesel exhaust, carbon black and titanium dioxide, *Inhal. Toxicol.*, 7, 533, 1995.

Herrero, M. C. et al., Nickel effects on hepatic amino acids, *Res. Commun. Chem. Pathol. Pharmacol.*, 79, 243, 1993.

Hiraishi, H. et al., Role for mucous glycoprotein in protecting cultured rat gastric mucosal cells against toxic oxygen metabolites, *J. Lab. Clin. Med.*, 121, 570, 1993.

Hirvonen, A., Polymorphism of xenobiotic-metabolizing enzymes and susceptibility to cancer, *Environ. Health Perspect.*, 107, 37, 1999.

Housley, D. G., Eccles, R., and Richards, R. J., Gender difference in the concentration of the antioxidant uric acid in human nasal lavage, *Acta Otolaryngol. (Stockh)*, 116, 751, 1996.

Huang, X., Klein, C. B., and Costa, M., Crystalline Ni3S2 specifically enhances the formation of oxidants in the nuclei of CHO cells as detected by dichlorofluorescein, *Carcinogenesis*, 15, 545, 1994.

Huang, X. et al., The role of nickel and nickel-mediated reactive oxygen species in the mechanism of nickel carcinogenesis, *Environ. Health Perspect.*, 102, 281, 1994.

Hunt, J. F. et al., Endogenous airway acidification. Implications for asthma pathophysiology, *Am. J. Respir. Crit. Care Med.*, 161, 694, 2000.

Ichinose, T. A. et al., Biological effects of diesel exhaust particles (DEP). II. Acute toxicity of DEP introduced into lung by intratracheal instillation, *Toxicology*, 99, 153, 1985.

Jenkinson, S. G., Black, R. D., and Lawrence, R. A., Glutathione concentrations in rat lung bronchoalveolar lavage fluid: effect of hyperoxia, *J. Lab. Clin. Med.*, 112, 345, 1988.

Kadiiska, M. B. et al., *In vivo* evidence of free radical formation in the rat lung after exposure to an emission source air pollution particle, *Chem. Res. Toxicol.*, 10, 1104, 1997.

Kane, A. B., Mechanisms of mineral fiber carcinogenesis, In *Mechanisms of Fiber Carcinogenesis*, Kane, A. B., Boffetta, P., Saracci, R., Wilbourn, J. D., eds., IARC Scientific Publications, No 140, International Agency for Research on Cancer, Lyon, 11, 1996.

Kasprzak, K. S., The role of oxidative damage in metal carcinogenesis, *Chem. Res. Toxicol.*, 4, 604, 1991.

Kawasaki, S. et al., Benzene-extracted components are important for the major activity of diesel exhaust particles: effect on interleukin-8 gene expression in human bronchial epithelial cells, *Am. J. Respir. Cell Mol. Biol.*, 24, 419, 2001.

Kelly, F. J., Oxidative stress; its role in air pollution and adverse health effects, *Occup. Environ. Med.*, 60, 612, 2003.

Kelly, F. J. and Richards, R., Antioxidant defenses in the human lung, In *Air Pollution and Health*, Holgate, S., Samet, J., Koren, H. S., and Maynard, R. L., eds., Academic Press, New York, 1999.

Kelly, F. J., Buhl, R., and Sandström, T., Measurement of antioxidants, oxidants and oxidant products in bronchoalveolar lavage fluid, *Eur. Respir. Rev.*, 9, 93, 1999.

Kelly, F. J. et al., The free radical basis of air pollution: focus on ozone, *Respir. Med.*, 89, 647, 1995.

Kelly, F. J. et al., Altered lung antioxidant status in patients with mild asthma, *Lancet*, 354, 482, 1999.

Klein, C. B., Frenkel, K., and Costa, M., The role of oxidative processes in metal carcinogenesis, *Chem. Res. Toxicol.*, 4, 592, 1991.

Koc, M. et al., Levels of some acute-phase proteins in the serum of patients with cancer during radiotherapy, *Biol. Pharm. Bull.*, 26, 1494, 2003.

Kodavanti, U. P. et al., Acute lung injury from intratracheal exposure to fugitive residual oil fly ash and its constituent metals in normo- and spontaneously hypertensive rats, *Inhal. Toxicol.*, 13, 37, 2001.

Kumagai, Y. et al., Oxidation of proximal protein sulfhydryls by phenanthraquinone, a component of diesel exhaust particles, Chem, *Res. Toxicol.*, 15, 483, 2002.

Lawton, L. J. and Donaldson, W., Lead-induced tissue fatty acid alterations and lipid peroxidation, *Biol. Trace Elem. Res.*, 28, 83, 1991.

Lee, K. P., Trochimowicz, H. J., and Reinhardt, C. F., Pulmonary response of rats exposed to titanium dioxide (TiO2) by inhalation for two years, *Toxicol. Appl. Pharmacol.*, 79, 179, 1985.

Li, N. et al., Use of a stratified oxidative stress model to study the biological effects of ambient concentrated and diesel exhaust particulate matter, *Inhal. Toxicol.*, 14, 459, 2002.

Li, W., Zhao, Y., and Chou, I. N., Alterations in cytoskeletal protein sulfhydryls and cellular glutathione in cultured cells exposed to cadmium and nickel ions, *Toxicology*, 77, 65, 1993.

Li, X. Y. et al., Free radical activity and pro-inflammatory effects of particulate air pollution (PM10) *in vivo* and *in vitro*, *Thorax*, 51, 1216, 1996.

Li, X. Y. et al., Short-term inflammatory responses following intratracheal instillation of fine and ultrafine carbon black in rats, *Inhal. Toxicol.*, 11, 709, 1996a.

Li, Y. et al., ESR evidence for the generation of reactive oxygen species from the copper-mediated oxidation of the benzene metabolite, hydroquinone: role in DNA damage, *Chem. Biol. Interact.*, 94, 101, 1995.

Lin, X., Zhuang, Z., and Costa, M., Analysis of residual amino acid—DNA crosslinks induced in intact cells by nickel and chromium compounds, *Carcinogenesis*, 13, 1763, 1992.

Linden, M. et al., Glutathione in bronchoalveolar lavage fluid from smokers is related to humoral markers of inflammatory cell activity, *Inflammation*, 13, 651–658, 1989.

Lynn, S. et al., Reactive oxygen species are involved in nickel inhibition of DNA repair, *Environ. Mol. Mutagen.*, 29, 208, 1997.

McCay, P. B., Vitamin E: interactions with free radicals and ascorbate, *Ann. Rev. Nutr.*, 5, 323, 1985.

Manoli, E. et al., Profile analysis of ambient and source emitted particle-bound polycyclic aromatic hydrocarbons from three sites in northern Greece, *Chemosphere*, 56, 867, 2004.

Menon, P. et al., Passive cigarette smoke-challenge studies: increase in bronchial hyper-reactivity, *J. Allergy Clin. Immunol.*, 89, 560, 1992.

Monn, C. et al., Ambient PM(10) extracts inhibit phagocytosis of defined inert model particles by alveolar macrophages, *Inhal. Toxicol.*, 14, 369, 2002.

Mudway, I. S. and Kelly, F. J., Ozone and the lung: a sensitive issue, *Mol. Aspects Med.*, 21, 1, 2000.

Mudway, I. S. et al., Antioxidant consumption and repletion kinetics in nasal lavage fluid following exposure of healthy human volunteers to ozone, *Eur. Respir. J.*, 13, 1429, 1999.

Mudway, I. S. et al., Differences in basal airway antioxidant concentrations are not predictive of individual responsiveness to ozone: a comparison of healthy and mild asthmatic subjects, *Free Radic. Biol. Med.*, 31, 962, 2001.

Mudway, I. S. et al., An *in vitro* and *in vivo* investigation of the effects of diesel exhaust on human airway lining fluid antioxidants, *Arch. Biochem. Biophys.*, 423, 200, 2004.

Mueller-Anneling, L. et al., Ambient Endotoxin Concentrations in PM10 from Southern California, *Environ. Health Perspect.*, 112, 583, 2004.

Nakamura, H. et al., Effect of vitamin E on the response of lung antioxidant enzymes in young rats exposed to hyperoxia, *Kobe J. Med. Sci.*, 33, 53, 1987.

Nauss, K., and Diesel Working Group, In *Diesel Exhaust: A Critical Analysis of Emissions, Exposure and Health Effects*, 1995.

Niki, E., Yamamoto, Y., and Kamiya, Y., Effect of the phytyl tail of vitamin E on its antioxidant activity, *Life Chem. Reps.*, 3, 35, 1985.

Niki, E., Takahashi, M., and Komuro, E., Antioxidant activity of vitamin E in liposomal membranes, *Chem. Let.*, 1573, 1986.

Nordenhall, C. et al., Airway inflammation following exposure to diesel exhaust: a study of time kinetics using induced sputum, *Eur. Respir. J.*, 15, 1046, 2000.

Nualart, F. J. et al., Recycling of vitamin C by a bystander effect, *J. Biol. Chem.*, 278, 10128, 2003.

Oberdörster, G., Ferin, J., and Lehnert, B. E., Correlation between particle size, *in vivo* particle persistence, and lung injury, *Environ. Health Perspect.*, 102, 173, 1994.

Oberdörster, G., Oberdörster, E., and Oberdörster, J., Nanotoxicology: an emerging discipline evolving from studies of ultrafine particles, *Environ. Health Perspect.*, 113, 823, 2005.

Oberdörster, G. et al., Thermal degradation events as health hazards: particle vs gas phase effects, mechanistic studies with particles, *Acta Astronaut.*, 27, 251, 1992.

Ollier, W. E. R. et al., Association of homozygosity for glutathione *S*-transferase: *GSTM1* null alleles is associated with the Ro+/La− autoantibody profile in patients with systemic lupus erythematosus, *Arthritis Rheum.*, 39, 1763, 1996.

Oury, T. D. et al., Extracellular superoxide dismutase in vessels and airways of humans and baboons, *Free Radic. Biol. Med.*, 20, 957, 1996.

Pacht, E. R. and Davis, W. B., Role of transferrin and ceruloplasmin in antioxidant activity of lung epithelial lining fluid, *J. Appl. Physiol.*, 64, 2092, 1988.

Packer, J. E., Slater, T. F., and Wilson, R. L., Direct observation of a free radical interaction between vitamin E and vitamin C, *Nature*, 278, 737, 1979.

Park, J. B. and Levine, M., Purification, cloning and expression of dehydroascorbic acid-reducing activity from human neutrophils: identification as glutaredoxin, *Biochem. J.*, 315, 931, 1996.

Park, J. H. et al., House dust endotoxin and wheeze in the first year of life, *Am. J. Respir. Crit. Care Med.*, 163, 322, 2001.

Peden, D. B. et al., Uric acid is a major antioxidant in human nasal airway secretions, *Proc. Natl. Acad. Sci. U.S.A.*, 87, 7638, 1990.

Peden, D. B. et al., Nasal secretion of the ozone scavenger uric acid, *Am. Rev. Respir. Dis.*, 148, 455, 1993.

Peto, R. et al., Mortality from tobacco in developed countries: indirect estimation from national vital statistics, *Lancet*, 339, 1268, 1992.

Pope, C. A. et al., Particulate air pollution as a predictor of mortality in a prospective study of U.S. adults, *Am. J. Respir. Crit. Care Med.*, 151, 669, 1995.

Pourazar, J. et al., Diesel exhaust activates redox-sensitive transcription factors and kinases in human airways, *Am. J. Physiol. Lung Cell. Mol. Physiol.*, 289, L724, 2005.

Quinlan, G. J., Evans, T. W., and Gutteridge, J. M., Iron and the redox status of the lungs, *Free Radic. Biol. Med.*, 33, 1306, 2002.

Rahman, I. and MacNee, W., Oxidant/antioxidant imbalance in smokers and chronic obstructive pulmonary disease, *Thorax*, 51, 348, 1996.

Rahman, I., Skwarska, E., and MacNee, W., Attenuation of oxidant/antioxidant imbalance during treatment of exacerbations of chronic obstructive pulmonary disease, *Thorax*, 52, 565, 1997.

Rahman, I. et al., Systemic oxidative stress in asthma, COPD, and smokers, *Am. J. Respir. Crit. Care Med.*, 154, 1055, 1996a.

Rahman, I. et al., Induction of gamma-glutamylcysteine synthetase by cigarette smoke is associated with AP-1 in human alveolar epithelial cell, *FEBS Lett.*, 396, 21, 1996b.

Rahman, I. et al., 4-Hydroxy-2-nonenal, a specific lipid peroxidation product, is elevated in lungs of patients with chronic obstructive pulmonary disease, *Am. J. Respir. Crit. Care Med.*, 166, 490, 2002.

Reif, D. W., Ferritin as a source of iron for oxidative damage, *Free Radic. Biol. Med.*, 12, 417, 1992.

Renwick, L. C. et al., Increased inflammation and altered macrophage chemotactic responses caused by two ultrafine particle types, *Occup. Environ. Med.*, 61, 442, 2004.

Repine, J. E., Bast, A., and Lankhorst, I., Oxidative stress in chronic obstructive pulmonary disease. Oxidative Stress Study Group, *Am. J. Respir. Crit. Care Med.*, 156, 341, 1997.

Ribarov, S. R., Benov, L. C., and Bechev, I. C., The effect of lead on haemoglobin-catalyzed lipid peroxidation, *Biochem. Biophys. Acta.*, 664, 453, 1981.

Rodriguez, R. E. et al., Nickel-induced lipid peroxidation in the liver of different strains of mice and its relation to nickel effects on antioxidant systems, *Toxicol. Lett.*, 57, 269, 1991.

Rumsey, S. C. et al., Glucose transporter isoforms GLUT1 and GLUT3 transport dehydroascorbic acid, *J. Biol. Chem.*, 272, 1898, 1997.

Rustow, B. et al., Type II pneumocytes secrete vitamin E together with surfactant lipids, *Am. J. Physiol.*, 265, L133, 1993.

Sagai, M. et al., Biological effects of diesel exhaust particles. I. *In vitro* production of superoxide and *in vivo* toxicity in mouse, *Free Radic. Biol. Med.*, 14, 37, 1993.

Salvi, S. et al., Acute inflammatory responses in the airways and peripheral blood after short-term exposure to diesel exhaust in healthy human volunteers, *Am. J. Respir. Crit. Care Med.*, 159, 702, 1999.

Samet, J. M., Fine particulate air pollution and mortality in 20 U.S. cities, 1987–1994, *N. Engl. J. Med.*, 343, 1742, 2000.

Scheepers, P. T. and Bos, R. P., Combustion of diesel fuel from a toxicological perspective. II. Toxicity, *Int. Arch. Occup. Environ. Health*, 64, 163, 1992.

Sherman, C. B. et al., Firefighting acutely increases airway responsiveness, *Am. Rev. Respir. Dis.*, 140, 185, 1989.

Shi, X. and Dalal, N. S., Superoxide-independent reduction of vanadate by rat liver microsomes/NAD(P)H: vanadate reductase activity, *Arch. Biochem. Biophys.*, 295, 70, 1992.

Shi, X. and Dalal, N. S., Vanadate-mediated hydroxyl radical generation from superoxide radical in the presence of NADH: Haber–Weiss vs Fenton mechanism, *Arch. Biochem. Biophys.*, 307, 336, 1993.

Shvedova, A. A. et al., Unusual inflammatory and fibrogenic pulmonary responses to single- walled carbon nanotubes in mice, *Am. J. Physiol. Lung Cell. Mol. Physiol.*, 289 (5), 698, 2005.

Sies, H., Oxidative stress. *Oxidants and Antioxidants*, Academic Press, London, 1991, chap. II, 1.

Skoza, L., Snyder, A., and Kikkawa, Y., Ascorbic acid in bronchoalveolar wash, *Lung*, 161, 99, 1983.

Smith, L. J., Anderson, J., and Shamsuddin, M., Gluthathione localization and distribution after intratrachial installation. Implications for treatment, *Am. Rev. Respir. Dis.*, 145, 153, 1992.

Squadrito, G. L. et al., Quinoid redox cycling as a mechanism for sustained free radical generation by inhaled airborne particulate matter, *Free Radic. Biol. Med.*, 31, 1132, 2001.

Stenfors, N., Nordenhall, C., Salvi, S. S., Mudway, I., Soderberg, M., Blomberg, A., Helleday, R., Levin, J. O., Holgate, S. T., Kelly, F. J., Frew, A. J., Sandstrom, T., Different airway inflammatory responses in asthmatic and healthy humans exposed to diesel, *Eur. Respir. J.*, 23, 82, 2004.

Stinson, T. J. et al., The relationship between nickel chloride-induced peroxidation and DNA strand breakage in rat liver, *Toxicol. Appl. Pharmacol.*, 117, 98, 1992.

Stites, S. W. et al., Increased concentrations of iron and isoferritins in the lower respiratory tract of patients with stable cystic fibrosis, *Am. J. Respir. Crit. Care Med.*, 160, 796, 1999.

Stohs, S. J. and Bagchi, D., Oxidative mechanisms in the toxicity of metal ions, *Free Radic. Biol. Med.*, 18, 321, 1995.

Strange, R. C. and Fryer, A. A., The glutathione *S*-transferases: influence of polymorphism on susceptibility to nonfamilial cancers, In *Metabolic Polymorphisms and Cancer*, Boffetta, P., Caporaso, N., Cuzick, J., Lang, M., and Vineis, P., eds., IARC Scientific Publications, Lyon, 231, 1999.

Takebe, G. et al., A comparative study on the hydroperoxide and thiol specificity of the glutathione peroxidase family and selenoprotein P, *J. Biol. Chem.*, 277, 41254, 2002.

Tang, L. X., Yang, J. L., and Shen, X., Effects of additional iron-chelators on $Fe(2+)$-initiated lipid peroxidation: evidence to support the $Fe2+ \ldots Fe3+$ complex as the initiator, *J. Inorg. Biochem.*, 68, 265, 1997.

Thompson, A. B. et al., Lower respiratory tract lactoferrin and lysozyme arise primarily in the airways and are elevated in association with chronic bronchitis, *J. Lab. Clin. Med.*, 115, 148, 1990.

Toth, K. M. et al., Erythrocytes from cigarette smokers contain more glutathione and catalase and protect endothelial cells from hydrogen peroxide better than erythrocytes from nonsmokers, *Am. Rev. Respir. Dis.*, 134, 281, 1986.

Tran, C. L. et al., Inhalation of poorly soluble particles. II. Influence of particle surface area on inflammation and clearance, *Inhal. Toxicol.*, 12, 101, 2000.

Turi, J. L. et al., The iron cycle and oxidative stress in the lung, *Free Radic. Biol. Med.*, 36, 850, 2003.

Ulfvarson, U. et al., Effects of exposure to vehicle exhaust on health, *Scand. J. Work Environ. Health*, 13, 505, 1987.

Vallyathan, V. et al., Changes in bronchoalveolar lavage indices associated with radiographic classification in coal miners, *Am. J. Respir. Crit. Care Med.*, 162, 958, 2000.

van den Dobbelsteen, D. J. et al., Rapid and specific efflux of reduced glutathione during apoptosis induced by anti-Fas/APO-1 antibody, *J. Biol. Chem.*, 271, 15420, 1996.

van der Vliet, A. et al., Determination of low-molecular-mass antioxidant concentrations in human respiratory tract lining fluids, *Am. J. Physiol.*, 276, L289, 1999.

van Strien, R. T. et al., Microbial exposure of rural school children, as assessed by levels of N-acetyl-muramic acid in mattress dust, and its association with respiratory health, *J. Allergy Clin. Immunol.*, 113, 860, 2004.

van Vliet, P. et al., Motor vehicle exhaust and chronic respiratory symptoms in children living near freeways, *Environ. Res.*, 74, 122, 1997.

Verstraeten, S. V. et al., Myelin is the preferential target for Al-mediated oxidative damage, *Arch. Biochem. Biophys.*, 344, 289, 1997.

Vistal, J. J., Health effects of diesel exhaust particulate emissions, *Bull. NY Acid Med.*, 56, 914, 1980.

Wang, X. and Quinn, P. J., Vitamin E and its function in membranes, *Prog. Lipid Res.*, 38, 309, 1999.

Willis, R. J. and Kratzing, C. C., Ascorbate acid in rat lung, *Biochem. Biophys. Res. Commun.*, 59, 1250, 1974.

Wilson, M. R. et al., Interactions between ultrafine particles and transition metals *in vivo* and *in vitro*, *Toxicol. Appl. Pharmacol.*, 184, 172, 2002.

Winkler, B. S., Orselli, S. M., and Rex, T. S., The redox couple between glutathione and ascorbic acid: a chemical and physiological perspective, *Free Radic. Biol. Med.*, 17, 333, 1994.

Winterbourn, C. C., Comparative reactivities of various biological compounds with myeloperoxidase-hydrogen peroxide-chloride, and similarity of the oxidant to hypochlorite, *Biochim. Biophys. Acta.*, 840, 204, 1985.

Winterbourn, C. C., Toxicity of iron and hydrogen peroxide: the Fenton reaction, *Toxicol. Lett.*, 82, 969, 1995.

Witting, L. A., Vitamin E and lipid antioxidants in free radical-initiated reactions, In *Free Radicals in Biology*, Pryor, W. A., ed., Academic Press, New York, p. 259, 1980.

Yang, F. et al., Cellular expression of ceruloplasmin in baboon and mouse lung during development and inflammation, *Am. J. Respir. Cell Mol. Biol.*, 14, 161, 1996.

Ye, J. et al., Critical role of glass fiber length in TNF_ production and transcription factor activation in macrophages, *Am. J. Physiol.*, 276, L426, 1999.

Yiin, S. J. and Lin, T. H., Lead catalyzed peroxidation of essential unsaturated fatty acid, *Biol. Trace Elem. Res.*, 50, 167, 1995.

Zago, M. P. and Oteiza, P. I., The antioxidant properties of zinc: interactions with iron and antioxidants, *Free Rad. Biol. Med.*, 31, 266, 2001.

Zatta, P. et al., Aluminium (III) as a promoter of cellular oxidation, *Coord. Chem. Rev.*, 228, 271, 2002.

Zhang, Q. et al., Comparative toxicity of standard nickel and ultrafine nickel in lung after intratracheal instillation, *J. Occup. Health*, 45, 23, 2003.

Zhuang, Z., Huang, X., and Costa, M., Protein oxidation and amino acid-DNA crosslinking by nickel compounds in intact cultured cells, *Toxicol. Appl. Pharmacol.*, 126, 319, 1994.

Zielinski, H. et al., Modeling the interactions of particulates with epithelial lining fluid antioxidants, *Am. J. Physiol.*, 277, L719, 1999.

6 Particles and Cellular Oxidative and Nitrosative Stress

Dale W. Porter, Stephen S. Leonard, and Vincent Castranova
Health Effects Laboratory Division,
National Institute for Occupational Safety and Health

CONTENTS

6.1 INTRODUCTION

There are three types of reactive oxygen species (ROS): oxygen-containing free radicals, reactive anions containing oxygen atoms, or molecules containing oxygen atoms that can either produce free radicals or are chemically activated by them. Examples are hydroxyl radical ($^{\bullet}OH$), superoxide radical ($^{\bullet}O_2^-$), and hydrogen peroxide (H_2O_2). Similar to ROS, reactive nitrogen species (RNS) can be nitrogen-containing free radicals, reactive anions containing nitrogen atoms, or molecules containing nitrogen atoms that can either produce free radicals or are chemically activated by them. Examples of RNS include nitric oxide ($NO^{\bullet}$) and peroxynitrite ($ONOO^-$). Under normal conditions, an equilibrium exists between ROS and RNS generation, and antioxidant defenses. This equilibrium can be disturbed by a number of factors, many of which are organ, tissue, and/or cell specific. In the lung, inhaled particles can induce an inflammatory response, a component of which is an increase in ROS and RNS production. This increase in ROS/RNS generation can be the result of oxidants being generated from inhaled particles, or from lung phagocytes or epithelial cells, which have been stimulated to produce oxidants. In this review, we describe the sources and mechanisms of particle-induced oxidative stress.

6.2 SOURCES OF CELLULAR ROS

6.2.1 MITOCHONDRIA

One source of cellular ROS is the mitochondria. Oxidative phosphorylation is the process by which adenosine-5′-triphosphate (ATP) is formed as electrons are transferred from an electron donor (i.e., nicotinamide adenine dinucleotide (NADH) or flavin adenine dinucleotide (FADH), to the terminal electron acceptor, oxygen, by a series of electron carrying complexes located within the inner mitochondrial membrane. It has been estimated that 2–4% of the oxygen consumed by oxidative phosphorylation produces superoxide as a result of unpaired electrons "leaking" from the electron transport chain (Kirkinezos and Moraes 2001). The most likely sites of superoxide radical formation during oxidative phosphorylation are at complexes I and II of the electron transport chain, because these complexes can exist as semiquinones with unpaired electrons (Ohnishi 1998; Magnitsky et al. 2002; Muller, Crofts, and Kramer 2002). These unpaired electrons can be donated to molecular oxygen, forming superoxide radical.

6.2.2 NADPH OXIDASE

Another source of cellular ROS is the "respiratory burst," a term first used in 1933 to describe an increase in oxygen consumption when phagocytic cells were exposed to microorganisms (Balridge and Gerad 1933). Since this initial report, studies have determined that a multi-subunit enzyme complex, called nicotinamide adenine dinucleotide phosphate (NADPH) oxidase, is responsible for the respiratory burst (Patriarca et al. 1971; Suh et al. 1999; De Deken et al. 2000). Active NADPH oxidase is a membrane-bound, five sub-unit complex. At rest, three of these sub-units ($p40^{phox}$, $p47^{phox}$, and $p67^{phox}$) are complexed in the cytosol, while $p22^{phox}$ and $gp91^{phox}$ are membrane bound. Upon stimulation, all subunits are brought together into one macromolecular complex by mechanisms involving phosphoinositide, produced by activated PI3 kinase, and phosphorylation of $p47^{phox}$ by protein kinase C, and activation of mitogen-activated kinases (MAPKs), protein kinase A, and p21-activated kinases (PAK) result in membrane assembly of the active five sub-unit NADPH oxidase (Chen and Castranova 2004). This active NADPH oxidase produces superoxide radical, which in turn can generate other forms of ROS, such as hydrogen peroxide and hydroxyl radical.

6.3 NON-CELLULAR PARTICLE-MEDIATED ROS GENERATION

6.3.1 SILICA

As early as 1966, it was proposed that the toxicity of α-quartz (silica) was due to silanol groups (SiOH) on the surface of silica particles acting as hydrogen donors, forming hydrogen bonds with biological membranes, and disrupting their normal functioning (Nash, Allison, and Harington 1966). Later studies, which examined freshly fractured silica produced by milling or grinding silica, determined that the surface of freshly fractured silica had cleavage planes characterized by the presence of various siloxyl groups (e.g., $Si^{\cdot}$, $SiO^{\cdot}$, Si^{+}, and SiO^{-}) on its surface (Vallyathan et al. 1988; Fubini et al. 1990; Castranova, Dalal, and Vallyathan 1996; Fubini 1998). In an aqueous environment, silica can generate hydrogen peroxide, hydroxyl and superoxide radicals, and singlet oxygen ($^1O_2$) (Vallyathan et al. 1988; Konecny et al. 2001). In addition, there is a positive correlation between the amount of ROS generated and the distribution and quantity of silanol groups on the silica particle surface (Fubini et al. 2001). In cell-free systems, hydroxyl radical generated from silica can interact with membrane lipids, causing lipid peroxidation in proportion to the amount of ROS produced (Dalal, Shi, and Vallyathan 1990; Shi et al. 1994), and also can produce DNA strand breaks (Shi et al. 1994). Electron spin resonance (ESR) has been used to detect siloxyl radicals on the surface of silica particles (Figure 6.1, panels a and b) and also the generation of hydroxyl radical in aqueous medium (Figure 6.1, panels c and d). Furthermore, radical signals produced by freshly ground silica are larger compared to aged silica, which is consistent with freshly fractured silica being more toxic than aged silica (Vallyathan et al. 1988; Vallyathan et al. 1995; Castranova, Dalal, and Vallyathan 1996).

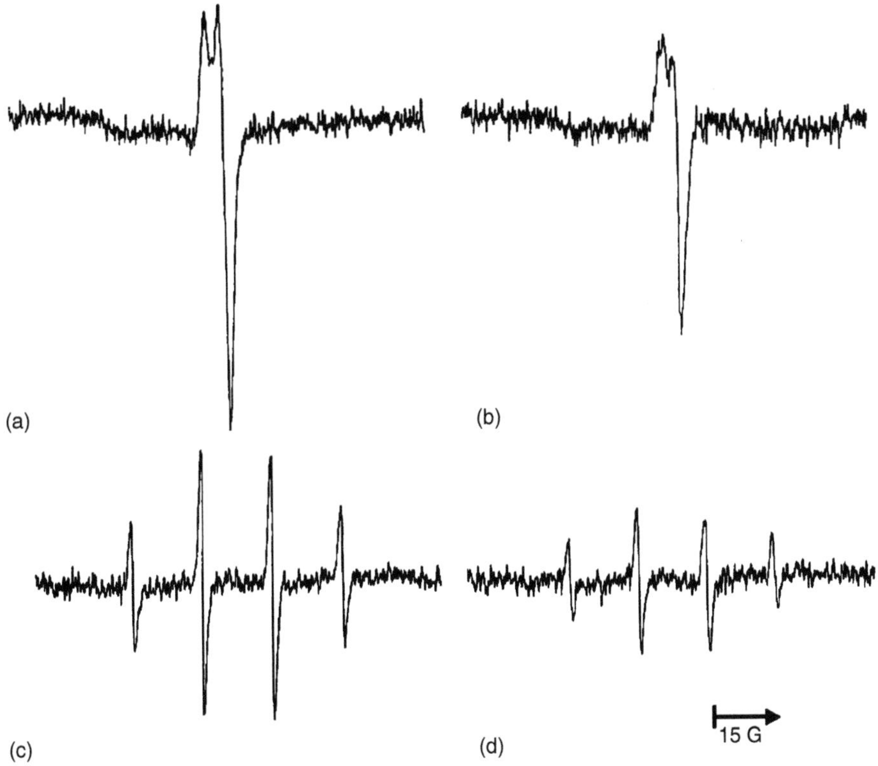

FIGURE 6.1 ESR spectra of freshly fractured and aged silica. Spectra (panel a and panel b) were recorded from 100 mg of dry silica placed in a quartz NMR tube and scanned using the following parameters: receiver gain, 5.02×10^4; time constant, 0.08 s; modulation amplitude, 1 G; scan time, 83 s; number of scans, 5; magnetic field, 3505 ± 50 G. Spectra (panel c and panel d) were recorded 3 min after reaction initiation from a pH 7.4 phosphate buffered saline containing 100 mM DMPO, 10 mM $H_2 0_2$, and the following reactants: (panel c) fresh silica (10 mg/mL); (panel d) aged silica (l0 mg/mL). The ESR spectrometer settings were: receiver gain, 6.32×10^4 time constant, 0.04 s; modulation amplitude, 1 G; scan time, 41 s; number of scans, 2; magnetic field, 3490 ± 100 G.

Agents that modify the surface of silica can alter its ability to generate ROS. For example, polyvinylpyridine-*N*-oxide (PVPNO), the organosilane Prosil 28, and aluminum lactate, all decrease non-cellular ROS generation from silica (Wallace et al. 1985; Vallyathan et al. 1991; Mao et al. 1995; Duffin et al. 2001; Knaapen et al. 2002). Iron contamination of silica also affects non-cellular ROS generation. The trace iron contamination of silica may not be soluble iron, but actually iron complexed into the crystal lattice (Donaldson et al. 2001; Fubini et al. 2001). *In vitro*, hydrogen peroxide and trace iron contamination can significantly increase hydroxyl radical production, and this can be inhibited by catalase, suggesting that a Fenton mechanism is responsible for the hydroxyl radical generation (Ghio et al. 1992; Shi et al. 1995). However, the presence of extractable iron is not absolutely required for hydroxyl radical generation, because iron chelation (Fubini et al. 2001) and iron-free or iron-depleted silica (Fenoglio et al. 2001) are still capable of generating $^{\cdot}OH$, albeit at lower levels.

6.3.2 Coal Dust

As determined by ESR, coal dust can produce carbon-centered radicals (Figure 6.2). ESR studies of coal dust samples, obtained from autopsied lymph nodes from asymptomatic miners and patients with Coal Workers' Pneumoconiosis (CWP), determined that coal dust obtained from CWP patients had higher amounts of stable carbon radicals, and the amount of these radicals was related to disease severity (Dalal et al. 1991). In addition, coal dust can generate hydroxyl radical and hydrogen peroxide (Dalal et al. 1995). Coal dust-mediated hydroxyl radical generation is inhibited by deferoxamine and catalase, and is partially inhibited by superoxide dismutase, indicating Fenton chemistry may be responsible for hydroxyl radical generation (Dalal et al. 1995). ESR studies conducted in our laboratory have determined that bituminous coal, which has a high iron contamination, produces more hydroxyl radicals, as measured by ESR, than lignite coal, which has a lower amount of iron in comparison to bituminous coal (data not shown). These determinations add further support to the role of iron in the generation of ROS from coal dust.

6.3.3 Asbestos

All forms of asbestos contain iron, either as a component of their crystalline structure, or as a surface impurity. For example, crocidolite and amosite contain high amounts of iron within their

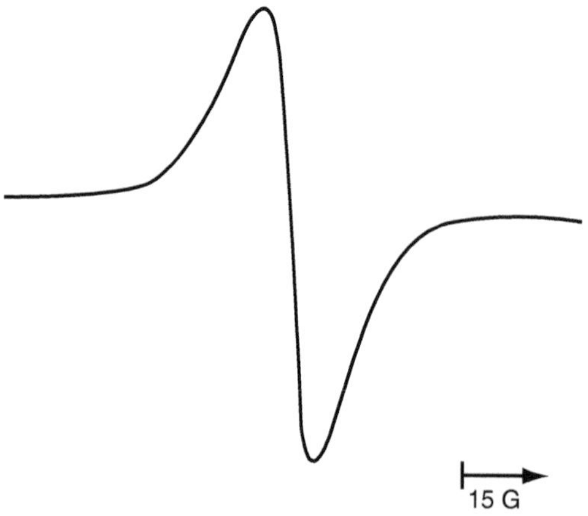

FIGURE 6.2 ESR spectrum of bituminous coal. ESR spectra were recorded from 40 mg of dry bituminous coal placed in a quartz NMR tube and scanned using the following parameters: receiver gain, 5.02×10^4; time constant, 0.08 s; modulation amplitude, 1 G; scan time, 83 s; magnetic field, 3505 ± 50 G.

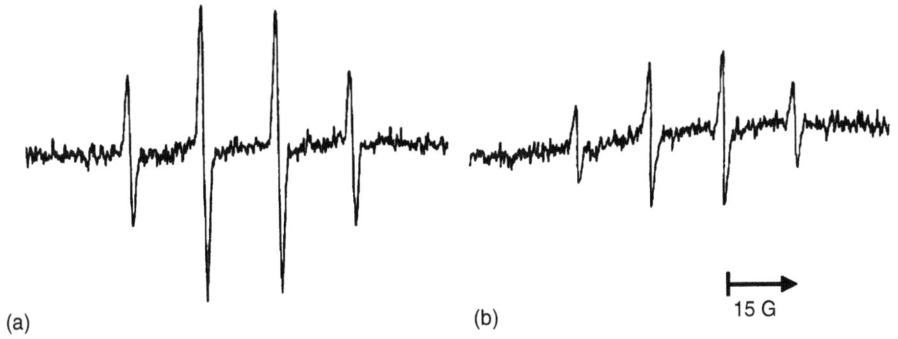

FIGURE 6.3 ESR spectra of crocidolite and chrysotile asbestos. ESR spectra were recorded 3 min after reaction initiation from a pH 7.4 phosphate buffered saline containing 100 mM DMPO, 10 mM H_2O_2, and the following reactants: (panel a) crocidolite (10 mg/mL); (panel b) chrysotile (10 mg/mL). The ESR spectrometer settings were: receiver gain, 6.32×10^4; time constant, 0.04 s; modulation amplitude, 1 G; scan time, 41 s; number of scans, 2; magnetic field, 3490 ± 100 G.

crystal lattice, whereas chrysotile contains trace iron as a contaminant (Harrington 1965; Timbrell 1970; Zussman 1978; Hodgson 1979; Pooley 1981; DeWaele and Adams 1988). The chemical properties of asbestos, especially their iron content, made it likely that they may cause the formation of hydroxyl radicals through iron-catalyzed reactions. This hypothesis was confirmed in a study which reported that chrysotile, amosite, and crocidolite asbestos all generate hydroxyl radical, detected by ESR spectroscopy, in the presence of hydrogen peroxide (Weitzman and Graceffa 1984). As seen in Figure 6.3, hydroxyl radical generation from both crocidolite (panel a) and chrysotile (panel b) is easily detected using ESR, with crodidolite producing a larger signal in comparison to chrysotile. This relates to the fact that Fe is part of the crocidolite crystal structure, whereas Fe is a contaminate of chrysotile. The pivotal role of iron was further established using the iron chelator deferoxamine. Deferoxamine inhibited asbestos-induced $^{\bullet}OH$ radical generation when it was added to the incubation mixture, or when the asbestos was pretreated with desferrioxamine, then washed to remove the extractable iron (Weitzman and Graceffa 1984). Lastly, fibers coated with a passivating material which resisted dissolution, making the iron inaccessible to react with oxygen, exhibited little ability to generate ROS. When the passivating material was removed by grinding or chemical reduction, the asbestos fibers were able to generate ROS (Pezerat et al. 1989).

6.3.4 Other Particles

Residual oil fly ash (ROFA) is a particulate pollutant produced by the combustion of fossil fuels, and is composed of soluble and insoluble metals. In one study (Antonini et al. 2004), ROFA (ROFA-total) was resuspedend in phosphate buffered saline (PBS) for 24 h, and then the particle-free supernatant (ROFA-sol) sample was separated from the insoluble component (ROFA-insol). Elemental analysis of the ROFA-total sample found it to contain greater amounts of Fe and other transition metals than ROFA-insol sample. ESR studies obtained a spectrum representative of hydroxyl radical when each of the samples was treated with H_2O_2 (Figure 6.4). The response was much stronger for the ROFA-total than the ROFA-insol sample, which correlated with the higher amounts of Fe and other transition metals in the ROFA-total sample compared to ROFA-insol sample. This association was further supported by the observation that deferoxamine significantly reduced hydroxyl radical signal from ROFA (Antonini et al. 2004).

Welding is another source of particulates that can generate ROS. Arc welding joins pieces of metal that have been made liquid by the heat produced as electricity passes from one conductor to another. The extremely high temperatures (>4,000°C) of this process heat both the base metal

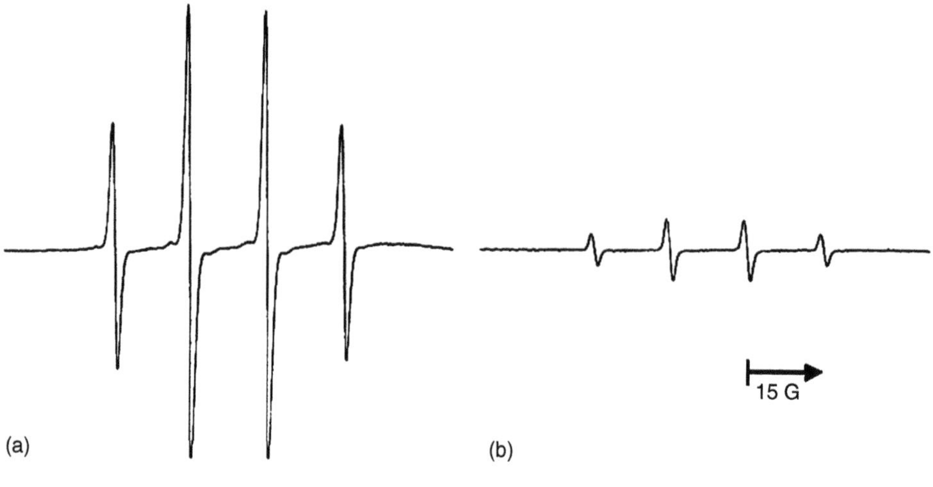

FIGURE 6.4 ESR spectra of ROFA-total and ROFA-insol. ESR spectra were recorded 3 min after reaction initiation from a pH 7.4 phosphate buffered saline containing 100 mM DMPO, 10 mM H_2O_2, and the following reactants: (panel a) ROFA-total (10 mg/mL); (panel b) ROFA-insol (10 mg/mL). The ESR spectrometer settings were: receiver gain, 6.32×10^4; time constant, 0.04 s; modulation amplitude, 1 G; scan time, 41 s; number of scans, 2; magnetic field, 3490 ± 100 G.

pieces to be joined and a consumable electrode fed into the weld. Fumes are formed by the evaporation of the metals, primarily at the tip of the electrode. The metal vapors are oxidized on contact with the air and form small particulates of different complexes of metal oxides. The fumes produced by welding can vary greatly. For example, welding fumes collected from manual metal arc (MMA) welding using a stainless steel (SS) electrode contains soluble metals, in particular chromium. In contrast, welding fume from gas metal arc (GMA) with a SS electrode, or GMA with a mild steel (MS) electrode produces low levels of soluble metals (Taylor et al. 2003). ESR was used to assess the ability of the fumes to produce free radicals in cell-free systems, and only MMA–SS fume produced a spectra characteristic of hydroxyl radical. Furthermore, when the total MMA–SS was compared with its insol fraction, the soluble metals in total MMA–SS were found to be most responsible for the production of hydroxyl radicals (Taylor et al. 2003). The ability of welding fumes to produce ROS decays with time after collection, and is highest in freshly generated welding fume, as measured by dichlorofluorescein fluorescence (Antonini et al. 1998).

Wood smoke, produced by the combustion of wood, has been identified as a source of particles that can generate free radicals (Leonard et al. 2000). Wood smoke particulate, collected on a filter, has been determined by ESR to have carbon-centered radicals based on the spectral line shape and position. These carbon-centered radicals are relatively stable, with a half-life of several days, depending on environmental conditions. In addition to carbon centered radicals, filters treated with H_2O_2 exhibited an ESR spectra indicative of hydroxyl radical generation. This generation of hydroxyl radicals was associated with the ability of wood smoke to cause DNA damage and induce lipid peroxidation, nuclear factor kappa B (NF-κB) activation, and tumor necrosis factor-alpha (TNF-α) production in macrophages (Leonard et al. 2000).

6.4 PARTICLE-MEDIATED CELLULAR ROS GENERATION

6.4.1 Silica

In vitro silica exposure has been shown to significantly increase alveolar macrophage (AM) intracellular superoxide radical and hydrogen peroxide levels in comparison to controls

(Zeidler et al. 2003). Silica exposure also has been shown to activate NADPH oxidase, resulting in increased oxygen consumption and extracellular secretion of superoxide and hydrogen peroxide from AMs (Castranova, Pailes, and Li 1990), polymorphonuclear leukocytes (PMNs) (Kang et al. 1991), and alveolar type II cells (Kanj, Kang, and Castranova 2005).

Extensive data exist regarding ROS production by lung pneumocytes after *in vivo* quartz exposure. Exposure of rats to silica results in potentiation of particle-stimulated ROS and RNS production in harvested AMs ex vivo. AMs isolated from silica-exposed animals have increased hydrogen peroxide production (Castranova 1994). Chemiluminescence, which is an indicator of ROS production, has also been shown to be increased in silica-exposed rats (Castranova et al. 1985; Porter et al. 2002a), and exposure to freshly fractured silica stimulates AM chemiluminescence to an even greater extent than aged silica (Castranova et al. 1996; Porter et al. 2002a).

The impact of trace iron contamination on toxicity *in vivo* is unclear. In one study, silica with surface associated iron caused greater pulmonary inflammation in comparison to iron-free silica (Ghio et al. 1992). However, another study which compared the amount of iron, ROS generation, and toxicity between different silica samples, found that silica-induced toxicity and iron contamination were not correlated (Donaldson et al. 2001).

Similar to the animal study results, human pneumocytes isolated from silica-exposed subjects exhibit increased ROS production. Specifically, human AMs, obtained from patients with silicosis, a disease caused by inhalation of silica, have increased production of superoxide (Rom et al. 1987; Wallaert et al. 1990), hydrogen peroxide (Rom et al. 1987), and AM chemiluminescence (Goodman et al. 1992; Castranova et al. 1998), in comparison to healthy controls.

6.4.2 Coal Dust

AMs obtained from rats 24 h after intratracheal (IT) instillation of coal dust have significantly increased ROS generation, as measured by zymosan-stimulated chemiluminescence (Blackford et al. 1997). In another study, rats exposed by inhalation to 2 mg/m^3 coal dust (6 h/day) for two years also had significantly increased chemiluminescence (Castranova et al. 1985). AMs obtained from human patients with CWP have increased AM chemiluminescence (Goodman et al. 1992; Castranova et al. 1998) and superoxide production (Rom et al. 1987; Wallaert et al. 1990) in comparison to healthy controls.

6.4.3 Asbestos

Oxidant release, specifically hydrogen peroxide and superoxide, have been determined to occur after *in vitro* exposure of alveolar and peritoneal macrophages to asbestos (Donaldson et al. 1985; Hansen and Mossman 1987; Petruska et al. 1990). Comparison of the ability long and short crocidolite fibers to stimulate the release of hydrogen peroxide and induce cytotoxicity found no differences (Goodglick and Kane 1990). However, with respect to superoxide production, fibrous asbestos (length:diameter ratio greater than 3:1) caused a significant increase in superoxide release from rat AMs in comparison to non-fibrous dusts, suggesting the geometry of the particles does effect superoxide generation (Hansen and Mossman 1987). In an earlier study (Goodglick and Kane 1986), mouse peritoneal macrophages exposed to crocidolite asbestos *in vitro* were found to release ROS and experience increased cytotoxicity. This crocidolite-induced cytotoxicity was prevented by incubation in a hypoxic environment, by addition of superoxide dismutases (SOD) and catalase, or if the crocidolite fibers were pretreated with deferoxamine, suggesting that oxygen and asbestos-associated iron play a role in asbestos-induced ROS and cytotoxicity.

In vivo exposure to asbestos has been shown to enhance the capacity of lung inflammatory cells to release oxidants. Bronchoalveolar lavage (BAL) cells obtained from sheep exposed to chrysotile asbestos did not have an increased basal level of superoxide production, but did release significantly higher amounts when stimulated with phorbol myristate acetate, in compassion to BAL cells obtained from controls (Cantin, Dubois, and Begin 1988). Chemiluminescence, measured from peritoneal macrophages obtained from asbestos-exposed and control mice, demonstrated that chemiluminescence was higher for asbestos-exposed mice (Donaldson and Cullen 1984). ROS production from human AMs, obtained from patients with asbestosis, a disease linked to asbestos exposure, and healthy controls, has also been studied. AMs obtained from asbestosis patients had increased release of superoxide and hydrogen peroxide, in comparison to healthy controls (Rom et al. 1987). Thus, the *in vitro* data and animal studies are consistent with the results obtained from humans, suggesting a role of asbestos-induced ROS in disease initiation and progression.

6.4.4 Other Particles

In vitro exposure of RAW 264.7 macrophages to lead chromate ($PbCrO_4$) particles has been reported to cause a respiratory burst and increase hydrogen peroxide production by 7-fold. This ROS production has been associated with activation of NF-κB and activator protein-1 (AP-1) (Leonard et al. 2004).

In vivo exposure of rats to a variety of environmentally or occupationally relevant particles has been reported to potentiate the production of ROS by AM, as measured by stimulant-induced chemiluminescence (Table 6.1). In general, the potency of a particle to stimulate ROS production has been associated with its inflammatory potential. For example, MMA/SS electrode welding fume has been shown to generate more ROS than GMA/MS or SS welding fumes and cause greater lung damage (BAL fluid LDH and albumin) and oxidant injury (lung lipid peroxidation), respectively (Taylor et al. 2003). In addition, oxidant stress was reported in welders as increased serum isoprostane and antioxidant levels, with the degree of oxidant stress being associated with years of welding in a shipyard (Han et al. 2005).

TABLE 6.1
Stimulation of ROS Production by Alveolar Macrophages Harvested from Particle Exposed Rats

Particle	Exposure	Chemiluminescence (Increase From Control)	Reference
Diesel exhaust	IT (5 mg/kg BW); 3 days post	2.3-Fold zymosan-stimulated	Yang et al. (2001)
Carbon black	IT (5 mg/kg BW); 3 days post	2.3-Fold zymosan-stimulated	Yang et al. (2001)
Titanium dioxide	IT (5 mg/100 g BW); 1 day post	3.0-Fold zymosan-stimulated	Blackford et al. (1997)
Carbonyl iron	IT (5 mg/100 g BW); 1 days post	1.6-Fold zymosan-stimulated	Blackford et al. (1997)
Residual oil fly ash (ROFA)	IT (1 mg/100 g BW); 1 days post	9–11-Fold PMA-stimulated	Antonini et al. (2004), Lewis et al. (2003)
Residual oil fly ash (ROFA)	IT (2 mg/rat); 1 days post	3.0-Fold zymosan-stimulated	Nurkiewicz et al. (2004)
Welding fume (manual metal arc/stainless sleet electrode)	IT (5 mg/rat BW); 3 days post	2.5-Fold zymosan-stimulated	Antonini et al. (2004)

6.5 CELLULAR RNS GENERATION

NO synthase catalyzes the formation of $NO^{\cdot}$ using L-arginine as a substrate. Three isoforms of NO synthase, two of which are constitutively expressed and one which is inducible, have been described. The inducible isoform of nitric oxide synthase (NOS), commonly referred to as inducible nitric oxide synthase (iNOS or NOS2), is the isoform important with respect to particle-induced toxicity because its expression can be induced in various pneumocytes by particle exposure.

$NO^{\cdot}$ is a free radical, but despite this, it is not particularly toxic (Beckman and Koppenol 1996). However, the conditions that stimulate pneumocyte $NO^{\cdot}$ production from iNOS (i.e., particle exposure), also stimulate ROS production by many of these same cells. One of these forms of ROS, superoxide, can react in a rapid isostoichiometric reaction with $NO^{\cdot}$, forming peroxynitrite in a near-diffusion-limited reaction (Beckman and Koppenol 1996). Peroxynitrite is a potent oxidant, reacting with and disrupting the normal functions of proteins via nitrosation of tyrosine residues (Beckman 1996; Beckman and Koppenol 1996; van der Vliet and Cross 2000), and has also been associated with enhanced lipid peroxidation and DNA damage (Rubbo et al. 1994; Eiserich, Patel, and O'Donnell 1998; Hofseth et al. 2003).

6.6 PARTICLE-MEDIATED CELLULAR RNS GENERATION

6.6.1 Silica

The mouse macrophage cell line, IC-21, when exposed to silica *in vitro*, has a 12-fold increase in $NO^{\cdot}$ production at 4 h post-exposure (Srivastava et al. 2002). In contrast, rat primary AMs exposed to silica *in vitro* do not exhibit increased production of $NO^{\cdot}$ (Huffman, Judy, and Castranova 1998; Kanj, Kang, and Castranova 2005). However, naïve primary rat AM, when cultured in media previously conditioned by BAL cells obtained from silica-exposed rats, do produce $NO^{\cdot}$ in response to *in vitro* silica exposure, suggesting that extracellular mediators are critical to the induction of iNOS (Huffman, Judy, and Castranova 1998). Neither primary alveolar type II cells, nor the rat type II cell line RLE-6TN, releases $NO^{\cdot}$ after *in vitro* exposure to silica (Kanj, Kang, and Castranova 2005).

Many studies have been conducted that report that *in vivo* silica exposure results in increased $NO^{\cdot}$ production from various lung cells. Silica administered by IT instillation to rats results in 3-fold increase in mRNA for iNOS and a 5-fold increase in $NO^{\cdot}$ production from BAL cells 24 h after exposure (Blackford et al. 1994; Huffman, Judy, and Castranova 1998). A silica time course inhalation study reported that the BAL fluid level of NO products, nitrite and nitrate (NO_x), was elevated 1.8-fold after 10 days of exposure, while NO-dependent chemiluminescence was elevated 15-fold at this exposure time, and these levels remained relatively constant throughout the first 41 days of exposure (Porter et al. 2002b). Continued exposure after 41 days inhalation resulted in a rapid rise in $NO^{\cdot}$ production (i.e., BAL fluid NO_x levels increased 22-fold and NO-dependent chemiluminescence 151-fold) after 116 days of silica exposure (Porter et al. 2002b). Immunohistochemical evidence of iNOS induction in AMs and alveolar type II epithelial cells suggested these cells were the source of the $NO^{\cdot}$ production (Porter et al. 2002b). There was a temporal and spatial relationship between induction of $NO^{\cdot}$ production and pulmonary inflammation, in this study. Consistent with these observations was the determination that iNOS expression was induced in AMs in response to silica inhalation and that silica- induced pathology was significantly decreased in iNOS knockout mice (Srivastava et al. 2002).

Increased $NO^{\cdot}$ production has also been reported in humans with silica-induced lung disease. Specifically, iNOS mRNA levels and $NO^{\cdot}$ production from BAL cells were determined from a silica-exposed coal miner with an abnormal chest x-ray, a silica-exposed coal miner with a normal chest x-ray, and an unexposed control. iNOS mRNA from BAL cells isolated from the two coal miners demonstrated that both were higher than the unexposed control, and that the miner with the abnormal chest x-ray had more iNOS mRNA than that from the miner with a normal chest x-ray

(Castranova et al. 1998). AM NO˙ production was measured by NO-dependent chemiluminescence, and in comparison to the unexposed control, the coal miners with normal and abnormal chest x-rays had 15- and 31-fold higher NO-dependent chemiluminescence, respectively (Castranova et al. 1998).

6.6.2 Coal Dust

Rat BAL cells, obtained 24 h after IT instillation of coal dust, express iNOS and have increased NO˙ production, measured as NO-dependent chemiluminescence, in comparison to saline-exposed controls (Blackford et al. 1997).

6.6.3 Asbestos

Exposure of rat AMs and the mouse peritoneal monocyte-macrophage cell line, RAW 264.7, to crocidolite induces activation of the iNOS promoter gene and transcription of iNOS mRNA (Quinlan et al. 1998). The mouse AM cell line MH-S, when exposed to crocidolite, exhibited a 4-fold increase in NOS activity and NO˙ production at 24 h post-exposure (Aldieri et al. 2001). Increased mRNA levels for iNOS and NO˙ production have also been reported for A549 cells, a human alveolar type II cell line, in response to asbestos (Chao, Park, and Aust 1996). *In vitro* exposure of rat AMs to asbestos fibers results in a significant increase in NO˙ production 48 h after exposure, with chrysotile being a more potent stimulant than crocidolite on an equal mass basis (Thomas et al. 1994).

IT instillation of rats with asbestos has been shown to increase NOS activity of lung tissue 48 h post-IT exposure (Iguchi, Kojo, and Ikeda 1996). In mice, IT instillation of crocidolite causes induction of iNOS mRNA in lung tissue and increased immunohistochemical staining for iNOS protein and nitrotyrosine residues in bronchial epithelial cells, alveolar epithelial cells, and AMs (Dorger et al. 2002a). In rats, 24 h after IT instillation of crocidolite, increased iNOS mRNA and protein have been observed in lung tissue, as well as positive staining for iNOS and nitrotyrosine in AMs and alveolar epithelial cells (Dorger et al. 2002b). Inhalation exposure of rats to crocidolite or chrysotile asbestos results in a more than a 2-fold increase in NO˙ production from AMs, and was temporally correlated with pulmonary inflammation (Quinlan et al. 1998).

6.6.4 Other Particles

Exposure of rats by IT instillation to fine titanium dioxide or carbonyl iron significantly increased NO-dependent chemiluminescence from harvested AMs 24 h post-exposure (Blackford et al. 1997). However, as with inflammatory potency, these nuisance dusts were significantly less stimulatory than silica or coal dust. Intratracheal instillation of rats with diesel exhaust particles or ultrafine carbon black also caused a small (2-fold increase) but significant increase in NO production by AMs (Yang et al. 2001). Exposure of rats to MMA/SS electrode welding fumes resulted in a 3-fold increase in nitrate/nitrite levels in bronchoalveolar lavage fluid three days after IT instillation (Antonini et al. 2004). Additionally, iNOS protein was found by immunohistochemical staining of the lung to be associated anatomically with areas of welding fume-induced inflammation. A recent study also reported that NO-dependent chemiluminescence from AMs was elevated 6.8-fold 24 h after IT instillation of ROFA (Nurkiewicz et al. 2004).

6.7 PARTICLE-INDUCED ACTIVATION OF NUCLEAR FACTOR-κB

6.7.1 ROS and RNS Regulation of Nuclear Factor-κB

NF-κB is a transcription factor found in many different cell types, and functions in the molecular signaling between the cytoplasm and nucleus. In resting cells, NF-κB is retained in the cytoplasm in an inactive form by binding to inhibitory proteins IκBα and IκBβ. Upon activation by extracellular stimuli, these inhibitory proteins dissociate from the complex, and NF-κB translocates to the nucleus in an activated form. Once in the nucleus, activated NF-κB binds to specific binding sequences found in the promoter region of many genes, contributing to the regulation of these genes. Specifically, with respect to particles and cellular oxidative stress, NF-κB regulates a variety of genes involved in inflammatory or acute phase responses, including several proinflammatory or fibrogenic cytokines (i.e., IL-1, IL-2, IL-6, and TNF-α).

It is well established by numerous studies, reviewed elsewhere, that cellular oxidative stress induced by ROS can cause activation of NF-κB (Kabe et al. 2005). In contrast, the effect of $NO^{\cdot}$ on NF-κB activation is controversial. Studies using the RAW 264.7 cells have indicated that exogenous $NO^{\cdot}$ inhibits NF-κB activation (Chen et al. 1995), whereas a later study demonstrated that exogenous $NO^{\cdot}$ can stimulate NF-κB activation (Kang et al. 2000a). The contradictory results of these studies may reflect differences in $NO^{\cdot}$ concentrations, the duration of $NO^{\cdot}$ exposure, and/or the basal activity of macrophages (Kang et al. 2000a), since these variables have been shown to alter the effect of $NO^{\cdot}$ on NF-κB activation in other studies (Umansky et al. 1998; Diaz-Cazorla, Perez-Sala, and Lamas 1999).

6.7.2 Particle-Induced Activation of NF-κB

Particle-induced activation of NF-κB has been demonstrated in *in vitro* and *in vivo* studies. Exposure of RAW 264.7 cells to silica *in vitro* results in activation of NF-κB (Chen et al. 1995). Furthermore, NF-κB activation is stimulated in rat primary AM exposed to silica *in vitro*, and ROS scavengers and iNOS inhibitors reduced this NF-κB activation (Kang et al. 2000a; Kang, Lee, and Castranova 2000b). BAL cells isolated from rats after IT instillation of silica demonstrated NF-κB activation, which was decreased by pre-treatment with the anti-inflammatory agent dexamethasone (Sacks et al. 1998). In a silica inhalation study, progressive increase in NF-κB activation during a 116-day time course study was demonstrated (Porter et al. 2002c). Crocidolite fibers have also been shown to activate NF-κB in hamster tracheal epithelia cells (Janssen et al. 1995a). Fiberglass has been reported to activate NF-κB in cultured AMs (Ye et al. 1999). This NF-κB activation, and the resultant TNF-α production, were shown to depend on fiber length and be inhibited by antioxidant treatment. Exposure of macrophages to lead chromate particles has also been reported to activate NF-κB in a ROS-dependent manner (Leonard et al. 2004).

6.7.3 Particle-Induced Activation of AP-1

AP-1 is a transcription factor composed of homodimers and/or heterodimers of Jun (*c*-Jun, Jun B, and Jun D) and Fos (*c*-Fos, Fos B, Fra-1, Fra-2, and FosB2) gene families (Angel and Karin 1991). It interacts with DNA sequences known as TPA response elements, or AP-1 sites that govern inflammation, proliferation, and apoptosis. AP-1 activation is controlled by MAPK (Bernstein et al. 1994). ROS have been shown to act as MAPK activators.

Crystalline silica has been reported to induce the phosphorylation of MAPK, specifically p38 and extracellular signal regulated protein kinases (ERK1 and ERK 2), in an epidermal cell line (Ding et al. 1999). This MAPK activation was associated with activation of AP-1 in cultured cells as well as in AP-1 luciferase reporter transgenic mice. This silica-induced AP-1 activation was mediated through ROS (Ding et al. 2001). Asbestos has been shown to elevate expression of *c*-Fos and *c*-Jun in mesothelial cells (Janssen, Heintz, and Mossman 1995b). Asbestos activated ERKs in cell culture systems via an ROS-dependent mechanism (Ding et al. 1999; Buder-Hoffmann et al. 2001). This asbestos-induced MAPK activation resulted in activation of AP-1 both *in vitro* and in lung and bronchiolar tissue from AP-1 luciferase transgenic mice (Ding et al. 1999). ERK and p38 MAPK and AP-1 activation have also been demonstrated in macrophages in response to *in vitro* treatment with glass fibers (Ye et al. 2001). Likewise, lead chromate particles have been shown to activate AP-1 in macrophages via an ROS-dependent mechanisms (Leonard et al. 2004).

6.8 PARTICLE-INDUCED APOPTOSIS

The molecular mechanisms regulating apoptosis have been reviewed elsewhere (Granville et al. 1998; Green and Reed 1998), but it is clear that ROS generation and/or alteration in cellular redox state may stimulate apoptosis. Since many particulates induce ROS and RNS, and thus disturb the cellular redox status, they might be expected to induce apoptosis.

In the normal rat lung, apoptosis is very low, but asbestos exposure causes a significant increase in apoptosis in bronchiolar and alveolar epithelial cells (Mossman and Churg 1998). ROS has been suggested to play a role in asbestos-induced apoptosis, because asbestos-induced apoptosis can be ameliorated by exogenous antioxidants, catalase, superoxide dismutase, and deferoxamine (BéruBé et al. 1996; Broaddus et al. 1996). Asbestos-induced apoptosis is observed in human cell systems, as well as in rats; indeed chrysotile and crocidolite have been reported to cause apoptosis in human AMs 48 h after exposure (Hamilton, Iyer, and Holian 1996).

Apoptosis has been observed in BAL cells isolated from rats 10 days after IT instillation of silica (Leigh et al. 1997). At two months post-IT exposure, apoptotic cells were present in granulomatous lesions (Leigh et al. 1997). A silica inhalation time course study, conducted using rats, also reported that apoptotic cells were located in the airspaces and increased significantly in number as silica lung burden increased during a 116-day exposure (Porter et al. 2002b). *In vitro* exposure of human AMs to crystalline silica for 6 h or 24 h induces apoptosis, whereas exposures to amorphous silica or titanium dioxide did not; these observations suggest the human AM apoptotic response appears to be specific to crystalline silica (Iyer et al. 1996). Incubation of human AMs with a caspase inhibitor (Z-VAD-FMK) prevented silica-induced apoptosis, indicating silica induces apoptosis via activation of the caspase system (Iyer et al. 1996).

Generation of ROS appears to play a role in silica-induced apoptosis. The role of ROS was established in an *in vitro* study using silica exposed rat AMs. The determination that ROS generation preceded caspase activation and subsequent apoptosis suggested ROS generation was an initiating step (Shen et al. 2001). The antioxidant ebselen prevented silica-induced apoptosis in this *in vitro* system, thus adding support to the role of ROS in initiating apoptosis.

Another mechanism that may participate in silica-induced apoptosis is RNS generation. The mouse macrophage cell line IC21, when exposed to crystalline silica *in vitro*, exhibited increased apoptosis (Srivastava et al. 2002). However, silica-induced apoptosis can be prevented by the NOS inhibitor N(G)-nitro-L-arginine-methyl ester, suggesting that $NO^{\cdot}$ also contributes to the initiation of apoptosis in silica-exposed AMs (Srivastava et al. 2002). To confirm this observation *in vivo*, wild-type and iNOS knockout mice were exposed to either air (controls) or silica by inhalation exposure (Srivastava et al. 2002). Silica-exposed wild-type mice had much more apoptosis in comparison air-exposed wild-type mice, indicating silica did induce apoptosis. Comparison of the silica-exposed wild-type mice and iNOS knockout mice demonstrated that iNOS knockout mice had much less apoptosis. These results indicate that silica induces apoptosis, and $NO^{\cdot}$ participated in initiating apoptosis induced by silica exposure.

The role of apoptosis in the pathogenesis of silicosis is not without controversy. Since apoptosis, unlike necrosis, is cell death without the release of pro-inflammatory stimuli and apoptotic bodies are rapidly phagocytized, it is considered to play a role in the resolution of inflammation (Savill et al. 1993). However, free apoptotic bodies have been observed in rat lungs after a threshold burden of silica was achieved. Evidence indicated that such free apoptotic bodies induce production of TNF-α and TGF-β, causing inflammation, and lead to pulmonary fibrosis (Wang et al. 2003).

6.9 ANTIOXIDANT DEFENSES AND PARTICULATE EXPOSURE

6.9.1 Antioxidant Defenses

The respiratory system has numerous nonenzymatic and enzymatic antioxidant defense systems which are present both in intracellular and extracellular compartments. These systems will be discussed here briefly, but have been extensively reviewed elsewhere (Heffner and Kensler 1989). Enzymatic defense mechanisms include SOD, catalase, glutathione peroxidase, and heme oxygenase. SOD catalyzes the reaction of superoxide to hydrogen peroxide. There are three types of SODs in the lung: extracellular superoxide dismutases (ECSOD), Mn-SOD, and Cu,Zn-SOD. These enzymes are located in the epithelial lining fluid and interstitial space in the mitochondria, and in the cytosol and peroxisomes, respectively. Catalase and glutathione peroxidase participate in scavenging hydrogen peroxide. Catalase detoxifies hydrogen peroxide by converting it to oxygen and water, and is located primarily in peroxisomes. Glutathione peroxidase is part of the glutathione oxidation–reduction system, which also includes glutathione reductase and glucose-6-phosphate dehydrogenase. The enzyme heme oxygenase is also believed to be an important cell-associated antioxidant in the lung and other tissues. The primary function of this enzyme is the degradation of heme-containing proteins, which is thought to prevent iron-mediated hydroxyl radical production. The epithelial lining and interstitial fluid, representing the extracellular compartment of the lung, contains many low-molecular-weight antioxidants. These include α-tocopherol (vitamin E), uric acid, glutathione, and ascorbic acid, all of which function by scavenging free radicals.

6.9.2 Particulate Exposure Induces Antioxidant Defenses

As previously discussed in this chapter, particulate exposure results in increased ROS and RNS production, and this in turn contributes to particulate-induced oxidative stress. Antioxidant defenses can respond to particulate-induced cellular oxidative stress in an attempt to limit the oxidative damage.

The effect of *in vitro* particulate exposure on glutathione levels is unclear. In one study, *in vitro* exposure of rat AMs to silica or crocidolite resulted in an increase in glutathione production and release (Boehme, Maples, and Henderson 1992), whereas asbestos exposure of rat pleural mesothelial cells caused a depletion of total cellular glutathione (Janssen, Heintz, and Mossman 1995b).

Effects of particulate exposure on enzymatic antioxidant defenses have been more consistent. Crocidolite- and chrysotile-exposed hamster tracheal epithelial cells have increased SOD activity (Mossman, Marsh, and Shatos 1986). This up-regulation of SOD appears to be a specific response, because non-pathogenic glass fibers did not cause increased SOD activity in tracheal epithelial cells (Shatos et al. 1987). Small increases in heme oxygenase and Mn-SOD activities have been reported for crocidolite- and chrysotile-exposed human mesothelial cells and human adult lung fibroblasts (Janssen et al. 1994). This response also appears to be particle-specific, because expression of MnSOD or heme oxygenase did not occur after exposure to polystyrene beads or riebeckite, a nonfibrous analog of crocidolite (Janssen et al. 1994).

In vivo studies in rodents have suggested that SOD is produced in proportion to the oxidant stress present in the lung, and this has been suggested to be a defensive response (Vallyathan et al. 1995). This hypothesis was supported by data obtained from a rat silica inhalation time course study. In this study, BAL fluid SOD increased steadily throughout the 116 days of exposure, and paralleled increases in lung lipid peroxidation levels, a marker of oxidative stress (Porter et al. 2002b). Asbestos exposure has also been shown to alter enzymatic antioxidant levels. Specifically, rats exposed to asbestos have increased mRNA and enzymatic activity for Mn-SOD, Cu,Zn-SOD, and glutathione peroxidase (Janssen et al. 1992).

Changes in antioxidant defenses in response to particle-exposure have also been determined to occur in humans. PMA-induced ROS production and also increased SOD and Mn-SOD levels in AMs obtained from subjects with CWP (Voisin et al. 1985). Coal miners with early stage CWP have decreased total glutathione levels in erythrocytes, but levels are increased in those with more severe fibrosis, suggesting an up-regulation of glutathione-dependent enzymes in late stage CWP (Engelen et al. 1990; Perrin-Nadif et al. 1998). A longitudinal study, which evaluated disease progression and antioxidant status, reported that SOD activity increased with progression of CWP (Schins, Keman, and Borm 1997). In addition, this study found that individuals with increased levels of nonenzymatic antioxidants had less risk, while individuals with elevated enzymatic antioxidants were at higher risk of developing CWP (Schins, Keman, and Borm 1997). Another study (Vallyathan et al. 2000) reported that bronchoalveolar lavage levels of catalase, superoxide dismutase, and glutathione peroxidase were unchanged in asymptomatic coal miners, but increased in miners with simple CWP. This induction in antioxidant levels was associated with oxidative stress (lipid peroxidation) in these symptomatic miners. Lastly, elevated serum antioxidant levels have been associated with oxidant stress (i.e., increased isoprostanes), in shipyard welders (Han et al. 2005).

6.10 SUMMARY

Oxidant stress has been suggested as a causative factor in the initiation and progression of pneumoconiosis (Castranova 2000). Under normal physiological conditions, an equilibrium exists between ROS and RNS generation and antioxidant defenses. However, the inhalation of particles can induce oxidant stress in the lung, disrupting the equilibrium existing between ROS and RNS generation and antioxidant defenses. Specifically, at high particle burdens, the oxidant–antioxidant balance shifts to an excess of oxidant production, causing oxidant injury and initiating the disease process.

As reviewed in this chapter, particle exposure can induce oxidant stress by two distinct mechanisms: (1) noncellular particle-mediated ROS generation, and (2) particle-mediated cellular ROS and RNS generation. Non-cellular ROS generation results from, or is enhanced by, Fenton-like reaction(s) involving iron or other transition metals. Particle-mediated cellular ROS and RNS generation results from the stimulation of lung cells, in particular AMs, PMNs, and type II cells. These cells produce ROS via NADPH oxidase, as well as from the mitochondrial electron transport chain. In addition, particle-exposure can induce the expression of the enzyme iNOS, which results in these cells producing nitric oxide, a form of RNS. Both ROS and RNS can induce radical-specific cell damage, such as lipid peroxidation and nitration of tyrosine residues in proteins. Additionally, oxidant stress activates transcription factors (i.e., NF-κB and AP-1), which contribute to the transcriptional regulation of genes for inflammatory chemokines and cytokines, as well as growth and fibrogenic factors. Oxidant stress also can induce apoptosis, which at high dust burdens may reach levels which exceed the ability of AMs to clear apoptotic bodies. Free apoptotic bodies may further enhance particle-induced inflammatory and fibrotic processes.

REFERENCES

Aldieri, E., Ghigo, D., Tomatis, M., Prandi, L., Fenoglio, I., Costamagna, C., Pescarmona, G., Bosia, A., and Fubini, B., Iron inhibits the nitric oxide synthesis elicited by asbestos in murine macrophages, *Free Radic. Biol. Med.*, 31, 412–417, 2001.

Angel, P. and Karin, M., The role of Jun Fos and the AP-1 complex in cell-proliferation and transformation, *Biochim. Biophys. Acta*, 1072, 129–157, 1991.

Antonini, J. M., Clarke, R. W., Krishna Murthy, G. G., Sreekanthan, P., Jenkins, N., Eagar, T. W., and Brain, J. D., Freshly generated stainless steel welding fume induces greater lung inflammation in rats as compared to aged fume, *Toxicol. Lett.*, 98, 77–86, 1998.

Antonini, J. M., Taylor, M. D., Leonard, S. S., Lawryk, N. J., Shi, X., Clarke, R. W., and Roberts, J. R., Metal composition and solubility determine lung toxicity induced by residual oil fly ash collected from different sites within a power plant, *Mol. Cell Biochem.*, 255, 257–265, 2004.

Balridge, C. W. and Gerad, R. W., The extra respiration of phagocytosis, *Am. J. Physiol.*, 103, 235–236, 1933.

Beckman, J. S., Oxidative damage and tyrosine nitration from peroxynitrite, *Chem. Res. Toxicol.*, 9, 836–844, 1996.

Beckman, J. S. and Koppenol, W. H., Nitric oxide, superoxide, and peroxynitrite: the good, the bad, and ugly, *Am. J. Physiol.*, 271, C1424–C1437, 1996.

Bernstein, L. R., Ferris, D. K., Colburn, N. H., and Sobel, M. E., A family of mitogen-activated protein kinase-related proteins interacts *in vivo* with activator protein-1 transcription factor, *J. Biol. Chem.*, 269, 9401–9404, 1994.

BéruBé, K. A., Quinlan, T. R., Fung, H., Magae, J., Vacek, P., Taatjes, D. J., and Mossman, B. T., Apoptosis is observed in mesothelial cells after exposure to crocidolite asbestos, *Am. J. Respir. Cell Mol. Biol.*, 15, 141–147, 1996.

Blackford, J. A., Jr., Antonini, J. M., Castranova, V., and Dey, R. D., Intratracheal instillation of silica up-regulates inducible nitric oxide synthase gene expression and increases nitric oxide production in alveolar macrophages and neutrophils, *Am. J. Respir. Cell Mol. Biol.*, 11, 426–431, 1994.

Blackford, J. A., Jr., Jones, W., Dey, R. D., and Castranova, V., Comparison of inducible nitric oxide synthase gene expression and lung inflammation following intratracheal instillation of silica, coal, carbonyl iron, or titanium dioxide in rats, *J. Toxicol. Environ. Health Part A*, 51, 203–218, 1997.

Boehme, D. S., Maples, K. R., and Henderson, R. F., Glutathione release by pulmonary alveolar macrophages in response to particles *in vitro*, *Toxicol. Lett.*, 60, 53–60, 1992.

Broaddus, V. C., Yang, L., Scavo, L. M., Ernst, J. D., and Boylan, A. M., Asbestos induces apoptosis of human and rabbit pleural mesothelial cells via reactive oxygen species, *J. Clin. Invest.*, 98, 2050–2059, 1996.

Buder-Hoffmann, S., Palmer, C., Vacek, P., Taatjes, D., and Mossman, B., Different accumulation of activated extracellular signal-regulated kinases (ERK 1/2) and role in cell-cycle alterations by epidermal growth factor, hydrogen peroxide, or asbestos in pulmonary epithelial cells, *Am. J. Respir. Cell Mol. Biol.*, 24, 405–413, 2001.

Cantin, A., Dubois, F., and Begin, R., Lung exposure to mineral dusts enhances the capacity of lung inflammatory cells to release superoxide, *J. Leukoc. Biol.*, 43, 299–303, 1988.

Castranova, V., Generation of oxygen radicals and mechanisms of injury prevention, *Environ. Health Perspect.*, 102 (suppl. 10), 65–68, 1994.

Castranova, V., From coal mine dust to quartz: mechanisms of pulmonary pathogenicity, *Inhal. Toxicol.*, 12, 7–14, 2000.

Castranova, V., Bowman, L., Reasor, M. J., Lewis, T., Tucker, J., and Miles, P. R., The response of rat alveolar macrophages to chronic inhalation of coal dust and/or diesel exhaust, *Environ. Res.*, 36, 405–419, 1985.

Castranova, V., Pailes, W. H., and Li, C., Effects of silica exposure on alveolar macrophages: action of tetrandrine, In *Proceedings of the International Conference on Pneumoconiosis*, Li, Y., Yao, P., Schlipkoter, H. W., Idel, H., and Rosenbrach, M., eds., Stefan Walbers Verlag, Dusseldorf, 256–260, 1990.

Castranova, V., Dalal, N. S., and Vallyathan, V., Role of surface free radicals in the pathogenicity of silica, In *Silica and Silica-Induced Lung Diseases*, Castranova, V., Vallyathan, V., and Wallace, W. E., eds., CRC Press, Boca Raton, FL, 91–105, 1996.

Castranova, V., Pailes, W. H., Dalal, N. S., Miles, P. R., Bowman, L., Vallyathan, V., Pack, D. et al., Enhanced pulmonary response to the inhalation of freshly fractured silica as compared with aged silica dust exposure, *Appl. Occup. Environ. Hyg.*, 11, 937–941, 1996.

Castranova, V., Huffman, L. J., Judy, D. J., Bylander, J. E., Lapp, L. N., Weber, S. L., Blackford, J. A., and Dey, R. D., Enhancement of nitric oxide production by pulmonary cells following silica exposure, *Environ. Health Perspect.*, 106 (suppl. 5), 1165–1169, 1998.

Chao, C. C., Park, S. H., and Aust, A. E., Participation of nitric oxide and iron in the oxidation of DNA in asbestos-treated human lung epithelial cells, *Arch. Biochem. Biophys.*, 326, 152–157, 1996.

Chen, F. and Castranova, V., Reactive oxygen species in the activation and regulation of intracellular signaling events, In *Reactive Oxygen/Nitrogen Species: Lung Injury and Disease*, Vallyathan, V., Shi, X., and Castranova, V., eds., Marcel Decker, New York, 59–90, 2004.

Chen, F., Sun, S. C., Kuh, D. C., Gaydos, L. J., and Demers, L. M., Essential role of NF-κ B activation in silica-induced inflammatory mediator production in macrophages, *Biochem. Biophys. Res. Commun.*, 214, 985–992, 1995.

Dalal, N. S., Shi, X. L., and Vallyathan, V., Role of free radicals in the mechanisms of hemolysis and lipid peroxidation by silica: comparative ESR and cytotoxicity studies, *J. Toxicol. Environ. Health*, 29, 307–316, 1990.

Dalal, N. S., Jafari, B., Petersen, M., Green, F. H., and Vallyathan, V., Presence of stable coal radicals in autopsied coal miners' lungs and its possible correlation to coal workers' pneumoconiosis, *Arch. Environ. Health*, 46, 366–372, 1991.

Dalal, N. S., Newman, J., Pack, D., Leonard, S., and Vallyathan, V., Hydroxyl radical generation by coal mine dust: possible implication to coal workers' pneumoconiosis (CWP), *Free Radic. Biol. Med.*, 18, 11–20, 1995.

De Deken, X., Wang, D., Many, M. C., Costagliola, S., Libert, F., Vassart, G., Dumont, J. E., and Miot, F., Cloning of two human thyroid cDNAs encoding new members of the NADPH oxidase family, *J. Biol. Chem.*, 275, 23227–23233, 2000.

DeWaele, J. K. and Adams, F. C., The surface characterization of modified chrysotile asbestos, *Scanning Electon Microsc.*, 2, 209–228, 1988.

Diaz-Cazorla, M., Perez-Sala, D., and Lamas, S., Dual effect of nitric oxide donors on cyclooxygenase-2 expression in human mesangial cells, *J. Am. Soc. Nephrol.*, 10, 943–952, 1999.

Ding, M., Dong, Z., Chen, F., Pack, D., Ma, W. Y., Ye, J., Shi, X., Castranova, V., and Vallyathan, V., Asbestos induces activator protein-1 transactivation in transgenic mice, *Cancer Res.*, 59, 1884–1889, 1999a.

Ding, M., Shi, X., Dong, Z., Chen, F., Lu, Y., Castranova, V., and Vallyathan, V., Freshly fractured crystalline silica induces activator protein-1 activation through ERKs and p38 MAPK, *J. Biol. Chem.*, 274, 30611–30616, 1999b.

Ding, M., Shi, X., Lu, Y., Huang, C., Leonard, S., Roberts, J., Antonini, J., Castranova, V., and Vallyathan, V., Induction of activator protein-1 through reactive oxygen species by crystalline silica in JB6 cells, *J. Biol. Chem.*, 276, 9108–9114, 2001.

Donaldson, K. and Cullen, R. T., Chemiluminescence of asbestos-activated macrophages, *Br. J. Exp. Pathol.*, 65, 81–90, 1984.

Donaldson, K., Slight, J., Hannant, D., and Bolton, R. E., Increased release of hydrogen peroxide and superoxide anion from asbestos-primed macrophages. Effect of hydrogen peroxide on the functional activity of alpha 1-protease inhibitor, *Inflammation*, 9, 139–147, 1985.

Donaldson, K., Stone, V., Duffin, R., Clouter, A., Schins, R., and Borm, P., The quartz hazard: effects of surface and matrix on inflammogenic activity, *J. Environ. Pathol. Toxicol. Oncol.*, 20 (suppl. 1), 109–118, 2001.

Dorger, M., Allmeling, A. M., Kiefmann, R., Munzing, S., Messmer, K., and Krombach, F., Early inflammatory response to asbestos exposure in rat and hamster lungs: role of inducible nitric oxide synthase, *Toxicol. Appl. Pharmacol.*, 181, 93–105, 2002a.

Dorger, M., Allmeling, A. M., Kiefmann, R., Schropp, A., and Krombach, F., Dual role of inducible nitric oxide synthase in acute asbestos-induced lung injury, *Free Radic. Biol. Med.*, 33, 491–501, 2002b.

Duffin, R., Gilmour, P. S., Schins, R. P., Clouter, A., Guy, K., Brown, D. M., MacNee, W., Borm, P. J., Donaldson, K., and Stone, V., Aluminium lactate treatment of DQ12 quartz inhibits its ability to cause inflammation, chemokine expression, and nuclear factor-κ B activation, *Toxicol. Appl. Pharmacol.*, 176, 10–17, 2001.

Eiserich, J. P., Patel, R. P., and O'Donnell, V. B., Pathophysiology of nitric oxide and related species: free radical reactions and modification of biomolecules, *Mol. Aspects. Med.*, 19, 221–357, 1998.

Engelen, J. J., Borm, P. J., van Sprundel, M., and Leenaerts, L., Blood anti-oxidant parameters at different stages of pneumoconiosis in coal workers, *Environ. Health Perspect.*, 84, 165–172, 1990.

Fenoglio, I., Prandi, L., Tomatis, M., and Fubini, B., Free radical generation in the toxicity of inhaled mineral particles: the role of iron speciation at the surface of asbestos and silica, *Redox Rep.*, 6, 235–241, 2001.

Fubini, B., Health effect of silica, In *The Surface Properties of Silicas*, Legrand, J. P., Ed., John Wiley & Sons, Chichester, U.K., 415–464, 1998.

Fubini, B., Giamello, E., Volante, M., and Bolis, V., Chemical functionalities at the silica surface determining its reactivity when inhaled. Formation and reactivity of surface radicals, *Toxicol. Ind. Health*, 6, 571–598, 1990.

Fubini, B., Fenoglio, I., Elias, Z., and Poirot, O., Variability of biological responses to silicas: effect of origin, crystalinity, and state of surface on generation of reactive oxygen species and morphological transformation of mammalian cells, *J. Environ. Pathol. Toxicol. Oncol.*, 20 (suppl. 1), 95–108, 2001.

Ghio, A. J., Kennedy, T. P., Whorton, A. R., Crumbliss, A. L., Hatch, G. E., and Hoidal, J. R., Role of surface complexed iron in oxidant generation and lung inflammation induced by silicates., *Am. J. Physiol.*, 263, L511–L518, 1992.

Goodglick, L. A. and Kane, A. B., Role of reactive oxygen metabolites in crocidolite asbestos toxicity to mouse macrophages, *Cancer Res.*, 46, 5558–5566, 1986.

Goodglick, L. A. and Kane, A. B., Cytotoxicity of long and short crocidolite asbestos fibers in vitro and *in vivo*, *Cancer Res.*, 50, 5153–5163, 1990.

Goodman, G. B., Kaplan, P. D., Stachura, I., Castranova, V., Pailes, W. H., and Lapp, N. L., Acute silicosis responding to corticosteroid therapy, *Chest*, 101, 366–370, 1992.

Granville, D. J., Carthy, C. M., Hunt, D. W., and McManus, B. M., Apoptosis: molecular aspects of cell death and disease, *Lab. Invest.*, 78, 893–913, 1998.

Green, D. R. and Reed, J. C., Mitochondria and apoptosis, *Science*, 281, 1309–1312, 1998.

Hamilton, R. F., Iyer, L. L., and Holian, A., Asbestos induces apoptosis in human alveolar macrophages, *Am. J. Physiol.*, 271, L813–L819, 1996.

Han, S. G., Kim, Y., Kashon, M. L., Pack, D., Castranova, V., and Vallyathan, V., Correlates of oxidative stress and free radical activity in asymptomatic shipyard welders' Serum, *Am. J. Respir. Crit. Care Med.*, 172, 1541–1548, 2005.

Hansen, K. and Mossman, B. T., Generation of superoxide (O_2^-) from alveolar macrophages exposed to asbestiform and nonfibrous particles, *Cancer Res.*, 47, 1681–1686, 1987.

Harrington, J. S., Chemical studies of asbestos, *Ann. NY. Acad. Sci.*, 132, 31–47, 1965.

Heffner, J. and Kensler, T., Pulmonary strategies of antioxidant defense, *Ann. Rev. Respir. Dis.*, 140, 531–554, 1989.

Hodgson, A. A., Chemistry and physics of asbestos, In *Asbestos*, Michaels, L. and Chissick, S. S., eds., Wiley, New York, 67–114, 1979.

Hofseth, L. J., Hussain, S. P., Wogan, G. N., and Harris, C. C., Nitric oxide in cancer and chemoprevention, *Free. Radic. Biol. Med.*, 34, 955–968, 2003.

Huffman, L. J., Judy, D. J., and Castranova, V., Regulation of nitric oxide production by rat alveolar macrophages in response to silica exposure, *J. Toxicol. Environ. Health Part A*, 53, 29–46, 1998.

Iguchi, H., Kojo, S., and Ikeda, M., Nitric oxide (NO) synthase activity in the lung and NO synthesis in alveolar macrophages of rats increased on exposure to asbestos, *J. Appl. Toxicol.*, 16, 309–315, 1996.

Iyer, R., Hamilton, R. F., Li, L., and Holian, A., Silica-induced apoptosis mediated via scavenger receptor in human alveolar macrophages, *Toxicol. Appl. Pharmacol.*, 141, 84–92, 1996.

Janssen, Y. M., Marsh, J. P., Absher, M. P., Hemenway, D., Vacek, P. M., Leslie, K. O., Borm, P. J., and Mossman, B. T., Expression of antioxidant enzymes in rat lungs after inhalation of asbestos or silica, *J. Biol. Chem.*, 267, 10625–10630, 1992.

Janssen, Y. M., Marsh, J. P., Absher, M. P., Gabrielson, E., Borm, P. J., Driscoll, K., and Mossman, B. T., Oxidant stress responses in human pleural mesothelial cells exposed to asbestos, *Am. J. Respir. Crit. Care Med.*, 149, 795–802, 1994.

Janssen, Y. M., Barchowsky, A., Treadwell, M., Driscoll, K. E., and Mossman, B. T., Asbestos induces nuclear factor kappa B (NF-κ B) DNA-binding activity and NF-kappa B-dependent gene expression in tracheal epithelial cells, *Proc. Natl. Acad. Sci. U.S.A.*, 92, 8458–8462, 1995a.

Janssen, Y. M., Heintz, N. H., and Mossman, B. T., Induction of *c*-Fos and *c*-Jun proto-oncogene expression by asbestos is ameliorated by *N*-acetyl-L-cysteine in mesothelial cells, *Cancer Res.*, 55, 2085–2089, 1995b.

Kabe, Y., Ando, K., Hirao, S., Yoshida, M., and Handa, H., Redox regulation of NF-κB activation: distinct redox regulation between the cytoplasm and the nucleus, *Antioxid. Redox Signal*, 7, 395–403, 2005.

Kang, J. H., Van Dyke, K., Pailes, W. H., and Castranova, V., Potential role of platelet-activating factor in development of occupational lung disease: action as an activator or potentiator of pulmonary phagocytes, *Proc. Resp. Dust Miner. Ind.*, 183–190, 1991.

Kang, J. L., Go, Y. H., Hur, K. C., and Castranova, V., Silica-induced nuclear factor-kappaB activation: involvement of reactive oxygen species and protein tyrosine kinase activation, *J. Toxicol. Environ. Health Part A*, 60, 27–46, 2000a.

Kang, J. L., Lee, K., and Castranova, V., Nitric oxide up-regulates DNA-binding activity of nuclear factor-kappaB in macrophages stimulated with silica and inflammatory stimulants, *Mol. Cell Biochem.*, 215, 1–9, 2000b.

Kanj, R. S., Kang, J. L., and Castranova, V., Measurement of the release of inflammatory mediators from rat alveolar macrophages and alveolar type II cells following lipopolysaccharide or silica exposure: a comparative study, *J. Toxicol. Environ. Health Part A*, 68, 185–207, 2005.

Kirkinezos, I. G. and Moraes, C. T., Reactive oxygen species and mitochondrial diseases, *Semin. Cell Dev. Biol.*, 12, 449–457, 2001.

Knaapen, A. M., Albrecht, C., Becker, A., Hohr, D., Winzer, A., Haenen, G. R., Borm, P. J., and Schins, R. P., DNA damage in lung epithelial cells isolated from rats exposed to quartz: role of surface reactivity and neutrophilic inflammation, *Carcinogenesis*, 23, 1111–1120, 2002.

Konecny, R., Leonard, S., Shi, X., Robinson, V., and Castranova, V., Reactivity of free radicals on hydroxylated quartz surface and its implications for pathogenicity experimental and quantum mechanical study, *J. Environ. Pathol. Toxicol. Oncol.*, 20 (suppl. 1), 119–132, 2001.

Leigh, J., Wang, H., Bonin, A., Peters, M., and Ruan, X., Silica-induced apoptosis in alveolar and granulomatous cells *in vivo*, *Environ. Health Perspect. 105*, 105 (suppl. 5), 1241–1245, 1997.

Leonard, S. S., Wang, S., Shi, X., Jordan, B. S., Castranova, V., and Dubick, M. A., Wood smoke particles generate free radicals and cause lipid peroxidation, DNA damage, NF-kappa B activation and TNF-α release in macrophages, *Toxicology*, 150, 147–157, 2000.

Leonard, S. S., Roberts, J. R., Antonini, J. M., Castranova, V., and Shi, X., $PbCrO_4$ mediates cellular responses via reactive oxygen species, *Mol. Cell Biochem.*, 255, 171–179, 2004.

Magnitsky, S., Toulokhonova, L., Yano, T., Sled, V. D., Hagerhall, C., Grivennikova, V. G., Burbaev, D. S., Vinogradov, A. D., and Ohnishi, T., EPR characterization of ubisemiquinones and iron-sulfur cluster N2, central components of the energy coupling in the NADH-ubiquinone oxidoreductase (complex I) in situ, *J. Bioenerg. Biomembr.*, 34, 193–208, 2002.

Mao, Y., Daniel, L. N., Knapton, A. D., Shi, X., and Saffiotti, U., Protective effects of silanol group binding agents on quartz toxicity to rat lung alveolar cells, *Appl. Occup. Environ. Hyg.*, 10, 1–6, 1995.

Mossman, B. T. and Churg, A., Mechanisms in the pathogenesis of asbestosis and silicosis, *Am. J. Respir. Crit. Care Med.*, 157, 1666–1680, 1998.

Mossman, B. T., Marsh, J. P., and Shatos, M. A., Alteration of superoxide dismutase activity in tracheal epithelial cells by asbestos and inhibition of cytotoxicity by antioxidants, *Lab. Invest.*, 54, 204–212, 1986.

Muller, F., Crofts, A. R., and Kramer, D. M., Multiple Q-cycle bypass reactions at the Qo site of the cytochrome bc1 complex, *Biochemistry*, 41, 7866–7874, 2002.

Nash, T., Allison, A. C., and Harington, J. S., Physico-chemical properties of silica in relation to its toxicity, *Nature*, 210, 259–261, 1966.

Nurkiewicz, T. R., Porter, D. W., Barger, M., Castranova, V., and Boegehold, M. A., Particulate matter exposure impairs systemic microvascular endothelium-dependent dilation, *Environ. Health Perspect.*, 112, 1299–1306, 2004.

Ohnishi, T., Iron–sulfur clusters/semiquinones in complex I, *Biochim. Biophys. Acta*, 1364, 186–206, 1998.

Patriarca, P., Cramer, R., Moncalvo, S., Rossi, F., and Romeo, D., Enzymatic basis of metabolic stimulation in leucocytes during phagocytosis: the role of activated NADPH oxidase, *Arch. Biochem. Biophys.*, 145, 255–262, 1971.

Perrin-Nadif, R., Bourgkard, E., Dusch, M., Bernadac, P., Bertrand, J. P., Mur, J. M., and Pham, Q. T., Relations between occupational exposure to coal mine dusts, erythrocyte catalase and Cu^{++}/Zn^{++} superoxide dismutase activities, and the severity of coal workers' pneumoconiosis, *Occup. Environ. Med.*, 55, 533–540, 1998.

Petruska, J. M., Marsh, J., Bergeron, M., and Mossman, B. T., Brief inhalation of asbestos compromises superoxide production in cells from bronchoalveolar lavage, *Am. J. Respir. Cell Mol. Biol.*, 2, 129–136, 1990.

Pezerat, H., Zalma, R., Guignard, J., and Javrand, M. C., Production of oxygen radicals by the reduction of oxygen arising from the surface activity of mineral fibers, In *Non-Occupational Exposure to Mineral Fibers*, Bignon, J., Peto, J., and Seracci, R., eds., IARC Scientific Publications, Lyon, France, 100–111, 1989.

Pooley, F. D., Mineralogy of asbestos: the physical and chemical properties of the dusts they form, *Semin. Oncol.*, 8, 243–249, 1981.

Porter, D. W., Barger, M., Robinson, V. A., Leonard, S. S., Landsittel, D., and Castranova, V., Comparison of low doses of aged and freshly fractured silica on pulmonary inflammation and damage in the rat, *Toxicology*, 175, 63–71, 2002a.

Porter, D. W., Millecchia, L., Robinson, V. A., Hubbs, A., Willard, P., Pack, D., Ramsey, D. et al., Enhanced nitric oxide and reactive oxygen species production and damage after inhalation of silica, *Am. J. Physiol. Lung. Cell Mol. Physiol.*, 283, L485–L493, 2002b.

Porter, D. W., Ye, J., Ma, J., Barger, M., Robinson, V. A., Ramsey, D., McLaurin, J. et al., Time course of pulmonary response of rats to inhalation of crystalline silica: NF-kappa B activation, inflammation, cytokine production, and damage, *Inhal. Toxicol.*, 14, 349–367, 2002c.

Quinlan, T. R., BéruBé, K. A., Hacker, M. P., Taatjes, D. J., Timblin, C. R., Goldberg, J., Kimberley, P. et al., Mechanisms of asbestos-induced nitric oxide production by rat alveolar macrophages in inhalation and in vitro models, *Free Radic. Biol. Med.*, 24, 778–788, 1998.

Rom, W. N., Bitterman, P. B., Rennard, S. I., Cantin, A., and Crystal, R. G., Characterization of the lower respiratory tract inflammation of nonsmoking individuals with interstitial lung disease associated with chronic inhalation of inorganic dusts, *Am. Rev. Respir. Dis.*, 136, 1429–1434, 1987.

Rubbo, H., Radi, R., Trujillo, M., Telleri, R., Kalyanaraman, B., Barnes, S., Kirk, M., and Freeman, B. A., Nitric oxide regulation of superoxide and peroxynitrite-dependent lipid peroxidation. Formation of novel nitrogen-containing oxidized lipid derivatives, *J. Biol. Chem.*, 269, 26066–26075, 1994.

Sacks, M., Gordon, J., Bylander, J., Porter, D., Shi, X. L., Castranova, V., Kaczmarczyk, W., Van Dyke, K., and Reasor, M. J., Silica-induced pulmonary inflammation in rats: activation of NF-κ B and its suppression by dexamethasone, *Biochem. Biophys. Res. Commun.*, 253, 181–184, 1998.

Savill, J., Fadok, V., Henson, P., and Haslett, C., Phagocyte recognition of cells undergoing apoptosis, *Immunol. Today*, 14, 131–136, 1993.

Schins, R., Keman, S., and Borm, P., Blood antioxidant status in coal dust-induced respiratory disorders: a longitudinal evaluation of multiple biomarkers, *Biomarkers*, 2, 45–50, 1997.

Shatos, M. A., Doherty, J. M., Marsh, J. P., and Mossman, B. T., Prevention of asbestos-induced cell death in rat lung fibroblasts and alveolar macrophages by scavengers of active oxygen species, *Environ. Res.*, 44, 103–116, 1987.

Shen, H. M., Zhang, Z., Zhang, Q. F., and Ong, C. N., Reactive oxygen species and caspase activation mediate silica-induced apoptosis in alveolar macrophages, *Am. J. Physiol. Lung Cell Mol. Physiol.*, 280, L10–L17, 2001.

Shi, X., Mao, Y., Daniel, L. N., Saffiotti, U., Dalal, N. S., and Vallyathan, V., Silica radical-induced DNA damage and lipid peroxidation, *Environ. Health Perspect.*, 102 (suppl. 10), 149–154, 1994.

Shi, X., Mao, Y., Daniel, L. N., Saffiotti, U., Dalal, N. S., and Vallyathan, V., Generation of reactive oxygen species by quartz particles and its implication for cellular damage, *Appl. Occup. Environ. Hyg.*, 10, 1138–1144, 1995.

Srivastava, K. D., Rom, W. N., Jagirdar, J., Yie, T. A., Gordon, T., and Tchou-Wong, K. M., Crucial role of interleukin-1beta and nitric oxide synthase in silica-induced inflammation and apoptosis in mice, *Am. J. Respir. Crit. Care Med.*, 165, 527–533, 2002.

Suh, Y. A., Arnold, R. S., Lassegue, B., Shi, J., Xu, X., Sorescu, D., Chung, A. B., Griendling, K. K., and Lambeth, J. D., Cell transformation by the superoxide-generating oxidase Mox1, *Nature*, 401, 79–82, 1999.

Taylor, M. D., Roberts, J. R., Leonard, S. S., Shi, X., and Antonini, J. M., Effects of welding fumes of differing composition and solubility on free radical production and acute lung injury and inflammation in rats, *Toxicol. Sci.*, 75, 181–191, 2003.

Thomas, G., Ando, T., Verma, K., and Kagan, E., Asbestos fibers and interferon-gamma up-regulate nitric oxide production in rat alveolar macrophages, *Am. J. Respir. Cell Mol. Biol.*, 11, 707–715, 1994.

Timbrell, V., Characteristics of the international union against cancer standard reference samples of asbestos, In *Proceedings of the International Conference*, Shapiro, H. A., ed., Oxford University Press, Oxford, England, 28–36, 1970.

Umansky, V., Hehner, S. P., Dumont, A., Hofmann, T. G., Schirrmacher, V., Droge, W., and Schmitz, M. L., Co-stimulatory effect of nitric oxide on endothelial NF-kappa B implies a physiological self-amplifying mechanism, *Eur. J. Immunol.*, 28, 2276–2282, 1998.

Vallyathan, V., Shi, X. L., Dalal, N. S., Irr, W., and Castranova, V., Generation of free radicals from freshly fractured silica dust. Potential role in acute silica-induced lung injury, *Am. Rev. Respir. Dis.*, 138, 1213–1219, 1988.

Vallyathan, V., Kang, J. H., Van Dyke, K., Dalal, N. S., and Castranova, V., Response of alveolar macrophages to in vitro exposure to freshly fractured versus aged silica dust: the ability of Prosil 28, an organosilane material, to coat silica and reduce its biological reactivity, *J. Toxicol. Environ. Health Part A*, 33, 303–315, 1991.

Vallyathan, V., Castranova, V., Pack, D., Leonard, S., Shumaker, J., Hubbs, A. F., Shoemaker, D. A. et al., Freshly fractured quartz inhalation leads to enhanced lung injury and inflammation. Potential role of free radicals, *Am. J. Respir. Crit. Care Med.*, 152, 1003–1009, 1995.

Vallyathan, V., Goins, M., Lapp, L. N., Pack, D., Leonard, S., Shi, X., and Castranova, V., Changes in bronchoalveolar lavage indices associated with radiographic classification in coal miners, *Am. J. Respir. Crit. Care Med.*, 162, 958–965, 2000.

van der Vliet, A. and Cross, C. E., Oxidants, nitrosants, and the lung, *Am. J. Med.*, 109, 398–421, 2000.

Voisin, C., Wallaert, B., Aerts, C., and Grosbois, J., Bronchoalveolar lavage in coal workers' pneumoconiosis: oxidant and antioxidant activities of alveolar macrophages, In *In Vitro Effects of Mineral Dusts*, Beck, E. G. and Bignon, J., eds., Springer-Verlag, Berlin, 93–100, 1985.

Wallace, W. E., Jr., Vallyathan, V., Keane, M. J., and Robinson, V., In vitro biologic toxicity of native and surface-modified silica and kaolin, *J. Toxicol. Environ. Health Part A*, 16, 415–424, 1985.

Wallaert, B., Lassalle, P., Fortin, F., Aerts, C., Bart, F., Fournier, E., and Voisin, C., Superoxide anion generation by alveolar inflammatory cells in simple pneumoconiosis and in progressive massive fibrosis of nonsmoking coal workers, *Am. Rev. Respir. Dis.*, 141, 129–133, 1990.

Wang, L., Antonini, J. M., Rojanasakul, Y., Castranova, V., Scabilloni, J. F., and Mercer, R. R., Potential role of apoptotic macrophages in pulmonary inflammation and fibrosis, *J. Cell Physiol.*, 194, 215–224, 2003.

Weitzman, S. A. and Graceffa, P., Asbestos catalyzes hydroxyl and superoxide radical generation from hydrogen peroxide, *Arch. Biochem. Biophys.*, 228, 373–376, 1984.

Yang, H. M., Antonini, J. M., Barger, M. W., Butterworth, L., Roberts, B. R., Ma, J. K., Castranova, V., and Ma, J. Y., Diesel exhaust particles suppress macrophage function and slow the pulmonary clearance of Listeria monocytogenes in rats, *Environ. Health Perspect.*, 109, 515–521, 2001.

Ye, J., Shi, X., Jones, W., Rojanasakul, Y., Cheng, N., Schwegler-Berry, D., Baron, P., Deye, G. J., Li, C., and Castranova, V., Critical role of glass fiber length in TNF-alpha production and transcription factor activation in macrophages, *Am. J. Physiol.*, 276, L426–L434, 1999.

Ye, J., Zeidler, P., Young, S. H., Martinez, A., Robinson, V. A., Jones, W., Baron, P., Shi, X., and Castranova, V., Activation of mitogen-activated protein kinase p38 and extracellular signal-regulated kinase is involved in glass fiber-induced tumor necrosis factor-alpha production in macrophages, *J. Biol. Chem.*, 276, 5360–5367, 2001.

Zeidler, P. C., Roberts, J. R., Castranova, V., Chen, F., Butterworth, L., Andrew, M. E., Robinson, V. A., and Porter, D. W., Response of alveolar macrophages from inducible nitric oxide synthase knockout or wild-type mice to an in vitro lipopolysaccharide or silica exposure, *J. Toxicol. Environ. Health Part A*, 66, 995–1013, 2003.

Zussman, J., The crystal structures of amphibole and serpentine minerals, *National Bureau of Standards Special Publication*, 506, 1978.

7 Interaction of Particles with Membranes

Barbara Rothen-Rutishauser
Institute of Anatomy, University of Bern

Samuel Schürch
Institute of Anatomy, University of Bern

Department of Physiology and Biophysics, University of Calgary

Peter Gehr
Institute of Anatomy, University of Bern

CONTENTS

7.1 INTRODUCTION

The concept of membrane includes two differing interfacial entities in biology: tissue barriers such as the mucus barrier for the inner or luminal tissue lining of the airways on the one hand, and cell layers, the cell membrane itself, lipid air–water films, and lipid bilayers, as biophysical models for the cell interfacial layers on the other hand. The tissue membrane lining the airways forms a barrier that protects the underlying tissue layers. However, increasing scientific evidence has demonstrated, for instance, that nanoparticles cannot only cross easily tissue membranes, they also can pass subsequently into secondary organs by crossing cell boundaries, including the outer cellular membranes as well as internal cellular membranes like the nuclear membrane and membranes of organelles. These organelles encompass structural and functional units including mitochondria, endoplasmic reticulum, the Golgi apparatus, lysosomes, and others. Cellular membranes are composed of phospholipids, glycolipids, and numerous membrane proteins.

7.1.1 Tissue Membranes

A series of structural and functional barriers (Figure 7.1) protect the respiratory system against harmful and innocuous particulate material (for a review, see Nicod 2005). This is important, as the internal surface area of the lungs is vast (alveoli plus airways approximately 140 m^2, Gehr, Bachofen, and Weibel 1978), facilitating easy access to the lung tissue, including immunocompetent cells such as dendritic cells. The particle may enter blood capillaries and then pass to other organs. The series of structural barriers includes:

1. The surfactant film (Gil and Weibel 1971; Gehr et al. 1990; Schürch et al. 1990)
2. The aqueous surface lining layers including the mucociliary escalator (Kilburn 1968)
3. A population of macrophages (professional phagocytes) in the airways and in the alveoli (Brain 1988; Lehnert 1992)
4. The epithelium cellular layer endowed with tight junctions between the cells (Schneeberger and Lynch 1984; Godfrey 1997)
5. A network of dendritic cells inside and underneath the epithelium (Holt and Schon-Hegrad 1987; McWilliam, Holt, and Gehr 2000)
6. The basal lamina (basement membrane) (Timpl and Dzladek 1986; Yurchenco et al. 1986; Maina and West 2005)
7. The connective tissue (Dunsmore and Rannels 1996)
8. The capillary endothelium (Schneeberger 1977; Dudek and Garcia 2001)

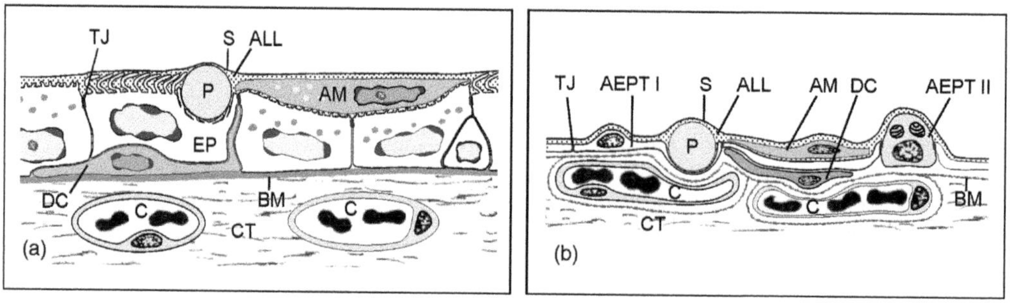

FIGURE 7.1 Tissue barrier system of airways (a) and alveoli (b). Abbreviations: alveolar epithelium type I (AEPT I), alveolar epithelium type II (AEPT II), aqueous lining layer (ALL), airway/alveolar macrophage (AM), basal membrane (BM), capillary (C), connective tissue (CT), dendritic cell (DC), epithelial cell (EP), particle (P), surfactant (S), tight junction (TJ).

However, despite the existence of these barriers, described in 7.1.1, respiratory diseases are frequent and increasing (Peters et al. 1997; Wichmann et al. 2000; Schulz et al. 2005) and more attention has been directed towards elucidating how and when the antigens evade these barriers. Insoluble particles deposited in the airways are largely cleared by the mucociliary action, but not all deposited particles are removed by this mechanism. The fate of these particles depends on their physical–chemical characteristics and the nature of their interaction with the surfactant film at the air–liquid interface. All inhaled particles (PM_{10}: particulate matter with a diameter equal or smaller than 10 μm) deposited in the airways are displaced into the subphase below the surfactant film and may be modified by surfactant components or coated with surfactant during the displacement process (Gehr et al. 1990; Schürch et al. 1990; Geiser, Schürch, and Gehr 2003a, Geiser et al. 2003b). As a result of the displacement, particles come into contact with the epithelium and alveolar macrophages (Gehr et al. 1990; Schürch et al. 1990; Gehr et al. 1996; Geiser, Schürch, and Gehr 2003a). Of particular importance and interest is to clarify the mechanisms involved in the particle translocation to the dendritic cells. These cells are located at the base of the epithelium and are the most competent antigen presenting cells in the lung (Holt, Schon-Hegrad, and McMenamin 1990; Nicod 1997). One of the most crucial specialized functions of dendritic cells is the capture and delivery of antigen to local lymphoid tissues. Their unique responsibility is to decide whether to present sampled antigen in an immunogenic or tolerogenic way (for a review, see Vermaelen and Pauwels 2005).

In vivo, alveolar macrophages occupy the luminal aspect of the epithelium (Brain, Gehr, and Kavet 1984; Geiser et al. 1994) while immature dendritic cells occupy the basal aspect of the epithelium. Dendritic cells are located within the basement lamina and reach the tight junctions of the epithelium with fine cytoplasmic processes pushed between the epithelial cells (Figure 7.1) (Holt and Schon-Hegrad 1987; McWilliam, Holt, and Gehr 2000).

By using of laser scanning and transmission electron microscopy, it has been demonstrated that dendritic cells are efficiently phagocytic for a variety of particles, such as polystyrene particles (Matsuno et al. 1996; Kiama et al. 2001; Thiele et al. 2001), puffball spores, biodegradable microspheres (Walter et al. 2001), and *Salmonella typhimurium* (Dreher et al. 2001). Although both macrophages and dendritic cells are derived from circulating blood monocytes, macrophages are twice as phagocytic as immature dendritic cells (Kiama et al. 2001). Whereas much is known on the interaction of particles with dendritic cells and with macrophages, nothing is known on how the antigens reach the dendritic cells and if macrophages and epithelial cells are involved. Transport of the particles to the dendritic cells presupposes their passage across the epithelium either through the epithelial cells or between the epithelial cells, through the tight junctions. However, the mechanism in which dendritic cells cross the tight junctions, which "seal" the airway epithelium at the apical side, is not clear yet (Holt 2005). It has been shown in gut mucosa that subepithelial dendritic cells are capable of capturing antigens outside the epithelium by extending fine cytoplasmic processes through the tight junctions (Rescigno et. al. 2001; Niess et al. 2005). We have hints that dendritic cells behave this way at the airway epithelium (data not published). Vermaelen and colleagues (Vermaelen et al. 2001) found in an *in vivo* study that fluorescein isothiocyanate conjugated macromolecules are transported to the tracheal lymph nodes by airway dendritic cells after intratracheal instillation. The mechanism by which the macromolecules pass through the epithelium to reach the dendritic cells was not addressed. Takano and colleagues (Takano et al. 2005) showed that dendritic cells easily access antigens beyond epithelial tight junctions in human nasal mucosa, but only in allergic rhinitis. Another *in vitro* model using mouse tracheal epithelial cells and mouse bone marrow dendritic cells showed impaired migration of metalloproteinase-9-deficient dendritic cells through tracheal epithelial tight junctions (Ichiyasu et al. 2004), but the mechanism of dendritic cell migration through the lung epithelium still remains largely unknown. We have obtained evidence from preliminary *in vitro* studies that dendritic cells collect particles on the luminal side of the epithelium. Few publications describe this kind of dendritic cell behavior in disease. There is evidence that dendritic cells play an important role in the pathogenesis of allergic asthma, and there are an increased numbers of dendritic cells in the airway mucosa of patients with chronic obstructive pulmonary disease (Vermaelen and Pauwels 2005).

After antigens have passed through the epithelial barrier, they may pass through the basement membrane and subsequently through the subepithelial connective tissue layer and eventually come into contact with endothelial cells lining the capillaries. Since endothelial cells play an important role in inflammation processes (Michiels 2003), particles can have some effect on endothelial cell function and viability inducing proinflammatory stimuli. It has been proposed that the permeability of the lung tissue barrier to nanoparticles is controlled at the epithelial and the endothelial level (Meiring et al. 2005). Particulate antigens $<0.1\ \mu m$ are able to cross the air–blood barrier of the lung, and thus can enter the circulatory system (Nemmar et al. 2002). Antigens that have passed through the epithelial and endothelial barrier are transported by the blood circulation and reach other organs including the liver, the heart (Brown, Zeman, and Bennett 2002; Kreyling et al. 2002), the brain (Oberdörster et al. 2004), or other organs.

7.1.2 Cellular Membranes

The surfactant air–liquid film in airways and alveoli and phospholipid bilayers with different types of proteins are considered actual membranes. These membranes also include the nuclear membrane and the membranes of the different organelles.

We have observed that particle uptake *in vitro* into cells does not occur by any of the expected endocytic processes, but rather by other mechanisms including adhesive interactions (see also Section 7.5) (Rimai et al. 2000). Particles within cells have been found not to be membrane-bound (Kapp et al. 2004; Geiser et al. 2005), hence they have direct access to cytoplasmic proteins, important biochemical molecules in organelles, and DNA in the nucleus, which may greatly enhance their hazardous potential. *In vitro* experiments revealed penetration of nanoparticles into mitochondria of macrophages and epithelial cells, associated with oxidative stress and mitochondrial damage (Li et al. 2003). This penetration of nanoparticles led to a loss of christae in the mitochondria. The inner mitochondrial membrane was destroyed by nanoparticles. Regardless of the mechanisms causing this mitochondrial damage, there is evidence that it is the organic substances attached to the particle that are responsible for this damage (Xia et al. 2004). The inner mitochondrial membrane is the structure carrying the respiratory chain that converts molecular oxygen into energy stored as ATP. One wonders what consequences a loss of christae may have.

Penetration of particles into the nucleus has been shown in a number of studies (Liu et al. 2003; Geiser et al. 2005; Gurr et al. 2005; Tsoli et al. 2005). Tsoli and colleagues (Tsoli et al. 2005) have shown that Au_{55} clusters interact with DNA in a way that may be the reason for the strong toxicity of these tiny 1.4 nm particles promoting human cancer.

7.1.3 Adverse Health Effects of Nanoparticles

In addition to the generation of nanoparticles from combustion processes in large amounts, there are progressively more nanoparticles released into the air, into water, and into soil every year from other sources, i.e., nanotechnology (Mazzola 2003; Paull et al. 2003). One example of nanotechnology is to deliver and target pharmaceutical, therapeutic, and diagnostic agents and has recently been referred to as nanomedicine (for a review see Moghimi, Hunter, and Murray 2005). All of these artificial nanoparticles have unique chemical, physical, and electrical properties. They behave neither like solids and liquids, nor like gases of the same materials. New properties emerge that are not exhibited by larger particles having the same chemical composition. These properties include different colors, different electric, magnetic, and mechanical properties, any or all of which may be altered at the nanoscale.

The series of these barriers described in 7.1.1 may not be as effective to protect the body against particles $<0.1\ \mu m$ in size. It has been shown that these small particles may eventually reach the capillaries and nerves embedded in the subepithelial connective tissue layer (Figure 7.1). Since junctions between epithelial cells may be reached by nerve fibres, nanoparticles may already gain access to the epithelia via nerve fibres (as they do in the olfactory region of the upper nose cavity). If nanoparticles once reach the capillaries, they are translocated into other organs where they may penetrate into or through cells (Kreyling et al. 2002; Oberdörster et al. 2002). The nanoparticles may clear the blood–brain barrier (since they can obviously pass through endothelial cells and astrocytes) and penetrate into the neurons of the brain (Oberdörster et al. 2004). Once they enter nerve cells, they may be rapidly transported by axonal transport mechanisms into any region of the brain.

There is evidence from a number of epidemiological studies that particulate matter causes adverse health effects associated with increased pulmonary and cardiovascular mortality (Pope, Dockery, and Schwartz 1995; Peters et al. 1997; Schulz et al. 2005). Recent studies indicate a specific toxicological role of inhaled nanoparticles (Borm and Kreyling 2004). Thus particles with a few nm in diameter are of particular interest, but hardly any information is available today on their effects on cells, tissues, and organs. It has been described that inhaled combustion-derived nanoparticles provoke oxidative stress causing inflammation as well as oxidative adducts in the epithelium that can contribute to carcinogenesis (for a review see Donaldson et al. 2005). Particle-induced pulmonary and systemic inflammation, accelerated atherosclerosis, and altered cardiac autonomic function may be part of the patho-physiological pathways, linking particulate air pollution to cardiovascular mortality (Kunzli and Tager 2005). Oxidative stress related pathways might mediate the effect of air pollutants to the cardiovascular system through the efflux of inflammatory cytokines and other mediators of oxidative stress into the blood system (Donaldson et al. 2001; Brook, Brook, and Rajagopalan 2003). Studies with cellular and animal models suggest a variety of possible mechanisms, including direct effects of particle components on the intracellular sources of reactive oxygen species, indirect effects due to proinflammatory mediators released from macrophages stimulated by particulate matter, and neural stimulation after particle deposition in the lungs (reviewed in Gonzalez-Flecha 2004). Reactive oxygen species in low concentration mediated mechanisms may cause inflammation, but also increase calcium concentrations, which can activate the transcription factors nuclear factor-κB and activator protein-1 after exposure of cells to nanoparticles (Brown et al. 2004).

It is very important to collect risk data—in particular, health risk data—so that questions can be answered and problems can be addressed early in the stage of the new technology developments (Hoet, Bruske-Hohlfeld, and Salata 2004). The rapid proliferation of many different engineered nanomaterials, i.e., artificially produced nanoparticles and nanotubes, as well as nanoparticles generated by combustion processes, requires defined screening strategies for the characterization of the potential human health effects from exposure to nanomaterial (Oberdörster et al. 2005). Most of the concerns regarding nanomaterials stem from the experiences with nanoparticles being substantially more inflammatory and toxic than fine particles (Ferin, Oberdörster, and Penney 1992; Li et al. 2003).

7.2 VISUALIZATION OF NANOPARTICLES

As nanoparticles have the size of small cell components (for example, ribosomes, which have a diameter of approximately 30 nm), their identification in cells is difficult. We have already established a number of techniques, including the application of laser scanning microscopy in combination with immunofluorescence methods, which provide an important tool to optimize and characterize cell culture models. Several fluorescent markers can be used simultaneously with laser scanning microscopy (Rothen-Rutishauser et al. 1998a, 2000; Lamprecht and Lehr 2002). The resolution is the limiting factor in light microscopy. If, for instance, green light with a wavelength of 500 nm is used, the theoretical resolution according to the Abbe equation is approximately 200 nm. In laser scanning microscopy, the resolution is limited to 200 nm and

500–900 nm in the lateral and axial directions respectively. Mathematical restoration of the images (Shaw 1994; van der Voort and Strasters 1995; Rothen-Rutishauser et al. 1998b, 2006) can improve the resolution up to a factor of 2–3, provided that the signal-to-noise ratio is high and the effective point-spread function of the microscope is well known. However, when electrons instead of light are used, the theoretical resolution is much better and is in the range of 0.2 nm.

7.2.1 Laser Scanning Microscopy

Using a 100×/1.3 N.A. objective combined with a digital zoom (resulting in voxel dimensions of 50×50×150 nm), fluorescently labeled nanoparticles with a size of 20 nm can be detected within the cells (Figure 7.2a). However, single particles can only be detected after applying a deconvolution algorithm (Huygens Professional from SVI, Netherlands) to reduce the noise (Figure 7.2a).

7.2.2 Transmission Electron Microscopy

7.2.2.1 Conventional Transmission Electron Microscopy

Electron dense colloidal gold particles with a diameter of 0.025 μm (25 nm), a routinely used marker in electron microscopy (Bendayan 1984), could be identified in red blood cells, which do not contain other cytoplasmic structures inside (Figure 7.2b).

7.2.2.2 Analytical Transmission Electron Microscopy

The identification of small objects in transmission electron microscopy is not adequate for the ultrastructural analysis of nanoparticles, since their size and electron density are similar to those of other cellular components and with the contrast reagents also unspecific electron dense aggregates can also be produced. Electron energy loss spectroscopy and related methods have been applied to analyze nanoparticles, e.g., 20–30 nm titanium dioxide particles on ultrathin sections of lung tissue (Kapp et al. 2004) and in cell cultures (Figure 7.2c) (Rothen-Rutishauser et al. 2006).

7.2.3 New Visualization Tools

Nanoparticles are very suitable for manipulations at the molecular level and nanotechnology offers new tools for the delivery, imaging, and sensing in cancer research. Nanoparticle vectors circulating *in vivo*, assembled under molecular control, are capable of selective tumor targeting and potent delivery of therapeutics (Portney and Ozkan 2006). The development of quantum dots that are colloid nanocrystals with a few nanometers in diameter offers new nanoscale visualization tools for biological applications. Beyond biotechnology and cell imaging applications, quantum dots may be used for intravenous injection to target cellular markers of diseased tissue and organs in the human body. In addition to the visualization of quantum dots with fluorescence microscopy, quantum dots can be used as contrast agents in bioimaging with a combination of magnetic resonance imaging, positron emission tomography, and computed tomography (Michalet et al. 2005).

7.3 ANIMAL STUDIES

Animal studies have shown the potential of particles to cause pulmonary inflammation, blood changes, and alterations of specific cardiac endpoints. Increased pulmonary and cardiovascular mortality might be associated with high concentrations of airborne particles. Indications of a specific toxicological role for nanoparticles were found (Peters et al. 1997). Nanoparticles may induce inflammatory and prothrombotic responses, promoting atherosclerosis, thrombogenesis, and

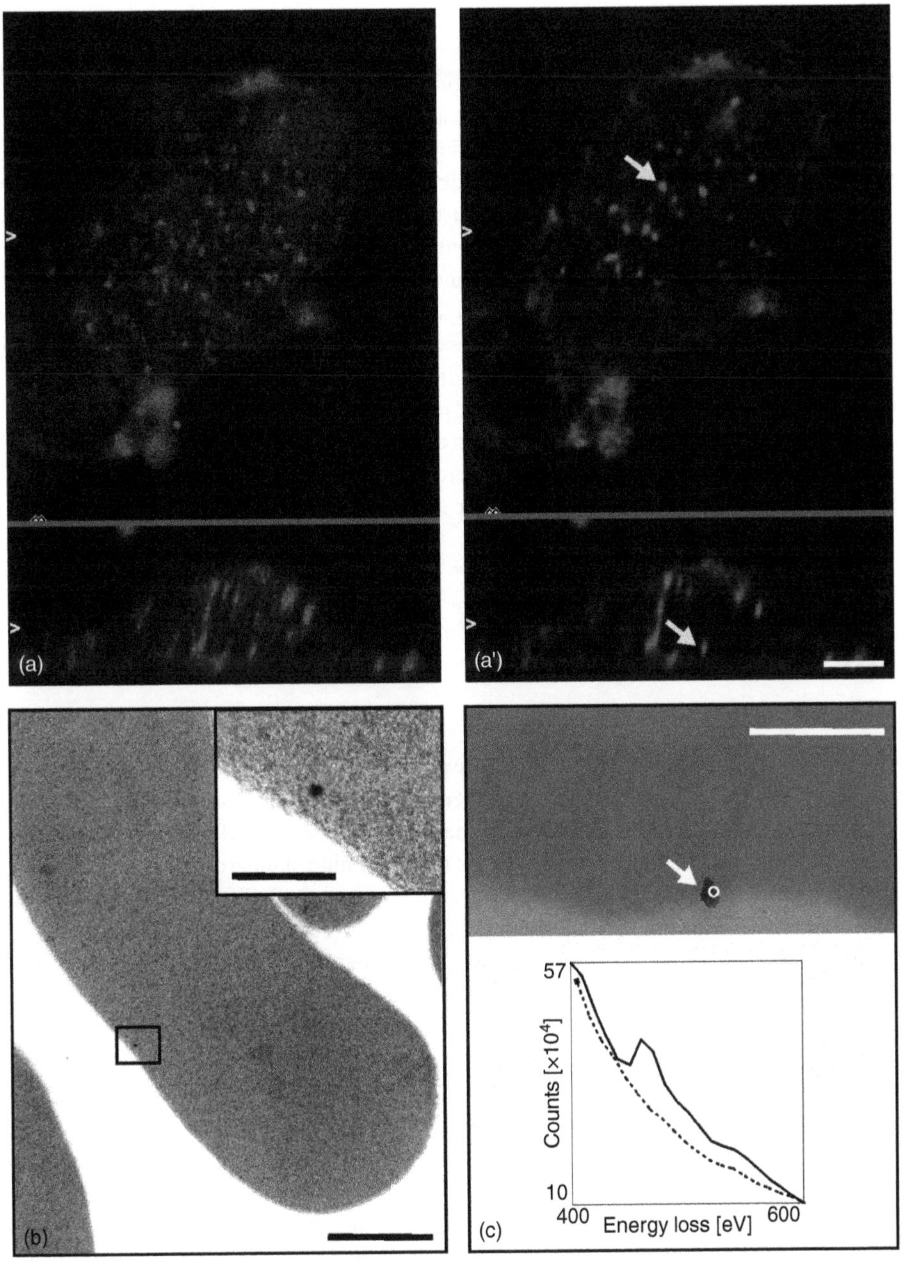

FIGURE 7.2 Visualization of nanoparticles by laser scanning microscopy. (a, a′) Micrograph of 0.02 μm fluorescent polystyrene spheres within a cultured macrophage before (a) and after deconvolution (a′). Single particles can only be detected after deconvolution (a′, arrows). Yellow open arrowheads mark the position of the projections. (b) Transmission electron microscopy image of a 0.025 μm gold particle within a red blood cell. The particle is not membrane-bound (b, inset). (c) Electron energy loss spectroscopy image of titanium dioxide (arrow). The white open circles mark the positions were the energy loss analysis was performed. The corresponding energy loss spectra (black lines) is shown, the dotted lines represent the background. (a, a′) Scale bar is 2 μm; (b) scale bar is 1 μm; (b, inset) scale bar is 0.2 μm; (c) scale bar is 0.5 μm.

the occurrence of other cardiovascular events. Moreover, inhaled nanoparticles may affect human lung physiology (Pietropaoli et al. 2004). In addition, there is evidence that nanoparticles may affect the autonomic nervous system or act directly on cells in various organs, and induce mutations (Samet, DeMarini, and Malling 2004; Schulz et al. 2005).

In lungs of rats that inhaled titanium dioxide nanoparticles of a count median diameter of 22 nm, particles were found on the luminal side of airways and alveoli as well as within each tissue compartment (Figure 7.3). The relative distributions of the nanoparticles in the lung compartments, airspace, epithelium and endothelium, connective tissue, and capillary lumen correlate with the volume fractions of these compartments, and interestingly enough, the relative distribution among the different compartments at 1 h and at 24 h after inhalation was the same (Geiser et al. 2005). This may be interpreted that nanoparticles move around freely within the different pulmonary tissue compartments and between them within 24 h. As nanoparticles were found in the capillary lumen, including in erythrocytes, it is not surprising that in other studies nanoparticles could be shown to appear in many compartments of the body, including the liver, the heart, and the nervous system within a few hours after exposure of the respiratory system (Brown, Zeman, and Bennett 2002; Kreyling et al. 2002; Oberdörster et al. 2004).

7.4 CELL CULTURE STUDIES

7.4.1 GENERAL

It is not possible to solve all problems and to answer all questions by conducting studies on animal models. Often cell cultures offer a much better system of investigation to study cellular and subcellular functions. Cultured human and animal cells can be controlled better and more reproducibly than *in vivo* systems, but *in vitro* systems require a high standardization to be reproducible. Guidelines for good cell culture practice (GCCP) are required, including the control of the starting material, e.g., the cultured cells, the culture medium, and the culture substratum (Gstraunthaler and Hartung 2002). The cells may be used from freshly isolated tissue (primary cultures) or may be a continuous cell line (secondary cultures). Both systems have advantages and disadvantages. Primary cultures represent a heterogeneous population of different cell types and they are extremely difficult to standardize. Cell lines are homogenous and more stable and hence are more reproducible. The disadvantage of cell lines, however, is that they retain little phenotypic differentiation. Nevertheless, if cell cultures are used properly, they represent a sophisticated and reproducible system with which basic questions can be answered and which may help to understand what happens *in vivo*.

The ability to characterize properly cell cultures is an important requirement for the work with cell cultures. The application of laser scanning microscopy, in combination with immunofluorescence methods, is an important tool to optimize and characterize cell culture models. Several fluorescent markers can be used simultaneously with the laser scanning microscopy (Rothen-Rutishauser et al. 1998a, 2000; Lamprecht and Lehr 2002), and, in combination with digital data restoration, the resolution of the images can be increased and the accuracy of the data improved (Rothen-Rutishauser et al. 1998b). The availability of fluorescent probes as indicators used in studies with living cells has increased during the last years (Lamprecht and Lehr 2002; Rothen-Rutishauser et al. 2002).

It is with cell culture experiments that particle cell interactions can be investigated. These experiments are necessary to study the interaction of nanoparticles with the membranes of a cellular system, the cell membrane, the nuclear membrane, and the organelle membrane (see Section 7.5 for a detailed description of particle–cell membrane interaction).

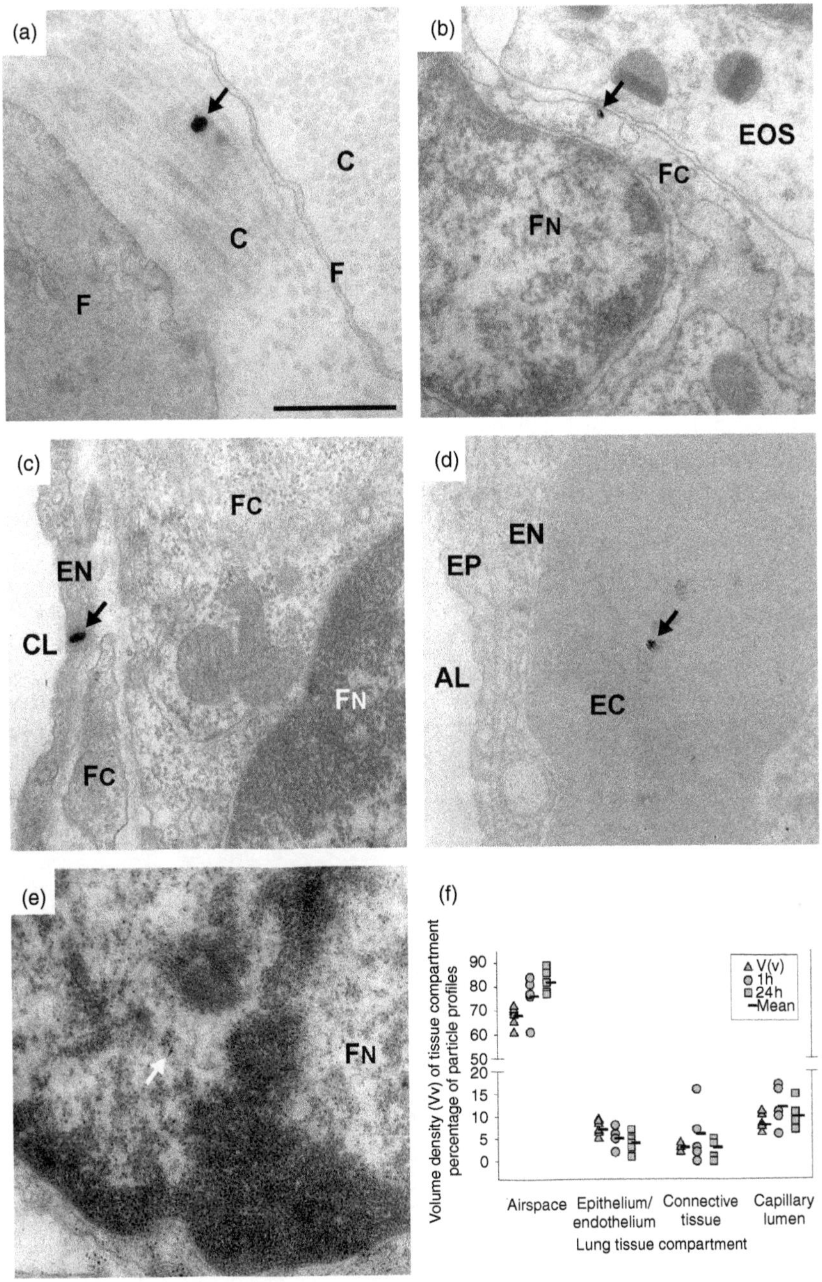

FIGURE 7.3 Analytical transmission electron microscopy micrographs of particles (arrows) in the lung parenchyma. (a) In the connective tissue, between collagen fibrils (C); fibrocytes (F). (b) In the cytoplasm of a fibrocyte (F_C), near its nucleus (F_N); eosinophil granulocyte (EOS). (c) In the cytoplasm of a capillary-endothelial cell (EN). (d) Within an erythrocyte (EC) in the capillary lumen (CL); alveolar lumen (AL), epithelium (EP). (e) Within the nucleus of a fibroblast (F_N). Scale bar is 500 nm. (f) Relative frequencies of particle profiles localized in the different lung compartments at 1 h and 24 h after inhalation. The volume densities of the different lung tissue compartments were taken from published data. (Burri et al. 1973; Tschanz et al. 1995, 2003; Pinkerton et al. 1992). (Reproduced from Geiser, M., Rothen-Rutishauser, B., Kapp, N., Schürch, S., Kreyling, W., Schulz, H., Semmler, M., Im Hof, V., Heyder, J., and Gehr, P., *Environ. Health Perspect.*, 113, 1555–1560, 2005. With permission.)

7.4.2 Macrophages (As a Model for Phagocytic Cells)

Cultured porcine lung macrophages have been used as a model for phagocytic cells. The uptake of uncharged fine (1 and 0.2 μm), ultrafine (0.078 or 78 nm), and negatively charged ultrafine (0.02 or 20 nm) fluorescently labeled polystyrene particles was investigated comparatively (Geiser et al. 2005). We found particles of all sizes and charges within macrophages, whereby the fraction of cells containing particles was differing (Figure 7.3). On average, 73%–77% (SD 11%–17%) of the macrophages contained uncharged as well as negatively charged nanoparticles, while 21% (SD 11%) contained 0.2 μm particles, and 56% (SD 30%) the 1 μm particles (Figure 7.4). The presence of cytochalasin D, a fungal metabolite that is known to block phagocytosis due to depolymerization of the actin filament network, did not inhibit the uptake of ultrafine (negatively charged and uncharged) or 0.2 μm particles by macrophages, whereas it blocked the uptake of 1 μm particles (Figure 7.3 and Figure 7.4). None of the endocytic pathways, which all include vesicle formation, are likely to account for the translocation of nanoparticles into the cells. Moreover, since cytochalasin D treatment of macrophages did not prevent nanoparticle translocation into these cells, particle uptake by any actin-based mechanism can also be excluded.

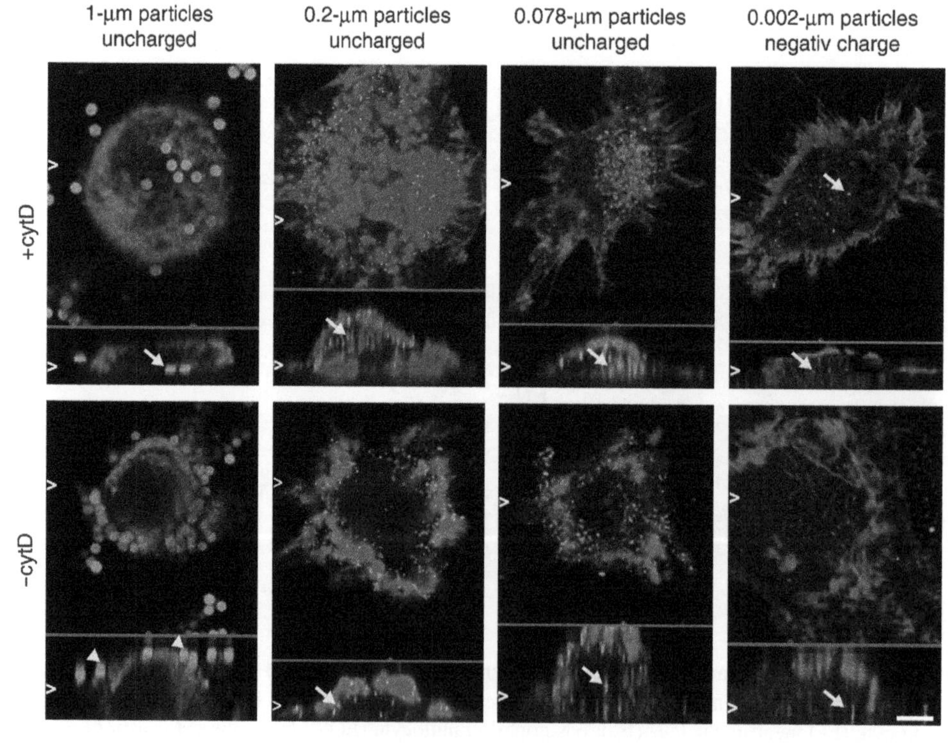

FIGURE 7.4 (See color insert) Laser scanning microscopy micrographs of fluorescent polystyrene spheres (1, 0.2, 0.078 and 0.02 μm) taken up by macrophages in the absence or presence of cytochalasin D (cytD). F-Actin is shown in red, particles are green. The *xy* and *xz* projections allow the clear differentiation between intracellular (arrows) and extracellular (arrowheads) particles. Yellow open arrowheads mark the position of projections. Scale bar is 2 μm. (Pictures with 1, 0.2, and 0.078 μm particles are reproduced from Geiser, M., Rothen-Rutishauser, B., Kapp, N., Schürch, S., Kreyling, W., Schulz, H., Semmler, M., Im Hof, V., Heyder, J., and Gehr, P., *Environ. Health Perspect.*, 113, 1555–1560, 2005. With permission.)

7.4.3 Red Blood Cells (As a Model for Nonphagoctic Cells)

Human red blood cells have been used as a model for nonphagocytic cells as they do not have any receptors on their surface and the cytoskeleton of the red cell is not arranged for endocytosis. The red blood cell membrane consists of the phospholipid bilayer containing integral membrane proteins and an underlying membrane cytoskeleton typical for these cells (for a review, see Gratzer 1981). We found ultrafine (uncharged and negatively charged) and 0.2 μm particles within red blood cells, but not larger ones (Figure 7.5) (Geiser et al. 2005; Rothen-Rutishauser et al. 2006). Other particles with different surface properties, i.e., gold and ultrafine titanium dioxide particles, were also included in the *in vitro* studies with red blood cells. We also showed by transmission electron microscopy and electron energy loss spectroscopic analysis that 0.025 μm gold and ultrafine titanium dioxide particles, after their incubation with red blood cells, were inside the cells and not membrane bound (Rothen-Rutishauser et al. 2006). In conclusion, we found by microscopic analyses that different types of nanoparticles are taken up by red blood cells as long as they are equal or smaller than 0.2 μm in diameter. We did not see any difference in particle uptake

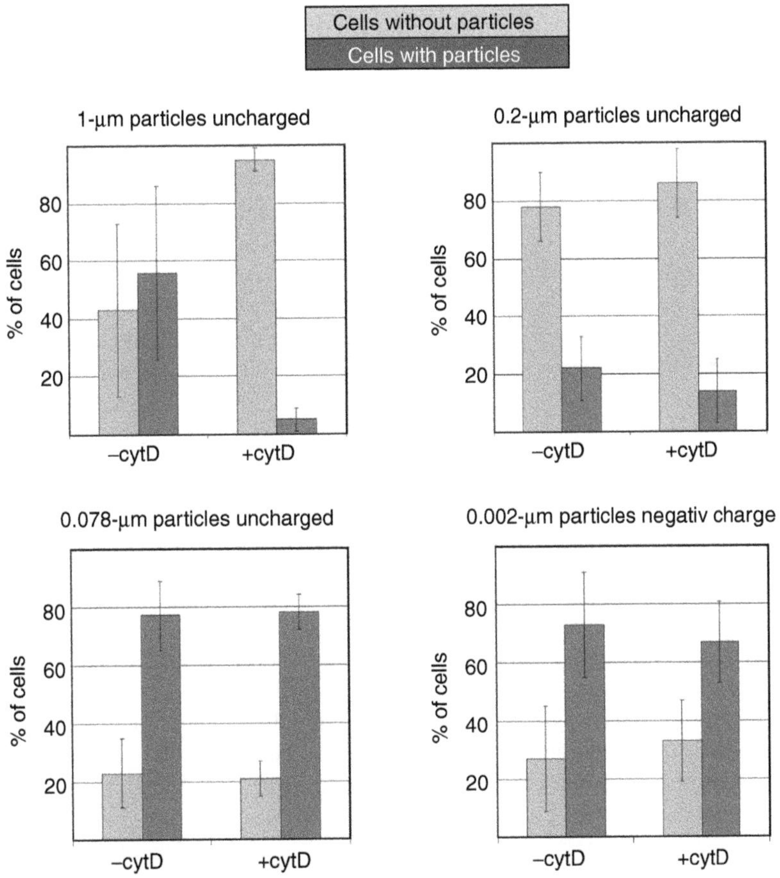

FIGURE 7.5 Particle uptake by macrophages in the absence and presence of cytochalasin D (cytD). Data are expressed as mean ± SD of 3–4 experiments with scanning of 30–50 cells each by laser scanning microscopy. Blue bars represent cells with particles, yellow bars those without particles. (Graphs with 1, 0.2, and 0.078 μm particles are reproduced from Geiser, M., Rothen-Rutishauser, B., Kapp, N., Schürch, S., Kreyling, W., Schulz, H., Semmler, M., Im Hof, V., Heyder, J., and Gehr, P., *Environ. Health Perspect.*, 113, 1555–1560, 2005. With permission.)

with respect to differing surface charges or surface chemistry, as we used three different particle types, metal, metal oxide, and synthetic polymeric materials uncharged or with negative or positive charges. All particles found within the red blood cells were not membrane-bound. Thus, it appears that particle size is the most important factor for their translocation into cells. However, the translocation mechanism is still not clear.

7.4.4 *In Vitro* Models of Airway and Alveolar Epithelium

The cultivation of epithelial cells on permeable supports allows keeping the solutions on both sides of the cultured epithelium and differentiation is increased. After the introduction of this method by Handler, Green, and Steele (1989), many cell culture models of the human air–blood barrier have been described in the literature (Fuchs et al. 2002; Meaney et al. 2002; Steimer, Haltner, and Lehr 2005).

It has been suggested that the alveolar epithelium in particular is the appropriate lung surface to target for delivery of macromolecules. Therefore, *in vitro* models of the alveolar epithelium represent valuable research tools. The use of primary cultures of isolated alveolar type II cells has been described extensively (Fuchs et al. 2002). There is evidence that isolated alveolar type II cells in culture loose their characteristic alveolar type-II phenotype and acquire some morphological and biochemical markers of the alveolar type I cell *in vivo* (Elbert et al. 1999). It is known from *in vivo* studies that alveolar type II cells are able to repair damaged type I epithelium (Bachofen and Weibel 1974).

Among primary cultures, the use of cell lines is also very common. For example, the A549 human epithelial cell line is proposed to be a standardized model for any studies of lung epithelium (Lieber et al. 1976; Foster et al. 1998). It has been shown that the A549 cells have many important biological properties of alveolar epithelial type II cells, e.g., membrane-bound inclusions, which resemble lamellar bodies of type II cells (Shapiro et al. 1978). Other ultrastructural characteristics common to type II cells have also been described, as for example, distinct polarization, tight junctions, and extensive cytoplasmic extensions (Stearns, Paulauskis, and Godleski 2001).

Only recently, some drug transport studies involving airway epithelial cell cultures appeared in the literature (for a review, see Steimer, Haltner, and Lehr 2005). These studies involved primary cultures of different species (Meaney et al. 2002) and also established cell lines, such as the well-defined human 16HBE14o-cells (Forbes and Lansey 1998) and Calu-3 (Cavet, West, and Simmons 1997), both representing bronchial epithelial cells.

7.4.5 Triple Cell Co-Culture System of the Airway Epithelial Barrier

Recently, an *in vitro* system has been introduced and characterized as a model to simulate the human airway epithelial barrier to study the interactions of the airway cells with deposited particles (Rothen-Rutishauser, Kiama, and Gehr 2005; Blank et al. in press).

A triple co-culture model system composed of cuboidal epithelial cells, macrophages, and dendritic cells was established and evaluated in terms of its functional relevance to the *in vivo* tissue. For this purpose, the human (alveolar) epithelial cell line A549, which originated from human lung carcinoma (Lieber et al. 1976), was chosen. The A549 cell line is routinely used as an *in vitro* model for pulmonary cuboidal epithelium to study the interaction of environmental particles (Stringer, Imrich, and Kobzik 1996) and nanoparticles (Stearns, Paulauskis, and Godleski 2001) with these cells. Bilayer models with A549 and endothelial cells to study bacteria infections have also been described (Birkness et al. 1999; Bermudez et al. 2002), as well as co-cultures of A549 cells with an alveolar macrophage cell line to study the effect of lung surfactant phospholipids during the uptake of microspheres (Jones et al. 2002). In our study, human macrophages and dendritic cells derived from human blood monocytes (Thiele et al. 2001) were combined with A549 cells. The cultures of single cells as well as the triple co-cultures were characterized in terms of their typical features, e.g., morphology of the cells, integrity of the epithelial layer, and expression of specific cell surface markers (CD14 for macrophages, and CD86 for dendritic cells) (Figure 7.5).

The interplay of epithelial cells with macrophages and dendritic cells during the uptake of polystyrene particles was investigated with laser scanning microscopy and transmission electron microscopy. With this triple co-culture system, it could clearly be shown that particles ≤ 1 μm reach dendritic cells (Figure 7.6). The fact that transepithelial electrical resistance did not decrease during the experiment suggests that particles are transported through the epithelial cells. It was confirmed that particles were within all three cell types, but to a different extent. Whereas macrophages and dendritic cells had taken up numerous particles, only few particles were seen in the epithelial cells (Rothen-Rutishauser, Kiama, and Gehr 2005). The 3D model will be used to investigate particle uptake and translocation by the three cell types and their possible interplay

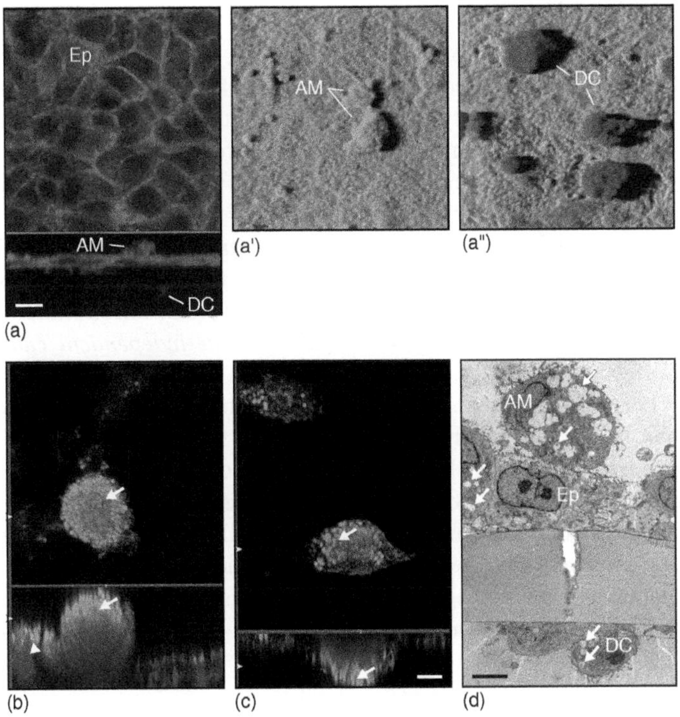

FIGURE 7.6 (See color insert) Laser scanning microscopy and transmission electron microscopy images of triple co-cultures with epithelial cells, macrophages, and DCs grown on a filter insert. (a, a′, a″) Epithelial cells are shown in yellow, macrophages (AM) in turquoise, and dendritic cells (DC) in pink. Macrophages reside on the surface of the epithelial cells, whereas dendritic cells were localized at the bottom side of the insert. (a) Represents *xy* and *xz* projections; yellow arrowheads mark the position of projections. (a′ and a″) are 3D reconstructions from the data set, (a′) from the upper side, (a″) from the lower side. (b, c, d) 1 μm Particles uptake in triple cell co-cultures. Triple cell co-cultures were incubated for 24 h, fixed and stained afterwards for F-Actin (red) and CD14 (b, turquoise) or CD86 (c, turquoise). Particles (green) were found in CD14 positive (b, arrows), in epithelial cells (b, arrowhead), and in CD86 positive dendritic cells (c, arrows). Images represent *xy* and *xz* projections; yellow arrowheads mark the position of projections. Transmission electron microscopy images of triple co-cultures exposed 24 h to 1 μm particles (d). Particles (d, arrows) were found within epithelial cells (Ep), macrophages (AM), and dendritic cells (DC). (a, a′, a″) Scale bar is 10 μm; (b, c) scale bar is 5 μm; (d) scale bar is 2 μm. (Images b, c and d are reproduced from Rothen-Rutishauser, B. M., Kiama, S. G., and Gehr, P., *Am. J. Respir. Cell Mol. Biol.*, 32, 281–289, 2005. With permission.)

during this process. The two-chamber system with the possibility of cell-to-cell communication is a more realistic human tissue model than the standard monolayers. More investigations are needed, including quantification of the particle transfer from the air to the dendritic cells, to understand the pathway of antigens and the interplay of the different cell types involved.

7.5 PARTICLE–CELL MEMBRANE INTERACTIONS: MECHANISM AND SPECULATIONS

The microscopic analyses of phagocytic and nonphagocytic cells incubated with different particle types has shown that macrophages take up fine and ultrafine polystyrene microspheres and that treatment with cytochalasin D inhibits the uptake of 1 μm particles, but not of the smaller ones by these cells. Ultrafine polystyrene and gold particles were also found to enter red blood cells, and they were not membrane-bound either.

7.5.1 Uptake by Endocytosis

The mechanisms of intracellular uptake of macromolecules, particles, and even cells are subsumed as *endocytosis*. Material to be ingested is progressively enclosed by the plasma membrane, which eventually detaches to form an endocytic vesicle. Phagocytosis and pinocytosis are distinguished by the size of endocytic vesicles formed. *Phagocytosis*, a receptor-mediated, actin-based process is characteristic for neutrophils, macrophages, and dendritic cells. It is the main mechanism for the clearance of insoluble 1–3 μm particles from the alveoli. *Pinocytosis* involves the ingestion of fluid and solutes via vesicles of about 100 nm in diameter. There are at least four basic mechanisms, most of which can be demonstrated in lungs; they involve specific receptor–ligand interactions: *Macropinocytosis*, *clathrin-mediated actin-based endocytosis*, *caveolae-mediated endo- or transcytosis*, and *clathrin- and caveolae-independent endocytosis* (Conner and Schmid 2003).

Kruth et al. (1999) described an additional special endocytic process called *patocytosis,* in which hydrophobic <0.5 μm polystyrene particles are transported through induced plasma membrane channels into an extensive labyrinth of interconnected membrane-bound compartments. None of these endocytic pathways, which all include vesicle formation, is likely to account for the translocation of nanoparticles in our study, as intracellularly localized particles were not membrane-bound. Moreover, since red blood cells contained ultrafine particles (Figure 7.7) and cytochalasin D treatment of macrophages did not prevent nanoparticle translocation into these cells (Figure 7.4), uptake of individual nanoparticles by any actin-based mechanism can also be excluded.

7.5.2 Transport via Pores

Transport via pores as suggested for lung–blood substance exchange (Conhaim et al. 1988; Hermans and Bernard 1999) is a potential mechanism for nanoparticle translocation. Titanium dioxide particles may diffuse through such pores. Transport mechanism by diffusion is consistent with the observed spatial distribution of nanoparticles in our inhalation study. Signal-mediated transport via pores has as of yet been demonstrated only for ultrafine gold particles of up to 39 nm in diameter through the nuclear pore complex in Xenopus oocytes, where transport velocities depended on particle size (Panté and Kann 2002).

7.5.3 Passive Uptake

Passive uptake (not triggered by receptor–ligand interactions) may occur by electrostatic, including hydrogen bond, van der Waals, and hydrophobic interactions. In addition, steric interactions may have to be considered for particles with aqueous surface polymers. These interactions are subsumed

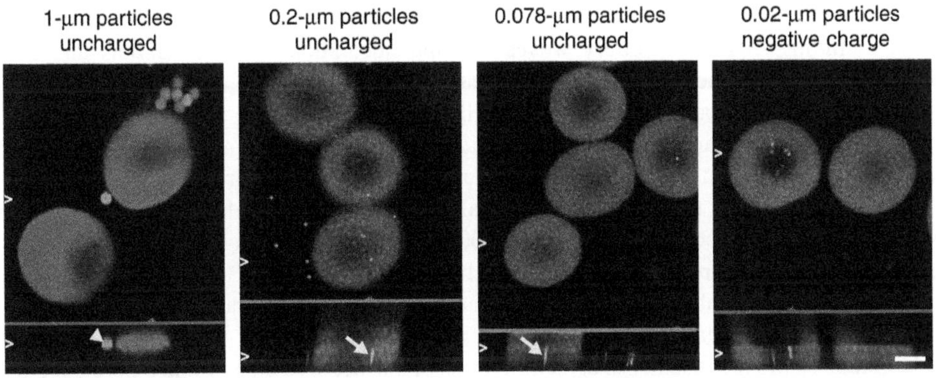

FIGURE 7.7 (See color insert) Laser scanning microscopy micrographs of fluorescent polystyrene spheres (1, 0.2, 0.078 and 0.02 μm) taken up by red blood cells. Autofluorescence of the cells is shown in red, particles are green. The *xy* and *xz* projections allow the clear differentiation between internalized (arrows) and extracellular (arrowheads) particles. Yellow open arrowheads mark the position of the projections. Scale bar is 2 μm. (Pictures with 1, 0.2, and 0.078 μm particles are reproduced from Geiser, M., Rothen-Rutishauser, B., Kapp, N., Schürch, S., Kreyling, W., Schulz, H., Semmler, M., Im Hof, V., Heyder, J., and Gehr, P., *Environ. Health Perspect.*, 113, 1555–1560, 2005. With permission.)

under "adhesive interactions" (Rimai et al. 2000). Rimai and coworkers showed that 8 μm glass particles were engulfed by a polystyrene substrate by approximately 90%, while 22 μm particles only by 30%. However, the influence of particle size on their engulfment was not clarified in these publications.

7.5.4 Thermodynamic Considerations at Vapor–Liquid and Fluid–Liquid Interfaces

Several concepts for the nonspecific engulfment of particles through interfacial structures (including cell membranes) have been suggested. A thermodynamic model using the "wettability criterion" was successful in predicting passive particle uptake, although it did not take into account the elastic properties of the cell membrane (Chen, Langer, and Edwards 1997). Another thermodynamic analysis combined with a molecular dynamics simulation showed that line tension influences the wetting behavior of nanoparticles at liquid–vapor and liquid–liquid interfaces (Bresme and Quirke 1999). The authors found negative line tension values for particles of few nanometers in diameter, but positive values for those an order of magnitude larger. A negative line tension favors the initial wetting of a spherical particle following its approach to a vapor–fluid interface.

The engulfment of solid surface heterogeneities equivalent to particles in the nanometer range by unbalanced capillary forces (free energy perturbations) was studied by Shanahan (Shanahan 1990).

Experimental results demonstrate consistently greater immersion of smaller particles than larger ones into a liquid substrate covered by surfactant film *in vitro*, as well as in situ in airways. This supports the concept that line tension plays a significant role in particle displacement into a liquid substrate (Schürch et al. 1999; Geiser et al. 2000).

7.5.5 Adhesive Interactions, Membrane Stability, and Dynamics

Biomembrane structure is mainly fluid and held together by hydrophobic interactions, and in addition, there are intimate covalent links within proteins and lipid molecules (Evans and Ludwig 2000). Different types of lipids, cholesterol, and proteins play an important role for the membrane stability.

Nanoparticles may cross biomembranes through holes created by a thermally activated process, including thermal capillary waves and the nucleation of pores (Evans and Ludwig 2000). Adhesive interaction between the molecules of the membrane and the nanoparticle may cause the nucleation of pores. Small, unstable holes may have a life-time of only a few nanoseconds. Pore size can range from nano- to micrometer in radius. For phosphatidyl bilayers, a membrane area expansion of only 2%–4% can cause membrane rupture, which is caused by the nucleation of pores. The corresponding external tension is in the order of 10^{-3}–10^{-2} N/m as shown by Evans and associates (Evans et al. 2003). In the lung, during inhalation, the epithelial cells will increase their area during normal tidal breathing by 5%–10%, and during a big breath up to 20% (Bachofen et al. 1987). This likely causes sufficient membrane stress to cause nucleation of cell membrane pores.

Membrane stability is usually analyzed in terms of surface free energy and a pore edge energy that replaces the line energy or line tension of a 3-phase line of, for example, an oil droplet sitting an a water–vapor surfactant film (Schürch, Bachofen, and Weibel 1985).

As mentioned above, membranes are held together mainly by hydrophobic interactions, as the exposure to water of the fatty acid chains is not favored thermodynamically. However, under lateral stress caused by membrane stretching or by adhesion driven tension (adhesive interactions), holes may be created. Evans and Heinrich (2003) have explored steps of tension in time to analyze the kinetic process of hole nucleation and the dynamics of membrane failure (Evans and Heinrich 2003). Thus, the passage of nanoparticles through the cell membrane may be initiated by a membrane tension caused by cell stretching or by adhesive interaction between the cell membrane and the particle. A nanoscale hole will open the cell membrane facilitating particle uptake by the cell.

The mechanics of membrane stability as outlined above may offer a way to further explore which chemical and physical properties of membranes and particles are responsible for the translocation of nanoparticles *in vivo*. Interestingly, we did not see any difference in particle uptake *in vitro* with respect to differing surface charges or surface chemistry as we used three different particle types, a metal, a metal oxide, and a synthetic polymeric material (data not shown). However, these particles were added to the cells in suspension and did not approach the cells from the air and require passage through a surfactant film first, as in the *in vivo* inhalation experiments. In these *in vivo* experiments, electrostatic interactions are likely important for particle deposition and subsequent retention.

7.6 CONCLUSION

Nanoparticles of various materials can cross any cellular membrane, but neither endocytosis, which is based on vesicle formation, nor any actin-based mechanisms are likely to account for individual nanoparticle translocation into the cell. Our results from the inhalation experiments with titanium dioxide particles suggest a transport mechanism that includes a nonspecific entering of particles of cells caused by adhesive interactions and perhaps the creation of temporary membrane pores by such interactions. After particle deposition in the lung, nanoparticles may cross the tissue and cellular barriers and may reach different tissue compartments with various cell types. They may also enter capillaries and ultimately end up in many other organs. It may be impossible to prevent, influence, or direct their entering on the cellular level. Moreover, the adverse potential of nanoparticles is greatly enhanced by their free location and movement within cells, which promotes interactions with intracellular proteins and organelles and even the nuclear DNA.

ACKNOWLEDGMENTS

We are grateful to Sandra Frank for her help to prepare the manuscript.

This work was supported by the Swiss National Science Foundation (Nr. 3200-065352.01), the Swiss Agency for the Environment, Forests and Landscape, and the Silva Casa Foundation.

REFERENCES

Bachofen, M. and Weibel, E. R., Basic pattern of tissue repair in human lungs following unspecific injury, *Chest*, 65, 14S–19S, 1974.

Bachofen, H., Schurch, S., Urbinelli, M., and Weibel, E. R., Relations among alveolar surface tension, surface area, volume and recoil pressure in rabbit lungs, *J. Appl. Physiol.*, 62, 1878–1887, 1987.

Bendayan, M., Protein-A-gold electron microscopic cytochemistry: methods, applications, and limitations, *J. Electron Microsc. Technol.*, 1, 243–270, 1984.

Bermudez, L. E., Sangari, F. J., Kolonoski, P., Petrofsky, M., and Goodman, J., The efficiency of the translocation of Mycobacterium tuberculosis across a bilayer of epithelial and endothelial cells as a model of the alveolar wall is a consequence of transport within mononuclear phagocytes and invasion of alveolar epithelial cells, *Infect. Immun.*, 70, 140–146, 2002.

Birkness, K. A., Deslauriers, M., Bartlett, J. H., White, E. H., King, C. H., and Quinn, F. D., An in vitro tissue culture bilayer model to examine early events in Mycobacterium tuberculosis infection, *Infect. Immun.*, 67, 653–658, 1999.

Blank, F., Rothen-Rutishauser, B., Schurch, S., and Gehr, P., An optimized in vitro model of the respiratory tract wall to study particle cell interactions, *J. Aerosol Med.* in press, 2006.

Borm, P. J. and Kreyling, W., Toxicological hazards of inhaled nanoparticles—potential implications for drug delivery, *J. Nanosci. Nanotechnol.*, 4 (5), 521–531, 2004.

Brain, J. D., Lung macrophages: how many kinds are there? What do they do? *Am. Rev. Respir. Dis.*, 137, 507–509, 1988.

Brain, J. D., Gehr, P., and Kavet, R. I., Airway macrophages. The importance of the fixation method, *Am. Rev. Respir. Dis.*, 129, 823–826, 1984.

Bresme, F. and Quirke, N., Nanoparticulates at liquid/liquid interfaces, *Phys. Chem. Chem. Phys.*, 1, 2149–2155, 1999.

Brook, R. D., Brook, J. R., and Rajagopalan, S., Air pollution: the "Heart" of the problem, *Curr. Hypertens. Rep.*, 5, 32–39, 2003.

Brown, J. S., Zeman, K. L., and Bennett, W. D., Ultrafine particle deposition and clearance in the healthy and obstructed lung, *Am. J. Respir. Crit. Care Med.*, 166, 1240–1247, 2002.

Brown, D. M., Donaldson, K., Borm, P. J., Schins, R. P., Dehnhardt, M., Gilmour, P., Jimenez, L. A., and Stone, V., Calcium and ROS-mediated activation of transcription factors and TNF-alpha cytokine gene expression in macrophages exposed to ultrafine particles, *Am. J. Physiol. Lung Cell Mol. Physiol.*, 286, L344–L353, 2004.

Burri, H. P., Dbaly, J., and Weibel, E. W., The postnatal growth of the rat lung I. Morphometry, *Anat. Rec.*, 178, 711–730, 1973.

Cavet, M. E., West, M., and Simmons, N. L., Transepithelial transport of the fluoroquinolone ciprofloxacin by human airway epithelial Calu-3 cells, *Antimicrob. Agents Chemother.*, 41, 2693–2698, 1997.

Chen, H., Langer, R., and Edwards, D. A., A film tension theory of phagocytosis, *J. Colloid Interface Sci.*, 190, 118–133, 1997.

Conhaim, R. L., Eaton, A., Staub, N. C., and Heath, T. D., Equivalent pore estimate for the alveolar-airway barrier in isolated dog lung, *J. Appl. Physiol.*, 64, 1134–1142, 1988.

Conner, S. D. and Schmid, S. L., Regulated portals of entry into the cell, *Nature*, 422, 36–44, 2003.

Donaldson, K., Stone, V., Seaton, A., and MacNee, W., Ambient particle inhalation and the cardiovascular system: potential mechanisms, *Environ. Health Perspect.*, 109, 523–527, 2001.

Donaldson, K., Tran, L., Jimenez, L. A., Duffin, R., Newby, D. E., Mills, N., MacNee, W., and Stone, V., Combustion-derived nanoparticles: a review of their toxicology following inhalation exposure, *Part. Fibre Toxicol.*, 2, 10, 2005.

Dreher, D., Kok, M., Cochand, L., Kiama, S. G., Gehr, P., Pechere, J. C., and Nicod, L. P., Genetic background of attenuated Salmonella typhimurium has profound influence on infection and cytokine patterns in human dendritic cells, *J. Leukoc. Biol.*, 69, 583–589, 2001.

Dudek, S. M. and Garcia, J. G., Cytoskeletal regulation of pulmonary vascular permeability, *J. Appl. Physiol.*, 91, 1487–1500, 2001.

Dunsmore, S. E. and Rannels, D. E., Extracellular matrix biology in the lung, *Am. J. Physiol.*, 270, L3–L27, 1996.

Elbert, K. J., Schafer, U. F., Schafers, H. J., Kim, K. J., Lee, V. H., and Lehr, C. M., Monolayers of human alveolar epithelial cells in primary culture for pulmonary absorption and transport studies, *Pharm. Res.*, 16, 601–608, 1999.

Evans, E. and Heinrich, V., Hydrodynamics and physics of soft objects. Dynamic strength of fluid membranes, *C.R. Physique*, 4, 265–274, 2003.

Evans, E. and Ludwig, F., Dynamic strength of molecular anchoring and material cohesion in fluid biomembranes, *J. Phys. Condens. Matter*, 12, A315–A320, 2000.

Evans, E., Heinrich, V., Ludwig, F., and Rawicz, W., Dynamic tension spectroscopy and strength of biomembranes, *Biophys. J.*, 85, 2342–2350, 2003.

Ferin, J., Oberdörster, G., and Penney, D. P., Pulmonary retention of ultrafine and fine particles in rats, *Am. J. Respir. Cell Mol. Biol.*, 6, 535–542, 1992.

Forbes, B. and Lansey, A. B., Transport characteristics of formoterol and salbutamol across a bronchial epithelial drug absorption model, *Eur. J. Pharm. Sci.*, 6, S24, 1998.

Foster, K. A., Oster, C. G., Mayer, M. M., Avery, M. L., and Audus, K. L., Characterization of the A549 cell line as a type II pulmonary epithelial cell model for drug metabolism, *Exp. Cell Res.*, 243, 359–366, 1998.

Fuchs, S., Gumbleton, M., Schäefer, U. F., and Lehr, C. M., Models of the alveolar epithelium, In *Cell Culture Models of Biological Barriers. In-Vitro Test Systems for Drug Absorption and Delivery*, Lehr, C.M., ed., Taylor and Francis, London, New York, 13, 189–210, 2002.

Gehr, P., Bachofen, M., and Weibel, E. R., The normal human lung: ultrastructure and morphometric estimation of diffusion capacity, *Respir. Physiol.*, 32, 121–140, 1978.

Gehr, P., Schürch, S., Berthiaume, Y., Im Hof, V., and Geiser, M., Particle retention in airways by surfactant, *J. Aerosol Med.*, 3, 27–43, 1990.

Gehr, P., Green, F. H., Geiser, M., Im Hof, V., Lee, M. M., and Schürch, S., Airway surfactant, a primary defense barrier: mechanical and immunological aspects, *J. Aerosol Med.*, 9, 163–181, 1996.

Geiser, M., Baumann, M., Cruz-Orive, L. M., Im Hof, V., Waber, U., and Gehr, P., The effect of particle inhalation on macrophage number and phagocytic activity in the intrapulmonary conducting airways of hamsters, *Am. J. Respir. Cell Mol. Biol.*, 10, 594–603, 1994.

Geiser, M., Schürch, S., Im Hof, V., and Gehr, P., Structural and interfacial aspects of particle retention, In *Particle-Lung Interaction*, Gehr, P., and Heyder, J., eds., vol. 143, Lung Biology in Health and disease, Lenfant C, exec. ed., Marcel Dekker Inc., New York, 291–322, 2000.

Geiser, M., Schürch, S., and Gehr, P., Influence of surface chemistry and topography of particles on their immersion into the lung's surface-lining layer, *J. Appl. Physiol.*, 94, 1793–1801, 2003a.

Geiser, M., Matter, M., Maye, I., Im Hof, V., Gehr, P., and Schurch, S., Influence of airspace geometry and surfactant on the retention of man-made vitreous fibers (MMVF 10a), *Environ. Health Perspect.*, 111, 895–901, 2003b.

Geiser, M., Rothen-Rutishauser, B., Kapp, N., Schürch, S., Kreyling, W., Schulz, H., Semmler, M., Im Hof, V., Heyder, J., and Gehr, P., Ultrafine particles cross cellular membranes by non-phagocytic mechanisms in lungs and in cultured cells, *Environ. Health Perspect.*, 113, 1555–1560, 2005.

Gil, J. and Weibel, E. R., Extracellular lining of bronchioles after perfusion-fixation of rat lungs for electron microscopy, *Anat. Rec.*, 169, 185–200, 1971.

Godfrey, R. W., Human airway epithelial tight junctions, *Microsc. Res. Tech.*, 38, 488–499, 1997.

Gonzalez-Flecha, B., Oxidant mechanisms in response to ambient air particles, *Mol. Aspects Med.*, 25, 169–182, 2004.

Gratzer, W. B., The red cell membrane and its cytoskeleton, *Biochem. J.*, 198, 1–8, 1981.

Gstraunthaler, G. and Hartung, T., Good cell culture practice: good laboratory practice in the cell culture laboratory for the standardization and quality assurance of *in vitro* studies, In *Cell Culture Models of Biological Barriers. In-Vitro Test Systems for Drug Absorption and Delivery*, Lehr, C.-M., ed., vol. 7, Taylor and Francis, New York, 112–120, 2002.

Gurr, J. R., Wang, A. S., Chen, C. H., and Jan, K. Y., Ultrafine titanium dioxide particles in the absence of photoactivation can induce oxidative DNA damage to human bronchial epithelial cells, *Toxicology*, 15, 66–73, 2005.

Handler, J. S., Green, N., and Steele, R. E., Cultures as epithelial models: porous-bottom culture dishes for studying transport and differentiation, *Methods Enzymol.*, 171, 736–744, 1989.

Hermans, C. and Bernard, A., Lung epithelium-specific proteins: characteristics and potential applications as markers, *Am. J. Respir. Crit. Care Med.*, 159, 646–678, 1999.

Hoet, P. H., Bruske-Hohlfeld, I., and Salata, O. V., Nanoparticles—known and unknown health risks, *J. Nanobiotechnol.*, 2, 12, 2004.

Holt, P. G., Pulmonary dendritic cells in local immunity to inert and pathogenic antigens in the respiratory tract, *Proc. Am. Thorac. Soc.*, 2, 116–120, 2005.

Holt, P. G. and Schon-Hegrad, M. A., Localization of T cells, macrophages and dendritic cells in rat respiratory tract tissue: implications for immune function studies, *Immunology*, 62, 349–356, 1987.

Holt, P. G., Schon-Hegrad, M. A., and McMenamin, P. G., Dendritic cells in the respiratory tract, *Int. Rev. Immunol.*, 6, 139–149, 1990.

Ichiyasu, H., McCormack, J. M., McCarthy, K. M., Dombkowski, D., Preffer, F. I., and Schneeberger, E. E., Matrix metalloproteinase-9-deficient dendritic cells have impaired migration through tracheal epithelial tight junctions, *Am. J. Respir. Cell Mol. Biol.*, 30, 761–770, 2004.

Jones, B. G., Dickinson, P. A., Gumbleton, M., and Kellaway, I. W., Lung surfactant phospholipids inhibit the uptake of respirable microspheres by the alveolar macrophage NR8383, *J. Pharm. Pharmacol.*, 54, 1065–1072, 2002.

Kapp, N., Kreyling, W., Schulz, H., Im Hof, V., Gehr, P., Semmler, M., and Geiser, M., Electron energy loss spectroscopy for analysis of inhaled ultrafine particles in rat lungs, *Microsc. Res. Tech.*, 63, 298–305, 2004.

Kiama, S. G., Cochand, L., Karlsson, L., Nicod, L. P., and Gehr, P., Evaluation of phagocytic activity in human monocyte-derived dendritic cells, *J. Aerosol Med.*, 14, 289–299, 2001.

Kilburn, K. H., A hypothesis for pulmonary clearance and its implications, *Am. Rev. Respir. Dis.*, 98, 449–463, 1968.

Kreyling, W. G., Semmler, M., Erbe, F., Mayer, P., Takenaka, S., and Schulz, H., Translocation of ultrafine insoluble iridium particles from lung epithelium to extrapulmonary organs is size dependent but very low, *J. Toxicol. Environ. Health*, 65, 1513–1530, 2002.

Kruth, H. S., Chang, J., Ifrim, I., and Zhang, W. Y., Characterization of patocytosis: endocytosis into macrophage surface- connected compartments, *Eur. J. Cell. Biol.*, 78, 91–99, 1999.

Kunzli, N. and Tager, I. B., Air pollution: from lung to heart, *Swiss Med. Wkly*, 135 (47–48), 697–702, 2005.

Lamprecht, A. and Lehr, C.-M., Confocal and two-photon fluorescence microscopy, In *Cell Culture Models of Biological Barriers. In-Vitro Test Systems for Drug Absorption and Delivery*, Lehr, C.-M., ed., vol. 7, Taylor and Francis, London, New York, 378–391, 2002.

Lehnert, B. E., Pulmonary and thoracic macrophage subpopulations and clearance of particles from the lung, *Environ. Health Perspect.*, 97, 17–46, 1992.

Li, N., Sioutas, C., Cho, A., Schmitz, D., Misra, C., Sempf, J., Wang, M., Oberley, T., Froines, J., and Nel, A., Ultrafine particulate pollutants induce oxidative stress and mitochondrial damage, *Environ. Health Perspect.*, 111, 455–460, 2003.

Lieber, M., Smith, B., Szakal, A., Nelson-Rees, W., and Todaro, G., A continuous tumor-cell line from a human lung carcinoma with properties of type II alveolar epithelial cells, *Int. J. Cancer*, 17, 62–70, 1976.

Liu, Y., Meyer-Zalka, W., Franzka, S., Schmid, G., Leis, M., and Kuhn, H., Gold-cluster degradation by the transition of B-DNA into A-DNA and the formation of nanowires, *Angew. Chem. Int. Ed.*, 42, 2853–2857, 2003.

Maina, J. N. and West, J. B., Thin and strong! The bioengineering dilemma in the structural and functional design of the blood–gas barrier, *Physiol. Rev.*, 85, 811–844, 2005.

Matsuno, K., Ezaki, T., Kudo, S., and Uehara, Y., A life stage of particle-laden rat dendritic cells in vivo: their terminal division, active phagocytosis, and translocation from the liver to the draining lymph, *J. Exp. Med.*, 183, 1865–1878, 1996.

Mazzola, L., Commercializing nanotechnology, *Nat. Biotechnol.*, 21, 1137–1143, 2003.

McWilliam, A. S., Holt, P. G., and Gehr, P., Dendritic cells as sentinels of immune surveillance in the airways, In *Particle-Lung Interaction*, Gehr, P., Heyder, J., eds., vol. 143, Lung Biology in Health and Disease Series, Lenfant C exec. ed., Marcel Dekker Inc., New York, 473–489, 2000.

Meaney, C., Florea, B. I., Ehrhardt, C., Schaefer, U. F., Lehr, C.-M., Junginger, H. E., and Borchard, G., Bronchial epithelial cell cultures, In *Cell Culture Models of Biological Barriers In-Vitro Test Systems for Drug Absorption and Delivery*, Lehr, C.M., ed., Taylor and Francis, London, New York, vol. 13, 211–227, 2000.

Meiring, J. J., Borm, P. J., Bagate, K., Semmler, M., Seitz, J., Takenaka, S., and Kreyling, W. G., The influence of hydrogen peroxide and histamine on lung permeability and translocation of iridium nanoparticles in the isolated perfused rat lung, *Part. Fibre Toxicol.*, 2 (3), 2005.

Michalet, X., Pinaud, F. F., Bentolila, L. A., Tsay, J. M., Doose, S., Li, J. J., Sundaresan, G., Wu, A. M., Gambhir, S. S., and Weiss, S., Quantum dots for live cells, *in vivo* imaging, and diagnostics, *Science*, 307, 538–544, 2005.

Michiels, C., Endothelial cell functions, *J. Cell Physiol.*, 196, 430–443, 2003.

Moghimi, S. M., Hunter, A. C., and Murray, J. C., Nanomedicine: current status and future prospects, *FASEB J.*, 19 (3), 311–330, 2005.

Nemmar, A., Hoet, P. H., Vanquickenborne, B., Dinsdale, D., Thomeer, M., Hoylaerts, M. F., Vanbilloen, H., Mortelmans, L., and Nemery, B., Passage of inhaled particles into the blood circulation in humans, *Circulation*, 105, 411–414, 2002.

Nicod, L. P., *Function of Human Lung Dendritic Cells*, Marcel Dekker Inc., New York, 311–334, 1997

Nicod, L. P., Lung defenses: an overview, *Eur. Respir. Rev.*, 95, 45–50, 2005.

Niess, J. H., Brand, S., Gu, X., Landsman, L., Jung, S., McCormick, B. A., Vyas, J. M. et al., CX3CR1-mediated dendritic cell access to the intestinal lumen and bacterial clearance, *Science*, 307, 254–258, 2005.

Oberdörster, G., Sharp, Z., Atudorei, V., Elder, A., Gelein, R., Lunts, A., Kreyling, W., and Cox, C., Extrapulmonary translocation of ultrafine carbon particles following whole body inhalation exposure of rats, *J. Toxicol. Environ. Health A*, 65, 1531–1543, 2002.

Oberdörster, G., Sharp, Z., Atudorei, V., Elder, A., Gelein, R., Kreyling, W., and Cox, C., Translocation of inhaled ultrafine particles to the brain, *Inhal. Toxicol.*, 16, 437–445, 2004.

Oberdörster, G., Maynard, A., Donaldson, K., Castranova, V., Fitzpatrick, J., Ausman, K., Carter, J. et al., ILSI research foundation/risk science institute nanomaterial toxicity screening working group, principles for characterizing the potential human health effects from exposure to nanomaterials: elements of a screening strategy, *Part. Fibre Toxicol.*, 6 (2), 8, 2005.

Panté, N. and Kann, M., Nuclear pore complex is able to transport macromolecules with diameters of ~39 nm, *Mol. Biol. Cell*, 13, 425–434, 2002.

Paull, R., Wolfe, J., Hebert, P., and Sinkula, M., Investing in nanotechnology, *Nat. Biotechnol.*, 21, 1144–1147, 2003.

Peters, A., Wichmann, H. E., Tuch, T., Heinrich, J., and Heyder, J., Respiratory effects are associated with the number of ultrafine particles, *Am. J. Respir. Crit. Care Med.*, 155, 1376–1383, 1997.

Pietropaoli, A. P., Frampton, M. W., Hyde, R. W., Morrow, P. E., Oberdörster, G., Cox, C., Speers, D. M. et al., Pulmonary function, diffusing capacity, and inflammation in healthy and asthmatic subjects exposed to ultrafine particles, *Inhal. Toxicol.*, 16, 59–72, 2004.

Pinkerton, K. E., Gehr, P., and Crapo, J. D., Architecture and cellular composition of the air–blood barrier, In *Comparative Biology of the Normal Lung. Volume I. Treatise on Pulmonary Toxicology*, Parent, R.A., ed., CRC Press Inc., Boca Raton, FL, 121–144, 1992.

Pope III, C. A., Dockery, D. W., and Schwartz, J., Review of epidemiological evidence of health effects of particulate air pollution, *Inhal. Toxicol.*, 7, 1–18, 1995.

Portney, N. G. and Ozkan, M., Nano-oncology: drug delivery, imaging, and sensing., *Anal. Bioanal. Chem.*, 384 (3), 620–630, 2006.

Rescigno, M., Urbano, M., Valzasina, B., Francolini, M., Rotta, G., Bonasio, R., Granucci, F., Kraehenbuhl, J. P., and Ricciardi-Castagnoli, P., Dendritic cells express tight junction proteins and penetrate gut epithelial monolayers to sample bacteria, *Nat. Immunol.*, 2, 361–367, 2001.

Rimai, D. S., Quesnel, D. J., and Busnaia, A. A., The adhesion of dry particles in the nanometer to micrometer-size range, *Colloids Surf. A. Physicochem. Eng. Aspects*, 165, 3–10, 2000.

Rothen-Rutishauser, B., Kramer, S. D., Braun, A., Gunthert, M., and Wunderli-Allenspach, H., MDCK cell cultures as an epithelial *in vitro* model: cytoskeleton and tight junctions as indicators for the definition of age-related stages by confocal microscopy, *Pharm. Res.*, 15, 964–971, 1998a.

Rothen-Rutishauser, B. M., Messerli, M. J., van der Voort, H., Günthert, M., and Wunderli-Allenspach, H., Deconvolution combined with digital colocalization analysis to study the spatial distribution of tight and adherens junction proteins, *J. Comput. Assist. Microsc.*, 10, 103–111, 1998b.

Rothen-Rutishauser, B., Braun, A., Gunthert, M., and Wunderli-Allenspach, H., Formation of multilayers in the caco-2 cell culture model: a confocal laser scanning microscopy study, *Pharm. Res.*, 17, 460–465, 2000.

Rothen-Rutishauser, B., Riesen, F. K., Braun, A., Gunthert, M., and Wunderli-Allenspach, H., Dynamics of tight and adherens junctions under EGTA treatment, *J. Membr. Biol.*, 188, 151–162, 2002.

Rothen-Rutishauser, B. M., Kiama, S. G., and Gehr, P., A 3D cellular model of the human respiratory tract to study the interaction with particles, *Am. J. Respir. Cell Mol. Biol.*, 32, 281–289, 2005.

Rothen-Rutishauser, B. M., Schürch, S., Haenni, B., Kapp, N., and Gehr, P., Interaction of fine and nanoparticles with red blood cells visualized with advanced microscopic techniques, *Environ. Sci. Technol.*, 40, 4353–4359, 2006.

Samet, J. M., DeMarini, D. M., and Malling, H. V., Biomedicine Do airborne particles induce heritable mutations? *Science*, 304, 971–972, 2004.

Schneeberger, E. E., Ultrastructure of intercellular junctions in the freeze fractured alveolar-capillary membrane of mouse lung, *Chest*, 71, 299–300, 1977.

Schneeberger, E. E. and Lynch, R. D., Tight junctions. Their structure, composition, and function, *Circ. Res.*, 55, 723–733, 1984.

Schulz, H., Harder, V., Ibald-Mulli, A., Khandoga, A., Koenig, W., Krombach, F., Radykewicz, R., Stampfl, A., Thorand, B., and Peters, A., Cardiovascular effects of fine and ultrafine particles, *J. Aerosol Med.*, 18, 1–22, 2005.

Schürch, S., Bachofen, H., and Weibel, E. R., Alveolar surface tension in excised rabbit lungs: effect of temperature, *Respir. Physiol.*, 62, 3–45, 1985.

Schürch, S., Gehr, P., Im Hof, V., Geiser, M., and Green, F., Surfactant displaces particles toward the epithelium in airways and alveoli, *Respir. Physiol.*, 80, 17–32, 1990.

Schürch, S., Geiser, M., Lee, M. M., and Gehr, P., Particles at the airway interfaces of the lung, *Colloids Surf. B: Biointerfaces*, 15, 339–353, 1999.

Shanahan, M. E. R., Capillary movement of nearly axisymmetric sessile drops, *J. Phys. D. Appl. Phys.*, 23, 321–327, 1990.

Shapiro, D. L., Nardone, L. L., Rooney, S. A., Motoyama, E. K., and Munoz, J. L., Phospholipid biosynthesis and secretion by a cell line (A549) which resembles type II aleveolar epithelial cells, *Biochim. Biophys. Acta*, 530, 197–207, 1978.

Shaw, P., Deconvolution in 3D optical microscopy, *Histochem. J.*, 26, 687–694, 1994.

Stearns, R. C., Paulauskis, J. D., and Godleski, J. J., Endocytosis of ultrafine particles by A549 Cells, *Am. J. Respir. Cell Mol. Biol.*, 24, 108–115, 2001.

Steimer, A., Haltner, E., and Lehr, C. M., Cell culture models of the respiratory tract relevant to pulmonary drug delivery, *J. Aerosol Med.*, 18, 137–182, 2005.

Stringer, B., Imrich, A., and Kobzik, L., Lung epithelial cell (A549) interaction with unopsonized environmental particulates: quantitation of particle-specific binding and IL-8 production, *Exp. Lung Res.*, 22, 495–508, 1996.

Takano, K., Kojima, T., Go, M., Murata, M., Ichimiya, S., Himi, T., and Sawada, N., HLA-DR- and CD11c-positive dendritic cells penetrate beyond well-developed epithelial tight junctions in human nasal mucosa of allergic rhinitis, *J. Histochem. Cytochem.*, 53, 611–619, 2005.

Thiele, L., Rothen-Rutishauser, B., Jilek, S., Wunderli-Allenspach, H., Merkle, H. P., and Walter, E., Evaluation of particle uptake in human blood monocyte-derived cells in vitro. Does phagocytosis activity of dendritic cells measure up with macrophages? *J. Control Release*, 76, 59–71, 2001.

Timpl, R. and Dzladek, M., Structure, development and molecular pathology of basement membranes, *Int. Rev. Exp. Pathol.*, 29, 1–112, 1986.

Tschanz, S. A., Damke, B. M., and Burri, H. P., Influence of postnatally administrated glucocorticoids on rat lung growth, *Biol. Neonate.*, 68, 229–245, 1995.

Tschanz, S. A., Makanya, A. N., Haenni, B., and Burri, H. P., Effects of neonatal high-dose short-term glucocorticoid treatment on the lung: a morphologic and morphometric study in the rat, *Pediatr. Res.*, 53, 72–80, 2003.

Tsoli, M., Kuhn, H., Brandau, W., Esche, H., and Schmid, G., Cellular uptake and toxicity of Au_{55} clusters, *Small*, 8–9, 841–844, 2005.

van der Voort, H. T. M. and Strasters, K. C., Restoration of confocal images for quantitative image analysis, *J. Microsc.*, 178, 165–181, 1995.

Vermaelen, K. and Pauwels, R., Pulmonary dendritic cells, *Am. J. Respir. Crit. Care Med.*, 172, 530–551, 2005.

Vermaelen, K. Y., Carro-Muino, I., Lambrecht, B. N., and Pauwels, R. A., Specific migratory dendritic cells rapidly transport antigen from the airways to the thoracic lymph nodes, *J. Exp. Med.*, 193, 51–60, 2001.

Walter, E., Dreher, D., Kok, M., Thiele, L., Kiama, S. G., Gehr, P., and Merkle, H. P., Hydrophilic poly(DL-lactide-co-glycolide) microspheres for the delivery of DNA to human-derived macrophages and dendritic cells, *J. Control Release*, 76, 149–168, 2001.

Wichmann, H. E., Spix, C., Tuch, T., Wolke, G., Peters, A., Heinrich, J., Kreyling, W. G., and Heyder, J., Daily mortality and fine and ultrafine particles in Erfurt, Germany part I: role of particle number and particle mass, *Res. Rep. Health Eff. Inst.*, 98, 5–86, 2000.

Xia, T., Korge, P., Weiss, J. N., Li, N., Venkatesen, M. I., Sioutas, C., and Nel, A., Quinones and aromatic chemical compounds in particulate matter induce mitochondrial dysfunction: implications for ultrafine particle toxicity, *Environ. Health Perspect.*, 112, 1347–1358, 2004.

Yurchenco, P. D., Tsilibary, E. C., Charonis, A. S., and Furthmayr, H., Models for the self-assembly of basement membrane, *J. Histochem. Cytochem.*, 34, 93–102, 1986.

8 Particle-Associated Metals and Oxidative Stress in Signaling

James M. Samet and Andrew J. Ghio
U.S. Environmental Protection Agency

CONTENTS

8.1 METALS IN PM

Metals have long been recognized as toxic components of particles found in occupational settings. There are a number of well-characterized pathological entities caused by the inhalation of particles of specific metallic compounds (Table 8.1). While no single physicochemical property has emerged as the common etiological factor, a substantial body of work carried out over the last decade supports a role for metals in the toxicity of ambient air pollution. Epidemiological studies have correlated health outcomes with the content of specific metallic components in a particle. Studies on Utah Valley particulate matter (PM) suggest that soluble ions of Cu, V, and Zn participate in its biological effects (Kennedy et al. 1998). Metal depletion and reconstitution experiments strengthen these findings and provide some insight on the speciation of the metals involved, pointing to soluble metal salts rather than insoluble oxides. A series of studies using the fugitive emission PM residual oil fly ash (ROFA) as a model particle has also yielded a significant amount of information demonstrating the biological effects of PM metal compounds. ROFA can be distinguished from other combustion PM by its high content of V, which has been demonstrated to be the major, but not sole, contributor to its toxicity.

Metals carry out a wide range of essential structural and catalytic functions in living systems. Consequently, life evolved with a dependency on metal availability and all living organisms require

TABLE 8.1
Pulmonary Diseases Associated with Metal Exposures

Disease	Metal
Shaver's disease	Aluminum
Potroom asthma	Aluminum
Berylliosis	Beryllium
Pulmonary emphysema	Cadmium and others
Lung cancer	Chromium and others
Hard metal disease	Cobalt
Vineyard workers' lung	Copper
Siderosis	Iron
Asthma	Nickel and others
Stannosis	Tin
Boilermakers' bronchitis	Vanadium
Metal fume fever	Zinc and others

metals, including Fe, Cu, Zn, Co, Mn, Cr, Mo, V, Se, Ni, and Sn. Some of these metals have electron-transfer capabilities that present toxicity by generating an oxidative stress. Metals that exist in more than one stable valence state can directly participate in electron transfer reactions and, therefore, possess the potential to generate oxidants. Those metals that assume only one stable valence state can also present an oxidative stress through depletion of sulfhydryls or by interfering with the metabolism of other metals. Consequently, cells are presented with a conundrum: an absolute requirement to procure metals for homeostatic and synthetic functions, and the inherent peril posed by oxidative stress associated with the handling of metals.

The atmosphere constitutes a prime vehicle for the movement and distribution of metals. Metal compounds in the atmosphere are generally nonvolatile and thus are associated with particles (exceptions include elemental mercury and organomercury compounds, tetraethyl and tetramethyl lead, arsenic and selenium oxides, elemental selenium, arsine and arsenic alkyls, and species such as nickel carbonyls) (Schroeder et al. 1987). Worldwide, the total particle production is in the order of $1–4 \times 10^9$ tonnes/year, with approximately 5%–20% of that being anthropogenic (Schroeder et al. 1987). Metals comprise a fraction of the total PM burden but, given its magnitude, the absolute mass of metals is quite significant. The residence time for metals associated with particles in the atmosphere is typically less than 40 days.

Particles originating from natural sources can be derived from terrestrial dust dispersed by wind and automobiles, sea spray, biological processes, volcanic emissions, emissions from fires, pollen, plant debris, and reactions between natural gaseous emissions. The metals in crustal dust (in order of abundance) are Al, Fe, Mn, Zn, Pb, V, Cr, Ni, Cu, Co, Hg, and Cd. Similarly, volcanic ash can have great amounts of Fe, Mn, V, Zn, Co, As, and Sb (in decreasing order of concentration). However, the content of these metals in any one particle can vary greatly.

Metals can be preferentially associated with particles that have a diameter less than 2.5 μm and are derived from anthropogenic processes. These particles emanate from the incomplete combustion of carbon-containing materials at power plants, smelters, incinerators, cement kilns, home furnaces, fireplaces, and motor vehicles. These fine particles contain acid sulfates, soot, and fly ash. Sulfur in particles is present as ammonium sulfate rather than sulfite and can make up 40%–67% of the mass in the fine particle fraction. The incomplete oxidation of carbonaceous materials generates soot, a mixture of particulate carbon (C_8H; 90%–98% carbon by weight) with organic and inorganic components. The residue that remains after burning a substance (fuel or waste) is largely an inorganic mixture that is termed fly ash (e.g., coal fly ash, oil fly ash, and incinerator fly ash).

The composition of fly ash is not uniform but generally contains Fe_2O_3, Fe_3O_4, Al_2O_3, SiO_2, and a small amount of carbonaceous compounds with varying degrees of oxidation (Greenberg, Zoller, and Gordon 1978).

Iron is the metal found in highest concentration in the atmosphere. Its concentration is frequently 10-fold higher than that of other transition metals (Stevens et al. 1978; Vossler et al. 1989; Mamane 1990; Dodd et al. 1991) and Fe often accounts for the majority of particle-associated metal. Iron is most frequently released into the atmosphere as insoluble oxides (i.e., FeO, Fe_2O_3, and Fe_3O_4). Fe_2O_3 makes up 5%–15% by mass of the fuel ash produced by power stations. Therefore, Fe is ubiquitous in the atmosphere and it concentrates in the air of urban settings and fogwater, where its concentration can approximate 1.0 mM (Waldman et al. 1982; Munger et al. 1983). However, a majority of atmospheric iron is found to be in a soluble form rather than the highly insoluble oxides and oxyhydroxides that are actually discharged from individual sources. A relatively large fraction (12%–56%) of the iron in fuel ash can solubilize within 20 min in atmospheric water at acidic pH values (Weschler, Mandich, and Graedel 1986). The rate of this reaction is dependent on the presence of a suitable ligand for iron in the atmosphere such as sulfates, nitrates, oxalate, humic-like substances, and surface functional groups on oxides.

Fly ash from coal-fired power plants have high concentrations of Fe, Zn, Pb, V, Mn, Cr, Cu, Ni, As, Co, Cd, Sb, and Hg (Schroeder et al. 1987). As already mentioned, oil fly ash is unique in its content of V, Ni, Fe, Zn, Pb, Cu, As, Co, Cr, Mn, and Sb (Henry and Knapp 1980; Schroeder et al. 1987). The incineration of municipal waste releases Zn, Fe, Hg, Pb, Sn, As, Cd, Co, Cu, Mn, Ni, and Sb (Schroeder et al. 1987). Open-hearth furnaces in steel mills are a notable source of Fe, Zn, Cr, Cu, Mn, Ni, and Pb (Schroeder et al. 1987). Vehicle emissions contain smaller concentrations of both Zn and Fe. Lead was previously derived predominantly from leaded gasoline; this metal now originates primarily from resuspended soil, oil burning, and small scale smelting.

8.2 METAL-DEPENDENT GENERATION OF REACTIVE OXYGEN AND NITROGEN SPECIES

Those metals that exist in more than one stable valence state can directly participate in electron transfer reactions and consequently possess a potential to generate reactive oxidants. A labile or incomplete coordination of the metal allows its direct participation in electron transfer and leads to the generation of highly reactive radicals. Iron and Cu in particles complexed to sulfate, nitrate, oxalate, incompletely oxidized carbon fragments, and functional groups at the surface of an oxide can directly catalyze oxidant generation. In addition to the metal, peroxide and reductants are available in both the atmosphere and in the lower respiratory tract in humans (Jacob, Gottlieb, and Prather 1989). The resultant cycling through reduced and oxidized states to generate radical species has been shown to occur in the atmosphere (Logan et al. 1981; Behra and Sigg 1990). Metals which can assume only one stable valence state (e.g., Zn) can also present an oxidative stress to cells and tissues through depletion of sulfhydryls or disruption of Fe homeostasis (Stohs and Bagchi 1995).

The rate of oxidant generation after particle exposure correlates positively with the concentration of several metals (Pritchard 1996). Deferoxamine inhibits oxidant production by these particles, consistent with metal participation in catalysis of radicals (Ghio et al. in press). It has also been demonstrated that exposure to particles results in hydroxylation of salicylate (Ghio et al. in press). Oxidative stress after *in vivo* exposure of the lung to metal-containing particles has been demonstrated using several different methodologies including electron spin resonance and analysis for aldehydes (Madden, Thomas, and Ghio 1999). However, in many particles, oxidant production may show no relationship to metals but will instead correlate with organic content, ultrafines, and endotoxin.

Comparable to reactive oxygen species, there are numerous interactions between nitric oxide (NO) and its metabolites with metals. NO can function as a ligand for numerous metals (e.g., Cu and Fe) (Cooper 1999; Torres and Wilson 1999). Such binding is predicted to affect the state of activation of numerous proteins. Exposure of cells to particles laden with metals has demonstrated that particles function to modulate NO production by affecting expression of inducible NO synthase (Chauhan et al. 2004). Similarly, human exposure to a metal-abundant particle was accompanied by a decrease in expired NO (Kim et al. 2003). It was hypothesized that, given the protective role of NO in the environment of the lower respiratory tract, decrements in its concentration could potentially be associated with detrimental health effects including airway inflammation and pulmonary hypertension. However, there have also been several investigations demonstrating an increase in expired NO levels following exposures to particles.

8.3 DISRUPTION OF NORMAL CELL AND TISSUE IRON HOMEOSTASIS BY METALS IN PARTICLES AS A MECHANISM OF OXIDANT INJURY

Living systems demonstrate an absolute requirement for Fe availability. Consequently, its transport and storage are pivotal in determining the success of the system. In addition to the biological effects of metals included in the particle itself, exposure to any particle will disrupt normal host iron homeostasis and induce oxidative stress. Non-ferrous metals that interfere with the normal transport of Fe can also potentially create an oxidative stress. Decrements in cellular uptake of Fe following V exposure contribute to an oxidative stress and biological effects (Huang 2003; Ghio et al. 2005). Similar potential effects of non-ferrous metals on the numerous mechanisms responsible for Fe homeostasis can be proposed. Among these would be effects of metals on ferrireductases and ferroxidases included in Fe uptake and release, on storage in ferritin, and on release by ferroportin.

In addition to the effects of individual metals on Fe homeostasis, every particle can induce an accumulation of Fe at their surface. All retained particles present a solid–liquid interface. In an aqueous environment, oxides will be covered with surface hydroxyl groups (e.g., Al–OH and Ti–OH groups at the surface of Al_2O_3 and TiO_2, respectively). Consequently, all surfaces of oxide particles in a cell or tissue will have some concentration of oxygen-containing functional groups. Similarly, the incomplete oxidation of carbonaceous materials will produce oxygen-containing functional groups (e.g., C–OH, C=O, and COOH) that occur in greatest concentration at the surface of the particle. An open network of negatively charged functional groups on a surface presents spaces large enough to accommodate adsorbed metal cations. As a result of its electropositivity, Fe^{3+} has a high affinity for oxygen-donor ligands. Consequently, this metal will react with the oxygen-containing functional groups at the surfaces of particles. Therefore, following exposure to any insoluble particle, cells may accumulate Fe. This disrupts normal homeostasis and is conducive to Fe-catalyzed generation of reactive oxygen species.

8.4 PATHOPHYSIOLOGICAL EFFECTS OF METAL-INDUCED SIGNALING

An introduction of reactive metal into a tissue is associated with neutrophilic influx. Disparities between the inflammatory potency of different PM have been associated with their metal content (Pritchard et al. 1996). Such dissimilarities in the acute inflammatory response in rats have also been observed to correlate with the content of other PM constituents such as endotoxin. An association between metal and inflammation can be exemplified by the introduction of either the Fe chelate bleomycin or V in the lower respiratory tract. Exposure to either is followed by a marked influx of neutrophils and injury (i.e., a pneumonitis). This response has been associated with resultant oxidative stress in the lung in which reactive oxygen and nitrogen species are considered pivotal mediators.

Metals can also effect an extensive fibrotic response in the lung (Bonner et al. 2000). Exposures to organic or inorganic particles, which initiate a diffuse fibrotic response in the lung (i.e., pneumoconiosis), are similarly associated with an accumulation of metal and an oxidative stress.

The role of metals is supported by the inhibition of fibrosis in experiments with chelators and dietary depletion (Kennedy et al. 1986; Chandler et al. 1988).

Finally, carcinogenesis can be a consequence of exposure of host tissues to increased concentrations of available metal. Accumulation of metal can damage DNA through the generation of oxidative stress (Hutchinson 1985; Imlay and Linn 1988). DNA strand nicks and mutations resulting from oxidant exposure are assumed to be events in tumor induction. Unlike As, Cr, or Ni, the carcinogenicity of Fe is still under debate. However, Fe dextran given intramuscularly can induce cancers (Weinbren, Salm, and Greenberg 1978). In addition, depletion of Fe in the diet will diminish the incidence of certain malignancies and neoplastic cell growth can be inhibited using deferoxamine (Hann, Stahlhut, and Menduke 1991; Tabor and Kim 1991).

The sections that follow will briefly review current information on major signaling events associated with exposure to metals commonly found in PM in ambient or occupational settings. Whenever possible, a distinction is drawn between the direct effects of metals and those mediated by oxidant stress that results from metal exposure.

8.5 SIGNALING EFFECTS OF COPPER

Copper (Cu) is a pervasive contaminant of ambient PM of natural and anthropogenic origin. Smelting and ore processing are leading sources of Cu in ambient air PM. Therefore, Cu concentrations in ambient air PM are largely influenced by proximity to these industrial activities and atmospheric transport (Goforth and Christoforou 2005).

Given its physiologic essentiality and ubiquitous presence in everything from coins and wiring to food utensils and construction materials, the relatively high toxicity of Cu is somewhat surprising (Prieditis and Adamson 2002; Uriu-Adams and Keen 2005; Valko, Morris, and Cronin 2005). Instillation of soluble Cu salts produces lung injury and inflammation in rodent models, and there is evidence that Cu ions are mutagenic. Cu is known to interact non-specifically with cysteine, methionine, and histidine residues in proteins which can cause displacement of functional metal centers and steric deformation in proteins (Koch, Pena, and Thiele 1997). There is evidence that Cu^{2+} can activate signaling through both oxidant- and non-oxidant mechanisms. Cu is also a potent and reversible inhibitor of at least some protein tyrosine phosphatases (PTPs) which, as discussed below, may account for its signaling effects (Hanahisa and Yamaguchi 1998; Kim et al. 2000).

Unlike Zn and Cd, Cu is an efficient redox metal capable of generating ROS under physiological conditions (see Mattie and Freedman 2004). In addition to direct ROS generation, Cu^{2+} exposure reduces total glutathione concentrations and increases oxidized glutathione levels. Cu^{2+}-induced cytoxicity and PKC-dependent transcription are blunted by antioxidants. Cu^{2+}-induced metallothionein-1 transcription is PKC- and MAPK-dependent in COS-7 cells. Furthermore, metallothionein-1 expression induced by Cu^{2+} is protein kinase C- and MAPK-dependent and requires the transcription factor MTF-1 (Mattie and Freedman 2004).

Exposure to Cu^{2+} has been shown to increase phosphorylation levels and activate the MAPKs ERK, p38 and JNK, as well as to increase phosphorylation of the transcription factors ATF-2 and c-Jun in a human airway epithelial cell line (Samet et al. 1998a), and to induce MEK1/2 and epidermal growth factor receptor (EGFR) phosphorylation (Wu et al. 1999a). Cu^{2+} has also been shown to inhibit the formation of GTP-G alpha s, leading to a decrease in adenylyl cyclase activity in S49 cell membrane preparations (Gao, Du, and Patel 2005).

Ostrakhovich et al. showed that exposure of HeLa cells or human skin fibroblasts to relatively low doses of Cu^{2+}, but not Cu^{+}, induced PI3 kinase activation of Akt and phosphorylation of its downstream target glycogen synthase kinase-3 (Ostrakhovitch et al. 2002). The same researchers subsequently reported that activation of p53 is a prerequisite for Cu-induced Akt activation and apoptosis in MF-7 cells (Ostrakhovitch and Cherian 2004). Although Cu^{2+}-induced generation of ROS was reported, the latter was dissociated from the activation of Akt (Ostrakhovitch et al. 2002).

Kennedy and coworkers showed increased IL-8, IL-6, and ICAM1 expression in human bronchial epithelial cells treated with an aqueous extract prepared from PM collected from an area close to a smelter (Kennedy et al. 1998). These effects were preceded by activation of NFκB. The IL-8-induced effect could be reproduced using a solution of Cu^{2+} equivalent in concentration to that found in the aqueous extract. In contrast, equimolar solutions of Fe, Zn, or Pb failed to induce the effect of the aqueous extract of PM. Extract induced IL-8 secretion was blunted by the inclusion of superoxide dismutase, deferoxamine, or *N*-acetylcysteine in the media, but not by the addition of the hydroxyl radical scavenger dimethylthiourea (Kennedy et al. 1998).

Additional evidence of Cu^{2+}-induced activation of NFκB has been reported by Chen et al. who examined NFκB-dependent reporter activity in JB6 mouse epidermal cells treated with a number of relevant PM metals (Chen et al. 2001). Using their system, these workers found an approximate potency rank order of $Cu^{II}=Co>As^{III}=As^{V}=V^{V}>Cr^{VI}=Ni^{II}$ for NFκB activation. In the same study, the effect of exposure to these metals on AP-1-dependent reporter activity was found to closely track their activation of NFκB (Chen et al. 2001).

8.6 SIGNALING EFFECTS OF IRON

Iron participates at three specific levels of signaling: sensing of its own available concentrations, mitogen-activated protein (MAP) kinase activation, and transcription factor activation. Figure 8.1 summarizes the role of bioavailable iron in signaling and oxidant injury.

Iron is necessary for the transcription of numerous genes that play critical roles in cell cycle regulation, homeostasis, and development. The cell has developed mechanisms to ensure a sufficient supply of iron for its metabolic and growth needs. Iron homeostasis in living systems is regulated in response to iron availability at the transcriptional level. Iron deficiency induces an accumulation of mRNAs that are then translated into Fe transport proteins. Cellular iron sensing is activated in iron deficiency by disassembly of the 4Fe–4S cluster in aconitase to produce iron response protein-1 (IRP-1) (Eisenstein and Blemings 1998). IRP function as a sensor of Fe status. Under conditions of

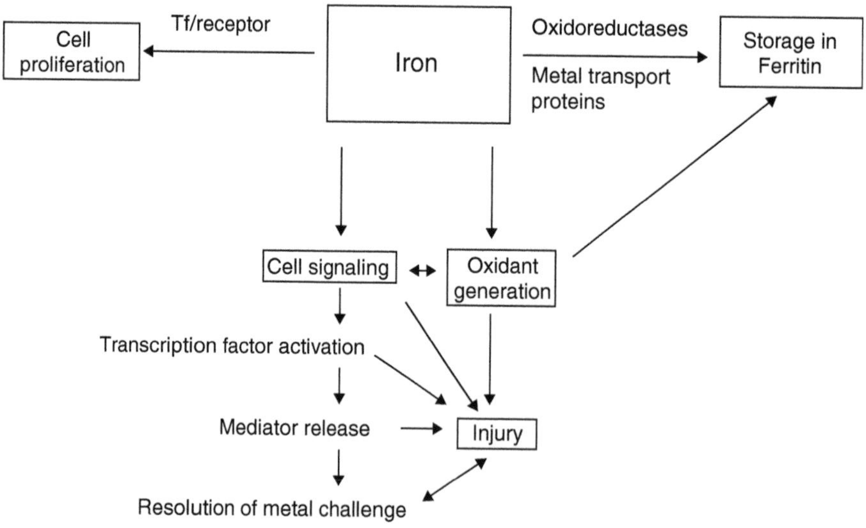

FIGURE 8.1 Role of bioavailable iron in signaling leading to lung injury. The transferrin/transferrin receptor pathway facilitates uptake of Fe required to meet metabolic and proliferation needs. Excess Fe can be stored in ferritin, where its catalytical activity is diminished. Unstored ("free") Fe can activate cell signaling through oxidant generation or other mechanisms. Signaling leads to transcription factor activation, mediator release, and either resolution of the metal challenge or injury.

Fe deprivation, an IRP will bind to the iron-responsive element (IRE), a stem-loop structure present on either 3′ or 5′ untranslated regions of several pivotal mRNAs. The increased binding of IRP to the 3″ end of transferrin receptor (TfR) mRNA stabilizes the mRNA, resulting in increased translation and TfR synthesis. The IRP/IRE interaction at the 5′ end of ferritin mRNA prevents its translation. When cellular Fe is in excess, IRP/IRE binding is decreased leading to a degradation of TfR mRNA and to increased translation of ferritin mRNA.

The IRP, therefore, couples Fe availability with TfR and ferritin expression. There are two IRPs: IRP-1 and IRP-2 (molecular weights of 90–95 and 105 kDa, respectively) with the former being the more abundant. In resting cells, IRP-1 is a cytosolic aconitase that converts citrate into isocitrate and contains a fully assembled iron sulfur cluster [4Fe–4S]. It becomes a post-transcriptional regulator of Fe metabolism when present as an apoprotein (3Fe–4S). Comparable to IRP-1, IRP-2 is a monomeric cytosolic protein. However, IRP-2 does not keep aconitase activity nor does it possess an Fe–S cluster. IRP-2 is regulated post-translationally by proteosome-mediated degradation, a process that involves an Fe-dependent oxidation mechanism and requires a unique amino acid sequence containing three cysteine residues. Reactive oxygen species (including H_2O_2) and the activation of NO synthase can also stimulate IRP to bind to an IRE and modulate expression of TfR and ferritin (Boldt 1999). This data demonstrates a regulatory loop between Fe metabolism and reactive oxygen species and the NO pathway, both recognized as important physiologic mediators of signal transduction. Finally, proteins other than TfR and ferritin have been recognized to have an IRE and these include proteins that function in the uptake of Fe (divalent metal transporter, DMT-1), the synthesis of heme (δ-amino-laevulinate synthetase, which is the rate limiting enzyme in protoporphyrin) and the tricarboxylic acid cycle (mitochondrial aconitase and succinic dehydrogenase). Therefore, signaling properties of environmental iron complexes are not limited to regulating metal availability but also influence fundamental processes of the cell.

Iron has been shown to stimulate the MAP kinase signaling pathways. Soluble Fe showed a greater activity than insoluble iron compounds in activating the extracellular signal-regulated kinases (ERKs) and p38 MAP kinase (but not Jun N-terminal kinases or JNKs) (Huang 2003). Pretreatment with inhibitors of the MAP kinase pathways significantly decreased the cellular release of proinflammatory cytokines. Similarly, coals containing bioavailable iron also induced phosphorylation of ERKs and p38 MAP kinase (but not JNK) (Huang 2003). The response of cells to ultraviolet irradiation demonstrated that Fe-driven generation of reactive oxygen species was a key event in the signaling pathway. This resulted in a 15-fold increase in the activity of JNK2. Iron has also been demonstrated to stimulate a differentiation of cells that were ERK-dependent.

Iron can induce early signaling pathways that modulate activities of several transcription factors. Activator protein-1 (AP-1) consists of a family of Jun/Fos dimers that include different Jun proteins (c-Jun, JunB, and JunD) and Fos proteins (c-Fos, FosB, Fra-1, and FosB2). This transcription factor participates in signaling following environmental stress and regulates complex biological processes. It is characterized by a capacity to alter gene expression in response to growth factors and cytokines. In a study of coal particles, Fe was demonstrated to activate AP-1 (as well as nuclear factor of activated T cells). However, there is conflicting evidence that suggest that there is no correlation between AP-1 activation and iron availability (Huang et al. 2002).

NF-κB is the prototypic transcription factor in eukaryotic cells recognized to play a role in transactivation of promoters for genes involved in inflammation. In the cytoplasm, NF-κB exists as an inactive form bound to its inhibitor IκB. Upon activation, following phosphorylation of IkB, the p50–p65 heterodimer migrates to the nucleus where it binds to κB binding sites in the promoter sequences of a variety of genes. Numerous gene products are associated with NF-κB activation including cytokines (e.g., the genes for interleukins 6 and 8 have functional AP-1 and NF-κB binding sites), cell adhesion molecules, and acute response proteins. Fe exposure can activate NF-κ B. Addition of ferrous ion to isolated cells was shown to increase electron paramagnetic resonance-detected radicals, IκB activity, and NF-kappa B binding while treatment with an Fe chelator inhibited the latter two of these effects (Xiong, She, and Tsukamoto 2004). The Fe^{2+}-exposed cells developed increased TNFα-promoter activity and later released elevated concentrations of TNFα. Therefore, ferrous iron served as a direct and independent inducer of IKK, NF-κB, and TNF-α promoter activity as well as for the release of TNF-α protein by cells, most likely through an oxidant mechanism. Finally, there was a transient rise in intracellular labile level of Fe in cells following exposure to LPS or TNFα. This elevation in metal concentration was associated with both IKK and NF-kB activation. These results support the existence of Fe-dependent signaling for the activation of IKK/NF-κB.

8.7 SIGNALING EFFECTS OF NICKEL

Nickel (Ni) is one of the metals frequently found associated with PM derived from fossil fuel combustion. Ni is also an occupational toxicant in Ni refining operations (Salnikow and Costa 2000; Magari et al. 2002; Goforth and Christoforou 2005; Sorensen et al. 2005). Major toxic effects associated with Ni inhalation include: carcinogenicity, allergic sensitization, and inflammation/fibrosis.

With the exception of metallic Ni, all forms of Ni are considered carcinogenic—nasal and pulmonary tumors are the primary concern. The mechanism responsible for the carcinogenic effect of Ni is not completely understood. However, Ni is classified as an epigenetic carcinogen because even its most reactive compounds exhibit weak mutagenicity and minimal reactivity towards DNA (Salnikow and Costa 2000). As reviewed by Salnikow and Costa, there is now considerable evidence that Ni induces transformation by altering signaling processes in the cell (Salnikow and Costa 2000). Moreover, Ni exposure has been shown to affect multiple signal transduction cascades simultaneously. Activation of all 3 MAPKs (ERK, p38, and JNK) has been demonstrated in dendritic cells, a pivotal cell in the sensitization of T-cells, which are induced to undergo migration and maturation by Ni exposure (Boisleve et al. 2004; Boisleve, Kerdine-Romer, and Pallardy 2005). Interestingly, increased expression of the tumor suppressor transcription factor p53 has also been reported following exposure to Ni, although its role in Ni-induced carcinogenesis is not clear (Salnikow and Costa 2000).

Ni is a transition metal with two adjacent valence states, Ni^{+} and Ni^{2+}, and redox cycling by Ni ions is known to occur (Oller, Costa, and Oberdörster 1997; Yu et al. 2001; Oller 2002; Seilkop and Oller 2003). Increases in intracellular ROS have been reported in cells treated with Ni. Furthermore, oxidative damage to DNA can be seen in cells treated with $NiCl_2$ (Huang et al. 1994; Salnikow et al. 1994; Huang et al. 1995). The primary mechanism for these effects appears to be a marked decrease in levels of the intracellular antioxidants glutathione (Misra, Rodriguez, and Kasprzak 1990; Rodriguez et al. 1991) and ascorbate in cells treated with Ni (Maxwell and Salnikow 2004; Salnikow et al. 2004; Salnikow and Kasprzak 2005).

The pro-oxidative effect of Ni is also remarkable in view of the ability of this metal to act as a hypoxia mimetic. Ni ions have been shown to be potent inducers of the expression of hypoxia-inducible factor 1 (HIF-1) (Andrew, Klei, and Barchowsky 2001; Salnikow, Davidson, and Costa 2002; Salnikow et al. 2003; Maxwell and Salnikow 2004). This transcription factor is a heterodimer of two proteins: the constitutively expressed HIF-1β, and HIF-1α, which undergoes rapid proteolytic degradation under normoxic conditions. Hypoxia and Ni exposure result in stabilization of HIF-1α, allowing for the active HIF-1 to induce the expression of genes that control cell survival, angiogenesis, glucose metabolism, and tumor invasion (Maxwell and Salnikow 2004).

The mechanism through which Ni exposure results in stabilization of HIF-1α is not known, but it may involve substitution of Ni for Fe in a putative "oxygen sensor," possibly a kinase, which may be responsible for responding to oxygen tensions in the cell by phosphorylating, and thereby stabilizing, HIF-1α (Ruas and Poellinger 2005; Ryter and Choi 2005). Interestingly,

Salnikow et al. have shown that Ni-induced HIF-1 activation occurs through an oxidant-independent mechanism (Salnikow et al. 2000).

Oxidant-dependent activation of the NF-AT (Huang et al. 2001b) and NF-κ B (Goebeler et al. 1995) transcription factors has been reported following exposure to Ni^{2+} in lung cells and endothelial cells, respectively. However, Barchoswsky and colleagues showed that NFκB activation does not occur during Ni-induced expression of IL-8 in human airway epithelial cells (Barchowsky et al. 2002). In contrast, Ni^{2+} exposure increases plasminogen activator inhibitor 1 (PAI-1) in airway epithelial cells through an oxidant-independent activation of c-Jun and AP-1 (Andrew, Klei, and Barchowsky 2001, #4240). Thus, there appears to be substantial cell-type specificity regarding the oxidative effects of Ni on signaling pathways (reviewed in Barchowsky and O'Hara 2003).

8.8 SIGNALING EFFECTS OF VANADIUM

V is a redox-active transition metal that can be a significant contributor to the ambient air PM metal burden in areas where residual fuel oil combustion emissions are prevalent. Soluble V has been shown to be a major active component of ROFA, which is essentially an inorganic ash characterized by its high content of metals (Henry and Knapp 1980). Power plant maintenance workers exposed to ROFA develop "boilermaker's bronchitis," an inflammatory disease of the airways associated with exposure to vanadium pentoxide (Levy, Hoffman, and Gottsegen 1984; Hauser et al. 1995a, 1995b, 1995c; Hauser et al. 1996).

Soluble V salts are potent inducers of protein phosphorylation and, as such, they have been used as an experimental stimulus for many years. As a result, there are numerous studies showing the effects of V on a broad array of signal transduction processes. Similarly, the use of ROFA as a model PM in studies on the adverse effects of metals of PM in ambient and occupational settings has produced a substantial literature that is relevant to the effects of V exposure on signaling (Ghio et al. 2002).

The initiating signaling effect of V is believed to be inhibition of PTPs (Gordon 1991). Inhibition of PTPs by V leads to unopposed kinase activity and a progressive accumulation of multiple tyrosine phosphoproteins (Samet et al. 1997), which would be expected to result in a simultaneous activation of multiple signaling cascades in the cell. Multiple mechanisms have been proposed for the inhibitory effect of V on PTPs. The vanadate ion has been proposed as a phosphate analog that acts as a competitive (and, therefore, reversible) inhibitor of PTPs (Gordon 1991). Vanadyl (V_{IV}, VO^{2+}) and vanadate (V_V, VO^{3-}) species occur in equilibrium in solution (Gordon 1991) and solutions of either form appear equally potent in inducing phosphotyrosine accumulation in BEAS cells (Samet et al. 1997).

Oxidized forms of vanadium, such as pervanadate, produced by the reaction of vanadate with H_2O_2 (Krejsa et al. 1997b; Krejsa and Schieven 1998), act as potent irreversible PTP inhibitors by oxidizing the catalytic cysteine in the active site (Huyer et al. 1997). A third potential mechanism of V-induced PTP inhibition is V-catalyzed formation of ROS through redox cycling. H_2O_2 is a direct inhibitor of PTP activity that functions by oxidizing the reactive cysteine (thiolate, $R–S^-$) to form the sulfenyl derivative (R–S–OH) of the cysteinyl thiol (Groen et al. 2005). PTP oxidation is now recognized as an essential event in physiological signaling (Kamata et al. 2000; Kim et al. 2000; Meng, Fukada, and Tonks 2002; Meng and Tonks 2003; Salmeen et al. 2003; Meng et al. 2004; Persson et al. 2004; Groen et al. 2005), and as discussed later in this section, may prove to be a common signaling initiation mechanism by toxicants (Samet et al. 1999a); (T. Tal, in preparation).

Some studies have attempted to evaluate the direct versus oxidant-mediated roles of V exposure on intracellular signaling. Jaspers et al. showed that overexpression of catalase, but not Cu/zinc superoxide dismutase, prevented V-induced NFκB nuclear translocation and NFκB-dependent transcriptional activity in BEAS cells, thus implicating the production of H_2O_2 as essential in this process. Interestingly, V-induced activation of NFκB in this study was dependent on the phosphorylation of the MAPK p38, which was demonstrated to require H_2O_2 production in these cells. In contrast, increases in protein tyrosine phosphorylation in response to V treatment were not altered by catalase overexpression. These results suggest that V exposure activates signaling processes through both direct and oxidant-dependent mechanisms (Jaspers et al. 2000). Another study showed that pervanadate, but not vanadate or vanadyl ions, could induce Stat1 alpha activation in HeLa cells, further supporting a role for oxidants in V-induced signaling (Haque et al. 1995).

However, significant cell type-dependent differences appear to exist in the role of redox status in the effect of V compounds on NFκB activation. In a separate study conducted in lymphocytic cell lines, Krejsa and coworkers compared the effects of orthovanadate (V_v) to those induced by two stabilized peroxovanadium compounds on NFκB activation. They showed that V compounds can trigger NFκB activation by PTP inhibition in a redox-independent manner in these cells (Krejsa et al. 1997a).

The contribution of oxidants to the effects of V on insulin receptor signaling was examined by Lu et al., who showed that V-induced enhancement of insulin receptor sensitivity could be blocked by *N*-acetyl cysteine and is associated with reduction of V^{5+} to V^{4+} and decreased glutathione levels in adipocytes (Lu et al. 2001).

Activation of the MAP kinases JNK, ERK, p38, and MEK by V ions has been shown in the BEAS 2B human airway epithelial cell line (Samet et al. 1998b; Wu et al. 1999b). EGFR phosphorylation and Ras activation in response to exposure to V has been demonstrated in the same cells (Wu et al. 1999b). Similarly, it has been reported that vanadium pentoxide (V_2O_5) induces MAPK activation in human lung fibroblasts (Ingram et al. 2003) and rat lung myofibroblasts (Wang et al. 2003). These studies showed that V_2O_5-induced activation of p38 and EGFR was linked to Stat-1 activation through a mechanism that requires the generation of ROS (Wang et al. 2003). HB-EGF expression induced by exposure to V_2O_5 was linked to ERK and p38 activation and could be blocked by antioxidants *N*-acetyl-cysteine (NAC) or catalase (Ingram et al. 2003).

A separate study reported that exposure to V ions increases AP-1-dependent transcriptional activity in mouse epidermal JB6 cells through a PKC-dependent process, and that this effect was inhibited by superoxide dismutase (SOD) as well as catalase and NAC (Ding et al. 1999). Another report using the same cells confirmed vanadate-induced activation of PKC, specifically PKC lambda and zeta, and not PKC alpha (Li et al. 2004). The PKC lambda activation was accompanied by its translocation to the nucleus in vanadate-exposed cells, and was further shown to be necessary for phosphorylation of Akt, but dispensable in the phosphorylation of p70S6K induced by vanadate (Li et al. 2004). Also in JB6 cells, Huang and coworkers demonstrated that vanadate induces transactivation of P53 that could be blocked by pre-treatment with NAC, catalase, and deferoxamine, and enhanced by SOD, implicating H_2O_2 as the ROS involved (Huang et al. 2000). In a subsequent study, the same authors described activation of the transcription factor NFAT in vanadate exposed mouse embryo fibroblasts and mouse epidermal CI 41 cells (Huang et al. 2001a). Again, NAC, catalase, and deferoxamine blocked NFAT activation but SOD synergized its activation by vanadate. Interestingly, a calcium channel blocker was effective in blocking NFAT activation, while calcium ionophores potentiated the effect of vanadate in these cells (Huang et al. 2001a).

8.9 SIGNALING EFFECTS OF ZINC

Zn compounds are ubiquitous environmental contaminants derived from fossil fuel combustion, metallurgical operations, mining operations, and tire erosion (Kuschner et al. 1995; Cyrys et al. 2003; Kodavanti et al. 2003, Fosmire 1990, #4101; Councell et al. 2004; Gualtieri et al. 2005). Next to iron, Zn is the most abundant and common metal associated with ambient PM (Adamson et al. 2000).

Zn is also an air contaminant in occupational settings, with inhalation of Zn particles being the primary cause of metal fume fever in welding operations (Kuschner et al. 1995).

Zn^{2+}, the only soluble Zn species encountered biologically, does not redox cycle. Therefore, direct ROS production during Zn^{2+} can be ruled out as a contributing mechanism in Zn-induced signaling. Whether exposure to Zn^{2+} alters intracellular redox potential or induces the formation of oxidant species intracellularly is less clear, however. Zn^{2+} has been ascribed antioxidant properties due to its ability to "protect" protein sulfhydryls from oxidation (Bray and Bettger 1990a). On the other hand, Zn^{2+} exposure has also been shown to reduce intracellular glutathione levels (May and Contoreggi 1982; Walther et al. 2000).

Another interesting aspect of Zn^{2+} signaling is the functional description of a Zn-sensing plasma membrane receptor that triggers the release of intracellular calcium and is involved in the proliferation and survival of epithelial cells (Hershfinkel et al. 2001). Zn is also reported to induce P2X-dependent entry of extracellular calcium and to restore chloride secretion in airway epithelial cells derived from cystic fibrosis patients (Zsembery et al. 2004; Liang, Zsembery, and Schwiebert 2005). In addition, Zn^{2+} exposure has been shown to inhibit the formation of GTP-G alpha s, leading to a decrease in adenylyl cyclase activity in S49 cell membranes (Gao, Du, and Patel 2005).

Controlled human exposure, animal, and *in vitro* studies confirm that exposure to Zn-containing particles results in inflammatory responses that involve increased expression of inflammatory mediators (Fine et al. 1997; Samet et al. 1998b; Fine et al. 2000; Kodavanti et al. 2003; Beckett et al. 2005). Consequently, a number of studies have characterized the effects of Zn exposure on signal transduction pathways that regulate the expression of inflammatory proteins.

In particulate form, exposure to the Cu- and Zn-laden PM derived from the Utah Valley increases EGFR phosphorylation and Ras activation (Kennedy et al. 1998). Exposure to soluble Zn^{2+} has been shown to activate a number of signaling intermediates in a variety of cell types. Kinases whose activities have been reported to be activated after cellular exposure to Zn^{2+} include EGFR (Wu et al. 1999a, 2001, 2002b), Src (Wu et al. 2002a; Samet et al. 2003), Akt (Wu et al. 2003), and the MAPKs p38, ERK, and JNK (Samet et al. 1998a). In one study, Zn^{2+} was second only to V^{4+} among a list of combustion metal ions that included Cu^{2+}, Ni^{2+}, Fe^{2+}, Fe^{3+}, Cr^{3+}, and Cr^{6+} in inducing MAPK phosphorylation (Samet et al. 1998a). Functional activation of MAPKs in BEAS 2B cells by Zn^{2+} exposure has also been demonstrated, inasmuch as the activities of p38 and ERK were also elevated in relative proportion to the phosphorylation increases. Interestingly, in spite of the induction of phosphorylation, Zn^{2+} exposure did not increase JNK in the human bronchial epithelial cell line (Samet et al. 1998a). As discussed below, this may reflect the relative inhibitory effect of Zn^{2+} towards some kinases. Upstream in the MAPK cascade, MEK1/2 phosphorylation and activity were increased in response to Zn^{2+} exposure of BEAS 2B human airway epithelial cells (Wu et al. 1999a). Activation of these MEK1/2 and ERK were shown to occur through the EGFR in a process that requires the formation of Ras-GTP (Wu et al. 2002c). Increased phosphorylation of the transcription factors ATF-2 and c-Jun has also been documented in human bronchial epithelial cells exposed to Zn^{2+}, lending additional support for functional signaling in response to Zn^{2+}-initiated signaling (Samet et al. 1998a). However, IkBalpha degradation and NFκB DNA binding are not induced by Zn^{2+} exposure of HAEC (Wu et al. 2002c).

HAEC, BEAS 2B human bronchial epithelial cells, A431 skin carcinoma cells, and B82 mouse lung fibroblasts exposed acutely to high concentrations (500 μm) of Zn^{2+} ions show multi-site phosphorylation of EGFR (Wu et al. 1999c, 2002a; Samet et al. 2003). Functional activation of EGFR, assessed as elevated GTP-Ras concentrations, have also been described (Wu et al. 2002a). In B82 cells and A431 cells, Zn^{2+}-induced phosphorylation of EGFR is dependent on Src activity and activating phosphorylation of Src (Y416) accompanies it (Wu et al. 2002a; Samet et al. 2003). Moreover, Src-EGFR co-localization is induced by Zn^{2+} and phosphorylation of EGFR at a Src-specific transphosphorylation site (Y845) is required for Zn^{2+}-mediated Ras activation in B82 cells

(Wu et al. 2002a). These findings strongly implicate Src activation in Zn^{2+}-induced EGFR activation in some cell types.

However, the involvement of Src in Zn^{2+}-induced EGFR activation is not universal because cell-type specific differences are known to exist (Tal, in preparation). In primary HAEC, exposure to Zn^{2+} results in EGFR phosphorylation that does not require Src activity or involve activating phosphorylation of Src. Further arguing against a role for Src in Zn^{2+}induced EGFR phosphorylation is the fact that Zn^{2+} is actually an effective inhibitor of tyrosine kinase activity, including that of Src. The Zn^{2+} ion is known to act as a competitive inhibitor of kinase activity by occupying the second metal binding site (normally occupied by Mg^{2+} or Mn^{2+}) present in tyrosine kinases (Sun and Budde 1999). In fact, Zn^{2+} inhibits Src activity *in vitro*, and HAEC exposed to Zn^{2+} show a dose-dependent reduction in Src activity while displaying increased EGFR phosphorylation (Tal, in preparation).

Studies on the mechanism of signal transduction initiation by Zn have followed the origin of Zn^{2+}-induced EGFR phosphorylation in an attempt to identify the first event triggered by cellular exposure to Zn^{2+}. Using EGF as a positive control, dimerization of EGFR by Zn exposure could not be demonstrated in membrane preparations or intact A431 cells (Samet et al. 2003). Experiments using the Zn^{2+}-specific ionophore pyrithione indicate that the target for Zn^{2+}-induced EGFR is intracellular (Bromberg, unpublished). Recent attention has been placed on the role of inhibition of PTPs in Zn^{2+} induced initiation of signals.

Zn^{2+} is a known inhibitor of PTP activity in a variety of cell types (Tonks, Diltz, and Fischer 1990; Samet et al. 1999b; Haase and Maret 2003). In HAEC, the potency of Zn^{2+} as a PTP inhibitor rivals that of V ions (all PTPs are inhibited by V). As discussed above, V-mediated inhibition of PTPs is thought to occur through competitive inhibition by the vanadate (VO_4^{3-}) ion, which is structurally similar to phosphate (PO_4^{3-}) (Gordon 1991). In contrast, the mechanism of Zn^{2+}-induced inhibition of PTPs may involve coordination of the divalent Zn^{2+} ion with a conserved cysteine that is present in the active site of all PTPs. The catalytic cysteine in PTPs has an unusually low pK_a, which results in the thiol (–RSH) group being ionized to the thiolate anion ($-RS^-$) at physiological pH. The low pK_a of this cysteine is believed to render tyrosine phosphatases highly susceptible to oxidant mediated inactivation, since cysteinyl thiolates are readily oxidized to the sulfenyl (–RSOH) form, and further (irreversibly in most cases) to the sulfinyl (–RSOOH) and even the sulfonyl (–RSOOOH) species. It is the thiolate anion that initiates the attack on the phosphate group of the substrate. Therefore, oxidation of the active site cysteine sulfenyl form in tyrosine phosphatases effectively inactivates the activity of the enzyme. The sulfenyl-, but not the sulfinyl- or sulfonyl-, oxidation derivatives can be reduced back to the thiolate form by intracellular reducing agents such as glutathione (via formation of the mixed disulfide) (Lee et al. 1998; Tonks 2005). Thus, it would seem that parallels exist between redox- and Zn^{2+}-mediated inhibition of PTPs. However, the exact mechanism through which Zn^{2+} inhibits PTPs is not established. In addition to interacting with the catalytic cysteine, Zn^{2+} could also impair PTP activity by coordinating with conserved histidine or aspartate residues in the active site (Haase and Maret 2003).

8.10 TOWARDS A UNIFIED MECHANISM OF INJURY BY PM: INHIBITION OF TYROSINE PHOSPHATASES BY METALS AND OXIDANTS

Oxidative stress has recently emerged as a common theme in the toxicology of multiple PM components. Organic components in PM, specifically quinones recovered as polar extracts from diesel exhaust PM, have been shown to impose an oxidative stress and lead to decreased mitochondrial membrane potential, reduced mitochondrial membrane mass, and the induction of apoptosis (Xia et al. 2004). Basis on their work, Nel and colleagues have proposed a stratified oxidative stress model as an approach to the study of PM health effects. This model monitors the expression of several markers of cellular stress, including expression of the redox-sensitive gene HO-1 and levels of reduced glutathione (Li et al. 2002).

Similarly, ultrafine carbon particles with very high surface area are also oxidative. Beck-Speier et al. recently demonstrated that elemental carbon ultrafine particles oxidize methionine and induce oxidant-dependent responses in canine alveolar macrophages (Beck-Speier et al. 2005). These authors showed that LTB4 and 8-isoprostane production by macrophages were strongly correlated with the oxidative potential of the particle, while c-PLA2 activation, PGE2 release, and ROS formation appeared to be induced in an oxidant-independent manner. The oxidant-dependent effects of the particles correlated with their physicochemical properties (Beck-Speier et al. 2005). These results are similar to those previously reported by Dick et al., in which ultrafine PM induced free radical formation, as assessed using a plasmid scission assay, was correlated with the inflammogenicity of the particle when instilled intratracheally into rats (Dick et al. 2003).

Metals associated with PM can present an oxidative stress to cells through direct and indirect mechanisms (Valko, Morris, and Cronin 2005). Direct production of ROS is understood to occur through the Haber–Weiss reaction and is specific to certain metals with two stable adjacent valence states such as Fe, Cr, Cu, Ni, and V, although the efficiency for redox cycling among these metals can vary considerably in biological systems. It is apparent that metal exposure can also induce cellular production of oxidants by inducing a respiratory or oxidative burst in some cells (Conde et al. 1995; Noh and Koh 2000; Zhang et al. 2001).

While oxidative stress induced by metals through indirect mechanisms is less well characterized, it is believed to involve depletion of protein thiols and reduced glutathione (Grabowski, Paulauskis, and Godleski 1999; Zhang et al. 2001; Valko, Morris, and Cronin 2005). Depletion of protein thiols can occur through an oxidation that involves the addition of oxygen to the sulfur as in the formation of sulfenyl, sulfinyl, or sulfonyl derivatives. Alternatively, two thiols can be condensed into a disulfide, which is an oxidation of the cysteinyl sulfur atoms as well. Coordination by metal ions may represent another form of protein thiol depletion in which the sulfur atom involved is neither oxidized nor reduced, but is otherwise made functionally unavailable by its electrostatic interaction with the metal. It has been argued that such coordination of sulfhydryl groups constitutes a protective antioxidant effect of certain metals, notably Zn (Bray and Bettger 1990b; Bettger 1993). This argument is supported by observations of increased susceptibility to oxidant damage in dietary Zn deficiency (Bray and Bettger 1990b). Similarly, oxidation of glutathione has been reported in cells exposed to Cd, Ni, Pb, and Hg ions (Stohs and Bagchi 1995), and susceptibility to metal injury has been associated with reduced glutathione levels (Gochfeld 1997; Norwood et al. 2001).

Thus, each of the three leading physicochemical properties of PM that have been implicated in its toxicity, organic carbon, surface area reactivity, and metals have been related to oxidative stress. Since oxidant species are known activators of numerous signal transduction pathways, it is perhaps not surprising that exposure to PM and PM constituents has also been linked with multiple signaling events. A single mechanism of action for oxidant-induced signaling has yet to be proposed. However, a growing body of evidence points to oxidant mediated inhibition of protein phosphatases as an essential event in cellular signaling. Signal initiation by growth factor stimulated activation of the insulin (Meng et al. 2004), epidermal growth factor (Lee et al. 1998), and platelet derived growth factor (Meng, Fukada, and Tonks 2002) receptors has been shown to involve intracellular generation of reactive oxygen species and a transient inhibition of tyrosine phosphatase activity. The redox regulation of PTPs that accompanies growth factor receptor kinase activation is now well characterized. It involves the reversible oxidation of thiol group of the catalytic cysteine found in the active site of all tyrosine phosphatases, resulting in the loss of enzymatic activity. The sulphenyl derivative formed by the oxidation of the cysteinyl residue is reversible by intracellular reductants such as glutathione (coupled with glutathione reductase to nicotinamide adenine dinucleotide phosphate (NADPH)), which restores the enzymatic activity of the phosphatase and, as such, is a mechanism for the termination of receptor-mediated signaling (Lee et al. 1998; Meng, Fukada, and Tonks 2002; Meng et al. 2004). Thus, during ligand stimulation there is concurrent kinase activation and suppression of phosphatase activity, which allows for unopposed kinase activity and results in potentiation of target protein phosphorylation.

Exogenous oxidative stress such as that presented by PM organics, metals, or ultrafine carbon particles may be able to inhibit phosphatase activity without inducing an increase in kinase activity. A basal receptor kinase activity can be demonstrated to exist even in unstimulated cells. The phosphorylation carried out by this basal kinase activity is normally opposed by an excess of phosphatase activity. In a scenario in which phosphatase activity is suppressed in the absence of receptor kinase activation, it would be expected that the loss of phosphatase activity would permit phosphoproteins to accumulate and reach levels sufficient to effect initiation of signaling. In this manner, oxidant stress caused by exposure to PM may exert a non-specific inhibition of tyrosine phosphatases, which could lead to activation of multiple signaling pathways in the cell. Such a mechanism could account for the effect of PM on signaling activation that is reported in the literature, wherein diverse signaling pathways seem to be affected simultaneously. PM metals fit in this paradigm as direct and indirect inducers of oxidant stress that inhibit tyrosine phosphatase activity, possibly in a reversible manner. In addition, V compounds and Zn^{2+} are broad inhibitors of tyrosine phosphatases that can act without the involvement of oxidants. The reversibility of these metal inhibitors of tyrosine phosphatases is not entirely clear.

Thus, at this stage of the research effort on the toxicology of PM, oxidant stress has emerged as a common theme in the search for the etiological agent responsible for the health effects of PM of widely varying physicochemical composition. On this basis and with new understanding of the pivotal role of redox regulation in cell signaling, it is plausible that inhibition of tyrosine phosphatases is a unifying mechanism that explains the broad disregulation of signal transduction induced by PM exposure (Figure 8.2).

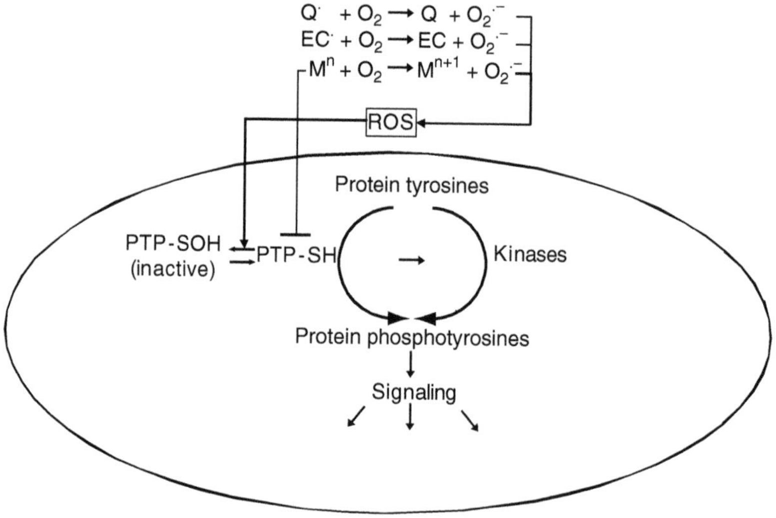

FIGURE 8.2 Proposed mechanism of cell signaling activation by PM metals and oxidants. PM organic compounds such as quinones (Q) can exist as radicals and form reactive oxygen species (ROS) in biological systems by donating an electron to oxygen. Elemental carbon (EC) particles can carry carbon-centered radical species on their surface, which can react with oxygen in solution to form superoxide and ROS. Some transition metals redox cycle between two adjacent valence states to produce superoxide anion through the Haber–Weiss reaction. Protein tyrosine phosphate levels in resting cells are kept at low levels through the activity of the protein tyrosine phosphatases (PTP), which normally grossly exceeds the basal activity of tyrosine kinases, serving to maintain signaling quiescence. ROS react with the catalytic cysteine of PTP, inhibiting their activity. Certain metal ions (e.g., V^{4+}, Zn^{2+}) can also attack the catalytic cysteine of PTP directly. In the absence of PTP activity, the basal activity of tyrosine kinases works unopposed, resulting in a progressive accumulation of protein tyrosine phosphates and signal transduction.

Finally, it is possible to identify a number of gaps that remain in our knowledge about the role of metals and oxidants in the toxicology of PM. Much is yet to be known of the speciation of bioactive metals associated with PM and of the oxidants that they produce. A related issue is that the physical chemistry of metal coordination is not well understood, and this limits our ability to discern between bioavailable and biologically inert metal species in PM. An important topic that has not received sufficient attention to date is the role of endogenous metals that are mobilized following exposure to PM. There is an extensive body of knowledge on physiology of the uptake and distribution of essential metals. However, there has been very little application or integration of this information into the toxicology of metals. Additionally, there is a pressing need for dosimetry data that can be used to reconcile experimental exposures with the concentrations of metals to which lung cells are exposed under real world conditions.

ACKNOWLEDGMENTS

The authors are indebted to Drs. Philip Bromberg, Donald Graff, and Alejandro Molinelli for their critical review of this chapter.
Disclaimer: This chapter has been reviewed by the National Health and Environmental Effects Research Laboratory, U.S. Environmental Protection Agency, and approved for publication. Approval does not signify that the contents necessarily reflect the views and policies of the Agency, nor does the mention of trade names constitute endorsement or recommendation for use.

REFERENCES

Adamson, I. Y., Prieditis, H., Hedgecock, C., and Vincent, R., Zinc is the toxic factor in the lung response to an atmospheric particulate sample, *Toxicol. Appl. Pharmacol.*, 166, 111–119, 2000.

Andrew, A. S., Klei, L. R., and Barchowsky, A., Nickel requires hypoxia-inducible factor-1 alpha, not redox signaling, to induce plasminogen activator inhibitor-1, *Am. J. Physiol. Lung Cell Mol. Physiol.*, 281, L607–L615, 2001.

Barchowsky, A. and O'Hara, K. A., Metal-induced cell signaling and gene activation in lung diseases, *Free Radic. Biol. Med.*, 34, 1130–1135, 2003.

Barchowsky, A., Soucy, N. V., O'Hara, K. A., Hwa, J., Noreault, T. L., and Andrew, A. S., A novel pathway for nickel-induced interleukin-8 expression, *J. Biol. Chem.*, 277, 24225–24231, 2002.

Beckett, W. S., Chalupa, D. F., Pauly-Brown, A., Speers, D. M., Stewart, J. C., Frampton, M. W., Utell, M. J. et al., Comparing inhaled ultrafine versus fine zinc oxide particles in healthy adults: a human inhalation study, *Am. J. Respir. Crit. Care Med.*, 171, 1129–1135, 2005.

Beck-Speier, I., Dayal, N., Karg, E., Maier, K. L., Schumann, G., Schulz, H., Semmler, M. et al., Oxidative stress and lipid mediators induced in alveolar macrophages by ultrafine particles, *Free Radic. Biol. Med.*, 38, 1080–1092, 2005.

Behra, P. and Sigg, L., Evidence for redox cycling of iron in atmospheric water droplets, *Nature*, 344, 419–421, 1990.

Bettger, W. J., Zinc and selenium, site-specific versus general antioxidation, *Can. J. Physiol. Pharmacol.*, 71, 721–724, 1993 Review.

Boisleve, F., Kerdine-Romer, S., Rougier-Larzat, N., and Pallardy, M., Nickel and DNCB induce CCR7 expression on human dendritic cells through different signaling pathways: role of TNF-alpha and MAPK, *J. Investig. Dermatol.*, 123, 494–502, 2004.

Boisleve, F., Kerdine-Romer, S., and Pallardy, M., Implication of the MAPK pathways in the maturation of human dendritic cells induced by nickel and TNF-alpha, *Toxicology*, 206, 233–244, 2005.

Boldt, D. H., New perspectives on iron: an introduction, *Am. J. Med. Sci.*, 318, 207–212, 1999.

Bonner, J. C., Rice, A. B., Moomaw, C. R., and Morgan, D. L., Airway fibrosis in rats induced by vanadium pentoxide, *Am. J. Physiol. Lung Cell Mol. Physiol.*, 278, L209–L216, 2000.

Bray, T. M. and Bettger, W. J., The physiological role of zinc as an antioxidant, *Free Radic. Biol. Med.*, 8, 281–291, 1990a.

Bray, T. M. and Bettger, W. J., The physiological role of zinc as an antioxidant, *Free Radic. Biol. Med.*, 8, 281–291, 1990b Review.

Chandler, D. B., Barton, J. C., Briggs, D. D. III, Butler, T. W., Kennedy, J. I., Grizzle, W. E., and Fulmer, J. D., Effect of iron deficiency on bleomycin-induced lung fibrosis in the hamster, *Am. Rev. Respir. Dis.*, 137, 85–89, 1988.

Chauhan, V., Breznan, D., Goegan, P., Nadeau, D., Karthikeyan, S., Brook, J. R., and Vincent, R., Effects of ambient air particles on nitric oxide production in macrophage cell lines, *Cell Biol. Toxicol.*, 20, 221–239, 2004.

Chen, F., Ding, M., Castranova, V., and Shi, X., Carcinogenic metals and NF-kappaB activation, *Mol. Cell Biochem.*, 222, 159–171, 2001.

Conde, M., Chiara, M. D., Pintado, E., and Sobrino, F., Modulation of phorbol ester-induced respiratory burst by vanadate, genistein, and phenylarsine oxide in mouse macrophages, *Free Radic.Biol. Med.*, 18, 343–348, 1995.

Cooper, C. E., Nitric oxide and iron proteins, *Biochim. Biophys. Acta.*, 1411, 290–309, 1999.

Councell, T. B., Duckenfield, K. U., Landa, E. R., and Callender, E., Tire-wear particles as a source of zinc to the environment, *Environ. Sci. Technol.*, 38, 4206–4214, 2004.

Cyrys, J., Stolzel, M., Heinrich, J., Kreyling, W. G., Menzel, N., Wittmaack, K., Tuch, T., and Wichmann, H. E., Elemental composition and sources of fine and ultrafine ambient particles in Erfurt, Germany, *Sci. Total Environ.*, 305, 143–156, 2003.

Dick, C. A., Brown, D. M., Donaldson, K., and Stone, V., The role of free radicals in the toxic and inflammatory effects of four different ultrafine particle types, *Inhal. Toxicol.*, 15, 39–52, 2003.

Ding, M., Li, J. J., Leonard, S. S., Ye, J. P., Shi, X., Colburn, N. H., Castranova, V., and Vallyathan, V., Vanadate-induced activation of activator protein-1: role of reactive oxygen species, *Carcinogenesis*, 20, 663–668, 1999.

Dodd, J., Ondov, J. M., Tuncel, G., Dzubat, T. G., and Stevens, R. K., Multimodal size spectra of submicrometer particles bearing various elements in rural air, *Environ. Sci. Technol.*, 25, 890–903, 1991.

Eisenstein, R. S. and Blemings, K. P., Iron regulatory proteins, iron responsive elements, and iron homeostasis, *J. Nutr.*, 128, 2295–2298, 1998.

Fine, J. M., Gordon, T., Chen, L. C., Kinney, P., Falcone, G., and Beckett, W. S., Metal fume fever: characterization of clinical and plasma IL-6 responses in controlled human exposures to zinc oxide fume at and below the threshold limit value, *J. Occup. Environ. Med.*, 39, 722–726, 1997.

Fine, J. M., Gordon, T., Chen, L. C., Kinney, P., Falcone, G., Sparer, J., and Beckett, W. S., Characterization of clinical tolerance to inhaled zinc oxide in naive subjects and sheet metal workers, *J. Occup. Environ. Med.*, 42, 1085–1091, 2000.

Fosmire, G. J., Zinc toxicity, *Am. J. Clin. Nutr.*, 51, 225–227, 1990.

Gao, X., Du, Z., and Patel, T. B., Copper and zinc inhibit Galphas function: a nucleotide-free state of Galphas induced by Cu2+ and Zn2+, *J. Biol. Chem.*, 280, 2579–2586, 2005.

Ghio, A. J., Stonehuerner, J., Pritchard, R. J., Quigley, D. R., Dreher, K. L., and Costa, D. L., Humic-like substances in air pollution particulates correlate with concentrations of transition metals and oxidant generation, *Inhal. Toxicol.*, 8, 479–494, 1996.

Ghio, A. J., Silbajoris, R., Carson, J. L., and Samet, J. M., Biologic effects of oil fly ash, *Environ. Health Perspect.*, 110 (suppl. 1), 89–94, 2002.

Ghio, A. J., Piantadosi, C. A., Wang, X., Dailey, L. A., Stonehuerner, J. D., Madden, M. C., Yang, F., Dolan, K. G., Garrick, M. D., and Garrick, L. M., Divalent metal transporter-1 decreases metal-related injury in the lung, *Am. J. Physiol. Lung Cell Mol. Physiol.*, 289, L460–L467, 2005.

Gochfeld, M., Factors influencing susceptibility to metals, *Environ. Health Perspect.*, 105, 817–822, 1997 Review.

Goebeler, M., Roth, J., Brocker, E. B., Sorg, C., and Schulze-Osthoff, K., Activation of nuclear factor-kappa B and gene expression in human endothelial cells by the common haptens nickel and cobalt, *J. Immunol.*, 155, 2459–2467, 1995.

Goforth, M. R. and Christoforou, C. S., Particle size distribution and atmospheric metals measurements in a rural area in the South Eastern USA, *Sci. Total Environ.*, 356, 217–227, 2005.

Gordon, J. A., Use of vanadate as protein-phosphotyrosine phosphatase inhibitor, *Methods Enzymol.*, 201, 477–482, 1991.

Grabowski, G. M., Paulauskis, J. D., and Godleski, J. J., Mediating phosphorylation events in the vanadium-induced respiratory burst of alveolar macrophages, *Toxicol. Appl. Pharmacol.*, 156, 170–178, 1999.

Greenberg, R., Zoller, W. H., and Gordon, G. E., Composition and size distributions of particles released in refuse incineration, *Environ. Sci. Technol.*, 12, 566–573, 1978.

Groen, A., Lemeer, S., van der Wijk, T., Overvoorde, J., Heck, A. J., Ostman, A., Barford, D., Slijper, M., and den Hertog, J., Differential oxidation of protein-tyrosine phosphatases, *J. Biol. Chem.*, 280, 10298–10304, 2005.

Gualtieri, M., Andrioletti, M., Vismara, C., Milani, M., and Camatini, M., Toxicity of tire debris leachates, *Environ. Int.*, 31, 723–730, 2005.

Haase, H. and Maret, W., Intracellular zinc fluctuations modulate protein tyrosine phosphatase activity in insulin/insulin-like growth factor-1 signaling, *Exp. Cell Res.*, 291, 289–298, 2003.

Hanahisa, Y. and Yamaguchi, M., Inhibitory effect of zinc and copper on phosphatase activity in the brain cytosol of rats: involvement of SH groups, *Biol. Pharm. Bull.*, 21, 1222–1225, 1998.

Hann, H. W., Stahlhut, M. W., and Menduke, H., Iron enhances tumor growth. Observation on spontaneous mammary tumors in mice, *Cancer*, 68, 2407–2410, 1991.

Haque, S. J., Flati, V., Deb, A., and Williams, B. R., Roles of protein-tyrosine phosphatases in Stat1 alpha-mediated cell signaling, *J. Biol. Chem.*, 270, 25709–25714, 1995.

Hauser, R., Elreedy, S., Hoppin, J. A., and Christiani, D. C., Airway obstruction in boilermakers exposed to fuel oil ash, *Am. J. Respir. Crit. Care Med.*, 152, 1478–1484, 1995a.

Hauser, R., Elreedy, S., Hoppin, J. A., and Christiani, D. C., Airway obstruction in boilermakers exposed to fuel oil ash. A prospective investigation, *Am. J. Respir. Crit. Care Med.*, 152, 1478–1484, 1995b.

Hauser, R., Elreedy, S., Hoppin, J. A., and Christiani, D. C., Upper airway response in workers exposed to fuel oil ash: nasal lavage analysis, *Occup. Environ. Med.*, 52, 353–358, 1995c.

Hauser, R., Daskalakis, C., and Christiani, D. C., A regression approach to the analysis of serial peak flow among fuel oil ash exposed workers, *Am. J. Respir. Crit. Care Med.*, 154, 974–980, 1996.

Henry, W. M. and Knapp, K. T., Compound forms of fossil fuel fly ash emissions, *Environ. Sci. Technol.*, 14, 450–456, 1980.

Hershfinkel, M., Moran, A., Grossman, N., and Sekler, I., A zinc-sensing receptor triggers the release of intracellular $Ca2+$ and regulates ion transport, *Proc. Natl Acad. Sci. USA*, 98, 11749–11754, 2001.

Huang, X., Iron overload and its association with cancer risk in humans: evidence for iron as a carcinogenic metal, *Mutat. Res.*, 533, 153–171, 2003.

Huang, X., Zhuang, Z., Frenkel, K., Klein, C. B., and Costa, M., The role of nickel and nickel-mediated reactive oxygen species in the mechanism of nickel carcinogenesis, *Environ. Health Perspect.*, 102 (suppl. 3), 281–284, 1994.

Huang, X., Kitahara, J., Zhitkovich, A., Dowjat, K., and Costa, M., Heterochromatic proteins specifically enhance nickel-induced 8-oxo-dG formation, *Carcinogenesis*, 16, 1753–1759, 1995.

Huang, C., Zhang, Z., Ding, M., Li, J., Ye, J., Leonard, S. S., Shen, H. M. et al., Vanadate induces p53 transactivation through hydrogen peroxide and causes apoptosis, *J. Biol. Chem.*, 275, 32516–32522, 2000.

Huang, C., Ding, M., Li, J., Leonard, S. S., Rojanasakul, Y., Castranova, V., Vallyathan, V., Ju, G., and Shi, X., Vanadium-induced nuclear factor of activated T cells activation through hydrogen peroxide, *J. Biol. Chem.*, 276, 22397–22403, 2001a.

Huang, C., Li, J., Costa, M., Zhang, Z., Leonard, S. S., Castranova, V., Vallyathan, V., Ju, G., and Shi, X., Hydrogen peroxide mediates activation of nuclear factor of activated T cells (NFAT) by nickel subsulfide, *Cancer Res.*, 61, 8051–8057, 2001b.

Huang, C., Li, J., Zhang, Q., and Huang, X., Role of bioavailable iron in coal dust-induced activation of activator protein-1 and nuclear factor of activated T cells: difference between Pennsylvania and Utah coal dusts, *Am. J. Respir. Cell Mol. Biol.*, 27, 568–574, 2002.

Hutchinson, F., Chemical changes induced in DNA by ionizing radiation, *Prog. Nucleic Acid Res. Mol. Biol.*, 32, 115–154, 1985.

Huyer, G., Liu, S., Kelly, J., Moffat, J., Payette, P., Kennedy, B., Tsaprailis, G., Gresser, M. J., and Ramachandran, C., Mechanism of inhibition of protein-tyrosine phosphatases by vanadate and pervanadate, *J. Biol. Chem.*, 272, 843–851, 1997.

Imlay, J. A. and Linn, S., DNA damage and oxygen radical toxicity, *Science*, 240, 1302–1309, 1988.

Ingram, J. L., Rice, A. B., Santos, J., Van Houten, B., and Bonner, J. C., Vanadium-induced HB-EGF expression in human lung fibroblasts is oxidant dependent and requires MAP kinases, *Am. J. Physiol. Lung Cell Mol. Physiol.*, 284, L774–L782, 2003.

Jacob, D., Gottlieb, E. W., and Prather, M. J., Chemistry of a polluted cloudy boundary layer, *J. Geophys. Res.*, 94, 12975–13002, 1989.

Jaspers, I., Samet, J. M., Erzurum, S., and Reed, W., Vanadium-induced kappaB-dependent transcription depends upon peroxide- induced activation of the p38 mitogen-activated protein kinase, *Am. J. Respir. Cell Mol. Biol.*, 23, 95–102, 2000.

Kamata, H., Shibukawa, Y., Oka, S. I., and Hirata, H., Epidermal growth factor receptor is modulated by redox through multiple mechanisms. Effects of reductants and H_2O_2, *Eur. J. Biochem.*, 267, 1933–1944, 2000.

Kennedy, J. I., Chandler, D. B., Jackson, R. M., and Fulmer, J. D., Reduction in bleomycin-induced lung hydroxyproline content by an iron-chelating agent, *Chest*, 89, 123S–125S, 1986.

Kennedy, T., Ghio, A. J., Reed, W., Samet, J., Zagorski, J., Quay, J., Carter, J., Dailey, L., Hoidal, J. R., and Devlin, R. B., Copper-dependent inflammation and nuclear factor-kappaB activation by particulate air pollution, *Am. J. Respir. Cell Mol. Biol.*, 19, 366–378, 1998.

Kim, J. H., Cho, H., Ryu, S. E., and Choi, M. U., Effects of metal ions on the activity of protein tyrosine phosphatase VHR: highly potent and reversible oxidative inactivation by Cu2+ ion, *Arch. Biochem. Biophys.*, 382, 72–80, 2000.

Kim, J. Y., Hauser, R., Wand, M. P., Herrick, R. F., Amarasiriwardena, C. J., and Christiani, D. C., The association of expired nitric oxide with occupational particulate metal exposure, *Environ. Res.*, 93, 158–166, 2003.

Koch, K. A., Pena, M. M., and Thiele, D. J., Copper-binding motifs in catalysis, transport, detoxification and signaling, *Chem. Biol.*, 4, 549–560, 1997.

Kodavanti, U. P., Moyer, C. F., Ledbetter, A. D., Schladweiler, M. C., Costa, D. L., Hauser, R., Christiani, D. C., and Nyska, A., Inhaled environmental combustion particles cause myocardial injury in the Wistar Kyoto rat, *Toxicol. Sci.*, 71, 237–245, 2003.

Krejsa, C. M. and Schieven, G. L., Impact of oxidative stress on signal transduction control by phosphotyrosine phosphatases, *Environ. Health Perspect.*, 106 (suppl. 5), 1179–1184, 1998.

Krejsa, C. M., Nadler, S. G., Esselstyn, J. M., Kavanagh, T. J., Ledbetter, J. A., and Schieven, G. L., Role of oxidative stress in the action of vanadium phosphotyrosine phosphatase inhibitors. Redox independent activation of NF-kappaB, *J. Biol. Chem.*, 272, 11541–11549, 1997a.

Krejsa, C. M., Nadler, S. G., Esselstyn, J. M., Kavanagh, T. J., Ledbetter, J. A., and Schieven, G. L., Role of oxidative stress in the action of vanadium phosphotyrosine phosphatase inhibitors. Redox independent activation of NF-kappaB, *J. Biol. Chem.*, 272, 11541–11549, 1997b.

Kuschner, W. G., Alessandro, D. A., Wintermeyer, S. F., Wong, H., Boushey, H. A., and Blanc, P. D., Pulmonary responses to purified zinc oxide fume, *J. Investig. Med.*, 43, 371–378, 1995.

Lee, S. R., Kwon, K. S., Kim, S. R., and Rhee, S. G., Reversible inactivation of protein-tyrosine phosphatase 1B in A431 cells stimulated with epidermal growth factor, *J. Biol. Chem.*, 273, 15366–15372, 1998.

Levy, B. S., Hoffman, L., and Gottsegen, S., Boilermakers' bronchitis, *J. Occup. Med.*, 26, 567–570, 1984.

Li, N., Kim, S., Wang, M., Froines, J., Sioutas, C., and Nel, A., Use of a stratified oxidative stress model to study the biological effects of ambient concentrated and diesel exhaust particulate matter, *Inhal. Toxicol.*, 14, 459–486, 2002.

Li, J., Dokka, S., Wang, L., Shi, X., Castranova, V., Yan, Y., Costa, M., and Huang, C., Activation of aPKC is required for vanadate-induced phosphorylation of protein kinase B (Akt), but not p70S6k in mouse epidermal JB6 cells, *Mol. Cell Biochem.*, 255, 217–225, 2004.

Liang, L., Zsembery, A., and Schwiebert, E. M., RNA interference targeted to multiple P2X receptor subtypes attenuates zinc-induced calcium entry, *Am. J. Physiol. Cell Physiol.*, 289, C388–C396, 2005.

Logan, J., Prather, M. J., Wofsy, S. C., and McElroy, M. B., Tropospheric chemistry: a global perspective, *J. Geophys. Res.*, 86, 7210–7254, 1981.

Lu, B., Ennis, D., Lai, R., Bogdanovic, E., Nikolov, R., Salamon, L., Fantus, C., Le-Tien, H., and Fantus, I. G., Enhanced sensitivity of insulin-resistant adipocytes to vanadate is associated with oxidative stress and decreased reduction of vanadate (+5) to vanadyl (+4), *J. Biol. Chem.*, 276, 35589–35598, 2001.

Madden, M. C., Thomas, M. J., and Ghio, A. J., Acetaldehyde (CH_3CHO) production in rodent lung after exposure to metal-rich particles, *Free Radic. Biol. Med.*, 26, 1569–1577, 1999.

Magari, S. R., Schwartz, J., Williams, P. L., Hauser, R., Smith, T. J., and Christiani, D. C., The association of particulate air metal concentrations with heart rate variability, *Environ. Health Perspect.*, 110, 875–880, 2002.

Mamane, Y., Estimate of municipal refuse incinerator contribution to Philadelphia aerosol using single particle analysis. Ambient measurements, *Atmos. Environ.*, 24B, 127–135, 1990.

Mattie, M. D. and Freedman, J. H., Copper-inducible transcription: regulation by metal- and oxidative stress-responsive pathways, *Am. J. Physiol. Cell Physiol.*, 286, C293–C301, 2004.

Maxwell, P. and Salnikow, K., HIF-1: an oxygen and metal responsive transcription factor, *Cancer Biol. Ther.*, 3, 29–35, 2004.

May, J. M. and Contoreggi, C. S., The mechanism of the insulin-like effects of ionic zinc, *J. Biol. Chem.*, 257, 4362–4368, 1982.

Meng, T. C. and Tonks, N. K., Analysis of the regulation of protein tyrosine phosphatases in vivo by reversible oxidation, *Methods Enzymol.*, 366, 304–318, 2003.

Meng, T. C., Fukada, T., and Tonks, N. K., Reversible oxidation and inactivation of protein tyrosine phosphatases in vivo, *Mol. Cell*, 9, 387–399, 2002.

Meng, T. C., Buckley, D. A., Galic, S., Tiganis, T., and Tonks, N. K., Regulation of insulin signaling through reversible oxidation of the protein-tyrosine phosphatases TC45 and PTP1B, *J. Biol. Chem.*, 279, 37716–37725, 2004.

Misra, M., Rodriguez, R. E., and Kasprzak, K. S., Nickel induced lipid peroxidation in the rat: correlation with nickel effect on antioxidant defense systems, *Toxicology*, 64, 1–17, 1990.

Munger, J., Jacob, D. J., Waldman, L. M., and Hoffman, M. R., Fogwater chemistry in an urban atmosphere, *J. Geophys. Res.*, 88, 5109–5121, 1983.

Noh, K. M. and Koh, J. Y., Induction and activation by zinc of NADPH oxidase in cultured cortical neurons and astrocytes, *J Neurosci.*, 20, RC111, 2000.

Norwood, J., Jr., Ledbetter, A. D., Doerfler, D. L., and Hatch, G. E., Residual oil fly ash inhalation in guinea pigs: influence of absorbate and glutathione depletion, *Toxicol. Sci.*, 61, 144–153, 2001.

Oller, A. R., Respiratory carcinogenicity assessment of soluble nickel compounds, *Environ. Health Perspect.*, 110 (suppl. 5), 841–844, 2002.

Oller, A. R., Costa, M., and Oberdörster, G., Carcinogenicity assessment of selected nickel compounds, *Toxicol. Appl. Pharmacol.*, 143, 152–166, 1997.

Ostrakhovitch, E. A. and Cherian, M. G., Differential regulation of signal transduction pathways in wild type and mutated p53 breast cancer epithelial cells by copper and zinc, *Arch. Biochem. Biophys.*, 423, 351–361, 2004.

Ostrakhovitch, E. A., Lordnejad, M. R., Schliess, F., Sies, H., and Klotz, L. O., Copper ions strongly activate the phosphoinositide-3-kinase/Akt pathway independent of the generation of reactive oxygen species, *Arch. Biochem. Biophys.*, 397, 232–239, 2002.

Persson, C., Sjoblom, T., Groen, A., Kappert, K., Engstrom, U., Hellman, U., Heldin, C. H., den Hertog, J., and Ostman, A., Preferential oxidation of the second phosphatase domain of receptor-like PTP-alpha revealed by an antibody against oxidized protein tyrosine phosphatases, *Proc. Natl Acad. Sci. USA*, 101, 1886–1891, 2004.

Prieditis, H. and Adamson, I. Y., Comparative pulmonary toxicity of various soluble metals found in urban particulate dusts, *Exp. Lung Res.*, 28, 563–576, 2002.

Pritchard, R. J., Ghio, A. J., Lehman, J. R., Winsett, D. W., Tepper, J. S., Park, P., Gilmour, M. I., Dreher, K. L., and Costa, D., Oxidant generation and lung injury after particulate air pollutant exposure increase with the concentrations of associated metals, *Inhal. Toxicol.*, 8, 457–477, 1996.

Rodriguez, R. E., Misra, M., North, S. L., and Kasprzak, K. S., Nickel-induced lipid peroxidation in the liver of different strains of mice and its relation to nickel effects on antioxidant systems, *Toxicol. Lett.*, 57, 269–281, 1991.

Ruas, J. L. and Poellinger, L., Hypoxia-dependent activation of HIF into a transcriptional regulator, *Semin. Cell Dev. Biol.*, 16, 514–522, 2005.

Ryter, S. W. and Choi, A. M., Heme oxygenase-1: redox regulation of a stress protein in lung and cell culture models, *Antioxid. Redox Signal*, 7, 80–91, 2005.

Salmeen, A., Andersen, J. N., Myers, M. P., Meng, T. C., Hinks, J. A., Tonks, N. K., and Barford, D., Redox regulation of protein tyrosine phosphatase 1B involves a sulphenyl-amide intermediate, *Nature*, 423, 769–773, 2003.

Salnikow, K. and Costa, M., Epigenetic mechanisms of nickel carcinogenesis, *J. Environ. Pathol. Toxicol. Oncol.*, 19, 307–318, 2000.

Salnikow, K. and Kasprzak, K. S., Ascorbate depletion: a critical step in nickel carcinogenesis? *Environ. Health Perspect.*, 113, 577–584, 2005.

Salnikow, K., Gao, M., Voitkun, V., Huang, X., and Costa, M., Altered oxidative stress responses in nickel-resistant mammalian cells, *Cancer Res.*, 54, 6407–6412, 1994.

Salnikow, K., Blagosklonny, M. V., Ryan, H., Johnson, R., and Costa, M., Carcinogenic nickel induces genes involved with hypoxic stress, *Cancer Res.*, 60, 38–41, 2000.

Salnikow, K., Davidson, T., and Costa, M., The role of hypoxia-inducible signaling pathway in nickel carcinogenesis, *Environ. Health Perspect.*, 110 (suppl. 5), 831–834, 2002.

Salnikow, K., Davidson, T., Zhang, Q., Chen, L. C., Su, W., and Costa, M., The involvement of hypoxia-inducible transcription factor-1-dependent pathway in nickel carcinogenesis, *Cancer Res.*, 63, 3524–3530, 2003.

Salnikow, K., Donald, S. P., Bruick, R. K., Zhitkovich, A., Phang, J. M., and Kasprzak, K. S., Depletion of intracellular ascorbate by the carcinogenic metals nickel and cobalt results in the induction of hypoxic stress, *J. Biol. Chem.*, 279, 40337–40344, 2004.

Samet, J. M., Stonehuerner, J., Reed, W., Devlin, R. B., Dailey, L. A., Kennedy, T. P., Bromberg, P. A., and Ghio, A. J., Disruption of protein tyrosine phosphate homeostasis in bronchial epithelial cells exposed to oil fly ash, *Am. J. Physiol. Lung Cell. Mol. Physiol.*, 272, L426–L432, 1997.

Samet, J. M., Graves, L. M., Quay, J., Dailey, L. A., Devlin, R. B., Ghio, A. J., Bromberg, P. A., and Reed, W., Activation of map kinases in human airway epithelial cells exposed to metals, *Am. J. Physiol. Lung Cell. Mol. Physiol.*, 275, L551–L558, 1998a.

Samet, J. M., Graves, L. M., Quay, J., Dailey, L. A., Devlin, R. B., Ghio, A. J., Wu, W., Bromberg, P. A., and Reed, W., Activation of MAPKs in human bronchial epithelial cells exposed to metals, *Am J Physiol.*, 275, L551–L558, 1998b.

Samet, J. M., Silbajoris, R., Wu, W., and Graves, L. M., Tyrosine phosphatases as targets in metal-induced signaling in human airway epithelial cells, *Am. J. Respir. Cell. Mol. Biol.*, 21, 357–364, 1999a.

Samet, J. M., Silbajoris, R., Wu, W., and Graves, L. M., Tyrosine phosphatases as targets in metal-induced signaling in human airway epithelial cells, *Am. J. Respir. Cell. Mol. Biol.*, 21, 357–364, 1999b.

Samet, J. M., Dewar, B. J., Wu, W., and Graves, L. M., Mechanisms of Zn(2+)-induced signal initiation through the epidermal growth factor receptor, *Toxicol. Appl. Pharmacol.*, 191, 86–93, 2003.

Schroeder, W. H., Dobson, M., Kane, D. M., and Johnson, N. D., Toxic trace elements associated with airborne particulate matter: a review, *JAPCA*, 37, 1267–1285, 1987.

Seilkop, S. K. and Oller, A. R., Respiratory cancer risks associated with low-level nickel exposure: an integrated assessment based on animal, epidemiological, and mechanistic data, *Regul. Toxicol. Pharmacol.*, 37, 173–190, 2003.

Sorensen, M., Schins, R. P., Hertel, O., and Loft, S., Transition metals in personal samples of PM2.5 and oxidative stress in human volunteers, *Cancer Epidemiol. Biomarkers Prev.*, 14, 1340–1343, 2005.

Stevens, R. K., Dzubay, T. G., Russwurm, G., and Rickel, D., Sampling and analysis of atmospheric sulfates and related species, *Atmos. Environ.*, 12, 55–68, 1978.

Stohs, S. J. and Bagchi, D., Oxidative mechanisms in the toxicity of metal ions, *Free Radic. Biol. Med.*, 18, 321–336, 1995.

Sun, G. and Budde, R. J., Substitution studies of the second divalent metal cation requirement of protein tyrosine kinase CSK, *Biochemistry*, 38, 5659–5665, 1999.

Tabor, E. and Kim, C. M., Inhibition of human hepatocellular carcinoma and hepatoblastoma cell lines by deferoxamine, *J. Med. Virol.*, 34, 45–50, 1991.

Tonks, N. K., Redox redux: revisiting PTPs and the control of cell signaling, *Cell*, 121, 667–670, 2005.

Tonks, N. K., Diltz, C. D., and Fischer, E. H., CD45, an integral membrane protein tyrosine phosphatase. Characterization of enzyme activity, *J. Biol. Chem.*, 265, 10674–10680, 1990.

Torres, J. and Wilson, M. T., The reactions of copper proteins with nitric oxide, *Biochim. Biophys. Acta*, 1411, 310–322, 1999.

Uriu-Adams, J. Y. and Keen, C. L., Copper, oxidative stress, and human health, *Mol. Aspects Med.*, 26, 268–298, 2005.

Valko, M., Morris, H., and Cronin, M. T., Metals, toxicity and oxidative stress, *Curr. Med. Chem.*, 12, 1161–1208, 2005.

Waldman, J., Munger, J. W., Jacob, D. J., Flagan, R. C., Morgan, J. J., and Hoffmann, M. R., Chemical composition of acid fog, *Science*, 218, 677–680, 1982.

Walther, U. I., Wilhelm, B., Walther, S. C., Muckter, H., and Forth, W., Effect of zinc chloride on GSH synthesis rates in various lung cell lines, *In Vitro Mol. Toxicol.*, 13, 145–152, 2000.

Wang, Y. Z., Ingram, J. L., Walters, D. M., Rice, A. B., Santos, J. H., Van Houten, B., and Bonner, J. C., Vanadium-induced STAT-1 activation in lung myofibroblasts requires H_2O_2 and p38 MAP kinase, *Free Radic. Biol. Med.*, 35, 845–855, 2003.

Weinbren, K., Salm, R., and Greenberg, G., Intramuscular injections of iron compounds and oncogenesis in man, *Br. Med. J.*, 1, 683–685, 1978.

Weschler, C., Mandich, M. L., and Graedel, T. E., Speciation, photosensitivity, and reactions of transition metal ions in atmospheric droplets, *J. Geophys. Res.*, 91, 5189–5204, 1986.

Wu, W., Graves, L. M., Jaspers, I., Devlin, R. B., Reed, W., and Samet, J. M., Activation of the EGF receptor signaling pathway in human airway epithelial cells exposed to metals, *Am. J. Physiol. Lung Cell. Mol. Physiol.*, 277, L924–L931, 1999a.

Wu, W., Graves, L. M., Jaspers, I., Devlin, R. B., Reed, W., and Samet, J. M., Activation of the EGF receptor-signaling pathway in human airway epithelial cells exposed to metals, *Am. J. Physiol.*, 277, L924–L931, 1999b.

Wu, W., Graves, L. M., Jaspers, I., Devlin, R. B., Reed, W., and Samet, J. M., Activation of the EGF receptor-signaling pathway in human airway epithelial cells exposed to metals, *Am. J. Physiol.*, 277, L924–L931, 1999c.

Wu, W., Samet, J. M., Ghio, A. J., and Devlin, R. B., Activation of the EGF receptor signaling pathway in airway epithelial cells exposed to Utah Valley PM, *Am. J. Physiol. Lung Cell Mol. Physiol.*, 281, L483–L489, 2001.

Wu, W., Graves, L. M., Gill, G. N., Parsons, S. J., and Samet, J. M., Src-dependent phosphorylation of the epidermal growth factor receptor on tyrosine 845 is required for zinc-induced ras activation, *J. Biol. Chem.*, 277, 24252–24257, 2002a.

Wu, W., Jaspers, I., Zhang, W., Graves, L., and Samet, J. M., Role of Ras in metal-induced EGF receptor signaling and NFKB activation in human airway epithelial cells, *Am. J. Physiol.: Lung Cell Mol. Physiol.*, 282, L924–L931, 2002b.

Wu, W., Jaspers, I., Zhang, W., Graves, L. M., and Samet, J. M., Role of Ras in metal-induced EGF receptor signaling and NF-kappaB activation in human airway epithelial cells, *Am. J. Physiol. Lung Cell Mol. Physiol.*, 282, L1040–L1048, 2002c.

Wu, W., Wang, X., Zhang, W., Reed, W., Samet, J. M., Whang, Y. E., and Ghio, A. J., Zinc-induced PTEN protein degradation through the proteasome pathway in human airway epithelial cells, *J. Biol. Chem.*, 278, 28258–28263, 2003.

Xia, T., Korge, P., Weiss, J. N., Li, N., Venkatesen, M. I., Sioutas, C., and Nel, A., Quinones and aromatic chemical compounds in particulate matter induce mitochondrial dysfunction: implications for ultrafine particle toxicity, *Environ. Health Perspect.*, 112, 1347–1358, 2004.

Xiong, S., She, H., and Tsukamoto, H., Signaling role of iron in NF-kappa B activation in hepatic macrophages, *Comp. Hepatol.*, 3 (suppl. 1), S36, 2004.

Yu, C. P., Hsieh, T. H., Oller, A. R., and Oberdörster, G., Evaluation of the human nickel retention model with workplace data, *Regul. Toxicol. Pharmacol.*, 33, 165–172, 2001.

Zhang, Z., Huang, C., Li, J., Leonard, S. S., Lanciotti, R., Butterworth, L., and Shi, X., Vanadate-induced cell growth regulation and the role of reactive oxygen species, *Arch. Biochem. Biophys.*, 392, 311–320, 2001.

Zsembery, A., Fortenberry, J. A., Liang, L., Bebok, Z., Tucker, T. A., Boyce, A. T., Braunstein, G. M. et al., Extracellular zinc and ATP restore chloride secretion across cystic fibrosis airway epithelia by triggering calcium entry, *J. Biol. Chem.*, 279, 10720–10729, 2004.

9 Proinflammatory Effects of Particles on Macrophages and Epithelial Cells

Vicki Stone
School of Life Sciences, Napier University

Peter G. Barlow
Queen's Medical Research Institute, University of Edinburgh

Gary R. Hutchison
Medical Research Council, Queens Medical Research Institute

David M. Brown
School of Life Sciences, Napier University

CONTENTS

This chapter aims to compare the ability of a number of potentially pathogenic particles in terms of their ability to induce inflammation and disease. The article will focus on the effects of α-quartz, asbestos, and environmental particulate air pollution particles and how these different particles activate epithelial cells and macrophages leading to inflammation.

9.1 PARTICLE-INDUCED INFLAMMATION AND DISEASE

As described in the previous chapter, inflammation is considered to play a key role in driving disease induced by a number of pathogenic respirable particles. For example, carcinogenic and pro-fibrotic particles such as α-quartz and asbestos have both been shown to induce a chronic inflammatory response involving a number of cell types in the lung.[1,2] The chronic inflammatory response coupled with the surface, chemical, or physical reactivity of the particles is thought to activate neutrophils and macrophages allowing inflammation to persist.[3,4] This prolonged inflammation can result in a number of processes that contribute to the induction of fibrosis and carcinogenesis. For example, the inflammatory process results in the production of mitogenic stimuli that induce epithelial cell proliferation (Figure 9.1).[5] As cell proliferation rates increase, the chance of successful repair to damaged deoxyribonucleic acid (DNA) prior to cell division is diminished and hence the risk of passing on a mutation to daughter cells is enhanced. In addition, such particles have been demonstrated to generate reactive oxygen species (ROS) due to activity at the particle surface,[6,7] as well as from phagocytic cells (Figure 9.1). Both of these processes can lead to oxidative stress and damage to macromolecules such as DNA, again increasing the potential for mutations. A more detailed discussion can be found in the Chapter on Genotoxicity of Particles (Schins & Hei).

Particulate air pollution particles, PM_{10} (particulate matter collected through a size selective inlet with 50% efficiency for particles of 10 μm aerodynamic diameter), are associated with short-term effects such as increased hospital admissions or deaths due to respiratory and cardiovascular causes.[8] Again, inflammation is postulated to play a key role in the mechanism by which PM_{10} exacerbates pre-existing inflammatory diseases in susceptible individuals.[9] The composition of PM_{10} is complex and variable with location and time, however a number of studies have suggested that various components such as the ultrafine or nanoparticle fraction (definitions provided below),[10,11] metals,[12–14] and endotoxin[15,16] play a key role in driving the proinflammatory effects of PM_{10}.

Much of the information that is available relating to the toxicology of ultrafine particles (diameter less than 100 nm) has been obtained from the study of low toxicity, low-solubility nanoparticles (one diameter less than 100 nm) such as carbon black,[17,18] polystyrene[19] and TiO_2.[20] These studies and many more all suggest that as particle size decreases, potential to induce oxidative stress and inflammation increases. Such results are likely to be relevant to the toxicology of engineered nanoparticles, although this may vary if the nanoparticles are soluble, preventing biopersistance, if they are coated (e.g., polyethyleneglycol; PEG),[21] allowing avoidance of macrophage phagocytosis and cellular uptake, or if the particles are extremely toxic (e.g., α-quartz). With the recent expansion of nanotechnology, engineered nanoparticles are used in a wide variety of applications including sunscreens, cosmetics, food, and medicine, as well as

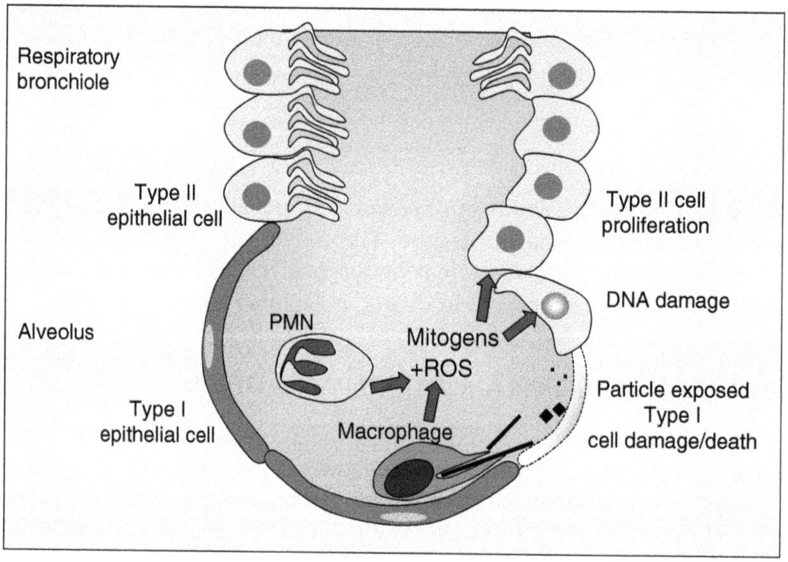

FIGURE 9.1 A schematic diagram of the respiratory regions of the lung depicting the effects of particles on cell proliferation and mutation. In the presence of pathogenic particles, reactive oxygen species (ROS) can damage the DNA of dividing cells leading to mutation. The ROS can also activate intracellular factors that drive cell proliferation (AP-1). These factors combined with mitogenic factors released by the inflammatory cells (neutrophils and macrophages) stimulate type II cell proliferation to replace the damaged type I cells. An increase in cell proliferation rate increased the possibility that mutations are not corrected and hence become permanent.

technical applications such as electronics. This means that exposure to a wider range of nanoparticles is likely to occur for workers, consumers, and patients via a number of exposure routes including inhalation, ingestion, injection, and dermal adsorption. Hence, the variety of target tissues affected by nanoparticle exposure is likely to increase.[22,23]

In a study by Brown et al.[20] using polystyrene beads, and by Duffin et al.[24] using carbon black, polystyrene beads and TiO_2 of varying sizes, it was demonstrated that a clear linear relationship exists between the surface area dose of the particle instilled into the rat lung and the ability to induce inflammation, as indicated by neutrophil influx 18 h after instillation. Stoeger et al. recently identified a link between surface area and the ability of six different carbon particles to induce inflammation in the mouse lung.[25] In this study, they identified a threshold dose for the instilled particles of 20 cm^2 surface area, below which an acute proinflammatory responses could not be detected. A similar study was reported by Tran et al. using inhalation of TiO_2 and barium sulphate particles into the rat lung, but in this study the threshold dose for inducing inflammation was between 200 and 300 cm^2.[26] This tenfold difference in threshold could be due to animal size and species differences. The subsequent translation of the threshold dose, identified by Tran et al. into particle surface area per surface area of lung took into consideration that deposition of the particles would be most prevalent in the centroacinar region of the lung, and found that the particle dose that initiated inflammation equated to 1 cm^2 of particle surface area/1 cm^2 of lung surface area.[27] This represents a useful way of considering dose assessment in relation to toxicological impact. A simple explanation for the link between surface area and inflammation is that the surface/volume ratio changes dramatically at low radius and that the surface area is directly proportional to the number of atoms at the particle surface.

Many papers have described the roles of macrophages and epithelial cells in driving the inflammatory response to ultrafine particles;[9,28] α-quartz,[3] and asbestos.[4] This review, however, will concentrate on the signaling mechanisms elicited in these two cell types on exposure to a variety of particles.

9.2 THE ROLE OF EPITHELIAL CELLS IN DRIVING PARTICLE-INDUCED INFLAMMATION

When inhaled particles deposit in the upper airways, they interact with epithelial cells that are ciliated and covered in thick, sticky mucus. This allows the particles to be cleared by the wafting actions of the mucociliary escalator, causing the particles to be transported out of the lung airways to be swallowed into the stomach, or to be blown from the nose. In contrast, particles depositing in the respiratory parts of the lung, including the alveoli and terminal respiratory bronchioles, must be cleared by phagocytic cells such as macrophages. In this region of the lung, the epithelium is not ciliated—instead the type I epithelial cells of the alveolus are large, thin, flat structures that are designed for gaseous exchange. The type I epithelial cells are incapable of division due to their specialized nature. If damaged by toxins or particles, the type I cells must be replaced by division of type II epithelial cells (type II cell hyperplasia) that subsequently differentiate into type I cells.

Type II cells are functionally very different to type I cells, with functions that include synthesis of surfactant for the reduction of alveolar surface tension.[29] These cells are also very different in structure, possessing a cuboidal structure with microvilli that extend into the alveolar lumen.[30] The type II cells are also capable of synthesizing a range of inflammatory mediators, so they play a key role in particle induced inflammation in the lung. Particle deposition in the alveolus results in interaction with these epithelial cells, which generate chemotactic factors (e.g., IL8) to stimulate the recruitment of phagocytic cells. For example, PM_{10} has been shown to increase IL8 expression by the human type II epithelial cell line A549,[31,32] as has α-quartz,[33,34] and nanoparticle carbon black.[35] Hence, this would suggest that activation of epithelial cells by particles is an essential part of the clearance process. A more detailed description of the chemotactic actions of epithelial products generated by particle treatment is provided below.

Few studies have demonstrated an ability of ultrafine or nanoparticles to induce cytokine production by epithelial cells *in vitro*, unless studied at very high mass doses. This is because the surface of the particles is very adsorbent, and hence binds the proteins released by the cells, resulting in an underestimation of the cytokine production. Seagrave et al. have also demonstrated that IL8 can adsorb onto diesel exhaust particulate matter.[36] This IL8 appeared to remain biologically active, as it was able to induce neutrophil shape change. We have recently found that this effect is not limited to cytokines such as IL8 and TNFα, but also includes the cytotoxicity marker lactate dehydrogenase (LDH), resulting in an underestimation of the particle-induced toxicity.[37] Hohr et al. also demonstrated that ultrafine TiO_2 particles adsorb myeloperoxidase protein, preventing its accurate determination in BAL fluid of instilled animals.[38] The consequence of protein adsorption onto the particle surface in terms of the protein function and the particle toxicity requires further investigation; perceivable effects range from protein dysfunction to hyper-reactivity of cells on receiving a bolus dose concentrated on the particle surface.[39]

Despite the lack of evidence that nanoparticles stimulate epithelial cells to generate cytokine proteins, there is, however, sufficient evidence to suggest that the messenger ribonucleic acid (mRNA) for cytokines such as IL8 is upregulated by these particles (e.g., Schins et al.[28]).

The gene expression of many proinflammatory cytokines such as IL8 and TNFα is under the control of the transcription factor nuclear factor kappa B (NF-κB). NF-κB, when not activated, is retained in the cytoplasm of the cell by the binding of inhibitor kappa B (IκB). Phosphorylation of IκB by IκB kinase (IKK) results in the release of NF-κB and its subsequent nuclear localization.[40] Schins et al.[28] demonstrated that treatment of the human alveolar type II epithelial cell line (A549) with α-quartz led to a persistent depletion of IκB allowing proinflammatory signaling and IL8 expression to continue. Treatment of A549 cells with asbestos has also been shown to induce DNA binding of NF-κB and NF-κB dependent gene expression.[41] In a different study, PM_{10} particles were also found to induce NF-κB nuclear localization, DNA binding, and transcriptional activation in A549 cells.[42] However, this effect occurred in the absence of IκB degradation, a phenomenon that has been observed for hydrogen peroxide.[43]

Ramage et al.[44] demonstrated that air pollution particles (PM_{10}) and 14 nm carbon black particles activate the expression of heat shock protein 70 (HSP70) by the A549 lung epithelial cell line. HSP70 can prevent NF-κB activation by the stabilization of IκB kinase. This means that HSP70 is a cytoprotective protein that exists in cells as a molecular chaperone, and during oxidative stress and inflammation it can up-regulate processes to protect the cell from damage.[45–47]

HSP70 secretion has also been shown to be increased in response to pathogenic particles such as asbestos.[48] In fact, in the study by Ramage et al.[31] both PM_{10} and nanoparticle carbon black stimulated increased HSP70 secretion by A549 cells—an effect that was significantly inhibited by the addition of antioxidants, suggesting a role for ROS in the particle induced upregulation and release of this molecule. The role of the released HSP70 is not fully understood, but extracellular HSP70 has been shown to activate macrophages leading to calcium influx, NF-κB activation and TNFα production,[49] and to be elevated during cardiovascular disease.[50]

The late activation of HSP70 (as seen by Ramage et al.[31] at 6 h after cell treatment) may be associated with a proinflammatory effect, since release of HSP70 into the blood has been associated with a proinflammatory status. We hypothesize that the differential activation of NF-κB and inhibition via HSP70 may be treatment dependent, with relatively low dose or low toxicity materials allowing HSP70 activation and upregulation of antioxidant defenses, while higher dose or toxicity materials bypass this protective mechanism leading to NF-κB activation and inflammation (Figure 9.2). As with many signaling pathways, it is likely that the two pathways suggested are not mutually exclusive and that there is either overlap or that they form part of a continuum.

Many of the cell signaling events described above include a role for oxidative stress or ROS. For example, nanoparticle carbon black has been shown to deplete the antioxidant glutathione in the epithelial cell line A549.[51] While the effects of PM_{10} on the glutathione content of epithelial cells *in vitro* have not been published, PM_{10} instillation into the rat lung was found to deplete both lung

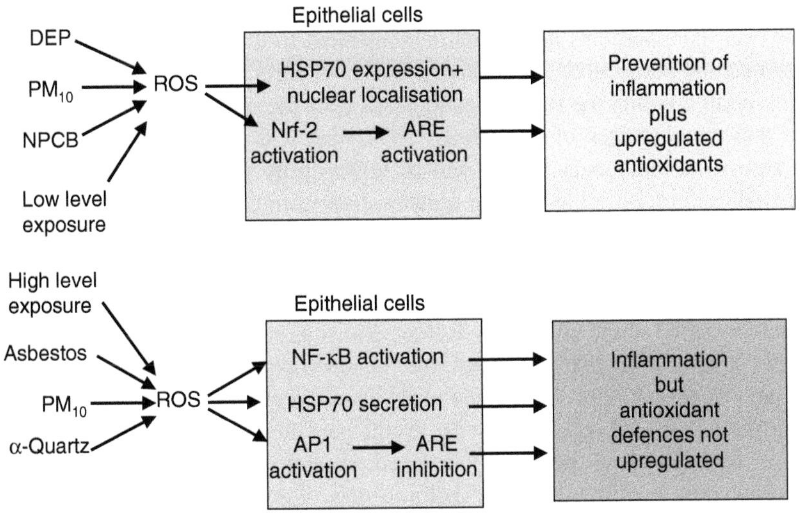

FIGURE 9.2 Hypothetical mechanisms involving epithelial cells by which particles may differentially regulate proinflammatory and antioxidant defense mechanisms. PM_{10} (respirable particulate air pollution) and nanoparticle carbon black (NPCB) have all been demonstrated to activate HSP70 nuclear localization, while diesel exhaust particles (DEP) have also been shown to activate the antioxidant response element (ARE), allowing antioxidant upregulation and prevention of inflammation. Conversely, asbestos, PM_{10} and α-quartz have all been shown to activate NF-κB leading to cytokine gene expression and inflammation. The two pathways are not suggested to be mutually exclusive, but may be a continuum or overlap.

tissue and bronchoalveolar lavage fluid glutathione content.[52] The pathology of asbestos has also been linked to oxidative stress as indicated by its ability to upregulate HO-1[53] and to deplete glutathione from lung lining fluid.[7]

The role of ROS in driving the proinflammatory effects of many particle types is evidenced by the ability of antioxidants to inhibit a variety of particle induced signaling events. For example, as described above, nacystelyn prevented HSP70 nuclear localization in A549 cells treated with PM_{10} or nanoparticle carbon black.[31] Brown et al.[54] also demonstrated that nacystelyn, along with the other antioxidants such as curcumin, could prevent activation of the transcription factor NF-κB on exposure of A549 cells to pathogenic fibres. Mannitol, a hydroxyl radical scavenger, has also been demonstrated to prevent nanoparticle carbon black induced cytotoxicity to A549 cells.[51]

9.3 THE EFFECTS OF PARTICLES ON MACROPHAGE SIGNALING MECHANISMS

Not all particles entering the body are pathogenic; this is because macrophages play a major role in the clearance of foreign particles. Alveolar macrophage make up approximately 5% of the total lung cell population[55] and occur at a frequency of approximately one macrophage per alveolus.[56] Interstitial and intravascular macrophages are also present in the lung, but these are less well studied due to their inaccessibility by bronchoalveolar lavage.

In order for macrophages to respond to a chemotactic stimulus, to migrate, to phagocytose a particle, and then to clear that particle from the tissue, a complex interaction of extracellular and intracellular signaling pathways is required to control the cellular response (Figure 9.1 and Figure 9.2). However, these pathways are open to modulation by environmental factors and by the toxic effects of the particle. An alteration in such signaling pathways can lead to decreased particle clearance and hence increased risk of inflammation and disease.

For most particles, the site from which particles must be cleared is usually the respiratory system. However, nanoparticles have been demonstrated to translocate to other organs.[57,58] It is well documented that intravenous injection of a variety of particles, including nanoparticles, results in their accumulation within the reticulo-endothelial cells including the Kupffer macrophage cells of the liver and macrophages of the spleen.[59,60] Hence it is possible that nanoparticles which translocate across the pulmonary barrier will be taken up by this system of tissue macrophages. Indeed, the studies demonstrating particle translocation from the lung do exhibit accumulation in organs such as the liver.[61] Furthermore, due to the diverse applications devised for nanoparticles, exposure routes will also include ingestion as components of foods and injection as components of medicines. A description of cellular uptake and translocation mechanisms for nanoparticles are described and discussed in the chapter by Rothen-Rutishauser et al.

Many studies have demonstrated a key role for macrophages in driving the proinflammatory response to pathogenic particles. Of these studies, a large proportion have been conducted *in vitro* using both primary cells and cell lines, so the results generated—although focused initially on the lung—are now relevant to any potentially exposed tissue type. Since nanoparticles are known to activate macrophages *in vitro* (described below), it is likely that regardless of the tissue type, macrophages can be activated by nanoparticle exposure, but that clearance will not be fully effective, leading to a proinflammatory status. However, this response may not be a universal reaction to nanoparticles, since modification with agents such as PEG is known to prevent uptake by the reticulo-endothelial system.[62] Moghimi and colleagues have published a number of studies demonstrating how drug delivery particles such as liposomes and polystyrene beads can be modified to avoid macrophage uptake.[63–66] Non-ionic surfactants and polymeric macromolecules have proven to be very successful for this purpose. Surfactants decrease the van der Waals forces that are responsible for particle aggregation, and actually increase the repulsive forces. Polymer-coated particles are thought to avoid macrophage uptake by prevention of opsinization of the particle surface. While this field of research is fairly well advanced for nanomedicine, this data requires examination in order to allow application to other types of nanoparticles.

In vitro studies with macrophages have demonstrated that nanoparticle carbon black and PM_{10} activate the expression of proinflammatory mediators such as tumor necrosis factor alpha (TNFα).[67–69] Furthermore, these studies have investigated the signaling mechanisms activated by the nanoparticles and PM_{10}, identifying that both particle types activate calcium signaling[70,71] via a mechanism involving ROS. This would suggest that oxidative stress or ROS are important in both the epithelial cell (as described above) and macrophage responses to pathogenic particles. The signals induced then activate transcription factors such as NF-κB, leading to the subsequent increased production of TNFα protein production.[15,50]

To date, most of the studies relating to the induction of inflammation and nanoparticles have concentrated on the upregulation of inflammation. However, a number of signaling pathways are activated by oxidative stress that can protect the cell. For example, activation of the antioxidant response element (ARE) by Nrf-2 leads to the upregulation of antioxidant defenses in response to ROS production. ARE is a genetic sequence found in the promoter of many genes that controls the expression of antioxidant defense pathways, including enzymes such as heme-oxygenase-1 (HO-1) and glutathione S-transferase (GST).[72,73] NF-κB activation leads to increased expression of proinflammatory cytokines, while Nrf-2 activation leads to the induction of antioxidant defense mechanisms, including HO-1. Li et al.[74] demonstrated that both organic and inorganic extracts of diesel exhaust particulates induced the expression of HO-1 and GST in macrophages via a transcription factor Nrf-2. In other studies, the oxidant *tert*-butyl hydroperoxide and lipopolysaccharide (LPS) have both been demonstrated to activate the ARE in macrophages.[75] The ARE actually contains a binding site for the transcription factor AP-1. We have previously demonstrated that nanoparticle carbon black increases AP-1 DNA binding. In the study by Ng et al.[57] AP-1 binding to the ARE inhibited its activation thus preventing the upregulation of antioxidant defenses.

This would suggest that in macrophages exposed to environmental particles or nanoparticles there is the potential for activation of pathways that both increase inflammation (via NF-κB and AP-1) and increase antioxidant defense mechanisms (via Nrf-2). Furthermore, these pathways have the potential to interact with AP-1 inhibiting ARE and therefore preventing upregulation of antioxidant defense mechanisms by Nrf-2.

One of the primary roles of macrophages is to phagocytose foreign material. Nanoparticle uptake by macrophages has been observed in a number of studies[76–78] and in these studies the uptake of the nanopartices has been associated with a subsequent decrease in the ability of the macrophages to take up either fluorescent 2 μm polystyrene beads,[76] yeast,[79] or fluorescent E.Coli.[62] Similarly, macrophages recovered from the lungs of rats instilled with PM_{10} particles exhibited a decrease in the uptake of fluorescent polystyrene beads ex vivo.[80] These studies would suggest that particle clearance is not efficient in the presence of nanoparticles, allowing the particles to persist in the lung, drive inflammation, and perhaps cross the epithelial barrier and gain access to the lung interstitium[20] and blood.[57]

However, macrophages obviously have a limited capacity to phagocytose particles of any type. In the future, it would be interesting to try to estimate the relative ability of macrophages to phagocytose particles of different sizes, and then to identify whether the maximum tolerated volume of particle uptake with particles of different size is comparable. With the data generated thus far, it is difficult to determine whether the nanoparticles per se specifically inhibit the further uptake of larger particles, or simply "fill" or "overload" the cells preventing further particle uptake. It is worth noting that in the study by Renwick et al.[58] the lower doses of nanoparticles actually induced a small increase rather than a decrease in bead uptake.[81]

The effects of asbestos on macrophage are also well studied. The ability of asbestos to induce proinflammatory cytokine production by macrophages has been related to the fibre length, with longer fibres being more effective.[82] The length of fibres also impacts the ability of macrophages to take up asbestos by phagocytosis. Fibers of longer than 15 μm are not easily ingested and lead to a process known as frustrated phagocytosis.[83] Asbestos also induces an increased influx of calcium into the cytoplasm of macrophages, although this could be as a consequence of cell death rather than a specific signaling event that controls cytokine production.[84] Treatment of macrophages with asbestos fibres has also been shown to deplete glutathione and to activate NF-κB DNA, an effect that was inhibited by antioxidants.[85]

Particles of α-Quartz are thought to be highly cytotoxic to macrophages, such that uptake of α-quartz leads to inhibition of macrophage function, macrophage cell death, and subsequent release of the α-quartz particle load back into the lung tissue.[3] For instance, α-quartz has been shown to induce calcium elevation in macrophages, but as a part of the cell death process rather than as a specific signaling event.[86,87] Several studies have demonstrated an ability of α-quartz to induce TNFα production by macrophages,[88,89] suggesting that in addition to simply killing the macrophages, these particles must also be able to activate the cells leading to a proinflammatory response.

9.4 INTERACTIONS BETWEEN MACROPHAGES AND EPITHELIAL CELLS

Of course, in the body, particles never encounter one cell type at a time, and instead the response is a culmination of complex interactions between particles and many cell types. There are a number of ways in which such interactions can be studied, including:

(i) Animal models
(ii) Lung slices or isolated organs
(iii) Co-cultures
(iv) Conditioned media from individual cell types and transferring these to other cells in culture

With options (i)–(iv), attributing the inflammatory or signaling responses observed to any particular cell types can be difficult, but they have the advantage of providing a more physiologically relevant response.

Our own studies using conditioned media suggest that exposure of epithelial cells to nanoparticle carbon black results in the generation of chemotactic factors that stimulate the migration of macrophages.[90] These results indicate that the effects of nanoparticles are not limited to macrophages and that a complex interaction between macrophages and epithelial cells plays a key role in driving the inflammatory response. Once the macrophages migrate to and locate the nanoparticles, it might be expected that these particles would be taken up by phagocytosis and subsequently cleared from the lung tissue by migration of the macrophage onto the mucociliary escalator, or by migration to the lymph nodes. However, there may be a discord in the effectiveness of this mechanism when nanoparticles are involved. Hypothetically, if particles can induce epithelial cells to secrete macrophage chemoattractants, a prolonged exposure of epithelial cells to nanoparticles may result in hypersecretion of chemoattractants into the alveolar space. It is possible that this may disrupt the normal chemotactic gradient within the lung and result in particle-laden macrophages remaining within the respiratory region instead of migrating to the mucociliary escalator for clearance. The impact of nanoparticles on the chemotactic gradient within the lung clearly requires further investigation.

Our recent studies also suggest that in response to PM_{10} macrophages make substances that activate epithelial cells, resulting in an upregulation of the adhesion molecule ICAM-1.[91] These adhesion molecules are important *in vivo* as they allow interaction with migrating leukocytes, facilitating their traffic through the tissue.

Other methods of macrophage recruitment to sites of particle deposition could involve activation of the alternative complement cascade, resulting in the local generation of the chemotactic protein C5a. Barlow et al.[60] demonstrated that carbon black nanoparticles could induce the generation of macrophage chemotactic factors in blood serum. Although serum is not generally present in the alveolar space, in times of prolonged inflammation, the lung vasculature may be compromised, allowing seepage of serum proteins into the alveolar space. If nanoparticles were to come into contact with these proteins, it is feasible that an excess generation of chemotactic substances could be generated, resulting in heightened inflammation. This observation may also have ramifications with regard to particle translocation from the lung into the blood stream and systemic activation of complement could have serious effects on the cardiovascular system.

Our own studies have also shown that macrophages from rats instilled with PM_{10} show a decreased potential to migrate ex vivo[80]. This could prove deleterious to particle clearance in the lung as particle-laden macrophages may be unable to migrate towards the mucociliary escalator to be removed from the lung. This would again result in prolonged macrophage retention and increased inflammation as a result.

Hutchison et al.[92] treated either an alveolar epithelial cell line (A549) or a monocytic cell line (MM6) with PM_{10}. The PM_{10} was collected in near vicinity of a steel plant in the U.K. The PM_{10} samples were found to activate proinflammatory (interleukin 1(IL-1B) and TNFα) and pro-fibrotic (TGFβ) cytokine gene expression by macrophages, with those samples highest in metal content being most effective. The same PM_{10} samples were not effective at inducing expression of proinflammatory cytokines such as IL8, IL6, or TNFα by epithelial cells.

In the study by Hutchison et al.[92] the secretory products of the epithelial cells were recovered and used to treat the macrophage cell line in order to ascertain whether these products would elicit changes in macrophage activity. Despite the lack of IL8, IL6, or TNFα production measured in this experiment, the supernatant from these cells was very potent at inducing the expression of a number of proinflammatory mediators (IL8 and granulocyte-macrophage colony-stimulating factor) by the macrophages. In fact, the supernatant from the epithelial cells treated with the PM_{10} of highest metal content was so potent that it inhibited macrophage cytokine production, probably due to toxicity. These findings suggest cellular interactions are taking place, however the factors driving these responses by epithelial cells are currently unknown. Previous studies using conditioned media have highlighted the importance of cytokine/chemokine release and the role these molecules have in driving effects both locally and systemically, particularly in the field of particle toxicology. Schmidt et al. published a study in which macrophages were treated with silica, a know agent of pulmonary fibrosis.[93] The silica treatment stimulated release of IL-1B, which was subsequently found to modulate proliferation of fibroblasts, providing evidence that the particles induced release of cytokines from one cell type to impact on the function of another cell type. Albrecht et al. also reported similar effects attributed to unknown factors in media created from lavage of quartz treated animals.[94] The results of this study indicate that stimulation of NF-κB was partially TNFα independent, suggesting other mediators were involved in NF-κB activation and the associated inflammation. Due to the complex overlapping and redundant nature of inflammatory signaling, it is often difficult to attribute specific effects to one molecule; however, while studies using conditioned media are not always useful to determine exactly which factors are responsible for the cellular response, they are useful for providing information on the combined and overall effects of the particle induced mediators.

9.5 USING MACROPHAGES AND EPITHELIAL CELLS AS A MODEL FOR STUDYING PARTICLE-INDUCED INFLAMMATION—CONCLUSION

As outlined above, a wide variety of studies indicate that different pathogenic particles induce oxidative stress and signaling mechanisms in both epithelial cells and macrophages that drive the inflammatory response leading to disease. Many of the studies conducted to investigate the role of these two cell types in driving the inflammatory response have been conducted *in vitro*. In many cases, the *in vitro* cultured cells reflect closely the potency of the particles observed using *in vivo* models with respect to oxidative stress and the activation of proinflammatory signaling events.

With the rapid expansion of nanotechnology and the potential for increased exposure to a wide variety of particles of unknown toxicity, it will be essential to exploit such *in vitro* protocols in order to prevent excessive animal testing. It is therefore essential that a systematic assessment of the relevance and reliability of a number of *in vitro* models be conducted. Such models are likely to include single cell types, cell lines, primary cells, and mixed cultures. With such models, it is essential to prevent artifacts due to the use of particles (e.g., protein adsorption). Relevant controls (e.g., low-toxicity particles) must also be included, and of course, it will be necessary to compare the results to *in vivo* responses, using historical data where appropriate and possible. Finally, the validation of such models will result in a valuable tool for the toxicity assessment of any potentially pathogenic particle in the future.

REFERENCES

1. Brown, G. M., Donaldson, K., and Brown, D. M., Bronchoalveolar leukocyte response in experimental silicosis: Modulation by a soluble aluminum compound, *Toxicol. Appl. Pharmacol.*, 101, 95, 1989.
2. Donaldson, K., Brown, G. M., Brown, D. M., Bolton, R. E., and Davis, J. M., Inflammation generating potential of long and short fibre amosite asbestos samples, *Br. J. Ind. Med.*, 46, 271, 1989.
3. Donaldson, K. and Borm, P. J., The quartz hazard: A variable entity, *Ann. Occup. Hyg.*, 42, 287, 1998.
4. Mossman, B. T., Faux, S., Janssen, Y., Jimenez, L. A., Timblin, C., Zanella, C., Goldberg, J., Walsh, E., Barchowsky, A., and Driscoll, K., Cell signaling pathways elicited by asbestos, *Environ. Health Perspect.*, 105 (suppl. 5), 1121, 1997.
5. BéruBé, K. A., Quinlan, T. R., Moulton, G., Hemenway, D., O'Shaughnessy, P., Vacek, P., and Mossman, B. T., Comparative proliferative and histopathologic changes in rat lungs after inhalation of chrysotile or crocidolite asbestos, *Toxicol. Appl. Pharmacol.*, 137, 67, 1996.

6. Duffin, R., Gilmour, P. S., Schins, R. P., Clouter, A., Guy, K., Brown, D. M., MacNee, W., Borm, P. J., Donaldson, K., and Stone, V., Aluminum lactate treatment of DQ12 quartz inhibits its ability to cause inflammation, chemokine expression, and nuclear factor-kappa B activation, *Toxicol. Appl. Pharmacol.*, 176, 10, 2001.
7. Brown, D. M., Beswick, P. H., Bell, K. S., and Donaldson, K., Depletion of glutathione and ascorbate in lung lining fluid by respirable fibres, *Ann. Occup. Hyg.*, 44, 101, 2000.
8. Pope, C. A. III, Dockery, D. W., Spengler, J. D., and Raizenne, M. E., Respiratory health and PM10 pollution. A daily time series analysis, *Am. Rev. Respir. Dis.*, 144, 668, 1991.
9. Donaldson, K., Stone, V., Seaton, A., and MacNee, W., Ambient particle inhalation and the cardiovascular system: potential mechanisms, *Environ. Health Perspect.*, 109 (suppl. 4), 523, 2001.
10. Donaldson, K., Stone, V., Gilmour, P. S., Brown, D. M., and MacNee, W., Ultrafine particles: Mechanisms of lung injury, *Phil. Trans. R. Soc. Lond.*, 358, 2741–2749, 2000. Ref Type: Journal (Full).
11. Peters, A., Wichmann, H. E., Tuch, T., Heinrich, J., and Heyder, J., Respiratory effects are associated with the number of ultrafine particles, *Am. J. Respir. Crit. Care Med.*, 155, 1376, 1997.
12. Carter, J. D., Ghio, A. J., Samet, J. M., and Devlin, R. B., Cytokine production by human airway epithelial cells after exposure to an air pollution particle is metal-dependent, *Toxicol. Appl. Pharmacol.*, 146, 180, 1997.
13. Gilmour, P. S., Brown, D. M., Lindsay, T. G., Beswick, P. H., MacNee, W., and Donaldson, K., Adverse health effects of PM_{10} particles: involvement of iron in generation of hydroxyl radical, *Occup. Environ. Med.*, 53, 817, 1996.
14. Hutchison, G. R., Brown, D. M., Hibbs, L. R., Heal, M. R., Donaldson, K., Maynard, R. L., Monaghan, M., Nicholl, A., and Stone, V., The effect of refurbishing a UK steel plant on PM10 metal composition and ability to induce inflammation, *Respir. Res.*, 6, 43, 2005.
15. Becker, S., Soukup, J. M., Gilmour, M. I., and Devlin, R. B., Stimulation of human and rat alveolar macrophages by urban air particulates: effects on oxidant radical generation and cytokine production, *Toxicol. Appl. Pharmacol.*, 141, 637, 1996.
16. Schins, R. P., Lightbody, J. H., Borm, P. J., Shi, T., Donaldson, K., and Stone, V., Inflammatory effects of coarse and fine particulate matter in relation to chemical and biological constituents, *Toxicol. Appl. Pharmacol.*, 195, 1, 2004.
17. Li, X. Y., Brown, D., Smith, S., MacNee, W., and Donaldson, K., Short-term inflammatory responses following intratracheal instillation of fine and ultrafine carbon black in rats, *Inhal. Toxicol.*, 11, 709, 1999.
18. Oberdörster, G., Sharp, Z., Atudorei, V., Elder, A., Gelein, R., Lunts, A., Kreyling, W., and Cox, C., Extrapulmonary translocation of ultrafine carbon particles following whole-body inhalation exposure of rats, *J. Toxicol. Environ. Health A*, 65, 1531, 2002.
19. Brown, D. M., Wilson, M. R., MacNee, W., Stone, V., and Donaldson, K., Size-dependent proinflammatory effects of ultrafine polystyrene particles: a role for surface area and oxidative stress in the enhanced activity of ultrafines, *Toxicol. Appl. Pharmacol.*, 175, 191, 2001.
20. Ferin, J., Oberdörster, G., and Penney, D. P., Pulmonary retention of ultrafine and fine particles in rats, *Am. J. Respir. Cell Mol. Biol.*, 6, 535, 1992.
21. Calvo, P., Gouritin, B., Chacun, H., Desmaele, D., D'Angelo, J., Noel, J. P., Georgin, D., Fattal, E., Andreux, J. P., and Couvreur, P., Long-circulating PEGylated polycyanoacrylate nanoparticles as new drug carrier for brain delivery, *Pharm. Res.*, 18, 1157, 2001.
22. Donaldson, K., Stone, V., Tran, C. L., Kreyling, W., and Borm, P. J., Nanotoxicology, *Occup. Environ. Med.*, 61, 727, 2004.
23. Oberdörster, G., Oberdörster, E., and Oberdörster, J., Nanotoxicology: an emerging discipline evolving from studies of ultrafine particles, *Environ. Health Perspect.*, 113, 823, 2005.
24. Duffin, R., Tran, C. L., Clouter, A., Brown, D. M., MacNee, W., Stone, V., and Donaldson, K., The importance of surface area and specific reactivity in the acute pulmonary inflammatory response to particles, *Ann. Occup. Hyg.*, 46 (suppl. 1), 242–245, 2002.
25. Stoeger, T., Reinhard, C., Takenaka, S., Schroeppel, A., Karg, E., Ritter, B., Heyder, J., and Schulz, H., Instillation of six different ultrafine carbon particles indicates a surface area threshold dose for acute lung inflammation in mice, *Environ. Health Perspect.*, 114, 328, 2006.

26. Tran, C. L., Buchanan, D., Cullen, R. T., Searl, A., Jones, A. D., and Donaldson, K., Inhalation of poorly soluble particles. II. Influence of particle surface area on inflammation and clearance, *Inhal. Toxicol.*, 12, 1113, 2000.
27. Faux, S. P., Tran, C. L., Miller, B., Montellier, C., and Donaldson, K., In vitro determinants of particulate toxicity: the dose-metric for poorly soluble dusts, In *HSE Books Research Report*, 2001.
28. Beck-Speier, I., Dayal, N., Karg, E., Maier, K. L., Schumann, G., Schulz, H., Semmler, M. et al., Oxidative stress and lipid mediators induced in alveolar macrophages by ultrafine particles, *Free Radic. Biol. Med.*, 38, 1080, 2005.
29. Burri, P. H., Morphology and respiratory function of the alveolar unit, *Int. Arch. Allergy Appl. Immunol.*, 76 (suppl. 1), 2, 1985.
30. Nemery, B. and Hoet, P. H., Use of isolated lung cells in pulmonary toxicology, *Toxicol. In Vitro*, 7, 359, 1993.
31. Dick, C. A., Dennekamp, M., Howarth, S., Cherrie, J. W., Seaton, A., Donaldson, K., and Stone, V., Stimulation of IL-8 release from epithelial cells by gas cooker PM(10): A pilot study, *Occup. Environ. Med.*, 58, 208, 2001.
32. Hetland, R. B., Refsnes, M., Myran, T., Johansen, B. V., Uthus, N., and Schwarze, P. E., Mineral and/or metal content as critical determinants of particle-induced release of IL-6 and IL-8 from A549 cells, *J. Toxicol. Environ. Health A*, 60, 47, 2000.
33. Schins, R. P., McAlinden, A., MacNee, W., Jimenez, L. A., Ross, J. A., Guy, K., Faux, S. P., and Donaldson, K., Persistent depletion of I kappa B alpha and interleukin-8 expression in human pulmonary epithelial cells exposed to quartz particles, *Toxicol. Appl. Pharmacol.*, 167, 107, 2000.
34. Hetland, R. B., Schwarze, P. E., Johansen, B. V., Myran, T., Uthus, N., and Refsnes, M., Silica-induced cytokine release from A549 cells: Importance of surface area versus size, *Hum. Exp. Toxicol.*, 20, 46, 2001.
35. Montellier, C., Tran, C. L., MacNee, W., Faux, S., Miller, B., and Donaldson, K., The proinflammatory effects of low solubility low toxicity particles, nanoparticles and fine particles on epithelial cells in vitro: The role of surface area and surface reactivity, *Occup. Environ. Med.*, 2006, in press.
36. Seagrave, J., Knall, C., McDonald, J. D., and Mauderly, J. L., Diesel particulate material binds and concentrates a proinflammatory cytokine that causes neutrophil migration, *Inhal. Toxicol.*, 16 (suppl. 1), 93, 2004.
37. Stone, V., Al-Attili, F., Duncan, P., Dickson, C., and Brown, D. M., Protein adsorption onto nanoparticle surface—impact on protein detection and function, *Nanoletters*, 2006. (submitted).
38. Hohr, D., Steinfartz, Y., Schins, R. P., Knaapen, A. M., Martra, G., Fubini, B., and Borm, P. J., The surface area rather than the surface coating determines the acute inflammatory response after instillation of fine and ultrafine TiO_2 in the rat, *Int. J. Hyg. Environ. Health*, 205, 239, 2002.
39. Borm, P. J. and Kreyling, W., Toxicological hazards of inhaled nanoparticles—potential implications for drug delivery, *J. Nanosci. Nanotechnol.*, 4, 521, 2004.
40. Rahman, I., Oxidative stress, chromatin remodeling and gene transcription in inflammation and chronic lung diseases, *J. Biochem. Mol. Biol.*, 36, 95, 2003.
41. Janssen, Y. M., Barchowsky, A., Treadwell, M., Driscoll, K. E., and Mossman, B. T., Asbestos induces nuclear factor kappa B (NF-kappa B) DNA-binding activity and NF-kappa B-dependent gene expression in tracheal epithelial cells, *Proc. Natl Acad. Sci. USA*, 92, 8458, 1995.
42. Jimenez, L. A., Thompson, J., Brown, D. A., Rahman, I., Antonicelli, F., Duffin, R., Drost, E. M., Hay, R. T., Donaldson, K., and MacNee, W., Activation of NF-kappaB by PM(10) occurs via an iron-mediated mechanism in the absence of IkappaB degradation, *Toxicol. Appl. Pharmacol.*, 166, 101, 2000.
43. Janssen-Heininger, Y. M., Macara, I., and Mossman, B. T., Cooperativity between oxidants and tumor necrosis factor in the activation of nuclear factor (NF)-kappaB: Requirement of Ras/mitogen-activated protein kinases in the activation of NF-kappaB by oxidants, *Am. J. Respir. Cell Mol. Biol.*, 20, 942, 1999.
44. Ramage, L. and Guy, K., Expression of C-reactive protein and heat-shock protein-70 in the lung epithelial cell line A549, in response to PM10 exposure, *Inhal. Toxicol.*, 16, 447, 2004.
45. Jacquier-Sarlin, M. R., Fuller, K., Dinh-Xuan, A. T., Richard, M. J., and Polla, B. S., Protective effects of hsp70 in inflammation, *Experientia*, 50, 1031, 1994.

46. Madamanchi, N. R., Li, S., Patterson, C., and Runge, M. S., Reactive oxygen species regulate heat-shock protein 70 via the JAK/STAT pathway, *Arterioscler. Thromb. Vasc. Biol.*, 21, 321, 2001.
47. Kohn, G., Wong, H. R., Bshesh, K., Zhao, B., Vasi, N., Denenberg, A., Morris, C., Stark, J., and Shanley, T. P., Heat shock inhibits tnf-induced ICAM-1 expression in human endothelial cells via I kappa kinase inhibition, *Shock*, 17, 91, 2002.
48. Timblin, C. R., Janssen, Y. M., Goldberg, J. L., and Mossman, B. T., GRP78, HSP72/73, and cJun stress protein levels in lung epithelial cells exposed to asbestos, cadmium, or H_2O_2, *Free Radic. Biol. Med.*, 24, 632, 1998.
49. Asea, A., Kraeft, S. K., Kurt-Jones, E. A., Stevenson, M. A., Chen, L. B., Finberg, R. W., Koo, G. C., and Calderwood, S. K., HSP70 stimulates cytokine production through a CD14-dependant pathway, demonstrating its dual role as a chaperone and cytokine, *Nat. Med.*, 6, 435, 2000.
50. Pockley, A. G., Georgiades, A., Thulin, T., de, F. U., and Frostegard, J., Serum heat shock protein 70 levels predict the development of atherosclerosis in subjects with established hypertension, *Hypertension*, 42, 235, 2003.
51. Stone, V., Shaw, J., Brown, D. M., MacNee, W., Faux, S. P., and Donaldson, K., The role of oxidative stress in the prolonged inhibitory effect of ultrafine carbon black on epithelial cell function, *Toxicol. In Vitro*, 12, 649–659, 1998.
52. Li, X. Y., Gilmour, P. S., Donaldson, K., and MacNee, W., In vivo and in vitro proinflammatory effects of particulate air pollution (PM10), *Environ. Health Perspect.*, 105 (suppl. 5), 1279, 1997.
53. Nagatomo, H., Morimoto, Y., Oyabu, T., Hirohashi, M., Ogami, A., Yamato, H., Kuroda, K., Higashi, T., and Tanaka, I., Expression of heme oxygenase-1 in the lungs of rats exposed to crocidolite asbestos, *Inhal. Toxicol.*, 17, 293, 2005.
54. Brown, D. M., Beswick, P. H., and Donaldson, K., Induction of nuclear translocation of NF-kappaB in epithelial cells by respirable mineral fibres, *J. Pathol.*, 189, 258, 1999.
55. Crapo, J. D., Barry, B. E., Gehr, P., Bachofen, M., and Weibel, E. R., Cell number and cell characteristics of the normal human lung, *Am. Rev. Respir. Dis.*, 125, 740, 1982.
56. Gordon, S. B. and Read, R. C., Macrophage defenses against respiratory tract infections, *Br. Med. Bull.*, 61, 45, 2002.
57. Oberdörster, G., Sharp, Z., Atudorei, V., Elder, A., Gelein, R., Lunts, A., Kreyling, W., and Cox, C., Extrapulmonary translocation of ultrafine carbon particles following whole-body inhalation exposure of rats, *J. Toxicol. Environ. Health A*, 65, 1531, 2002.
58. Nemmar, A., Vanbilloen, H., Hoylaerts, M. F., Hoet, P. H., Verbruggen, A., and Nemery, B., Passage of intratracheally instilled ultrafine particles from the lung into the systemic circulation in hamster, *Am. J. Respir. Crit. Care Med.*, 164, 1665, 2001.
59. Armstrong, T. I., Moghimi, S. M., Davis, S. S., and Illum, L., Activation of the mononuclear phagocyte system by poloxamine 908: Its implications for targeted drug delivery, *Pharm. Res.*, 14, 1629, 1997.
60. Khandoga, A., Stampfl, A., Takenaka, S., Schulz, H., Radykewicz, R., Kreyling, W., and Krombach, F., Ultrafine particles exert prothrombotic but not inflammatory effects on the hepatic microcirculation in healthy mice in vivo, *Circulation*, 109 (10), 1320–1325, 2004.
61. Kreyling, W. G., Semmler, M., Erbe, F., Mayer, P., Takenaka, S., Schulz, H., Oberdörster, G., and Ziesenis, A., Translocation of ultrafine insoluble iridium particles from lung epithelium to extrapulmonary organs is size dependent but very low, *J. Toxicol. Environ. Health A*, 65, 1513, 2002.
62. Calvo, P., Gouritin, B., Chacun, H., Desmaele, D., D'Angelo, J., Noel, J. P., Georgin, D., Fattal, E., Andreux, J. P., and Couvreur, P., Long-circulating PEGylated polycyanoacrylate nanoparticles as new drug carrier for brain delivery, *Pharm. Res.*, 18, 1157, 2001.
63. Moghimi, S. M. and Szebeni, J., Stealth liposomes and long circulating nanoparticles: Critical issues in pharmacokinetics, opsonization and protein-binding properties, *Prog. Lipid Res.*, 42, 463, 2003.
64. Moghimi, S. M., Modulation of lymphatic distribution of subcutaneously injected poloxamer 407-coated nanospheres: The effect of the ethylene oxide chain configuration, *FEBS Lett.*, 540, 241, 2003.
65. Moghimi, S. M., Hunter, A. C., and Murray, J. C., Nanomedicine: Current status and future prospects, *FASEB J.*, 19, 311, 2005.
66. Moghimi, S. M., The effect of methoxy-PEG chain length and molecular architecture on lymph node targeting of immuno-PEG liposomes, *Biomaterials*, 27, 136, 2006.

67. Brown, D. M., Donaldson, K., and Stone, V., Role of calcium in the induction of TNFa expression by macrophages on exposure to ultrafine particles, *Ann. Occup. Hyg.*, 46 (suppl. 1), 219–222, 2002.
68. Brown, D. M., Donaldson, K., Borm, P. J., Schins, R. P., Denhart, M., Gilmour, P., Jimenez, L. A., and Stone, V., Calcium and reactive oxygen species-mediated activation of transcription factors and TNFa cytokine gene expression in macrophages exposed to ultrafine particles, *Am. J. Physiol. Lung Cell. Mol. Physiol.*, 286, L344–L353, 2004.
69. Brown, D., Donaldson, K., and Stone, V., Effects of PM10 in human peripheral blood monocytes and J774 macrophages, *Respir. Res.*, 5, 29, 2004.
70. Stone, V., Brown, D. M., Watt, N., Wilson, M., Donaldson, K., Ritchie, H., and MacNee, W., Ultrafine particle-mediated activation of macrophages: Intracellular calcium signaling and oxidative stress, *Inhal. Toxicol.*, 12 (suppl. 3), 345–351, 2000.
71. Stone, V., Tuinman, M., Vamvakopoulos, J. E., Shaw, J., Brown, D., Petterson, S., Faux, S. P. et al., Increased calcium influx in a monocytic cell line on exposure to ultrafine carbon black, *Eur. Respir. J.*, 15, 297, 2000.
72. Ishii, T., Itoh, K., Sato, H., and Bannai, S., Oxidative stress-inducible proteins in macrophages, *Free Radic. Res.*, 31, 351, 1999.
73. Ishii, T., Itoh, K., Takahashi, S., Sato, H., Yanagawa, T., Katoh, Y., Bannai, S., and Yamamoto, M., Transcription factor Nrf2 coordinately regulates a group of oxidative stress-inducible genes in macrophages, *J. Biol. Chem.*, 275, 16023, 2000.
74. Li, N., Alam, J., Venkatesan, M. I., Eiguren-Fernandez, A., Schmitz, D., Di, S. E., Slaughter, N. et al., Nrf2 is a key transcription factor that regulates antioxidant defense in macrophages and epithelial cells: Protecting against the proinflammatory and oxidizing effects of diesel exhaust chemicals, *J. Immunol.*, 173, 3467, 2004.
75. Ng, D., Kokot, N., Hiura, T., Faris, M., Saxon, A., and Nel, A., Macrophage activation by polycyclic aromatic hydrocarbons: Evidence for the involvement of stress-activated protein kinases, activator protein-1, and antioxidant response elements, *J. Immunol.*, 161, 942, 1998.
76. Renwick, L. C., Donaldson, K., and Clouter, A., Impairment of alveolar macrophage phagocytosis by ultrafine particles, *Toxicol. Appl. Pharmacol.*, 172, 119, 2001.
77. Wilson, M. R., Barlow, P. G., Hutchison, G. R., Sales, J., Simpson, R., and Stone, V., Nanoparticle interactions with zinc and iron: Implications for toxicology and inflammation, *Environ. Health Perspect.*, 2006, (submitted).
78. Geiser, M., Kapp, N., Schurch, S., Kreyling, W., Schulz, H., Semmler, M., Im Hof, V., Heyder, J., and Gehr, P., Ultrafine particles cross cellular membranes by non-phagocytic mechanisms in lungs and in cultured cells, *Environ. Health Perspect*, 113, 1555–1560, 2005. doi:101289/ehp8006.
79. Lundborg, M., Dahlen, S. E., Johard, U., Gerde, P., Jarstrand, C., Camner, P., and Lastbom, L., Aggregates of ultrafine particles impair phagocytosis of microorganisms by human alveolar macrophages, *Environ. Res.*, 100, 197, 2006.
80. Barlow, P., Brown, D. M., Donaldson, K., Maccallum, J., and Stone, V., Acute ambient particle (PM10) exposure reduces alveolar macrophage clearance potential ex vivo, *Particle Fibre Toxicol.*, 2006, (submitted).
81. Hoet, P. H. and Nemery, B., Stimulation of phagocytosis by ultrafine particles, *Toxicol. Appl. Pharmacol.*, 176, 203, 2001.
82. Donaldson, K., Li, X. Y., Dogra, S., Miller, B. G., and Brown, G. M., Asbestos-stimulated tumour necrosis factor release from alveolar macrophages depends on fibre length and opsonization, *J. Pathol.*, 168, 243, 1992.
83. Goodglick, L. A. and Kane, A. B., Role of reactive oxygen metabolites in crocidolite asbestos toxicity to mouse macrophages, *Cancer Res.*, 46, 5558, 1986.
84. Kalla, B., Hamilton, R. F., Scheule, R. K., and Holian, A., Role of extracellular calcium in chrysotile asbestos stimulation of alveolar macrophages, *Toxicol. Appl. Pharmacol.*, 104, 130, 1990.
85. Gilmour, P. S., Brown, D. M., Beswick, P. H., MacNee, W., Rahman, I., and Donaldson, K., Free radical activity of industrial fibers: Role of iron in oxidative stress and activation of transcription factors, *Environ. Health Perspect.*, 105 (suppl. 5), 1313, 1997.
86. Chen, J., Armstrong, L. C., Liu, S. J., Gerriets, J. E., and Last, J. A., Silica increases cytosolic free calcium ion concentration of alveolar macrophages in vitro, *Toxicol. Appl. Pharmacol.*, 111, 211, 1991.

87. Tarnok, A., Schluter, T., Berg, I., and Gercken, G., Silica induces changes in cytosolic free calcium, cytosolic pH, and plasma membrane potential in bovine alveolar macrophages, *Anal. Cell. Pathol.*, 15, 61, 1997.
88. Mohr, C., Gemsa, D., Graebner, C., Hemenway, D. R., Leslie, K. O., Absher, P. M., and Davis, G. S., Systemic macrophage stimulation in rats with silicosis: Enhanced release of tumor necrosis factor-alpha from alveolar and peritoneal macrophages, *Am. J. Respir. Cell Mol. Biol.*, 5, 395, 1991.
89. Driscoll, K. E., Hassenbein, D. G., Carter, J. M., Kunkel, S. L., Quinlan, T. R., and Mossman, B. T., TNF alpha and increased chemokine expression in rat lung after particle exposure, *Toxicol. Lett.*, 82–83, 483, 1995.
90. Barlow, P. G., Clouter-Baker, A., Donaldson, K., Maccallum, J., and Stone, V., Carbon black nanoparticles induce type II epithelial cells to release chemotaxins for alveolar macrophages, *Part Fibre Toxicol.*, 2, 11, 2005.
91. Brown, D. M., Hutchison, L., Donaldson, K., and Stone, V., The effects of PM10 particles and oxidative stress on macrophages and lung epithelial cells: Modulating effects of calcium signaling antagonists, *Am. J. Physiol. Lung Cell. Mol. Physiol.*, 2006, (submitted).
92. Hutchison, G. R., Hibbs, L., Heal, M., Maynard, R. L., Donaldson, K., and Stone, V., The relationship between macrophage and epithelial cell signaling interactions and the metal content of PM10 collected near a U.K. Steel Plant., *Toxicol. Appl. Pharmacol.*, 2006, (submitted).
93. Schmidt, J. A., Oliver, C. N., Lepe-Zuniga, J. L., Green, I., and Gery, I., Silica-stimulated monocytes release fibroblast proliferation factors identical to interleukin 1. A potential role for interleukin 1 in the pathogenesis of silicosis, *J. Clin. Invest.*, 73, 1462, 1984.
94. Albrecht, C., Schins, R. P., Hohr, D., Becker, A., Shi, T., Knaapen, A. M., and Borm, P. J., Inflammatory time course after quartz instillation: role of tumor necrosis factor-alpha and particle surface, *Am. J. Respir. Cell Mol. Biol.*, 31, 292, 2004.

10 Cell-Signaling Pathways Elicited by Particulates

Jamie E. Levis and Brooke T. Mossman
University of Vermont College of Medicine, University of Vermont

CONTENTS

10.1 PREFACE

Inhaled particles impinge upon epithelial cells of the respiratory tract after inhalation, facilitating an inflammatory response. In addition to causing epithelial cell injury through mechanisms involving DNA damage, pathogenic particles such as silica or asbestos elicit toxic and proliferative responses in lung cells through cell signaling pathways that can be triggered by direct interactions of fibers with the plasma membrane (Rom et al. 1991; Adamson 1997; Mossman and Churg 1998) or indirectly via reactive oxygen species (ROS) (Shukla et al. 2003a). At high concentrations of particles, exposures result in cell death and repair or compensatory proliferation of surrounding epithelial cells. If this phenomenon occurs subsequent to DNA damage, a situation could arise whereby the replicating population, including initiated cells that have an increased propensity towards further genetic instability, could continue on the route towards malignancy, i.e., lung cancers. The elucidation of the molecular mechanisms of cell injury and proliferation by inhaled particles is therefore critically important for understanding mechanisms of lung cancer and mesothelioma, a tumor unique to asbestos fibers, as well as pulmonary or pleural fibrosis. In these diseases, proliferation of epithelial cells or mesothelial cells may play dual roles: (1) repair of damaged epithelium, and (2) production of cytokines and chemokines that encourage inflammation and proliferation. In this chapter, we focus on cell signaling pathways controlling these processes. Although these cascades were first characterized in epithelial and mesothelial cells after exposure to asbestos or silica, several of these pathways have now been documented in various cell types after exposure to airborne particulate matter (PM), diesel exhaust, and/or ultrafine particles from a variety of sources. Because cell-signaling pathways initiated by particulates are studied in an effort to understand how to control proliferative and inflammatory alterations intrinsic to particulate-associated lung diseases, we first present the relevance of these processes to the pathogenesis of fibrogenic, carcinogenic, and inflammatory diseases such as asthma. We then describe relevant signaling cascades impinging upon the activator protein-1 (AP-1) and nuclear factor-κB (NF-κB) transcription factors and what is known about their activation by various particulates. Lastly, we provide a perspective on how these pathways can be verified in lung tissue after inhalation or instillation of particles for screening and therapy of particle-associated diseases.

10.2 RELEVANCE OF CELL PROLIFERATION IN LUNG TO DISEASE

In asbestosis and idiopathic pulmonary fibrosis (IPF), the histological sequence leading to disease is believed to occur in the following fashion: an initial alveolitis, which may involve polymorphonuclear (PMN) leukocytes but is predominately monocytic, occurs before fibrotic changes become evident (Rom et al. 1987; Spurzem et al. 1987; Mossman and Gee 1989). Proliferation is noted in alveolar macrophages, fibroblasts, and epithelial cells of the bronchioles. Importantly, there is evidence to suggest that smooth muscle cells, as well as endothelial cells of the arterioles near alveolar duct bifurcations, undergo proliferation in response to inhalation of chrysotile asbestos (McGavran et al. 1990). This initial inflammatory response is followed by an accumulation of PMNs in the alveoli and lung interstitium, followed by an influx of interstitial macrophages and fibroblast proliferation, which leads to interstitial thickening and eventual irreversible architectural distortion, particularly in the terminal bronchioles and alveolar ducts (Brody et al. 1981). Damage to the basement membrane occurs, with loss of endothelial and type I alveolar epithelial cells and epithelial integrity, allowing access of growth factors, cytokines, and chemokines into the interstitium (Rom et al. 1987; Chang et al. 1988; Mossman and Marsh 1989). Type II epithelial cell hyperplasia develops along with interstitial fibrosis as typified by deposition of collagen and other extracellular matrix proteins. Finally, fibrosis of the peribronchiolar and interstitial tissues develops and becomes the hallmark of advanced asbestosis (Becklake 1976; Craighead et al. 1982; Rom et al. 1987, 1991; Mossman and Churg 1998).

Injury to cells is often followed by compensatory cell proliferation. Alveolar type II and bronchiolar epithelial cells in rat lungs undergo proliferation in response to high exposures to crocidolite and chrysotile asbestos (BéruBé et al. 1996). It is known that proliferation of these cell types is a prominent repercussion of lung injury, such as that occurring in pulmonary fibrosis (Crouch 1990). Studies using 5′-bromo-2′-deoxyuridine (BrdU) and ^{3}H thymidine incorporation, as well as immunodetection of proliferating cell nuclear antigen (PCNA), have shown that areas of developing fibrotic foci in lung in response to chrysotile asbestos are characterized by proliferation of alveolar type II as well as bronchiolar epithelial cells (Dixon et al. 1995; Quinlan et al. 1995; BéruBé et al. 1996; Robledo et al. 2000). As noted above, the degree of injury in asbestos exposed animals is dose-dependant and followed by epithelial cell proliferation with a more intense and protracted inflammatory response, and eventually, fibrosis. Observations suggest that the increases in epithelial cell proliferation may be important in lung remodeling following injury, but if allowed to proceed unchecked and unregulated, can culminate in fibrogenesis or carcinogenesis. A logical conclusion stemming from these data would be that early responses of lung epithelial cells are instrumental to the development of fibrogenesis.

Other data support this view. For example, a study using bleomycin instillation into lung (a well characterized model of fibrosis in rodents) shows that early injury and repair of epithelial cells can govern whether fibrosis develops (Nomoto et al. 1997). Interestingly, this work provides evidence that programmed cell death, i.e., apoptosis of epithelial cells, is sustained during fibrogenesis, and that glucocorticoids administered to rodents block the apoptotic response of these cells and the accompanying fibrogenesis. A further study by this group has demonstrated that inhalation of an anti-Fas antibody (mimicking Fas/Fas-ligand interaction) induces apoptosis of epithelial cells and results in fibrogenesis (Kuwano et al. 1999).

As cited above, the proliferative responses of epithelial cells to asbestos are well documented, but is there evidence that asbestos really causes apoptosis in epithelial cells *in vivo*? Recent studies have in fact demonstrated apoptotic effects of asbestos on epithelial cells *in vitro* (Aljandali et al. 2001; Yuan et al. 2004; Upadhyay et al. 2005) and *in vivo* following intratracheal instillation of asbestos (Aljandali et al. 2001), but apoptosis has not been reported after inhalation of asbestos. Taken together, these data certainly suggest that functional responses of epithelial cells are crucial in the development of fibrosis, carcinogenesis, and lung remodeling. Epithelial cell injury is also a prominent feature of asthma, a disease often associated with the development of airway fibrosis (Comhair et al. 2005).

When compared with normal subjects, asbestos-exposed individuals demonstrate increased numbers of macrophages undergoing mitosis (Takemura et al. 1989), and the surfaces of alveolar macrophages from individuals with fibrosis show a striking increase in blebs, ruffling, and filopodia, presumably reflecting the enhanced phagocytic capability of these cells (Bitterman et al. 1984). In summary, understanding the cell signaling pathways controlling death and cell proliferation of epithelial cells and macrophages is critical to modulation of these processes which may be important in both disease prevention and therapy.

10.3 IMPORTANCE OF UNDERSTANDING CELL-SIGNALING PATHWAYS LEADING TO INFLAMMATORY ALTERATIONS IN LUNG DISEASE

The initial and protracted inflammatory response, which characterizes a number of models of pulmonary fibrosis, is believed to be important in asbestosis as well. In a study using Fisher 344 rats, lower-dose exposure to crocidolite asbestos resulted in a transient inflammation in bronchoalveolar lavage fluid and reversible inflammatory foci in lung with a maintenance of normal lung architecture (Quinlan et al. 1994). At higher concentrations of asbestos, neutrophil infiltration into lung and focal fibrotic lesions were noted, along with increased levels of the collagen marker, hydroxyproline. Interestingly, we also noted that changes in levels of expression of genes involved in antioxidant defense (*manganese superoxide dismutase* and *copper–zinc superoxide dismutase*) as well as cell proliferation (*ornithine decarboxylase* and *c-jun*) correlated with histopathologic findings, inflammatory cell influx, and lung hydroxyproline levels. The increase in *c-jun* levels in response to asbestos inhalation in this fibrosis model is particularly significant in light of the changes in the expression of this gene in response to asbestos and its association with altered cellular proliferation in carcinogenesis (Schutte et al. 1989).

The development of asbestosis has been linked to oxidants which are either generated directly from asbestos fibers induced by cells contacting asbestos fibers or are associated with inflammation (Robledo and Mossman 1999). On high iron-containing particles or fibers, ROS generated by the Fenton reaction can produce reactive oxygen intermediates, which directly participate in cell damage at high concentrations or cell proliferation at low concentrations. Generation of ROS during frustrated phagocytosis, i.e., an oxidative burst, can also initiate cell signaling and inflammation (Shukla et al. 2003a). More recently, attention has been focused on the interaction of ROS and reactive nitrogen species (RNS). This interaction can result in the generation of peroxynitrite, which has been shown to nitrate macromolecules, including proteins *in vitro*, thereby critically altering their function (MacMillan-Crow et al. 1998). Inhalation of asbestos induces RNS in rat lungs (Tanaka et al. 1998), and tyrosine nitration resulting from asbestos inhalation is associated with increased activation of signaling pathways in rat lungs (Iwagaki et al. 2003). It is conceivable that RNS, acting alone or with ROS, contribute to cell death and proliferation seen following asbestos exposure, thereby contributing to the development of fibrosis.

The inflammatory cascade, involving paracrine and autocrine events, is believed to be crucial in the pathology of asbestos-induced lung injury (Robledo and Mossman 1999). The protracted pulmonary inflammation noted in animal models of asbestosis can be correlated with the fibroproliferative responses, and cytokines, a major class of inflammatory modulators, are implicated in clinical asbestosis and animal models of this disease (Mossman and Churg 1998). Tumor necrosis factor α (TNFα) and its interaction with cytokines and growth factors has been the most extensively studied factor in the pathogenesis of asbestosis (Mossman and Churg 1998; Robledo and Mossman 1999). For example, crocidolite and chrysotile asbestos cause increased production of TNFα in alveolar macrophages (Driscoll et al. 1995b). Transgenic mice that overexpress TNFα in alveolar type II epithelial cells develop pulmonary fibrosis independent of pathogenic stimuli (Miyazaki et al. 1995). Conversely, mice that lack the TNF receptor produce TNF in response to a fibrogenic dose of chrysotile, but do not demonstrate markers for cellular proliferation nor develop fibrotic lesions (Liu et al. 1998). Increased expression and production of TNF was noted in the lungs of inducible nitric oxide synthase (iNOS) knockout mice exposed to asbestos, and this increase was correlated with an increase in neutrophil influx into the alveolar space (Dorger et al. 2002). Interestingly, this study provides evidence that iNOS-derived nitric oxide exerts a dual role in this model—it results in an exacerbated inflammatory response but attenuates oxidant-promoted tissue damage.

An exhaustive elucidation of the inflammatory mediators downstream from TNFα in asbestos-induced fibrosis is beyond the scope of this review. It should be noted, however, that TNFα is not directly chemotactic for neutrophils and macrophages (Robledo and Mossman 1999), thus work has focused on TNF-inducible chemotactic cytokines as effectors of asbestos induced lung damage, or fibrosis. These include interleukins 1, 6, and 8 (IL-1, IL-6, and IL-8), and transforming growth factor α and β (TGFα and TGFβ). These factors may be of particular importance in fibrogenesis, as they induce production of extracellular matrix proteins, induce epithelial cell proliferation, and are chemotactic for lung fibroblasts (Robledo and Mossman 1999). There is evidence to show that TGFβ is produced in the lungs following exposure to asbestos, and that macrophages showing strong positive staining for this peptide are found at sites of developing fibrotic lesions (Perdue and Brody 1994). A recent study has shown that expression of TGFβ-1 is noticeably absent in the lungs of TNFα receptor mice, and, importantly, these mice do not develop fibrosis following asbestos exposure (Liu and Brody 2001). This finding supports the contention that TNFα is an integral part of a pathway that is important in the fibrotic process resulting from asbestos exposure, and that it is exerting at least part of its effect through inducing the expression of downstream effectors and signaling pathways (see below).

10.4 SIGNALING PATHWAYS ACTIVATED BY PARTICULATES

10.4.1 Mitogen-Activated Protein Kinases, Fos/Jun Family Members, and Activator Protein-1

The mitogen-activated protein kinases (MAPK) cascades consist of a series of phosphorylated serine threonine kinases that are divided into three major pathways: extracellular signal-regulated kinases (ERKs), of which ERKs1 and 2 represent the major mammalian kinases of this group; c-Jun-NH_2-terminal kinases (JNKs 1, 2, and 3), also known as stress-activated protein kinases (SAPKs); and p38 kinases (Karin 1995; Shukla et al. 2003b). MAPK cascades can be initiated by receptor tyrosine kinases or factors stimulating phosphorylation of upstream MAPKKK or MAPKK. Alternatively, factors inhibiting the phosphatases that normally check these pathways will also cause net increases in phosphorylation of these proteins.

Specific MAPKs control the activation of *fos* and *jun* family proto-oncogene and their protein products that have been implicated in both apoptotic and proliferative responses to asbestos (Manning et al. 2002). In mesothelial and pulmonary epithelial cells, asbestos preferentially

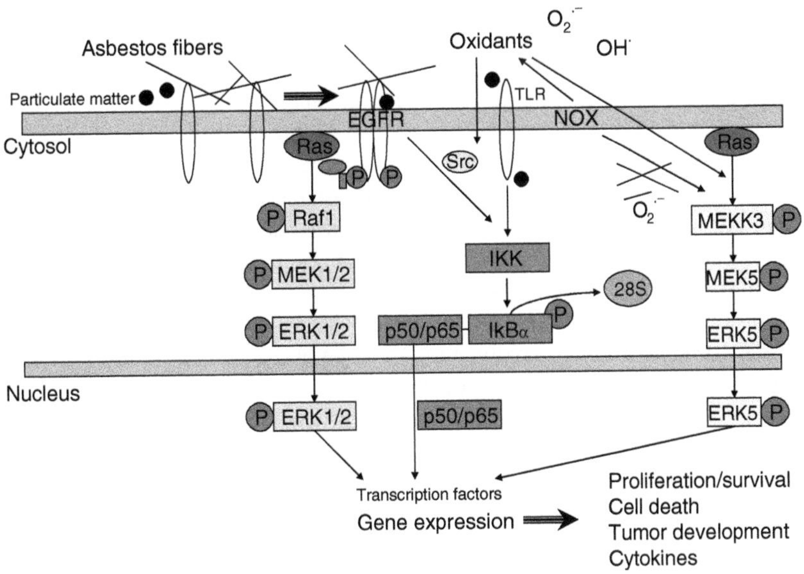

FIGURE 10.1 A diagram illustrating the primary signaling pathways stimulated by particulates such as asbestos fibers and airborne particulate matter in lung epithelium and mesothelium. All abbreviations and definitions are provided in the text.

activates the ERK1/2 pathway via an oxidant-dependant mechanism involving phosphorylation of the epidermal growth factor receptor (EGFR) (Figure 10.1) (Zanella et al. 1996; Jimenez et al. 1997). In rat pleural mesothelial (RPM) cells, addition of either chrysotile or crocidolite asbestos, in contrast to a number of other particles and synthetic fibers, induces phosphorylation and increased kinase activity of ERK1 and ERK2. Asbestos induced activation can be blocked by treating these cells with tyrphostin AG1478, a specific inhibitor of the tyrosine kinase activity of EGFR (Zanella et al. 1996). Treatment with this inhibitor prevents the induction of *c-fos* and apoptosis in these cell types (Zanella et al. 1999), further strengthening the case for interaction of asbestos fibers with the EGFR (Pache et al. 1998a). These finding are of particular relevance regarding the pathobiology of mesothelioma, as EGF is a growth factor required by human mesothelial cells (Gabrielson et al. 1988). EGFR and ERK1/2 activation by asbestos have also been associated with initiating cell cycle alterations in a murine alveolar type II epithelial cell line (C10), suggesting that EGFR and ERK may play a role in aberrant proliferation in lung epithelial cells (Buder-Hoffmann et al. 2001). ERK1 and 2 phosphorylation by crocidolite asbestos can also be inhibited by administration of catalase in RPM cells, suggesting that this is an oxidant-dependent process. Moreover, integrins appear to be integral to stimulation of ERK1/2 by asbestos in mesothelial cells (Berken et al. 2003).

ERK5 is also induced in C10 alveolar epithelial cells by crocidolite asbestos fibers through an oxidant-dependent process that is not dependent on EGFR activation, unlike ERK1/2 (Scapoli et al. 2004). Moreover, both ERK1/2 and ERK5 activation by asbestos involves Src activation, and activation of all three pathways are essential for initiation of cell proliferation. An intriguing line of investigation regarding fiber length and activation of cellular pathways which can lead to cell proliferation, apoptosis, and cell survival has shown that EGFR activity in human mesothelial cells exposed to crocidolite is greatest in areas where the cell contacts the fiber, and that fibers longer than 60 μm are associated with increased EGFR immunoreactivity in contrast to shorter fibers (Pache et al. 1998b). Shorter fibers are also less apt to cause frustrated phagocytosis, a process releasing large amounts of oxidants from cells due to a phagocytic burst, and these reactive species are known to alter EGFR activation (Goldkorn et al. 2005).

It has been known for over a decade that asbestos fibers activate the early response protooncogenes, *c-fos* and *c-jun,* in rodent mesothelial and tracheal epithelial cells *in vitro* (Heintz et al. 1993; Janssen et al. 1994). Activation is not seen with nonpathogenic synthetic fibers or particles, suggesting a link to the pathobiology of lung cancers and mesothelioma. This viewpoint has been reinforced with observations that erionite, the most potent mesotheliomagenic fiber in man and rodents, causes potent and prolonged *c-fos/c-jun* activation in mesothelial cells (Janssen et al. 1994). Moreover, ultrafine airborne particles (uPM) cause increases in *c-jun, junB, fra-1,* and *fra-2* at proliferative concentrations in C10 epithelial cells whereas increased concentrations of uPM-causing apoptosis are associated with upregulation of genes involved in Fas-associated and TNFR-associated death pathways (Timblin et al. 1998b).

Early response genes encode proteins that form AP-1, a redox sensitive transcription factor that activates a variety of genes that are involved in DNA synthesis. AP-1 also has been shown to be of paramount importance in tumor promotion in skin carcinogenesis (Young et al. 1999). The induction of these protooncogenes in response to asbestos is persistent in *in vitro* models (Heintz et al. 1993; Janssen et al. 1994), and may be a chronic source of aberrant cell proliferation in asbestos exposed lung via activation of EGFR-mediated signaling (Timblin et al. 1995). Although overexpression of *c-jun* has been shown to cause proliferative changes in tracheal epithelial cells (Reddy and Mossman 2002), the function of other AP-1 family members in carcinogenesis is unclear, and may in fact be cell type- and AP-1 partner type-specific (Reddy and Mossman 2002).

We have also shown that a signature of asbestos inhalation and coal dust instillation is increased expression of phosphorylated ERK1/2 using immunohistochemistry (IHC) (Robledo et al. 2000; Albrecht et al. 2002; Cummins et al. 2002). This is most striking in distal bronchiolar epithelium and the alveolar duct region, sites of asbestos fiber and particle impaction after inhalation. Phospho-ERK1/2 is translocated to the nucleus of C10 alveolar epithelial cells after addition of crocidolite asbestos *in vitro*, which eventually determines cell fate after exposure. At low concentrations of asbestos fibers, there is initial nuclear accumulation of phospho-ERK1/2, which diminishes over time and results in expression of cyclin D1, an AP-1 regulated gene, and entry of cells into S phase. At higher concentrations of fibers, phospho-ERK1/2 accumulates in the nucleus where apoptosis-inducing factor (AIF) is detected and precedes apoptosis (Yuan et al. 2004). These events correlate with nuclear accumulation of Fos (Burch et al. 2004), whereas we have linked ERK1/2 dependent Fra-1 expression to proliferation and transformation of RPM cells (Ramos-Nino et al. 2002, 2003).

Most recently, we have linked asbestos-induced EGFR activation, *fra-1* transactivation, expression of AP-1 family members, and AP-1 to DNA binding cells to intracellular levels of glutathione and y-glutamylcysteine synthetase levels, suggesting again a critical role of particle-induced oxidative stress (Shukla et al. 2004). The recent observation that diesel exhaust, a known source of particles and other agents inducing oxidative stress, activates redox-sensitive transcription factors, and kinases in human airways (Pourazar et al. 2005), confirms the relevance of these signaling pathways to human lung responses. Using gene profiling, we have confirmed that expression of more than 38 signal transduction genes and oxidative-stress genes, including the AP-1 regulated gene, *heme oxygenase*, is altered in mouse lungs after inhalation of chrysotile asbestos over a 40-day period (Sabo-Attwood et al. 2005).

10.4.2 Nuclear Factor-κB

Of the many signaling cascades activated in airway epithelium in response to oxidant or particle stimulation, NF-κB has been implicated as one of the most important in both regulation of inflammation and cell survival. NF-κB is a ubiquitous transcription factor that can be activated by cytokines, ROS, growth factors, bacteria and viruses, ultraviolet irradiation, airborne PM and inorganic minerals such as asbestos or silica (Janssen et al. 1995, 1997; Ghosh et al. 1998; Janssen-Heininger et al. 2000; Shukla et al. 2000; Ding et al. 2002). NF-κB activity is tightly controlled by the inhibitory protein, IκBα, that is normally present in the cytosol complexed to NF-κB dimers, thereby preventing the nuclear localization of NF-κB and ensuring low basal transcriptional activity (Figure 10.1). Upon cellular stimulation of this signaling pathway, IκBα becomes phosphorylated at serines 32 and 36 by the activity of the IκB kinase (IKK) complex, then is ubiquinated and degraded through the 26S proteasome pathway. This exposes the nuclear localization sequence of NF-κB, allowing its entry into the nucleus and thus facilitating DNA binding and the transcriptional up-regulation of NF-κB regulated genes. The regulation of NF-κB and its degradation products are topics of contemporary interest, as many NF-κB inducible genes encode inflammatory chemokines and cytokines, adhesion molecules, growth factors, enzymes, and transcription factors (Sanceau et al. 1995). For example, interleukin-6 (IL-6) (Harant et al. 1996), interleukin-8 (IL-8) (Driscoll et al. 1995a), and macrophage inflammatory protein-2 (MIP-2) (Poynter et al. 1999), three putative mediators of inflammation and fibrogenesis in lung, have NF-κB binding sequences in their promoter regions which are critical to their transcriptional activation.

We have shown previously that asbestos and silica fibers cause activation of the NF-κB signaling pathway *in vitro* (Hubbard et al. 2002) and in lung epithelium after inhalation of crocidolite asbestos by rats (Hubbard et al. 2001). *In vivo*, striking increases in nuclear translocation of p65 (Rel A), the subunit causing transcriptional activation of NF-κB, occur in distal bronchiolar and alveolar epithelial cells after brief exposures to fibers (Hubbard et al. 2001). Thus, the induction of NF-κB in airway epithelium by asbestos or other particles may be a critical initial event promoting epithelial cell alterations, inflammation, and lung disease.

In a collaborative study, we have also demonstrated that brief inhalation of $PM_{2.5}$ from Sterling Forest, NY (a 6 h exposure to 300 μg/mm^3 particles) caused upregulation of a number of NF-κB regulated genes in lung homogenates, including TNFα (Shukla et al. 2000). Transcriptional activation of NF-κB-dependent gene expression was also observed by PM in an alveolar epithelial NF-κB luciferase reporter cell line and was inhibited by catalase administration.These findings support the concept that NF-κB is redox-sensitive transcription factor, like AP-1 (Janssen-Heininger et al. 2000). A recent report establishes that Ottawa Urban Air Particles or iron-loaded fine TiO_2 causes NF-κB activation in the absence of epithelial particle uptake by rat tracheal explants *in vitro* (Churg et al. 2005). Both dusts and an iron-containing citrate extract from them caused phosphorylation of the EGFR and activated NF-κB through a pathway involving oxidative stress and Src activation. These studies imply that extracellular stimulation of NF-κB by oxidants elaborated by particles occurs through the EGFR (see Figure 10.1). We have shown previously that NF-κB activation in C10 lung epithelial cells by asbestos fibers does not require EGFR phosphorylation by crocidolite asbestos fibers (Ramos-Nino et al. 2002). However, frustrated phagocytosis involving stimulation of NADPH oxidases (NOX) and elaboration of intracellular oxidants occurs in response to iron-containing asbestos types, such as crocidolite, in these and other cell types (Shukla et al. 2003a), and these processes might activate NF-κB.

Cell signaling and cytokine production by ambient and diesel sources of PM have been studied extensively in human alveolar macrophages (HAM) and human airway epithelial cells (NHBE) *in vitro* (Becker et al. 2005). These studies reveal that oxidant-induced stress plays a major role in production of cytokines by both coarse and fine particles in HAM, can be blocked by a toll like receptor 4 (TLR4) agonist involved in the recognition of LPS, and Gram negative bacteria-exposure to PM decreases the expression of TLR4 associated with hyperesponsiveness to LPS, i.e., tolerance. NHBE also recognize PM through TLR2, a receptor with preference for recognition of Gram-positive bacteria. TLRs have been linked to LPS-stimulation of the NF-κB signaling pathways, and it is highly likely that they modulate PM-induced NF-κB signaling responses and cytokine production.

NF-κB activation is also induced by silica in various cell types (Ding et al. 2002), and unlike asbestos, PM and silica induce JNK activation by lung epithelial cells *in vitro* (Timblin et al. 1998a; Shukla et al. 2001). Although JNK activation is classically associated with cell death (Yanase et al. 2005), crosstalk mediated between the JNK signaling pathway and NF-κB, a transcription factor promoting survival as opposed to cell death (Wang et al. 2005), may dictate eventual proliferative or apoptotic responses to particulates. For example, inhibition of JNK activation may occur through NF-κB target genes, GADD45β, and c-IAP (an inhibitor of apoptosis protein) (Tang et al. 2001).

10.4.3 Other Signaling Pathways Induced by Particulates

Other signaling pathways that impact upon the MAPK/AP-1 and or NF-κB pathways have been shown to be activated by asbestos in a variety of cell types (Shukla et al. 2003b). These include members of the Protein Kinase C family (Lounsbury et al. 2002; Shukla et al. 2003c), nuclear factor of activated T cells (NFAT) (Li et al. 2002), calcium-dependent pathways leading to activation of the CREB transcription factor (Barlow et al. 2006), and the phosphatidylinositol-3 kinase (PI3-K)/AKT pathway leading to mTOR activation (Swain et al. submitted). The interplay between these pathways will likely prove critical in determining phenotypic and inflammatory outcomes of particulate exposures in epithelial and other lung cell types.

10.5 CONCLUSIONS

In vitro studies have shed light on mechanisms of cell signaling by pathogenic particulates, including asbestos fibers, silica particles and most recently, airborne PM from a variety of sources. While initial work has shown that redox-associated transcription factors are activated by these particulates in several cell types, the significance of these pathways in terms of lung responses and remodeling remains to be determined *in vivo*. The fact that many of these pathways can be demonstrated in rodent and human lungs *in vivo* using cell imaging after inhalation of particulates (Poynter et al. 1999; Taatjes and Mossman 2005) is encouraging and validates *in vitro* investigations. Moreover, microarray analysis and *in situ* hybridization studies now allow profiling and localization of genes regulated by AP-1 and NF-κB transcription factors in rodent and human cells of the lung (Sabo-Attwood et al. 2005). Transgenic targeting of genes and proteins modulating cell signaling using lung epithelial and other cell-specific promoters are exciting developments that will verify the functional ramifications of signaling pathways in animal models of particulate-induced lung diseases.

REFERENCES

Adamson, I. Y., Early mesothelial cell proliferation after asbestos exposure: *in vivo* and *in vitro* studies, *Environ. Health Perspect.*, 105 (suppl. 5), 1205–1208, 1997.

Albrecht, C., Borm, P. J., Adolf, B., Timblin, C. R., and Mossman, B. T., *In vitro* and *in vivo* activation of extracellular signal-regulated kinases by coal dusts and quartz silica, *Toxicol. Appl. Pharmacol.*, 184, 37–45, 2002.

Aljandali, A., Pollack, H., Yeldandi, A., Li, Y., Weitzman, S. A., and Kamp, D. W., Asbestos causes apoptosis in alveolar epithelial cells: role of iron-induced free radicals, *J. Lab. Clin. Med.*, 137, 330–339, 2001.

Barlow, C. A., Shukla, A., Mossman, B. T., and Lounsbury, K. M., Oxidant-mediated cAMP response element binding protein activation: calcium regulation and role in apoptosis of lung epithelial cells, *Am. J. Respir. Cell Mol. Biol.*, 34, 7–14, 2006.

Becker, S., Mundandhara, S., Devlin, R. B., and Madden, M., Regulation of cytokine production in human alveolar macrophages and airway epithelial cells in response to ambient air pollution particles: further mechanistic studies, *Toxicol. Appl. Pharmacol.*, 207, 269–275, 2005.

Becklake, M. R., Asbestos-related diseases of the lung and other organs: their epidemiology and implications for clinical practice, *Am. Rev. Respir. Dis.*, 114, 187–227, 1976.

Berken, A., Abel, J., and Unfried, K., beta1-integrin mediates asbestos-induced phosphorylation of AKT and ERK1/2 in a rat pleural mesothelial cell line, *Oncogene*, 22, 8524–8528, 2003.

BéruBé, K. A., Quinlan, T. R., Moulton, G., Hemenway, D., O'Shaughnessy, P., Vacek, P., and Mossman, B. T., Comparative proliferative and histopathologic changes in rat lungs after inhalation of chrysotile or crocidolite asbestos, *Toxicol. Appl. Pharmacol.*, 137, 67–74, 1996.

Bitterman, P. B., Saltzman, L. E., Adelberg, S., Ferrans, V. J., and Crystal, R. G., Alveolar macrophage replication one mechanism for the expansion of the mononuclear phagocyte population in the chronically inflamed lung, *J. Clin. Invest.*, 74, 460–469, 1984.

Brody, A. R., Hill, L. H., Adkins, B., Jr., and O'Connor, R. W., Chrysotile asbestos inhalation in rats: deposition pattern and reaction of alveolar epithelium and pulmonary macrophages, *Am. Rev. Respir. Dis.*, 123, 670–679, 1981.

Buder-Hoffmann, S., Palmer, C., Vacek, P., Taatjes, D., and Mossman, B., Different accumulation of activated extracellular signal-regulated kinases (ERK 1/2) and role in cell-cycle alterations by epidermal growth factor, hydrogen peroxide, or asbestos in pulmonary epithelial cells, *Am. J. Respir. Cell. Mol. Biol.*, 24, 405–413, 2001.

Burch, P. M., Yuan, Z., Loonen, A., and Heintz, N. H., An extracellular signal-regulated kinase 1- and 2-dependent program of chromatin trafficking of c-Fos and Fra-1 is required for cyclin D1 expression during cell cycle reentry, *Mol. Cell Biol.*, 24, 4696–4709, 2004.

Chang, L. Y., Overby, L. H., Brody, A. R., and Crapo, J. D., Progressive lung cell reactions and extracellular matrix production after a brief exposure to asbestos, *Am. J. Pathol.*, 131, 156–170, 1988.

Churg, A., Xie, C., Wang, X., Vincent, R., and Wang, R. D., Air pollution particles activate NF-κB on contact with airway epithelial cell surfaces, *Toxicol. Appl. Pharmacol.*, 208, 37–45, 2005.

Comhair, S. A., Xu, W., Ghosh, S., Thunnissen, F. B., Almasan, A., Calhoun, W. J., Janocha, A. J., Zheng, L., Hazen, S. L., and Erzurum, S. C., Superoxide dismutase inactivation in pathophysiology of asthmatic airway remodeling and reactivity, *Am. J. Pathol.*, 166, 663–674, 2005.

Craighead, J. E., Abraham, J. L., Churg, A., Green, F. H., Kleinerman, J., Pratt, P. C., Seemayer, T. A., Vallyathan, V., and Weill, H., The pathology of asbestos-associated diseases of the lungs and pleural cavities: diagnostic criteria and proposed grading schema. Report of the Pneumoconiosis Committee of the College of American Pathologists and the National Institute for Occupational Safety and Health, *Arch. Pathol. Lab. Med.*, 106, 544–596, 1982.

Crouch, E., Pathobiology of pulmonary fibrosis, *Am. J. Physiol.*, 259, L159–L184, 1990.

Cummins, M. M., O'Mullane, L. M., Barden, J. A., Cook, D. I., and Poronnik, P., Antisense co-suppression of G(alpha)(q) and G(alpha)(11) demonstrates that both isoforms mediate M(3)-receptor-activated Ca(2+) signaling in intact epithelial cells, *Pflugers. Arch.*, 444, 644–653, 2002.

Ding, M., Chen, F., Shi, X., Yucesoy, B., Mossman, B., and Vallyathan, V., Diseases caused by silica: mechanisms of injury and disease development, *Int. Immunopharmacol.*, 2, 173–182, 2002.

Dixon, D., Bowser, A. D., Badgett, A., Haseman, J. K., and Brody, A. R., Incorporation of bromodeoxyuridine (BrdU) in the bronchiolar-alveolar regions of the lungs following two inhalation exposures to chrysotile asbestos in strain A/J mice, *J. Environ. Pathol. Toxicol. Oncol.*, 14, 205–213, 1995.

Dorger, M., Allmeling, A. M., Kiefmann, R., Munzing, S., Messmer, K., and Krombach, F., Early inflammatory response to asbestos exposure in rat and hamster lungs: role of inducible nitric oxide synthase, *Toxicol. Appl. Pharmacol.*, 181, 93–105, 2002.

Driscoll, K. E., Hassenbein, D. G., Howard, B. W., Isfort, R. J., Cody, D., Tindal, M. H., Suchanek, M., and Carter, J. M., Cloning, expression, and functional characterization of rat MIP-2: a neutrophil chemoattractant and epithelial cell mitogen, *J. Leukoc. Biol.*, 58, 359–364, 1995.

Driscoll, K. E., Maurer, J. K., Higgins, J., and Poynter, J., Alveolar macrophage cytokine and growth factor production in a rat model of crocidolite-induced pulmonary inflammation and fibrosis, *J. Toxicol. Environ. Health*, 46, 155–169, 1995.

Gabrielson, E. W., Gerwin, B. I., Harris, C. C., Roberts, A. B., Sporn, M. B., and Lechner, J. F., Stimulation of DNA synthesis in cultured primary human mesothelial cells by specific growth factors, *Faseb J.*, 2, 2717–2721, 1988.

Ghosh, S., May, M. J., and Kopp, E. B., NF-κB and Rel proteins: evolutionarily conserved mediators of immune responses, *Annu. Rev. Immunol.*, 16, 225–260, 1998.

Goldkorn, T., Ravid, T., and Khan, E. M., Life and death decisions: ceramide generation and EGF receptor trafficking are modulated by oxidative stress, *Antioxid. Redox. Signal.*, 7, 119–128, 2005.

Harant, H., de Martin, R., Andrew, P. J., Foglar, E., Dittrich, C., and Lindley, I. J., Synergistic activation of interleukin-8 gene transcription by all-trans-retinoic acid and tumor necrosis factor-alpha involves the transcription factor NF-kappa B, *J. Biol. Chem.*, 271, 26954–26961, 1996.

Heintz, N. H., Janssen, Y. M., and Mossman, B. T., Persistent induction of c-fos and c-jun expression by asbestos, *Proc. Natl. Acad. Sci. U S A*, 90, 3299–3303, 1993.

Hubbard, A. K., Timblin, C. R., Rincon, M., and Mossman, B. T., Use of transgenic luciferase reporter mice to determine activation of transcription factors and gene expression by fibrogenic particles, *Chest*, 120, 24S–25S, 2001.

Hubbard, A. K., Timblin, C. R., Shukla, A., Rincon, M., and Mossman, B. T., Activation of NF-kappa B-dependent gene expression by silica in lungs of luciferase reporter mice, *Am. J. Physiol. Lung Cell Mol. Physiol.*, 282, L968–L975, 2002.

Iwagaki, A., Choe, N., Li, Y., Hemenway, D. R., and Kagan, E., Asbestos inhalation induces tyrosine nitration associated with extracellular signal-regulated kinase 1/2 activation in the rat lung, *Am. J. Respir. Cell Mol. Biol.*, 28, 51–60, 2003.

Janssen, Y. M., Heintz, N. H., Marsh, J. P., Borm, P. J., and Mossman, B. T., Induction of c-fos and c-jun proto-oncogenes in target cells of the lung and pleura by carcinogenic fibers, *Am. J. Respir. Cell Mol. Biol.*, 11, 522–530, 1994.

Janssen, Y. M., Barchowsky, A., Treadwell, M., Driscoll, K. E., and Mossman, B. T., Asbestos induces nuclear factor κB (NF-κB) DNA-binding activity and NF-κB-dependent gene expression in tracheal epithelial cells, *Proc. Natl. Acad. Sci. U S A*, 92, 8458–8462, 1995.

Janssen, Y. M., Driscoll, K. E., Howard, B., Quinlan, T. R., Treadwell, M., Barchowsky, A., and Mossman, B. T., Asbestos causes translocation of p65 protein and increases NF-κB DNA binding activity in rat lung epithelial and pleural mesothelial cells, *Am. J. Pathol.*, 151, 389–401, 1997.

Janssen-Heininger, Y. M., Poynter, M. E., and Baeuerle, P. A., Recent advances towards understanding redox mechanisms in the activation of nuclear factor kappaB, *Free Radic. Biol. Med.*, 28, 1317–1327, 2000.

Jimenez, L. A., Zanella, C., Fung, H., Janssen, Y. M., Vacek, P., Charland, C., Goldberg, J., and Mossman, B. T., Role of extracellular signal-regulated protein kinases in apoptosis by asbestos and H2O2, *Am. J. Physiol.*, 273, L1029–L1035, 1997.

Karin, M., The regulation of AP-1 activity by mitogen-activated protein kinases, *J. Biol. Chem.*, 270, 16483–16486, 1995.

Kuwano, K., Hagimoto, N., Kawasaki, M., Yatomi, T., Nakamura, N., Nagata, S., Suda, T. et al., Essential roles of the Fas–Fas ligand pathway in the development of pulmonary fibrosis, *J. Clin. Invest.*, 104, 13–19, 1999.

Li, J., Huang, B., Shi, X., Castranova, V., Vallyathan, V., and Huang, C., Involvement of hydrogen peroxide in asbestos-induced NFAT activation, *Mol. Cell Biochem.*, 234–235, 161–168, 2002.

Liu, J. Y. and Brody, A. R., Increased TGF-beta1 in the lungs of asbestos-exposed rats and mice: reduced expression in TNF-alpha receptor knockout mice, *J. Environ. Pathol. Toxicol. Oncol.*, 20, 97–108, 2001.

Liu, J. Y., Brass, D. M., Hoyle, G. W., and Brody, A. R., TNF-alpha receptor knockout mice are protected from the fibroproliferative effects of inhaled asbestos fibers, *Am. J. Pathol.*, 153, 1839–1847, 1998.

Lounsbury, K. M., Stern, M., Taatjes, D., Jaken, S., and Mossman, B. T., Increased localization and substrate activation of protein kinase C delta in lung epithelial cells following exposure to asbestos, *Am. J. Pathol.*, 160, 1991–2000, 2002.

MacMillan-Crow, L. A., Crow, J. P., and Thompson, J. A., Peroxynitrite-mediated inactivation of manganese superoxide dismutase involves nitration and oxidation of critical tyrosine residues, *Biochemistry*, 37, 1613–1622, 1998.

Manning, C. B., Cummins, A. B., Jung, M. W., Berlanger, I., Timblin, C. R., Palmer, C., Taatjes, D. J., Hemenway, D., Vacek, P., and Mossman, B. T., A mutant epidermal growth factor receptor targeted to lung epithelium inhibits asbestos-induced proliferation and proto-oncogene expression, *Cancer Res.*, 62, 4169–4175, 2002.

McGavran, P. D., Moore, L. B., and Brody, A. R., Inhalation of chrysotile asbestos induces rapid cellular proliferation in small pulmonary vessels of mice and rats, *Am. J. Pathol.*, 136, 695–705, 1990.

Miyazaki, Y., Araki, K., Vesin, C., Garcia, I., Kapanci, Y., Whitsett, J. A., Piguet, P. F., and Vassalli, P., Expression of a tumor necrosis factor-alpha transgene in murine lung causes lymphocytic and fibrosing alveolitis. A mouse model of progressive pulmonary fibrosis, *J. Clin. Invest.*, 96, 250–259, 1995.

Mossman, B. T. and Churg, A., Mechanisms in the pathogenesis of asbestosis and silicosis, *Am. J. Respir. Crit. Care Med.*, 157, 1666–1680, 1998.

Mossman, B. T. and Gee, J. B., Asbestos-related diseases, *N. Engl. J. Med.*, 320, 1721–1730, 1989.

Mossman, B. T. and Marsh, J. P., Evidence supporting a role for active oxygen species in asbestos-induced toxicity and lung disease, *Environ. Health Perspect.*, 81, 91–94, 1989.

Nomoto, Y., Kuwano, K., Hagimoto, N., Kunitake, R., Kawasaki, M., and Hara, N., Apoptosis and Fas/Fas ligand mRNA expression in acute immune complex alveolitis in mice, *Eur. Respir. J.*, 10, 2351–2359, 1997.

Pache, J. C., Christakos, P. G., Gannon, D. E., Mitchell, J. J., Low, R. B., and Leslie, K. O., Myofibroblasts in diffuse alveolar damage of the lung, *Mod. Pathol.*, 11, 1064–1070, 1998a.

Pache, J. C., Janssen, Y. M., Walsh, E. S., Quinlan, T. R., Zanella, C. L., Low, R. B., Taatjes, D. J., and Mossman, B. T., Increased epidermal growth factor-receptor protein in a human mesothelial cell line in response to long asbestos fibers, *Am. J. Pathol.*, 152, 333–340, 1998b.

Perdue, T. D. and Brody, A. R., Distribution of transforming growth factor-beta 1, fibronectin, and smooth muscle actin in asbestos-induced pulmonary fibrosis in rats, *J. Histochem. Cytochem.*, 42, 1061–1070, 1994.

Pourazar, J., Mudway, I. S., Samet, J. M., Helleday, R., Blomberg, A., Wilson, S. J., Frew, A. J., Kelly, F. J., and Sandstrom, T., Diesel exhaust activates redox-sensitive transcription factors and kinases in human airways, *Am. J. Physiol. Lung Cell Mol. Physiol.*, 289, L724–L730, 2005.

Poynter, M. E., Janssen-Heininger, Y. M., Buder-Hoffmann, S., Taatjes, D. J., and Mossman, B. T., Measurement of oxidant-induced signal transduction proteins using cell imaging, *Free Radic. Biol. Med.*, 27, 1164–1172, 1999.

Quinlan, T. R., Marsh, J. P., Janssen, Y. M., Leslie, K. O., Hemenway, D., Vacek, P., and Mossman, B. T., Dose-responsive increases in pulmonary fibrosis after inhalation of asbestos, *Am. J. Respir. Crit. Care Med.*, 150, 200–206, 1994.

Quinlan, T. R., BéruBé, K. A., Marsh, J. P., Janssen, Y. M., Taishi, P., Leslie, K. O., Hemenway, D., O'Shaughnessy, P. T., Vacek, P., and Mossman, B. T., Patterns of inflammation, cell proliferation, and related gene expression in lung after inhalation of chrysotile asbestos, *Am. J. Pathol.*, 147, 728–739, 1995.

Ramos-Nino, M. E., Haegens, A., Shukla, A., and Mossman, B. T., Role of mitogen-activated protein kinases (MAPK) in cell injury and proliferation by environmental particulates, *Mol. Cell Biochem.*, 234–235, 111–118, 2002.

Ramos-Nino, M. E., Heintz, N., Scappoli, L., Martinelli, M., Land, S., Nowak, N., Haegens, A., et al., Gene profiling and kinase screening in asbestos-exposed epithelial cells and lungs, *Am. J. Respir. Cell Mol. Biol.*, 29, S51–S58, 2003.

Reddy, S. P. and Mossman, B. T., Role and regulation of activator protein-1 in toxicant-induced responses of the lung, *Am. J. Physiol. Lung Cell Mol. Physiol.*, 283, L1161–L1178, 2002.

Robledo, R. and Mossman, B., Cellular and molecular mechanisms of asbestos-induced fibrosis, *J. Cell Physiol.*, 180, 158–166, 1999.

Robledo, R. F., Buder-Hoffmann, S. A., Cummins, A. B., Walsh, E. S., Taatjes, D. J., and Mossman, B. T., Increased phosphorylated extracellular signal-regulated kinase immunoreactivity associated with proliferative and morphologic lung alterations after chrysotile asbestos inhalation in mice, *Am. J. Pathol.*, 156, 1307–1316, 2000.

Rom, W. N., Bitterman, P. B., Rennard, S. I., Cantin, A., and Crystal, R. G., Characterization of the lower respiratory tract inflammation of nonsmoking individuals with interstitial lung disease associated with chronic inhalation of inorganic dusts, *Am. Rev. Respir. Dis.*, 136, 1429–1434, 1987.

Rom, W. N., Travis, W. D., and Brody, A. R., Cellular and molecular basis of the asbestos-related diseases, *Am. Rev. Respir. Dis.*, 143, 408–422, 1991.

Sabo-Attwood, T., Ramos-Nino, M., Bond, J., Butnor, K. J., Heintz, N., Gruber, A. D., Steele, C., Taatjes, D. J., Vacek, P., and Mossman, B. T., Gene Expression Profiles Reveal Increased mClca3 (Gob5) Expression and Mucin Production in a Murine Model of Asbestos-Induced Fibrogenesis, *Am. J. Pathol.*, 167, 1243–1256, 2005.

Sanceau, J., Kaisho, T., Hirano, T., and Wietzerbin, J., Triggering of the human interleukin-6 gene by interferon-gamma and tumor necrosis factor-alpha in monocytic cells involves cooperation between interferon regulatory factor-1, NF-κB, and Sp1 transcription factors, *J. Biol. Chem.*, 270, 27920–27931, 1995.

Scapoli, L., Ramos-Nino, M. E., Martinelli, M., and Mossman, B. T., Src-dependent ERK5 and Src/EGFR-dependent ERK1/2 activation is required for cell proliferation by asbestos, *Oncogene*, 23, 805–813, 2004.

Schutte, J., Minna, J. D., and Birrer, M. J., Deregulated expression of human c-jun transforms primary rat embryo cells in cooperation with an activated c-Ha-ras gene and transforms rat-1a cells as a single gene, *Proc. Natl. Acad. Sci. USA*, 86, 2257–2261, 1989.

Shukla, A., Timblin, C., BéruBé, K., Gordon, T., McKinney, W., Driscoll, K., Vacek, P., and Mossman, B. T., Inhaled particulate matter causes expression of nuclear factor (NF)-κB-related genes and oxidant-dependent NF-κB activation in vitro, *Am. J. Respir. Cell Mol. Biol.*, 23, 182–187, 2000.

Shukla, A., Timblin, C. R., Hubbard, A. K., Bravman, J., and Mossman, B. T., Silica-induced activation of c-Jun-NH2-terminal amino kinases, protracted expression of the activator protein-1 proto-oncogene, fra-1, and S-phase alterations are mediated via oxidative stress, *Cancer Res.*, 61, 1791–1795, 2001.

Shukla, A., Gulumian, M., Hei, T. K., Kamp, D., Rahman, Q., and Mossman, B. T., Multiple roles of oxidants in the pathogenesis of asbestos-induced diseases, *Free Radic. Biol. Med.*, 34, 1117–1129, 2003a.

Shukla, A., Ramos-Nino, M., and Mossman, B., Cell signaling and transcription factor activation by asbestos in lung injury and disease, *Int. J. Biochem. Cell Biol.*, 35, 1198–1209, 2003b.

Shukla, A., Stern, M., Lounsbury, K. M., Flanders, T., and Mossman, B. T., Asbestos-induced apoptosis is protein kinase C delta-dependent, *Am. J. Respir. Cell Mol. Biol.*, 29, 198–205, 2003c.

Shukla, A., Flanders, T., Lounsbury, K. M., and Mossman, B. T., The gamma-glutamylcysteine synthetase and glutathione regulate asbestos-induced expression of activator protein-1 family members and activity, *Cancer Res.*, 64, 7780–7786, 2004.

Spurzem, J. R., Saltini, C., Rom, W., Winchester, R. J., and Crystal, R. G., Mechanisms of macrophage accumulation in the lungs of asbestos-exposed subjects, *Am. Rev. Respir. Dis.*, 136, 276–280, 1987.

Taatjes, D. J. and Mossman, B. T., *Cell Imaging Techniques: Methods and Protocols, (Methods in Molecular Biology)*, Humanan Press, Totowa, NJ, p. 319, 2005.

Takemura, T., Rom, W. N., Ferrans, V. J., and Crystal, R. G., Morphologic characterization of alveolar macrophages from subjects with occupational exposure to inorganic particles, *Am. Rev. Respir. Dis.*, 140, 1674–1685, 1989.

Tanaka, S., Choe, N., Hemenway, D. R., Zhu, S., Matalon, S., and Kagan, E., Asbestos inhalation induces reactive nitrogen species and nitrotyrosine formation in the lungs and pleura of the rat, *J. Clin. Invest.*, 102, 445–454, 1998.

Tang, G., Minemoto, Y., Dibling, B., Purcell, N. H., Li, Z., Karin, M., and Lin, A., Inhibition of JNK activation through NF-κB target genes, *Nature*, 414, 313–317, 2001.

Timblin, C. R., Janssen, Y. W., and Mossman, B. T., Transcriptional activation of the proto-oncogene c-jun by asbestos and H2O2 is directly related to increased proliferation and transformation of tracheal epithelial cells, *Cancer Res.*, 55, 2723–2726, 1995.

Timblin, C., BéruBé, K., Churg, A., Driscoll, K., Gordon, T., Hemenway, D., Walsh, E., Cummins, A. B., Vacek, P., and Mossman, B., Ambient particulate matter causes activation of the c-jun kinase/stress-activated protein kinase cascade and DNA synthesis in lung epithelial cells, *Cancer Res.*, 58, 4543–4547, 1998a.

Timblin, C. R., Guthrie, G. D., Janssen, Y. W., Walsh, E. S., Vacek, P., and Mossman, B. T., Patterns of c-fos and c-jun proto-oncogene expression, apoptosis, and proliferation in rat pleural mesothelial cells exposed to erionite or asbestos fibers, *Toxicol. Appl. Pharmacol.*, 151, 88–97, 1998b.

Upadhyay, D., Panduri, V., and Kamp, D. W., Fibroblast growth factor-10 prevents asbestos-induced alveolar epithelial cell apoptosis by a mitogen-activated protein kinase-dependent mechanism, *Am. J. Respir. Cell Mol. Biol.*, 32, 232–238, 2005.

Wang, L., Reinach, P., and Lu, L., TNF-alpha promotes cell survival through stimulation of K+ channel and NFkappaB activity in corneal epithelial cells, *Exp. Cell Res.*, 311, 39–48, 2005.

Yanase, N., Hata, K., Shimo, K., Hayashida, M., Evers, B. M., and Mizuguchi, J., Requirement of c-Jun NH2-terminal kinase activation in interferon-alpha-induced apoptosis through upregulation of tumor necrosis factor-related apoptosis-inducing ligand (TRAIL) in Daudi B lymphoma cells, *Exp. Cell Res.*, 310, 10–21, 2005.

Young, M. R., Li, J. J., Rincon, M., Flavell, R. A., Sathyanarayana, B. K., Hunziker, R., and Colburn, N., Transgenic mice demonstrate AP-1 (activator protein-1) transactivation is required for tumor promotion, *Proc. Natl. Acad. Sci. USA*, 96, 9827–9832, 1999.

Yuan, Z., Taatjes, D. J., Mossman, B. T., and Heintz, N. H., The duration of nuclear extracellular signal-regulated kinase 1 and 2 signaling during cell cycle reentry distinguishes proliferation from apoptosis in response to asbestos, *Cancer Res.*, 64, 6530–6536, 2004.

Zanella, C. L., Posada, J., Tritton, T. R., and Mossman, B. T., Asbestos causes stimulation of the extracellular signal-regulated kinase 1 mitogen-activated protein kinase cascade after phosphorylation of the epidermal growth factor receptor, *Cancer Res.*, 56, 5334–5338, 1996.

Zanella, C. L., Timblin, C. R., Cummins, A., Jung, M., Goldberg, J., Raabe, R., Tritton, T. R., and Mossman, B. T., Asbestos-induced phosphorylation of epidermal growth factor receptor is linked to c-fos and apoptosis, *Am. J. Physiol.*, 277, L684–L693, 1999.

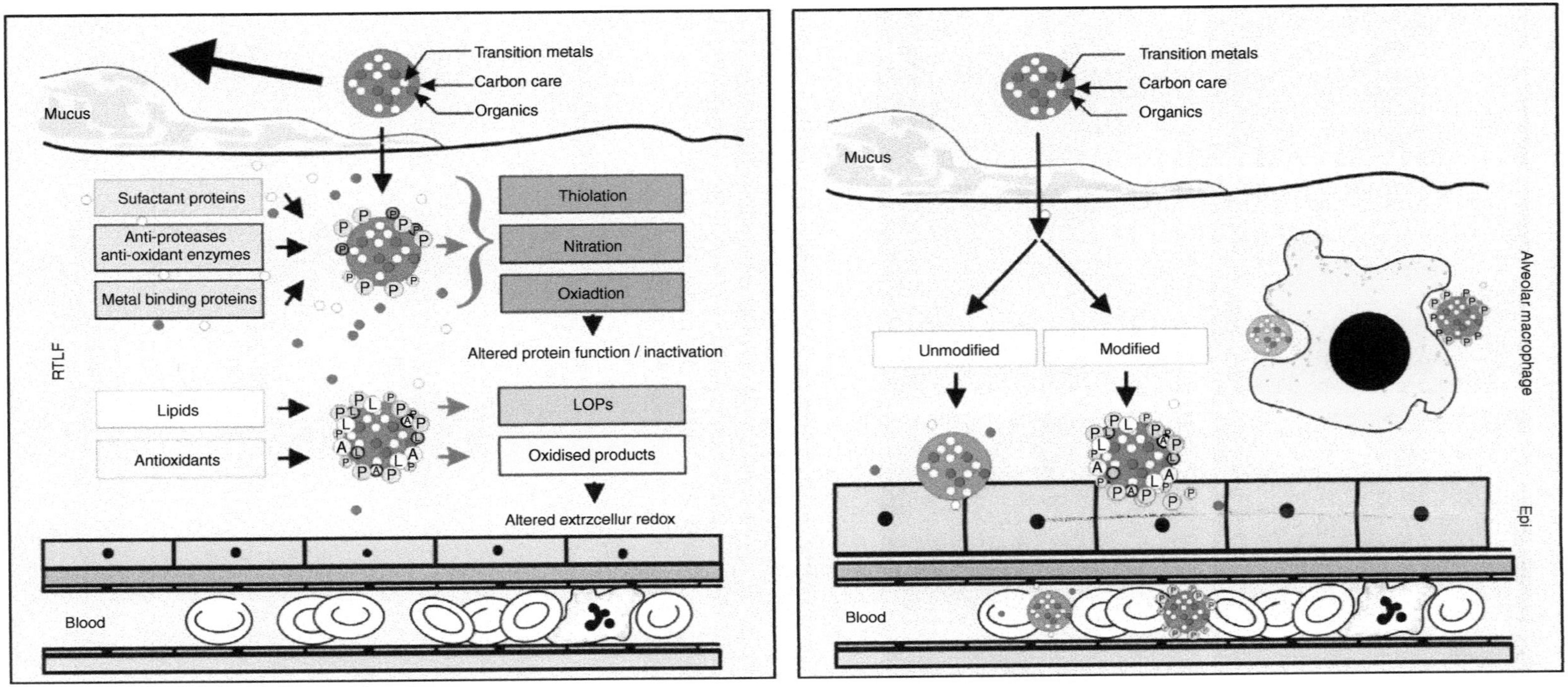

FIGURE 5.4 Particle-RTLF interactions. Inhaled particles depositing in the airways may become trapped by the layer of mucus and transported from the airways by the mucocillary elevator. Those particles that are retained in the airways can either leach soluble components into the RTLF that oxidize antioxidants, lipid, and protein components of the RTLF, or absorb the oxidized or native RTLF components onto their surface. These interactions therefore alter both the redox state of the RTLF, which will impact upon the underlying cells, as well as modifying the particle surface that ultimately reaches the underlying epithelium, or that is intercepted by alveolar macrophages within the extracellular compartment.

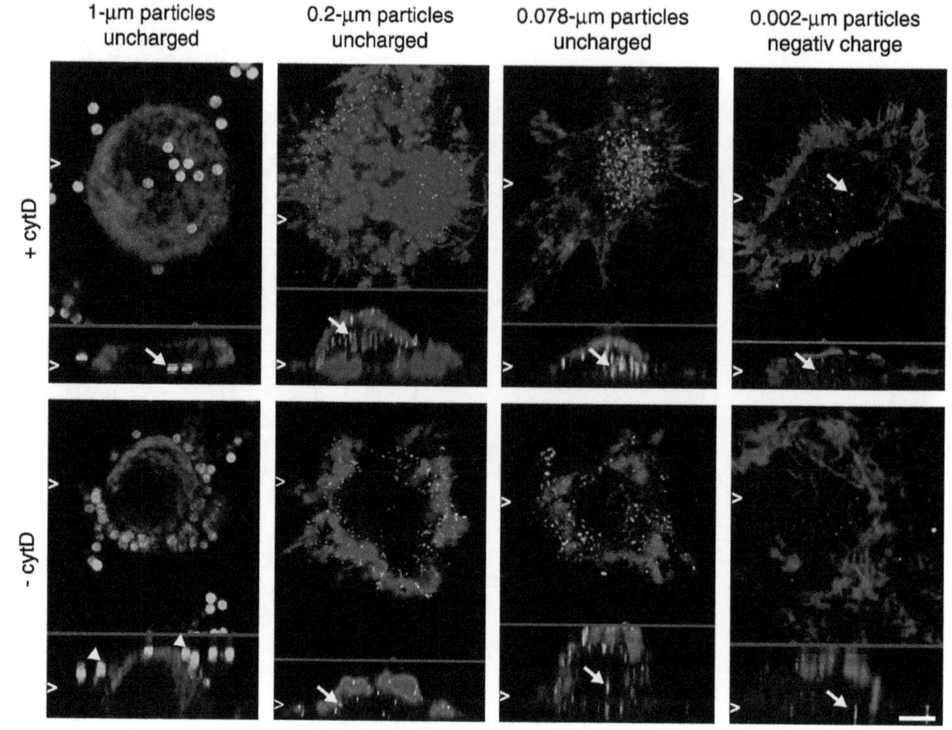

FIGURE 7.4 Laser scanning microscopy micrographs of fluorescent polystyrene spheres (1, 0.2, 0.078 and 0.02 μm) taken up by macrophages in the absence or presence of cytochalasin D (cytD). F-Actin is shown in red, particles are green. The *xy* and *xz* projections allow the clear differentiation between intracellular (arrows) and extracellular (arrowheads) particles. Yellow open arrowheads mark the position of projections. Scale bar is 2 μm. (Pictures with 1, 0.2, and 0.078 μm particles are reproduced from Geiser, M., Rothen-Rutishauser, B., Kapp, N., Schürch, S., Kreyling, W., Schulz, H., Semmler, M., Im Hof, V., Heyder, J., and Gehr, P., *Environ. Health Perspect.*, 2005, (accepted, doi:10.1289/ehp.8006). With permission.)

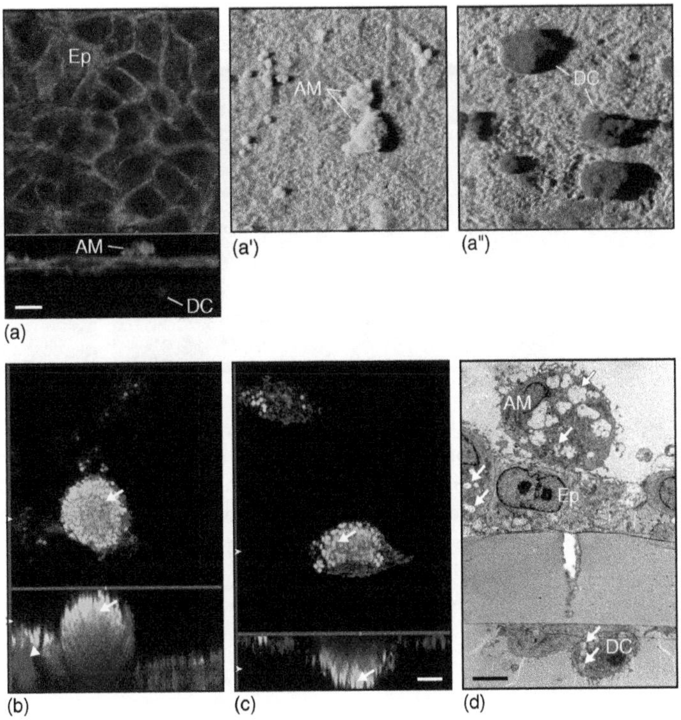

FIGURE 7.6 Laser scanning microscopy and transmission electron microscopy images of triple co-cultures with epithelial cells, macrophages, and DCs grown on a filter insert. (a, a′, a″) Epithelial cells are shown in yellow, macrophages (AM) in turquoise, and dendritic cells (DC) in pink. Macrophages reside on the surface of the epithelial cells, whereas dendritic cells were localized at the bottom side of the insert. (a) Represents *xy* and *xz* projections; yellow arrowheads mark the position of projections. (a′ and a″) are 3D reconstructions from the data set, (a′) from the upper side, (a″) from the lower side. (b, c, d) 1 μm particles uptake in triple cell co-cultures. Triple cell co-cultures were incubated for 24 h, fixed and stained afterwards for F-Actin (red) and CD14 (b, turquoise) or CD86 (c, turquoise). Particles (green) were found in CD14 positive (b, arrows), in epithelial cells (b, arrowhead), and in CD86 positive dendritic cells (c, arrows). Images represent *xy* and *xz* projections; yellow arrowheads mark the position of projections. Transmission electron microscopy images of triple co-cultures exposed 24 h to 1 μm particles (d). Particles (d, arrows) were found within epithelial cells (Ep), macrophages (AM), and dendritic cells (DC). (a, a′, a″) Scale bar is 10 μm; (b, c) scale bar is 5 μm; (d) scale bar is 2 μm. (Images b, c and d are reproduced from Rothen-Rutishauser, B. M., Kiama, S. G., and Gehr, P., *Am. J. Respir. Cell Mol. Biol.*, 32, 281–289, 2005. With permission.)

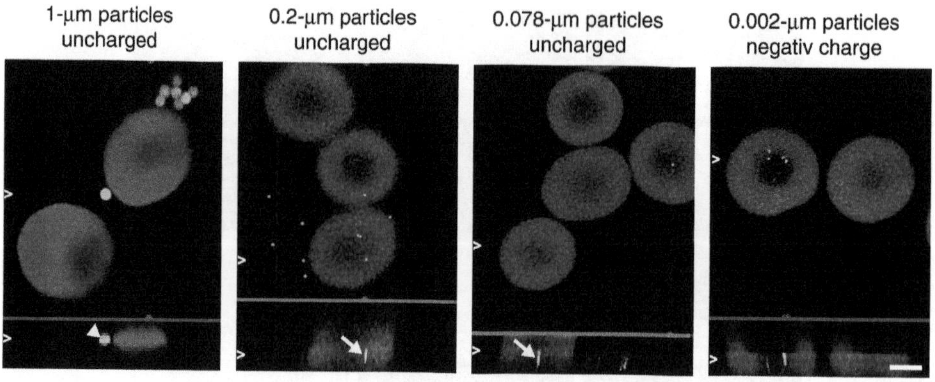

FIGURE 7.7 Laser scanning microscopy micrographs of fluorescent polystyrene spheres (1, 0.2, 0.078 and 0.02 μm) taken up by red blood cells. Autofluorescence of the cells is shown in red, particles are green. The *xy* and *xz* projections allow the clear differentiation between internalized (arrows) and extracellular (arrowheads) particles. Yellow open arrowheads mark the position of the projections. Scale bar is 2 μm. (Pictures with 1, 0.2, and 0.078 μm particles are reproduced from Geiser, M., Rothen-Rutishauser, B., Kapp, N., Schürch, S., Kreyling, W., Schulz, H., Semmler, M., Im Hof, V., Heyder, J., and Gehr, P., *Environ. Health Perspect.*, 2005, (accepted, doi:10.1289/ehp.8006). With permission.)

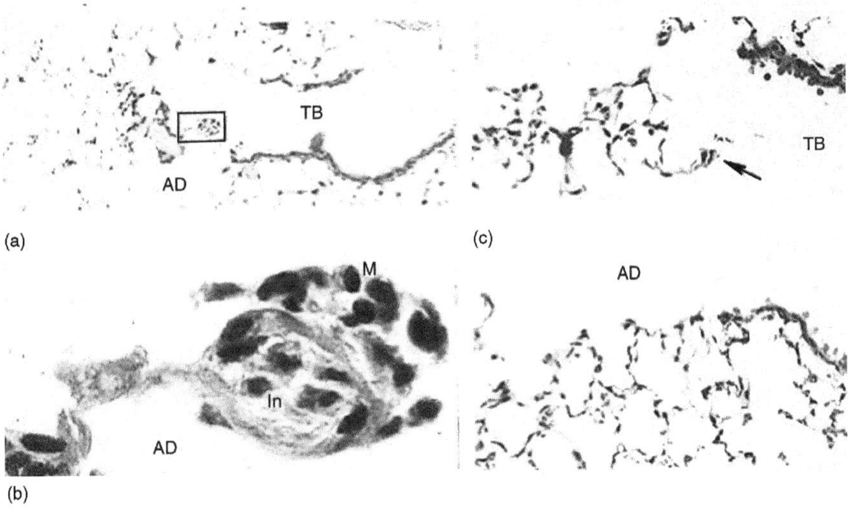

FIGURE 12.2 Histopathology. (a) Lung of a mouse (Bl6/129) 48 h after exposure to chrysotile asbestos. The lesion (box) located at the alveolar duct (AD) bifurcation near the ends of the terminal bronchioles (TB) is hypercellular and hypertrophic, with increased extracellular matrix, macrophages (M) and interstitial cells (IN). (b) Higher magnification of lesion (X40). (c) TNF-aRKO mice fail to develop fibroproliferative lesions in response to asbestos. This knockout animal exhibits a normal duct bifurcation (arrow) 48 h after asbestos exposure. Photographs reproduced with permission.

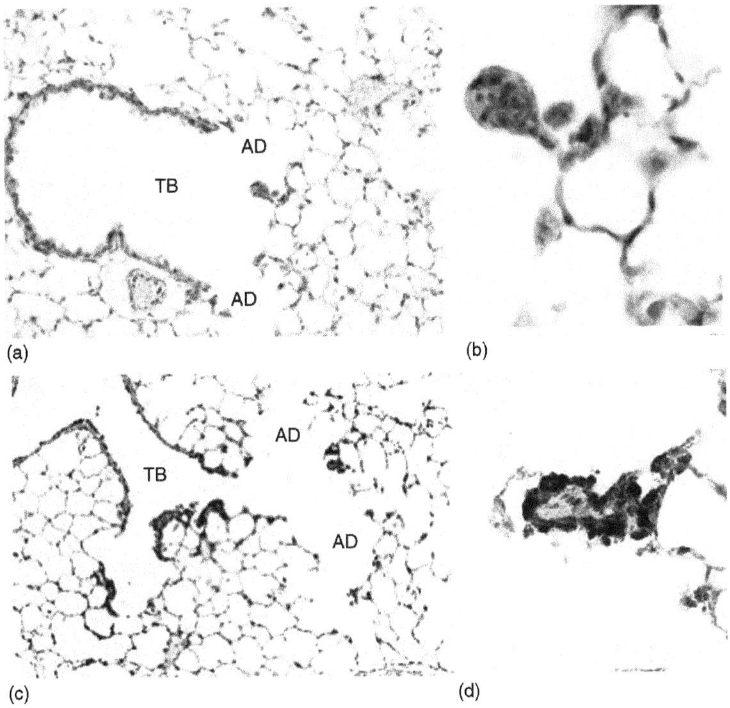

FIGURE 12.3 TGF-β1 expression in lungs of mice 48 h after exposure to chrysotile asbestos. (a) Immunohistochemistry using an antibody to TGF-β1 latent associated peptide demonstrates expression of TGF-β1 protein in epithelial cells and macrophages. (b) Higher magnification from 3A. Developing lesion exhibits robust staining of epithelium, macrophages and interstitial cells (X40). Photograph (a & b) courtesy of Dr. Derek Pociask, Tulane University Health Science Center. (c) TGF-β1 in situ hybridization (ISH) of a typical bronchiolar alveolar duct region. Numerous epithelial cells and macrophages express TGF-β1 mRNA. (d) Developing lesion with epithelium (arrow) and macrophages (arrow heads) positive for TGF-β1 ISH (X40).

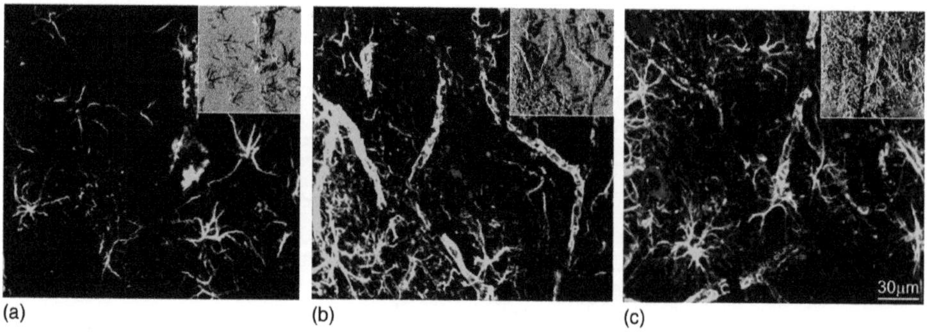

FIGURE 19.3 Reactive gliosis and astrocytic proliferation in the frontal cortex white matter of healthy Mexico City dogs. Glial cells and proliferating astrocytes were localized in paraffin sections of frontal cortex of Mexico City dogs by immunohistochemistry using fluorescein-labeled anti-glial fibrillary acidic protein (GFAP, green) and phycoerythrin-labeled anti-bromodeoxyuridine (BrdU, red), respectively, and examined by confocal microscopy: (a) 3-year-old male, (b) 5-year-old female, and (c) 14-year-old female. Gliosis worsens with age. Images represent maximum intensity projections, showing the maximum intensity of all layers along the viewing direction. The inserts represent 3D reconstructions from the same data sets. (Pictures were taken by Dr. Barbara Rothen-Rutishauser Ph.D. Institute of Anatomy, University of Bern, Bern, Switzerland.)

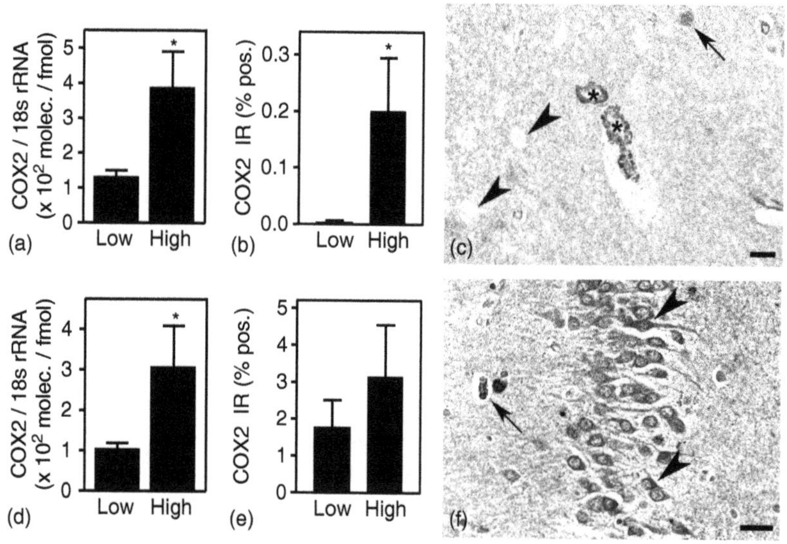

FIGURE 19.4 COX2 expression in frontal cortex (a–c) and hippocampus (d–f). COX2 mRNA abundance was measured by RT-PCR and normalized for 18s rRNA levels. COX2 protein was localized in sections of paraffin-embedded tissues by IHC and its abundance was measured by quantitative image analysis. COX2 mRNA was significantly elevated in the high exposure group in both frontal cortex (a, $p = 0.009$), and hippocampus (d, $p = 0.04$). COX2 immunoreactivity (IR) was significantly elevated in the high exposure group in frontal cortex (b, $p = 0.01$), but not in hippocampus (e). Means ± SEMs are shown in A, B, D, and F. (c) Representative COX IHC in frontal cortex from a subject in the high exposure group showing strong staining of endothelial cells in the capillaries (*), and pyramidal neurons (arrow), while other neurons were negative (arrowheads). Scale = 20 μm. (f) Representative COX IHC in dentate gyrus from a subject in the high exposure group showing COX2 positive neurons (arrowheads) and capillaries (short arrow). Scale = 15 μm. (Calderón-Garcidueñas, L. et al., Brain inflammation and Alzheimer's-like pathology in individuals exposed to severe air pollution, *Toxicol. Pathol.*, 32, 650–658, 2004. With permission.)

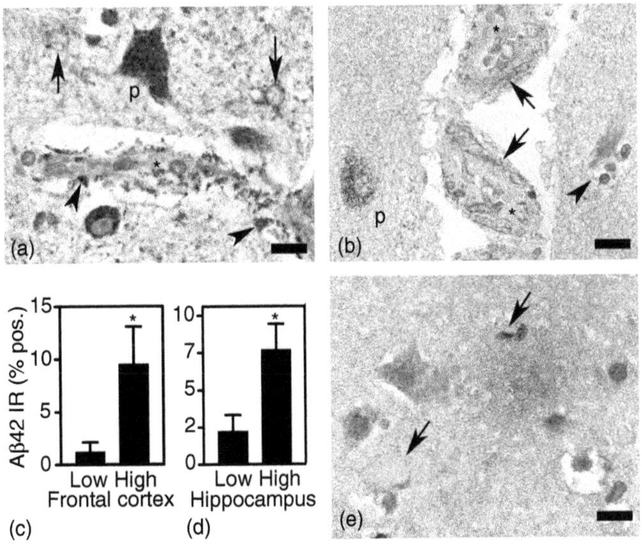

FIGURE 19.5 Aβ42 accumulation in frontal cortex and hippocampus. Aβ42 was localized in sections of paraffin-embedded tissues by IHC. (a) Anti-Aβ42 stained pyramidal neurons (p), astrocytes (arrows) and astrocytic processes (arrowheads) around blood vessels (*). (b) In addition to accumulation in pyramidal neurons, (p) Aβ42 was deposited in smooth muscle cells (arrows) in cortical arterioles (*). A dead neuron surrounded by glial cells is indicated (arrowhead). (c and d) Quantitative image analysis of Aβ42 IHC showed a significant increase in Aβ42 immunoreactivity (Aβ42 IR) in both frontal cortex (c, * $p = 0.04$) and hippocampus (d, * $p = 0.001$) in the high exposure group. Means ± SEMs are shown. (e) Aβ42 IHC of frontal cortex from a 38-year-old subject from Mexico City showing diffuse plaque-like staining with surrounding reactive astrocytes (arrows). Scale = 20 μm. (Calderón-Garcidueñas, L. et al., Brain inflammation and Alzheimer's-like pathology in individuals exposed to severe air pollution, *Toxicol. Pathol.*, 32, 650–658, 2004. With permission.)

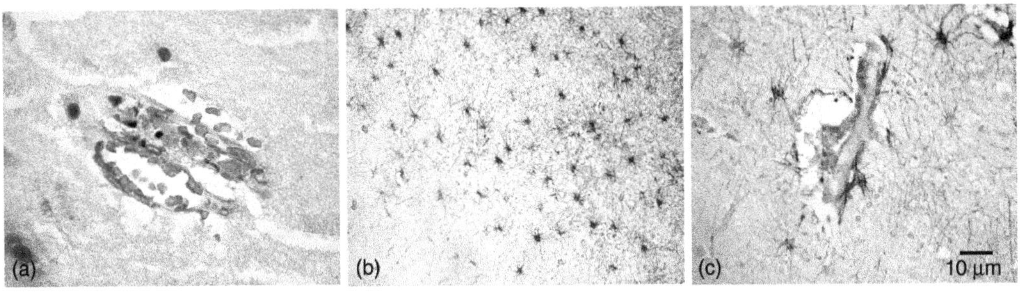

FIGURE 19.8 Reactive pathology and breakdown of the BBB in frontal cortex of an 11-year-old Mexico City male. Paraffin embedded frontal cortex was stained with hematoxylin and eosin (a) and anti-GFAP (b and c). (a) There are numerous perivascular macrophages with hemosiderin-like granules and free RBC surrounding a blood vessel. A neuron is seen in the lower left corner. (b) Same 11-year-old child showing reactive gliosis in subcortical frontal white matter (GFAP). (c) A 17-year-old Mexico City male with reactive astrocytes around cortical frontal blood vessels. The vessel is surrounded by macrophages loaded with hemosiderin-like pigment (GFAP). Leaking blood vessels are indirect evidence of a breakdown of the BBB.

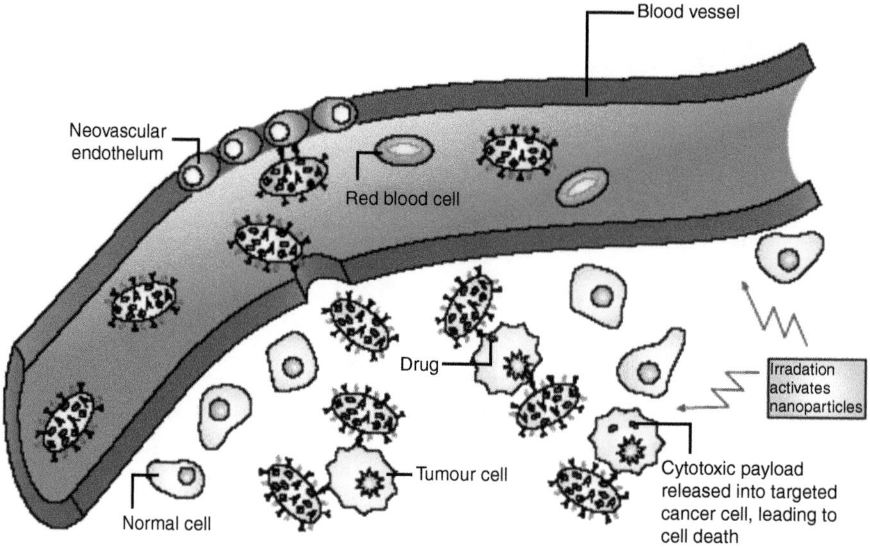

FIGURE 21.3 Multi-component targeting strategies using nanoparticles in treating cancer. Nanoparticles extravasate into the tumor stroma through the fenestrations of the angiogenic vasculature, demonstrating targeting by enhanced permeation and retention. The particles carry multiple antibodies, which further target them to epitopes on cancer cells, and direct antitumour action. Nanoparticles are activated and release their cytotoxic action when irradiated by external energy. Not shown: nanoparticles might preferentially adhere to cancer neovasculature and cause it to collapse, providing anti-angiogenic therapy. The red blood cells are not shown to scale; the volume occupied by a red blood cell would suffice to host 1–10 million nanoparticles of 10 nm diameter. (Reproduced from Ferrari, M., *Nat. Rev.*, 5, 161–171, 2005. With permission.)

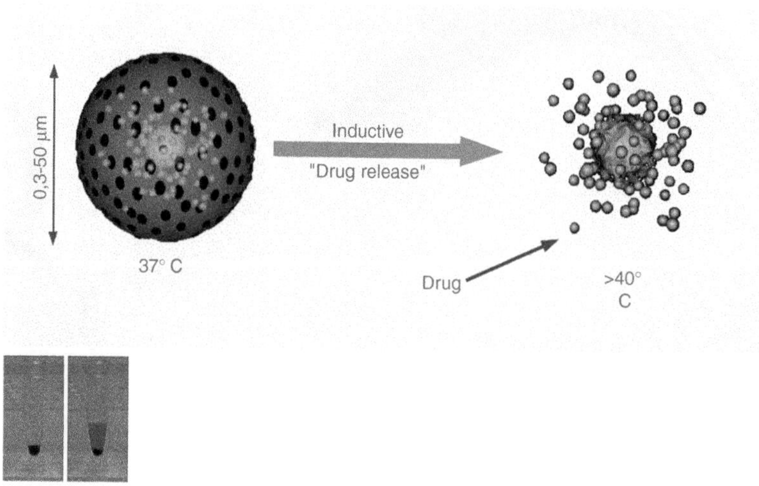

FIGURE 21.5 Graph illustrating a contactless controllable drug carrying system based on thermosensitive magnetic nano- and micro particles. The insert shows the application of the system with Rhodamine B encapsulated beads that is released after heating up to 45°C. By encapsulating ferro (i)magnetic colloids into the polymer matrix and exposing these magnetic beads to an external high frequency magnetic field (induction coil, magnetic amplitude 10–50 kA/m, 0.3–0.5 MHz), the subsequent shrink process can be remotely induced.

11 Particle-Associated Organics and Proinflammatory Signaling

Francelyne Marano, Sonja Boland, and Armelle Baeza-Squiban
Laboratoire de Cytophysiologie et Toxicologie Cellulaire, Université Paris 7 – Denis Didèrot

CONTENTS

11.1 INTRODUCTION

The burning of fossil fuels generates fine and ultrafine airborne particles, which contain a large amount of organic compounds including polyaromatic hydrocarbons (PAH). These particles must be taken into account as they are considered to be among the most abundant components of particulate matter 2.5 μm ($PM_{2.5}$) in urban areas. Most of them are produced by diesel engine-powered cars and diesel exhaust particles (DEP) constitute, on average, 40% of the PM_{10} in a city such as Los Angeles (Diaz-Sanchez 1997), and in a kerbside station in Paris more than 50% of particles were close to the ultrafine range ($\leq$0.26 μm), likely due to the influence of the traffic (Baulig et al. 2004). Chemical analysis between $PM_{2.5}$ collected in a kerbside and a background station in Paris revealed that PAH are twice as important in the kerbside station. The results are more relevant with heavy PAH than light PAH, due to their higher stability. We have also observed variations of PAH according to the seasons, probably due to chemical reactions with atmospheric oxidants. However, PAH are only a part of the organic component and they do not greatly influence the soluble organic fraction (SOF) measured after dichloromethane extraction that appear to be between 10 and 12% of the mass of the particles whatever the station and approximately 45% lower than the SOF of DEP (20%) (Baulig et al. 2004).

11.2 WHAT IS THE ROLE OF THESE ORGANIC COMPOUNDS IN THE EFFECTS OF PM?

In this chapter, we do not consider the genotoxic and carcinogenic effects of these organic compounds. The dramatic increase of human allergic airway diseases in the last century has followed the increase in the use of fossil fuels and many epidemiological studies have provided indirect evidence for a correlation between particulate pollution and increased incidence of asthma and allergic rhinitis. Numerous experimental studies in animals, in human volunteers, and *in vitro* were performed to provide a causal explanation for these observations. Diaz-Sanchez et al. have published numerous studies on the role of DEP and their associated PAH in the induction of allergic airway diseases (Riedl and Diaz-Sanchez 2005). *In vivo* nasal provocation studies, using amounts of DEP equivalent to the total exposure in 1–3 days in Los Angeles, showed enhanced immunoglobulin E (IgE) production in the human upper airway. This response appears to be linked to organic compounds of DEP since it has been established that PAH–DEP could significantly increase IgE mRNA and protein production in IgE-secreting Epstein–Barr-virus transformed human B cells *in vitro* (Tsien et al. 1997). The effects of DEP were specific since they do not increase IgG, IgA, or IgM (Diaz-Sanchez et al. 1994). The ability of DEP to act as an adjuvant was tested by performing nasal provocation challenges with DEP, ragweed antigen *Amb a 1*, or both simultaneously in ragweed-sensitive subjects (Diaz-Sanchez 1997). Ragweed-specific IgE was 16 times higher following ragweed plus DEP challenge compared with ragweed alone. However, IgG levels remained constant after challenge with DEP plus antigen. Further experiments have shown an increase in cytokine mRNA levels such as Interleukin-2 (IL-2), IL-4, IL-5, IL-6, and Interferon γ. The susceptibility to the adjuvant effect of DEP is an intrinsic trait, as there is a high intraindividual reproducibility of the nasal allergic responses in human exposure studies (Bastain et al. 2003). Several candidate genes could be involved in this susceptibility to particles, such as antioxidant enzymes or Toll Like Receptor 4, CD14 or Tumor Necrosis Factor α (reviewed in Granum and Lovik 2002). This adjuvant effect of DEP for allergic sensitization is a delayed response and does not explain acute PM effects on airway hyperreactivity. It has been recently demonstrated by Hao (Hao et al. 2003), using BALB/c mouse model sensitized by ovalbumin, that aerosolized DEP could induce increased airway hyperreactivity even if DEP delivery is delayed after the peak inflammatory response. It was concluded that DEP induced airway hyperreactivity independently of adjuvant effects. Interestingly, DEP co-administration with a neoallergen such as keyhole limpet hemocyanin shows that the particles could synergize with the neoallergen and drive the de novo production of antigen-specific IgE (Diaz-Sanchez et al. 1999). These results suggest that DEP exposure with a neoallergen leads to sensitization IgE mucosal production. Thus particulate air pollution may influence both—the sensitization and the provocation phase of allergy by inducing oxidative and inflammatory reactions in the respiratory mucosa (Granum and Lovik 2002).

Moreover, various animal experiments suggest that DEP may alter both innate and acquired cellular immunity. Beside their adjuvant effect, these particles have also an immunosuppressive effect in animals. The increased susceptibility of the lung to infection in rats exposed to DEP was related to the inhibition of the functions of alveolar macrophages by organic compounds, but not the carbonaceous core (Castranova et al. 2001). Using nitric oxide (NO) production as a marker of macrophage function, it was shown that crude DEP organic extracts inhibit both Lipopolysaccharide and Bacillus Calmette–Guérin (BCG) induced NO production by a murine macrophage cell line explaining the impaired bacterial clearance noticed in a BCG mouse lung infection model (Saxena et al. 2003b). By fractionation of the organic extract, it appears that this inhibitory effect was mainly due to PAH and resin fractions (Saxena et al. 2003a).

The identification of the chemical components involved in these biological effects and the understanding of the underlying mechanisms are still imperfect. Such studies are difficult, as there exists a great variability in the chemical composition of PM according to their emission sources, age, and site of sampling.

11.2.1 The Particles Organic Fraction

The organic fraction of particles comprises a countless quantity of compounds (such as aliphatic hydrocarbons, PAH, nitroaromatics hydrocarbons, quinones, aldehydes, and heterocyclics), some of which are still unidentified. This fraction can represent up to 50% of the mass of the particle and may contain toxic compounds. At the present time, PAH are quantitatively and qualitatively the best known family of organic compounds adsorbed on particles, although they only represent a few percent of the organic fraction. The interest in these compounds lies in their known genotoxic and inflammatory properties and their use as a tracer of source. They have been shown to be in higher concentrations in submicron particles (De Kok et al. 2005; Rehwagen et al. 2005), which can be explained by the fact that soot from combustion sources consist primarily of fine particles with high PAH content and that the smaller particles have a relatively high surface area for PAH adsorption (Ravindra, Mittal, and Van Grieken 2001).

Another category of organic compounds that has held the attention of biologists are quinones, due to their ability to induce various hazardous effects *in vivo* such as acute cytotoxicity, immunotoxicity, and carcinogenesis (Bolton et al. 2000). Four quinones (1,2-naphthoquinone, 1,4-naphthoquinone, 9,10-phenanthraquinone, 9,10-anthraquinone) have been identified and quantified in DEP (7.9–4.04 μg/g) and in Los Angeles $PM_{2.5}$ (5–730 pg/m^3) (Cho et al. 2004).

11.2.2 Bioavailability of Organic Compounds

The presence on particles of organic compounds exhibiting a potential biological effect raises the question of their bioavailablity. To understand the processes whereby particles deliver and transfer toxic components to target cells, experiments have been done using radio-labeled benzo(a)pyrene (B(a)P)-bound denuded particles. After their administration to dogs, the extent and rate of release as well as their metabolic fate were investigated (Gerde et al. 2001b). It reveals that in the alveolar region, B(a)P was adsorbed mostly unaltered into the blood and was systematically metabolized (Gerde et al. 2001a). In the conducting airways, a smaller fraction of B(a)P was slowly deposited but metabolized in the airway epithelium (Gerde et al. 2001b). Nevertheless, a large fraction of B(a)P remained bound to particles even 6 months after the exposure (Gerde et al. 2001b).

In cells, foreign substances are detoxified by two sequential reaction processes, namely, Phase I and Phase II. In Phase I reactions, xenobiotics are mainly oxidized by cytochrome P450 (CYP) enzymes to become more polarized metabolites. Phase II metabolism, catalyzed by enzymes such as glutathione *S*-transferase (GST) and NADP(P)H:quinone oxidoreductase (NQO1), converts the reactive Phase I metabolites to more hydrophilic substances, allowing their elimination.

Among the members of the *CYP* gene family, *CYP1* is known to be induced by PAH through a receptor-dependent mechanism. The cytosolic aryl hydrocarbon receptor (AhR), when bound by PAH, translocates to the nucleus, heterodimerizes with another partner, and activates the transcription of CYP1 family genes through binding to the xenobiotic response element. Native DEP, PM and their respective extracts act as activators of the AhR, inducing CYP1A1 expression and activity (Meek 1998; Bonvallot et al. 2001; Baulig et al. 2003a). As shown in Figure 11.1a, DEP and their organic extract induce a transient CYP1A1 mRNA expression in human bronchial epithelial cells (HBE) similar to B(a)P whereas carbon black particles have not such an effect (Baulig et al. 2003a).

The genes of Phase II metabolism (GST, NQO-1) are regulated in a concerted manner at the transcriptional level through the antioxidant-responsive element (ARE)/electrophile-responsive element. The transcription factor NF-E2-related factor-2 (Nrf2) is central to ARE-mediated gene expression (Itoh et al. 1997) and Nrf2 (−/−) mice exhibit significant reduction of phase II enzymes (Cho et al. 2002). DEP induce the translocation of Nrf2 to the nucleus of HBE cells, increase nuclear protein binding to the ARE (Baulig et al. 2003a), as well as NQO1 expression as shown in Figure 11.1b.

Whereas this biotransformation process aimed to detoxify xenobiotics, a bioactivation may occur and reactive metabolites are produced especially during the Phase I. By this way, PAH give rise to electrophilic metabolites responsible for their genotoxicity.

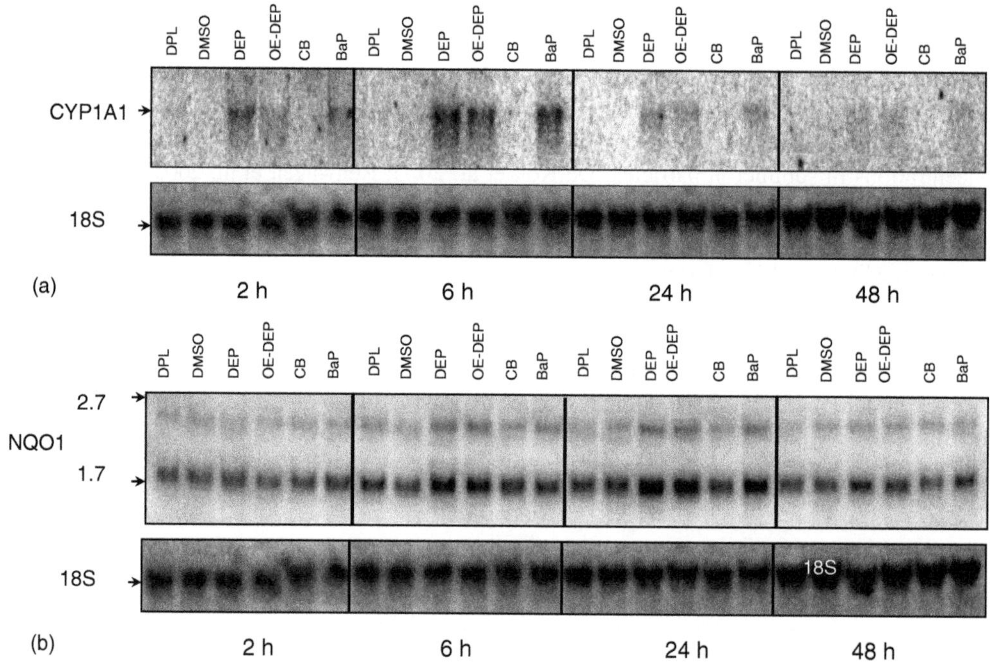

FIGURE 11.1 Induction of cytochrome P-450 1A1 (CYP1A1) and NADPH: quinone oxidoreductase 1 (NQO-1) gene expression in HBE cells (a and b respectively). Cells were treated or not with DEP (10 μg/cm^2), carbon black (10 μg/cm^2) organic extracts of DEP (OE-DEP, 10 μg/mL) or benzo(a)pyrene (B(a)*P*, 3 μM). RNA (30 μg) were extracted from cells after 2, 6, 24, or 48 h of treatment, electrophoresed, Northern-blotted, and then incubated with a ^{32}P-labeled cDNA probe for CYP1A mRNA, NQO-1 mRNA, or 18S RNA. (From Baulig, A., Garlatti, M., Bonvallot, V., Marchand, A., Barouki, R., Marano, F., and Baeza-Squiban, A., *Am. J. Physiol. Lung. Cell. Mol. Physiol.*, 285, L671–L679, 2003a. With permission).

The observation of the CYP1A1 gene induction in lung homogenates of Big Blue rats exposed, by inhalation, to whole DEP fumes supports the argument that the leaching of organic compounds from particles can occur (Sato et al. 2000). Moreover, in another study, it was shown that this transient CYP1A1 induction in lungs of rats only occurs with DEP and not with carbon black (Ma and Ma 2002). The mechanisms of transfer of organic compounds from particles to the target cells can involve the uptake of particles (Bonvallot et al. 2001). However, a recent study using fresh butadiene soots suggests that the transfer to cells occurs by the direct contact between soots and the plasma membrane, likely involving a partitioning mechanism (Penn et al. 2005).

Moreover, it is not known how effectively biological media (e.g., serum or interstitial fluids) can solubilize the organic compounds (Keane et al. 1991). It has been shown that the addition of surfactant in an aqueous suspension of DEP or carbon black particles on which a PAH mixture has been previously adsorbed doesn't favor the leaching of PAH (Borm et al. 2005). In addition, the PAH bioavailability is negligible when the PAH content is low relative to the particle monolayer surface (Borm et al. 2005).

11.2.3 Organic Compounds and Oxidative Stress

Evidence for the involvement of oxidative stress in the effects of organic compounds came from the initial observation that the mortality resulting from lung edema after intratracheal administration of whole DEP into mice was suppressed by pretreatment with polyethylene glycol-modified superoxide dismutase (Sagai et al. 1993) and that it was limited with methanol-washed DEP. In this same study, it was shown that in acellular conditions, whole DEP produced oxygen radicals (superoxide anion radical $O_2^{\cdot-}$ and hydroxyl radical $^{\cdot}OH$) identified by electron paramagnetic resonance, which were not produced with methanol-washed DEP (Sagai et al. 1993). Quinones have been reported to be responsible for this radical production due to their ability to undergo enzymatic (P450/P450 reductase) and non enzymatic redox cycling with their corresponding semiquinone radical giving rise to $O_2^{\cdot-}$ (Bolton et al. 2000). Further enzymatic or spontaneous dismutation of $O_2^{\cdot-}$ produces hydrogen peroxide, which in presence of trace amounts of transition metals such as iron gives $^{\cdot}OH$ by the Fenton reaction. A methanol extract of DEP has been shown to cause a significant formation of $O_2^{\cdot-}$ in the presence of Cyt P450 reductase (Kumagai et al. 1997), an enzyme which activity is increased in DEP-treated mice (Lim et al. 1998). More recently, $PM_{2.5}$ have been found to contain abundant and stable semiquinone radicals detected by EPR and to induce DNA damage (Dellinger et al. 2001; Squadrito et al. 2001). To these semiquinone radicals directly present on particles, others can be produced during an alternative PAH-metabolization involving dihydrodiol dehydrogenase leading to the generation of PAH *o*-quinones (Penning et al. 1999). Furthermore, the CYP1A1 catalytic activity generates reactive oxygen species (ROS) (Perret and Pompon 1998). Redox-active transition metals, redox cycling quinones, and PAH present on PM can act synergistically to produce ROS.

Taken altogether, these data reveal that organic compounds are a source of ROS. It explains the pro-oxidant status measured using various specific fluorescent probes in airway epithelial cells and macrophages treated either with DEP, PM, or their corresponding organic extract, whereas carbon black particles or solvent-extracted particles do not have such an effect (Hiura et al. 1999; Li et al. 2002; Baulig et al. 2003a; Baulig et al. 2004). For example, increased ROS production determined by the dichlorofluorescein fluorescence was observed in HBE cells exposed for 4 h to DEP, urban $PM_{2.5}$ sampled in Paris, and their respective extracts. The extracts gave a fluorescence signal similar to native particles (Figure 11.2). A pro-oxidant status is known to induce cellular specific responses in the order that cells face oxidant insult. Various studies have shown that such responses occur in DEP-treated cells. By a genomic approach, the expression profiles of genes induced by organic extracts of DEP in rat alveolar macrophages reveals the increased expression of anti-oxidant enzymes (heme oxygenase (HO-1), thioredoxin peroxidase 2, NADPH dehydrogenase) (Koike et al. 2002; Koike et al. 2004). Similarly, the overexpression of HO-1 was observed in a murine macrophages cell line (RAW264.7) exposed to organic extracts of DEP as well as in epithelial cells (Li et al. 2002). These data were completed by observation of a change in the proteome of RAW264.7 exposed to DEP organic extracts 51 proteins were newly expressed but were suppressed by *N*-acetylcysteine, a thiol antioxidant (Xiao et al. 2003).

Furthermore, from crude DEP extracts, Li and collaborators (Li et al. 2000) have shown that only the polar fraction that is enriched in quinones and the aromatic fraction enriched in PAH were able to decrease the cellular GSH/GSSG ratio in macrophages as well as to induce HO-1 expression, both indicative of a situation of oxidative stress. In other respects, the comparison of coarse, fine, and ultrafine PM revealed that ultrafine PM have the highest redox activity (Cho et al. 2005), in agreement with the observation that ultrafines were the most potent towards inducing HO-1 expression and depleting intracellular GSH (Li et al. 2003). However, the susceptibility to an adverse health effect of DEP is linked to the functional variation in natural antioxidant defenses. The polymorphism of GST genes has been associated with atopy and experimental studies provide evidence that the GSTM1 and GSTP1 genotypes can play a role in the susceptibility to the adjuvant effect of DEP (Gilliland et al. 2004).

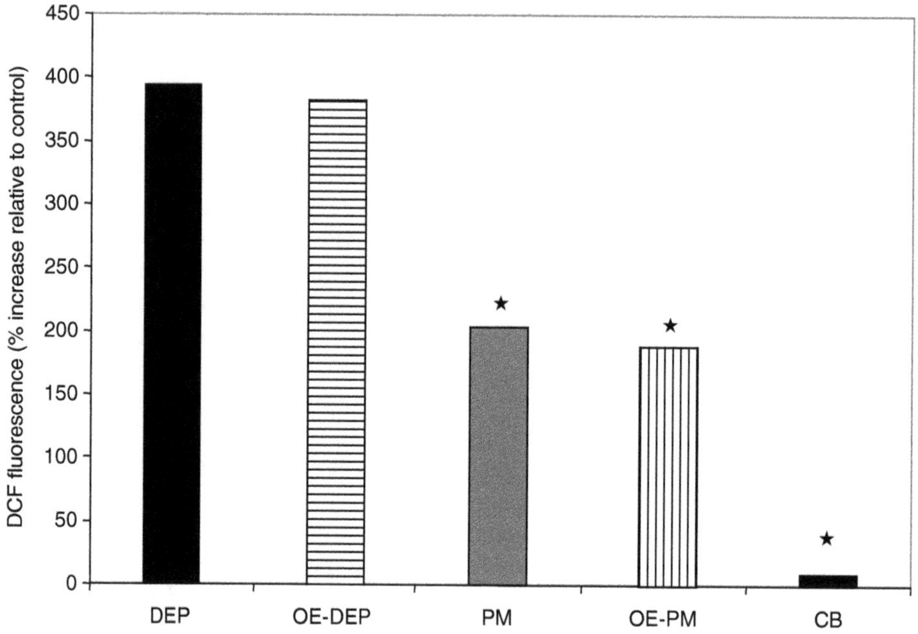

FIGURE 11.2 Dichlorofluorescein (DCF) fluorescence intensity in human bronchial epithelial cells treated with diesel exhaust particles (DEP, 10 μg/cm^2) or their corresponding organic extract (OE-DEP), Paris urban $PM_{2.5}$ (PM, 10 μg/cm^2) or their corresponding organic extract (OE-PM), or carbon black particles (CB, 10 μg/cm^2). The cells were loaded with 2′,7′-dichlorofluorescein-diacetate (H2DCF-DA) at 20 μM for 20 min and then treated or not with the toxics for 4 h. The DCF fluorescence was measured by cytometry. Results are expressed in % of increase of DCF fluorescence relative to control.

11.2.4 Organic Compounds and Inflammation

In vivo human exposure studies show that phenantrene, in contrast to carbon black, increase IgE production, but did not cause inflammatory cell infiltration (Saxon and Diaz-Sanchez 2000). In animal studies however PAH not only increased IgE production (Heo, Saxon, and Hankinson 2001) but also the recruitment of inflammatory cells (Hiyoshi et al. 2005). Furthermore, *in vitro* studies have shown that organic compounds are involved in the proinflammatory response induced by particles in the two respiratory target cells (airway epithelial cells and macrophages). Several studies using HBE cell lines (BEAS-2B, 16HBE), normal human airway epithelial cells, and macrophages have shown that an inflammatory mediator release (IL-8, GM-CSF, RANTES, TNF-α) can be induced by exposure to DEP extract (Boland et al. 2000; Fahy et al. 2000; Li et al. 2002; Vogel et al. 2005). From the comparison of native DEP with their organic extracts obtained with benzene (Kawasaki et al. 2001) or dichloromethane extraction with extracted DEP and carbon black particles (Figure 11.3a) (Boland et al. 2000), it was concluded that organic compounds mimic native DEP and that the carbonaceous core is not involved in the proinflammatory response. Moreover, the role of organic compounds was strengthened by the observation that DEP from vehicles equipped with a catalytic converter exhibiting a SOF of 8.3% induce a lower GM-CSF release by HBE cells than DEP from non-equipped vehicles having a 35% SOF (Figure 11.3b) (Boland et al. 2000). Concerning PM, until now, few studies have addressed the involvement of organic compounds. By chemical fractionation (organic vs. aqueous fraction) of Paris urban $PM_{2.5}$, it was shown that the GM-CSF secretion induced by native $PM_{2.5}$ in HBE cells was mimicked by their organic extracts

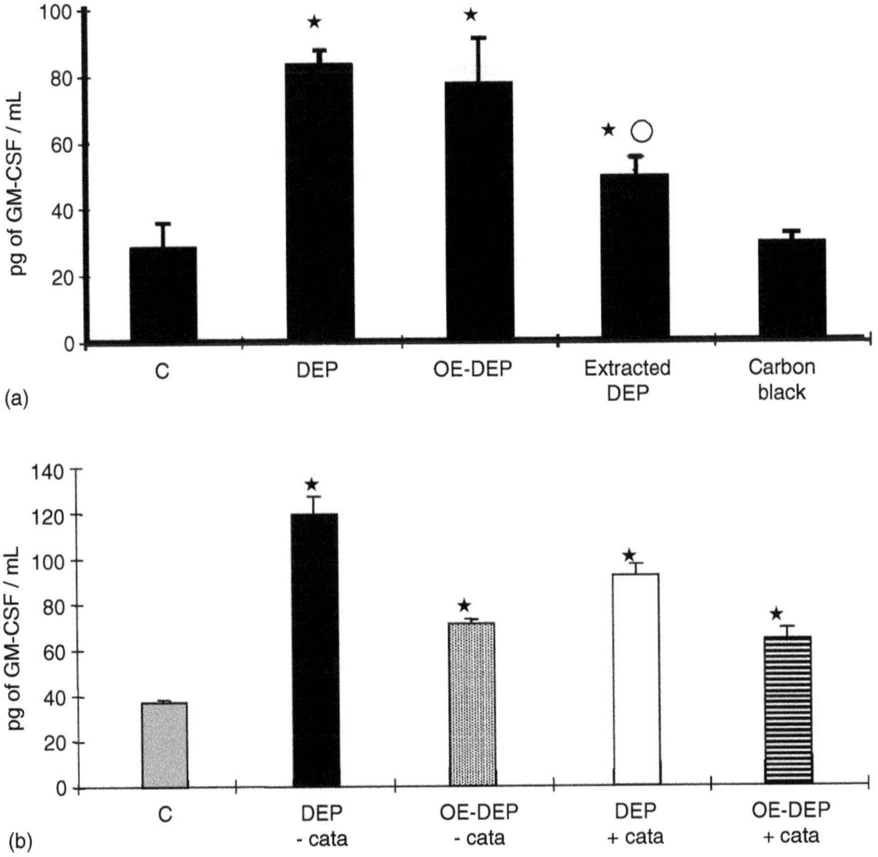

FIGURE 11.3 GM-CSF release by HBE cells treated for 24 h with (a) DEP, extracted DEP, or carbon black particles at 10 μg/cm^2 and dichloromethane extracts of DEP (OE-DEP) at 20 μg/mL, (b) DEP collected from a diesel engine with and without an oxidation catalyst and their corresponding organic extracts (OE-DEP). $*P<0.05$ compared with control value, $^{\circ}P<0.05$ compared with DEP-treated culture. (From Boland, S., Bonvallot, V., Fournier, T., Baeza-Squiban, A., Aubier, M., and Marano, F., *Am. J. Physiol. Lung Cell. Mol. Physiol.* 278, L25–L32, 2000. With permission).

whereas the aqueous extract had a slight effect (Baeza-Squiban et al. 2005). The absence of effect of the soluble fraction was also observed with PM_{10} (EHC-93) in normal HBE cells (Fujii et al. 2001). The effect of urban particles on the cytokine production of macrophages was also due to the organic fraction (Vogel et al. 2005) (Table 11.1).

The induction of chemokine release may be responsible for the inflammatory cell infiltration observed in the *in vivo* studies (Hiyoshi et al. 2005). Figure 11.4 shows the inflammatory response, which may result from leukocyte recruitment and activation. Beside the stimulation of these inflammatory cells by locally secreted cytokines, it has also been shown that PAH have direct effects on leucocytes. Pyrene increase the production of IL-4 by T lymphocytes (Bommel et al. 2000) and organic extracts of DEP increase CD1a and costimulatory molecule expression on monocyte derived dendritic cells (Koike and Kobayashi 2005), IgE production of B lymphocytes (Takenaka et al. 1995; Tsien et al. 1997), IL-4 and histamine release from basophils (Devouassoux et al. 2002) as well as mast cell (Diaz-Sanchez, Penichet-Garcia, and Saxon 2000) and eosinophil degranulation (Terada et al. 1997). This release of granulocyte mediators

TABLE 11.1
Inflammatory Effects of Organic Compounds

In vivo studies in humans		
	CB increase the number of inflammatory cells but NOT IgE	Saxon and Diaz-Sanchez (2000)
	Phenanthrene (with or without allergen) increase IgE but NOT inflammatory cells	
In vivo studies in animals		
Mice	Phenanthraquinone increase neutrophils, eosinophils, IL-5 and eotaxin	Hiyoshi et al. (2005)
	CB and organic extracts of DEP enhance ovalbumin specific IgE and IgG1	Heo, Saxon, and Hankinson (2001)
	Organic extracts of DEP increase neutrophil number in response to LPS but did NOT effect LPS induced cytokine secretion	Yanagisawa et al. (2003)
	PAH enhance IgE production in response to allergen	Kanoh et al. (1996)
In vitro studies		
Epithelial cells	Organic extracts of DEP increase histamine receptor in nasal epithelial cells as well as histamine-induced IL-8 and GM-CSF release	Terada et al. (1999)
	Organic extracts of DEP increase GM-CSF	Boland et al. (1999)
	DEP but not CB stimulate amphiregulin secretion	Blanchet et al. (2004)
	Organic extracts of DEP increase IL-8 and HO-1	Li et al. (2002)
	Pyrene increase IL-8 but NOT eotaxin expression in A549 cells	Bommel et al. (2003)
	B(a)P and organic extracts of DEP increase IL-8, GM-CSF and RANTES	Kawasaki et al. (2001)
Alveolar macrophages	Organic extracts of DEP and PM increase IL-8, TNF and COX2	Vogel et al. (2005)
	Organic extracts of DEP increase IL-8	Li et al. (2002)
	Organic extracts of DEP did NOT alter costimulatory molecules (B7) and MHC class II (CDIa) expression and antigen presentation	Koike and Kobayashi (2005)
	Organic extracts of DEP decrease PGE2 production by blocking COX-2 enzyme activity	Rudra-Ganguly et al. (2002)
	Organic extracts of DEP reduce production of IL-1and TNF-α in response to inflammatory agents	Siegel et al. (2004)
	Organic extracts of DEP increase IL-1 but not TNF-α	Yang et al. (1997)
Peripheral blood mononuclear cells (PBMC)	Organic extracts of DEP and PAH in presence of LPS increase or decrease IL-10 production depending on the order of exposure	Pacheco et al. (2001)
	Organic extracts of DEP decrease MCP-1 but increase IL-8, RANTES and chemotactic activity for neutrophils and eosinophils	Fahy et al. (1999)
	Organic extracts of DEP decrease IP10 and increase MDC production induced by allergen	Fahy et al. (2002)
	Organic extracts of DEP increase IL-8, RANRES and TNFα in PBMC of allergic persons	Fahy et al. (2000)
MonoDC	Organic extracts of DEP increase the expression of costimulatory molecules (B7) and MHC class II (CDIa) expression and enhance allergen presentation	Koike and Kobayashi (2005)
	PAH during monoDC differentiation decrease expression of CD1a, B7.1 and CD40 as well as DC function	Laupeze et al. (2002)
Lymph node cells	B(a)P induce IL-4 and IL-6 production	Fujimaki et al. (1997)
T lymphocyte	Pyrene increases IL-4 production	Bommel et al. (2000)

(continued)

Table 11.1 ***(Continued)***

B lymphocyte	Organic extracts of DEP increase IgE production	Tsien et al. (1997)
	Organic extracts of DEP increase IgE production only in the presence of IL-4 and CD40Ab	Takenaka et al. (1995)
Basophils	Organic extracts of DEP increase IL-4 and histamine release in cells from allergic and non-allergic subjects	Devouassoux et al. (2002)
	PAH increase IL-4 and histamine release only in presence of IgE	Kepley et al. (2003)
Mast cell	Organic extracts of DEP increase histamine release in the presence of IgE	Diaz-Sanchez et al. (2000)
Eosinophils	Organic extracts of DEP increase degranulation and adhesion to epithelial cells	Terada et al. (1997)

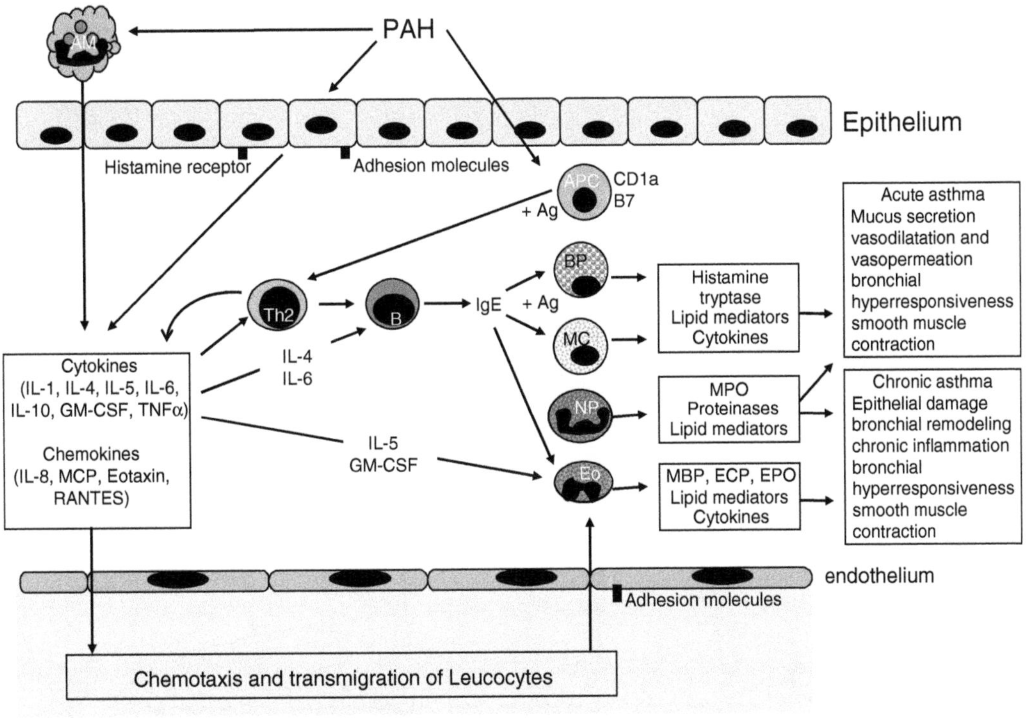

FIGURE 11.4 Scheme of the inflammatory response induced by PAH. PAH stimulate alveolar macrophages (AM) and epithelial cells to release cytokines and chemokines, inducing the recruitment of peripheral blood leukocytes on which PAH have also direct effects. The cytokines and activation of antigen presenting cells (APC: dendritic cells or macrophages/monocytes) stimulate the differentiation of T lymphocytes into a Th2 phenotype that is able to induce, in combination with cytokines, the activation and isotype switching of B lymphocytes (B), resulting in IgE production. IgE could induce the release of mediators by mast cells (MC) and basophils (BP) leading to symptoms of acute asthma. Furthermore, the infiltration of eosinophils (Eo) and their prolonged survival and activation by cytokines and IgE conduces to the release of mediators involved in the development of chronic asthma. Neutrophil (NP) degranulation may lead to both acute and chronic asthma.

could lead to symptoms of asthma, which have been shown to be aggravated after an increase in PM_{10} levels (von Klot et al. 2002).

This cytokine and chemokine expression is associated with the activation of upstream signaling cascades among which mitogen activated protein kinases (MAPKs) pathways have been shown to be activated by particles. DEP and their organic extracts increase the ERK 1/2 phosphorylation correlated to the GM-CSF release by 16HBE cell line as well as that of p38 (Bonvallot et al. 2001). p38 activation has also been implicated in the IL-8 mRNA expression induced by DEP and their benzene organic extracts in BEAS-2B cell line (Kawasaki et al. 2001) and in the release of IL-8 and RANTES induced by DEP extracts in peripheral blood mononuclear cells from allergic patients (Fahy et al. 2000). Finally, JNK phosphorylation was observed in both HBE cells and macrophages exposed to DEP extracts (Li et al. 2002). A more global overview of signaling pathways activation was obtained combining proteomic and phosphoproteins detection in HBE cells and macrophages exposed to crude or fractionated DEP extracts (Wang et al. 2005). The p38 MAPK, JNK, and ERK cascades are activated mainly by a quinone-containing polar fraction and to a lesser extend by PAH-containing aromatic fraction.

The expression of many inflammatory mediators is regulated by transcription factors among which the redox sensitive transcription factors NF-κB and AP-1. Whereas many studies have shown the activation of these transcription factors in particles treated-cells, few of them address the role of organic compounds. In HBE cells, DEP and their extracts induced NF-κB (Bonvallot et al. 2001; Kawasaki et al. 2001), but not AP-1 activation (Bonvallot et al. 2000).

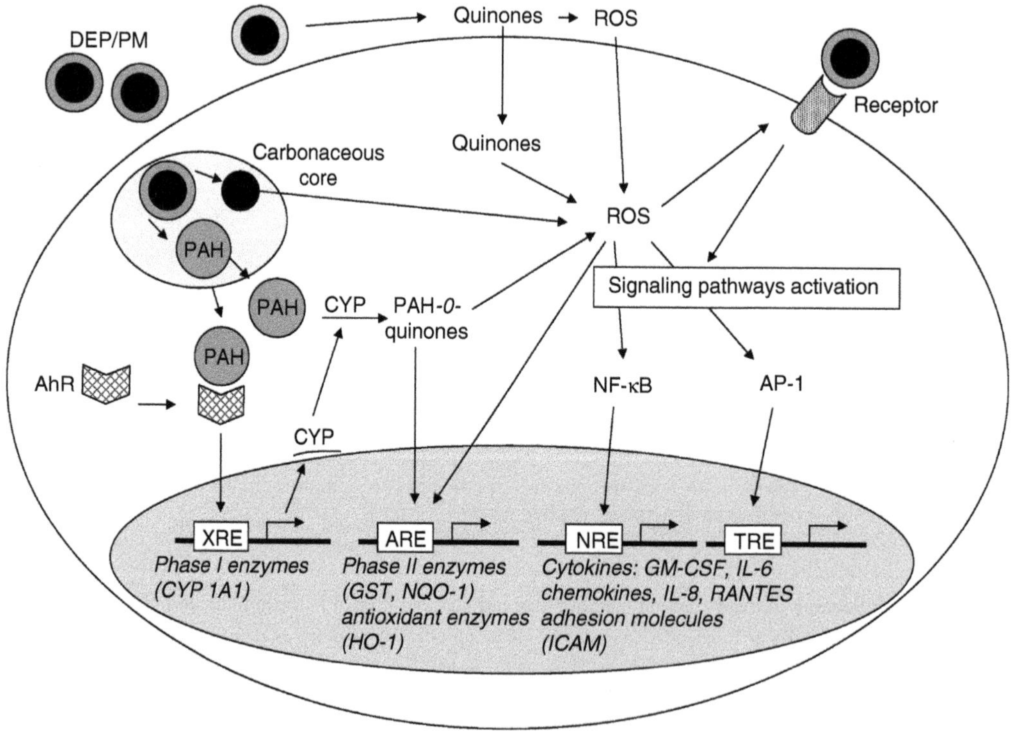

FIGURE 11.5 Scheme of the metabolic pathways activated by particles and the involvement of their organic component. DEP, diesel exhaust particles; PAH, polyaromatic hydrocarbons; ROS, reactive oxygen species; AhR, aryl hydrocarbon receptor; CYP, cytochrome P450; XRE, xenobiotic responsive element; ARE, antioxidant responsive element; NRE, NF-kB responsive element; TRE, TPA responsive element.

11.3 CONCLUSION

It is now clear that organic compounds play an essential role in the adverse effects of atmospheric particles. Most of the studies were performed on DEP, which were used as a model for PM containing xenobiotics such as PAH and quinones. The comparison of the effects induced by DEP, organic extracts of DEP, and extracted DEP revealed that organic compounds are mainly involved in the proinflammatory response. The extend of the response depends on the bioavailability of these compounds that could be extracted by broncho-alveolar fluid or after the internalization of particles by the airway epithelium and the macrophages. The metabolism of these molecules generates ROS that could be responsible for the activation of signaling pathways and redox-sensitive transcription factors such as NF-κB and AP1. Thus, cytokine genes under the control of these factors could initiate transcription. Moreover, recent data have shown that DEP and PM could upregulate the expression of growth factors such as amphiregulin, a ligand of the epidermal growth factor receptor in human HBE, and increase its secretion (Blanchet et al. 2004). Amphiregulin, which could have autocrine as well as paracrine effects, was demonstrated to be involved in cytokine secretion and would participate to a more general cellular response for sustaining the proinflammatory effects of particles. All of these results, which are summarized in Figure 11.5, provide a mechanistic explanation for the *in vivo* observations of animal or human responses to DEP or PM exposures and to epidemiological studies.

ABBREVIATIONS

AhR: Aryl hydrocarbon receptor
ARE: Antioxidant-responsive element
BCG: Bacillus Calmette–Guérin
B(a)P: Benzo(a)pyrene
CYP: Cytochrome P
DEP: Diesel exhaust particles
GST: Glutathione *S*-transferase
HO-1: Heme oxygenase
HBE: Human bronchial epithelial cells
Ig: Immunoglobulin
IL: Interleukin
MAPK: Mitogen activated protein kinase
NQO1: NADP(P)H: quinone oxido-reductase
Nrf2: NF-E2-related factor-2
NO: Nitric oxide
$PM_{2.5}$: Particulate matter
PAH: Polyaromatic hydrocarbons
ROS: Reactive oxygen species
SOF: Soluble organic fraction

REFERENCES

Baeza-Squiban, A., Baulig, A., Schins, R., Garlatti, M., Barouki, R., and Marano, F., Role of the different Paris PM2.5 components in the proinflammatory response induced in airway epithelial cells, *Am. J. Respir. Cell Mol. Biol.*, 2, A169, 2005.

Bastain, T. M., Gilliland, F. D., Li, Y. F., Saxon, A., and Diaz-Sanchez, D., Intraindividual reproducibility of nasal allergic responses to diesel exhaust particles indicates a susceptible phenotype, *Clin. Immunol.*, 109, 130–136, 2003.

Baulig, A., Garlatti, M., Bonvallot, V., Marchand, A., Barouki, R., Marano, F., and Baeza-Squiban, A., Involvement of reactive oxygen species in the metabolic pathways triggered by diesel exhaust particles in human airway epithelial cells, *Am. J. Physiol. Lung Cell Mol. Physiol.*, 285, L671–L679, 2003a.

Baulig, A., Sourdeval, M., Meyer, M., Marano, F., and Baeza-Squiban, A., Biological effects of atmospheric particles on human bronchial epithelial cells. Comparison with diesel exhaust particles, *Toxicol. In Vitro*, 17, 567–573, 2003b.

Baulig, A., Poirault, J. J., Ausset, P., Schins, R., Shi, T., Baralle, D., Dorlhene, P., et al., Physicochemical characteristics and biological activities of seasonal atmospheric particulate matter sampling in two locations of Paris, *Environ. Sci. Technol.*, 38, 5985–5992, 2004.

Blanchet, S., Ramgolam, K., Baulig, A., Marano, F., and Baeza-Squiban, A., Fine particulate matter induces amphiregulin secretion by bronchial epithelial cells, *Am. J. Respir. Cell Mol. Biol.*, 30, 421–427, 2004.

Boland, S., Baeza-Squiban, A., Fournier, T., Houcine, O., Gendron, M. C., Chevrier, M., Jouvenot, G., Coste, A., Aubier, M., and Marano, F., Diesel exhaust particles are taken up by human airway epithelial cells *in vitro* and alter cytokine production, *Am. J. Physiol.*, 276, L604–L613, 1999.

Boland, S., Bonvallot, V., Fournier, T., Baeza-Squiban, A., Aubier, M., and Marano, F., Mechanisms of GM-CSF increase by diesel exhaust particles in human airway epithelial cells, *Am. J. Physiol. Lung Cell Mol. Physiol.*, 278, L25–L32, 2000.

Bolton, J. L., Trush, M. A., Penning, T. M., Dryhurst, G., and Monks, T. J., Role of quinones in toxicology, *Chem. Res. Toxicol.*, 13, 135–160, 2000.

Bommel, H., Li-Weber, M., Serfling, E., and Duschl, A., The environmental pollutant pyrene induces the production of IL-4, *J. Allergy Clin. Immunol.*, 105, 796–802, 2000.

Bommel, H., Haake, M., Luft, P., Horejs-Hoeck, J., Hein, H., Bartels, J., Schauer, C., Poschl, U., Kracht, M., and Duschl, A., The diesel exhaust component pyrene induces expression of IL-8 but not of eotaxin, *Int. Immunopharmacol.*, 3, 1371–1379, 2003.

Bonvallot, V., Baeza-Squiban, A., Boland, S., and Marano, F., Activation of transcription factors by diesel exhaust particles in human bronchial epithelial cells *in vitro*, *Inhal. Toxicol.*, 12, 359–364, 2000.

Bonvallot, V., Baeza-Squiban, A., Baulig, A., Brulant, S., Boland, S., Muzeau, F., Barouki, R., and Marano, F., Organic compounds from diesel exhaust particles elicit a proinflammatory response in human airway epithelial cells and induce cytochrome p450 1A1 expression, *Am. J. Respir. Cell Mol. Biol.*, 25, 515–521, 2001.

Born, P. J. A., Cakmak, G., Termann, E., Weishaupt, C., van Schooten, F. J., Oberdörster, G., and Schins, R. P. F., Formation of PAH-DNA adducts after in vivo and in vitro exposure of rats and lung cells to different commercial carbon black, *Toxicol. Appl. Pharmacol.*, 205, 157–167, 2005.

Castranova, V., Ma, J. Y., Yang, H. M., Antonini, J. M., Butterworth, L., Barger, M. W., Roberts, J., and Ma, J. K., Effect of exposure to diesel exhaust particles on the susceptibility of the lung to infection, *Environ. Health Perspect.*, 109 (suppl. 4), 609–612, 2001.

Cho, H. Y., Jedlicka, A. E., Reddy, S. P., Kensler, T. W., Yamamoto, M., Zhang, L. Y., and Kleeberger, S. R., Role of NRF2 in protection against hyperoxic lung injury in mice, *Am. J. Respir. Cell Mol. Biol.*, 26, 175–182, 2002.

Cho, A., Di Stefano, E., You, Y., Rodriguez, C., Schmitz, D., Kumagai, Y., Miguel, A., et al., Determination of four quinones in diesel exhaust particles, SRM 1649a, and atmospheric PM2.5, *Aerosol Sci. Technol.*, 38, 68–81, 2004.

Cho, A. K., Sioutas, C., Miguel, A. H., Kumagai, Y., Schmitz, D. A., Singh, M., Eiguren-Fernandez, A., and Froines, J. R., Redox activity of airborne particulate matter at different sites in the Los Angeles Basin, *Environ. Res.*, 99, 40–47, 2005.

De Kok, T. M., Hogervorst, J. G., Briede, J. J., van Herwijnen, M. H., Maas, L. M., Moonen, E. J., Driece, H. A., and Kleinjans, J. C., Genotoxicity and physicochemical characteristics of traffic-related ambient particulate matter, *Environ. Mol. Mutagen.*, 46, 71–80, 2005.

Dellinger, B., Pryor, W. A., Cueto, R., Squadrito, G. L., Hegde, V., and Deutsch, W. A., Role of free radicals in the toxicity of airborne fine particulate matter, *Chem. Res. Toxicol.*, 14, 1371–1377, 2001.

Devouassoux, G., Saxon, A., Metcalfe, D. D., Prussin, C., Colomb, M. G., Brambilla, C., and Diaz-Sanchez, D., Chemical constituents of diesel exhaust particles induce IL-4 production and histamine release by human basophils, *J. Allergy Clin. Immunol.*, 109, 847–853, 2002.

Diaz-Sanchez, D., Dotson, A. R., Takenaka, H., and Saxon, A., Diesel exhaust particles induce local IgE production *in vivo* and alter the pattern of IgE messenger RNA isoforms, *J. Clin. Invest.*, 94, 1417–1425, 1994.

Diaz-Sanchez, D., The role of diesel exhaust particles and their associated polyaromatic hydrocarbons in the induction of allergic airway disease, *Allergy*, 52, 52–56, 1997 (discussion 57–58).

Diaz-Sanchez, D., Garcia, M. P., Wang, M., Jyrala, M., and Saxon, A., Nasal challenge with diesel exhaust particles can induce sensitization to a neoallergen in the human mucosa, *J. Allergy Clin. Immunol.*, 104, 1183–1188, 1999.

Diaz-Sanchez, D., Penichet-Garcia, M., and Saxon, A., Diesel exhaust particles directly induce activated mast cells to degranulate and increase histamine levels and symptom severity, *J. Allergy Clin. Immunol.*, 106, 1140–1146, 2000.

Fahy, O., Tsicopoulos, A., Hammad, H., Pestel, J., Tonnel, A. B., and Wallaert, B., Effects of diesel organic extracts on chemokine production by peripheral blood mononuclear cells, *J. Allergy Clin. Immunol.*, 103, 1115–1124, 1999.

Fahy, O., Hammad, H., Senechal, S., Pestel, J., Tonnel, A. B., Wallaert, B., and Tsicopoulos, A., Synergistic effect of diesel organic extracts and allergen Der p 1 on the release of chemokines by peripheral blood mononuclear cells from allergic subjects: involvement of the map kinase pathway, *Am. J. Respir. Cell Mol. Biol.*, 23, 247–254, 2000.

Fahy, O., Senechal, S., Pene, J., Scherpereel, A., Lassalle, P., Tonnel, A. B., Yssel, H., Wallaert, B., and Tsicopoulos, A., Diesel exposure favors Th2 cell recruitment by mononuclear cells and alveolar macrophages from allergic patients by differentially regulating macrophage-derived chemokine and IFN-gamma-induced protein-10 production, *J. Immunol.*, 168, 5912–5919, 2002.

Fujimaki, H., Shiraishi, F., Aoki, Y., and Saneyoshi, K., Modulated cytokine production from cervical lymph node cells treated with B[a]P and PCB, *Chemosphere*, 34, 1487–1493, 1997.

Fujii, T., Hayashi, S., Hogg, J. C., Vincent, R., and Van Eeden, S. F., Particulate matter induces cytokine expression in human bronchial epithelial cells, *Am. J. Respir. Cell Mol. Biol.*, 25, 265–271, 2001.

Gerde, P., Muggenburg, B. A., Lundborg, M., and Dahl, A. R., The rapid alveolar absorption of diesel soot-adsorbed benzo[a]pyrene: bioavailability, metabolism and dosimetry of an inhaled particle-borne carcinogen, *Carcinogenesis*, 22, 741–749, 2001.

Gerde, P., Muggenburg, B. A., Lundborg, M., Tesfaigzi, Y., and Dahl, A. R., Respiratory epithelial penetration and clearance of particle-borne benzo[a]pyrene, *Res. Rep. Health Eff. Inst.*, 5–25, 2001b (discussion 27–32).

Gilliland, F. D., Li, Y. F., Saxon, A., and Diaz-Sanchez, D., Effect of glutathione-*S*-transferase M1 and P1 genotypes on xenobiotic enhancement of allergic responses: randomized, placebo-controlled crossover study, *Lancet*, 363, 119–125, 2004.

Granum, B. and Lovik, M., The effect of particles on allergic immune responses, *Toxicol. Sci.*, 65, 7–17, 2002.

Hao, M., Comier, S., Wang, M., Lee, J. J., and Nel, A., Diesel exhaust particles exert acute effects on airway inflammation and function in murine allergen provocation models, *J. Allergy Clin. Immunol.*, 112, 905–914, 2003.

Heo, Y., Saxon, A., and Hankinson, O., Effect of diesel exhaust particles and their components on the allergen-specific IgE and IgG1 response in mice, *Toxicology*, 159, 143–158, 2001.

Hiura, T. S., Kaszubowski, M. P., Li, N., and Nel, A. E., Chemicals in diesel exhaust particles generate reactive oxygen radicals and induce apoptosis in macrophages, *J. Immunol.*, 163, 5582–5591, 1999.

Hiyoshi, K., Takano, H., Inoue, K., Ichinose, T., Yanagisawa, R., Tomura, S., Cho, A. K., Froines, J. R., and Kumagai, Y., Effects of a single intratracheal administration of phenanthraquinone on murine lung, *J. Appl. Toxicol.*, 25, 47–51, 2005.

Itoh, K., Chiba, T., Takahashi, S., Ishii, T., Igarashi, K., Katoh, Y., Oyake, T. et al., An Nrf2/small Maf heterodimer mediates the induction of phase II detoxifying enzyme genes through antioxidant response elements, *Biochem. Biophys. Res. Commun.*, 236, 313–322, 1997.

Kanoh, T., Suzuki, T., Ishimori, M., Ikeda, S., Ohasawa, M., Ohkuni, H., and Tunetoshi, Y., Adjuvant activities of pyrene, anthracene, fluoranthene, and benzo(a)pyrene in production of anti-IgE antibody to Japanese cedar pollen allergen in mice, *J. Clin. Lab. Immunol.*, 48, 133–147, 1996.

Kawasaki, S., Takizawa, H., Takami, K., Desaki, M., Okazaki, H., Kasama, T., Kobayashi, K. et al., Benzene-extracted components are important for the major activity of diesel exhaust particles: effect on interleukin-8 gene expression in human bronchial epithelial cells, *Am. J. Respir. Cell Mol. Biol.*, 24, 419–426, 2001.

Keane, M. J., Xing, S. G., Harrison, J. C., Ong, T., and Wallace, W. E., Genotoxicity of diesel-exhaust particles dispersed in simulated pulmonary surfactant, *Mutat. Res.*, 260, 233–238, 1991.

Kepley, C. L., Lauer, F. T., Oliver, J. M., and Burchiel, S. W., Environmental polycyclic aromatic hydrocarbons, benzo(a) pyrene (BaP), and BaP–quinones, enhance IgE-mediated histamine release and IL-4 production in human basophils, *Clin. Immunol.*, 107, 10–19, 2003.

Koike, E. and Kobayashi, T., Organic extract of diesel exhaust particles stimulates expression of la and costimulatory molecules associated with antigen presentation in rat peripheral blood monocytes but not in alveolar macrophages, *Toxicol. Appl. Pharmacol.*, 209, 277–285, 2005.

Koike, E., Hirano, S., Shimojo, N., and Kobayashi, T., cDNA microarray analysis of gene expression in rat alveolar macrophages in response to organic extract of diesel exhaust particles, *Toxicol. Sci.*, 67, 241–246, 2002.

Koike, E., Hirano, S., Furuyama, A., and Kobayashi, T., cDNA microarray analysis of rat alveolar epithelial cells following exposure to organic extract of diesel exhaust particles, *Toxicol. Appl. Pharmacol.*, 201, 178–185, 2004.

Kumagai, Y., Arimoto, T., Shinyashiki, M., Shimojo, N., Nakai, Y., Yoshikawa, T., and Sagai, M., Generation of reactive oxygen species during interaction of diesel exhaust particle components with NADPH-cytochrome P450 reductase and involvement of the bioactivation in the DNA damage, *Free Radic. Biol. Med.*, 22, 479–487, 1997.

Laupeze, B., Amiot, L., Sparfel, L., Le Ferrec, E., Fauchet, R., and Fardel, O., Polycyclic aromatic hydrocarbons affect functional differentiation and maturation of human monocyte-derived dendritic cells, *J. Immunol.*, 168, 2652–2658, 2002.

Li, N., Venkatesan, M. I., Miguel, A., Kaplan, R., Gujuluva, C., Alam, J., and Nel, A., Induction of heme oxygenase-1 expression in macrophages by diesel exhaust particle chemicals and quinones via the antioxidant-responsive element, *J. Immunol.*, 165, 3393–3401, 2000.

Li, N., Wang, M., Oberley, T. D., Sempf, J. M., and Nel, A. E., Comparison of the pro-oxidative and proinflammatory effects of organic diesel exhaust particle chemicals in bronchial epithelial cells and macrophages, *J. Immunol.*, 169, 4531–4541, 2002.

Li, N., Sioutas, C., Cho, A., Schmitz, D., Misra, C., Sempf, J., Wang, M., Oberley, T., Froines, J., and Nel, A., Ultrafine particulate pollutants induce oxidative stress and mitochondrial damage, *Environ. Health Perspect.*, 111, 455–460, 2003.

Lim, H. B., Ichinose, T., Miyabara, Y., Takano, H., Kumagai, Y., Shimojyo, N., Devalia, J. L., and Sagai, M., Involvement of superoxide and nitric oxide on airway inflammation and hyperresponsiveness induced by diesel exhaust particles in mice, *Free Radic. Biol. Med.*, 25, 635–644, 1998.

Ma, J. Y. and Ma, J. K., The dual effect of the particulate and organic components of diesel exhaust particles on the alteration of pulmonary immune/inflammatory responses and metabolic enzymes, *J. Environ. Sci. Health C Environ. Carcinog. Ecotoxicol. Rev.*, 20, 117–147, 2002.

Meek, M. D., Ah receptor and estrogen receptor-dependent modulation of gene expression by extracts of diesel exhaust particles, *Environ. Res.*, 79, 114–121, 1998.

Pacheco, K. A., Tarkowski, M., Sterritt, C., Negri, J., Rosenwasser, L. J., and Borish, L., The influence of diesel exhaust particles on mononuclear phagocytic cell-derived cytokines: IL-10, TGF-beta and IL-1 beta, *Clin. Exp. Immunol.*, 126, 374–383, 2001.

Penn, A., Murphy, G., Barker, S., Henk, W., and Penn, L., Combustion-derived ultrafine particles transport organic toxicants to target respiratory cells, *Environ. Health Perspect.*, 113, 956–963, 2005.

Penning, T. M., Burczynski, M. E., Hung, C. F., McCoull, K. D., Palackal, N. T., and Tsuruda, L. S., Dihydrodiol dehydrogenases and polycyclic aromatic hydrocarbon activation: generation of reactive and redox active *o*-quinones, *Chem. Res. Toxicol.*, 12, 1–18, 1999.

Perret, A. and Pompon, D., Electron shuttle between membrane-bound cytochrome P450 3A4 and b5 rules uncoupling mechanisms, *Biochemistry*, 37, 11412–11424, 1998.

Ravindra, Mittal, A. K., and Van Grieken, R., Health risk assessment of urban suspended particulate matter with special reference to polycyclic aromatic hydrocarbons: a review, *Rev. Environ. Health*, 16, 169–189, 2001.

Rehwagen, M., Muller, A., Massolo, L., Herbarth, O., and Ronco, A., Polycyclic aromatic hydrocarbons associated with particles in ambient air from urban and industrial areas, *Sci. Total Environ.*, 348, 199–210, 2005.

Riedl, M. and Diaz-Sanchez, D., Biology of diesel exhaust effects on respiratory function, *J. Allergy Clin. Immunol.*, 115, 221–228, 2005 (quiz 229).

Rudra-Ganguly, N., Reddy, S. T., Korge, P., and Herschman, H. R., Diesel exhaust particle extracts and associated polycyclic aromatic hydrocarbons inhibit Cox-2-dependent prostaglandin synthesis in murine macrophages and fibroblasts, *J. Biol. Chem.*, 277, 39259–39265, 2002.

Sagai, M., Saito, H., Ichinose, T., Kodama, M., and Mori, Y., Biological effects of diesel exhaust particles. I. *In vitro* production of superoxide and *in vivo* toxicity in mouse, *Free Radic. Biol. Med.*, 14, 37–47, 1993.

Sato, H., Sone, H., Sagai, M., Suzuki, K. T., and Aoki, Y., Increase in mutation frequency in lung of Big Blue rat by exposure to diesel exhaust, *Carcinogenesis*, 21, 653–661, 2000.

Saxena, Q. B., Saxena, R. K., Siegel, P. D., and Lewis, D. M., Identification of organic fractions of diesel exhaust particulate (DEP) which inhibit nitric oxide (NO) production from a murine macrophage cell line, *Toxicol. Lett.*, 143, 317–322, 2003a.

Saxena, R. K., Saxena, Q. B., Weissman, D. N., Simpson, J. P., Bledsoe, T. A., and Lewis, D. M., Effect of diesel exhaust particulate on Bacillus Calmette–Guerin lung infection in mice and attendant changes in lung interstitial lymphoid subpopulations and IFNgamma response, *Toxicol. Sci.*, 73, 66–71, 2003.

Saxon, A. and Diaz-Sanchez, D., Diesel exhaust as a model xenobiotic in allergic inflammation, *Immunopharmacology*, 48, 325–327, 2000.

Siegel, P. D., Saxena, R. K., Saxena, Q. B., Ma, J. K., Ma, J. Y., Yin, X. J., Castranova, V., Al-Humadi, N., and Lewis, D. M., Effect of diesel exhaust particulate (DEP) on immune responses: contributions of particulate versus organic soluble components, *J. Toxicol. Environ. Health A*, 67, 221–231, 2004.

Squadrito, G. L., Cueto, R., Dellinger, B., and Pryor, W. A., Quinoid redox cycling as a mechanism for sustained free radical generation by inhaled airborne particulate matter, *Free Radic. Biol. Med.*, 31, 1132–1138, 2001.

Takenaka, H., Zhang, K., Diaz-Sanchez, D., Tsien, A., and Saxon, A., Enhanced human IgE production results from exposure to the aromatic hydrocarbons from diesel exhaust: direct effects on B-cell IgE production, *J. Allergy Clin. Immunol.*, 95, 103–115, 1995.

Terada, N., Maesako, K., Hiruma, K., Hamano, N., Houki, G., Konno, A., Ikeda, T., and Sai, M., Diesel exhaust particulates enhance eosinophil adhesion to nasal epithelial cells and cause degranulation, *Int. Arch. Allergy Immunol.*, 114, 167–174, 1997.

Terada, N., Hamano, N., Maesako, K. I., Hiruma, K., Hohki, G., Suzuki, K., Ishikawa, K., and Konno, A., Diesel exhaust particulates upregulate histamine receptor mRNA and increase histamine-induced IL-8 and GM-CSF production in nasal epithelial cells and endothelial cells, *Clin. Exp. Allergy*, 29, 52–59, 1999.

Tsien, A., Diaz-Sanchez, D., Ma, J., and Saxon, A., The organic component of diesel exhaust particles and phenanthrene, a major polyaromatic hydrocarbon constituent, enhances IgE production by IgE-secreting EBV-transformed human B cells in vitro, *Toxicol. Appl. Pharmacol.*, 142, 256–263, 1997.

Vogel, C. F., Sciullo, E., Wong, P., Kuzmicky, P., Kado, N., and Matsumura, F., Induction of proinflammatory cytokines and C-reactive protein in human macrophage cell line U937 exposed to air pollution particulates, *Environ. Health Perspect.*, 113, 1536–1541, 2005.

von Klot, S., Wolke, G., Tuch, T., Heinrich, J., Dockery, D. W., Schwartz, J., Kreyling, W. G., Wichmann, H. E., and Peters, A., Increased asthma medication use in association with ambient fine and ultrafine particles, *Eur. Respir. J.*, 20, 691–702, 2002.

Wang, M., Xiao, G. G., Li, N., Xie, Y., Loo, J. A., and Nel, A. E., Use of a fluorescent phosphoprotein dye to characterize oxidative stress-induced signaling pathway components in macrophage and epithelial cultures exposed to diesel exhaust particle chemicals, *Electrophoresis*, 26, 2092–2108, 2005.

Xiao, G. G., Wang, M., Li, N., Loo, J. A., and Nel, A. E., Use of proteomics to demonstrate a hierarchical oxidative stress response to diesel exhaust particle chemicals in a macrophage cell line, *J. Biol. Chem.*, 278, 50781–50790, 2003.

Yanagisawa, R., Takano, H., Inoue, K., Ichinose, T., Sadakane, K., Yoshino, S., Yamaki, K., et al., Enhancement of acute lung injury related to bacterial endotoxin by components of diesel exhaust particles, *Thorax*, 58, 605–612, 2003.

12 The Asbestos Model of Interstitial Pulmonary Fibrosis: TNF-α and TGF-β_1 as Mediators of Asbestos-Induced Lung Fibrogenesis

Deborah E. Sullivan and Arnold R. Brody
Tulane University Health Sciences Center, Tulane University

CONTENTS

12.1 INTRODUCTION

Inhalation of asbestos fibers has been known for more than a century as a potent inducer of lung cancer, mesothelioma, and scar tissue formation in the lung (Brody 1997). This latter disease, known as pulmonary "asbestosis," is a form of interstitial pulmonary fibrosis (IPF), and thus has all of the hallmarks of this disease; i.e., increased extra-cellular matrix production, varying degrees and types of inflammation, and proliferation of a variety of cells as the disease develops (Brody 1997; Mossman and Churg 1998). There currently are no effective treatments for any of the forms of IPF (Selman et al. 2004), despite the fact that millions of individuals worldwide are afflicted. The incidence of asbestosis has decreased dramatically over the past few decades as fewer individuals have been exposed to the heavy concentrations of fibers required to produce clinical disease. This reduced exposure has resulted from the banning of most commercial uses of asbestos in developed countries around the world, although developing countries still use tons of fibers annually, and the risk of contracting asbestosis and certain cancers remains high wherever asbestos remains in use.

IPF has multiple etiologies, including immune mechanisms, infectious agents, and toxic gases and particles (Morris and Brody 1998). The disease can also remain idiopathic in a large proportion of cases (Katzenstein and Myers 1998). Since so many individuals are affected and there are few reasonable therapeutic approaches, investigators have been using animal models of IPF for many years in attempt to understand the fundamental mechanisms of the disease process (Gharaee-Kermani, Ullenbruch, and Phan 2005). While every animal model has some limitations in our abilities to draw correlations between the model and the human disease, investigators have developed a number of highly useful model systems. The best known is the IPF caused by treatment with bleomycin, a potent anti-cancer drug (Chua, Gauldie, and Laurent 2005). Others have used such agents as ozone (Last et al. 1993), oxygen (Haschek et al. 1981), viruses (Sime et al. 1997a; Kolb et al. 2001), silica, and other inorganic particles (Castranova, Robinson, and Frazer 1996; Morris and Brody 1998). We and others have used inhalation of asbestos fibers as a useful model of IPF and have learned a great deal about the cellular and molecular mechanisms that mediate cell proliferation and scar tissue formation (Chang et al. 1988; Mossman et al. 1990; Brody 1997; Mossman and Churg 1998). The model system used and some of our more recent findings are described in the following review.

12.2 THE ASBESTOS-INDUCED MODEL OF IPF

Over the past years, we have used well-established rat and mouse models of IPF induced by the inhalation of fibrogenic asbestos fibers (Chang et al. 1988; McGavran and Brody 1989a; Brody 1992; Perdue and Brody 1994; Coin et al. 1996; Li et al. 2004). The models have been useful for determining the temporal and cellular events that culminate in fibrotic lesions in the lung interstitium. We have been using brief exposures (1–5 h) to fibers in rats and mice since asbestos serves as a paradigm for inducing the initial fibrogenic lesions that lead to diffuse IPF upon continued exposure (Rom, Travis, and Brody 1991). Our work has focused on the earliest temporal and anatomic events of cellular proliferation (Brody and Overby 1989; McGavran, Moore, and Brody 1990) and matrix production (Chang et al. 1988). A single one-hour exposure to chrysotile asbestos (at a concentration of ~1000 fibers/cc) induces highly significant increases in proliferation of mesenchymal (Brody and Overby 1989), epithelial (Brody and Overby 1989; Liu et al. 1996), and vascular cells (McGavran, Moore, and Brody 1990) within the first 24 h post-exposure. Some proliferative events begin as early as 12 h after the single exposure (Brody and Overby 1989). Proliferation of the various cell types results in a fibrotic lesion at the sites of initial fiber deposition and lung injury (Figure 12.1) (Chang et al. 1988; Brody and Overby 1989; McGavran, Butterick, and Brody 1989b; Brody 1992). This anatomic region, termed the bronchiolar-alveolar duct (BAD) junction, is well known to investigators interested in environmental lung disease (Crapo et al. 1989). This is so because the BAD junctions commonly are sites of early disease development after exposure to inhaled toxic agents (Crapo et al. 1989). After one hour of asbestos exposure, cell proliferation at these sites wanes by 96 h post-exposure (Brody and Overby 1989), leaving a fibrogenic lesion (Figure 12.2a and Figure 12.2b) that is prominent at least one month later (Chang et al. 1988). After three consecutive days of exposure (three hours per day), the proliferative events remain increased for at least two weeks, and the fibrotic lesion can be measured for six months post-exposure (Coin et al. 1996). There is no acute inflammation, and the fiber movement in our model post-exposure follows normal clearance curves as reported for a number of species, including man (Coin, Roggli, and Brody 1992). These two findings show that there is no "overload" of lung defense mechanisms (Coin et al. 1996). The longest duration of exposure using this model system in our laboratory was one day per week for eight weeks (Li et al. 2004). One year after cessation of exposure, the regimen resulted in diffuse IPF and several adenomas (Li et al. 2004). The diffuse fibrogenesis extended from the duct junctions, as predicted from our earlier findings using brief exposures (Li et al. 2004). An interesting feature of these developing fibrogenic lesions in both the short and long-term exposures is their similarity to the so-called "fibrotic foci" recognized as "hot spots" of fibrogenic activity in the lungs of individuals with progressive IPF (Selman, King, and Pardo 2001). The foci in humans and in this asbestos model exhibit little or no inflammation other than macrophage populations, clear epithelial injury and repair, and rapid elaboration of an extra-cellular matrix. The questions that must be answered relate to the mechanisms that cause

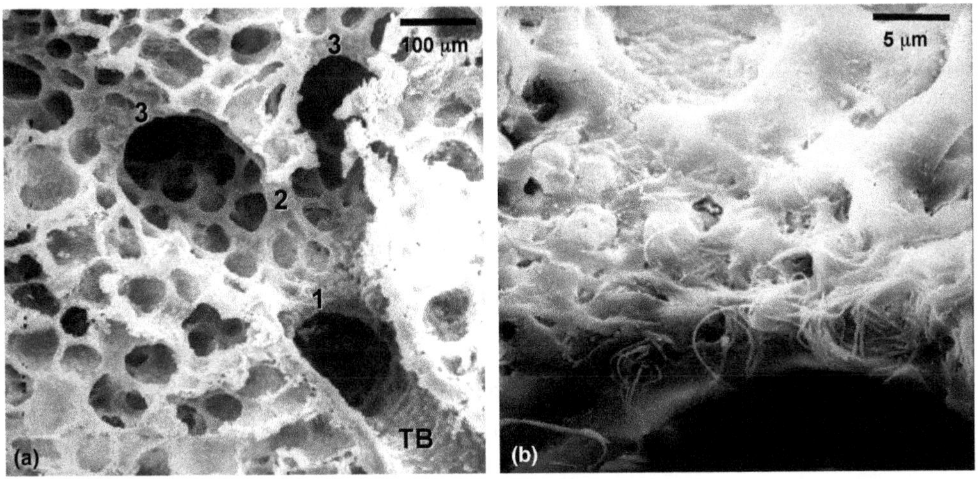

FIGURE 12.1 Scanning electron micrograph (SEM) of the left lung of a rat immediately after exposure to chrysotile asbestos for one hour. (a) A terminal bronchiole (Br) and three levels (1–3) of alveolar duct bifurcations are evident. (b) High magnification of the first bifurcation. Numerous thin asbestos fibers are lying upon the epithelial surface where they initially deposit. Reprinted from Brody, A. R. and Roe, M. N., *Am. Rev. Respir. Dis.*, 128, 724–729, 1983.

the initial epithelial injury, the mediators that control epithelial and mesenchymal cell proliferation and matrix production. The genes that code for TNF-α and TGF-β_1 and the mechanisms that control the signal transduction cascades controlling expression of the genes and protein synthesis are the focus of our ongoing work. Some of these studies are summarized below.

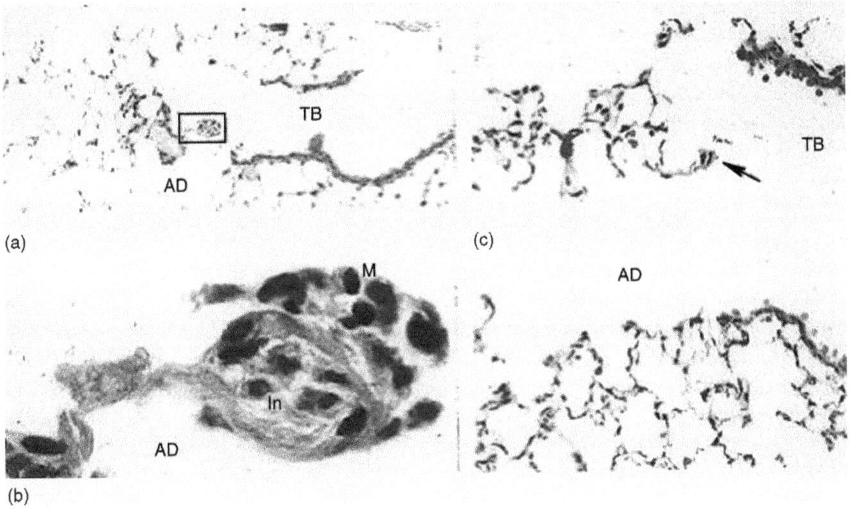

FIGURE 12.2 **(See color insert)** Histopathology. (a) Lung of a mouse (Bl6/129) 48 h after exposure to chrysotile asbestos. The lesion (box) located at the alveolar duct (AD) bifurcation near the ends of the terminal bronchioles (TB) is hypercellular and hypertrophic, with increased extracellular matrix, macrophages (M) and interstitial cells (IN). (b) Higher magnification of lesion (X40). (c) TNF-αRKO mice fail to develop fibroproliferative lesions in response to asbestos. This knockout animal exhibits a normal duct bifurcation (arrow) 48 h after asbestos exposure. Reprinted from Liu, J. -Y., Brass, D. M., Hoyle, G. W., and Brody, A. R., TNF-α receptor knockout mice are ptotected from the fibroproliferative effects of inhaled asbestos fibers, *Am. J. Pathol. Dis.*, 153, 1839–1847, 1998.

12.3 TNF-α AND TGF-β IN ASBESTOSIS

A number of peptide growth factors have been identified in the lungs of individuals with IPF or asbestosis and in animal models of the disease (Khalil et al. 1991b; Zhang et al. 1993; Brody et al. 1997; Zhang et al. 1999). Potent factors such as platelet-derived growth factor, insulin-like growth factor, transforming growth factor alpha and epidermal growth factor have all been demonstrated (Morris and Brody 1998). Two peptide growth factors, however, appear to stand out above the rest in their potential to mediate fibrogenesis. These are tumor necrosis factor alpha (TNF-α) (Table 12.1) and transforming growth factor beta1 (TGF-β_1) (Table 12.2). The reasons for this line of thinking are several: (1) Mice with both the 55–75 kDa receptors for TNF-α knocked out fail to develop significant bleomycin-induced fibrosis, silicosis, or asbestosis (Figure 12.2c) (Liu et al. 1998; Ortiz et al. 1998a). (2) Inbred mice (129 strain) that are genetically-resistant to the fibrogenic effects of asbestos exhibit reduced expression of TGF-β_1 subsequent to exposure (Brass et al. 1999) and less interstitial disease consequent to treatment with TGF-β_1 (Warshamana et al. 2002b). (3) TNF-α up-regulates the expression of TGF-β_1 *in vitro* (Warshamana, Corti, and Brody 2001; Sullivan et al. 2005) and *in vivo* (Sime et al. 1998) (the signal transduction mechanism mediating this phenomenon is detailed below).

The first evidence of a role for TGF-β_1 came from human lungs with IPF. Investigators showed immunohistochemical proof of TGF-β_1 throughout fibrotic regions of lung (Broekelmann et al. 1991). Khalil followed this concept with a series of studies on human lung tissues and macrophages (Khalil et al. 1989; Khalil et al. 1991b). In 1994, we showed the first immunohistochemical

TABLE 12.1
Evidence Suggesting TNF-α Plays a Significant Role in Development of IPF

Patients	
Increased expression of TNF-α in alveolar macrophages and proliferating type II pneumocytes of patients with IPF	Zhang et al. (1993) and Pan et al. (1996)
Anti-TNF-α agents have shown promise in treatment of IPF	Selman et al. (2004)
TNF-α promoter polymorphisms correlate with increased TNF-α expression and increased risk of developing IPF	Whyte et al. 2000 and Yucesoy et al. (2001)
Animal Models	
TNF-α transgenic mice develop pulmonary pathology similar to IPF	Miyazaki et al. (1995)
Overexpression of TNF-α in the lung results in diffuse fibrosis and increased expression of TGF-β1	Sime et al. (1998)
Agents that block TNF-α biological activity protect mice from fibrogenic effects of bleomycin or silica	Piguet et al. (1989), Piguet and Vesin (1994), and Phan and Kunkel (1992)
TNF-RKO mice are resistant to fibrogenic effects of asbestos, silica and bleomycin	Liu and Brody (2001) and Ortiz et al. (1999)
In vitro	
TNF-α induces expression of collagen and fibronectin in human fibroblasts	Paulsson et al. (1989)
TNF-α induces expression of TGF-β1 in:	
Lung fibroblasts	Warshamana, Corti, and Brody (2001) and Sullivan et al. (2005)
Type II pneumocytes	Warshamana, Corti, and Brody (2001)
Micro-glial cells	Chao et al. (1995)
Mature adipocytes	Samad et al. (1999)
Human proximal tubular cells	Phillips et al. (1996)
Rat pulmonary artery endothelial cells	Phan et al. (1992)

TABLE 12.2
Evidence Suggesting TGF-β_1 Plays a Significant Role in Development of IPF

Patients	
TGF-β_1 is up-regulated in bronchiolar and hyperplastic type II epithelial cells in lungs of patients with IPF	Kahlil et al. (1996)
Levels of TGF-β_1 in lung correlate with severity of IPF	Bartram and Speer (2004)
Animal Models	
Fibrogenic agents induce expression of TGF-β_1	Perdue and Brody (1994) and Santana et al. (1995)
Increased expression of TGF-β_1 correlates with elevated levels of ECM	Chang et al. (1988) and Coin et al. (1996)
Over expression of TGF-β_1 in the lung caused histological changes consistent with pulmonary fibrosis	Sime et al. (1997b) and Warshamana et al. (2002a & 2002b)
Agents that block TGF-β_1 biological activity attenuate pulmonary fibrosis after bleomycin	Giri, Hyde, and Hollinger (1993), Wang et al. (2002), and Nakao et al. (1999)
In vitro	
TGF-β_1 is a potent inducer of matrix proteins such as collagen, fibronectin and glycoproteins	Bertram and Speer (2004)
TGF-β_1 inhibits the expression of matrix metalloproteinases and increases synthesis of tissue inhibitor of metalloproteinase	Lasky and Brody (2000) (for review)
TGF-β_1 increases expression of CTGF as well as TGF-β_1 itself	Grotendorst (1997) and Van Obberghen-Schilling et al. (1988)

evidence of increased TGF-β_1 protein in the lungs of asbestos-exposed rats (Figure 12.3a and Figure 12.3b) (Perdue and Brody 1994). This was accompanied by increases in fibronectin and smooth muscle cells (Perdue and Brody 1994; Liu and Brody 2001). Increased protein should be preceded by expression of the gene that codes for TGF-β_1. If the peptide is playing any role in the disease process, gene expression and the consequent protein production should be exhibited most strongly at the sites of fiber deposition and ongoing fibrogenesis. This turned out to be the case inasmuch as studies employing *in situ* hybridization for the gene encoding TGF-β_1 showed a clear distribution of expression at the bronchiolar and alveolar duct bifurcations where fibers are initially deposited and the disease process is initiated and progresses (Figure 12.3c and Figure 12.3d) (Liu and Brody 2001). Since TGF-β_1 is, molecule for molecule, the most potent factor in the induction of extra-cellular matrix components, TGF-β_1 remains high on the list as a central mediator of interstitial fibrogenesis. This is consistent with the anatomic findings in human lungs, and with the temporal and anatomic distribution of the peptide in our asbestosis model. The question then remained as to whether TGF-β_1 was sufficient to induce fibrosis, or if other factors were required.

TNF-α had been shown to increase the expression of TGF-β_1 *in vitro* (Phan et al. 1992; Chao et al. 1995; Phillips et al. 1996; Samad et al. 1999; Warshamana, Corti, and Brody 2001). Elegant studies using a recombinant (AdV) to transduce the expression of TNF-α and TGF-β_1 *in vivo* were developed in rats (Sime et al. 1997b; Sime et al. 1998) and then mice (Warshamana et al. 2002a). TGF-β_1 over-expression caused severe inflammation and fibrosis as expected (Sime et al. 1997b; Warshamana et al. 2002a). TNF-α over-expression produced similar changes but also increased the expression of TGF-β_1 (Sime et al. 1998). Thus, we asked several questions that we hoped would shed light on the roles of TNF-α and TGF-β_1 in mediating control of cell growth and fibrogenesis: (1) Is TGF-β_1 expressed in the lungs of asbestos-exposed rats and mice and does expression correlate with disease development? YES- as described above (Liu et al. 1996; Liu and Brody 2001). (2) Is TNF-α expressed in the lungs of rats and mice and does it correlate with disease? YES (Ortiz et al. 1998a; Ortiz et al. 1998b; Brass et al. 1999). (3) Does TGF-β_1 over-expression induce

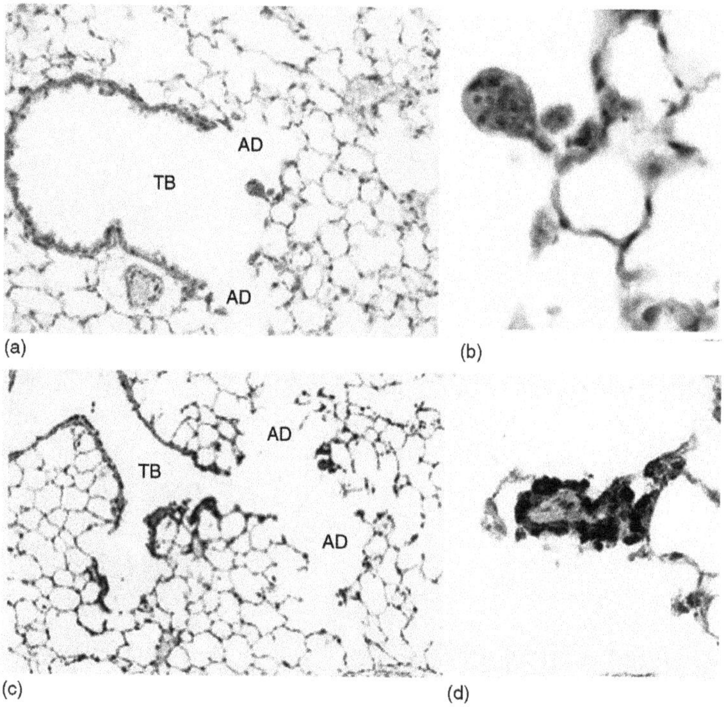

FIGURE 12.3 **(See color insert)** TGF-β1 expression in lungs of mice 48 h after exposure to chrysotile asbestos. (a) Immunohistochemistry using an antibody to TGF-β1 latent associated peptide demonstrates expression of TGF-β1 protein in epithelial cells and macrophages. (b) Higher magnification from 3A. Developing lesion exhibits robust staining of epithelium, macrophages and interstitial cells (X40). Photograph (a & b) courtesy of Dr. Derek Pociask, Tulane University Health Science Center. (c) TGF-β1 in situ hybridization (ISH) of a typical bronchiolar alveolar duct region. Numerous epithelial cells and macrophages express TGF-β1 mRNA. (d) Developing lesion with epithelium (arrow) and macrophages (arrow heads) positive for TGF-β1 ISH (X40).

fibrogenic lesions in the fibrogenic-resistant TNF-α-receptor knockout mice (TNF-αRKO)? YES (Liu et al. 2001). This latter set of experiments was interesting because, *a priori*, there was no way to know how the lungs of these knockout mice would respond to the potent fibrogenic growth factor. It turned out that TGF-β_1 over-expressed by the recombinant AdV caused diffuse IPF (Liu et al. 2001; Warshamana et al. 2002a, 2002b). Based on our understanding that TNF-α up-regulates TGF-β_1, it appears that in a natural setting where TNF-α can be rapidly expressed following injury (such as in the asbestos model), TGF-β_1 is consequently increased by TNF-α and fibrogenesis results. In an experimental setting where the TNF-α receptors are missing, TGF-β_1 alone is sufficient to induce fibrogenesis, thus shedding some light on its role. This does not mean that TGF-β_1 is, in fact, alone. TGF-β_1 itself induces the expression of other factors, including chemokines and cytokines (and TGF-β_1 itself), as evidenced by the severe inflammatory cell infiltrates that are present in the model systems. Further studies will be required to sort out the roles of each of these factors in this complex disease process.

12.3.1 Molecular and Cell Biology of TNF-α and TGF-β_1 in Disease Processes

TNF-α is a pleiotropic agent produced mostly by activated macrophages and monocytes, but also by many other cell types including fibroblasts and epithelial cells. Exposure to bacteria or parasitic proteins can induce production of high levels of TNF-α. However, a wide variety of other agents, such as viruses, tumor cells, complement, cytokines (interleukin-1β, IL-2, interferon-γ, granulocyte-macrophage colony-stimulating factor, and TNF-α itself), ischemia, trauma, irradiation, phorbol esters (Schottelius et al. 2004), and asbestos (Xie et al. 2000) can stimulate TNF-α gene expression.

TNF-α is synthesized as a 26-kDa proprotein that is displayed on the plasma membrane (Schottelius et al. 2004). The soluble 17 kDa mature monomer form of TNF-α is released by proteolytic cleavage of the proprotein by matrix metalloproteases. TACE (TNF-α-converting enzyme; an Adamalysin also known as ADAM 17) is the enzyme primarily responsible for processing membrane-associated to secreted TNF-α. Upon cleavage, the 17 kDa secreted protein oligomerizes to form the active homotrimer. Both soluble and membrane-bound forms of TNF-α are biologically active, however they have different affinities for cellular TNF-α receptors.

Biological responses to TNF-α are mediated by two structurally distinct receptors, TNF-αR1 (p55) and TNF-αR2 (p75) (Bazzoni and Beutler 1996). Both receptors are transmembrane glycoproteins and are present on all cell types except erythrocytes. TNF-R1 expression is constitutive in most cell types, whereas expression of TNF-R2 appears to be inducible. The extracellular domains of both receptors are subject to proteolytic cleavage and can be shed from the surface of cells. The resulting soluble receptors are capable of binding TNF-α and rendering it biologically inactive, and therefore they may act as natural inhibitors of TNF-α bioactivity. TNF-α signaling through either TNF-R1 or TNF-R2 requires dimerization or trimerization of the intracellular domains of the TNF receptors. Following recruitment of TNF receptor-associated factors (TRAFs) and adaptors to the ligand-activated receptor, numerous signaling pathways can be initiated simultaneously resulting in regulation of a large spectrum of cellular genes by both transcriptional and post-transcriptional mechanisms. Intracellular conditions determine which pathways dominate.

TNF-α was first characterized as a factor that produced tumor necrosis *in vivo* and exhibited antitumor activity by inducing cellular apoptosis *in vitro*. It has subsequently been recognized that TNF-α modulates growth, differentiation, and metabolism of a variety of cell types (Aggarwal 2003). Besides these effects, TNF-α is a well-known inducer of the inflammatory response and a regulator of immunity. In the lung, TNF-α increases fibroblast proliferation, differentiation, and extracellular matrix deposition, and promotes induction of matrix metalloproteinases that enhance basement membrane disruption and can facilitate fibroblast migration (Allen and Spiteri 2002). Accumulating evidence suggests that TNF-α expression in the lung plays a key role in the development of pulmonary fibrosis. Alveolar macrophages from patients with asbestosis and idiopathic pulmonary fibrosis display increased expression of TNF-α (Zhang et al. 1993). Proliferating type II pneumocytes of humans with acute fibrotic changes in the lung also contain high levels of TNF-α (Pan et al. 1996). Agents with anti-TNF-α properties such as pirfenidone and etanercept have shown promise in treatment of some patients with IPF (Selman et al. 2004). In addition, TNF-α promoter polymorphisms that cause higher levels of TNF-α production correlate with an increased risk of developing IPF (Whyte et al. 2000; Maier et al. 2001; Yucesoy et al. 2001). Studies in animal models of lung fibrosis also support the importance of TNF-α. Constitutive expression of TNF-α in the lung from the human secretory protein C promoter in transgenic mice produces pulmonary pathology that resembles fibrotic lung disease in humans (Miyazaki et al. 1995). Intratracheal instillation of TNF-α causes infiltration of inflammatory cells into the lung (Ulich et al. 1991), and concomitant stimulation of adhesion molecule expression and of chemokine release by TNF-α may further enhance pulmonary recruitment of inflammatory cells (Zhang et al. 1993). In addition to this proinflammatory role, TNF-α has a clear influence on the expression of other growth factors such as PDGF-A (platelet derived growth factor) (Sime et al. 1998) and TGF-β_1 (Phan et al. 1992; Sime et al. 1998; Sullivan et al. 2005). TNF-α has also been reported to stimulate production of types I and III collagen and fibronectin in human diploid fibroblasts (Paulsson et al. 1989) and in combination with IL-1β has been shown to up-regulate collagen transcription in lung fibroblasts (Elias et al. 1990). Administration of anti-TNF-α antibodies (Piguet et al. 1989; Phan and Kunkel 1992) or soluble TNF receptor (Piguet and Vesin 1994) protects mice from the fibrogenic effects of bleomycin or silica. Similarly, mice with both TNF receptors knocked-out (TNF-αRKO) are resistant to the fibrogenic effects of inhaled asbestos (Liu and Brody 2001), silica, and bleomycin (Ortiz et al. 1999). Furthermore, we found that the 129 strain of mice are resistant to the fibrogenic effects of asbestos and produced very little TNF-α (reviewed in detail above). Thus, it appears that TNF-α plays a key role in the development of interstitial inflammation and fibrosis. However, TNF-α is also known to exhibit anti-fibrotic properties. Specifically, overexpression of TNF-α has been reported to lessen pulmonary fibrosis in several animal models of IPF (Fujita et al. 2003), and several studies suggest that TNF-α represses the type I collagen promoter (Verrecchia, Wagner, and Mauviel 2002) and induces the expression of matrix metalloproteinases that participate in the degradation of extracellular matrix (Hozumi et al. 2001).

There is accumulating evidence that the fibrotic properties of TNF-α may be mediated at least in part by its ability to up-regulate TGF-β_1. TGF-β_1 plays important roles in normal lung morphogenesis and function (for review, see Bartram and Speer 2004). However, TGF-β_1 is the most potent profibrotic mediator known and it is present in high levels in asbestos-induced fibrosis (Khalil et al. 1991b). TGF-β_1 is the most abundant member of a cytokine family consisting of five known isoforms, three of which are expressed in mammalian cells, with similar but nonoverlapping activities (Massague 1996). TGF-β_1 is expressed by many cell types in the lung, including alveolar macrophages, lymphocytes, fibroblasts, and pulmonary epithelial cells. TGF-β_1 is secreted in a latent form as a high molecular weight complex that can be converted to the mature, active form of TGF-β_1 by a number of mechanisms including proteolytic cleavage (Harpel et al. 1992), interaction with the integrin alpha V beta 6 (Munger et al. 1999) or thrombin and oxidation by reactive oxygen species (ROS) (Pociask, Sime, and Brody 2004). Once activated, TGF-β_1 binds to specific type II receptors leading to recruitment and activation of type I receptors to form tetrameric complexes. Virtually all cells express the TGF-β_1 receptors and the level of their expression represents another mechanism for regulating the activity of TGF-β_1. Activated type I receptors propagate the intracellular signal through phosphorylation of particular Smads, a family of transcription factors (Varga 2002). There are three structurally and functionally distinct groups of Smads. Receptor-activated Smads (R-Smads, i.e. Smad 2/3) interact with a common mediator (Co-Smad, i.e., Smad 4) and transmit signals directly from activated TGF-β receptors. Inhibitory-Smads (I-Smads, i.e. Smad 6/7) also interact with the TGF-β receptor, but block Smad-dependent signaling. In most cells, R-Smads and Co-Smad are constitutively expressed, whereas I-Smads are regulated by TGF-β_1 and other stimuli. Activated Smad complexes translocate to the nucleus and bind to specific promoter elements of TGF-β_1 responsive genes. TGF-β_1 induces concentration-dependent responses of these genes, thus relatively small changes in active TGF-β_1 can significantly change the expression profile within the microenvironment.

Up-regulation of TGF-β_1 expression is a consistent feature of most fibrotic diseases. Kahlil and colleagues demonstrated intense TGF-β_1 expression in bronchiolar and hyperplastic type II epithelial cells in lungs of patients with advanced pulmonary fibrosis. While the TGF-β_2 and β_3 isoforms appeared to be constitutively expressed throughout the lung, TGF-β_1 localized in areas of fibrotic lung disease (Khalil et al. 1996). Moreover, the levels of TGF-β_1 in the lung correlate with the severity of lung disease (Bartram and Speer 2004). We showed that inhalation exposure of rats to asbestos produces enhanced TGF-β_1 and fibronectin expression at the sites of fiber deposition (Perdue and Brody 1994), which correlates with elevated levels of extra cellular matrix (ECM) detected by morphometry at times post-exposure (Chang et al. 1988; Coin et al. 1996). During the early stages of lung injury and repair after asbestos exposure, TGF-β_1 is found primarily in alveolar macrophages and in pulmonary epithelial cells (Perdue and Brody 1994; Liu and Brody 2001) (reviewed in detail above). Mice treated with bleomycin also had increased expression of TGF-β_1 in alveolar macrophages and bronchial and alveolar epithelial cells (Santana et al. 1995). Importantly, neutralizing antibodies to TGF-β_1 (Giri, Hyde, and Hollinger 1993) and TGF-β_1 soluble receptor (Wang et al. 2002), as well as upregulation of smad 7 protein expression (smad 7 is a natural inhibitor of the TGF-β-induced smad signaling pathway), have been used to block the development of bleomycin-induced fibrogenesis in animal models (Nakao et al. 1999). Moreover, introduction via the trachea of a recombinant AdV expressing TGF-β_1 caused the histopathological changes of pulmonary fibrosis in rats (Sime et al. 1997b) and mice (Warshamana et al. 2002a, 2002b). Consistent with these results, TNF-αRKO mice exhibit a fibrogenic-resistant phenotype (Liu et al. 1998); however re-establishment of TGF-β_1 expression in the TNF-αRKO mice is sufficient to induce fibroproliferative lung disease (Liu and Brody 2001). These observations suggest a pathogenic role for TGF-β_1 in animal models of particle-induced lung disease and agree with the observations of elevated TGF-β_1 in patients with silicosis (Jagirdar et al. 1996), asbestosis (Rom, Travis, and Brody 1991), and idiopathic pulmonary fibrosis (Khalil et al. 1991a; Bergeron et al. 2003).

The mechanisms underlying the fibrotic properties of TGF-β_1 are not entirely known, but TGF-β_1 appears to act at different levels to increase lung collagen deposition. It may be chemotactic for some fibroblasts, but it clearly promotes the transformation of lung fibroblasts to myofibroblasts. TGF-β_1 also promotes the accumulation and activation of myofibroblasts from the bone marrow (Desmouliere, Chaponnier, and Gabbiani 2005) and may induce epithelial to mesenchymal cell transition (EMT) (Kasai et al. 2005). Active TGF-β_1 inhibits proliferation of epithelial cells while producing variable effects on mesenchymal cell proliferation (Massague 1996). TGF-β_1 is a potent inducer of the synthesis of matrix proteins, such as collagen and fibronectin as well as glycoproteins (Bartram and Speer 2004). TGF-β_1 also increases the synthesis of integrins involved in matrix assembly at the cell surface (Heino et al. 1989). These effects are enhanced by TGF-β_1 inhibition of matrix degradation by decreasing synthesis of matrix metalloproteinases (MMP-1, MMP-3 and PA) and increasing synthesis of proteinase inhibitors, including TIMPs (tissue inhibitor of matrix metalloproteinases) and PAI-1 (plasminogen activator inhibitor) (for review, see Lasky and Brody 2000). TGF-β_1 also induces the expression of connective tissue growth factor (CTGF) (Grotendorst 1997; Duncan et al. 1999; Lasky and Brody 2000) as well as TGF-β_1 itself (Van Obberghen-Schilling et al. 1988), leading to an amplification of the original signal.

Despite intensive investigation into the biologic activities of TGF-β_1, the mechanisms by which expression of this growth factor are regulated are not yet fully understood. TGF-β_1 production may be controlled at the levels of transcription, mRNA stability, mRNA translation, secretion of preformed protein, activation of the latent protein to its active form (Phillips et al. 1995, 1996, 1997; Morrisey et al. 2001), and inactivation by circulating proteins and ECM (Bartram and Speer 2004). It is clear, however, that regulation of TGF-β_1 is cell-type-specific. Several factors such as differentiation (Loveridge et al. 1993), hypoxia (Falanga et al. 1991), members of the steroid hormone superfamily (Wakefield et al. 1990), retinoic acid receptors (Salbert et al. 1993), ECM components (Streuli et al. 1993; Varedi et al. 1997), high glucose concentration (Hoffman et al. 1998), angiotensin II (Wolf et al. 1993), ROS (Bellocq et al. 1999), IL-1β (Phillips et al. 1996), and PDGF (Fraser, Wakefield, and Phillips 2002) have been shown to influence TGF-β_1 expression. It has also been shown that TGF-β_1 regulates its own expression in an autocrine mechanism (Van Obberghen-Schilling et al. 1988). We showed for the first time that TNF-α rapidly activates the extracellular-stimulus regulated kinase (ERK)-specific mitogen-activated protein kinase (MAPK) pathway in fibroblasts, resulting in stabilization of TGF-β_1 mRNA and increased expression of TGF-β_1 protein in primary mouse lung fibroblasts (Sullivan et al. 2005) and mouse lung epithelial cells (unpublished data). Indeed, there is accumulating data indicating a role for TNF-α in the expression of TGF-β_1 in a variety of cell types. Significantly, a study *in vivo* showed that an anti-TNF-α antibody reduced bleomycin-induced lung injury and decreased expression of TGF-β_1 in the treated mice (Phan and Kunkel 1992). Importantly, Sime and colleagues showed that AdV transduction of TNF-α expression in lungs of normal rats induces TGF-β_1 production and interstitial fibrogenesis (Sime et al. 1998). *In vitro*, TNF-α rapidly up-regulates TGF-β_1 and collagen mRNA in both primary mesenchymal and epithelial cells of the lung (Warshamana, Corti, and Brody 2001). In addition, administration of TNF-α to microglial cells (Chao et al. 1995), mature adipocytes (Samad et al. 1999), human proximal tubular cells

(Phillips et al. 1996), and rat pulmonary artery endothelial cells (Phan et al. 1992) has been shown to induce significant increases in TGF-β_1 mRNA. Together these results suggest TNF-α may initiate or perpetuate a cytokine cascade known to be important in the development of fibrosis.

12.3.2 Molecular Mechanisms through Which TNF-α Controls TGF-β_1 Production and Consequent Fibrogenesis

The mechanisms by which TNF-α up-regulate TGF-β_1 appear to be complex. We observed significant stabilization of the TGF-β_1 mRNA in response to TNF-α (Sullivan et al. 2005). In untreated cells, the half-life of TGF-β_1 was ~15 h whereas in TNF-α treated cells, no degradation of the mRNA was detected after a 24 h exposure (Sullivan et al. 2005). Our yet unpublished findings also indicate that TNF-α doubles the rate of transcription of the TGF-β_1 gene. This combination of transcriptional and post-transcriptional events leads to a 3–4 fold increase in steady-state TGF-β_1 mRNA 6–12 h after exposure to TNF-α. Inhibition of the ERK with PD98059 or UO126 markedly reduced TNF-α induction of TGF-β_1, suggesting a critical role for this MAPK (Sullivan et al. 2005).

MAPKs, a group of evolutionarily conserved serine and threonine kinases, transduce extracellular signals to intracellular responses and play an important role in mediating cellular responses to changes in their environment. Activation of MAPKs may result in regulation of gene expression at multiple pre-transcriptional, transcriptional, and post-transcriptional steps. The most thoroughly studied MAPKs are the ERK, stress-activated protein kinases/c-Jun N-terminal kinases (SAPK/JNK), and the p38 kinases which respond to different upstream signals and can interact synergistically. The ERKs become active in response to cytokines, phorbol esters and growth factors. Phosphorylated ERKs may translocate to the nucleus where they phosphorylate and thus activate the AP-1 family member c-Jun (Pulverer et al. 1991; Cobb et al. 1994; Frost et al. 1994), as well as Ets family transcription factors, such as Elk-1 (Whitmarsh et al. 1995; Yang et al. 1998). The SAPK/JNKs are active primarily in response to IL-1β, TNF-α, osmotic stress and UV light (Minden and Karin 1997). Activated SAPK/JNKs phosphorylate c-Jun (Derijard et al. 1994) and Elk-1 (Whitmarsh et al. 1995), as well as the transcription factor ATF-2 (Gupta et al. 1996). The p38 kinases can also be activated by environmental stresses and cytokines. Several studies have established a role of p38 MAPK in regulation of gene expression through ATF-2 (activating transcription factor-2) and Elk-1 (Rincon, Flavell, and Davis 2000). ATF-2 has been shown to mediate transcription of the c-Jun gene (Karin 1995).

Regulation of the transcription factor activator protein-1 (AP-1), AP-1, that is comprised of a number of homodimeric and heterodimeric complexes of members of the jun family (c-jun, jun-B, and jun-D) and Fos (c-fos, fos-B, fra1, and fra2) family is an important function of MAPK signaling pathways. AP-1 (Kim et al. 1989; Weigert et al. 2000), AP-2 (Geiser et al. 1991) and Sp-1 (93) sites have been shown to be important in the regulation of both the human and mouse TGF-β_1 gene. A recent report demonstrated that high glucose stimulates TGF-β_1 promoter activity in mesangial cells and increases specific binding of AP-1 proteins to the corresponding consensus DNA sequences (Weigert et al. 2000). In another study, activation of the ERK pathway was shown to mediate stimulation of TGF-$\beta_1$expression by high glucose in mesangial cells (Isono et al. 2000). As discussed above, ERK activation is able to activate Elk-1, a member of the ternary complex factors that can enhance the expression of c-fos (Janknecht et al. 1993) and the subsequent DNA binding of the transcription factor AP-1 (Karin 1995). Our preliminary results suggest that TNF-α rapidly induces expression of c-fos and c-jun through the ERK pathway, suggesting a mechanism by which TNF-α increases TGF-β_1 gene transcription.

TNF-α also stabilizes TGF-β_1 mRNA in fibroblasts, but the precise mechanisms are not known. The decay rates of mRNAs are thought to be regulated by cis-acting sequence determinants, mRNA-binding proteins, endo- and exo-ribonucleases, and translation (Guhaniyogi and Brewer 2001). The presence of adenylate, uridylate-rich elements (AREs) in the three untranslated regions

(UTR) of mRNAs usually correlates with a short half-life. No ARE has been identified in the TGF-β_1 mRNA, which could account for TGF-β_1's relatively long half-life, even under unstimulated conditions. Sequences within the coding region (Wisdom and Lee 1991), or stem-loop structures formed by the 3'UTR (Guhaniyogi and Brewer 2001) can theoretically provide binding sites for regulatory proteins. TNF-α has been shown to up-regulate a number of genes by stabilizing mRNA through the modulation of RNA-binding proteins (Esnault and Malter 2003; Nabors et al. 2003). Furthermore, ERK signaling has been associated with increased expression of mRNA-binding proteins that regulate mRNA stability (Westmark and Malter 2001). Our current work is focused on identification of cis elements within TGF-β_1 mRNA and the corresponding regulatory proteins involved in TNF-α modulation of TGF-β_1 mRNA stability via the ERK pathway.

Our finding that TNF-α up-regulates TGF-β_1 expression through the ERK pathway is consistent with numerous studies implicating activation of the ERK pathway in the pathogenesis

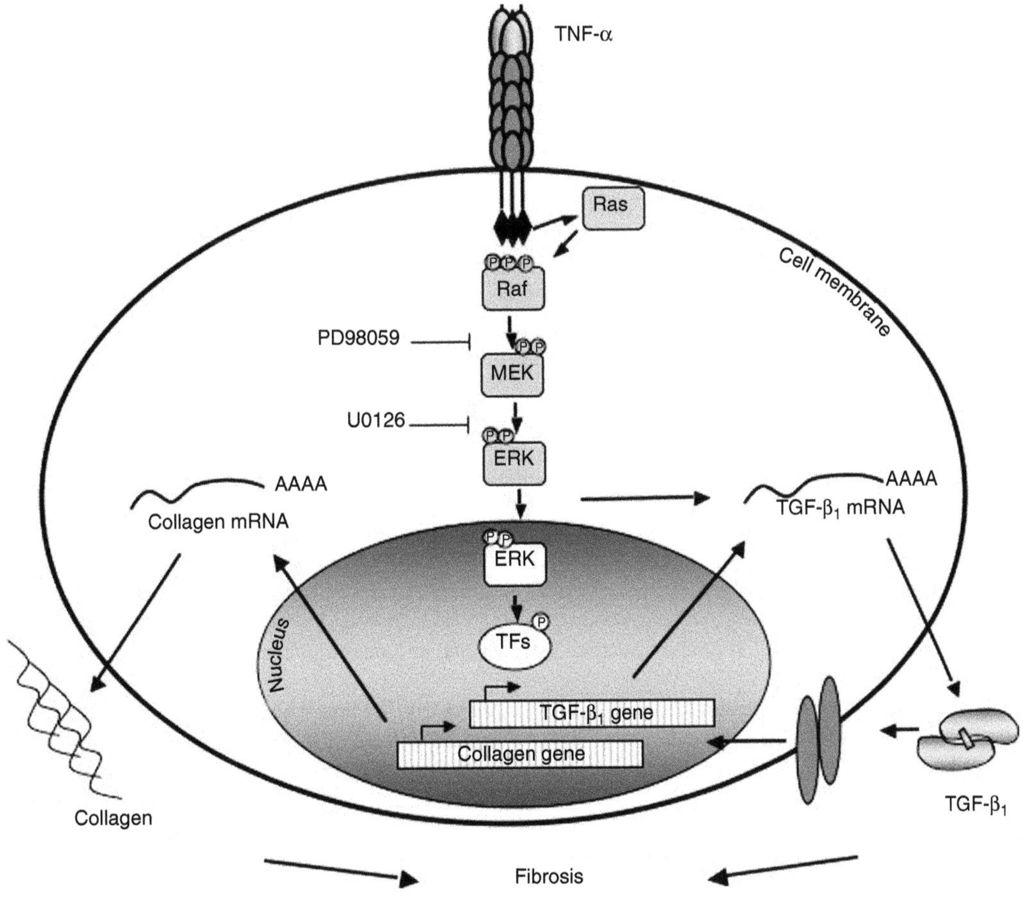

FIGURE 12.4 Proposed mechanism of TNF-α up-regulation of TGF-β_1 and induction of fibrosis: Exposure to asbestos leads to increased TNF-α expression by a variety of cell-types. TNF-α binds to its cellular receptor initiating a signal transduction pathway involving a cascade of phosphorylation events resulting in activation of the ERK pathway probably through Ras Raf and MEK (MAPK kinase). Activation of ERK leads to stabilization of TGF-β_1 mRNA and increased expression of the TGF-β_1 gene, presumably through activation of specific transcription factors (TFs). The net result is increased production of TGF-β_1. TGF-β_1 binding to its cellular receptor leads to increased production of collagen, a hallmark of fibrosis.

of particle-induced pulmonary fibrosis (Zanella et al. 1996; Cho et al. 1999). In a series of studies, Mossman and colleagues showed that asbestos exposure led to the production of ROS resulting in activation of ERK and increased expression of AP-1 complexes, including c-Jun, Jun-B and Fra-1, in rat pleural mesothelial cells (Mossman et al. 1997; Ramos-Nino, Timblin, and Mossman 2002). In an animal model, they showed that inhalation of chrysotile asbestos leads to significant chronic increases in ERK phosphorylation within pulmonary epithelial cells at the alveolar duct bifurcations where asbestos fibers initially deposit and accumulate (Robledo et al. 2000). Together, these results strongly suggest that signaling through the ERK pathway contributes to remodeling of the lung during the development of pulmonary fibrosis.

It is becoming clear that expression of TGF-β_1 plays a key role in the pathogenesis of IPF, and that TNF-α plays an important role in this process. The mechanism(s) involved in TNF-α up-regulation of TGF-β_1 appear to be complex, involving activation of ERK pathway resulting in both increased transcription of the TGF-β_1 gene and mRNA stabilization (Figure 12.4). Understanding the mechanisms involved in TNF-α regulation of TGF-β_1 may identify new targets for treatment of IPF.

12.4 SUMMARY

There are many models of IPF (Gharaee-Kermani, Ullenbruch, and, Phan 2005). We have settled on the asbestos model of lung injury because it recreates so many features of the human disease, allows extension of some of the findings to neoplastic processes, and offers hope for applying new therapeutic tools. It has become increasingly clear that investigators must focus on a central hypothesis that involves a few factors or mediators, their genes, and transduction pathways in order to progress in understanding the complex molecular mechanisms that control these diseases. Whether or not TNF-α and/or TGF-β_1 are central to disease development and should be the focus of new treatment regimens in animal models and clinical settings will be established.

REFERENCES

Aggarwal, B. B., Signaling pathways of the TNF superfamily: a double-edged sword, *Nat. Rev. Immunol.*, 3, 745–756, 2003.

Allen, J. T. and Spiteri, M. A., Growth factors in idiopathic pulmonary fibrosis: relative roles, *Respir. Res.*, 3, 13, 2002.

Bartram, U. and Speer, C. P., The role of transforming growth factor β in lung development and disease, *Chest*, 125, 754–765, 2004.

Bazzoni, F. and Beutler, B., The tumor necrosis factor ligand and receptor families, *N. Engl. J. Med.*, 334, 1717–1725, 1996.

Bellocq, A., Azoulay, E., Marullo, S., Flahault, A., Fouqueray, B., Philippe, C., Cadranel, J., and Baud, L., Reactive oxygen and nitrogen intermediates increase transforming growth factor-beta 1 release from human epithelial alveolar cells through two different mechanisms, *Am. J. Respir. Cell Mol. Biol.*, 21, 128–136, 1999.

Bergeron, A., Soler, P., Kambouchner, M., Loiseau, P., Milleron, B., Valeyre, D., Hance, A. J., and Tazi, A., Cytokine profiles in idiopathic pulmonary fibrosis suggest an important role for TGF-beta and IL-10, *Eur. Respir. J.*, 22, 69–76, 2003.

Brass, D., Hoyle, G., Poovey, H., Liu, J.-Y., and Brody, A., Reduced TNF-α and TGF-β1 expression in the lungs of inbred mice that fail to develop fibroproliferative lesions consequent to asbestos exposure, *Am. J. Pathol.*, 154, 853–862, 1999.

Brody, A. R., Asbestos exposure as a model of inflammation inducing interstitial pulmonary fibrosis, In *Inflammation: Basic Principles and Clinical Correlates*, Gallin, J. I., Goldstein, I. M., and Snyderman, R., eds., Raven Press, Ltd., New York, 1992.

Brody, A. R., Asbestos, In *Toxicology of the Respiratory System*, Roth, R., ed., Pergamon, New York, 393–413, 1997.

Brody, A. R. and Overby, L. H., Incorporation of tritiated thymidine by epithelial and interstitial cells in bronchiolar-alveolar regions of asbestos-exposed rats, *Am. J. Pathol.*, 134, 133–144, 1989.

Brody, A. R., Liu, J. Y., Brass, D., and Corti, M., Analyzing the genes and peptide growth factors expressed in lung cells in vivo consequent to asbestos exposure and in vitro, *Environ. Health Perspect.*, 105 (suppl. 5), 1165–1171, 1997.

Broekelmann, T. J., Limper, A. H., Colby, T. V., and McDonald, J. A., Transforming growth factor beta 1 is present at sites of extracellular matrix gene expression in human pulmonary fibrosis, *Proc. Natl. Acad. Sci. USA*, 88, 6642–6646, 1991.

Castranova, V., Robinson, V. A., and Frazer, D. G., Pulmonary reactions to organic dust exposures: development of an animal model, *Environ. Health Perspect.*, 104 (suppl. 1), 41–53, 1996.

Chang, L. Y., Overby, L. H., Brody, A. R., and Crapo, J. D., Progressive lung cell reactions and extracellular matrix production after a brief exposure to asbestos, *Am. J. Pathol.*, 131, 156–170, 1988.

Chao, C. C., Hu, S., Sheng, W. S., Tsang, M., and Peterson, P. K., Tumor necrosis factor-α mediates the release of bioactive transforming growth factor-β in murine microglial cell cultures, *Clin. Immunol. Immunopathol.*, 77, 358–365, 1995.

Cho, Y. J., Seo, M. S., Kim, J. K., Lim, Y., Chae, G., Ha, K. S., and Lee, K. H., Silica-induced generation of reactive oxygen species in Rat2 fibroblast: role in activation of mitogen-activated protein kinase, *Biochem. Biophys. Res. Commun.*, 262, 708–712, 1999.

Chua, F., Gauldie, J., and Laurent, G. J., Pulmonary fibrosis: Searching for model answers, *Am. J. Respir. Cell Mol. Biol.*, 33, 9–13, 2005.

Cobb, M. H., Hepler, J. E., Cheng, M., and Robbins, D., The mitogen-activated protein kinases, ERK1 and ERK2, *Semin. Cancer Biol.*, 5, 261–268, 1994.

Coin, P. G., Roggli, V. L., and Brody, A. R., Deposition, clearance, and translocation of chrysotile asbestos from peripheral and central regions of the rat lung, *Environ. Res.*, 58, 97–116, 1992.

Coin, P. G., Osornio-Vargas, A., Roggli, V. L., and Brody, A. R., Pulmonary fibrogenesis after three consecutive inhalation exposures to chrysotile asbestos, *Am. J. Respir. Crit. Care Med.*, 154, 1511–1519, 1996.

Crapo, J. D., Smolko, E. D., Miller, F. J., Graham, J. A., and Hayes, A. W., *Extrapolation of Dosimetric Relationships for Inhaled Articles and Gases*, Academic Press, San Diego, CA, 1989.

Derijard, B., Hibi, M., Wu, I. H., Barrett, T., Su, B., Deng, T., Karin, M., and Davis, R. J., JNK1: a protein kinase stimulated by UV light and Ha-Ras that binds and phosphorylates the c-Jun activation domain, *Cell*, 76, 1025–1037, 1994.

Desmouliere, A., Chaponnier, C., and Gabbiani, G., Tissue repair, contraction, and the myofibroblast, *Wound Repair Regen.*, 13, 7–12, 2005.

Duncan, M. R., Frazier, K. S., Abramson, S., Williams, S., Klapper, H., Huang, X., and Grotendorst, G. R., Connective tissue growth factor mediates transforming growth factor beta-induced collagen synthesis: down-regulation by cAMP, *FASEB J.*, 13, 1774–1786, 1999.

Elias, J. A., Freundlich, B., Adams, S., and Rosenbloom, J., Regulation of human lung fibroblast collagen production by recombinant interleukin-1, tumor necrosis factor, and interferon-gamma, *Ann. N. Y. Acad. Sci.*, 580, 233–244, 1990.

Esnault, S. and Malter, J. S., Hyaluronic acid or TNF-α plus fibronectin triggers granulocyte macrophage-colony-stimulating factor mRNA stabilization in eosinophils yet engages differential intracellular pathways and mRNA binding proteins, *J. Immunol.*, 171, 6780–6787, 2003.

Falanga, V., Qian, S. W., Danielpour, D., Katz, M. H., Roberts, A. B., and Sporn, M. B., Hypoxia upregulates the synthesis of TGF-beta 1 by human dermal fibroblasts, *J. Invest. Dermatol.*, 97, 634–637, 1991.

Fraser, D., Wakefield, L., and Phillips, A., Independent regulation of transforming growth factor-beta 1 transcription and translation by glucose and platelet-derived growth factor, *Am. J. Pathol.*, 161, 1039–1049, 2002.

Frost, J. A., Geppert, T. D., Cobb, M. H., and Feramisco, J. R., A requirement for extracellular signal-regulated kinase (ERK) function in the activation of AP-1 by Ha-Ras, phorbol 12-myristate 13-acetate, and serum, *Proc. Natl. Acad. Sci. USA*, 91, 3844–3848, 1994.

Fujita, M., Shannon, J. M., Morikawa, O., Gauldie, J., Hara, N., and Mason, R. J., Overexpression of tumor necrosis factor-alpha diminishes pulmonary fibrosis induced by bleomycin or transforming growth factor-β, *Am. J. Respir. Cell Mol. Biol.*, 29, 669–676, 2003.

Geiser, A. G., Kim, S. J., Roberts, A. B., and Sporn, M. B., Characterization of the mouse transforming growth factor-beta 1 promoter and activation by the Ha-ras oncogene, *Mol. Cell Biol.*, 11, 84–92, 1991.

Gharaee-Kermani, M., Ullenbruch, M., and Phan, S. H., Animal models of pulmonary fibrosis, *Methods Mol. Med.*, 117, 251–259, 2005.

Giri, S. N., Hyde, D. M., and Hollinger, M. A., Effect of antibody to transforming growth factor beta on bleomycin induced accumulation of lung collagen in mice, *Thorax*, 48, 959–966, 1993.

Grotendorst, G. R., Connective tissue growth factor: a mediator of TGF-beta action on fibroblasts, *Cytokine Growth Factor Rev.*, 8, 171–179, 1997.

Guhaniyogi, J. and Brewer, G., Regulation of mRNA stability in mammalian cells, *Gene*, 265, 11–23, 2001.

Gupta, S., Barrett, T., Whitmarsh, A. J., Cavanagh, J., Sluss, H. K., Derijard, B., and Davis, R. J., Selective interaction of JNK protein kinase isoforms with transcription factors, *Embo. J.*, 15, 2760–2770, 1996.

Harpel, J. G., Metz, C. N., Kojima, S., and Rifkin, D. B., Control of TGF-β activity: latency vs activation, *Prog. Growth Factor Res.*, 4, 321–335, 1992.

Haschek, W. M., Brody, A. R., Klein-Szanto, A. J., and Witschi, H., Animal model of human disease. Diffuse interstitial pulmonary fibrosis. Pulmonary fibrosis in mice induced by treatment with butylated hydroxytoluene and oxygen, *Am. J. Pathol.*, 105, 333–335, 1981.

Heino, J., Ignotz, R. A., Hemler, M. E., Crouse, C., and Massague, J., Regulation of cell adhesion receptors by transforming growth factor-beta. Concomitant regulation of integrins that share a common beta 1 subunit, *J. Biol. Chem.*, 264, 380–388, 1989.

Hoffman, B. B., Sharma, K., Zhu, Y., and Ziyadeh, F. N., Transcriptional activation of transforming growth factor-beta 1 in mesangial cell culture by high glucose concentration, *Kidney Int.*, 54, 1107–1116, 1998.

Hozumi, A., Nishimura, Y., Nishiuma, T., Kotani, Y., and Yokoyama, M., Induction of MMP-9 in normal human bronchial epithelial cells by TNF-alpha via NF-kappa B-mediated pathway, *Am. J. Physiol. Lung Cell. Mol. Physiol.*, 281, L1444–L1452, 2001.

Isono, M., Cruz, M. C., Chen, S., Hong, S. W., and Ziyadeh, F. N., Extracellular signal-regulated kinase mediates stimulation of TGF-beta 1 and matrix by high glucose in mesangial cells, *J. Am. Soc. Nephrol.*, 11, 2222–2230, 2000.

Jagirdar, J., Begin, R., Dufresne, A., Goswami, S., Lee, T. C., and Rom, W. N., Transforming growth factor-β in silicosis, *Am. J. Respir. Crit. Care Med.*, 154, 1076–1081, 1996.

Janknecht, R., Ernst, W. H., Pingoud, V., and Nordheim, A., Activation of ternary complex factor Elk-1 by MAP kinases, *Embo. J.*, 12, 5097–5104, 1993.

Karin, M., The regulation of AP-1 activity by mitogen-activated protein kinases, *J. Biol. Chem.*, 270, 16483–16486, 1995.

Kasai, H., Allen, J. T., Mason, R. M., Kamimura, T., and Zhang, Z., TGF-beta 1 induces human alveolar epithelial to mesenchymal cell transition (EMT), *Respir. Res.*, 6, 56, 2005.

Katzenstein, A. L. and Myers, J. L., Idiopathic pulmonary fibrosis: clinical relevance of pathologic classification, *Am. J. Respir. Crit. Care Med.*, 157, 1301–1315, 1998.

Khalil, N., Bereznay, O., Sporn, M., and Greenberg, A. H., Macrophage production of transforming growth factor beta and fibroblast collagen synthesis in chronic pulmonary inflammation, *J. Exp. Med.*, 170, 727–737, 1989.

Khalil, N., O'Connor, R. N., Unruh, H. W., Warren, P. W., Flanders, K. C., Kemp, A., Bereznay, O. H., and Greenberg, A. H., Increased production and immunohistochemical localization of transforming growth factor-beta in idiopathic pulmonary fibrosis, *Am. J. Respir. Cell Mol. Biol.*, 5, 155–162, 1991a.

Khalil, N., O'Connor, R. N., Unruh, H. W., Warren, P. W., Flanders, K. C., Kemp, A., Bereznay, O. H., and Greenberg, A. H., Increased production and immunohistochemical localization of transforming growth factor-beta in idiopathic pulmonary fibrosis, *Am. J. Respir. Cell Mol. Biol.*, 5, 155–162, 1991b.

Khalil, N., O'Connor, R. N., Flanders, K. C., and Unruh, H., TGF-β1, but not TGF-β2 or TGF-β3, is differentially present in epithelial cells of advanced pulmonary fibrosis: an immunohistochemical study, *Am. J. Respir. Cell Mol. Biol.*, 14, 131–138, 1996.

Kim, S. J., Denhez, F., Kim, K. Y., Holt, J. T., Sporn, M. B., and Roberts, A. B., Activation of the second promoter of the transforming growth factor-beta 1 gene by transforming growth factor-beta 1 and phorbol ester occurs through the same target sequences, *J. Biol. Chem.*, 264, 19373–19378, 1989.

Kolb, M., Margetts, P. J., Anthony, D. C., Pitossi, F., and Gauldie, J., Transient expression of IL-1beta induces acute lung injury and chronic repair leading to pulmonary fibrosis, *J. Clin. Invest.*, 107, 1529–1536, 2001.

Lasky, J. A. and Brody, A. R., Interstitial fibrosis and growth factors, *Environ. Health Perspect.*, 108 (suppl. 4), 751–762, 2000.

Last, J. A., Gelzleichter, T. R., Pinkerton, K. E., Walker, R. M., and Witschi, H., A new model of progressive pulmonary fibrosis in rats, *Am. Rev. Respir. Dis.*, 148, 487–494, 1993.

Li, J., Poovey, H. G., Rodriguez, J. F., Brody, A., and Hoyle, G. W., Effect of platelet-derived growth factor on the development and persistence of asbestos-induced fibroproliferative lung disease, *J. Environ. Pathol. Toxicol. Oncol.*, 23, 253–266, 2004.

Liu, J.-Y. and Brody, A. R., Increased TGF-β_1 in the lungs of asbestos-exposed rats and mice: reduced expression in TNF-α receptor knockout mice, *J. Environ. Pathol. Toxicol. Oncol.*, 20, 77–87, 2001.

Liu, J. Y., Morris, G. F., Lei, W. H., Corti, M., and Brody, A. R., Up-regulated expression of transforming growth factor-alpha in the bronchiolar-alveolar duct regions of asbestos-exposed rats, *Am. J. Pathol.*, 149, 205–217, 1996.

Liu, J.-Y., Brass, D. M., Hoyle, G. W., and Brody, A. R., TNF-α receptor knockout mice are protected from the fibroproliferative effects of inhaled asbestos fibers, *Am. J. Pathol.*, 153, 1839–1847, 1998.

Liu, J. Y., Sime, P. J., Wu, T., Warshamana, G. S., Pociask, D., Tsai, S. Y., and Brody, A. R., Transforming growth factor-beta (1) overexpression in tumor necrosis factor-alpha receptor knockout mice induces fibroproliferative lung disease, *Am. J. Respir. Cell Mol. Biol.*, 25, 3–7, 2001.

Loveridge, N., Farquharson, C., Hesketh, J. E., Jakowlew, S. B., Whitehead, C. C., and Thorp, B. H., The control of chondrocyte differentiation during endochondral bone growth in vivo: changes in TGF-beta and the proto-oncogene c-myc, *J. Cell Sci.*, 105 (Pt 4), 949–956, 1993.

Maier, L. A., Sawyer, R. T., Bauer, R. A., Kittle, L. A., Lympany, P., McGrath, D., Dubois, R., Daniloff, E., Rose, C. S., and Newman, L. S., High beryllium-stimulated TNF-alpha is associated with the −308 TNF-alpha promoter polymorphism and with clinical severity in chronic beryllium disease, *Am. J. Respir. Crit. Care Med.*, 164, 1192–1199, 2001.

Massague, J., TGF-beta signaling: receptors, transducers, and Mad proteins, *Cell*, 85, 947–950, 1996.

McGavran, P. D. and Brody, A. R., Chrysotile asbestos inhalation induces tritiated thymidine incorporation by epithelial cells of distal bronchioles, *Am. J. Respir. Cell Mol. Biol.*, 1, 231–235, 1989a.

McGavran, P. D., Butterick, C. J., and Brody, A. R., Tritiated thymidine incorporation and the development of an interstitial lesion in the bronchiolar-alveolar regions of the lungs of normal and complement deficient mice after inhalation of chrysotile asbestos, *J. Environ. Pathol. Toxicol. Oncol.*, 9, 377–391, 1989b.

McGavran, P. D., Moore, L. B., and Brody, A. R., Inhalation of chrysotile asbestos induces rapid cellular proliferation in small pulmonary vessels of mice and rats, *Am. J. Pathol.*, 136, 695–705, 1990.

Minden, A. and Karin, M., Regulation and function of the JNK subgroup of MAP kinases, *Biochim. Biophys. Acta*, 1333, F85–F104, 1997.

Miyazaki, Y., Araki, K., Vesin, C., Garcia, I., Kapanci, Y., Whitsett, J. A., Piguet, P. F., and Vassalli, P., Expression of a tumor necrosis factor-alpha transgene in murine lung causes lymphocytic and fibrosing alveolitis: a mouse model of progressive pulmonary fibrosis, *J. Clin. Invest.*, 96, 250–259, 1995.

Morris, G. and Brody, A., Molecular mechanisms of particle induced lung disease, In *Environmental and Occupational Medicine (Third Edition)*, Rom, W. N., ed., Lippincort-Raven, Philadelphia, 305–333, 1998.

Morrisey, K., Evans, R. A., Wakefield, L., and Phillips, A. O., Translational regulation of renal proximal tubular epithelial cell transforming growth factor-β_1 generation by insulin, *Am. J. Pathol.*, 159, 1905–1915, 2001.

Mossman, B. T. and Churg, A., Mechanisms in the pathogenesis of asbestosis and silicosis, *Am. J. Respir. Crit. Care Med.*, 157, 1666–1680, 1998.

Mossman, B. T., Marsh, J. P., Sesko, A., Hill, S., Shatos, M. A., Doherty, J., Petruska, J. et al., Inhibition of lung injury, inflammation, and interstitial pulmonary fibrosis by polyethylene glycol-conjugated catalase in a rapid inhalation model of asbestosis, *Am. Rev. Respir. Dis.*, 141, 1266–1271, 1990.

Mossman, B. T., Faux, S., Janssen, Y., Jimenez, L. A., Timblin, C., Zanella, C., Goldberg, J., Walsh, E., Barchowsky, A., and Driscoll, K., Cell signaling pathways elicited by asbestos, *Environ. Health Perspect.*, 105S, 1121–1125, 1997.

Munger, J. S., Huang, X., Kawakatsu, H., Griffiths, M. J., Dalton, S. L., Wu, J., Pittet, J. F. et al., The integrin alpha v beta 6 binds and activates latent TGF beta 1: a mechanism for regulating pulmonary inflammation and fibrosis, *Cell*, 96, 319–328, 1999.

Nabors, L. B., Suswam, E., Huang, Y., Yang, X., Johnson, M. J., and King, P. H., Tumor necrosis factor alpha induces angiogenic factor up-regulation in malignant glioma cells: a role for RNA stabilization and HuR, *Cancer Res.*, 63, 4181–4187, 2003.

Nakao, A., Fujii, M., Matsumura, R., Kumano, K., Saito, Y., Miyazono, K., and Iwamoto, I., Transient gene transfer and expression of Smad7 prevents bleomycin-induced lung fibrosis in mice, *J. Clin. Invest.*, 104, 5–11, 1999.

Ortiz, L. A., Lasky, J., Hamilton, R. F., Jr., Holian, A., Hoyle, G. W., Banks, W., Peschon, J. J., Brody, A. R., Lungarella, G., and Friedman, M., Expression of TNF and the necessity of TNF receptors in bleomycin-induced lung injury in mice, *Exp. Lung Res.*, 24, 721–743, 1998a.

Ortiz, L. A., Moroz, K., Liu, J. Y., Hoyle, G. W., Hammond, T., Hamilton, R. F., Holian, A., Banks, W., Brody, A. R., and Friedman, M., Alveolar macrophage apoptosis and TNF-alpha, but not p53, expression correlate with murine response to bleomycin, *Am. J. Physiol.*, 275, L1208–L1218, 1998b.

Ortiz, L. A., Lasky, J., Lungarella, G., Cavarra, E., Martorana, P., Banks, W. A., Peschon, J. J., Schmidts, H.-L., Brody, A. R., and Friedman, M., Upregulation of the p75 but not the p55 TNF-α receptor mRNA after silica and bleomycin exposure and protection from lung injury in double receptor knockout mice, *Am. J. Respir. Cell Mol. Biol.*, 20, 825–833, 1999.

Pan, L. H., Ohtani, H., Yamauchi, K., and Nagura, H., Co-expression of TNF alpha and IL-1 beta in human acute pulmonary fibrotic diseases: an immunohistochemical analysis, *Pathol. Int.*, 46, 91–99, 1996.

Paulsson, Y., Austgulen, R., Hofsli, E., Heldin, C. H., Westermark, B., and Nissen-Meyer, J., Tumor necrosis factor-induced expression of platelet-derived growth factor A-chain messenger RNA in fibroblasts, *Exp. Cell Res.*, 180, 490–496, 1989.

Perdue, T. D. and Brody, A. R., Distribution of transforming growth factor-β1, fibronectin, and smooth muscle actin in asbestos-induced pulmonary fibrosis in rats, *J. Histochem. Cytochem.*, 42, 1061–1070, 1994.

Phan, S. H. and Kunkel, S. L., Lung cytokine production in bleomycin-induced pulmonary fibrosis, *Exp. Lung Res.*, 18, 29–43, 1992.

Phan, S. H., Gharaee-Kermani, M., McGarry, B., Kunkel, S. L., and Wolber, F. W., Regulation of rat pulmonary artery endothelial cell transforming growth factor-β production by IL-1β and tumor necrosis factor-α, *J. Immunol.*, 149, 103–106, 1992.

Phillips, A. O., Steadman, R., Topley, N., and Williams, J. D., Elevated D-glucose concentrations modulate TGF-β1 synthesis by human cultured renal proximal tubular cells: the permissive role of platelet-derived growth factor, *Am. J. Pathol.*, 147, 362–374, 1995.

Phillips, A. O., Topley, N., Steadman, R., Morrisey, K., and Williams, J. D., Induction of TGF-β1 synthesis in D-glucose primed human proximal tubular cells: differential stimulation by the macrophage derived pro-inflammatory cytokines IL-1β and TNFα, *Kidney Int.*, 50, 1546–1554, 1996.

Phillips, A. O., Topley, N., Morrisey, K., Williams, J. D., and Steadman, R., Basic fibroblast growth factor stimulates the release of pre-formed TGF-β1 from human proximal tubular cells in the absence of de-novo gene transcription of mRNA translation, *Lab. Invest.*, 76, 591–600, 1997.

Piguet, P. F. and Vesin, C., Treatment by human recombinant soluble TNF receptor of pulmonary fibrosis induced by bleomycin or silica in mice, *Eur. Respir. J.*, 7, 515–518, 1994.

Piguet, P. F., Collart, M. A., Grau, G. E., Kapanci, Y., and Vassalli, P., Tumor necrosis factor/cachectin plays a key role in bleomycin-induced pneumopathy and fibrosis, *J. Exp. Med.*, 170, 655–663, 1989.

Pociask, D. A., Sime, P. J., and Brody, A. R., Asbestos-derived reactive oxygen species activate TGF-beta 1, *Lab. Invest.*, 84, 1013–1023, 2004.

Pulverer, B. J., Kyriakis, J. M., Avruch, J., Nikolakaki, E., and Woodgett, J. R., Phosphorylation of c-jun mediated by MAP kinases, *Nature*, 353, 670–674, 1991.

Ramos-Nino, M. E., Timblin, C. R., and Mossman, B. T., Mesothelial cell transformation requires increased AP-1 binding activity and ERK-dependent Fra-1 expression, *Cancer Res.*, 62, 6065–6069, 2002.

Rincon, M., Flavell, R. A., and Davis, R. A., The JNK and P38 MAP kinase signaling pathways in T cell-mediated immune responses, *Free Radic. Biol. Med.*, 28, 1328–1337, 2000.

Robledo, R. F., Buder-Hoffmann, S. A., Cummins, A. B., Walsh, E. S., Taatjes, D. J., and Mossman, B. T., Increased phosphorylated extracellular signal-regulated kinase immunoreactivity associated with proliferative and morphologic lung alterations after chrysotile asbestos inhalation in mice, *Am. J. Pathol.*, 156, 1307–1316, 2000.

Rom, W., Travis, W., and Brody, A., Cellular and molecular bases of the asbestos-related diseases, *Am. Rev. Respir. Dis.*, 143, 408–422, 1991.

Salbert, G., Fanjul, A., Piedrafita, F. J., Lu, X. P., Kim, S. J., Tran, P., and Pfahl, M., Retinoic acid receptors and retinoid X receptor-alpha down-regulate the transforming growth factor-beta 1 promoter by antagonizing AP-1 activity, *Mol. Endocrinol.*, 7, 1347–1356, 1993.

Samad, F., Uysal, K. T., Wiesbrock, S. M., Pandey, M., Hotamisligil, G. S., and Loskutoff, D. J., Tumor necrosis factor alpha is a key component in the obesity-linked elevation of plasminogen activator inhibitor 1, *Proc. Natl. Acad. Sci. USA*, 96, 6902–6907, 1999.

Santana, A., Saxena, B., Noble, N. A., Gold, L. I., and Marshall, B. C., Increased expression of transforming growth factor beta isoforms (β 1, β 2, β 3) in bleomycin-induced pulmonary fibrosis, *Am. J. Respir. Cell Mol. Biol.*, 13, 34–44, 1995.

Schottelius, A. J., Moldawer, L. L., Dinarello, C. A., Asadullah, K., Sterry, W., and Edwards, C. K., Biology of tumor necrosis factor-alpha implications for psoriasis, *Exp. Dermatol.*, 13, 193–222, 2004.

Selman, M., King, T. E., and Pardo, A., Idiopathic pulmonary fibrosis: prevailing and evolving hypotheses about its pathogenesis and implications for therapy, *Ann. Intern. Med.*, 134, 136–151, 2001.

Selman, M., Thannickal, V. J., Pardo, A., Zisman, D. A., Martinez, F. J., and Lynch, J. P., Idiopathic pulmonary fibrosis: pathogenesis and therapeutic approaches, *Drugs*, 64, 405–430, 2004.

Sime, P. J., Xing, Z., Foley, R., Graham, F. L., and Gauldie, J., Transient gene transfer and expression in the lung, *Chest*, 111, 89S–94S, 1997a.

Sime, P. J., Xing, Z., Graham, F. L., Csaky, K. G., and Gauldie, J., Adenovector-mediated gene transfer of active transforming growth factor-β1 induces prolonged severe fibrosis in rat lung, *J. Clin. Invest.*, 100, 768–776, 1997b.

Sime, P. J., Marr, R. A., Gauldie, D., Xing, Z., Hewlett, B. R., Graham, F. L., and Gauldie, J., Transfer of tumor necrosis factor-α to rat lung induces severe pulmonary inflammation and patchy interstitial fibrogenesis with induction of transforming growth factor-β1 and myofibroblasts, *Am. J. Pathol.*, 153, 825–832, 1998.

Streuli, C. H., Schmidhauser, C., Kobrin, M., Bissell, M. J., and Derynck, R., Extracellular matrix regulates expression of the TGF-beta 1 gene, *J. Cell Biol.*, 120, 253–260, 1993.

Sullivan, D. E., Ferris, M., Pociask, D., and Brody, A. R., Tumor necrosis factor-α induces transforming growth factor-β1 expression in lung fibroblasts through the extracellular signal-regulated kinase pathway, *Am. J. Respir. Cell Mol. Biol.*, 32, 342–349, 2005.

Ulich, T. R., Watson, L. R., Yin, S., Guo, K., Wang, P., Thang, H., and del Castillo, J., The intratracheal administration of endotoxin and cytokines: 1 Characterization of LPS-induced IL-1 and TNF mRNA expression and the LPS-, IL-1, and TNF-induced inflammatory infiltrate, *Am. J. Pathol.*, 138, 1485–1491, 1991.

Van Obberghen-Schilling, E., Roche, N. S., Flanders, K. C., Sporn, M. B., and Roberts, A. B., Transforming growth factor beta 1 positively regulates its own expression in normal and transformed cells, *J. Biol. Chem.*, 263, 7741–7746, 1988.

Varedi, M., Ghahary, A., Scott, P. G., and Tredget, E. E., Cytoskeleton regulates expression of genes for transforming growth factor-beta 1 and extracellular matrix proteins in dermal fibroblasts, *J. Cell Physiol.*, 172, 192–199, 1997.

Varga, J., TGF-beta, smads, and tissue fibrosis, *In Science and Medicine*, 298–307, 2002.

Verrecchia, F., Wagner, E. F., and Mauviel, A., Distinct involvement of the Jun-N-terminal kinase and NF-kappa B pathways in the repression of the human COL1A2 gene by TNF-alpha, *EMBO Rep.*, 3, 1069–1074, 2002.

Wakefield, L., Kim, S. J., Glick, A., Winokur, T., Colletta, A., and Sporn, M., Regulation of transforming growth factor-beta subtypes by members of the steroid hormone superfamily, *J. Cell Sci. Suppl.*, 13, 139–148, 1990.

Wang, Q., Hyde, D. M., Gotwals, P. J., and Giri, S. N., Effects of delayed treatment with transforming growth factor-beta soluble receptor in a three-dose bleomycin model of lung fibrosis in hamsters, *Exp. Lung Res.*, 28, 405–417, 2002.

Warshamana, G. S., Corti, M., and Brody, A. R., TNF-alpha PDGF, and TGF-beta (1) expression by primary mouse bronchiolar-alveolar epithelial and mesenchymal cells: TNF-alpha induces TGF-beta (1), *Exp. Mol. Pathol.*, 71, 13–33, 2001.

Warshamana, G. S., Pociask, D. A., Fisher, K. J., Liu, J. Y., Sime, P. J., and Brody, A. R., Titration of non-replicating adenovirus as a vector for transducing active TGF-beta 1 gene expression causing inflammation and fibrogenesis in the lungs of C57BL/6 mice, *Int. J. Exp. Pathol.*, 83, 183–201, 2002a.

Warshamana, G. S., Pociask, D. A., Sime, P., Schwartz, D. A., and Brody, A. R., Susceptibility to asbestos-induced and transforming growth factor-beta 1-induced fibroproliferative lung disease in two strains of mice, *Am. J. Respir. Cell Mol. Biol.*, 27, 705–713, 2002b.

Weigert, C., Sauer, U., Brodbeck, K., Pfeiffer, A., Haring, H. U., and Schleicher, E. D., AP-1 proteins mediate hyperglycemia-induced activation of the human TGF-β1 promoter in mesangial cells, *J. Am. Soc. Nephrol.*, 11, 2007–2016, 2000.

Westmark, C. J. and Malter, J. S., Up-regulation of nucleolin mRNA and protein in peripheral blood mononuclear cells by extracellular-regulated kinase, *J. Biol. Chem.*, 276, 1119–1126, 2001.

Whitmarsh, A. J., Shore, P., Sharrocks, A. D., and Davis, R. J., Integration of MAP kinase signal transduction pathways at the serum response element, *Science*, 269, 403–407, 1995.

Whyte, M., Hubbard, R., Meliconi, R., Whidborne, M., Eaton, V., Bingle, C., Timms, J. et al., Increased risk of fibrosing alveolitis associated with interleukin-1 receptor antagonist and tumor necrosis factor-alpha gene polymorphisms, *Am. J. Respir. Crit. Care Med.*, 162, 755–758, 2000.

Wisdom, R. and Lee, W., The protein-coding region of c-myc mRNA contains a sequence that specifies rapid mRNA turnover and induction by protein synthesis inhibitors, *Genes Dev.*, 5, 232–243, 1991.

Wolf, G., Mueller, E., Stahl, R. A., and Ziyadeh, F. N., Angiotensin II-induced hypertrophy of cultured murine proximal tubular cells is mediated by endogenous transforming growth factor-beta, *J. Clin. Invest.*, 92, 1366–1372, 1993.

Xie, C., Reusse, A., Dai, J., Zay, K., Harnett, J., and Churg, A., TNF-alpha increases tracheal epithelial asbestos and fiberglass binding via a NF-kappa B-dependent mechanism, *Am. J. Physiol. Lung Cell. Mol. Physiol.*, 279, L608–L614, 2000.

Yang, S. H., Yates, P. R., Whitmarsh, A. J., Davis, R. J., and Sharrocks, A. D., The Elk-1 ETS-domain transcription factor contains a mitogen-activated protein kinase targeting motif, *Mol. Cell Biol.*, 18, 710–720, 1998.

Yucesoy, B., Vallyathan, V., Landsittel, D. P., Sharp, D. S., Weston, A., Burleson, G. R., Simeonova, P., McKinstry, M., and Luster, M. I., Association of tumor necrosis factor-alpha and interleukin-1 gene polymorphisms with silicosis, *Toxicol. Appl. Pharmacol.*, 172, 75–82, 2001.

Zanella, C. L., Posada, J., Tritton, T. R., and Mossman, B. T., Asbestos causes stimulation of the extracellular signal-regulated kinase 1 mitogen-activated protein kinase cascade after phosphorylation of the epidermal growth factor receptor, *Cancer Res.*, 56, 5334–5338, 1996.

Zhang, Y., Lee, T. C., Guillemin, B., Yu, M. C., and Rom, W. N., Enhanced IL-1 beta and tumor necrosis factor-alpha release and messenger RNA expression in macrophages from idiopathic pulmonary fibrosis or after asbestos exposure, *J. Immunol.*, 150, 4188–4196, 1993.

Zhang, S., Smartt, H., Holgate, S. T., and Roche, W. R., Growth factors secreted by bronchial epithelial cells control myofibroblast proliferation: an in vitro co-culture model of airway remodeling in asthma, *Lab. Invest.*, 79, 395–405, 1999.

13 Effect of Particles on the Immune System

M. Ian Gilmour
National Health and Environmental Effects Research Laboratory,
U.S. Environmental Protection Agency

Tina Stevens[*]
Curriculum in Toxicology, University of North Carolina at Chapel Hill

Rajiv K. Saxena
School of Life Sciences, Jawaharlal Nehru University

CONTENTS

13.1 OVERVIEW

Respiratory allergies and infections are the most common form of illness in the United States and Europe, and together they account for more missed school and work days than any other types of disease (Akazawa, Sindelar, and Paltiel 2003; CDC 2004). From the well-documented air pollution episodes in London, England, and Donora, Pennsylvania (Holland et al. 1979) to the most recent time series analyses of multiple cities in the U.S. (Dominici et al. 2006), it is clear that elevated levels of airborne particles are associated with increased morbidity and mortality to respiratory infections, and increased hospital admissions for asthma (Koren 1995; Vigotti 1999).

A substantial body of experimental work has also shown that air pollutants such as tobacco smoke, ozone, diesel exhaust (DE), and other gases and particles can alter many aspects of the immune system to decrease resistance to infection and/or exacerbate respiratory allergies and asthma (Cohen, Zelikoff, and Schlesinger 2000). Inhaled pollutants affect a number of key host defenses, including mucociliary clearance activity in the airways, microbial killing in the lung

[*] UNC funded by US EPA training agreement EPA CT 829472

lining fluid, pulmonary macrophage function, and the development of specific immune responses such as antibody production and cell mediated immunity. In contrast, immune stimulation in the form of increased T cell activity and reaginic (IgE) antibody formation has also been shown to occur under some circumstances, resulting in increased incidence or severity of allergic lung disease.

These results continue to be confirmed in clinical, epidemiological, and experimental studies while basic research activities seek mechanistic explanations for the effects. This chapter will review recent research on the different ways that particles affect the immune system to either decrease resistance to infectious agents or increase allergic and inflammatory disorders in the respiratory tract. We conclude by summarizing research needs and future directions that will help in hazard identification and contribute to improved risk assessment of inhaled particles.

13.2 MODULATION OF PULMONARY RESPONSES TO PATHOGENS BY PM

In a seminal review article, Green and colleagues wrote (Green et al. 1977) that "despite the daily microbial assault that the respiratory tract experiences, the gas exchange area of the lung is maintained in a remarkably sterile condition by the combined antimicrobial activity of the mucociliary, phagocytic, and immune systems." Early studies using isolated macrophages from lung washes showed that exposure to various agents including ozone, nitrogen oxides, metal compounds, and tobacco smoke reduced the cells ability to ingest and/or kill bacteria through interfering with phagocytic uptake and intracellular anti-microbial activity (Gardner 1984). Later experiments revealed additional defects, including reduced anti-microbial activity of the lung lining fluid and impaired development of specific immune responses as measured by assessment of cellular effector function, antibody production after immunization, T cell phenotype changes, and cytokine production (Jakab et al. 1995).

Microorganisms are killed in phagocytes by an array of digestive enzymes, toxic oxygen species, and other anti-microbial agents. Rodents exposed to particulates such as carbon black (CB) (Jakab 1993), smoke (Moores, Janigan, and Hajela 1993), lead oxide (Zelikoff, Parsons, and Schlesinger 1993), titanium dioxide (Gilmour et al. 1989), and road dust (Ziegler et al. 1994) exhibit reduced alveolar macrophage (AM) phagocytosis and/or impaired pulmonary clearance of inhaled bacteria. A number of *in vitro* studies using both animal and human cells have also reported that particle matter (PM) or its toxic constituents reduces AM function. Exposure to acrolein or benzofuran adsorbed onto CB lowered rat AM phagocytosis (Jakab et al. 1990), while human peripheral blood monocytes had decreased phagocytic activity after incubation with a particulate air sample collected from an industrial area in Germany (Hadnagy and Seemayer 1994). Possible mechanisms for reduced macrophage phagocytosis include intracellular overloading of particulate, direct toxicity of internalized particles, and the co-production of suppressive mediators such as prostaglandins and corticosteroids (Canning et al. 1991). Uptake of ultra fine particles, in particular, may cause cytoskeletal dysfunction (Moller et al. 2005), and impair the phagocytic activity of AMs (Lundborg et al. 2001; Renwick, Donaldson, and Clouter 2001).

Less research has been conducted on the effect of particle exposure on the adaptive immune system (both local and systemic). It is known, however, that a significant proportion of particulate matter found in urban air is derived from combustion of fossil fuels and industrial discharges and that as a result, it contains varying amounts of metals, solvents, aromatic hydrocarbons, and other chemicals which modulate specific immune function. Of the metals investigated, cadmium, vanadium, chromium, lead, and nickel decrease antibody formation, antigen processing, and lymphocyte proliferation in experimental animals (Kowolenko et al. 1988; Newcombe 1992; Cohen, Zelikoff, and Schlesinger 2000). Organic compounds, which show immunotoxic properties such as benzene, trichloroethylene, dioxins, phenols, organotonins and diester phorbol compounds, are also found in the atmosphere at varying concentrations (including being part of the complex adsorbate on combustion particles) and have been reported under numerous experimental conditions to reduce immune function (Saboori and Newcombe 1992; Cohen, Zelikoff, and Schlesinger 2000).

The effect of particles on susceptibility to microbial pathogens has been most extensively studied with the streptococcus infectivity model. Coffin and Blommer (1967) first showed that mice exposed to irradiated automobile exhaust were more susceptible to a subsequent pulmonary infection with *Streptococcus zooepidemicus*. Later, it was also established that the same ranking of toxicity for a number of metal salts occurred whether the animals were exposed by inhalation or intratracheal instillation prior to infection (Hatch et al. 1981). With this validation in place between inhalation and instillation techniques, subsequent work then demonstrated that instillation of 100 μg of bentonite, oil fly ash, metal salts (CdO, ZnO, $NaAsO_2$, $SnCl_2$, and $CoNO_3$), and ambient air particles collected in Germany enhanced mortality of mice to infection by more than 50% (Hatch et al. 1985). Other particulates characterized as having intermediate potency (<50% excess mortality) included an ambient air particle sample from Washington DC, three coal fly ash samples, powdered latex, BeO, and Fe_2O_3, while low-potency particles that did not enhance mortality significantly at the 100 μg dose level included samples of coal fly ash, an ambient air sample from St. Louis and Mount St. Helens volcanic ash. More recent experiments with inhaled concentrated air particles (CAPs) from New York City air have indicated that exposure of normal healthy mice does not increase susceptibility to bacterial infection, however the exposures worsened existing pulmonary infections in aged rats (Zelikoff et al. 2003).

In addition to CAPs, a number of investigations have recently focused on the immunotoxic effects of emission particles derived from the combustion of diesel and wood. The streptococcus model has proved to be sensitive following exposure to relatively low levels (1.9 mg/m^3) of woodsmoke (Gilmour et al. 2001), and later studies demonstrated that exposure at even lower levels (0.75 mg/m^3) impairs pulmonary clearance of *S. aureus* (Zelikoff et al. 2002). While the streptococcus model has not been studied with diesel, a number of other infectivity systems have demonstrated effects with these particles. Harrod and colleagues reported that inhalation exposure to freshly generated DE (1 mg/m^3) reduced clearance of *Pseudomonas aeruginosa* (Harrod et al. 2005) and *Respiratory syncytial virus* (RSV) (Harrod et al. 2003). Other groups have similarly shown that inhalation or instillation exposure of re-entrained diesel particles at substantially higher concentrations increases lung burdens of *Mycobacterium tuberculosis* (Hiramatsu et al. 2005), *Bacillus Calmet Guerin* (Saxena et al. 2003b), *Listeria monocytogenes* (Yang et al. 2001).

It is thought that diesel exhaust particles (DEP) do not directly influence the bactericidal activity of AMs (Bonay et al. 2006), but rather suppress secretion of pro-inflammatory cytokines and oxidative processes involved in cellular activation (Saito et al. 2002; Mundandhara, Becker, and Madden 2005). Yang et al. (2001) demonstrated that the release of TH1 cytokines and reactive oxygen species (ROS) in response to *L. monocytogenes* was deficient in bronchoalveolar lavage (BAL) cells derived from DEP exposed rats. Exposure to CB did not produce a similar effect, indicating that the suppressive effect of DEP may be due to the absorbed organic molecules present in DEP preparations. Yin et al. (2004) further reported that the organic components isolated from DEP inhibited production of TNF-α and IL-12 by AMs and this effect may be secondary to the induction of oxidative stress. AMs from DEP treated rats were deficient in lipopolysaccharide (LPS) induced secretion of TNF-α and IL-1 as well as in the production of ROS in response to zymosan, and this inhibitory effect was due to adsorbed organic compounds on DEP (Castranova et al. 2001).

These immunomodulatory effects have also been recreated in *in vitro* systems, suggesting that cell based assays may have broad screening utility. For example, Bacillus Calmette-Guerin (BCG) and LPS-induced production of nitric oxide in a mouse macrophage cell line is inhibited by DEP, and this effect was found to reside in the more polar aromatic hydrocarbon and resin fractions of DEP with the most potent components occurring in the *n*-hexane soluble fraction (Saxena et al. 2003a; Shima et al. 2006).

Lung epithelial cells can secrete a variety of cytokines and this response is modulated by DEP or its associated extracts (Takizawa 2004). A gene array study showed that exposure to DEP extracts up-regulated about 50 genes in rat alveolar epithelial cells, with hemoxygenase-1 being the most prominent up-regulated (Koike et al. 2004), while bronchial epithelial cells obtained from human volunteers exposed to DEP have increased expression of the TH2 cytokine IL-13 (Pourazar et al. 2004). Induction or modulation of cytokine response as well as modulation of uptake and survival of pathogens in macrophages and epithelial cells by PM can have important consequences on the course of disease. Influenza virus and the RSV chiefly infect epithelial cells in airways (Harrod et al. 2003). DEP exposure augments the expression of receptors for many bacterial and viral pathogens on lung epithelial cells (Ito et al. 2006) and it has recently been demonstrated that pre-exposure to DEP increases influenza infection in human lung alveolar cells (Jaspers et al. 2005).

Cigarette smoke (CS) is another complex aerosol derived from the combustion of tobacco and its associated additives. It has been shown in numerous experimental and clinical studies that CS reduces mucociliary clearance, impairs macrophage function, reduces lymphocyte and antibody responses, and is associated with increased prevalence of respiratory infections (reviewed in Johnson et al. 1990). CS exposure has more recently been reported to inhibit the TH1 immune response to RSV infection in neonatal mice (Phaybouth et al. 2006) and promote TH2 priming in human dendritic cells (Vassallo et al. 2005).

Overall, it appears that exposure to DEP and CS, both of which consist of a complex mix of soot and organic condensate, can promote a TH2 cytokine pattern in lungs resulting in increased allergic and asthmatic-type responses (discussed below). As a consequence of the TH2 polarization and/or through other mechanisms, DEP and CS may also facilitate the growth of pathogens in lungs. Interestingly, the higher load of pathogens caused by PM exposure may eventually result in a stronger TH1 immune response, as seen in the BCG models of infection where bacterial load as well as IFNγ response are boosted by DEP (Saxena et al. 2003b). This has also been observed *in vitro* where cultured lung epithelial cells exposed to DEP and influenza display an increased viral infection accompanied with an augmented IFNγ response (Jaspers et al. 2005).

Unlike DEP exposure, silica particles enhance the clearance of *L. monocytogenes* from rat lungs (Antonini et al. 2000), indicating that particles may in some cases boost important parameters of the immune response. Whereas DEP exposure encourages a TH2 type of cytokine profile, silica exposure promotes a TH1 profile of cytokine release (Davis, Pfeiffer, and Hemenway 2000; Garn et al. 2000). In contrast to this apparent short term benefit, chronic exposure to silica still predisposes the host to pulmonary tuberculosis (TeWaternaude et al. 2006). Antibody responses in DEP and silica exposed animals are also different between these two types of particles. DEP exposed mice have a depressed systemic antibody response to sheep erythrocytes (Yang et al. 2003) while elevated serum levels of IgG and IgM have been reported in silicotic rats (Huang et al. 2001).

Taken together, most of these studies indicate that exposure to many airborne particulates and especially those containing toxic chemicals can affect immune function. The effect in general appears to be mediated by alteration in function of macrophages and epithelial cells and is accompanied with changes in the spectrum of cytokine release. Altered cytokine milieu may in turn modulate the subsequent adaptive immune responses.

13.3 TOLL LIKE RECEPTORS AND THEIR REGULATION IN THE RESPIRATORY TRACT

Toll like receptors (TLRs) constitute a family of structurally homologous receptors that recognize features common to many types of pathogens (pathogen-associated molecular patterns, or PAMPs, reviewed in Takeda and Akira 2005). The role of TLRs is to activate phagocytes and tissue dendritic cells in response to pathogens. This activation also leads to production of important mediators of innate immunity (cytokines and chemokines), as well as the promotion of surface expression of co-stimulatory molecules essential for the induction of adaptive immune responses. Since the respiratory tract constitutes a principal portal of entry for inhaled microbes, TLR bearing cells in the lung play an important role in responding to pathogens. Both pulmonary macrophages and epithelial cells express a spectrum of toll-like receptors that recognize virtually all classes of pathogens and can be stimulated to secrete a variety of immunomodulatory and chemotactic cytokines (Krutzik and Modlin 2004; Sha et al. 2004; Greene and McElvaney 2005). Several reports indicate that PM may modulate the expression of TLRs in lung. Expression of TLR2 that recognizes mycobacterial components is depressed in the lungs of smokers, indicating that the immune response to tuberculosis-like organisms may be sub-optimal (Droemann et al. 2005). DEP induced neutrophil influx in lungs and the release of MIP-1 was significantly lower in C3H/HeJ mice that have a point mutated and dysfunctional TLR4 molecule, as compared to C3H/HeN mice with a functional TLR4 molecule (Inoue et al. 2006).

TLRs may also be involved in the response of lung epithelial cells to PM. Becker et al. (2005) showed that IL-8 release by lung epithelial cells in response to ambient PM requires the participation of TLR2. This would indicate that modulation of expression and/or signaling through TLRs constitutes an important aspect of the biological effect of PM on the immune system in lungs. The link between innate immune responses modulated by Toll receptors and adaptive immune responses has received much attention recently through the observation that children exposed to the TLR4 ligand, bacterial endotoxin, show significant protection against developing allergies and asthma (Schaub, Lauener, and von Mutius 2006), and is discussed in Section 13.4.

13.4 EFFECT OF PARTICLES ON ALLERGIC IMMUNE RESPONSES

Over the last 30 years, the incidence of allergic disease has dramatically increased in industrialized countries. Currently, the prevalence ranges from 25 to 40% for allergic rhinitis and 6%–12% for allergic asthma (CDC 2004). Major environmental sensitizers, such as cockroach and dust mite feces, animal dander, molds, and seasonal pollens have been ubiquitous as long as people have lived in the world, but only recently has a significant percentage of the population (particularly in developed countries) developed an allergic response to these proteins. This would strongly indicate some environmental influence as opposed to a significant change in the gene pool. While changes in lifestyle—including alterations in diet, activity patterns, medication use, and housing conditions—have undoubtedly had an impact on the sensitization rate, epidemiology studies have also shown that increases in ambient PM correlate with increased hospitalizations due to respiratory illness, including asthma (Ostro 1993; Dockery and Pope 1994; Atkinson et al. 2001). Part of this association may be a result of allergenic pollens being bound to ambient particulates, as has been recently observed in four different European cities (Namork, Johansen, and Lovik 2006).

Most asthmatics experience exacerbations in airway inflammation and non-specific bronchial hyper-responsiveness to a wide range of inhaled substances, including CS, DE, hypertonic saline, and even cold air. These challenges are not antigenic in nature, but rather they behave as irritants in provoking inflammation and/or bronchoconstriction. Similar effects have been seen in allergic animals exposed to residual oil fly ash (ROFA) (Gavett et al. 1999; Hamada et al. 2002). A general increase in airway responsiveness and lung injury by exposure to PM may be an additive effect on top of pre-existing inflammation or through the development of increased tissue sensitivity. There are also indications that particles may stimulate the neuroimmune junction through the release of substance P (Wong et al. 2003), which in itself is a potent bronchoconstrictor and inflammatory mediator (Joos et al. 2003).

Particles can act directly on cells important in the effector phase of allergic reactions. Type I hypersensitivity reactions, such as those occurring in allergic asthma, are caused by the cross-linking of IgE molecules on the surface of mast cells. This signal induces the cells to degranulate and release preformed histamines, and synthesize prostaglandins, leukotrienes and immunomodulatory cytokines. An increase in the severity of allergic symptoms and histamine levels has been noted in dust mite-sensitive subjects when co-administered DEP and extract of house dust mite, compared to DEP or allergen extract alone. At the cellular level, DEP plus IgE antibody can also act directly on mast cells to secrete more histamine compared to DEP or anti IgE alone (Diaz-Sanchez, Penichet-Garcia, and Saxon 2000).

In addition to exacerbating existing allergic disease, there is epidemiological evidence that certain air pollutants including ozone and DE are associated with the development of new disease (Wade and Newman 1993; Rusznak, Devalia, and Davies 1994; D'Amato 1999; Hajat et al. 1999; Nicolai 1999), and recent associations have been specifically linked to proximity to highways (Brunekreef et al. 1997; Delfino et al. 2003; Gauderman et al. 2005). While these effects need to be confirmed with better personal exposure information, investigations in animals and in a few human clinical studies have reported that air pollutants may indeed contribute to the increased incidence of allergic disease and asthma.

Animal experiments have demonstrated that many types of particles, including ambient PM, DEP, ROFA, CB particles, and polystyrene particles (PSP), can act as immunologic adjuvants when administered with an antigen via intraperitoneal, intranasal, intratracheal, and inhalation routes of exposure (Takafuji et al. 1989; Fujimaki et al. 1997; Maejima et al. 1997; Lambert et al. 2000; Van Zijverden et al. 2000; de Haar et al. 2005; Nygaard, Aase, and Lovik 2005). In most cases the particles alone cause inflammation, but when administered during sensitization they also stimulate the development of allergic immune responses (in the form of increased IgE antibody, TH2 cytokines). Upon repeated challenge with antigen, these animals exhibit increased severity of allergic type disease (pulmonary eosinophils, airway hyperresponsiveness, increased mucus production, etc.) compared to control animals that received antigen exposure and vehicle control in the place of the pollutant.

The relationship between particle exposure and increased allergic symptoms has been examined in limited human studies with both allergic and non-allergic subjects. Individuals with allergic rhinitis and mild asthma exposed to 0.3 mg of DEP intra-nasally had significantly enhanced IgE antibody production in the nasal mucosa (Diaz-Sanchez et al. 1994). In a later study, atopic subjects given DEP prior to nasal immunization with a neoantigen, keyhole limpet hemocyanin (KLH), produced antigen-specific IgG, IgA, and IgE as well as IL-4 in nasal lavage fluid (Diaz-Sanchez et al. 1999), while subjects given KLH alone only produced IgG and IgA, indicating that the DEP acted as an adjuvant to promote primary allergic sensitization.

While these specific studies used a diesel particle highly enriched in organic constituents, another body of literature also shows that the carbonaceous core of the diesel, or more inert particles like CB and PSP, can similarly induce adjuvant-like effects (Granum and Lovik 2002). Rats instilled with 100 μg of fine (FCB) or ultrafine carbon black (UFCB) had some measure of allergic adjuvancy compared to DEP particles (Singh, Madden, and Gilmour 2005), while the adjuvant effects of PSP are directly related to increase in surface area of smaller particles instilled on the same mass basis as larger particles. Furthermore in another PSP study, smaller particles directly oxidized the oxidant-activated fluorophore dichlorofluoresein diacetate in a cell-free system compared to larger particles of the same chemical makeup (Brown et al. 2001), supporting the notion that the effect is related to surface area and particle number rather than mass.

13.5 MECHANISM OF ACTION OF PM ON THE IMMUNE RESPONSE

Particles come in a vast variety of shapes and chemical compositions and interact with different types of cells like macrophages and epithelial cells in the respiratory tract. It is therefore unlikely that a unified mechanism exists that can explain how PM exposure results in altered susceptibility to infections on the one hand, and augmented allergic and asthmatic responses on the other. Nonetheless, certain common features have been noted in the mechanism of action of many types of PM. One popular theme in this research area—and indeed in many diseases in general—is that of oxidative stress and the propensity for this phenomenon to cause tissue injury and dysfunction.

It is evident that inhaled particles and some of their components can injure and reduce the activity of pulmonary cells important to barrier function and the clearance of infectious agents. As such, these events offer pathogens a greater chance to colonize and cause disease (El-Etr and Cirillo 2001). At a secondary level, many particles and their components (transition and heavy metals, surface free radicals, organic moieties, etc.) also affect the development of specific immune responses through alterations in antigen processing and subsequent effector function. At the core of many of these effects are oxidative reactions, which are known to alter homeostatic balance and have dramatic effects on molecular, cellular, and tissue function, including host defenses.

There is also good evidence that ROS are involved in the adjuvant effect of diesel particles on the induction of TH2 type immune responses (Nel 2005). ROS are known to cause many forms of injury and inflammation and they are also produced by inflammatory cells induced during both sensitization and effector phases of allergic lung disease. Thiol antioxidants suppress DEP or DEP extract induced ROS in macrophage cell lines and inhibit DEP-enhanced allergic responses in mice (Whitekus et al. 2002). Nel and colleagues (Li and Nel 2006) have explained this paradigm in a three stage model. In the first tier, oxidative stress is at a low level and the induction of antioxidant enzymes such as NAD(P)H:quinone oxidoreductase (NQO1), glutathione-S-transferase M1 (GSTM1), and heme oxygenase-1 (HO-1) are able to restore cellular redox homeostasis. With continued oxidative stress these enzymes become overwhelmed and can no longer neutralize the effects of ROS. When this happens, (tier 2) activation of the MAPK and NF-kB cascade induces proinflammatory responses, including production of IL-4, IL-5, IL-8, IL-10, IL-13, RANTES, MIP-1α, MCP-3, GM-CSF, TNF-α, ICAM-1, and VCAM-1. At higher levels of oxidative stress (tier 3), the permeability of the mitochondria is compromised and disruption of the electron transfer chain results in cellular apoptosis and necrosis (Li and Nel 2006).

The mechanism by which more generic particles cause immune effects is also thought to be due to oxidative stress through the presence of surface free radicals that are generated by the interaction of PM with the aqueous milieu, as well as cellular elements in the respiratory tract (Shi et al. 2001; Aust et al. 2002). Ghio, Churg, and Roggli (2004) have suggested that oxygen containing functional groups present on the surface of PM through their capacity to coordinate iron result in the generation of radicals and activation of a variety of cell signaling pathways. Released reactive species and free radicals further activate and or interfere with the numerous cellular signaling pathways, resulting in a final expression of altered cytokine release profiles. Suppressive effects of anti-oxidants that neutralize the ROS and the general oxidative stress have been shown to mitigate the adverse effects of PM in many systems (Whitekus et al. 2002; Dick et al. 2003; Takizawa 2004; Kaimul Ahsan et al. 2005; McCunney 2005). It is interesting to note that respiratory burst and release of ROS is a normal consequence of interaction between macrophages and pathogens during allergic responses and possibly even during immunological priming. Clearly, augmentation of these processes by PM exposure may affect or exacerbate any intended outcome.

Mechanisms up or downstream of this oxidative injury are also important and noteworthy. A direct effect of ultrafine particles on the cytoskeleton of macrophages (Moller et al. 2005) and overwhelming of the cellular processes by an overload of PM (Oberdörster 1995) are illustrative examples. In the allergy/adjuvant models, a common theme has been that the particles cause some level of inflammation, which alters the cytokine balance in the lung. For example, allergic adjuvancy effects of ROFA can be replicated by direct administration of TNF-α, and are also reversed by treatment with anti-TNF-α antibodies (Lambert et al. 2000). Pulmonary injury also results in the recruitment of antigen presenting cells which may polarize subsequent immune responses to a different phenotype, while increased antigen trafficking to sub-mucosal tissue as occurs with ozone exposure (Koike and Kobayashi 2004), resulting in an amplification of allergic immune responses in susceptible individuals.

The interaction of innate and adaptive immunity is also a developing area of research inspired by the observation that low doses of bacterial endotoxin are associated with decreased allergies and asthma in children who live on farms (Ege et al. 2006). It is thought that stimulation through the Toll 4 receptor maintains mucosal immunity and a TH1 phenotype that suppresses the development of allergic type immune development.

13.6 CONCLUSIONS AND FUTURE DIRECTIONS

There are many reports showing that exposure to airborne particulates can adversely affect the immune system to increase the incidence and severity of respiratory disease. While these studies provide biological plausibility for the current epidemiological findings and offer clues for the mechanisms of these effects, additional information is needed in order to more effectively manage these risks. The chemical and physical characteristics of the particle that confer the observed toxicity, the shape of the dose response curve, and the impact of interactions with multiple air pollutants on the observed effects are far from clear. The potential for recovery versus permanent effects, factors contributing to susceptibility (particularly age and genetic predisposition), and effects of chronic low level exposures versus acute higher level exposures are all areas that require more study. A number of approaches are required to tackle these problems.

Firstly, there is a need to perform more immune testing in the many ongoing panel studies and epidemiology cohorts, and to analyze these measurements against personal exposure history and disease outcome. This is clearly a large and inherently complex task and interpretation will be confounded by many important parameters including infection, vaccination, and antigen exposure history, as well as other key immune modulators such as diet, activity patterns, alcohol use, etc. Nevertheless, general markers of immune competence such as total and antigen specific antibody levels, immune cell activity, and cytokine profiling in various cell types are needed to complement other health outcomes examined in epidemiology cohorts and will provide vital information for risk assessors.

To complement these studies, more intensive investigations in human clinical and animal studies will provide better information on the effects of individual and mixed inhaled particles on the immune system and subsequent development of disease. Inhalation studies, which either harness or create realistic pollution exposures, will identify the effect of both acute and chronic PM exposure on healthy and diseased animals to model real life situations. These can be achieved through CAPs technology and comparative testing of emission sources with the caveat that air pollution is a dynamic mix of aged particles and interactive gases. In addition, examination of PM samples from different areas taking into account seasonality and source apportionment models will provide a physicochemical basis to contrast and compare health effects of PM across various regions. To that end, instillation experiments like those of Steerenberg et al. (2006) assessing the immune effects in animals exposed to ambient PM from several different cities in Europe supplies crucial information about which chemical components are associated with PM health effects and provide mechanistic linkage to the epidemiology studies. *In vitro* studies can also be used as screening tools to provide relative toxicity data and mechanistic information in isolated and mixed cell systems. With these research activities in place, the comparative toxicity and mechanisms of action of PM can be studied and more meaningfully applied to hazard identification and risk assessment processes.

ACKNOWLEDGMENT

The authors appreciate the editorial comments of Dr. Maryjane Selgrade, U.S. EPA. This paper has been reviewed by the National Health and Environmental Effects Research Laboratory, U.S. Environmental Protection Agency, and approved for publication. Approval does not signify that the contents necessarily reflect the views and policies of the Agency, nor does the mention of trade names or commercial products constitute endorsement or recommendation for use.

REFERENCES

Akazawa, M., Sindelar, J. L., and Paltiel, A. D., Economic costs of influenza-related work absenteeism, *Value Health*, 6, 107–115, 2003.

Antonini, J. M., Roberts, J. R., Yang, H. M., Barger, M. W., Ramsey, D., Castranova, V., and Ma, J. Y., Effect of silica inhalation on the pulmonary clearance of a bacterial pathogen in Fischer 344 rats, *Lung*, 178, 341–350, 2000.

Atkinson, R. W., Anderson, H. R., Sunyer, J., Ayres, J., Baccini, M., Vonk, J. M., Boumghar, A. et al., Acute effects of particulate air pollution on respiratory admissions: results from APHEA 2 project. Air pollution and health: a European approach, *Am. J. Respir. Crit. Care Med.*, 164, 1860–1866, 2001.

Aust, A. E., Ball, J. C., Hu, A. A., Lighty, J. S., Smith, K. R., Straccia, A. M., Veranth, J. M., and Young, W. C., Particle characteristics responsible for effects on human lung epithelial cells, *Res. Rep. Health Eff. Inst.*, 110, 1–65, 2002, discussion 67–76.

Becker, S., Dailey, L., Soukup, J. M., Silbajoris, R., and Devlin, R. B., TLR-2 is involved in airway epithelial cell response to air pollution particles, *Toxicol. Appl. Pharmacol.*, 203, 45–52, 2005.

Bonay, M., Chambellan, A., Grandsaigne, M., Aubier, M., and Soler, P., Effects of diesel particles on the control of intracellular mycobacterial growth by human macrophages in vitro, *FEMS Immunol. Med. Microbiol.*, 46, 419–425, 2006.

Brown, D. M., Wilson, M. R., MacNee, W., Stone, V., and Donaldson, K., Size-dependent proinflammatory effects of ultrafine polystyrene particles: a role for surface area and oxidative stress in the enhanced activity of ultrafines, *Toxicol. Appl. Pharmacol.*, 175, 191–199, 2001.

Brunekreef, B., Janssen, N. A., de Hartog, J., Harssema, H., Knape, M., and van Vliet, P., Air pollution from truck traffic and lung function in children living near motorways, *Epidemiology*, 8, 298–303, 1997.

Canning, B. J., Hmieleski, R. R., Spannhake, E. W., and Jakab, G. J., Ozone reduces murine alveolar and peritoneal macrophage phagocytosis: the role of prostanoids, *Am. J. Physiol.*, 261, L277–L282, 1991.

Castranova, V., Ma, J. Y., Yang, H. M., Antonini, J. M., Butterworth, L., Barger, M. W., Roberts, J., and Ma, J. K., Effect of exposure to diesel exhaust particles on the susceptibility of the lung to infection, *Environ. Health Perspect.*, 109 (suppl. 4), 609–612, 2001.

CDC, Asthma prevalence and control characteristics by Race/Ethnicity: United States, *Morbidity Mortality Weekly Report*, 145–148, 2004.

Coffin, D. L. and Blommer, E. J., Acute toxicity of irradiated auto exhaust. Its indication by enhancement of mortality from streptococcal pneumonia, *Arch. Environ. Health*, 15, 36–38, 1967.

Cohen, M. D., Zelikoff, J. T., and Schlesinger, R. B., *Pulmonary Immunotoxicology*, Kluwer Academic Publishers, Norwell, MA, 2000.

D'Amato, G., Outdoor air pollution in urban areas and allergic respiratory diseases, *Monaldi Arch. Chest. Dis.*, 54, 470–474, 1999.

Davis, G. S., Pfeiffer, L. M., and Hemenway, D. R., Interferon-gamma production by specific lung lymphocyte phenotypes in silicosis in mice, *Am. J. Respir. Cell Mol. Biol.*, 22, 491–501, 2000.

de Haar, C., Hassing, I., Bol, M., Bleumink, R., and Pieters, R., Ultrafine carbon black particles cause early airway inflammation and have adjuvant activity in a mouse allergic airway disease model, *Toxicol. Sci.*, 87, 409–418, 2005.

Delfino, R. J., Gong, H., Jr., Linn, W. S., Pellizzari, E. D., and Hu, Y., Asthma symptoms in Hispanic children and daily ambient exposures to toxic and criteria air pollutants, *Environ. Health Perspect.*, 111, 647–656, 2003.

Diaz-Sanchez, D., Dotson, A. R., Takeaka, H., and Saxon, A., Diesel exhaust particles induce local IgE production in vivo and alter the pattern of messenger RNA isoforms, *J. Clin. Invest.*, 94, 1417–1425, 1994.

Diaz-Sanchez, D., Garcia, M. P., Wang, M., Jyrala, M., and Saxon, A., Nasal challenge with diesel exhaust particles can induce sensitization to a neoallergen in the human mucosa, *J. Allergy Clin. Immunol.*, 104, 1183–1188, 1999.

Diaz-Sanchez, D., Penichet-Garcia, M., and Saxon, A., Diesel exhaust particles directly induce activated mast cells to degranulate and increase histamine levels and symptom severity, *J. Allergy. Clin. Immunol.*, 106, 1140–1146, 2000.

Dick, C. A., Brown, D. M., Donaldson, K., and Stone, V., The role of free radicals in the toxic and inflammatory effects of four different ultrafine particle types, *Inhal. Toxicol.*, 15, 39–52, 2003.

Dockery, D. W. and Pope, C. A., Acute respiratory effects of particulate air pollution, *Annu. Rev. Public Health*, 15, 107–132, 1994.

Dominici, F., Peng, R. D., Bell, M. L., Pham, L., McDermott, A., Zeger, S. L., and Samet, J. M., Fine particulate air pollution and hospital admission for cardiovascular and respiratory diseases, *JAMA*, 295, 1127–1134, 2006.

Droemann, D., Goldmann, T., Tiedje, T., Zabel, P., Dalhoff, K., and Schaaf, B., Toll-like receptor 2 expression is decreased on alveolar macrophages in cigarette smokers and COPD patients, *Respir. Res.*, 6, 68, 2005.

Ege, M. J., Bieli, C., Frei, R., van Strien, R. T., Riedler, J., Ublagger, E., Schram-Bijkerk, D. et al., Prenatal farm exposure is related to the expression of receptors of the innate immunity and to atopic sensitization in school-age children, *J. Allergy Clin. Immunol.*, 117, 817–823, 2006.

El-Etr, S. H. and Cirillo, J. D., Entry mechanisms of mycobacteria, *Front. Biosci.*, 6, D737–D747, 2001.

Fujimaki, H., Saneyoshi, K., Shiraishi, F., Imai, T., and Endo, T., Inhalation of diesel exhaust enhances antigen-specific IgE antibody production in mice, *Toxicology*, 116, 227–233, 1997.

Gardner, D. E., Alterations in macrophage functions by environmental chemicals, *Environ. Health Perspect.*, 55, 343–358, 1984.

Garn, H., Friedetzky, A., Kirchner, A., Jager, R., and Gemsa, D., Experimental silicosis: a shift to a preferential IFN-gamma-based Th1 response in thoracic lymph nodes, *Am. J. Physiol. Lung Cell Mol. Physiol.*, 278, L1221–L1230, 2000.

Gauderman, W. J., Avol, E., Lurmann, F., Kuenzli, N., Gilliland, F., Peters, J., and McConnell, R., Childhood asthma and exposure to traffic and nitrogen dioxide, *Epidemiology*, 16, 737–743, 2005.

Gavett, S. H., Madison, S. L., Stevens, M. A., and Costa, D. L., Residual oil fly ash amplifies allergic cytokines, airway responsiveness, and inflammation in mice, *Am. J. Respir. Crit. Care Med.*, 160, 1897–1904, 1999.

Ghio, A. J., Churg, A., and Roggli, V. L., Ferruginous bodies: implications in the mechanism of fiber and particle toxicity, *Toxicol. Pathol.*, 32, 643–649, 2004.

Gilmour, M. I., Taylor, F. G., Baskerville, A., and Wathes, C. M., The effect of titanium dioxide inhalation on the pulmonary clearance of *Pasteurella haemolytica* in the mouse, *Environ. Res.*, 50, 157–172, 1989.

Gilmour, M. I., Daniels, M., McCrillis, R. C., Winsett, D., and Selgrade, M. K., Air pollutant-enhanced respiratory disease in experimental animals, *Environ. Health Perspect.*, 109 (suppl. 4), 619–622, 2001.

Granum, B. and Lovik, M., The effect of particles on allergic immune responses, *Toxicol. Sci.*, 65, 7–17, 2002.

Green, G. M., Jakab, G. J., Low, R. B., and Davis, G. S., Defense mechanisms of the respiratory membrane, *Am. Rev. Respir. Dis.*, 115, 479–514, 1977.

Greene, C. M. and McElvaney, N. G., Toll-like receptor expression and function in airway epithelial cells, *Arch. Immunol. Ther. Exp. (Warsz)*, 53, 418–427, 2005.

Hadnagy, W. and Seemayer, N. H., Inhibition of phagocytosis of human macrophages induced by airborne particulates, *Toxicol. Lett.*, 72, 23–31, 1994.

Hajat, S., Haines, A., Goubet, S. A., Atkinson, R. W., and Anderson, H. R., Association of air pollution with daily GP consultations for asthma and other lower respiratory conditions in London, *Thorax*, 54, 597–605, 1999.

Hamada, K., Goldsmith, C. A., Suzaki, Y., Goldman, A., and Kobzik, L., Airway hyperresponsiveness caused by aerosol exposure to residual oil fly ash leachate in mice, *J. Toxicol. Environ. Health A*, 65, 1351–1365, 2002.

Harrod, K. S., Jaramillo, R. J., Rosenberger, C. L., Wang, S. Z., Berger, J. A., McDonald, J. D., and Reed, M. D., Increased susceptibility to RSV infection by exposure to inhaled diesel engine emissions, *Am. J. Respir. Cell Mol. Biol.*, 28, 451–463, 2003.

Harrod, K. S., Jaramillo, R. J., Berger, J. A., Gigliotti, A. P., Seilkop, S. K., and Reed, M. D., Inhaled diesel engine emissions reduce bacterial clearance and exacerbate lung disease to *Pseudomonas aeruginosa* infection in vivo, *Toxicol. Sci.*, 83, 155–165, 2005.

Hatch, G. E., Slade, R., Boykin, E., Hu, P. C., Miller, F. J., and Gardner, D. E., Correlation of effects of inhaled versus intratracheally injected males on susceptibility to respiratory infection in mice, *Am. Rev. Respir. Dis.*, 124, 167–173, 1981.

Hatch, G. E., Boykin, E., Graham, J. A., Lewtas, J., Pott, F., Loud, K., and Mumford, J. L., Inhalable particles and pulmonary host defense: in vivo and in vitro effects of ambient air and combustion particles, *Environ. Res.*, 36, 67–80, 1985.

Hiramatsu, K., Saito, Y., Sakakibara, K., Azuma, A., Takizawa, H., and Sugawara, I., The effects of inhalation of diesel exhaust on murine mycobacterial infection, *Exp. Lung Res.*, 31, 405–415, 2005.

Holland, W. W., Bennett, A. E., Cameron, I. R., Florey, C. V., Leeder, S. R., Schilling, R. S., Swan, A. V., and Waller, R. E., Health effects of particulate pollution: reappraising the evidence, *Am. J. Epidemiol.*, 110, 527–659, 1979.

Huang, S. H., Hubbs, A. F., Stanley, C. F., Vallyathan, V., Schnabel, P. C., Rojanasakul, Y., Ma, J. K., Banks, D. E., and Weissman, D. N., Immunoglobulin responses to experimental silicosis, *Toxicol. Sci.*, 59, 108–117, 2001.

Inoue, K., Takano, H., Yanagisawa, R., Hirano, S., Ichinose, T., Shimada, A., and Yoshikawa, T., The role of toll-like receptor 4 in airway inflammation induced by diesel exhaust particles, *Arch. Toxicol.*, 80, 275–279, 2006.

Ito, T., Okumura, H., Tsukue, N., Kobayashi, T., Honda, K., and Sekizawa, K., Effect of diesel exhaust particles on mRNA expression of viral and bacterial receptors in rat lung epithelial L2 cells, *Toxicol. Lett.*, 2006.

Jakab, G. J., The toxicologic interactions resulting from inhalation of carbon black and acrolein on pulmonary antibacterial and antiviral defenses, *Toxicol. Appl. Pharmacol.*, 121, 167–175, 1993.

Jakab, G. J., Risby, T. H., Sehnert, S. S., Hmieleski, R. R., and Gilmour, M. I., Suppression of alveolar macrophage membrane-receptor-mediated phagocytosis by model particle–adsorbate complexes: physicochemical moderators of uptake, *Environ. Health Perspect.*, 89, 169–174, 1990.

Jakab, G. J., Spannhake, E. W., Canning, B. J., Kleeberger, S. R., and Gilmour, M. I., The effects of ozone on immune function, *Environ. Health Perspect.*, 103 (suppl. 2), 77–89, 1995.

Jaspers, I., Ciencewicki, J. M., Zhang, W., Brighton, L. E., Carson, J. L., Beck, M. A., and Madden, M. C., Diesel exhaust enhances influenza virus infections in respiratory epithelial cells, *Toxicol. Sci.*, 85, 990–1002, 2005.

Johnson, J. D., Houchens, D. P., Kluwe, W. M., Craig, D. K., and Fisher, G. L., Effects of mainstream and environmental tobacco smoke on the immune system in animals and humans: a review, *Crit. Rev. Toxicol.*, 20, 369–395, 1990.

Joos, G. F., De Swert, K. O., Schelfhout, V., and Pauwels, R. A., The role of neural inflammation in asthma and chronic obstructive pulmonary disease, *Ann. NY Acad. Sci.*, 992, 218–230, 2003.

Kaimul Ahsan, M., Nakamura, H., Tanito, M., Yamada, K., Utsumi, H., and Yodoi, J., Thioredoxin-1 suppresses lung injury and apoptosis induced by diesel exhaust particles (DEP) by scavenging reactive oxygen species and by inhibiting DEP-induced downregulation of Akt, *Free Radic. Biol. Med.*, 39, 1549–1559, 2005.

Koike, E. and Kobayashi, T., Ozone exposure enhances antigen-presenting activity of interstitial lung cells in rats, *Toxicology*, 196, 217–227, 2004.

Koike, E., Hirano, S., Furuyama, A., and Kobayashi, T., cDNA microarray analysis of rat alveolar epithelial cells following exposure to organic extract of diesel exhaust particles, *Toxicol. Appl. Pharmacol.*, 201, 178–185, 2004.

Koren, H. S., Associations between criteria air pollutants and asthma, *Environ. Health Perspect.*, 103 (suppl. 6), 235–242, 1995.

Kowolenko, M., Tracy, L., Mudzinski, S., and Lawrence, D. A., Effect of lead on macrophage function, *J. Leukoc. Biol.*, 43, 357–364, 1988.

Krutzik, S. R. and Modlin, R. L., The role of Toll-like receptors in combating mycobacteria, *Semin. Immunol.*, 16, 35–41, 2004.

Lambert, A. L., Dong, W., Selgrade, M. K., and Gilmour, M. I., Enhanced allergic sensitization by residual oil fly ash particles is mediated by soluble metal constituents, *Toxicol. Appl. Pharmacol.*, 165, 84–93, 2000.

Li, N. and Nel, A. E., Role of the Nrf2-mediated signaling pathway as a negative regulator of inflammation: implications for the impact of particulate pollutants on asthma, *Antioxid. Redox Signal.*, 8, 88–98, 2006.

Lundborg, M., Johard, U., Lastbom, L., Gerde, P., and Camner, P., Human alveolar macrophage phagocytic function is impaired by aggregates of ultrafine carbon particles, *Environ. Res.*, 86, 244–253, 2001.

Maejima, K., Tamura, K., Taniguchi, Y., Nagase, S., and Tanaka, H., Comparison of the effects of various fine particles on IgE antibody production in mice inhaling Japanese cedar pollen allergens, *J. Toxicol. Environ. Health*, 52, 231–248, 1997.

McCunney, R. J., Asthma, genes, and air pollution, *J. Occup. Environ. Med.*, 47, 1285–1291, 2005.

Moller, W., Brown, D. M., Kreyling, W. G., and Stone, V., Ultrafine particles cause cytoskeletal dysfunctions in macrophages: role of intracellular calcium, *Part. Fibre Toxicol.*, 2, 7, 2005.

Moores, H. K., Janigan, D. T., and Hajela, R. P., Lung injury after experimental smoke inhalation: particle-associated changes in alveolar macrophages, *Toxicol. Pathol.*, 21, 521–527, 1993.

Mundandhara, S. D., Becker, S., and Madden, M. C., Effects of diesel exhaust particles on human alveolar macrophage ability to secrete inflammatory mediators in response to lipopolysaccharide, *Toxicol. In Vitro,* 2005.

Namork, E., Johansen, B. V., and Lovik, M., Detection of allergens adsorbed to ambient air particles collected in four European cities, *Toxicol Lett.*, 165, 71–78, 2006.

Nel, A., Atmosphere. Air pollution-related illness: effects of particles, *Science*, 308, 804–806, 2005.

Newcombe, D. S., Immune surveillance, organophosphorus exposure, and lymphomagenesis, *Lancet*, 339, 539–541, 1992.

Nicolai, T., Environmental air pollution and lung disease in children, *Monaldi Arch. Chest. Dis.*, 54, 475–478, 1999.

Nygaard, U. C., Aase, A., and Lovik, M., The allergy adjuvant effect of particles—genetic factors influence antibody and cytokine responses, *BMC Immunol.*, 6, 11, 2005.

Oberdörster, G., Lung particle overload: implications for occupational exposures to particles, *Regul. Toxicol. Pharmacol.*, 21, 123–135, 1995.

Ostro, B., The association of air pollution and mortality: examining the case for inference, *Arch. Environ. Health*, 48, 336–342, 1993.

Phaybouth, V., Wang, S. Z., Hutt, J. A., McDonald, J. D., Harrod, K. S., and Barrett, E. G., Cigarette smoke suppresses Th1 cytokine production and increases RSV expression in a neonatal model, *Am. J. Physiol. Lung Cell Mol. Physiol.*, 290, L222–L231, 2006.

Pourazar, J., Frew, A. J., Blomberg, A., Helleday, R., Kelly, F. J., Wilson, S., and Sandstrom, T., Diesel exhaust exposure enhances the expression of IL-13 in the bronchial epithelium of healthy subjects, *Respir. Med.*, 98, 821–825, 2004.

Renwick, L. C., Donaldson, K., and Clouter, A., Impairment of alveolar macrophage phagocytosis by ultrafine particles, *Toxicol. Appl. Pharmacol.*, 172, 119–127, 2001.

Rusznak, C., Devalia, J. L., and Davies, R. J., The impact of pollution on allergic disease, *Allergy*, 49, 21–27, 1994.

Saboori, A. M. and Newcombe, D. S., *Environmental Chemicals with Immunotoxic Properties*, Raven Press Ltd, New York, 1992.

Saito, Y., Azuma, A., Kudo, S., Takizawa, H., and Sugawara, I., Long-term inhalation of diesel exhaust affects cytokine expression in murine lung tissues: comparison between low- and high-dose diesel exhaust exposure, *Exp. Lung Res.*, 28, 493–506, 2002.

Saxena, Q. B., Saxena, R. K., Siegel, P. D., and Lewis, D. M., Identification of organic fractions of diesel exhaust particulate (DEP) which inhibit nitric oxide (NO) production from a murine macrophage cell line, *Toxicol. Lett.*, 143, 317–322, 2003.

Saxena, R. K., Saxena, Q. B., Weissman, D. N., Simpson, J. P., Bledsoe, T. A., and Lewis, D. M., Effect of diesel exhaust particulate on bacillus Calmette–Guerin lung infection in mice and attendant changes in lung interstitial lymphoid subpopulations and IFN gamma response, *Toxicol. Sci.*, 73, 66–71, 2003.

Schaub, B., Lauener, R., and von Mutius, E., The many faces of the hygiene hypothesis, *J. Allergy Clin. Immunol.*, 117, 969–977, 2006.

Sha, Q., Truong-Tran, A. Q., Plitt, J. R., Beck, L. A., and Schleimer, R. P., Activation of airway epithelial cells by toll-like receptor agonists, *Am. J. Respir. Cell Mol. Biol.*, 31, 358–364, 2004.

Shi, X., Ding, M., Chen, F., Wang, L., Rojanasakul, Y., Vallyathan, V., and Castranova, V., Reactive oxygen species and molecular mechanism of silica-induced lung injury, *J. Environ. Pathol. Toxicol. Oncol.*, 20 (suppl. 1), 85–93, 2001.

Shima, H., Koike, E., Shinohara, R., and Kobayashi, T., Oxidative ability and toxicity of *n*-hexane insoluble fraction of diesel exhaust particles, *Toxicol. Sci.*, 91, 218–226, 2006.

Singh, P., Madden, M., and Gilmour, M. I., Effects of diesel exhaust particles and carbon black on induction of dust mite allergy in Brown Norway rats, *J. Immunotoxicol.*, 2, 41–49, 2005.

Steerenberg, P. A., van Amelsvoort, L., Lovik, M., Hetland, R. B., Alberg, T., Halatek, T., Bloemen, H. J. et al., Relation between sources of particulate air pollution and biological effect parameters in samples from four European cities: an exploratory study, *Inhal. Toxicol.*, 18, 333–346, 2006.

Takafuji, S., Suzuki, S., Koizumi, K., Tadokoro, K., Ohashi, H., Muranaka, M., and Miyamoto, T., Enhancing effect of suspended particulate matter on the IgE antibody production in mice, *Int. Arch. Allergy Appl. Immunol.*, 90, 1–7, 1989.

Takeda, K. and Akira, S., Toll-like receptors in innate immunity, *Int. Immunol.*, 17, 1–14, 2005.

Takizawa, H., Diesel exhaust particles and their effect on induced cytokine expression in human bronchial epithelial cells, *Curr. Opin. Allergy Clin. Immunol.*, 4, 355–359, 2004.

TeWaternaude, J. M., Ehrlich, R. I., Churchyard, G. J., Pemba, L., Dekker, K., Vermeis, M., White, N. W., Thompson, M. L., and Myers, J. E., Tuberculosis and silica exposure in South African gold miners, *Occup. Environ. Med.*, 63, 187–192, 2006.

Van Zijverden, M., van der Pijl, A., Bol, M., van Pinxteren, F. A., de Haar, C., Penninks, A. H., van Loveren, H., and Pieters, R., Diesel exhaust, carbon black, and silica particles display distinct Th1/Th2 modulating activity, *Toxicol. Appl. Pharmacol.*, 168, 131–139, 2000.

Vassallo, R., Tamada, K., Lau, J. S., Kroening, P. R., and Chen, L., Cigarette smoke extract suppresses human dendritic cell function leading to preferential induction of Th-2 priming, *J. Immunol.*, 175, 2684–2691, 2005.

Vigotti, M. A., Short-term effects of exposure to urban air pollution on human health in Europe. The APHEA projects (air pollution and health: a European approach), *Epidemiol. Prev.*, 23, 408–415, 1999.

Wade, J. F. and Newman, L. S., Diesel asthma. Reactive airways disease following overexposure to locomotive exhaust, *J. Occup. Med.*, 35, 149–154, 1993.

Whitekus, M. J., Li, N., Zhang, M., Wang, M., Horwitz, M. A., Nelson, S. K., Horwitz, L. D., Brechun, N., Diaz-Sanchez, D., and Nel, A. E., Thiol antioxidants inhibit the adjuvant effects of aerosolized diesel exhaust particles in a murine model for ovalbumin sensitization, *J. Immunol.*, 168, 2560–2567, 2002.

Wong, S. S., Sun, N. N., Keith, I., Kweon, C. B., Foster, D. E., Schauer, J. J., and Witten, M. L., Tachykinin substance P signaling involved in diesel exhaust-induced bronchopulmonary neurogenic inflammation in rats, *Arch. Toxicol.*, 77, 638–650, 2003.

Yang, H. M., Antonini, J. M., Barger, M. W., Butterworth, L., Roberts, B. R., Ma, J. K., Castranova, V., and Ma, J. Y., Diesel exhaust particles suppress macrophage function and slow the pulmonary clearance of *Listeria monocytogenes* in rats, *Environ. Health Perspect.*, 109, 515–521, 2001.

Yang, H. M., Butterworth, L., Munson, A. E., and Meade, B. J., Respiratory exposure to diesel exhaust particles decreases the spleen IgM response to a T cell-dependent antigen in female B6C3F1 mice, *Toxicol. Sci.*, 71, 207–216, 2003.

Yin, X. J., Ma, J. Y., Antonini, J. M., Castranova, V., and Ma, J. K., Roles of reactive oxygen species and heme oxygenase-1 in modulation of alveolar macrophage-mediated pulmonary immune responses to *Listeria monocytogenes* by diesel exhaust particles, *Toxicol. Sci.*, 82, 143–153, 2004.

Zelikoff, J. T., Parsons, E., and Schlesinger, R. B., Inhalation of particulate lead oxide disrupts pulmonary macrophage-mediated functions important for host defense and tumor surveillance in the lung, *Environ. Res.*, 62, 207–222, 1993.

Zelikoff, J. T., Chen, L. C., Cohen, M. D., and Schlesinger, R. B., The toxicology of inhaled woodsmoke, *J. Toxicol. Environ. Health B Crit. Rev.*, 5, 269–282, 2002.

Zelikoff, J. T., Chen, L. C., Cohen, M. D., Fang, K., Gordon, T., Li, Y., Nadziejko, C., and Schlesinger, R. B., Effects of inhaled ambient particulate matter on pulmonary antimicrobial immune defense, *Inhal. Toxicol.*, 15, 131–150, 2003.

Ziegler, B., Bhalla, D. K., Rasmussen, R. E., Kleinman, M. T., and Menzel, D. B., Inhalation of resuspended road dust, but not ammonium nitrate, decreases the expression of the pulmonary macrophage Fc receptor, *Toxicol. Lett.*, 71, 197–208, 1994.

14 Effects of Particles on the Cardiovascular System

Nicholas L. Mills
Centre for Cardiovascular Sciences, The University of Edinburgh

David E. Newby
Centre for Cardiovascular Sciences, The University of Edinburgh

William MacNee
MRC/University of Edinburgh Centre for Inflammation Research, Queen's Medical Research Institute

Ken Donaldson
MRC/University of Edinburgh Centre for Inflammation Research, Queen's Medical Research Institute

CONTENTS

14.1 INTRODUCTION

In recent years it has become widely appreciated that inhaled particles have adverse effects beyond the lungs and may exert an important influence on the cardiovascular system. This concept has come to the fore as a result of the burgeoning epidemiological literature documenting associations between exposure to environmental particles (PM_{10}) and cardiovascular morbidity and mortality. The expansion in our knowledge of the impact of particles on the cardiovascular system provides the focus of the present review. Because of the importance of the combustion-derived component of PM in determining adverse effects in the lungs (reviewed extensively by the authors elsewhere, Donaldson et al. 2005) and the link between nanoparticles and the cardiovascular system, this review focuses largely on the role of combustion-derived nanoparticles (CDNP).

14.2 ENVIRONMENTAL PM AND CARDIOVASCULAR EFFECTS

14.2.1 Epidemiological Evidence

The cardiovascular morbidity and mortality associated with increases in PM are well-documented (reviewed in Glantz 2002; Brook, Brook, and Rajagopalan 2003; Routledge, Ayres, and Townend 2003; Brook et al. 2004). Levels of PM vary temporally, fluctuating around an hourly or daily mean, with "time series" studies utilizing this temporal dimension to relate the moving average to a defined end-point, such as cardiovascular deaths. Using this approach, there is a clear relationship between PM_{10} and cardiovascular death in the hours and days following increases in the levels of PM_{10} (Table 14.1).

The composition of urban PM is dependent on local sources, with concentrations dependent on a number of variables including distance from source, wind speed, and direction. This spatial dimension is also utilized in environmental epidemiological studies and in this case, populations living in areas of high air pollution are compared to those in areas with low air pollution for a given health endpoint. A good example of this type of study is the Harvard Six Cities Study, which provided strong evidence that cardiovascular disease was more prevalent in areas where PM levels were higher (Table 14.2).

More detailed information that is helpful in understanding the likely mechanisms responsible for the cardiovascular effects of PM has been provided by panel studies, which have documented associations between elevated levels of particles and the onset of acute myocardial infarction (Peters et al. 2001a), increased heart rate (Peters et al. 1999), and decreased heart rate variability (Devlin et al. 2003). Chamber studies with concentrated airborne particles (CAPs) have also shown various effects, including lung inflammation (Ghio, Kim, and Devlin 2000), decreased heart rate variability (Devlin et al. 2003), and alterations in brachial artery diameter (Brook et al. 2002).

Further discussion of these observational studies are beyond the scope of this review that focuses on particle toxicology, but the following are useful and detailed reviews of the

TABLE 14.1
Percentage Change in Cardiovascular Outcomes for a 10 μg/m³ Increase in PM_{10}

Outcome	% Change per 10 μg/m³
Cardiovascular mortality	0.5 (0.4–0.7)
Cardiovascular admissions	0.5 (0.2–0.7)
Ischaemic heart disease admissions	0.8 (0.6–0.9)

Source: Adapted from *Committee on the Medical Effects of Air Pollutants*, 2006. With permission.

TABLE 14.2
Adjusted Mortality Rate Ratios for the Most versus the Least Polluted City in the U.S. 6 Cities Study

Outcome	Most vs. Least Polluted City
All mortality	1.26 (1.08–1.47)
Mortality due to cardiopulmonary disease	1.37 (1.11–1.68)

Source: Adapted from *Committee on the Medical Effects of Air Pollutants*, 2006. With permission.

epidemiological literature (Dockery et al. 1993; Brook, Brook, and Rajagopalan 2003; Goldberg et al. 2003; Brook et al. 2004; Schwartz 2004; Committee on the Medical Effects of Air Pollutants 2006).

14.2.2 Toxicology Studies

A number of interesting theories have been proposed to explain the association between increased PM and cardiovascular disease. Based on our existing knowledge, we present a summary of the potential pathways linking inhaled particles and adverse cardiovascular outcomes in Figure 14.1. According to this model, inhaled particles may cause pulmonary and systemic inflammation, which could indirectly impact the cardiovascular system. Smaller particles may also translocate into the bloodstream, exerting direct effects on the vasculature. In the blood, particles could affect thrombogenesis through effects on endothelial cells and platelets, or may they penetrate the vessel wall, promoting atherothrombosis or predisposing to plaque rupture.

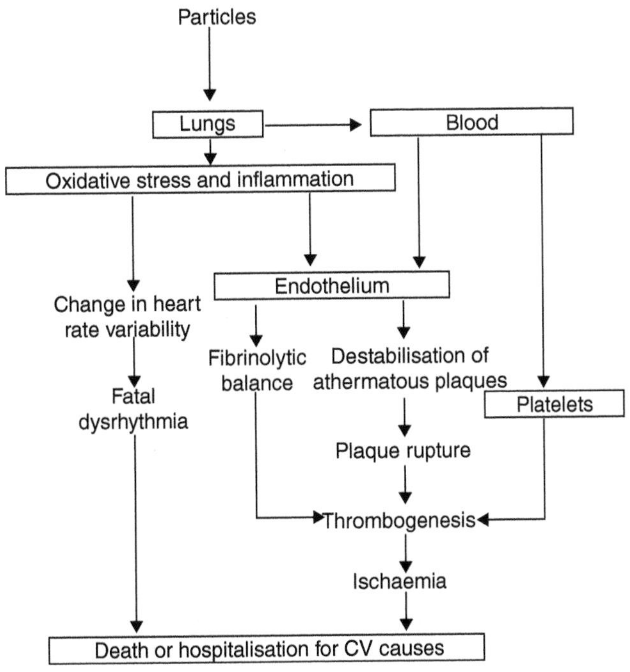

FIGURE 14.1 Hypothetical pathway for the effect of particles on the cardiovascular system based on toxicology studies.

14.2.2.1 The Inflammatory Effects of PM

14.2.2.1.1 Pulmonary Inflammation

Particles, in general, exert their pathogenic effects through oxidative stress and inflammation (Donaldson and Tran 2002). There are numerous reports of PM_{10} particles causes lung inflammation in animal models following intrapulmonary instillation (Li et al. 1996) and inhalation of concentrated ambient particles (CAPs) (Elder et al. 2004). In human studies, inflammatory effects have been demonstrated following inhalation of CAPs (Ghio, Kim, and Devlin 2000; Holgate et al. 2003) and instillation of aqueous extracts of PM (Ghio and Devlin 2001).

14.2.2.1.2 Systemic Inflammation

Increases in plasma or serum markers of systemic inflammation have been reported following exposure to particles. In panel and population studies, PM exposure is associated with evidence of an acute phase response with increased C-reactive protein (CRP) (Peters et al. 2001b) and plasma fibrinogen (Ghio, Kim, and Devlin 2000; Pekkanen et al. 2000; Schwartz 2001), enhanced plasma viscosity (Peters et al. 2001b), and altered hematological indices (Seaton et al. 1999). In animal studies, there are similar reports with increased fibrinogen in the blood of PM-exposed hypertensive (Cassee et al. 2002) and normal rats (Elder et al. 2002).

14.2.2.2 Potential Effects of PM on Vascular Inflammation, Atherosclerosis, and Plaque Stability

In health, the endothelial monolayer delicately balances regulatory pathways controlling vasomotion, thrombosis, cellular proliferation, inflammation, and oxidative stress. Atherosclerosis is widely recognized to be an inflammatory disease with endothelial dysfunction as one of the earliest pathological features (Ross 1999). Loss of endothelial integrity results in expression of leukocyte adhesion proteins, reduced anticoagulant activity and the release of growth factors, inflammatory mediators, and cytokines. Chronic inflammation results in leukocyte and monocyte recruitment, induction of atheroma formation, and further arterial damage. Plaque expansion and disruption can lead to angina, crescendo angina, and acute coronary syndromes, including myocardial infarction (Blum and Miller 1996; Brand et al. 1996; Ross 1999).

Biomarkers of systemic inflammation are elevated in patients with overt cardiovascular disease (Haverkate et al. 1997) and in those with established cardiovascular risk factors (Lee and Libby 1997). In apparently healthy individuals, elevated plasma concentrations of the acute phase reactant CRP have been shown to predict the development of ischemic heart disease (Kuller et al. 1996; Ridker et al. 2000a), and specifically the risk of a first myocardial or cerebral infarction, independent of other risk factors (Ridker et al. 2000b). Increases in markers of systemic inflammation such as CRP, serum amyloid A, and interleukin-6 accompany acute coronary events and correlate with short-term prognosis (Liuzzo et al. 1994). Elevation in inflammatory markers precedes myocardial necrosis, suggesting that inflammation may be the primary trigger of coronary plaque instability.

Short-term exposure to increased levels of PM_{10} induces changes that are indicative of systemic inflammation, such as increases in concentrations of white blood cells and platelets (Schwartz 2001), CRP (Seaton et al. 1999; Pope, III 2001), fibrinogen (Pekkanen et al. 2000; Prescott et al. 2000), and increased plasma viscosity (Peters et al. 1997), as well as alterations in coagulation Factor VII concentrations (Seaton et al. 1999). Experimental exposures mirror these clinical findings (van Eeden et al. 2001) and demonstrate direct evidence of combined systemic inflammation and endothelial dysfunction (Ross 1993; Vincent et al. 2001; Bonetti, Lerman, and Lerman 2003).

Repeated exposure to PM_{10} may, therefore, induce or exacerbate the vascular inflammation of atherosclerosis and promote plaque expansion or rupture. Indeed, using a Watanabe hereditary hyperlipidemic rabbit model, Suwa et al. (2002) described plaque progression and destabilization following instillation of high doses of PM_{10}. In the same model, PM_{10} exposure accelerated monocyte release from the bone marrow (Goto et al. 2004). The amount of particulate phagocytosed by alveolar macrophage correlated with both the bone marrow response (Goto et al. 2004) and plaque volume (Goto et al. 2004). ApoE mice that are predisposed to developing atheromatous plaques have been found to show more severe lesions on inhalation exposure to CAPs (Chen and Nadziejko 2005; Sun et al. 2005). These data together suggest that particulate-induced pulmonary inflammation is capable of systemic effects, which can contribute to the progression of atherosclerosis. A recent panel study in Los Angeles provided the first evidence of a link between chronic PM exposure and atherosclerosis in man (Kunzli et al. 2005). A 10 $\mu g/m^3$ increase in fine PM was associated with an increase in carotid intima-media thickness, an ultrasonic measure of atheroma, suggesting that long-term ambient PM exposure may affect the development of atherosclerosis in man.

14.2.2.3 Potential Effects of PM on Endothelial Function, Endogenous Fibrinolysis, and Thrombogenesis

Small areas of denudation and thrombus deposition are a common finding on the surface of atheromatous plaques and are usually subclinical. Endogenous fibrinolysis and "passification" of the lesion may therefore be able to prevent thrombus propagation and vessel occlusion (Davies 2000). However, in the presence of an adverse proinflammatory state or an imbalance in the fibrinolytic system, such microthrombi may propagate, ultimately leading to arterial occlusion and tissue infarction (Rosenberg and Aird 1999). Thus, the initiation, modification, and resolution of unstable and inflamed atheromatous plaques may be critically dependent on the cellular activation and function of the surrounding endothelium and vascular wall.

The endothelium plays a vital role in the control of blood flow, coagulation, fibrinolysis, and inflammation. Following the seminal work of Furchgott and Zawadski (1980), it is widely recognized that an array of mediators including cigarette smoking can influence vascular tone through endothelium-dependent actions, and there is now extensive evidence of abnormal endothelium-dependent vasomotion in patients with atherosclerosis (Ludmer et al. 1986; Celermajer et al. 1996; Newby et al. 1999). Mild systemic inflammation also causes a profound, but temporary, suppression of endothelium-dependent vasodilatation (Hingorani et al. 2000).

In a study of the mechanisms responsible for the association between short term increases in urban PM and acute myocardial infarction, we investigated the effects of diesel exhaust inhalation on vascular and endothelial function in humans (Mills et al. 2005). In a double-blind, randomized, crossover study, 30 healthy men were exposed to diluted diesel exhaust at 300 $\mu g/m^3$ particulate, or air, for 1 hour with intermittent exercise. Two and six hours after exposure, bilateral forearm blood flow was measured in response to intra-arterial infusions of the vasodilators bradykinin, acetylcholine, sodium nitroprusside, and verapamil using venous occlusion plethysmography—a well established clinical method of assessing endothelial function. Each vasodilator induced a dose-dependent increase in blood flow, but this vasomotor response was significantly attenuated 2 hours after exposure to diesel exhaust, and this impairment persisted at 6 hours (Figure 14.2).

Vasodilator drugs acetylcholine and bradykinin act on membrane bound G-protein coupled receptors on the vascular endothelium to stimulate calcium influx and nitric oxide synthase activation. This increases local nitric oxide (NO) levels in the adjacent smooth muscle cells, resulting in relaxation, vasodilatation, and an increase in forearm blood flow. The NO-donor sodium nitroprusside increases local NO concentrations via endothelial-independent pathways, while verapamil causes smooth muscle relaxation via a NO-independent pathway. The observed blunting of the vasomotor response to all vasodilators (except verapamil) suggests that NO consumption is central to the mechanism of CDNP induced vascular and endothelial dysfunction. Oxidative stress may result from pulmonary inflammation or directly from particles that gain access to the bloodstream. In this scenario, shown in Figure 14.3, excess superoxide produced as a result of enhanced oxidative

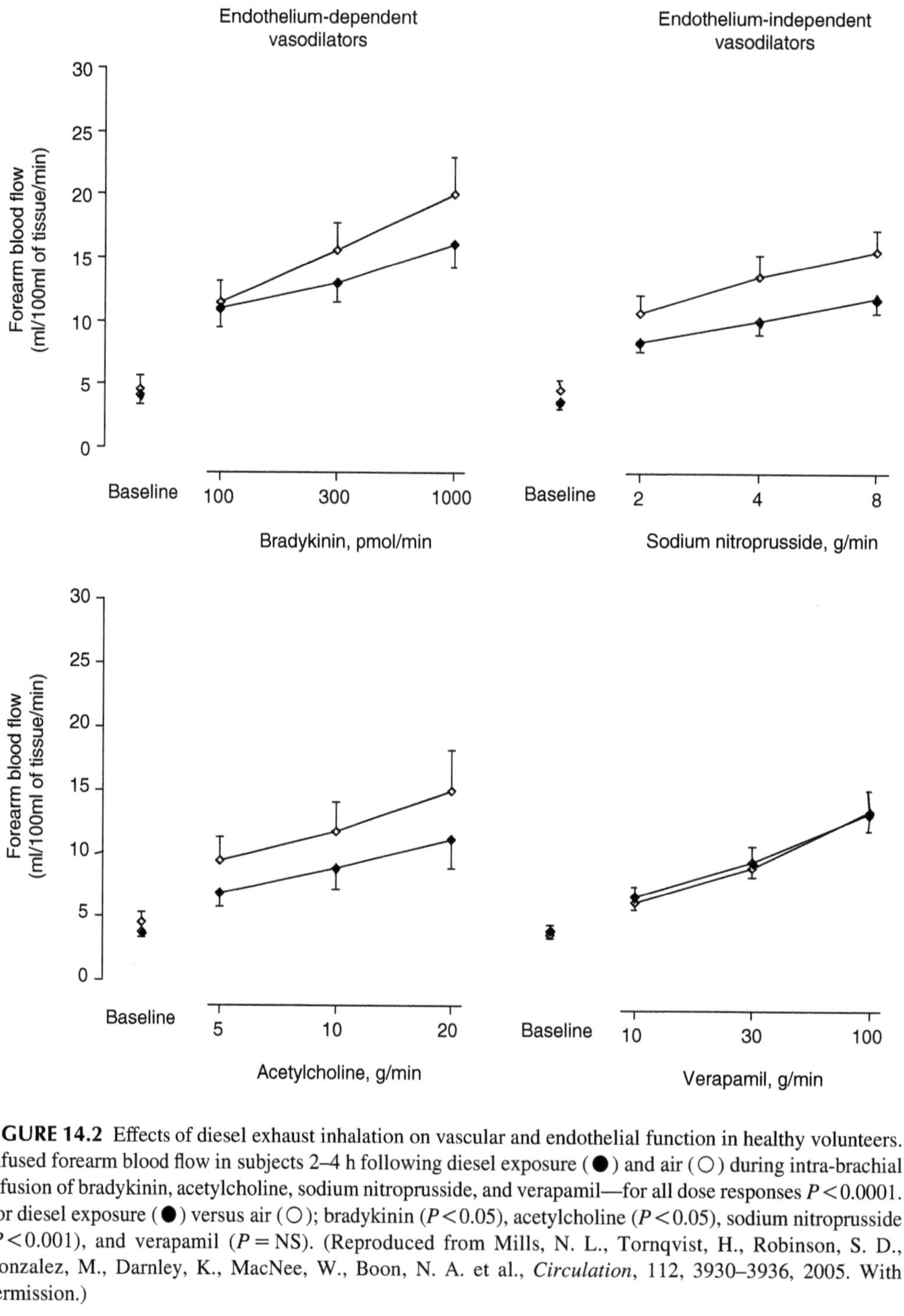

FIGURE 14.2 Effects of diesel exhaust inhalation on vascular and endothelial function in healthy volunteers. Infused forearm blood flow in subjects 2–4 h following diesel exposure (●) and air (○) during intra-brachial infusion of bradykinin, acetylcholine, sodium nitroprusside, and verapamil—for all dose responses $P<0.0001$. For diesel exposure (●) versus air (○); bradykinin ($P<0.05$), acetylcholine ($P<0.05$), sodium nitroprusside ($P<0.001$), and verapamil ($P=\text{NS}$). (Reproduced from Mills, N. L., Tornqvist, H., Robinson, S. D., Gonzalez, M., Darnley, K., MacNee, W., Boon, N. A. et al., *Circulation*, 112, 3930–3936, 2005. With permission.)

stress rapidly combines with NO in the vessel wall to form peroxynitrite, thus limiting NO availability and blunting smooth muscle relaxation.

However, whilst NO-dependent vasomotion is important, it may not be representative of other aspects of endothelial function, such as the regulation of fibrinolysis. The fibrinolytic factor tissue

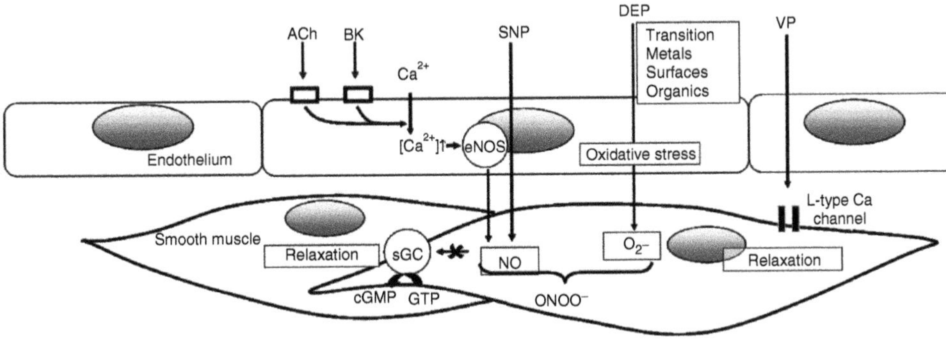

FIGURE 14.3 Hypothetical scheme to explain the observed effects of exposure to dilute diesel exhaust on vasomotor function. Enhanced vascular oxidative stress results in NO consumption and impaired smooth muscle relaxation (see text for detail). Ach, acetylcholine (endothelial-dependent vasodilator); BK, bradykinin (endothelial-dependent vasodilator); SNP, sodium nitroprusside (endothelial-independent vasodilator); DEP, diesel exhaust particles; VP, verapamil (endothelial-independent and nitric oxide-independent vasodilator); eNOS, endothelial nitric oxide synthase; sGC, soluble guanylyl cyclase; cGMP, guanosine cyclic monophosphate; GTP, guanosine triphosphate; NO, nitric oxide; O_2^-, superoxide anion.

plasminogen activator (t-PA) regulates the degradation of intravascular fibrin and is released from the endothelium through the translocation of a dynamic intracellular storage pool (Eijnden-Schrauwen et al. 1995; Hingorani et al. 2000). If endogenous fibrinolysis is to be effective, the rapid mobilization of t-PA from the endothelium is essential because thrombus dissolution is much more effective if t-PA is incorporated during, rather than after, thrombus formation (Brommer 1984; Fox et al. 1985). The efficacy of plasminogen activation and fibrin degradation is further determined by the relative balance between the acute local release of t-PA and its subsequent inhibition through formation of complexes with plasminogen activator inhibitor type 1 (PAI-1), as summarized in Figure 14.4. This dynamic aspect of endothelial function and fibrinolytic balance may be directly relevant to the pathogenesis of atherothrombosis.

Bradykinin is released during the contact phase of coagulation (Reddigari and Kaplan 1988), and results in a substantial dose-dependent release of t-PA from preformed granules within the endothelium (Labinjoh et al. 2000; Oliver, Webb and Newby 2005). In the human diesel exposure study outlined above (Mills et al. 2005), bradykinin caused a dose-dependent increase in plasma t-PA from the forearm vascular bed that was suppressed 6 hours after exposure to diesel ($P<0.001$; area under the curve decreased by 34%). We concluded that, at levels encountered in an urban environment, inhalation of dilute diesel exhaust impaired two important and complementary aspects of vascular function in man: the regulation of vascular tone and endogenous fibrinolysis. These findings combine the well-known inflammatory and pro-oxidant effects of combustion-derived nanoparticles with endothelial dysfunction and so provide a potential mechanism linking air pollution to the pathogenesis of atherothrombosis and acute myocardial infarction.

14.2.2.4 Effects of PM on the Autonomic Regulation of the Heart Rate

There is an important relationship between autonomic regulation of the cardiac cycle and cardiovascular mortality (Task Force of the European Society of Cardiology and the North American Society of Pacing and Electrophysiology 1996). Variation in the interval between consecutive heart beats, or heart rate variability (HRV), is controlled by the contrasting effects of the sympathetic and

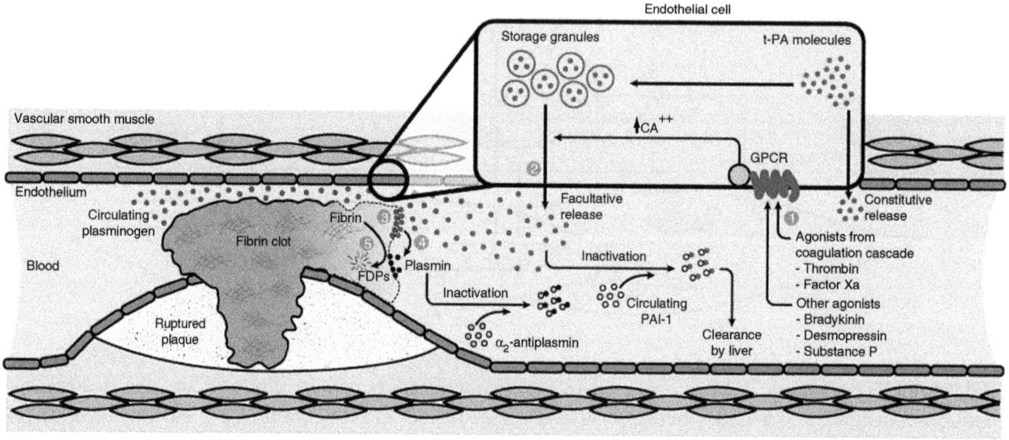

FIGURE 14.4 The endothelial fibrinolytic response to luminal thrombus. Agonists generated from the coagulation cascade act on endothelial cell surface G protein-coupled receptors (GPCRs) (1) to stimulate release of tissue plasminogen activator (t-PA) from storage granules, a step that requires an increase in intracellular calcium (Ca^{2+}) concentration (2). Free t-PA acts on thrombus-bound plasminogen (3) to produce plasmin (4) that, in turn, degrades cross-linked fibrin into fibrin degradation products (FDPs) (5), thus dissolving the thrombus. The fibrinolytic process is inhibited by inactivation of t-PA by plasminogen activator inhibitor-1 (PAI-1) and plasmin by alpha-2-antiplasmin. (Adapted from Oliver, J. J., Webb, D. J., and Newby, D. E., *Arterioscler. Thromb. Vasc. Biol.*, 25, 2470–2479, 2005. With permission.)

parasympathetic nervous systems. Reduction in HRV reflects either an increase in sympathetic drive or a decrease in vagal parasympathetic tone. Reduced HRV increases the risk of cardiovascular morbidity and mortality in both healthy individuals (Tsuji et al. 1996) and patients following myocardial infarction (Kleiger 1995).

Several panel studies have reported a consistent association between reduced HRV and high ambient PM (Liao et al. 1999; Pope, III et al. 1999; Gold et al. 2000; Creason et al. 2001; Magari et al. 2002). The finding of altered HRV in an elderly cohort exposed to concentrated ambient particles (CAPs) provides direct evidence of the effects of PM on autonomic activity (Devlin et al. 2003). The importance of this effect in the presence of cardiovascular pathology was demonstrated in a canine model of myocardial infarction, where exposure to residual oil fly ash (ROFA) reduced HRV and increased cardiac arrhythmia following infarction (Wellenius et al. 2002). Furthermore, in patients with implanted cardiac defibrillators, there appears to be a relationship between ambient PM and the incidence of ventricular fibrillation (Peters et al. 2000).

The influence of PM on the autonomic nervous system may result in hospitalization or cardiac death by triggering tachyarrhythmia or altering coronary vascular tone. How inhaled PM interacts with the nervous system is not clear. The effects could be mediated by interstitialized particles directly irritating the nerve endings or through the local release of inflammatory cytokines in response to the particles.

14.2.2.5 Direct Effects of Translocated Particles on the Endothelium and the Blood

The nanoparticle component of inhaled PM ($PM_{0.1}$) may influence the cardiovascular system through indirect effects mediated by pulmonary inflammation or through the direct action of particles that have become blood-borne. Translocation of inhaled nanoparticles across the alveolar-blood barrier has been demonstrated in animal studies for a range of nanoparticles delivered by inhalation and instillation (Nemmar et al. 2001; Kreyling et al. 2002a, 2002b; Nemmar et al. 2002; Oberdörster et al. 2002; Nemmar et al. 2004). Whether inhaled nanoparticulates can readily access the circulation in humans is currently the subject of intense research (Nemmar et al. 2002; Mills et al. 2006).

Once circulating, nanoparticles may interact with the vascular endothelium, or have direct effects on atherosclerotic plaques causing local oxidative stress and proinflammatory effects similar to those found in the lungs. Increased inflammation could destabilize the coronary plaque, resulting in rupture, thrombosis, and an acute coronary syndrome (Brook et al. 2004). Certainly, injured arteries can take up blood borne nanoparticles (Guzman et al. 2000), a fact exploited by the nanotechnology industry for both diagnostic and therapeutic purposes in cardiovascular medicine. The intra-arterial infusion of carbon black nanoparticles has a detrimental effect on the mouse microcirculation, with upregulation of von Willebrand factor expression and enhanced fibrin deposition on the endothelial surface (Khandoga et al. 2004). These prothrombotic effects are in keeping with toxicological evidence from inhalation studies, which suggest particle exposure may promote thrombogenesis (Nemmar et al. 2003b, 2003c).

14.2.2.5.1 Direct Effects of Particle on Endothelial Cells

Blood-borne particles could interact directly with vascular endothelial cells. In a study by the authors, endothelial cells were cultured and exposed to PM_{10} and expression of genes related to thrombogenesis and fibrinolysis were studied (Gilmour et al. 2005).

As shown in Figure 14.5, PM_{10} causes a direct, dose dependent and sustained impairment in endothelial cell production of t-PA. Impaired acute t-PA released in the event of intravascular thrombus formation would cause an imbalance between profibrinolytic factors and their endogenous inhibitors, preventing thrombus dissolution. Additionally, PM_{10} caused a concomitant upregulation of endothelial tissue factor expression, a potent stimulus for thrombus formation. Taken together, these observations suggest that PM_{10} promotes thrombogenesis in particle-exposed endothelium via two distinct but related pathways: enhanced endothelial tissue factor expression and impaired t-PA release.

14.2.2.5.2 Direct Effects of Particles on Platelets

Platelets are numerous in the bloodstream and are central to the process of intravascular thrombus formation. Few studies have addressed the direct effects of particles on platelets. In the experimental studies of Nemmar (Nemmar et al. 2003a), venous thrombosis is initiated by endothelial injury *in vivo*, using Rose-Bengal and green light from a xenon lamp to generate extreme oxidative stress in the femoral vein. In hamsters treated with intratracheal diesel particles, both arterial and venous thrombosis was increased in a dose-dependent manner. Platelets sampled from the instilled animals underwent more aggregation *in vitro*, although it is not clear whether enhance *in vivo* thrombus formation was due to inflammation or direct effects of the particles on platelet function. However, platelets isolated from normal hamsters and treated with DEP showed increased aggregation, suggesting that direct effects were at least plausible.

In a more comprehensive study (Radomski et al. 2005), a range of manufactured nanoparticles, including carbon nanotubes, fullerenes, urban PM, and carbon nanoparticles were incubated with platelets *in vitro*. Platelet aggregation was assessed and clear differences in the potency of these particles were evident. Fullerenes were without effect, urban dust and multiwalled nanotubes caused modest aggregation, whilst mixed carbon nanoparticles and single walled nanotubes were highly potent activators of platelets upregulating platelet glycoprotein IIb/IIIa receptor expression

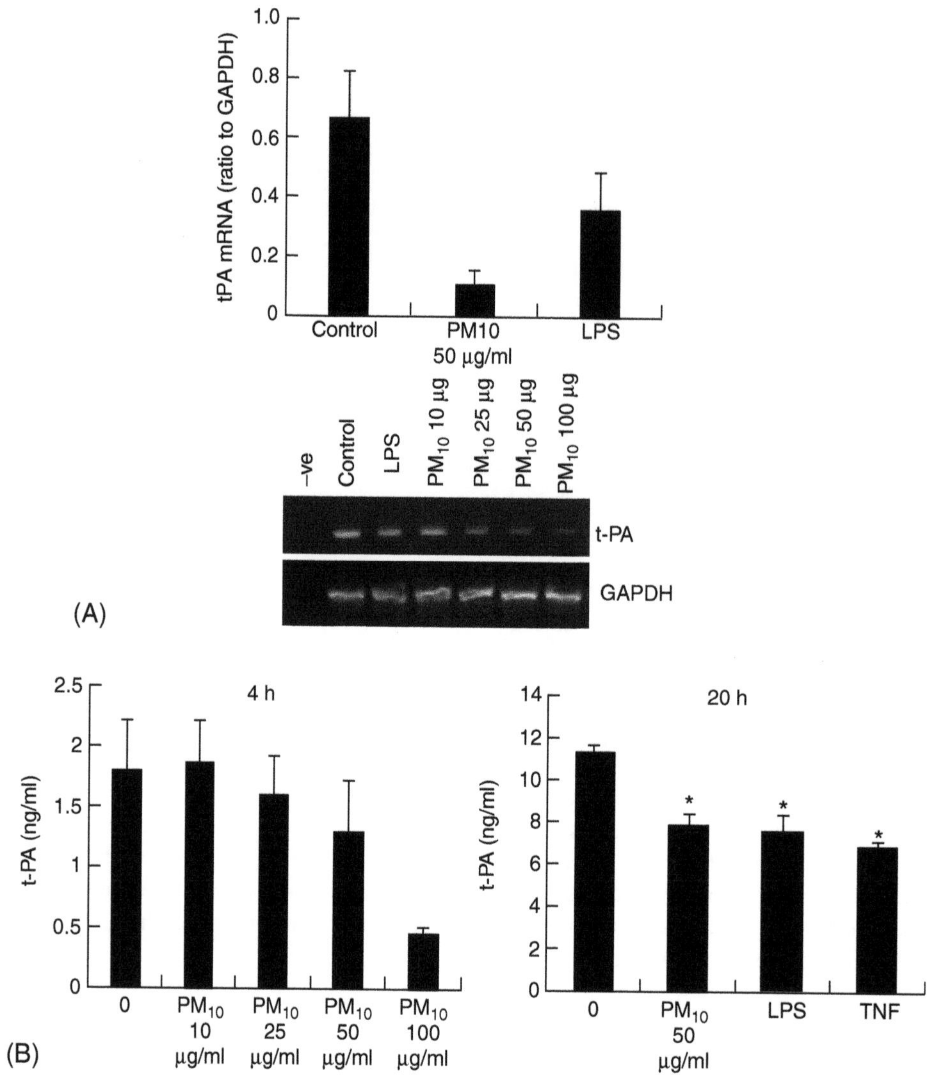

FIGURE 14.5 HUVEC t-PA mRNA expression as determined by real time-PCR following 6 h treatment with PM_{10} (10, 25, 50, and 100 mg/mL for dose response, 50 mg/mL for quantified histogram), TNF-α (10 ng/mL), and LPS (1 mg/mL). (A) Quantification of tissue factor RT-PCR bands by densitometry. (B) Protein released by HUVECs as determined by ELISA following 4 or 24 h exposure. Each bar represents the mean (SEM) from at least three separate experiments; $*p<0.05$, $**p<0.01$, $***p<0.001$. (Adapted from Gilmour, P. S., Morrison, E. R., Vickers, M. A., Ford, I., Ludlam, C. A., Greaves, M., Donaldson, K., and MacNee, W., *Occup. Environ. Med.*, 62, 164–171, 2005. With permission.)

and enhancing aggregate formation. In a ferric chloride induced rat carotid artery thrombosis model, a similar potency was observed with particles enhancing the rate of thrombus formation. Unfortunately, there was little characterization of these particles that might have allowed features such as surface area or chemistry to be linked to the variation in effect on platelet aggregation or thrombus formation (Figure 14.6).

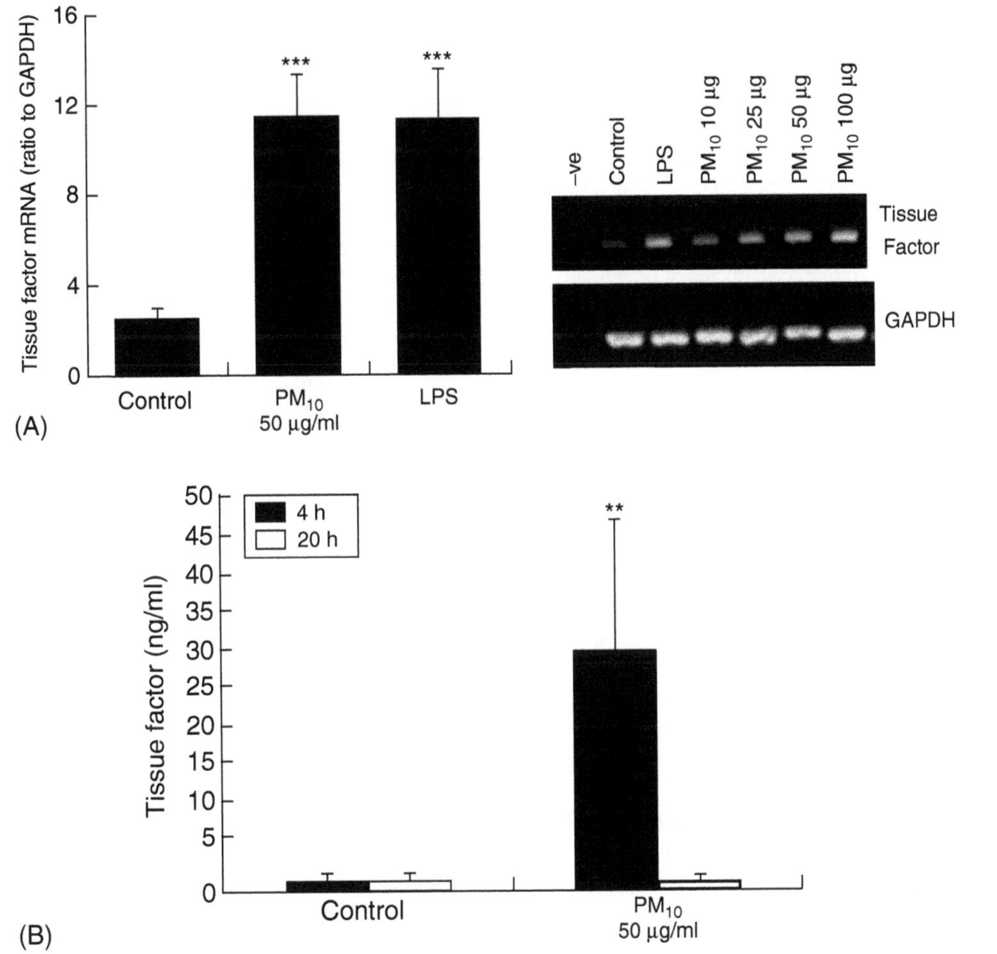

FIGURE 14.6 Human umbilical vein endothelial cell (HUVEC) tissue factor (TF) mRNA expression as determined by real time-PCR following 6 h treatment with PM_{10} (10, 25, 50, and 100 mg/mL for dose response, 50 mg/mL for quantified histogram), and LPS (1 mg/mL). (A) Quantification of tissue factor real time-PCR bands by densitometry. (B) Protein released by HUVECs as determined by ELISA following 4 or 24 h exposure. Each bar represents the mean (SEM) from at least three separate experiments; $^{**}p<0.01$, $^{***}p<0.001$. (Adapted from Gilmour, P. S., Morrison, E. R., Vickers, M. A., Ford, I., Ludlam, C. A., Greaves, M., Donaldson, K., and MacNee, W., *Occup. Environ. Med.*, 62, 164–171, 2005. With permission.)

14.3 CONCLUSIONS

We conclude that there are multiple plausible mechanistic pathways whereby increased PM deposition in the lungs may result in acute cardiovascular events. Central to this response is pulmonary and systemic inflammation, which may drive the pathogenesis and clinical manifestations of atherothrombosis. If particles or soluble components translocate into the bloodstream, we anticipate direct effects on the vascular endothelium, platelet function, and thrombosis. Ongoing research in this area is intense and will yield considerable insights into which of these pathways are most relevant, and the possibility of therapeutic intervention to reduce the impact of environmental PM on cardiovascular disease a realistic goal.

REFERENCES

Blum, A. and Miller, H. I., The role of inflammation in atherosclerosis, *Isr. J. Med. Sci.*, 32, 1059–1065, 1996.

Bonetti, P. O., Lerman, L. O., and Lerman, A., Endothelial dysfunction: a marker of atherosclerotic risk, *Arterioscler. Thromb. Vasc. Biol.*, 23, 168–175, 2003.

Brand, K., Page, S., Rogler, G., Bartsch, A., Brandl, R., Knuechel, R., Page, M., Kaltschmidt, C., Baeuerle, P.A., and Neumeier, D., Activated transcription factor nuclear factor-kappa B is present in the atherosclerotic lesion, *J. Clin. Invest.*, 97, 1715–1722, 1996.

Brommer, E. J., The level of extrinsic plasminogen activator (t-PA) during clotting as a determinant of the rate of fibrinolysis; inefficiency of activators added afterwards, *Thromb. Res.*, 34, 109–115, 1984.

Brook, R. D., Brook, J. R., Urch, B., Vincent, R., Rajagopalan, S., and Silverman, F., Inhalation of fine particulate air pollution and ozone causes acute arterial vasoconstriction in healthy adults, *Circulation*, 105, 1534–1536, 2002.

Brook, R. D., Brook, J. R., and Rajagopalan, S., Air pollution: the Heart of the problem, *Curr. Hypertens. Rep.*, 5, 32–39, 2003.

Brook, R. D., Franklin, B., Cascio, W., Hong, Y., Howard, G., Lipsett, M., Luepker, R. et al., Air pollution and cardiovascular disease: a statement for healthcare professionals from the Expert Panel on Population and Prevention Science of the American Heart Association, *Circulation*, 109, 2655–2671, 2004.

Cassee, F. R., Boere, A. J., Bos, J., Fokkens, P. H., Dormans, J. A., and van Loveren, H., Effects of diesel exhaust enriched concentrated $PM_{2.5}$ in ozone pre-exposed or monocrotaline-treated rats, *Inhal. Toxicol.*, 14, 721–743, 2002.

Celermajer, D. S., Adams, M. R., Clarkson, P., Robinson, J., McCredie, R., Donald, A., and Deanfield, J. E., Passive smoking and impaired endothelium-dependent arterial dilatation in healthy young adults, *N. Engl. J. Med.*, 334, 150–154, 1996.

Chen, L. C. and Nadziejko, C., Effects of subchronic exposures to concentrated ambient particles (CAPs) in mice. V. CAPs exacerbate aortic plaque development in hyperlipidemic mice, *Inhal. Toxicol.*, 17, 217–224, 2005.

Committee on the Medical Effects of Air Pollutants, Cardiovascular disease and air pollution, Department of Health, 2006, Ref Type: Generic.

Creason, J., Neas, L., Walsh, D., Williams, R., Sheldon, L., Liao, D., and Shy, C., Particulate matter and heart rate variability among elderly retirees: the Baltimore 1998 PM study, *J. Expo. Anal. Environ. Epidemiol.*, 11, 116–122, 2001.

Davies, M. J., Coronary disease: the pathophysiology of acute coronary syndromes, *Heart*, 83, 361–366, 2000. Ref Type: Generic

Devlin, R. B., Ghio, A. J., Kehrl, H., Sanders, G., and Cascio, W., Elderly humans exposed to concentrated air pollution particles have decreased heart rate variability, *Eur. Respir. J. Suppl.*, 40, 76s–80s, 2003.

Dockery, D. W., Pope, C. A., Xu, X. P., Spengler, J. D., Ware, J. H., Fay, M. E., Ferris, B. G., and Speizer, F. E., An association between air-pollution and mortality in 6 united-states cities, *N. Engl. J. Med.*, 329, 1753–1759, 1993.

Donaldson, K. and Tran, C. L., Inflammation caused by particles and fibres, *Inhal. Toxicol.*, 14, 2002.

Donaldson, K., Tran, L., Jimenez, L., Duffin, R., Newby, D. E., Mills, N., MacNee, W., and Stone, V., Combustion-derived nanoparticles: a review of their toxicology following inhalation exposure, *Part. Fibre Toxicol.*, 2, 10, 2005.

Eijnden-Schrauwen, Y., Kooistra, T., de Vries, R. E., and Emeis, J. J., Studies on the acute release of tissue-type plasminogen activator from human endothelial cells in vitro and in rats in vivo: evidence for a dynamic storage pool, *Blood*, 85, 3510–3517, 1995.

Elder A. P., Gelein, R., Azadniv, M., Frampton, M. W., Finkelstein, J., and Oberdörster, G., Systemic interactions between inhaled ultrafine particles and endotoxin, 2002, Ref Type: Generic.

Elder, A., Gelein, R., Finkelstein, J., Phipps, R., Frampton, M., Utell, M., Kittelson, D. B. et al., On-road exposure to highway aerosols. 2. Exposures of aged, compromised rats, *Inhal. Toxicol.*, 16 (suppl. 1), 41–53, 2004.

Fox, K. A., Robison, A. K., Knabb, R. M., Rosamond, T. L., Sobel, B. E., and Bergmann, S. R., Prevention of coronary thrombosis with subthrombolytic doses of tissue-type plasminogen activator, *Circulation*, 72, 1346–1354, 1985.

Furchgott, R. F. and Zawadzki, J. V., The obligatory role of endothelial cells in the relaxation of arterial smooth muscle by acetylcholine, *Nature*, 288, 373–376, 1980.

Ghio, A. J. and Devlin, R. B., Inflammatory lung injury after bronchial instillation of air pollution particles, *Am. J. Respir. Crit. Care Med.*, 164, 704–708, 2001.

Ghio, A. J., Kim, C., and Devlin, R. B., Concentrated ambient air particles induce mild pulmonary inflammation in healthy human volunteers, *Am. J. Respir. Crit. Care Med.*, 162, 981–988, 2000.

Gilmour, P. S., Morrison, E. R., Vickers, M. A., Ford, I., Ludlam, C. A., Greaves, M., Donaldson, K., and MacNee, W., The procoagulant potential of environmental particles (PM_{10}), *Occup. Environ. Med.*, 62, 164–171, 2005.

Glantz, S. A., Air pollution as a cause of heart disease. Time for action, *J. Am. Coll. Cardiol.*, 39, 943–945, 2002.

Gold, D. R., Litonjua, A., Schwartz, J., Lovett, E., Larson, A., Nearing, B., Allen, G., Verrier, M., Cherry, R., and Verrier, R., Ambient pollution and heart rate variability, *Circulation*, 101, 1267–1273, 2000.

Goldberg, M. S., Burnett, R. T., Valois, M. F., Flegel, K., Bailar, J. C., III, Brook, J., Vincent, R., and Radon, K., *Associations between ambient air pollution and daily mortality among persons with congestive heart failure Environ. Res.*, 91 2003.

Goto, Y., Hogg, J. C., Shih, C. H., Ishii, H., Vincent, R., and van Eeden, S. F., Exposure to ambient particles accelerates monocyte release from bone marrow in atherosclerotic rabbits, *Am. J. Physiol. Lung Cell Mol. Physiol.*, 287, L79–L85, 2004.

Guzman, M., Aberturas, M. R., Rodriguez-Puyol, M., and Molpeceres, J., Effect of nanoparticles on digitoxin uptake and pharmacologic activity in rat glomerular mesangial cell cultures, *Drug Deliv.*, 7, 215–222, 2000.

Haverkate, F., Thompson, S. G., Pyke, S. D., Gallimore, J. R., and Pepys, M. B., Production of C-reactive protein and risk of coronary events in stable and unstable angina. European Concerted Action on Thrombosis and Disabilities Angina Pectoris Study Group, *Lancet*, 349, 462–466, 1997.

Hingorani, A. D., Cross, J., Kharbanda, R. K., Mullen, M. J., Bhagat, K., Taylor, M., Donald, A. E. et al., Acute systemic inflammation impairs endothelium-dependent dilatation in humans, *Circulation*, 102, 994–999, 2000.

Holgate, S. T., Devlin, R. B., Wilson, S. J., and Frew, A. J., Health effects of acute exposure to air pollution. Part II: healthy subjects exposed to concentrated ambient particles, *Res. Rep. Health Eff. Inst.*, 31–50, 2003.

Khandoga, A., Stampfl, A., Takenaka, S., Schulz, H., Radykewicz, R., Kreyling, W., and Krombach, F., Ultrafine particles exert prothrombotic but not inflammatory effects on the hepatic microcirculation in healthy mice in vivo, *Circulation*, 109, 1320–1325, 2004.

Kleiger, R. E., Heart rate variability and mortality and sudden death post infarction, *J. Cardiovasc. Electrophysiol.*, 6, 365–367, 1995.

Kreyling, W., Semmler, M., Erbe, F., Mayer, P., Takenaka, S., Oberdörster, G., and Ziesenis, A., Minute translocation of inhlaed ultrafine insoluble iridium particles from lung epithelium to extrapulmonary tissues, *Ann. Occup. Hyg.*, 46 (suppl. 1), 223–226, 2002a.

Kreyling, W. G., Semmler, M., Erbe, F., Mayer, P., Takenaka, S., Schulz, H., Oberdörster, G., and Ziesenis, A., Translocation of ultrafine insoluble iridium particles from lung epithelium to extrapulmonary organs is size dependent but very low, *J. Toxicol. Environ. Health A*, 65, 1513–1530, 2002b.

Kuller, L. H., Tracy, R. P., Shaten, J., and Meilahn, E. N., Relation of C-reactive protein and coronary heart disease in the MRFIT nested case-control study. Multiple Risk Factor Intervention Trial, *Am. J. Epidemiol.*, 144, 537–547, 1996.

Kunzli, N., Jerrett, M., Mack, W. J., Beckerman, B., LaBree, L., Gilliland, F., Thomas, D., Peters, J., and Hodis, H. N., Ambient air pollution and atherosclerosis in Los Angeles, *Environ. Health Perspect.*, 113, 201–206, 2005.

Labinjoh, C., Newby, D. E., Dawson, P., Johnston, N. R., Ludlam, C. A., Boon, N. A., and Webb, D. J., Fibrinolytic actions of intra-arterial angiotensin II and bradykinin in vivo in man, *Cardiovasc. Res.*, 47, 707–714, 2000.

Lee, R. T. and Libby, P., The unstable atheroma, *Arterioscler. Thromb. Vasc. Biol.*, 17, 1859–1867, 1997.

Li, X. Y., Gilmour, P. S., Donaldson, K., and MacNee, W., Free radical activity and pro-inflammatory effects of particulate air pollution (PM_{10}) in vivo and in vitro, *Thorax*, 51, 1216–1222, 1996.

Liao, D., Creason, J., Shy, C., Williams, R., Watts, R., and Zweidinger, R., Daily variation of particulate air pollution and poor cardiac autonomic control in the elderly, *Environ. Health Perspect.*, 107, 521–525, 1999.

Liuzzo, G., Biasucci, L. M., Gallimore, J. R., Grillo, R. L., Rebuzzi, A. G., Pepys, M. B., and Maseri, A., The prognostic value of C-reactive protein and serum amyloid a protein in severe unstable angina, *N. Engl. J. Med.*, 331, 417–424, 1994.

Ludmer, P. L., Selwyn, A. P., Shook, T. L., Wayne, R. R., Mudge, G. H., Alexander, R. W., and Ganz, P., Paradoxical vasoconstriction induced by acetylcholine in atherosclerotic coronary arteries, *N. Engl. J. Med.*, 315, 1046–1051, 1986.

Magari, S. R., Schwartz, J., Williams, P. L., Hauser, R., Smith, T. J., and Christiani, D. C., The association between personal measurements of environmental exposure to particulates and heart rate variability, *Epidemiology*, 13, 305–310, 2002.

Mills, N. L., Tornqvist, H., Robinson, S. D., Gonzalez, M., Darnley, K., MacNee, W., Boon, N. A. et al., Diesel exhaust inhalation causes vascular dysfunction and impaired endogenous fibrinolysis, *Circulation*, 112, 3930–3936, 2005.

Mills, N. L., Amin, N., Robinson, S. D., Anand, A., Davies, J., Patel, D., de la Fuente, J. M. et al., Do inhaled carbon nanoparticles translocate directly into the circulation in humans?, *Am. J. Respir. Crit. Care Med.*, 173, 426–431, 2006.

Nemmar, A., Vanbilloen, H., Hoylaerts, M. F., Hoet, P. H., Verbruggen, A., and Nemery, B., Passage of intratracheally instilled ultrafine particles from the lung into the systemic circulation in hamster, *Am. J. Respir. Crit. Care Med.*, 164, 1665–1668, 2001.

Nemmar, A., Hoet, P. H., Vanquickenborne, B., Dinsdale, D., Thomeer, M., Hoylaerts, M. F., Vanbilloen, H., Mortelmans, L., and Nemery, B., Passage of inhaled particles into the blood circulation in humans, *Circulation*, 105, 411–414, 2002.

Nemmar, A., Hoet, P. H., Dinsdale, D., Vermylen, J., Hoylaerts, M. F., and Nemery, B., Diesel exhaust particles in lung acutely enhance experimental peripheral thrombosis, *Circulation*, 107, 1202–1208, 2003a.

Nemmar, A., Hoylaerts, M. F., Hoet, P. H., Vermylen, J., and Nemery, B., Size effect of intratracheally instilled particles on pulmonary inflammation and vascular thrombosis, *Toxicol. Appl. Pharmacol.*, 186, 38–45, 2003b.

Nemmar, A., Nemery, B., Hoet, P. H., Vermylen, J., and Hoylaerts, M. F., Pulmonary inflammation and thrombogenicity caused by diesel particles in hamsters: role of histamine, *Am. J. Respir. Crit. Care Med.*, 168, 1366–1372, 2003c.

Nemmar, A., Hoylaerts, M. F., Hoet, P. H., and Nemery, B., Possible mechanisms of the cardiovascular effects of inhaled particles: systemic translocation and prothrombotic effects, *Toxicol. Lett.*, 149, 243–253, 2004.

Newby, D. E., Wright, R. A., Labinjoh, C., Ludlam, C. A., Fox, K. A., Boon, N. A., and Webb, D. J., Endothelial dysfunction, impaired endogenous fibrinolysis, and cigarette smoking: a mechanism for arterial thrombosis and myocardial infarction, *Circulation*, 99, 1411–1415, 1999.

Oberdörster, G., Sharp, Z., Atudorei, V., Elder, A. P., Gelein, R., Lunts, A. K., Kreyling, W., and Cox, C., Extrapulmonary transocation of ultrafine carbon particles following inhalation exposure, 2002, Ref Type: Generic.

Oliver, J. J., Webb, D. J., and Newby, D. E., Stimulated tissue plasminogen activator release as a marker of endothelial function in humans 1, *Arterioscler. Thromb. Vasc. Biol.*, 25, 2470–2479, 2005.

Pekkanen, J., Brunner, E. J., Anderson, H. R., Tiittanen, P., and Atkinson, R. W., Daily concentrations of air pollution and plasma fibrinogen in London, *Occup. Environ. Med.*, 57, 818–822, 2000.

Peters, A., Doring, A., Wichmann, H. E., and Koenig, W., Increased plasma viscosity during an air pollution episode: a link to mortality?, *Lancet*, 349, 1582–1587, 1997.

Peters, A., Perz, S., Doring, A., Stieber, J., Koenig, W., and Wichmann, H. E., Increases in heart rate during an air pollution episode, *Am. J. Epidemiol.*, 150, 1094–1098, 1999.

Peters, A., Liu, E., Verrier, R. L., Schwartz, J., Gold, D. R., Mittleman, M., Baliff, J. et al., Air pollution and incidence of cardiac arrhythmia, *Epidemiology*, 11, 11–17, 2000.

Peters, A., Dockery, D. W., Muller, J. E., and Mittleman, M. A., Increased particulate air pollution and the triggering of myocardial infarction, *Circulation*, 103, 2810–2815, 2001a.

Peters, A., Frohlich, M., Doring, A., Immervoll, T., Wichmann, H. E., Hutchinson, W. L., Pepys, M. B., and Koenig, W., Particulate air pollution is associated with an acute phase response in men; results from the MONICA-Augsburg Study, *Eur. Heart J.*, 22, 1198–1204, 2001b.

Pope, C. A., III, Particulate air pollution, C-reactive protein, and cardiac risk, *Eur. Heart J.*, 22, 1149–1150, 2001.

Pope, C. A., III, Verrier, R. L., Lovett, E. G., Larson, A. C., Raizenne, M. E., Kanner, R. E., Schwartz, J., Villegas, G. M., Gold, D. R., and Dockery, D. W., Heart rate variability associated with particulate air pollution, *Am. Heart J.*, 138, 890–899, 1999.

Prescott, G. J., Lee, R. J., Cohen, G. R., Elton, R. A., Lee, A. J., Fowkes, F. G., and Agius, R. M., Investigation of factors which might indicate susceptibility to particulate air pollution, *Occup. Environ. Med.*, 57, 53–57, 2000.

Radomski, A., Jurasz, P., Alonso-Escolano, D., Drews, M., Morandi, M., Malinski, T., and Radomski, M. W., Nanoparticle-induced platelet aggregation and vascular thrombosis, *Br. J. Pharmacol.*, 146, 882–893, 2005.

Reddigari, S. and Kaplan, A. P., Cleavage of human high-molecular weight kininogen by purified kallikreins and upon contact activation of plasma 32, *Blood*, 71, 1334–1340, 1988.

Ridker, P. M., Hennekens, C. H., Buring, J. E., and Rifai, N., C-reactive protein and other markers of inflammation in the prediction of cardiovascular disease in women, *N. Engl. J. Med.*, 342, 836–843, 2000a.

Ridker, P. M., Rifai, N., Stampfer, M. J., and Hennekens, C. H., Plasma concentration of interleukin-6 and the risk of future myocardial infarction among apparently healthy men, *Circulation*, 101, 1767–1772, 2000b.

Rosenberg, R. D. and Aird, W. C., Vascular-bed-specific hemostasis and hypercoagulable states, *N. Engl. J. Med.*, 340, 1555–1564, 1999.

Ross, R., The pathogenesis of atherosclerosis: a perspective for the 1990s, *Nature*, 362, 801–809, 1993.

Ross, R., Atherosclerosis is an inflammatory disease, *Am. Heart J.*, 138, S419–S420, 1999.

Routledge, H. C., Ayres, J. G., and Townend, J. N., Why cardiologists should be interested in air pollution, *Heart*, 89, 1383–1388, 2003.

Schwartz, J., Air pollution and blood markers of cardiovascular risk, *Environ. Health Perspect.*, 109 (suppl. 3), 405–409, 2001.

Schwartz, J., The effects of particulate air pollution on daily deaths: a multi-city case crossover analysis, *Occup. Environ. Med.*, 61, 956–961, 2004.

Seaton, A., Soutar, A., Crawford, V., Elton, R., McNerlan, S., Cherrie, J., Watt, M., Agius, R., and Stout, R., Particulate air pollution and the blood, *Thorax*, 54, 1027–1032, 1999.

Sun, Q., Wang, A., Jin, X., Natanzon, A., Duquaine, D., Brook, R. D., Aguinaldo, J. G. et al., Long-term air pollution exposure and acceleration of atherosclerosis and vascular inflammation in an animal model, *JAMA*, 294, 3003–3010, 2005.

Suwa, T., Hogg, J. C., Quinlan, K. B., Ohgami, A., Vincent, R., and van Eeden, S. F., Particulate air pollution induces progression of atherosclerosis, *J. Am. Coll. Cardiol.*, 39, 935–942, 2002.

Task Force of the European Society of Cardiology and the North American Society of Pacing and Electrophysiology, Heart rate variability. Standards of measurement, physiological interpretation, and clinical use, *Eur. Heart J.*, 17, 354–381, 1996.

Tsuji, H., Venditti, F. J., Manders, E. S., Evans, J. C., Larson, M. G., Feldman, C. L., and Levy, D., Determinants of heart rate variability, *J. Am. Coll. Cardiol.*, 28, 1539–1546, 1996.

van Eeden, S. F., Tan, W. C., Suwa, T., Mukae, H., Terashima, T., Fujii, T., Qui, D., Vincent, R., and Hogg, J. C., Cytokines involved in the systemic inflammatory response induced by exposure to particulate matter air pollutants (PM_{10}), *Am. J. Respir. Crit. Care Med.*, 164, 826–830, 2001.

Vincent, R., Kumarathasan, P., Geogan, P., Bjarnason, S. G., Guenette, J., BéruBé, D., Adamson, I. Y. et al., Inhalation Toxicology of Urban Ambient Particulate Matter: Acute Cardiovascular Effects in Rats, Health Effects Research Report No. 104, 2001.

Wellenius, G. A., Saldiva, P. H., Batalha, J. R., Krishna Murthy, G. G., Coull, B. A., Verrier, R. L., and Godleski, J. J., Electrocardiographic changes during exposure to residual oil fly ash (ROFA) particles in a rat model of myocardial infarction, *J. Toxicol. Sci.*, 66, 327–335, 2002.

15 Susceptibility to Particle Effects

Steven R. Kleeberger and Reuben Howden
National Institute of Environmental Health Sciences,
National Institutes of Health

CONTENTS

15.1 INTRODUCTION

Linkage between exposure to particulate matter (PM) <10 μm in aerodynamic diameter (PM_{10}) and a number of acute and chronic health effects throughout the industrialized world has been well established in epidemiological studies (e.g., [1–3]). Acute effects include mortality, hospitalization, increased respiratory symptoms, decreased lung function, increased plasma viscosity, changes in heart rate and heart rate variability (HRV), and pulmonary inflammation [4]. Chronic effects associated with particulate exposures include increased mortality rates (e.g., cancer), chronic cardiopulmonary disease, and decreased lung function [4]. Understanding the mechanisms through which particulate exposures cause morbidity and mortality continues to be a critical public health concern.

It is also clear that susceptibility to particle effects is not the same from one individual to the next, i.e., interindividual differences in response to particulate exposures exist. The factors that determine interindividual susceptibility are almost certainly complex, and may include intrinsic (host) and extrinsic (environmental) factors (Figure 15.1). While considerable interest in genetic background as a host factor for susceptibility to pollutants has been generated [5–8], other host factors should also be considered [9]. These may include gender, age, pharmacokinetic and pharmacodynamic response parameters (including particle deposition and clearance [10]), pre-existing disease, and nutrition. Furthermore, these factors may, and probably do, interact to determine individual responsiveness. Numerous examples of gene X gene, gene X environment, and gene X gene X environment interaction in the pathogenesis of lung disease have been described (e.g., [11]).

Specific subpopulations may be particularly at risk to the toxic effects of particles. These subpopulations include the young and the elderly, patients with pre-existing diseases such as pulmonary hypertension, chronic obstructive pulmonary disease (COPD), asthma, diabetes, and compromised immune systems [4,12,13]. Environmental factors that may impact an individual's response to particle exposure include coexposures and the physical environment (e.g., temperature, altitude).

In this chapter, we identify and briefly discuss factors that have been found to contribute to interindividual susceptibility to particulate exposures. We include reports using animal models,

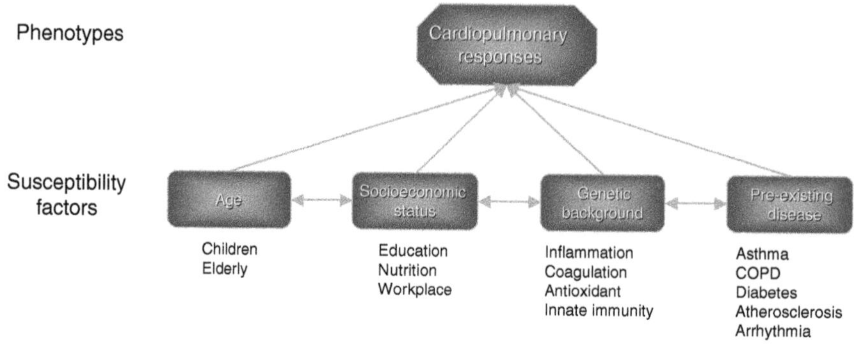

FIGURE 15.1 General susceptibility factors that may influence cardiopulmonary responses to particle exposures.

human subjects in clinical investigations, and epidemiological studies. Our overall objective is to provide the reader with a perspective on the relative importance of those factors that may impact on individual particle susceptibility and to suggest where further investigation is needed to clarify susceptibility factors.

15.2 SUSCEPTIBILITY FACTORS

15.2.1 Genetic Background

An extensive literature exists that describes interindividual differences in drug metabolism and susceptibility (e.g., [14–17]). Many of these differences have been attributed to polymorphisms in genes that encode metabolism and phase two enzymes (e.g., *N*-acetyltransferase and G6PD). For example, genetic background has an important role in determining susceptibility to infectious agents and pesticide exposures [18–20]. It is well known that many complex diseases cluster in families, and clustering can be explained by genetic background, shared environment, or a combination of the two. Two broad research strategies have been utilized to identify genes [or quantitative trait loci (QTLs)] that determine disease susceptibility. The first is positional cloning or linkage mapping, which exploits within-family associations between marker alleles and putative trait-influencing alleles that arise within families [21]. This approach is designed to identify association of a chromosomal interval(s) within the entire genome that may contain genes that are polymorphic and might account for the differential response phenotype under study. Linkage mapping is applicable to human populations and animal models, and has had considerable success in Mendelian (single gene) diseases. The second approach (candidate gene or association study) chooses loci, which a priori are likely to determine the phenotype of interest. Linkage is assessed between the phenotype of interest and markers flanking the candidate genes or the candidate genes themselves, and evaluates across-family associations. The principle underlying the association of genetic polymorphisms not directly involved in disease pathogenesis is that of linkage disequilibrium, which arises from the coinheritance of alleles at loci that are in close physical proximity on an individual chromosome [22]. Emergent technologies, including gene and protein expression arrays, have been important developments in the ability to identify candidate susceptibility genes (Figure 15.2).

A number of laboratories have used positional cloning in animal models to determine whether genetic background is an important determinant of pulmonary responses to particulates. Ohtsuka et al. [23] studied the interstrain variance of lung responses to particle-associated sulfate (acid coated particles, (ACP)) in inbred strains of mice. The ACP model was chosen for study

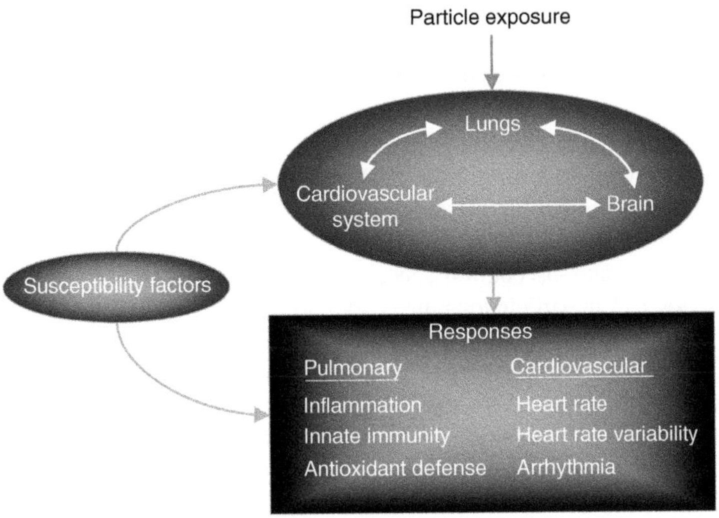

FIGURE 15.2 Schematic of the potential interactions between susceptibility factors (e.g., age, genetic background, pre-existing disease, socioeconomic status) and the route of particle entry into the body and physiological responses to the particles.

because the particles could be generated reproducibly in the laboratory and all mice could be exposed to the ACP under similar conditions. Nine strains of inbred mice were exposed to ACP for 4 h in a nose-only chamber, and inflammation was assessed after 1, 3, 7, or 14 days by bronchoalveolar lavage (BAL). Control exposures were carbon black (CB) or sulfur dioxide alone. Innate immune response was assessed by measuring Fc-receptor mediated phagocytosis of BAL alveolar macrophages (i.e., quantitation of ingested sensitized sheep red blood cells). ACP exposure did not cause an appreciable inflammatory response in any of the strains at any time postexposure. However, significant interstrain differences were found in Fc-receptor mediated phagocytosis by alveolar macrophages. Among the nine strains examined, C3H/HeJ (C3) mice were the most resistant and C57BL/6J (B6) were the most susceptible.

The significant interstrain variation in susceptibility indicated a strong genetic component contributed to ACP susceptibility. A genome-wide search for linkage of the Fc-receptor mediated phagocytosis phenotype was performed in segregant populations derived from B6 and C3 mice using informative simple sequence length polymorphisms (SSLPs) distributed at approximately 10 centi-Morgan (cM) intervals throughout the genome. Interval mapping by simple linear regression identified a susceptibility locus on chromosome 17 between approximately 16–22 cMs [24]. An additional QTL was detected on chromosome 11 between *D11Mit20* and *D11Mit12*, and no significant interaction between QTLs was detected. The presence of separate susceptibility loci for the phagocytosis phenotype was consistent with the cosegregation analysis for this parameter [24].

Within the chromosome 17 QTL are a number of candidate genes, including the proinflammatory cytokine *Tnf* (tumor necrosis factor-α, TNF-α), multiple histocompatibility loci, and heat shock proteins (HSP). The chromosome 11 QTL also contains a number of candidate genes, including a cluster of inducible cytokines and inducible nitric oxide synthase (iNOS). Interestingly, these QTLs nearly overlapped similar QTLs identified for ozone susceptibility. The common linkages suggest that similar genetic mechanisms may contribute to pulmonary responses to ozone-induced inflammation and macrophage phagocytic dysfunction induced by ACP, but further genetic analyses are required to confirm this hypothesis.

Tolerance to air pollutant-induced pulmonary inflammatory and hyperpermeability effects has been demonstrated in animal models and human subjects (e.g., [25,26]). Wesselkamper et al. [27] evaluated the interstrain variation in the ability of inbred mice to "become tolerant" to the toxic effects of repeated exposure to zinc oxide (ZnO). These investigators found significant interstrain variation in the inflammatory cell and hyperpermeability responses to single and multiple ZnO exposures. A genome-scan for susceptibility QTLs for the development of pulmonary tolerance to ZnO in a DBA/2J and Balb/cByJ intercross cohort from mice identified a significant QTL on chromosome 1, and suggestive QTLs on chromosomes 4 and 5 [28]. Toll-like receptor 5 (*Tlr5*) was identified as a candidate susceptibility gene in the chromosome 1 QTL, and functional analysis confirmed a role for *Tlr5* in particle tolerance [28]. This represents the first attempt to identify the genes responsible for the development of tolerance, and confirmation of candidate genes in additional models and human subjects may have important implications for understanding susceptibility and resistance to repeated exposures to pulmonary toxicants.

Wesselkamper et al. [29] developed another model of particle susceptibility and acute lung injury using nickel sulfate aerosol. Continuous exposure of mice to 150 mcg/m^3 of nickel sulfate causes death in a strain-dependent manner; A/J mice were significantly more susceptible to the aerosol than B6 mice [29]. A QTL analysis with backcross mice from A/J and B6 progenitors identified a significant QTL on chromosome 6, and suggestive QTLs on chromosomes 1 and 12. This study suggested that susceptibility to irritant-induced lung injury and subsequent survival was thus dependent on relatively few loci [30]. A number of interesting candidate genes were identified in these QTLs, including transforming growth factor alpha (*Tgfa*), and proof-of-concept testing in this model has begun. This group has also used gene expression arrays to identify candidate genes for particle-induced lung injury [31,32]. This approach led to the identification of transforming growth factor beta (*Tgfb*) and macrophage-stimulating 1 receptor (Mst1r) as candidate susceptibility genes, and functional analyses have confirmed an important role for both in the injury that follows exposure to nickel sulfate aerosol.

We have also compared the lung injury responses to residual oil fly ash (ROFA) exposure in inbred mouse strains [33]. Significant interstrain (genetic) variation was observed in ROFA-induced lung inflammation and hyperpermeability, and C3 mice were most resistant to the ROFA-induced injury responses, while B6 mice were among the most susceptible. Interestingly, ROFA-induced lung injury was significantly greater in C3H/HeOuJ mice compared to C3. C3H/HeOuJ and C3 mice differ only at a loss of function mutation in toll-like receptor 4 (*Tlr4*) that confers resistance to endotoxin and ozone in the C3 strain. ROFA also significantly enhanced transcript and protein levels of lung TLR4 in C3H/HeOuJ, but not in C3 mice. Furthermore, ROFA activated downstream TLR4 signaling molecules (i.e., MyD88, TRAF6, IRAK-1, NF-κB, MAPK, AP-1) to a greater extent in C3H/HeOuJ mice than in C3 mice before the development of pulmonary injury. These results support an important contribution of genetic background to particle-mediated lung injury and suggest that *Tlr4* is a candidate susceptibility gene. Gilmour et al. [34] also found that pulmonary responses to combustion source PM in hypertensive rats are mediated through TLR4 signaling. Interestingly, Hollingsworth et al. [35] found that the pulmonary inflammatory responses to ROFA were not different in B6 mice with targeted deletion of *Tlr4* compared to wild type mice, and may suggest that the interaction between *Tlr4* and genetic background (strain) is an important consideration in pulmonary response to ROFA.

It is interesting to note that similar QTLs have been identified for multiple independent models of susceptibility to pollutant-induced inflammation, injury, and immune dysfunction. For example, nearly identical QTLs on chromosomes 17 and 11 have been found to explain a significant portion of the genetic variance in susceptibility to ACP, ozone-induced inflammation, and ozone-induced lung injury and death. Further, these QTLs were also found to have an important role in bleomycin- and radiation-induced lung injury in the mouse [36,37]. The common linkage suggests that similar mechanisms control susceptibility to the various environmental agents [38].

Evidence for an important role of genetic background in susceptibility to particle effects in human populations has also emerged. For example, Schwartz et al. [39] found changes in the high frequency (HF) component of HRV associated with exposure to PM2.5 only in individuals without the glutathione-S-transferase M1 (*GSTM1*) allele. No particle effects on HRV were found in individuals with normal *GSTM1*. These investigators also found that use of statins reversed the particle effects on HRV in the *GSTM1* null subjects. This gene X drug X environment interaction on HRV demonstrates the complex nature of susceptibility to particle effects in human populations. Adonis et al. [40] found an association between a biomarker (1-OH-P) of exposure to PAHs in diesel exhaust and the presence of the *CYP1A1*2A* genotype, and may be useful in identifying individuals at higher risk among those exposed to diesel exhaust. Interestingly, the *GSTM1* null genotype was not associated with the exposure biomarker, though an interaction between *CYP1A1*2A* and *GSTM1* may be informative. These studies support the notion that oxidative stress may be an important component of particle toxicity, and that individuals with compromised antioxidant defenses may be at enhanced risk to the injurious effects of particles.

15.2.2 Nongenetic Factors

Factors other than genetic background that are involved in cardiopulmonary susceptibility to particle exposure are also of considerable importance in understanding the etiology of this public health concern (Figure 15.2). It is critical to understand which subsections of the general population are susceptible to particulate exposure in order to reduce the health risks. While it is beyond the scope of this chapter to discuss all nongenetic components associated with susceptibility, we will focus on three important subgroups: age; pre-existing disease; socioeconomic status (Table 15.1).

Age. Elderly individuals and developing infants are at particular risk from exposure to particulate air pollution. A number of epidemiological and laboratory studies have addressed this concern,

TABLE 15.1
Representative Investigations of the Effects of Age, Pre-Existing Disease, and Socioeconomic Status on Responsivity to Particle Effects

Subgroup	Study type	Conclusion	Reference
Age	Epidemiology	Elderly individuals with high airway hyperresponsiveness and IgE were susceptible to air pollution including PM_{10}	Boezen et al. [47]
	Mouse	The regulation of heart rate was altered when senescent AKR/J mice were exposed to carbon black	Tankersley [53]
	Epidemiology	The risk of respiratory mortality in postneonatal infants increased in relation to PM_{10} increases	Ha et al. [45]
Pre-existing disease	Epidemiology	Individuals with congestive heart failure, arrhythmia or atherosclerosis are at risk	Brook et al. [55]
	Epidemiology	COPD and asthma increase susceptibility via elevated oxidative stress	Li et al. [61]
	Human subjects	Higher fine particle deposition was found in subjects with obstructive lung disease	Kim and Kang [64]
Socioeconomic	Epidemiology	Higher particle associated mortality rates were found in low socioeconomic regions	Jerret et al. [67]
	Epidemiology	Low education levels and employment in manufacturing present additional particle related health risks	Levy et al. [68]
	Field study	Antioxidant supplementation attenuated lung function responses to particle exposure	Romieu et al. [70]

but the health effects associated with infant exposure to particulate matter has not been studied in sufficient detail. Pre- and perinatal infants can be affected by particulate air pollution, including increases in fetal mortality [41] and low birth weight [42,43]. These effects of particulate air pollution are of great concern since, for example, the relationship between low birth weight and infant mortality is well known. Evidence for infant health risk associated with particle exposure was also provided by Woodruff et al. [44]. In this study, increased levels of PM_{10} were linked to respiratory illness and sudden infant death syndrome in 1–11 month old children. Another epidemiological studies reported that postneonatal infants are highly susceptible to increases in respiratory mortality in relation to particulate exposure [45]. In this study, daily mortality records from 1995 to 1999 in Seoul were used to establish subsections in the Korean population that were susceptible to air pollution. Subsections of the population were divided into 3 age groups: 1 month–1 year; 2–64 years; and individuals 65 years and over. The authors reported that postneonates had the highest relative risk (1.142) of total or respiratory mortality upon exposure to PM_{10} when compared to elderly individuals (1.023) and the intermediate age group (1.008). This suggests that infants are at greater risk than the elderly, and that particle related illness at a young age could lead to important developmental consequences that may affect these individuals in later life.

The aged as a second subgroup of the general population are more susceptible to particles than younger adults. The elderly share many of the same responses to particle exposure as children and they include those already discussed above in addition to airway hyperresponsiveness (AHA) and allergic reaction mediated by immunoglobulin E (IgE). Furthermore, women may be at greater risk because they have higher AHA than men [46]. Concomitant AHA and high levels of IgE have been reported to result in higher susceptibility to PM_{10} exposure than either AHA or high levels of IgE alone [47]. However, the mechanisms involved in these responses to particle air pollution are currently not clear.

Recently, a number of epidemiological and observational experimental studies have investigated a possible association between particulate air pollution exposure and changes in the regulation of the heart as a risk phenotype. An association between changes in HRV and increased cardiovascular (CV) risk has been established for some time [48]. For example, myocardial infarction patients with low HRV are more likely to suffer complications than patients with normal HRV. Therefore, reduced HRV during or following exposure to particles could provide insight into the potential mechanisms involved in susceptibility. Epidemiological studies have shown changes in HRV that were associated with changes in ambient particulate matter. Such changes in HRV are believed to indicate a greater risk of life threatening arrhythmia or the occurrence of fatal CV events in those with pre-existing CV or cardiopulmonary disease. Reduced HRV in the elderly has been reported during or following periods of higher ambient PM [49,50]. It has been suggested that health status may be an important factor in determining the degree of PM induced HRV changes [51]. It is important to note that methods designed for HRV measurement require controlled laboratory conditions for ECG recording. Since epidemiologists are forced to use ambulatory ECG data for HRV, results from these studies can be difficult to interpret. Controlled studies using mice have shown reductions in heart rate (HR) and HRV following exposure to ultrafine particles [52]. These responses occurred within an hour of the bolus exposure but were transient and values returned to baseline rapidly. Moreover, in these young and healthy mice, baseline HR and HRV did not predict the response to PM exposure. In aged mice, carbon black (CB) induced changes in the autonomic nervous system that differed according to the degree of aging [53]. In older, but healthy mice, CB exposure resulted in changes in the sympathetic nervous system. Conversely, in terminally senescent mice changes in the parasympathetic nervous system were observed following CB exposure and these changes were related to HR regulation.

Pre-existing disease. Pre-existing disease increases susceptibility to environmental pollution such as particles. While many disease states are likely to be involved, there is some evidence available to suggest that CV and respiratory illnesses and diabetes are important. In the previous section responses to PM exposure were discussed, and these biologic systems may become overwhelmed if underlying disease exists. Moreover, early development of disease can be accelerated by frequent exposure to particles. For example, Sun et al. [54] demonstrated that long-term exposure to $PM_{2.5}$ altered vascular tone, induced vascular inflammation, and potentiated atherosclerosis in ApoE-/- mice.

PM has been associated with increased hospital admissions and CV mortality due to CV disease. Those with congestive heart failure, arrhythmia and atherosclerosis appear to be most at risk [55]. Park et al. [56] reported that a family history of ischaemic heart disease (IHD), hypertension, and diabetes are all associated with susceptibility to PM. Increases in blood coagulation factors are also known to occur with PM exposure, and could have fatal consequences in individuals with IHD, hypertension, and atherosclerosis. One study used a marker of potential myocardial ischaemia and found an increased risk of S-T segment depression during an exercise test that associated with ultra fine particles and $PM_{2.5}$ [57]. Furthermore, additional risk has been associated with a high concentration of plasma fibrinogen [58], giving the impression that susceptibility to particles may be partly dependant upon the number of these risk factors that are present together.

Cardiovascular injury or risk, in association with particle exposure, has also been linked to diabetes. Zanobetti and Schwartz [59] reported that the percentage increase in PM_{10}-related CV hospital admission in diabetics was twice as high as those of nondiabetics. The authors suggested that this observation might be associated with upregulation of inflammatory activity in diabetics, which may confer heightened susceptibility to PM. Patients with diabetes also present several risk factors associated independently with CV disease and exposure to PM. If compensatory mechanisms have a baseline reduction in capacity because of pre-existing disease, subsequent exposure to PM may overwhelm these systems causing severe illness or sudden death [60].

Pre-existing COPD and asthma are known to increase susceptibility to particle exposure perhaps due to a significant, additional oxidative stress [61]. In COPD patients, all cause associations with particle exposure were 10 times greater when compared to all subjects in a recent study [62]. Potential mechanisms for this high level of susceptibility in these patients include decreased antioxidant defenses [63] and higher fine particle deposition in the lungs of patients with obstructive airways disease [64]. Induction or exacerbation of asthmatic symptoms by exposure to particles has also been well established. In a recent study, fractional exhaled nitric oxide (FE_{NO}) was used as a noninvasive method of estimating airway inflammation [65]. FE_{NO} was measured over a 12-day period while local particle levels were monitored. FE_{NO} levels were associated with changes in PM_{10} and $PM_{2.5}$ in the elderly asthmatic patients. These data were in agreement with an earlier study by this group that showed similar results in asthmatic children [66].

Socioeconomic status. Subsections of a citywide population in an intraurban area with low socioeconomic status have been associated with higher mortality rates in relation to ambient air pollution [67]. Low education levels and employment in manufacturing present additional particle-related health risks [68]. Proposed explanations for this effect of socioeconomic status and education attainment include higher workplace particle exposures for those working in manufacturing material deprivation and poor material conditions [69]. Also, lower socioeconomic status has been associated with less exposure measurement error since there individuals are less mobile [69]. Since these groups must tolerate a disproportionate level of susceptibility to air pollution related health risks, they would account for a large percentage of emissions control benefits [68]. Low socioeconomic status may also affect the ability of people to achieve adequate nutrition. Poor nutrition status has been suggested as another particle susceptibility factor [69]. Inadequate nutrition could compromise antioxidant defenses thus increasing particle susceptibility—a recent study showed reduced particulate effects on lung function after dietary supplementation with antioxidants [70]. The importance of this factor becomes apparent when considering the fact that poor nutritional status is possible across all socioeconomic classes.

15.3 CONCLUSIONS

It is clear from the above discussion that genetics background is a critical component to inter-individual susceptibility to particle exposure. Further investigation is of great importance if our understanding of the genetic contribution to particle associated health risks is to be improved. For example, several QTLs are known to be involved, but few candidate genes have been rigorously tested. Little is known about how age, pre-existing disease, or socioeconomic status modifies the genetic components to particle susceptibility. Susceptibility to morbidity and mortality associated with particle exposure appears to be highly complex and it likely includes interactions between many of these components, both genetic and nongenetic. It is also clear that the mechanisms involved in many of these factors are poorly understood and much work is needed to reduce this important public health concern.

REFERENCES

1. Dockery, D. W. et al., An association between air pollution and mortality in six U.S. cities, *N. Engl. J. Med.*, 329, 1753, 1993.
2. Bell, M. L., Samet, J. M., and Dominici, F., Time-series studies of particulate matter, *Annu. Rev. Public Health*, 25, 247, 2004.
3. Dominici, F. et al., Fine particulate air pollution and hospital admission for cardiovascular and respiratory diseases, *JAMA*, 295, 1127, 2006.
4. Pope, C. A. III, et al., Cardiovascular mortality and long-term exposure to particulate air pollution: epidemiological evidence of general pathophysiological pathways of disease, *Circulation*, 109, 71, 2004.
5. Kleeberger, S. R., Genetic aspects of pulmonary responses to inhaled pollutants, *Exp. Toxicol. Pathol.*, 57 (suppl. 1), 147, 2005.
6. McCunney, R. J., Asthma, genes, and air pollution, *J. Occup. Environ. Med.*, 47, 1285, 2005.
7. Nebert, D. W., Inter-individual susceptibility to environmental toxicants—a current assessment, *Toxicol. Appl. Pharmacol.*, 207 (suppl. 2), 34, 2005.
8. Yang, I. A. et al., Association of tumor necrosis factor-alpha polymorphisms and ozone-induced change in lung function, *Am. J. Respir. Crit. Care Med.*, 171, 171, 2005.
9. Nemery, B. et al., Interstitial lung disease induced by exogenous agents: factors governing susceptibility, *Eur. Respir. J.*, 18 (suppl. 32), 30s, 2001.
10. Hattis, D. et al., Human interindividual variability in susceptibility to airborne particles, *Risk Anal.*, 21, 585, 2001.
11. Kleeberger, S. R. and Peden, D., Gene-environment interactions in asthma and other respiratory diseases, *Annu. Rev. Med.*, 56, 383, 2005.
12. Peters, A. et al., Short-term effects of particulate air pollution on respiratory morbidity in asthmatic children, *Eur. Respir. J.*, 10, 872, 1997.
13. Schwartz, J., PM10, ozone, and hospital admissions for the elderly in Minneapolis-St. Paul, Minnesota, *Arch. Environ. Health*, 49, 366, 1994.
14. Eichelbaum, M., Ingelman-Sundberg, M., and Evans, W. E., Pharmacogenomics and individualized drug therapy, *Annu. Rev. Med.*, 57, 119, 2006.
15. Kalow, W., Meyer, U. A., and Tyndale, R. F., eds., *Pharmacogenomics*, Marcel Dekker Inc., New York, 2002.
16. Nebert, D. W. and Vessell, E. S., Advances in pharmacogenomics and individualized drug therapy: exciting challenges that lie ahead, *Eur. J. Pharmacol.*, 500, 267, 2004.
17. Shastry, B. S., Pharmacogenetics and the concept of individualized medicine, *The Pharmacogenomics J.*, 6, 16, 2006.
18. Calabrese, E. F., *Ecogenetics: Genetic Variation in Susceptibility to Environmental Agents*, Wiley, New York, 1984.
19. Grandjean, P., ed., *Ecogenetics. Genetic Predisposition to the Toxic Effects of Chemicals*, Chapman and Hall, London, 1991.

20. Kellam, P. and Weiss, R. A., Infectogenomics: insights from the host genome into infectious diseases, *Cell*, 124, 695, 2006.
21. Schork, N. J., Fallin, D., and Launchbury, S., Single nucleotide polymorphisms and the future of genetic epidemiology, *Clin. Genet.*, 58, 250, 2000.
22. Palmer, L. J. and Cookson, W. O. C. M., Using single nucleotide polymorphisms as a means to understanding the pathophysiology of asthma, *Respir. Res.*, 2, 102, 2001.
23. Ohtsuka, Y. et al., Interstrain variation in murine susceptibility to inhaled acid-coated particles, *Am. J. Physiol. Lung Cell. Mol. Physiol.*, 278, L469, 2000.
24. Ohtsuka, Y. et al., Genetic linkage analysis of susceptibility to particle exposure in mice, *Am. J. Respir. Cell Mol. Biol.*, 22, 574, 2000.
25. Wiester, M. J. et al., Ozone adaptation in mice and its association with ascorbic acid in the lung, *Inhal. Toxicol.*, 12, 577, 2000.
26. Gong, H., Jr., McManus, M. S., and Linn, W. S., Attenuated response to repeated daily ozone exposures in asthmatic subjects, *Arch. Environ. Health*, 52, 34, 1997.
27. Wesselkamper, S. C. et al., Genetic variability in the development of pulmonary tolerance to inhaled pollutants in inbred mice, *Am. J. Physiol. Lung. Cell. Mol. Physiol.*, 281, L1200, 2001.
28. Wesselkamper, S. C., Chen, L. C., and Gordon, T., Quantitative trait analysis of the development of pulmonary tolerance to inhaled zinc oxide in mice, *Respir. Res.*, 6, 73, 2005.
29. Wesselkamper, S. C. et al., Genetic susceptibility to irritant-induced acute lung injury in mice, *Am. J. Physiol. Lung Cell. Mol. Physiol.*, 279, L575, 2000.
30. Prows, D. R. and Leikauf, G. D., Quantitative trait analysis of nickel-induced acute lung injury in mice, *Am. J. Respir. Cell Mol. Biol.*, 24, 740, 2001.
31. Mallakin, A. et al., Gene expression profiles of *Mst1r*-deficient mice during nickel-induced acute lung injury, *Am. J. Respir. Cell Mol. Biol.*, 34, 15, 2006.
32. Wesselkamper, S. C. et al., Gene expression changes during the development of acute lung injury. Role of transforming growth factor β, *Am. J. Respir. Crit. Care Med.*, 172, 1399, 2005.
33. Cho, H. Y. et al., Role of toll-like receptor-4 in genetic susceptibility to lung injury induced by residual oil fly ash, *Physiol. Genomics*, 22, 108, 2005.
34. Gilmour, P. S. et al., Hypertensive rats are more susceptible to TLR4-mediated signaling following exposure to combustion source particulate matter, *Inhal. Toxicol.*, 16(suppl. 1), 5, 2004.
35. Hollingsworth, J. W. II, et al., The role of toll-like receptor 4 in environmental airway injury in mice, *Am. J. Respir. Crit. Care Med.*, 170, 126, 2004.
36. Haston, C. K. et al., Bleomycin hydrolase and a genetic locus within the MHC affect risk for pulmonary fibrosis in mice, *Hum. Mol. Genet.*, 11, 1855, 2002.
37. Haston, C. K. et al., Universal and radiation-specific loci influence murine susceptibility to radiation-induced pulmonary fibrosis, *Cancer Res.*, 62, 3782, 2002.
38. Bauer, A. K., Malkinson, A. M., and Kleeberger, S. R., Susceptibility to neoplastic and non-neoplastic pulmonary diseases in mice: genetic similarities, *Am. J. Physiol. Lung Cell Mol. Physiol.*, 287, L685, 2004.
39. Schwartz, J. et al., Glutathione-S-transferase M1, obesity, statins, and autonomic effects of particles, *Am. J. Respir. Crit. Care Med.*, 172, 1529, 2005.
40. Adonis, M., Susceptibility and exposure biomarkers in people exposed to PAHs from diesel exhaust, *Toxicol. Lett.*, 144, 3, 2003.
41. Pereira, L. A. et al., Association between air pollution and intrauterine mortality in Sao Paulo, Brazil, *Environ. Health Perspect.*, 106, 325, 1998.
42. Ha, E. H. et al., Is air pollution a risk factor for low birth weight in Seoul?, *Epidemiology*, 12, 643, 2001.
43. Wang, X. et al., Association between air pollution and low birth weight: a community-based study, *Environ. Health Perspect.*, 105, 514, 1997.
44. Woodruff, T. J., Grillo, J., and Schoendorf, K. C., The relationship between selected causes of postneonatal infant mortality and particulate air pollution in the United States, *Environ. Health Perspect.*, 105, 608, 1997.
45. Ha, E. H. et al., Infant susceptibility of mortality to air pollution in Seoul, South Korea, *Pediatrics*, 111, 284, 2003.

46. Leynaert, B. et al., Is bronchial hyperresponsiveness more frequent in women than in men?, *Am. J. Respir. Crit. Care Med.*, 156, 1413, 1997.
47. Boezen, H. M. et al., Susceptibility to air pollution in elderly males and females, *Eur. Respir. J.*, 25, 1018, 2005.
48. Task Force, Heart rate variability: standards of measurement, physiological interpretation and clinical use. Task Force of the European Society of Cardiology and the North American Society of Pacing and Electrophysiology, *Circulation*, 93, 1043, 1996.
49. Creason, J., Particulate matter and heart rate variability among elderly retirees: the Baltimore 1998 PM study, *J. Expo. Anal. Environ. Epidemiol.*, 11, 116, 2001.
50. Gold, D. R. et al., Ambient pollution and heart rate variability, *Circulation*, 101, 1267, 2000.
51. Wheeler, A. et al., The relationship between ambient air pollution and heart rate variability differs for individuals with heart and pulmonary disease, *Environ. Health Perspect.*, 114, 560, 2006.
52. Howden, R. et al., Heart rate and heart rate variability responses to particulate matter exposure and the genetic contribution to susceptibility, *Am. J. Respir. Crit. Care Med.*, 171, A296, 2005.
53. Tankersley, C. G. et al., Particle effects on heart-rate regulation in senescent mice, *Inhal. Toxicol.*, 16, 381, 2004.
54. Sun, Q. et al., Long-term air pollution exposure and acceleration of atherosclerosis and vascular inflammation in an animal model, *JAMA*, 294, 3003, 2005.
55. Brook, R. D. et al., Air pollution and cardiovascular disease: a statement for healthcare professionals from the expert panel on population and prevention science of the American Heart Association, *Circulation*, 109, 2655, 2004.
56. Park, S. K. et al., Effects of air pollution on heart rate variability: the VA normative aging study, *Environ. Health Perspect.*, 113, 304, 2005.
57. Pakkanen, J. et al., Particulate air pollution and risk of ST-segment depression during repeated submaximal exercise tests among subjects with coronary heart disease, *Circulation*, 106, 933, 2002.
58. Prescott, G. J. et al., Investigation of factors which might indicate susceptibility to particulate air pollution, *J. Occup. Environ. Med.*, 57, 53, 2000.
59. Zanobetti, A. and Schwartz, J., Cardiovascular damage by airborne particles: are diabetics more susceptible?, *Epidemiology*, 13, 588, 2002.
60. Bateson, T. F. and Schwartz, J., Who is sensitive to the effects of particle air pollution on mortality? A case-crossover analysis of effect modifiers, *Epidemiology*, 15, 143, 2004.
61. Li, X. Y., Free radical activity and pro-inflammatory effects of particulate air pollution (PM10) in vivo and in vitro, *Thorax*, 51, 1216, 1996.
62. Sunyer, J. et al., Patients with chronic obstructive pulmonary disease are at increased risk of death associated with urban particle air pollution: a case-crossover analysis, *Am. J. Epidemiol.*, 151, 50, 2000.
63. Rahman, I. et al., Systemic oxidative stress in asthma, COPD and smokers, *Am. J. Respir. Crit. Care Med.*, 154, 1055, 1996.
64. Kim, C. S. and Kang, T. C., Comparative measurement of lung deposition of inhaled fine particles in normal subjects and patients with obstructive airway disease, *Am. J. Respir. Crit. Care Med.*, 155, 899, 1997.
65. Jansen, K. L. et al., Associations between health effects and particulate matter and black carbon in subjects with respiratory disease, *Environ. Health Perspect.*, 113, 1741, 2005.
66. Koenig, J. Q. et al., Measurement of offline exhaled nitric oxide in a study of community exposure of air pollution, *Environ. Health Perspect.*, 111, 1625, 2003.
67. Jerret, M. et al., Do socioeconomic characteristics modify the short term association between air pollution and mortality? Evidence from a zonal time series in Hamilton, Canada, *J. Epidemiol. Community Health*, 58, 31, 2004.
68. Levy, J. I., Greco, S. L., and Spengler, J. D., The importance of population susceptibility for air pollution risk assessment: a case study of power plants near Washington, DC, *Environ. Health Perspect.*, 110, 1253, 2002.
69. Villeneuve, P. J. et al., A time-series study of air pollution, socioeconomic status, and mortality in Vancouver, Canada, *J. Expo. Anal. Environ. Epidemiol.*, 13, 427, 2003.
70. Romieu, I. et al., Antioxidant supplementation and respiratory functions among workers exposed to high levels of ozone, *Am. J. Respir. Crit. Care Med.*, 158, 226, 1998.

16 Genotoxic Effects of Particles

Roel P. F. Schins
Institut für umweltmedizinische Forschung (IUF) an der Heinrich-Heine-Universität

Tom K. Hei
Center for Radiological Research, Columbia University

CONTENTS

16.1 INTRODUCTION

16.1.1 Implications of DNA Damage

It is nowadays well accepted that damage to the genomic DNA, both from endogenous and exogenous sources, may have major pathophysiological implications for cells, tissues, and organisms (Hoeijmakers 2001). Depending on its type, its extent, and its persistence, DNA damage may have various consequences for a cell, as shown in Figure 16.1. Of major importance is that DNA modifications can lead to mutations, altering the coding sequence of DNA. Mutations may involve a single gene, a block of genes, or even whole chromosomes. Damage to the DNA may also cause interference with various normal cellular processes, including DNA replication and DNA transcription, and this may result in cell death. To maintain the integrity of the DNA molecule before initiating DNA replication, transcription, and cell division, cells are provided with several efficient DNA repair mechanisms (Wood et al. 2001). In some circumstances, these repair processes are incorrect or incomplete, or otherwise deliberately by-passed (also referred to as damage tolerance).

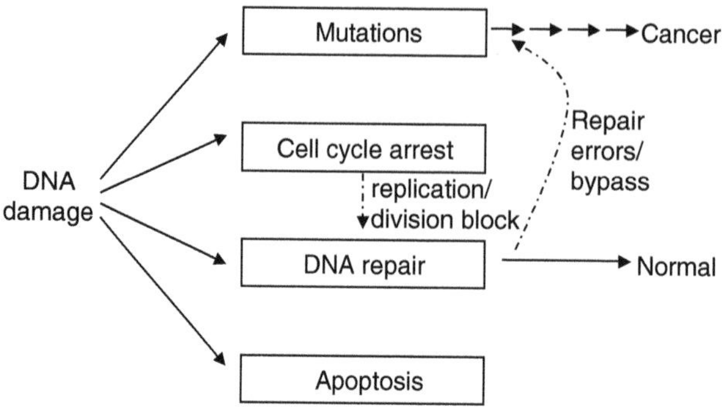

FIGURE 16.1 Cellular responses to DNA damage. DNA damage in a cell may lead to mutations that may involve single genes, a block of genes, or entire chromosomes. DNA damage can also trigger cell cycle arrest, which provides the cell an opportunity for prolonged DNA repair. DNA repair processes may be incorrect or incomplete and in certain circumstances, DNA lesions are deliberately by-passed. DNA damage also triggers signaling pathways that drive cells into apoptosis, thereby preventing the passing-on of genetic changes to subsequent cell generations.

The process of DNA damage in itself is known to trigger cell cycle arrest, whereby prolongation of G1 and G2 phases provides additional time for DNA repair prior to DNA synthesis and mitosis, respectively (Zhou and Elledge 2000). Finally, DNA damage is known to trigger signal transduction pathways, which promote cellular apoptosis (Figure 16.1).

16.1.2 Genotoxicity Testing of Particles and Carcinogenesis

Cancer is nowadays considered to result from a sequence of events that include genetic alterations involving activation of oncogenes, inactivation of suppressor genes, and aberrant controls in cell cycle checkpoint genes (Vogelstein and Kinzler 2004; Feinberg, Ohlsson, and Henikoff 2006). This concept has also been implicated for many years in fiber carcinogenesis (Barrett 1993; Kane 2000; Zhao et al. 2000). Due to its significance and relevance to the cancer process, genotoxicity testing has been introduced to identify potential carcinogenic and/or biologically active materials. Presently, a broad spectrum of such tests are available to screen for gene mutations, chromosomal mutations, and aneugenic effects (aneuploidy), as well as to detect DNA strand breakage, DNA adduct formation and DNA repair (e.g., reviewed in Vainio et al. 1992; McGregor, Rice, and Venitt 1999). Genotoxicity tests that have been used in particles research include DNA strand breakage assays (e.g., alkaline elution, comet assay), assays to detect DNA base damage or DNA adducts (e.g., HPLC, ^{32}P-postlabelling), chromosomal damage tests representing possible clastogenic activity of particles (e.g., sister chromatid exchanges, micronuclei), and finally, measurement of gene mutations (e.g., HPRT, Ames-test, DNA-sequencing) (Jaurand 1996; Schins 2002).

16.1.3 Genotoxicity Investigations with Particles

Various particles, both fibrous (e.g., asbestos) and nonfibrous, have been investigated for their genotoxic potential in various cell lines/types of both human and murine origins. In addition, various investigators have also addressed *in vivo* DNA damage and/or mutagenesis by particles. Selections of these investigations are shown in Table 16.1 and Table 16.2. In view of the importance of DNA damage in the carcinogenesis process, genotoxicity investigations have provided important (supportive) data for risk assessment purposes (Jaurand 1996; Greim et al. 2001). Beyond

TABLE 16.1
Examples of *In Vitro* Genotoxicity Studies by Particles

Particle	Effect	References
Asbestos fibers	CA (rat mesothelial cells)	Yegles et al. (1993)
	8-OHdG (A549)	Chao et al. (1996)
	Mutations (hgprt−/gpt+V79)	Park and Aust (1998)
	DSB (CHEF-electrophoresis) (xrs-5)	Okayasu et al. (1999)
	8-OHdG (ICC), multilocus deletions (Human-hamster hybrid (AL) cells)	Xu et al. (1999)
	SSB (comet) (Rat mesothelial)	Levresse et al. (2000)
	SSB (alkaline unwinding) (Rabbit mesothelial)	Liu et al. (2000)
Crystalline silica	CA, MN (Hel 299/V79-fibroblasts)	Nagalakshmi et al. (1995)
	MN, SSB (comet, V79)	Liu et al. (1996)
	SSB (comet, Hel 299/V79)	Zhong et al. (1997)
	SSB (comet), 8-OHdG (ICC) (A549/RLE)	Schins et al. (2002)
Amorphous silica	SSB (comet) (Hel 299/V79)	Zhong et al. (1997)
Carbon black	SSB (comet) (Hel 299/V79)	Zhong et al. (1997)
	SSB (comet) (A549/Monocytes)	Don Porto Carero et al. (2001)
	Bulky adducts (32P-postlabeling) (A549)	Borm et al. (2005)
Coal fly ashes	8-OHdG (HPLC) (RLE)	Van Maanen et al. (1999)
Diesel exhaust particles	SCE (V79)	Keane et al. (1991)
	SSB (A549)	Don Porto Carero et al. (2001)
	SSB (A549)	Dybdahl et al. (2004)
Cobalt/Tungsten carbide	SSB (Comet, Alkaline elution), MN (human lymphocytes/monocytes)	Anard et al. (1997); De Boeck et al. (2003)
Titanium dioxide	MN, SCE (Hamster ovary cells)	Lu et al. (1998)
	8-OHdG (HPLC) (human epithelial)	Prahalad et al. (2001)
	MN (Hamster fibroblasts)	Rahman et al. (2002)
	MN, oxidative DNA damage (fpg-comet) (BEAS-2B epithelial cells)	Gurr et al. (2005)
PM	SSB (Comet) (Lung epithelial cells)	Dellinger et al. (2001)
	8-OHdG (HPLC) (human epithelial)	Prahalad et al. (2001)
	SSB (Comet) (A549)	Knaapen et al. (2002a)
	SSB (Comet) (A549)	Don Porto Carero et al. (2001)
	8-OHdG (Comet) (A549)	Shi et al. (2003)

MN, Micronuclei; CA, chromosomal aberrations; SSB, single strand breaks; DSB, double strand breaks; ICC, immunocytochemistry; SCE, sister chromatid exchanges.

this, genotoxicity research has also contributed to our understanding of mechanisms involved in particle-induced diseases, including carcinogenesis (Hei 2000; Knaapen et al. 2004).

In view of risk assessment, one should be very cautious about the interpretation and extrapolation of genotoxicity data obtained with particles, especially with *in vitro* investigations. Firstly, particles have rather complex physicochemical properties (Fubini et al. 2001), which can modulate their biological activities when compared to soluble chemicals. Secondly, considerable mechanistic and/or dosimetrical discrepancies exist between *in vitro* studies and observations as obtained from *in vivo* genotoxicity and/or carcinogenicity studies. Typically, most available *in vivo* studies have been performed at high particle concentrations and/or long term exposures, which are associated with marked inflammatory and proliferative responses, and hence may obscure and/or modify genotoxicity readouts. This aspect will be further discussed in more detail later in this chapter.

TABLE 16.2
Examples of *In Vitro* Genotoxicity Studies by Particles

Particle	Effect (species)	References
Crystalline silica	HPRT-mutations in type II cells (Rats)	Driscoll et al. (1997)
	8-OHdG, p53 (IHC) (Rats)	Seiler et al. (2001)
	SSB (comet in lung epithelial cells, Rats)	Knaapen et al. (2002b)
	8-OHdG (IHC&HPLC) (Rats)	Albrecht et al. (2005)
Asbestos fibers	Strand breaks/tunel (Rats)	Jung et al. (2000)
	Mutation frequencies (transgenic mice)	Rihn et al. (2000)
	Mutation frequencies/spectra, 8-OHdG (transgenic rats)	Unfried et al. (2002)
Carbon black	Bulky DNA adducts (Rats)	Gallagher et al. (1994)
	HPRT mutations in type II cells (Rats)	Driscoll et al. (1997)
	8-OHdG (Rats) (HPLC)	Gallagher et al. (2003)
	Bulky DNA adducts (32P) (Rats)	Borm et al. (2005)
Diesel exhaust	Bulky DNA-adducts (32P- postlabelling) (Rats)	Wong et al. (1986)
	Bulky DNA-adducts (32P- postlabelling) (Rats)	Gallagher et al. (1994)
	Mutation frequencies/spectra, 8-OHdG (transgenic rats)	Sato et al. (2000)
Diesel exhaust particles (e.g., SRM1650)	8-OHdG (HPLC) (mice)	Ichinose et al. (1997)
	8-OHdG (fpg-comet) (mice)	Risom et al. (2003)
	8-OHdG (fpg-comet) (mice)	Dybdahl et al. (2004)
Titanium dioxide	Bulky DNA-adducts (32P-postlabelling) (Rats)	Gallagher et al. (1994)
	HPRT-mutations in type II cells (rats)	Driscoll et al. (1997)

IHC, immunocytochemistry.

DNA damage studies are highly relevant to the observed epidemiological associations between ambient particulate matter (PM) exposure and lung cancer (e.g., Pope et al. 2002). Because of the highly complex and variable composition of PM, it is by no means clear which component(s) is/are causally linked to the genotoxic and carcinogenic actions. Finally, there will always be a continuous demand for validated genotoxicity assays to screen novel particulate materials, such as engineered nanoparticles and nanotubes. There is recent concern in this issue because of (I) the intrinsically enhanced toxicity of ultrafine particles (TiO_2, carbon black) versus their nonultrafine counterparts in rodent studies (Donaldson et al. 2002; Oberdörster, Oberdörster, and Oberdörster 2005), and (II) the tumorigenicity of such ultrafine particles as observed in high-dose inhalation studies in rats (Borm, Schins, and Albrecht 2004).

16.2 MECHANISMS OF PARTICLE GENOTOXICITY

16.2.1 Effect of Particle-Specific Properties on DNA Damage

As discussed in several chapters throughout this book, toxic effects of particles are considered to be driven by a series of specific features and properties that are clearly distinguishable from nonparticulate chemicals. Based on current data, it appears that at least some of such particle-specific properties also may explain for the ability—or the lack of ability—of a certain particle to elicit DNA damage (Schins 2002). Major properties of particles that may impact on DNA damage are listed in Table 16.3.

First of all, in contrast to a soluble/nonparticulate genotoxic molecule, a single particle typically contains a multiplicity of intrinsic physicochemical properties that can act simultaneously or consecutively on various biological/toxicological endpoints and target sites (Fubini et al. 2001). The morphologies and aspect ratios of inhaled particles will determine their deposition, subsequent

TABLE 16.3
Properties of Particles Relevant for Their DNA Damaging Properties

Process/Mechanism	Particle Properties Implicated
Deposition site (inhalation): determines target cell for possible DNA damage induction	Size, shape, density, e.g., ultrafine particles, fibers
Cellular uptake and subcellular translocation:	Size/shapesurface properties, e.g., surface radicals, surface charge, adsorption of proteins (coating)
Association between particle uptake and DNA damage	
Nuclear penetration allows direct interaction between particle and DNA	
Interaction with mitochondria is associated with DNA damage	
Potential physical interaction with mitotic spindle during cell division, interaction with proteins, e.g., tubulin	Size, dimensions fibers, nanotubes, nanoparticles
Genotoxicity or (co)genotoxic effect of particle "contaminants" (carrier function)	Adsorbed mutagens, e.g., PAH, nitro-PAH DNA-damaging and/or DNA-repair inhibiting metals
Generation of ROS and oxidative stress by particles, activation of ROS generation by cells	Particle surface chemistry/composition, e.g., silanol groups iron leaching/mobilization activation of cellular ROS sources
DNA-damage/mutagenesis by products released during particle-elicited inflammation (i.e., ROS, RNS); Dependent on severity and persistency of inflammatory process	Any particle property which impacts on the severity and/or persistence of inflammation poorly soluble particles: implications of durability for inflammation and overload

translocation, and finally the primary cellular targets for an eventual genotoxic action. For particles of nano-size dimensions, target site may be systemic following inhalatory, dermal, or oral exposures (Oberdörster, Oberdörster, and Oberdörster 2005). In organs where nanoparticle translocation has already been demonstrated, such as the brain and the cardiovasculature, investigation of possible genotoxic effect represents a challenge.

On a cellular level, studies with asbestos and quartz indicate that DNA damage depends on their uptake (Jaurand 1997; Hei 2000; Liu, Ernst, and Broaddus 2000; Schins et al. 2002). Nuclear penetration of particles is also of major importance for genotoxicity, since this is a prerequisite for their direct interaction with the nuclear DNA. This aspect is likely to be of major importance for nanoparticles. Indeed, the presence of some of such particles in cellular nuclei has already been documented, partly because they are specifically engineered to reach this cell compartment (e.g., Thomas and Klibanov 2003; Chen and von Mikecz 2005; Geiser et al. 2005). An exception of this rule is during mitosis, where in the absence of a nuclear envelope, particles have been postulated to interact directly with condensated chromosomes and spindle-apparatus (Daniel et al. 1995). Furthermore, interactions with the spindle-apparatus and/or interference with the mitotic proteins, such as tubulins, have been specifically proposed as a genotoxic mechanism of asbestos fibers (Jaurand 1997), and may be relevant for nanoparticles and the "fibre-shaped" nanotubes.

In relation to particle genotoxicity, chemical composition and surface reactivity are known to play major modifying roles. Many particles present themselves as having an insoluble or poorly soluble core onto which various adsorbed mutagens (or carcinogens) can be carried from the environment into and throughout the human body. The mutagenicity of organic extracts of PM has been clearly established for several decades now. Significant contributors to this mutagenic pathway include the polycyclic aromatic hydrocarbons (PAH) and tobacco carcinogens, which upon biotransformation convert into reactive intermediates that can form bulky adducts to the

DNA (Hall and Grover 1990). Such adduct formation has also been linked to specific combustion-generated particles, including diesel exhaust particles (DEP) or commercial carbon black. Apart from organic constituents, PM often contain numerous metallic constituents, and many are known to illicit DNA damage and/or inhibit DNA repair (Hei, Liu, and Waldren 1998; Hartwig 2002; Kawanishi et al. 2002; Kessel et al. 2002; Tran et al. 2002). Obviously, the possible genotoxic and mutagenic effects of particle-associated organics or metals will crucially depend on their bioavailability (Hardy and Aust 1995; Greim et al. 2001). This concept of solubility in fact also holds true for core particles, that is, if the entire particle is readily soluble, any possible genotoxic effect will be of a nonparticulate chemical nature.

Of major importance for genotoxicity is the contribution by various oxidants including reactive oxygen species (ROS) such as hydrogen peroxide (H_2O_2), hydroxyl radicals ($^{\cdot}OH$), superoxide ($O_2^{\cdot -}$), and singlet oxygen ($^1O_2$), as well as reactive nitrogen species (RNS) including nitric oxide (NO•) and peroxynitrite (ONOO–) (Knaapen et al. 2004). As oxidants have been implicated in DNA damage and carcinogenesis (Marnett 2000), their generation by particles can be considered a major determinant of their genotoxic properties.

16.2.2 Oxidant Generation by Particles

In relation to its impact on DNA damage, the generation of oxidants by particles can be subdivided to occur in three principal manners (see Table 16.4). Firstly, oxidants may be intrinsically generated from the particles themselves, i.e., in the absence of cells. Herein, oxidants may be generated due to surface-associated reactive groups (e.g., $SiO^{\cdot}$ and $SiO_2^{\cdot}$ on quartz), or transitional metals such as iron, or due to the ability of particles to generate oxidants in aqueous suspension. Various metals (e.g., iron, copper) that are present in particles or fibers have been implicated in the generation of ROS via Haber-Weiss reactions (Hardy and Aust 1995; Donaldson et al. 1997). Besides, organic compounds associated with particles (e.g., PAH) can, upon metabolic activation, be converted into redox-cycling semiquinone radicals (Squadrito et al. 2001; Li et al. 2003). Importantly, the DNA-damaging ability of such particle-derived oxidants will depend on both the sub-cellular localization of the particle and the half-life of the generated oxidant species in this environment. Particle-generated ROS will likely have a significant impact on genotoxicity if they occur in direct

TABLE 16.4
Pathways of Particle-Mediated ROS Generation and Their Impact for Primary and Secondary Genotoxicity

Process/Mechanism	Genotoxicity
Intrinsic ROS generation from particles	Primary (direct or indirect)
Surface associated free radicals or oxidative groups (e.g., SiO• and SiO2• on crystalline silica, carbon-centred radicals on coal dusts)	
Oxidant generation by particles in aqueous suspension (e.g., Haber-Weiss reactions by available metals, semiquinone radical redox-cycling of biotransformed PAH	
High surface area of ultrafine particles/nanoparticles (possibly related to increased catalytic actions, quantum effects)	
ROS generation upon interaction of particle with target cells	Primary (indirect)
Damage to mitochondria/interaction with the electron transport chain	
Activation of NAD(P)H-like enzyme systems	
Disturbance of endogenous antioxidant defences	
Generation of ROS (and RNS) during particle-elicited inflammation	Secondary
Phagocytes: NADPHoxidase, Nitric oxides synthase, Myeloperoxidase	

proximity with the nuclear DNA, as indicated from naked-DNA damage assays (e.g., Daniel et al. 1995; Hardy and Aust 1995; Prahalad et al. 2001; Knaapen et al. 2002a).

A second route of oxidant generation is represented by the ability of particles to activate cells for (enhanced) intracellular ROS generation. Such cell-derived ROS generation may, for instance, involve interaction of particles with mitochondria and the electron transport chain or activation of NAD(P)H-like enzyme systems (Deshpande, Narayanan, and Lehnert 2002; Li et al. 2003). Given the extra-nuclear localization of these ROS sources, oxidative damage of the nuclear DNA would in such case result from stable, diffusible genotoxic oxidants such as hydrogen peroxide or other long-lived organic radicals. In this regard, there is supporting evidence that targeted cytoplasmic irradiation by alpha particles can result in mutations in the nuclei of the hit cells involving production of ROS (Wu et al. 1999).

Thirdly, ROS and RNS are known to be generated during particle-elicited inflammation, upon activation of (recruited) phagocytes (Kamp et al. 1994; Castranova et al. 1998; Albrecht et al. 2005), and they are considered to be implicated in the induction of oxidative DNA damage in target cells. Consequently, particle properties that are known to impact on inflammation and/or phagocyte-mediated ROS/RNS production, such as particle surface reactivity or transition metal content, may all indirectly impact on this pathway of DNA damage. Taken together, this pathway of ROS/RNS generation has been strongly implicated in the carcinogenic action of particles. Although phagocytosis has been implicated to be important in fibre- and particle-mediated genotoxicity (Hei 2000; Schins et al. 2002), the role of inflammatory response and the subsequent contributions from cytokines—often referred to as secondary genotoxicity and are the *in vivo* consequences of phagocytosis—are not well defined with animal studies.

16.2.3 The Concept of Primary and Secondary Genotoxicity in Particle Risk Assessment

Beyond the discrimination of the three oxidant-generating pathways by particles for mechanistic purposes (Table 16.4), a further subdivision is to be made between primary and secondary pathways (Schins 2002; Knaapen et al. 2004). Herein, secondary genotoxicity can be defined as genetic damage resulting from ROS/RNS (and possibly other mediators) that are generated during particle-elicited inflammation. Consequently, primary genotoxicity may be defined as genetic damage elicited by particles, direct or indirect, in the absence of inflammation (see Table 16.4). The aspect of secondary (versus primary) genotoxicity is defined in relation to its importance for risk assessment and originates from observations that various poorly soluble particles were tumorigenic in rat lungs after chronic high exposures associated with overloading and persistent inflammation (Greim et al. 2001; Borm, Schins, and Albrecht 2004). In support of this concept, alveolar epithelial cells obtained from rats that had been instilled with carbon black, TiO_2, or crystalline silica showed enhanced $HPRT^-$ mutagenicity, but only in those treatment conditions that resulted in a significant inflammation in the animals (Driscoll et al. 1997). As will be discussed in more detail in the next paragraph, such association between particle-elicited inflammation and genotoxicity/mutagenicity has been observed in several other studies.

Primary genotoxicity may result from various particle properties (as outlined in Section 16.2.1 and Section 16.2.2), and can be considered for particles that are genotoxic/mutagenic in the absence of inflammation. This genotoxicity may be either direct (e.g., DNA-binding of a PAH-metabolite) or indirect (e.g., ROS generation and redox cycling, interference with microtubule and induction of aneuploidy, inhibition of DNA synthesis, etc.). Thus, primary genotoxicity may, in principle, be identified from *in vitro* studies. In contrast, secondary genotoxic response may involve a threshold depending on the exposure concentrations that induce inflammation and overwhelm the antioxidant and DNA repair capacity in the lung (Greim et al. 2001). As mentioned before, the inflammatory effects of particles (extent, persistence) also depend on various particle properties, such as particle solubility and reactivity, and hence this threshold will likely vary with each particle type. [*Note*: Inflammatory cell-derived ROS and RNS are apart from their DNA-damaging properties also implicated in other processes involved in the carcinogenic process, including the induction of proliferation or apoptosis (Fubini and Hubbard 2003; Shukla et al. 2003; Knaapen et al. 2004)].

16.3 MAJOR TYPES OF DNA DAMAGE CAUSED BY PARTICLES

16.3.1 DNA Damage by Oxidants

ROS and RNS have been clearly implicated in the oxidative attack of DNA, and this can lead to base pair mutations, deletions, or insertions, all being commonly observed in mutated oncogenes and tumor suppressor genes (Wiseman and Halliwell 1996). Currently, more than 100 types of oxidative DNA lesions have been described, among which 8-hydroxydeoxyguanine (8-OHdG) is probably the best known and most investigated (Marnett 2000). Several particles have been shown to induce 8-OHdG, including asbestos, crystalline silica, coal fly ashes, and PM (Table 16.1). ROS and RNS can also elicit exocyclic etheno-adducts, which can be formed upon oxidation of poly-unsaturated fatty acid residues of phospholipids (i.e., lipid peroxidation), resulting in the formation of mutagenic electrophiles such as malondialdehyde and 4-hydroynonenal (Marnett 2000). Such lipid peroxidation-mediated DNA damage has been indicated for crystalline silica and asbestos (Shi et al. 1994; Howden and Faux 1996).

ROS are also well known to cause DNA backbone damage; i.e., by oxidation of deoxyribose sugar, which results in single strand breaks. Single-strand breakage has been shown for various particles including asbestos, quartz, DEP, carbon black, and TiO_2 (Table 16.1). Importantly, strand breakage may also occur spontaneously in DNA at abasic sites that are formed after depurination. If multiple single-strand breaks occur in close proximity, double-strand breaks may be formed. These are particularly dangerous, as they may lead to chromosomal aberrations. Double-strand breakage has been reported to occur with asbestos fibers (Okayasu et al. 1999). Further support for a role of ROS in particle-elicited DNA damage comes from various studies where genotoxic or mutagenic effects of particles were found to be reduced or abrogated in the presence of radical scavengers or antioxidants (e.g., Shi et al. 1994; Howden and Faux 1996; Xu et al. 1999; Dellinger et al. 2001).

Oxidant-mediated DNA damages by particles have also been demonstrated *in vivo* (Table 16.2). Several of these studies are indicative of a major role for inflammation-mediated ROS/RNS in these effects, and thus they are in support of a mechanism of secondary genotoxicity. Elevated 8-OHdG levels have been found in clear association with the extent of inflammation in rat lungs upon treatment with crystalline silica or DEP (Ichinose et al. 1997; Seiler et al. 2001). In *lacI* transgenic rats, enhanced 8-OHdG levels as observed in lungs following diesel exhaust inhalation, as well as in the omentum major upon intraperitoneal instillation of asbestos, were found to correlate with *in vivo* mutagenicity (Sato et al. 2000; Unfried, Schürkes, and Abel 2002). Further mutational spectrum analysis in both these studies demonstrated enhanced G to T transversions in concordance with the expected mutagenicity of 8-OHdG. Further support for inflammation-driven DNA damage and mutagenesis also comes from *in vitro* experiments; wherein activated neutrophils were found to induce DNA strand breakage and 8-OHdG in lung epithelial cells (Knaapen et al. 1999, 2004), and lavage cells from particle-treated rats were found to be mutagenic to lung epithelial cells (Driscoll et al. 1997), both in an antioxidant-dependent manner.

16.3.2 Bulky DNA Adduct Formation by Particles

Particles such as PM, DEP, and carbon black can carry surface-adsorbed components including PAH into the lung. The mutagenicity of PAH is closely associated with the ability of cells to metabolise these lipophilic compounds into electrophilic intermediates (Hall and Grover 1990).

These include epoxides, which form stable DNA adducts and radical cation intermediates, which result in readily depurinated = instable) DNA adducts. Diolepoxides contain high affinity for the exocyclic amino groups of deoxyguanine (dG) and deoxyadenosine (dA), which may lead to mutations following DNA replication. Lesions induced by PAH as well as by aromatic amines are referred to as bulky adducts, which lead to a strong local distortion of the DNA double helix and block transcription and DNA replication.

PAH are typically generated during combustion processes (e.g., traffic exhaust) and are emitted in semi-volatile as well as particle-associated forms, whereby the particle-associated PAH are predominantly found in the ultrafine mode (Bostrom et al. 2002). The gas/particle portioning of PAH depends on ambient temperature and humidity, as well as on the chemical properties and concentrations of both the PAH and the particles (Bostrom et al. 2002). The genotoxicity of organic extracts of ambient PM or DEP has been clearly established in several genotoxicity studies, and formation of bulky DNA adducts has been specifically shown using such extracts (Topinka et al. 2000; Borm et al. 2005). In line with this, chemical analysis of specified PM or DEP typically reveals various PAH, among which several are established mutagens/carcinogens. Further, major polycyclic compounds are nitro-PAH, such as 3-nitrobenzanthrone, which because of its extremely high mutagenicity has been referred to as "the devil in the diesel" (Arlt 2005). Importantly, however, genotoxicity studies with particle-extracts or specific constituents do not allow for a proper interpretation of genotoxicity of the particle in its entirety, as the effects of organics are dependent on bioavailability *in vivo* (Greim et al. 2001). Currently available studies in rodents following treatment with PAH-containing particles show rather contrasting outcomes (Borm, Schins, and Albrecht 2004). Apart from the bioavailability of PAH, which primarily relates to its surface area properties, PAH-load, and composition, various other aspects can account for these discrepancies. These include particle-deposition and retention patterns in relation to target cell and species-specific biotransformation activities, and obviously, the actual exposure levels. Although bulky adducts have been clearly identified in selected studies, several investigations with particles such as DEP and carbon black have revealed that oxidative DNA damage rather than bulky adduct formation may be implicated in the tumorigenesis (Gallagher et al. 1994; Ichinose et al. 1997; Sato et al. 2000; Gallagher et al. 2003; Borm et al. 2005). Such observations are in concordance with the ROS generating properties via semiquinone redox-cycling and/or (inflammatory) cell-derived ROS formation, as discussed earlier. Notably, inflammatory cell-derived ROS as well as quinones have been shown to play a role in the formation of DNA adducts by PAH, whereby ROS can convert PAH into DNA reactive intermediates (Knaapen et al. 2004).

16.4 EFFECTS OF PARTICLES ON CELL CYCLE ARREST, APOPTOSIS, AND DNA DAMAGE REPAIR

Apart from direct investigations of DNA damage, effects of particles on the genome integrity may also been derived from observations on damage associated responses (see Figure 16.1). The induction of DNA repair, and to a lesser extent, cell cycle arrest and apoptosis by particles may thus represent a specific cellular response to particle-elicited DNA damage. Importantly, however, it should be emphasized that such effects may also be triggered by particles in DNA damage-independent manner. In relation to cell cycle regulation and apoptosis, the tumor suppressor protein p53 represents a major component studied in particle research to date. DNA damage is known to up-regulate p53 and to induce apoptosis or cell cycle arrest, although the specific conditions that determine the fate of either of these processes are still poorly understood. Expression and/or mutation of p53 have, for instance, been addressed *in vivo* following treatment with silica, DEP, or carbon black (Swafford et al. 1995; Seiler et al. 2001; Ishihara et al. 2002). With regard to DNA repair induction by particles, two pathways are of importance. Oxidative DNA lesions, abasic sites, and strand breaks are principally repaired by base excision repair (BER), whereas bulky lesions are removed by nucleotide excision repair (NER) (Satoh and Lindahl 1994; Hartwig 2002). Double-strand breakage, as reported to occur with asbestos, is repaired via homologous recombination and nonhomologous end joining repair pathways. To date, only few studies have described effects of particles on DNA repair. Asbestos has been shown *in vitro* to up-regulate the expression or activity of the DNA glycosylase OGG-1 and apurinic/apyrimidic endonuclease (APE), which are specifically involved in the repair of oxidized bases (i.e., 8-OHdG) and their resulting apurinic/apyrimidic-sites (Fung et al. 1998; Kim et al. 2001). Asbestos treatment also results in increased mRNA expression of the human MTH1 gene, which is known to prevent mutagenicity via hydrolysation of 8-oxo-7,8-dihydrodeoxyguanosine triphosphate (Kim et al. 2001). Asbestos and silica particles have also been found to activate poly (ADP-ribose) polymerase (PARP) in relation to its function as a biomarker of DNA damage and/or DNA repair (Tsurudome et al. 1999; Kamp, Srinivasan, and Weitzman 2001). Enhanced expression of OGG1 or APE has been observed *in vivo*, i.e., in rat lungs, following instillation of, respectively, DEP (Tsurodome et al. 1999) or quartz (Albrecht et al. 2005). This indicates that BER may be induced *in vivo* by particles and/or by particle-elicited inflammatory effects. However, particles or their constituents may in fact also inhibit DNA repair and thereby enhance genotoxicity in an indirect manner. Specifically in the case of chemically heterogeneous particles such as PM, various metals are shown to be potent inhibitors of both BER and NER (Hartwig 2002). Finally, it is important to keep in mind that during inflammation, DNA damage outcome can be modulated by phagocyte-derived ROS such as HOCl and $NO^{\bullet}$, which are reported to be potent DNA repair inhibitors (Knaapen et al. 2004). In summary, in relation to particle-induced DNA damage investigations, more studies are needed to evaluate particle effects on DNA repair mechanisms as well as on particle properties that impact on inflammation, cell cycle regulatory signaling pathways, and apoptosis.

ACKNOWLEDGMENTS

Work supported in part by grants from the German Research Council DFG-SFB503 and IGK-738 [RS], and from the National Institutes of Health CA49062, ES05786, and Environmental Health Center grant P30 ES 09089 [TKH]. The authors thank Paul Borm, Ad Knaapen, and Catrin Albrecht for helpful discussions, and one of the authors [RS], for stimulating research collaborations over the years concerning the topic of this manuscript.

REFERENCES

Albrecht, C., Knaapen, A. M., Becker, A., Hoehr, D., Haberzettl, P., van Schooten, F. J., Borm, P. J., and Schins, R. P. F., The crucial role of particle surface-reactivity in respirable quartz-induced reactive oxygen/nitrogen species formation and APE/Ref-1 induction in rat lung, *Respir. Res.*, 6, 129, 2005.

Anard, D., Kirsch-Volders, M., Elhajouji, A., Belpaeme, K., and Lison, D., *In vitro* genotoxic effects of hard metal particles assessed by alkaline single cell gel and elution assays, *Carcinogenesis*, 18, 177–184, 1997.

Arlt, V. M., 3-Nitrobenzanthrone, a potential human cancer hazard in diesel exhaust and urban air pollution: a review of the evidence, *Mutagenesis*, 20, 399–410, 2005.

Barrett, J. C., Mechanisms of multi-step carcinogenesis and carcinogen risk assessment, *Environ. Health Perspect.*, 100, 9–20, 1993.

Borm, P. J. A., Schins, R. P. F., and Albrecht, C., Inhaled particles and lung cancer, part B: paradigms and risk assessment, *Int. J. Cancer.*, 110, 3–14, 2004.

Borm, P. J. A., Cakmak, G., Jermann, E., Weishaupt, C., Kempers, P., van Schooten, F. J., Oberdörster, G., and Schins, R. P. F., Formation of PAH-DNA adducts after *in vivo* and *in vitro* exposure of rats and lung cells to different commercial carbon blacks, *Toxicol. Appl. Pharmacol.*, 205, 157–167, 2005.

Bostrom, C. E., Gerde, P., Hanberg, A., Jernstrom, B., Johansson, C., Kyrklund, T., Rannug, A., Tornqvist, M., Victorin, K., and Westerholm, R., Cancer risk assessment, indicators, and guidelines for polycyclic aromatic hydrocarbons in the ambient air, *Environ. Health Perspect.*, 110 (suppl. 3), 451–488, 2002.

Castranova, V., Huffman, L. J., Judy, D. J., Bylander, J. E., Lapp, L. N., Weber, S. L., Blackford, J. A., and Dey, R. D., Enhancement of nitric oxide production by pulmonary cells following silica exposure, *Environ. Health Perspect.*, 106 (suppl. 5), 1165–1169, 1998.

Chao, C. C., Park, S. H., and Aust, A. E., Participation of nitric oxide and iron in the oxidation of DNA in asbestos-treated human lung epithelial cells, *Arch. Biochem. Biophys.*, 326, 152–157, 1996.

Chen, M. and von Mikecz, A., Formation of nucleoplasmic protein aggregates impairs nuclear function in response to SiO_2 nanoparticles, *Exp. Cell Res.*, 305, 51–62, 2005.

Daniel, L. N., Mao, Y., Williams, A. O., and Saffiotti, U., Direct interaction between crystalline silica and DNA—a proposed model for silica carcinogenesis, *Scand. J. Work Environ. Health*, 21 (suppl. 2), 22–26, 1995.

De Boeck, M., Lombaert, N., De Backer, S., Finsy, R., Lison, D., and Kirsch-Volders, M., In vitro genotoxic effects of different combinations of cobalt and metallic carbide particles, *Mutagenesis*, 18, 177–186, 2003.

Dellinger, B., Pryor, W. A., Cueto, R., Squadrito, G. L., Hegde, V., and Deutsch, W. A., Role of free radicals in the toxicity of airborne fine particulate matter, *Chem. Res. Toxicol.*, 14, 1371–1377, 2001.

Deshpande, A., Narayanan, P. K., and Lehnert, B. E., Silica-induced generation of extracellular factor(s) increases reactive oxygen species in human bronchial epithelial cells, *Toxicol. Sci.*, 67, 275–283, 2002.

Don Porto Carero, A., Hoet, P. H. M., Verschaeve, L., Schoeters, G., and Nemery, B., Genotoxic effects of carbon black particles, diesel exhaust particles, and urban air particulates and their extracts on a human alveolar epithelial cell line (A549) and a human monocytic cell line (THP-1), *Environ. Mol. Mutagen.*, 37, 155–163, 2001.

Donaldson, K., Brown, D. M., Mitchell, C., Dineva, M., Beswick, P. H., Gilmour, P., and MacNee, W., Free radical activity of PM_{10}: iron-mediated generation of hydroxyl radicals, *Environ. Health Perspect.*, 105 (suppl. 5), 1285–1289, 1997.

Donaldson, K., Brown, D., Clouter, A., Duffin, R., MacNee, W., Tran, L., and Stone, V., The pulmonary toxicology of ultrafine particles, *J. Aerosol Med.*, 15, 213–220, 2002.

Driscoll, K. E., Deyo, L. C., Carter, J. M., Howard, B. W., Hassenbein, D. G., and Bertram, T. A., Effects of particle exposure and particle-elicited inflammatory cells on mutation in rat alveolar epithelial cells, *Carcinogenesis*, 18, 423–430, 1997.

Dybdahl, M., Risom, L., Bornholdt, J., Autrup, H., Loft, S., and Wallin, H., Inflammatory and genotoxic effects of diesel particles *in vitro* and *in vivo*, *Mutat. Res.*, 562, 119–131, 2004.

Feinberg, A. P., Ohlsson, R., and Henikoff, S., The epigenetic progenitor origin of human cancer, *Nat. Rev. Genet.*, 7, 21–33, 2006.

Fubini, B. and Hubbard, A., Reactive oxygen species (ROS) and reactive nitrogen species (RNS) generation by silica in inflammation and fibrosis, *Free Radic. Biol. Med.*, 34, 1507–1516, 2003.

Fubini, B., Aust, A. E., Bolton, R. E., Borm, P. J. A., Bruch, J., Ciapetti, K., and Donaldson, K., Non-animal tests for evaluating the toxicity of solid xenobiotics. The report and recommendations of ECVAM Workshop 30, *Altern. Lab. Anim.*, 26, 579–617, 2001.

Fung, H., Kow, Y. W., Van Houten, B., Taatjes, D. J., Hatahet, Z., Janssen, Y. M., Vacek, P., Faux, S. P., and Mossman, B. T., Asbestos increases mammalian AP-endonuclease gene expression, protein levels, and enzyme activity in mesothelial cells, *Cancer Res.*, 58, 189–194, 1998.

Gallagher, J., Heinrich, U., George, M., Hendee, L., Phillips, D. H., and Lewtas, J., Formation of DNA adducts in rat following chronic inhalation of diesel emissions, carbon black and titanium dioxide, *Carcinogenesis*, 15, 1291–1299, 1994.

Gallagher, J., Sams, I. I. R., Inmon, J., Gelein, R., Elder, A., Oberdörster, G., and Prahalad, A. K., Formation of 8-oxo-7,8dihydro-2-deoxyguanosine in rat lung DNA following subchronic inhalation of carbon black, *Toxicol. Appl. Pharmacol.*, 190, 224–231, 2003.

Geiser, M., Rothen-Rutishauser, B., Kapp, N., Schurch, S., Kreyling, W., Schulz, H., Semmler, M., Im Hof, V., Heyder, J., and Gehr, P., Ultrafine particles cross cellular membranes by nonphagocytic mechanisms in lungs and in cultured cells, *Environ. Health Perspect.*, 113, 1555–1560, 2005.

Greim, H., Borm, P. J. A., Schins, R. P. F., Donaldson, K., Driscoll, K. E., Hartwig, A., Kuempel, E., Oberdörster, G., and Speit, G., Toxicity of fibers and particles. Report of the workshop held in Munich, Germany, *Inhal. Toxicol.*, 13, 101–119, 2001.

Gurr, J. R., Wang, A. S., Chen, C. H., and Jan, K. Y., Ultrafine titanium dioxide particles in the absence of photoactivation can induce oxidative damage to human bronchial epithelial cells, *Toxicology*, 213 (1–2), 66–73, 2005.

Hall, M. and Grover, P. L., Polycyclic aromatic hydrocarbons: metabolism, activation and tumor initiation, In *Chemical Carcinogenesis and Mutagenesis*, Cooper, C. S. and Grover, P. L., eds., Springer, Berlin, 227–272, 1990.

Hardy, J. A. and Aust, A. E., The effect of iron binding on the ability of crocidolite asbestos to catalyze DNA single-strand breaks, *Carcinogenesis*, 16, 319–325, 1995.

Hartwig, A., Role of DNA repair in particle- and fiber-induced lung injury, *Inhal. Toxicol.*, 14, 91–100, 2002.

Hei, T. K., Liu, S. X., and Waldren, C., Mutagenicity of arsenic in mammalian cells: role of reactive oxygen species, *Proc. Natl. Acad. Sci. USA*, 95, 8103–8107, 1998.

Hei, T. K., Genotoxicity versus carcinogenicity: implications from fiber toxicity studies, *Inhal. Toxicol.*, 12, 141–147, 2000.

Hoeijmakers, J. H. J., Genome maintenance mechanisms for preventing cancer, *Nature*, 411366–411374, 2001.

Howden, P. J. and Faux, S. P., Fiber-induced lipid peroxidation leads to DNA adduct formation in Salmonella typhimurium TA104 and rat lung fibroblasts, *Carcinogenesis*, 17, 413–419, 1996.

Ichinose, T., Yajima, Y., Nagashima, M., Takenoshita, S., Nagamachi, Y., and Sagai, M., Lung carcinogenesis and formation of 8-hydroxy-deoxyguanosine in mice by diesel exhaust particles, *Carcinogenesis*, 18, 185–192, 1997.

Ishihara, Y., Iijima, H., Matsunaga, K., Fukushima, T., Nishikawa, T., and Takenoshita, S., Expression and mutation of p53 gene in the lung of mice intratracheal injected with crystalline silica, *Cancer Lett.*, 177, 125–128, 2002.

Jaurand, M. C., Use of *in vitro* genotoxicity and cell transformation assays to evaluate the potential carcinogenicity of fibres, *IARC Sci. Publ.*, 140, 55–72, 1996.

Jaurand, M. C., Mechanisms of fiber-induced genotoxicity, *Environ. Health Perspect.*, 105, 1073–1084, 1997.

Jung, M., Davis, W. P., Taatjes, D. J., Churg, A., and Mossman, B. T., Asbestos and cigarette smoke cause increased DNA strand breaks and necrosis in bronchiolar epithelial cells *in vivo*, *Free Radic. Biol. Med.*, 28, 1295–1299, 2000.

Kamp, D. W., Dunn, M. M., Sbalchiero, J. S., Knap, A. M., and Weitzman, S. A., Contrasting effects of alveolar macrophages and neutrophils on asbestos-induced pulmonary epithelial cell injury, *Am. J. Physiol.*, 266, L84–L91, 1994.

Kamp, D. W., Srinivasan, M., and Weitzman, S. A., Cigarette smoke and asbestos activate poly-ADP-ribose polymerase in alveolar epithelial cells, *J. Invest. Med.*, 49, 68–76, 2001.

Kane, A. B., Oncogenes and tumor suppressor genes in the carcinogenicity of fibers and particles, *Inhal. Toxicol.*, 12, 131–140, 2000.

Kawanishi, S., Hiraku, Y., Murata, M., and Oikawa, S., The role of metals in site-specific DNA damage with reference to carcinogenesis, *Free Radic. Biol. Med.*, 32, 822–832, 2002.

Keane, M. J., Xing, S. G., Harrison, J. C., Ong, T., and Wallace, W. E., Genotoxicity of diesel-exhaust particles dispersed in simulated pulmonary surfactant, *Mutat. Res.*, 260, 233–238, 1991.

Kessel, M., Liu, S. X., Xu, A., Santella, R., and Hei, T. K., Arsenic induces oxidative DNA damage in mammalian cells, *Mol. Cell Biochem.*, 234–235 (1–2), 301–308, 2002.

Kim, H. N., Morimoto, Y., Tsuda, T., Ootsuyama, Y., Hirohashi, M., Hirano, T., Tanaka, I., Lim, Y., Yun, I. G., and Kasai, H., Changes in DNA 8-hydroxyguanine levels, 8-hydroxyguanine repair activity, and hOGG1 and hMTH1 mRNA expression in human lung alveolar epithelial cells induced by crocidolite asbestos, *Carcinogenesis*, 22, 265–269, 2001.

Knaapen, A. M., Seiler, F., Schilderman, P. A. E. L., Nehls, P., Bruch, J., Schins, R. P. F., and Borm, P. J. A., Neutrophils cause oxidative DNA damage in alveolar epithelial cells, *Free Radic. Biol. Med.*, 27, 234–240, 1999.

Knaapen, A. M., Shi, T., Borm, P. J. A., and Schins, R. P. F., Soluble metals as well as the insoluble particle fraction are involved in cellular DNA damage induced by particulate matter, *Mol. Cell Biochem.*, 234/235, 317–326, 2002a.

Knaapen, A. M., Albrecht, C., Becker, A., Höhr, D., Winzer, A., Haenen, G. R., Borm, P. J. A., and Schins, R. P. F., DNA damage in lung epithelial cells isolated from rats exposed to quartz: role of surface reactivity and neutrophilic inflammation, *Carcinogenesis*, 23, 1111–1120, 2002b.

Knaapen, A. M., Borm, P. J. A., Albrecht, C., and Schins, R. P. F., Inhaled particles and lung cancer, part A, *Mech. Int. J. Cancer*, 109, 799–809, 2004.

Levresse, V., Renier, A., Levy, F., Broaddus, V. C., and Jaurand, M., DNA breakage in asbestos-treated normal and transformed (TSV40) rat pleural mesothelial cells, *Mutagenesis*, 15, 239–244, 2000.

Li, N., Sioutas, C., Cho, A., Schmitz, D., Misra, C., Sempf, J., Wang, M., Oberley, T., Froines, J., and Nel, A., Ultrafine particulate pollutants induce oxidative stress and mitochondrial damage, *Environ. Health Perspect.*, 111, 455–460, 2003.

Liu, X., Keane, M. J., Zhong, B. Z., Ong, T. M., and Wallace, W. E., Micronucleus formation in V79 cells treated with respirable silica dispersed in medium and in simulated pulmonary surfactant, *Mutat. Res.*, 361(2–3), 89–94, 1996.

Liu, W., Ernst, J. D., and Broaddus, V. C., Phagocytosis of crocidolite asbestos induces oxidative stress, DNA damage, and apoptosis in mesothelial cells, *Am. J. Respir. Cell Mol. Biol.*, 23, 371–378, 2000.

Lu, P. J., Ho, I. C., and Lee, T. C., Induction of sister chromatid exchanges and micronuclei by titanium dioxide in Chinese hamster ovary-K1 cells, *Mutat. Res.*, 414, 15–20, 1998.

Marnett, L. J., Oxyradicals and DNA damage, *Carcinogenesis*, 21, 316–370, 2000.

McGregor, D.B., Rice, J.M., Venitt, S., eds., *The use of short- and medium-term tests for carcinogens and data on genetic effects in carcinogenic hazard evaluation*, IARC Scientific Publications No. 146, IARC, Lyon.

Nagalakshmi, R., Nath, J., Ong, T., and Whong, W. Z., Silica-induced micronuclei and chromosomal aberrations in Chinese hamster lung (V79) and human lung (Hel 299) cells, *Mutat. Res.*, 335, 27–33, 1995.

Oberdörster, G., Oberdörster, E., and Oberdörster, J., Nanotoxicology: an emerging discipline evolving from studies of ultrafine particles, *Environ. Health Perspect.*, 113, 823–839, 2005.

Okayasu, R., Takahashi, S., Yamada, S., Hei, T. K., and Ullrich, R. L., Asbestos and DNA double strand breaks, *Cancer Res.*, 59, 298–300, 1999.

Park, S. H. and Aust, A. E., Participation of iron and nitric oxide in the mutagenicity of asbestos in hgprt−, gpt+ Chinese hamster V79 cells, *Cancer Res.*, 58, 1144–1148, 1998.

Pope, C. A. I., Burnett, R. T., Thun, M. J., Calle, E. E., Krewski, D., Ito, K., and Thurston, G. D., Lung cancer, cardiopulmonary mortality, and long-term exposure to fine particulate air pollution, *JAMA*, 287 (9), 1132–1141, 2002.

Prahalad, A. K., Inmon, J., Dailey, L. A., Madden, M. C., Ghio, A. J., and Gallagher, J. E., Air pollution particles mediated oxidative DNA base damage in a cell free system and in human airway epithelial cells in relation to particulate metal content and bioreactivity, *Chem. Res. Toxicol.*, 14, 879–887, 2001.

Rahman, Q., Lohani, M., Dopp, E., Pemsel, H., Jonas, L., Weiss, D. G., and Schiffmann, D., Evidence that ultrafine titanium dioxide induced micronuclei and apoptosis in Syrian Hamster Embryo cells, *Environ. Health Perspect.*, 110, 797–800, 2002.

Rihn, B., Coulais, C., Kauffer, E., Bottin, M. C., Martin, P., Yvon, F., and Vigneron, J. C., Inhaled crocidolite mutagenicity in lung DNA, *Environ. Health Perspect.*, 108, 341–346, 2000.

Risom, L., Dybdahl, M., Bronholdt, J., Vogel, U., Wallin, H., Moller, P., and Loft, S., Oxidative DNA damage and defence gene expression in the mouse lung after short-term exposure to diesel exhaust particles by inhalation, *Carcinogenesis*, 24, 1847–1852, 2003.

Sato, H., Sone, H., Sagai, M., Suzuki, K. T., and Aoki, Y., Increase in mutation frequency in lung of Big Blue rat by exposure to diesel exhaust, *Carcinogenesis*, 21, 653–661, 2000.

Satoh, M. S. and Lindahl, T., Enzymatic repair of oxidative DNA damage, *Cancer Res.*, 54, 1899–1901, 1994.

Schins, R. P. F., Mechanisms of genotoxicity of particles and fibres, *Inhal. Toxicol.*, 14, 57–78, 2002.

Schins, R. P. F., Duffin, R., Höhr, D., Knaapen, A. M., Shi, T., Weishaupt, C., Stone, V., Donaldson, K., and Borm, P. J. A., Surface modification of quartz inhibits toxicity, particle uptake, and oxidative DNA damage in human lung epithelial cells, *Chem. Res. Toxicol.*, 15, 1166–1173, 2002.

Seiler, F., Rehn, B., Rehn, S., and Bruch, J., Quartz exposure of the rat lung leads to a linear dose response in inflammation but not in oxidative DNA damage and mutagenicity, *Am. J. Respir. Cell Mol. Biol.*, 24, 492–498, 2001.

Shi, X., Mao, Y., Daniel, L. N., Saffiotti, U., Dalal, N. S., and Vallyathan, V., Silica radical-induced DNA damage and lipid peroxidation, *Environ. Health Perspect.*, 02 (suppl. 10), 149–154, 1994.

Shi, T., Knaapen, A. M., Begerow, J., Birmili, W., Brom, P. J. A., and Schins, R. P. F., Temporal variation of hydroxyl radical generation and 8-hydroxy-20-deoxyguanosine formation by coarse and fine particulate matter, *Occup. Environ. Med.*, 69, 322–329, 2003.

Shukla, A., Gulumian, M., Hei, T. K., Kamp, D., Rahman, Q., and Mossman, B. T., Multiple roles of oxidants in the pathogenesis of asbestos-induced diseases, *Free Radic. Biol. Med.*, 34, 1117–1129, 2003.

Squadrito, G. L., Cueto, R., Dellinger, B., and Pryor, W. A., Quinoid redox cycling as a mechanism for sustained free radical generation by inhaled airborne particulate matter, *Free Radic. Biol. Med.*, 31, 1132–1138, 2001.

Swafford, D. S., Nikula, K. J., Mitchell, C. E., and Belinsky, S. A., Low frequency of alterations in p53, k-ras, and mdm2 in rat lung neoplasms induced by diesel exhaust or carbon black, *Carcinogenesis*, 16, 1215–1221, 1995.

Thomas, M. and Klibanov, A. M., Conjugation to gold nanoparticles enhances polyethylenimine's transfer of plasmid DNA into mammalian cells, *Proc. Natl. Acad. Sci. USA*, 100, 9138–9143, 2003.

Topinka, J., Schwarz, L. R., Wiebel, F. J., Cerna, M., and Wolff, T., Genotoxicity of urban air pollutants in the Czech Republic. Part II. DNA adduct formation in mammalian cells by extractable organic matter, *Mutat. Res.*, 469, 83–93, 2000.

Tran, H. P., Prakash, A. S., Barnard, R., Chiswell, B., and Ng, J. C., Arsenic inhibits the repair of DNA damage induced by benzo(a)pyrene, *Toxicol. Lett.*, 133, 59–67, 2002.

Tsurudome, Y., Hirano, T., Yamato, H., Tanaka, I., Sagai, M., Hirano, H., Nagata, N., Itoh, H., and Kasai, H., Changes in levels of 8-hydroxyguanine in DNA, its repair and OGG1 mRNA in rat lungs after intratracheal administration of diesel exhaust particles, *Carcinogenesis*, 20, 1573–1576, 1999.

Unfried, K., Schürkes, C., and Abel, J., Distinct spectrum of mutations induced by crocidolite asbestos: clue for 8-hydroxydeoguanosine dependent mutagenesis *in vivo*, *Cancer Res.*, 62, 99–104, 2002.

Vainio, H., Magee, P., McGregor, D., McMichael, A., eds., *Mechanisms of Carcinogenesis in Risk Identification*, IARC Scientific Publications No. 116, IARC, Lyon.

Van Maanen, J. M., Borm, P. J. A., Knaapen, A., van Herwijnen, M., Schilderman, P. A., Smith, K. R., Aust, A. E., Tomatis, M., and Fubini, B., In vitro effects of coal fly ashes: hydroxyl radical generation, iron release, and DNA damage and toxicity in rat lung epithelial cells, *Inhal. Toxicol.*, 11, 1123–1141, 1999.

Vogelstein, B. and Kinzler, K. W., Cancer genes and the pathways they control, *Nat. Med.*, 10, 789–799, 2004.

Wiseman, H. and Halliwell, B., Damage to DNA by reactive oxygen and nitrogen species: role in inflammatory disease and progression to cancer, *Biochem. J.*, 313, 17–29, 1996.

Wood, R. D., Mitchell, M., Sgouros, J., and Lindahl, T., Human DNA repair genes, *Science*, 191, 1284–1289, 2001.

Wong, D., Mitchell, C. E., Wolff, R. K., Mauderly, J. L., and Jeffrey, A. M., Identification of DNA damage as a result of exposure to rats to diesel exhaust, *Carcinogenesis*, 7, 1595–1597, 1986.

Wu, L. J., Randers-Pehrson, G., Xu, A., Waldren, C. A., Geard, C. R., and Hei, T. K., Targeted cytoplasmic irradiation with alpha particles induces mutations in mammalian cells, *Proc. Natl. Acad. Sci. USA*, 96, 4959–4964, 1999.

Yegles, M., Saint-Etienne, L., Renica, A., Janson, X., and Jaurand, M. C., Induction of metaphase and anaphase/telephase abnormalities by asbestos fibers in rat pleural mesothelial cells in vitro, *Am. J. Respir. Cell Mol. Biol.*, 9, 186–191, 1993.

Xu, A., Wu, L. J., Santella, R. M., and Hei, T. K., Role of oxyradicals in mutagenicity and DNA damage induced by crocidolite asbestos in mammalian cells, *Cancer Res.*, 59, 5922–5926, 1999.

Zhao, Y. L., Piao, C. Q., Wu, L. J., Suzuki, M., and Hei, T. K., Differentially expressed genes in asbestos-induced tumorigenic human bronchial epithelial cells: implication for mechanism, *Carcinogenesis*, 21, 2005–2010, 2000.

Zhong, B. Z., Whong, W. Z., and Ong, T. M., Detection of mineral-dust-induced DNA damage in two mammalian cell lines using the alkaline single cell gel/comet assay, *Mutat. Res.*, 393, 181–187, 1997.

Zhou, B. B. and Elledge, S. J., The DNA damage response: putting checkpoints in perspective, *Nature*, 408, 433–439, 2000.

17 Approaches to the Toxicological Testing of Particles

Ken Donaldson
MRC/University of Edinburgh Centre for Inflammation Research,
Queen's Medical Research Institute

Steve Faux
MRC/University of Edinburgh Centre for Inflammation Research,
Queen's Medical Research Institute

Paul J. A. Borm
Centre of Expertise in Life Sciences (CEL), Hogeschool Zuyd

Vicki Stone
School of Life Sciences, Napier University

CONTENTS

17.1 BACKGROUND TO TESTING OF PARTICLES

There is a wide spectrum of different approaches to the testing of particles that range from long-term inhalation studies, with the final endpoint of cancer, to short-term tests *in vitro* determining the ability of the particle to modulate cellular functions. The value of each of these tests in predicting pathogenicity varies; but in general, there is a playoff between the extended timescale and high cost of long-term pathogenicity experiments in animals which can be used in risk assessment and the short timescale and relative inexpensiveness of *in vitro* data, which can, at best, be used in hazard assessment. If epidemiological or clinical data were available on the pathogenic outcome of exposure to a novel particle, then this would form the basis of a rational test strategy. However, in the absence of such information, a strategy based on knowledge of structure, chemistry, and shape of the particle could allow benchmarking to similar particles of known pathogenicity in order to decide on the most appropriate endpoint.

This spectrum of possible pathology arising from particle exposure poses an immediate problem for a testing strategy. A strategy designed to detect a carcinogenic endpoint, for example, would be entirely different from one that would be chosen to detect the potential to cause asthma. So there is a need for some knowledge of the likely pathology that would arise. This could be provided by:

1. *A priori* knowledge from clinical observations or epidemiological studies indicating that a particular disease or manifestation of toxicity is associated with exposure to the particle.
2. In the absence of such information, there could be benchmarking to particles of known toxicity. For example, if the particle type was a mineral and contained some quartz, then the endpoints of fibrosis and cancer could be selected. If the particle was organic or contained heavy metals, then sensitization might be considered. Fibrous particles would be suspected of causing mesothelioma, etc.

17.2 FACTORS AFFECTING PARTICLE TOXICITY

17.2.1 Particle Size

Given the different diseases caused by particles, we assume that there are different target cells and tissues (e.g., airways, alveolar cells, immune cells) that are affected by different particles. This can be understood from the point of view of differences in dose to these different target cells caused by

TABLE 17.1
Anatomical Site of Deposition for the Different Deposition Fractions

Deposition Fraction	Approx Aerodynamic Diameter (μm)	Definition	Site of Deposition
Inhalable	~50–100	The fraction inhaled through the nose and mouth	Mouth, larynx, pharynx
Thoracic	~10–50	Fraction penetrating beyond the larynx	The above plus airways
Respirable	<~10	Fraction penetrating beyond the ciliated airways	The above plus terminal bronchioles and alveolar ducts

TABLE 17.2
The Adverse Effects Resulting from Different Sizes of Particles Depositing in Different Compartments

Deposition Site	Disease
Thoracic	COPD, asthma, central lung cancer
Respirable	Small airways disease, peripheral lung cancer, emphysema, interstitial fibrosis, cardiovascular effects

differences in the distribution of dose throughout the respiratory tract because of variation in particle size.

The fractional deposition of various sized particles for different pulmonary compartments is dictated by the aerodynamic diameter, D_{ae}, as shown Table 17.1. This is defined as the diameter of a particle of unit density with the same falling speed as the particle of interest.

Different-sized particles would be expected to elicit different types of pathogenic response because they deposit in different regions, as shown in Table 17.2.

From the foregoing, it can be concluded that the characterization of size of a novel particle is important in deciding what is the most likely endpoint to examine, since size will bear directly on the site of deposition and the subsequent response. When the site of the adverse health effect is not the local lung environment, it is more difficult to know which site of deposition is most important because there could be, in theory at least, translocation from any site of deposition—hence the question marks over the site of deposition of particles that are associated with mesothelioma and strokes/heart attacks.

In the case of nanoparticles (see later) additional targets, such as the blood and brain may be relevant, as a consequence of translocation.

17.2.1.1 Particle Shape

Particles come in a number of different shapes, e.g., compact, platy, and fibrous. The importance of particle shape is best understood for fibers, and fiber length is known to be a major factor in pathogenicity as reviewed in Donaldson and Tran (2004). Long fibers of amosite asbestos were much more pathogenic than a sample of the same material milled so that the fiber length was drastically shortened, with much of the sample being so short that it was classified as non-fibrous (Davis et al. 1986). For other particles, shape is less obvious as a parameter that mediates toxicity.

Shape may not be the only factor that dictates fiber pathogenicity—even amongst long and thin fibers there are differences in pathogenicity, especially in mesothelioma production. Erionite (Maltoni, Minardi, and Morisi 1982) and silicon carbide fibers (Davis et al. 1996), for example were much more active in causing mesothelioma following inhalation exposure than would be expected from their dimensions. For this reason it is likely that another factor, surface reactivity (see below), could be important in mediating some of their pathogenicity.

17.2.1.2 Particle-Derived Transition Metals

It has become evident that transition metals are key players in the pro-inflammatory effects of a range of particle types (Ghio et al. 2006). Iron is especially important because it has the ability to generate free radicals via Fenton chemistry that is well characterized. The state of the iron is all important; but in particular, the amount of Fe(II) is central since this is the directly harmful species (Ghio and Cohen 2005). Consequently, total iron is not necessarily informative as to the biologically active iron. The iron must redox-cycle to be capable of causing major injury to macromolecules. This is accomplished by a reluctant in the region of the particle, e.g., glutathione, ascorbate, NADH, or even superoxide anion. This means that the presence of anti-oxidants in the lungs is a "double-edged sword."

By the sequence of events shown in Figure 17.1, the highly toxic and reactive hydroxyl radical can be formed. The hydroxyl radical may be involved in diffusion-limited reactions which lead to the formation of various carbon-centered radicals, peroxyl, alkoxyl, and thiyl radicals, all of which have harmful consequences for cells. Because different reductants could have different potencies in causing reduction of Fe(III), and because of the known role of chelating agents, the micro-environment of the lung where the particle is present could be all-important in determining how much reactive iron is present at the surface. In addition, the particle may accumulate biological iron, which can also have free radical-generating activity (Ghio, Jaskot, and Hatch 1994). In various models, the biological effects of several different types of ambient particle, including PM_{10} and ROFA, have been suggested to be driven by their transition metal content (Dreher et al. 1997; Jimenez et al. 2000; Rice et al. 2001; Dai, Xie, and Churg 2002; Molinelli et al. 2002; McNeilly et al. in press). The measurement of the sum potential for a particle to generate oxidative stress has been advanced as a key parameter that might dictate overall toxicity.

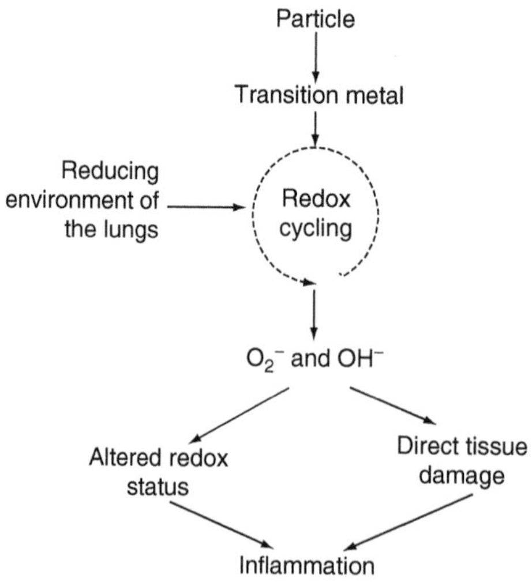

FIGURE 17.1 Generation of oxidative stress by transition metals.

17.2.1.3 Particle-Derived Free Radicals

Quartz is one of the most toxic particles and it is known to have a highly reactive surface. The quartz surface can generate reactive oxygen species (ROS) in several ways following interactions of the quartz particles with pulmonary cells or lung fluids (Castranova, Dalal, and Vallyathan 1995). PM_{10} has been demonstrated to generate free radicals and cause oxidative stress (Ghio et al. 1996; Gilmour et al. 1996; Squadrito et al. 2001; Aust et al. 2002) via transition metals and organics that redox cycle. Nanoparticles are also capable of generating ROS in cell-free systems (Wilson et al. 2002). In fact, almost all pathogenic particles studied, including asbestos (Lund and Aust 1991), glass fibers (Gilmour et al. 1997), and coalmine dust (Dalal et al. 1995) are capable of generating ROS in cell free systems. Oxidative stress has been suggested as a generic mechanism for the action of particles (Donaldson, Beswick, and Gilmour 1996), and more recently for nanoparticles (Donaldson et al. 2005; Oberdörster, Oberdörster, and Oberdörster 2005). Many systems are available for measuring the fee-radical-generating potential of a particle sample, including purely chemical methods such as HPLC (Brown, Fisher, and Donaldson 1998), the use of super-coiled plasmid DNA as a sensor of free radicals (Gilmour et al. 1997), the use of EPR and spin traps to capture these short-lived moieties (Shi et al. 2003), and calf thymus DNA to detect 8-hydroxydeoxyguanosine (8-OHdG).

17.2.1.4 Particle Biopersistence

Biopersistence is the capacity of particles to persist in the lungs. Biopersistence is limited by the potential of particles to dissolve or lose elements, break, or be mechanically cleared from the lungs by macrophages. The potential for a particle to dissolve in the lungs would seem intuitively to be an important factor, since the dose of particle would not be expected to build up in the case of soluble particles. However, little is know about the biochemical conditions that pertain in the lungs—the lung is largely a "black box" in this regard, although differences in pH and the impact of coating of the particles is to be anticipated.

The best case where this property is seen as being important is with fibers. Long fibers are not well cleared from the respiratory region of the lungs (Coin, Roggli, and Brody 1994; Searl et al. 1999), presumably because of the difficulties of the alveolar macrophage to successfully phagocytose and then move with them to the mucociliary escalator. Thus the ability of long fibers to persist in the lungs without being either dissolved away or weakened so that they break into smaller fibers which can be easily cleared is seen as an important factor contributing to pathogenicity (Hesterberg et al. 1994). For any particle, its ability to biopersist will be an important factor in modifying its pathogenicity.

TABLE 17.3
Important Particle Characteristics to Be Ascertained in Samples of Particles of Unknown Toxicity

Physical Characteristics	Chemical Characteristics
Dimensions	Transition metals
Surface area/unit mass	Quartz or cristobalite content
Biopersistence	Heavy metals
Free radical activity	PAH
Durability	Endotoxin

17.3 APPROACHES TO TESTING

17.3.1 Characterizing the Particles

An understanding of the nature of the test particle with regard to shape, size, elemental composition, transition metal content, endotoxin contamination, etc., is vital to the testing strategy since it will allow the particle to be benchmarked. There are a number of parameters that could be assessed. Once again, the concept of benchmarking is a useful one and the source of the particle, e.g., mineral-derived, man-made fiber, ash, etc., can be used to decide which is the most likely parameter that should be determined. Table 17.3 shows some particle characteristics that can be quantified, which might shed light on its likely pathogenicity.

Endotoxin is a potential confounder in all studies with particles and its presence should be rigorously monitored since it may explain all the toxicity of a dust sample (Brown and Donaldson 1996). Endotoxin can be measured by specific ELISA or functionally using the amoebocyte lysate assay.

17.4 ASSESSMENT OF TOXICITY *IN VITRO*

The European Centre for the Validation of Alternative Methods (ECVAM) published a report on "Nonanimal tests for evaluating the toxicity of solid xenobiotics," and this contains recommendations for *in vitro* tests with particles (Fubini et al. 1998). The majority of *in vitro* tests for detecting particle toxicity are aimed at detecting either direct or indirect pro-inflammatory effects. Acute and chronic inflammation is thought to be central to the etiology of many lung disorders, such as asthma and chronic obstructive pulmonary disease (COPD). The specific characteristics of the inflammatory response may be different, but all are characterized by the recruitment of inflammatory cells into the lung. These activated cells, such as alveolar macrophages and neutrophils, produce cytokines and ROS and many other mediators involved in inflammation. Once triggered, the inflammatory response will persist in these conditions leading to lung injury. The intracellular mechanism in the lung epithelium and the macrophages leading to lung injury in response to environmental particulates will involve the activation and upregulation of transcription factors, such as activator protein-1 (AP-1) and nuclear factor-κB (NF-κB), leading to increased gene expression and the biosynthesis of proinflammatory mediators. Particles may stimulate inflammation by a number of pathways such as:

1. Cell death, a potent stimulus for inflammation
2. Nonspecific stimulation of cell receptors
3. Oxidative stress
4. Calcium flux
5. Via the immune system (see Figure 17.2)

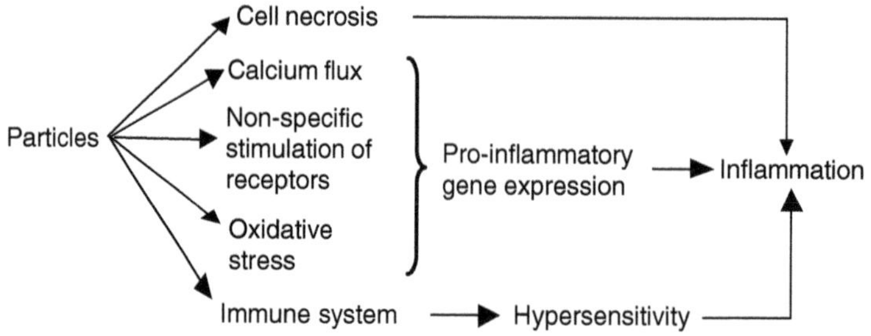

FIGURE 17.2 Pathways for particles to cause inflammation. There can be direct stimulation of target cells for pro-inflammatory gene expression, or inflammation can arise by more complex indirect routes that involve cytotoxicity or the immune system.

In vitro models are useful tools at two stages, namely the assessment of toxicity as the second tier of the testing system and the elucidation of the mechanism(s) of action. A reduction in the use of experimental animals is a clear advantage of such studies, but *in vitro* systems also provide a "simplified" model in which the details of the mechanism of action may be more readily examined.

The potential toxicity of particles is tested *in vitro* using either primary cells or cell lines in culture. There are a number of cell types that are of obvious interest when investigating the potential effects of particles, including type I and type II epithelial cells, Clara cells, alveolar macrophages, and neutrophils. Primary cells are obviously an advantage in that they are not transformed, hence they will correspond more closely to the cells found *in vivo*. Primary cells, however, often have a limited life span in culture, and for this reason, cell lines such as the A549 human type II cell line and THP-1 monocytes are widely used, due to their ready availability and the fact that the pathways under study are similar in these permanent cell lines to freshly derived cells of the same type. The acquisition of human primary cells remains difficult for many researchers, and for this reason primary rat cells are frequently used as an alternative. The use of cells has both benefits and drawbacks, both of which are well-known. The benefits include the ability to dissect out the sub-cellular pathways and responses and to isolate the responses specific to the cell type in question. The drawbacks are that there is no influence of the other cell types and the circulation that ordinarily plays an important role in the responses of any single cell type.

17.4.1 Cytotoxicity

A number of reliable and well-documented techniques are available to assess viability following particle exposure, including the MTT assay, which measures the metabolic competence of cells by assessing the activity of succinate dehydrogenase enzyme activity, a key enzyme in cellular respiration. Assessment of lactate dehydrogenase (LDH) enzyme leakage from the cells measures plasma membrane integrity. The MTT assay and measurement of LDH leakage are appropriate for death via necrosis. A number of particle types have been proposed to induce programmed cell death or apoptosis (BéruBé et al. 1996; Iyer and Holian 1997). The techniques available for the detection of apoptosis are numerous, from detection of DNA fragmentation to commercially available cell death ELISA kits and fluorescent dyes that label the DNA, such as propidium iodide and Hoechst 33342.

17.4.2 Cell Stimulation

Many particles are thought to induce effects on the lung by mechanisms other than toxicity. Some particles may in fact cause the stimulation of various cell types by, e.g., increasing entry of calcium or stimulating kinase activity, leading to cell proliferation or an increase in the production of cytokines and other pro-inflammatory mediators. Measurement of the production of ROS and cytokines, such as TNF-α and IL-1β by target cells, may form part of a testing strategy to discriminate between non-pathogenic and pathogenic particulates by the ability of various particulate preparations to differentially produce these mediators. There are a number of assays that can be used to assess cell stimulation *in vitro* and these are outlined below.

17.4.3 Cell Proliferation

Increased cell proliferation has been noted on exposure of various cell types to different particles. For example, treatment of primary rat type II epithelial cells with silica for 24 h has been shown to induce cell proliferation, as assessed by the incorporation of tritiated thymidine into the DNA of dividing cells. This simple technique has the disadvantage of using radioactivity, although at a low level. A similar technique involves the incorporation of 5-bromo-2′-deoxyuridine (BrdU) into DNA which is then assessed by immunostaining (Timblin, Janssen, and Mossman 1995). This non-radioactive technique has the advantage that BrdU can be used for observation by microscopy as well as quantification through either spectroscopy or fluorimetry.

17.4.4 Cytokine Measurement

One of the most obvious ways to assess the potential inflammogenic activity of a particle is by measuring the output of cytokines. Secreted cytokines are frequently assayed in the culture media through the use of ELISA. In addition, the quantification of specific mRNA sequences, through either Northern Blotting or RT-PCR, allows further investigation of gene regulation on exposure to particles. The same techniques are also applicable to other pro-inflammatory mediators.

17.4.5 Oxidative Stress Measurement

There is abundant data to suggest that many particle types induce their effects on the lungs in part through ROS, leading to oxidative stress (Tao, Gonzalez-Flecha, and Kobzik 2003). The free radicals produced at the surface of a variety of particle types along with the ROS released by leukocytes during phagocytosis and inflammation induce an oxidative stress within the lung leading to a range of events from oxidative damage to bio-molecules, such as DNA and protein, to activation of oxidative stress-responsive transcription factors that lead to transcription of pro-inflammatory genes (Rahman and MacNee 2000; Gilmour et al. 2003; Brown et al. 2004a). The measurement of intracellular glutathione in its reduced (GSH) and oxidized (GSSG) forms remains a sensitive means by which the induction of oxidative stress can be assessed. GSH is one of the major intracellular antioxidants (reviewed in Karin 1998). In acting as an antioxidant, two molecules of GSH are oxidized to form GSSG, which is then reduced back to GSH by the enzyme glutathione reductase using NADPH as a source of reductant. When the cell is exposed to high levels of oxidants, NADPH within the cell is decreased, allowing depletion of GSH and an increase in GSSG. The depletion of GSH is often used as a marker of oxidative stress in response to particles. Enzyme assays exist to measure the activity of enzymes such as γ-glutamyl transpeptidase (γ-GT), the enzyme responsible for the uptake of the components of GSH across the plasma membrane, and γ-glutamylcysteine synthetase (γ-GCS), the rate-limiting enzyme in the synthesis of GSH.

17.4.6 Alterations in Cell Signaling Cascades

17.4.6.1 Intracellular Calcium

Alterations in intracellular calcium homeostasis have been implicated following oxidative stress. In the resting nonstimulated cell, a Ca^{2+}ATPase pump in the plasma membrane actively extrudes Ca^{2+} from the cell while a different Ca^{2+}ATPase pump in the endoplasmic reticulum (ER) actively sequesters Ca^{2+} into this intracellular store. The ER Ca^{2+} store is released on activation of the cell by stimulants, which results in the production of inositol 1,4,5-trisphosphate (IP_3). This sharp increase in cytosolic calcium concentration stimulates the opening of Ca^{2+} channels in the plasma membrane (calcium release activated calcium channels; CRAC channels) allowing Ca^{2+} to enter the cell down it concentration gradient (calcium release activated calcium current, I_{CRAC}) resulting in a sustained increase in cytosolic Ca^{2+} concentration (Parekh and Penner 1997; Berridge, Bootman, and Lipp 1998; Berridge 2001). A number of the transport proteins involved in the maintenance of Ca^{2+} homeostasis are sensitive to oxidative stress. For example, the Ca^{2+}-ATPase of the ER contains a cysteine residue that is susceptible to oxidation, as are the IP_3 receptor calcium channels in the ER.

Cytosolic Ca^{2+} can be measured using fluorescent dyes such as fura-2 (Grynkiewicz, Poenie, and Tsien 1985) which alter their fluorescent properties on binding to Ca^{2+}. Fura-2 has the advantage of being a ratio dye, which permits alterations in background fluorescence, for example due to the introduction of particles, while measuring calcium.

Ultrafine carbon black (CB) has been shown to increase the resting cytosolic calcium concentration of a human monocytic cell line MonoMac 6 (MM6) (Stone et al. 2000). This effect was not observed with the same dose of larger, respirable CB particles or with pathogenic α-quartz (DQ12). PM_{10} was shown to produce the same type of calcium influx with associated TNFα gene expression (Brown, Donaldson, and Stone 2004b).

Thapsigargin is a useful tool to investigate the potential effects of particles on Ca^{2+} signaling (Thastrup et al. 1990). Thapsigargin works by inhibiting the ER Ca^{2+}ATPase, resulting in leak of the ER store contents into the cytosol. Treatment with thapsigargin results in a sharp increase in cytosolic Ca^{2+} (comparable to the effect of IP_3) followed by a stimulation of the I_{CRAC}. Treatment of a macrophage cell line (MonoMac 6) with ufCB for 30 min induced an increase in the I_{CRAC} observed on treatment with thapsigargin, through an increased opening of the plasma membrane Ca^{2+} channels (Stone et al. 2000). Similar effects were seen in response to PM_{10}—the increase in calcium was involved in TNFα gene expression (Brown, Donaldson, and Stone 2004b).

17.4.6.2 Mitogen-Activated Protein Kinase Cascade

The mitogen-activated protein kinase (MAPK) cascade includes the extra cellular signal-related kinase (ERK1, ERK2) activated in response to growth factors, oxidative stress or phorbol esters via a Ras-dependent mechanism, c-Jun amino terminal kinase/stress activated protein kinase (JNK1, JNK2) activated by TNF-α in a Ras-independent manner and p38 (Seger and Krebs 1995). Activation of the MAPK cascade involving phosphorylation and dephosphorylation of a number of proteins leads to the transactivation of c-fos and c-jun and a number of interrelated transcription factors (Seger and Krebs 1995). Moreover, in a number of cellular systems, the balance between the activation of several arms of this pathway appears to govern whether apoptosis or cell proliferation occurs (Xia et al. 1995).

Limited studies have been carried out investigating the influence of particulates on the MAPK pathway and these have been reviewed by the authors (Brown et al. 2004a; Donaldson et al. 2004). One recent study has shown that exposure of normal human bronchial epithelial (NHBE) cells to ultrafine elemental carbon particles induced the phosphorylation and activation of p38 MAPK. In addition, inhibition of p38 MAPK activity blocked the interleukin-8 mRNA expression in these cells.

17.4.6.3 Transcription Factor Activation

ROS and inflammatory cytokines both cause activation of the transcription factors NF-κB and AP-1 (Meyer, Schreck, and Baeuerle 1993). In addition, ROS have been suggested to act as second messenger molecules within the cell (Sen et al. 1997). The transcription factors NF-κB and AP-1 have been shown to be regulated by the intracellular redox status (Piette et al. 1997; Ginn-Pease and Whisler 1998). NF-κB is a transcription factor important in the regulation of a number of genes intrinsic to inflammation, proliferation, and lung defenses (Schins and Donaldson 2000) including cytokines, nitric oxide synthase, adhesion molecules, and protooncogenes, such as c-*myc*. The process of NF-κB activation involves the cytoplasmic phosphorylation, ubiquitination and subsequent proteolytic degradation of the IκB inhibitory subunits from NF-κB. Release of NF-κB from IκB allows uncovering of the nuclear localization site on the NF-κB subunits so that it can migrate to the nucleus. Once in the nucleus, the activated transcription factor complex, which include p65 protein subunits (Donaldson et al. 2004), then bind to promoter regions of genes that have consensus NF-κB DNA binding sequences. NF-κB has been shown to be activated by a number of particles (Jimenez et al. 2000; Shukla et al. 2000; Hubbard et al. 2002; McNeilly et al. 2005).

AP-1 is a family of accessory transcription factors that interact with other regulatory sequences called TPA-response element (TRE) or AP-1 sites (Donaldson et al. 2004). AP-1 transcription factors include homo- (Jun/Jun) and heterodimer (Fos/Jun) complexes encoded by various members of the c-fos and c-jun families of protooncogenes. The functional ramifications of c-fos and c-jun transactivation may be cell type specific, but Fos and Jun proteins may regulate the expression of other genes required for the progression through the cell cycle, apoptosis, or cell transformation (Angel and Karin 1991).

A number of different pathogenic particles have been found to activate AP-1 (Timblin, BéruBé, and Mossman 1998; Jimenez et al. 2002; Marwick et al. 2004; Shukla et al. 2004; Brown et al. 2004a; McNeilly et al. in press).

Activation of transcription factors can be investigated via a number of methods. These include the gel mobility shift or retardation assay of transcription factor DNA binding activity, immunohistochemical analysis of protein localization in cells, gene transactivation assays using reporter gene constructs measuring luciferase activity, and western blotting of protein levels.

17.4.7 Effects on Blood

With growing investigations into the effects of nanoparticles, there is concern that particles might gain access to the blood and thereby exert direct pathogenic effects on the cardiovascular system. It is therefore cogent to discuss testing systems for the effects of particles on the blood.

17.4.7.1 Endothelium

The endothelium is an extremely important cell type intimately involved in the regulation of vascular tone, clotting, fibrinolysis, and inflammation, so it is a highly relevant cell to study. HUVECs and variants are available that allow the measurement of the effects that particles gaining access to them might have, such as gene expression for clotting factors and inflammatory mediators (Gilmour et al. 2005).

17.4.7.2 Platelets

Platelets are a key cell in the initiation of thrombus formation. If particles gaining access to the blood could activate platelets, then there would likely be thrombus formation. Interactions between platelets and various nanoparticles have been reported (Nemmar et al. 2004; Radomski et al. 2005) with evidence that some particles can cause platelet aggregation and up-regulation of surface adhesion molecules on the platelets.

17.4.7.3 Atherosclerotic Plaques

If particles become blood borne, they are likely to be deposited in the vessel wall in the same way as LDL and by the same forces of turbulence. They could then interact directly with the cells in the atherosclerotic lesion. There are currently no specific *in vitro* models to investigate this aspect.

17.4.8 Effects on the Brain

Because of interest in the brain translocation of particles, studies of the effects of particles on neurons are warranted. These can address the penetration of the blood brain barrier, such as:

1. Evaluation of toxicity leading to increased permeability and opening of tight junctions between endothelial cells. If particles in the blood are able to have this effect, they may be able to leave the blood and traverse the BBB.
2. Testing the ability of particles to undergo endocytosis or transcytosis in endothelium of the BBB.
3. Assessment of the influence of NP on cell membrane fluidity, which may lead to inhibition of the brain efflux system.

They can also address the degenerative effect of particles on neural cells *in vitro* (Block et al. 2004).

17.4.9 *In Vitro* Tests for Genotoxicity

The recognition of carcinogenesis as the complex culmination of DNA damage, repair, mutations—both chromosomal and at the gene level—DNA methylation, signal transduction, translation, etc., means that there is a large battery of tests that can be used to detect the ability of particles to cause cancer. Indeed there is some overlap with pro-inflammatory tests described above, since inflammation may play a key role in carcinogenesis of some particles such as quartz (Donaldson and Borm 1998). The types of tests available include assays of mammalian cell transformation, DNA breaks, oxidative adducts of DNA, micronucleus assay, and inhibition of DNA repair. These assays are reviewed extensively (Fubini et al. 1998) with regard to their use for detecting particle effects. It is important to note that the carcinogenic effects detected for a number of particles in animal models, e.g., quartz and some non-toxic particles at overload, are a consequence of inflammation and so would not be detected by "classical" assays of genotoxicity.

17.5 Animal Studies

17.5.1 Bronchoalveolar Lavage

Inflammation lies at the center of the adverse effects of particles (Donaldson and Tran 2002). Bronchoalveolar lavage (BAL) analysis is an important method that is used to determine and quantify inflammation in the lungs as well as to derive mechanistic data.

17.5.2 Intratracheal Instillation

In intratracheal instillation, a mass of the particle sample is instilled into the lungs of rats in a small volume of saline. The dose rate is necessarily high in the case of instillation, as the entire dose is delivered to the lungs instantaneously; this contrasts with an inhalation exposure where small fractions of the total dose are delivered each day, i.e., at a low-dose rate. However, the dose in an instillation study can be low, and indeed should be low to circumvent the high-dose rate and the localized deposition in a small fraction of the total lung. Although instillation does have these drawbacks, it has value in comparing between particle samples. The technique is more problematic for fibers, which tend to clump and bridge small airways leading to granulomas in the airspaces. However, a comparison between instillation and inhalation for particles has demonstrated that the same types of qualitative response were seen (Henderson et al. 1995). The technique can be used for histopathology studies, but it is more often used in combination with BAL to study the short to medium-term inflammatory response to a suspect particle in relation to a known pathogenic particle such as quartz and a non-pathogenic particle such as pigment grade TiO_2 (Zhang et al. 1998). There is a distinct problem of "local overload" of lung defenses when non-toxic particles are used in this assay. Therefore, the test particles should be compared against control particles of known high toxicity and of low toxicity.

17.5.3 Inhalation Studies

Inhalation studies are commonly carried out in conventional toxicology protocols to determine the likely toxicity of particles. Standard protocols are available for inhalation studies from the OECD in Europe (see http://www.oecd.org/document/55/0,2340,en_2649_34377_2349687_1_1_1_1,00.html) and OPPTS in the U.S.A. (see http://www.epa.gov/opptsfrs/publications/OPPTS_Harmonized/870_Health_Effects_Test_Guidelines/Series/).

TABLE 17.4
Different Lengths of Toxicological Study

Study	Timescale	Endpoints Commonly Measured
Subacute	Up to 14 days	BAL, biochemistry
Subchronic	Up to 90 days	Biochemistry, histopathology, lung burden
Chronic	6 months to 2 years	Biochemistry, histopathology, lung burden
Carcinogenicity	24–30 months	Biochemistry, histopathology, lung burden
Mechanistic	Often short-term	BAL, special techniques, lung burden

Rats are maintained in chambers that deliver either whole body or nose-only exposure to clouds of particles conventionally for up to 7 h per day, 5 days per week (Wong 1999). At various points during exposure and post-exposure, rats are killed for various endpoints. Conventional protocols are shown in Table 17.4 (Wong 1999).

17.5.4 Monitoring Exposure, Dose, and Response

The key components of the toxicological paradigm, exposure, dose and response, require particular attention in inhalation toxicology in order to interpret results. The exposure needs to be carried out with a cloud of particles of the correct D_{ae} to ensure sufficient deposition of the material to the area of interest e.g. the airways or the respiratory zone (Schlesinger et al. 1997) and this is explicit in the OECD and OPPTS guidelines. The cutoff size for these regions can be very steep in the rat and so this requires due consideration and monitoring. Serial kills of rats need to be carried out for quantification of the lung tissue concentration of the particles to ensure that adequate dosing of the target tissues was achieved. This is usually achieved by ashing or digestion of the lung tissue and mass or chemical measures of the particle burden per lung. This can be refined with counts of particle sizes (e.g., fibers) or preliminary micro-dissection of lung to quantify distribution within different lung compartments (e.g., pleura or alveolar ducts). The response can be from extent of inflammation assessed by BAL, histological assessment of pathology or proportion of tumors.

17.5.5 Immunological Effects of Particles

In the case of particles that may be sensitizing or may elicit effects in sensitized individuals, special testing can be carried out (Kimber et al. 1996). Bioassays for detecting respiratory allergens are based on two approaches: (1) the conventional detection of IgE against the allergen and (2) lung lymph node cells from animals treated with the respiratory allergen are tested for Th2 cytokines, interleukins 4 and 10, and the Th1 cell cytokine interferon gamma; the former are generally increased in mice exposed to respiratory allergens. These assays can be expected to become more refined as the mechanism of allergic lung disease improves. For particles suspected of causing extrinsic allergic alveolitis, BAL profile, tests for complement activation and specific IgM and IgG and histological evidence of granulomatous disease should be carried out following conventional intratracheal or instillation exposure. For particles that might be expected to exacerbate asthma, Gilmour (1995) have described assays to assess the effect of co-exposure to pollutants on the immune and inflammatory response to common allergens.

17.5.6 Effects on the Microbicidal Activity of the Lungs

The ability of the lung to resist infection is an important function that could be compromised by exposure to particles, since particles could have effects on macrophages and epithelial cells that could compromise their microbicidal functions. This has been reviewed recently (Thomas and

Zelikoff 1999). Impairment of pulmonary defenses against bacteria is described following exposure to metals and to wood smoke particles. This is an important area where more information is needed and where specific particles may have considerable impact.

17.5.7 Animal Models to Study the Cardiovascular Effects of Particles

Animal models can be used to monitor effects of particles on the cardiovascular system and the brain. The dominant cardiovascular endpoint is atherosclerosis and models exist to investigate the effects of exposures on atherosclerosis development in rabbits (Suwa et al. 2002) and mice (Sun et al. 2005). In these models, the atherosclerotic lesions are assessed morphometrically or by biochemical analysis, as there is no standard method for detecting plaque rupture. Thrombosis is the other key aspect of plaque rupture that is important in dictating mortality and morbidity since the production of a thrombus is key to the production of infarction. The effect of particle exposure on thrombus formation has been studied in rose Bengal (a dye)-initiated thrombus formation in the femoral vein of rats (Nemmar et al. 2002a, 2002b). In this model, the small thrombus forming in response to the dye plus light is larger in hamsters exposed to particles, suggesting that there is a prothrombotic state in the blood after particle exposure. In a different model, vascular thrombosis was induced by ferric chloride in rat carotid arteries and the rate of thrombosis was measured with an ultrasonic flow probe. In the presence of particle exposure, there was an acceleration in the rate of vascular thrombosis (Radomski et al. 2005).

Human studies can illuminate the effects of particle exposures on the cardiovascular system, and such studies have demonstrated effects on heart rate variability (Devlin et al. 2003) and other markers of cardiovascular risk (Schwartz 2001). In unique studies, the authors were able to investigate the effect of diesel exhaust exposure on the endothelium. These studies demonstrated that exposure to diesel particles was associated with endothelial dysfunction that was very likely oxidative stress-mediated (Mills et al. 2005).

17.5.8 Animal Models of Neurological Effects of Particles

Following on from the report by Oberdörster that nanoparticles can translocate from the lungs to the brain (Oberdörster et al. 2004), reports of brain abnormalities in humans exposed to high pollution (Calderon-Garciduenas et al. 2004) and in mice exposed to CAPs (Veronesi et al. 2005), concern has arisen that particles should be tested for likely effects on the brain. Such tests should be based on direct transfer of particles via the olfactory neurons in the nose, but should also focus on the blood–brain barrier (BBB) and how particles in the blood might cross it. The following approaches may be used. After exposure to NP in animals/human subjects, assessment should be made of the following:

1. Assessment of brain function by use of non-invasive, functional changes (quantitative electroencephalographics, QEEG), or behavioral tests.
2. Assessment of histological markers of brain degeneration or inflammation.
3. Assessment of retention of NP in the blood brain capillaries due to adhesion, binding, or trapping that might play a role in crossing the BBB.

17.5.9 Conclusion—A Tiered Approach to Testing of a New Particle

Predictive testing of particles is a very difficult science. The variability between particles in size and composition and the spectrum of potentially adverse outcomes makes choices of assay difficult. The prohibitive cost and time needed for inhalation testing and the drive to reduce animal use means that there is an increasing emphasis on non-animal testing. We have advanced here the outline approach for the testing of particles of unknown toxicity. It is a layered system that can be entered at any point or terminated at any point, giving various degrees of confidence about the potential toxicity.

Tier 1: Consider the size of the particle and draw conclusions about its site of deposition and the likelihood of substantial deposition in the different compartments of the lungs. Determine the composition and biopersistence of the particle and benchmark the particle against other similarly structured particles to anticipate the likelihood and types of adverse effects. Tier 2: Armed with the data from benchmarking, carry out appropriate *in vitro* tests to determine whether there is any evidence for toxicity. Tier 3: Carry out instillation studies with due consideration of dose and inclusion of positive and negative control particles. Tier 4: Carry out inhalation studies. To understand the mechanism of toxic action of particles, further reductionist mechanistic studies on the cellular and molecular responses elicited by the particles can be carried out. At any point in the above tiered approach the particle may be exonerated and the subsequent testing tiers may not need to be carried out. Conversely, the system may also show up an important aspect of toxicity that deserves further study in specialized protocols.

REFERENCES

Aust, A. E., Ball, J. C., Hu, A. A., Lighty, J. S., Smith, K. R., Straccia, A. M., Veranth, J. M., and Young, W. C., Particle characteristics responsible for effects on human lung epithelial cells, *Res. Rep. Health Eff. Inst.*, 1–65, 2002.

Berridge, M. J., The versatility and complexity of calcium signaling, *Novartis Found. Symp.*, 239, 52–64, 2001.

Berridge, M. J., Bootman, M. D., and Lipp, P., Calcium—a life and death signal [News], *Nature*, 395, 645–648, 1998.

BéruBé, K. A., Quinlan, T. R., Fung, H., Magae, J., Vacek, P., Taatjes, D. J., and Mossman, B. T., Apoptosis is observed in mesothelial cells after exposure to crocidolite asbestos, *Am. J. Respir. Cell Mol. Biol.*, 15, 141–147, 1996.

Block, M. L., Wu, X., Pei, Z., Li, G., Wang, T., Qin, L., Wilson, B., Yang, J., Hong, J. S., and Veronesi, B., Nanometer size diesel exhaust particles are selectively toxic to dopaminergic neurons: the role of microglia, phagocytosis, and NADPH oxidase, *FASEB J.*, 18, 1618–1620, 2004.

Brown, D. M. and Donaldson, K., Wool and grain dusts stimulate TNF secretion by alveolar macrophages in vitro, *Occup. Environ. Med.*, 53, 387–393, 1996.

Brown, D. M., Fisher, C., and Donaldson, K., Free radical activity of synthetic vitreous fibers: iron chelation inhibits hydroxyl radical generation by refractory ceramic fiber, *J. Toxicol. Environ. Health*, 53, 545–561, 1998.

Brown, D. M., Donaldson, K., and Stone, V., Effects of PM10 in human peripheral blood monocytes and J774 macrophages, *Respir. Res.*, 5, 29, 2004.

Brown, D. M., Donaldson, K., Borm, P. J., Schins, R. P., Dehnhardt, M., Gilmour, P., Jimenez, L. A., and Stone, V., Calcium and ROS-mediated activation of transcription factors and TNF-alpha cytokine gene expression in macrophages exposed to ultra-fine particles, *Am. J. Physiol. Lung Cell Mol. Physiol.*, 286, L344–L353, 2004.

Calderon-Garciduenas, L., Reed, W., Maronpot, R. R., Henriquez-Roldan, C., Gado-Chavez, R., Calderon-Garciduenas, A., Dragustinovis, I. et al., Brain inflammation and Alzheimer's-like pathology in individuals exposed to severe air pollution, *Toxicol. Pathol.*, 32, 650–658, 2004.

Castranova, V., Dalal, N. S., and Vallyathan, V., The role of surface free Radicals in the pathogenicity of silicosis in silica and silica-induced lung diseases, CRC Press, Boca Raton, FL, 1995 Ref Type: Generic.

Coin, P. G., Roggli, V. L., and Brody, A. R., Persistence of long, thin chrysotile asbestos fibers in the lungs of rats, *Environ. Health Perspect.*, 102 (suppl. 5), 197–199, 1994 Review.

Dai, J., Xie, C., and Churg, A., Iron loading makes a non-fibrogenic model air pollutant particle fibrogenic in rat tracheal explants, *Am. J. Respir. Cell Mol. Biol.*, 26, 685–693, 2002.

Dalal, N. S., Newman, J., Pack, D., Leonard, S., and Vallyathan, V., Hydroxyl radical generation by coal mine dust: Possible implication to coal worker's pneumoconiosis (CWP), *Free Radic. Biol. Med.*, 18, 11–20, 1995.

Davis, J. G., Addison, J., Bolton, R. E., Donaldson, K., Jones, A. D., and Smith, T., The pathogenicity of long versus short fiber samples of amosite asbestos administered to rats by inhalation and intraperitoneal injection, *Br. J. Exp. Pathol.*, 67, 415–430, 1986.

Davis, J. G., Brown, D. M., Cullen, R. T., Donaldson, K., Jones, A. D., Miller, B. G., Mcintosh, C., and Searl, A., A comparison of methods of determining and predicting the pathogenicity of mineral fibers, *Inhal. Toxicol.*, 8, 747–770, 1996.

Devlin, R. B., Ghio, A. J., Kehrl, H., Sanders, G., and Cascio, W., Elderly humans exposed to concentrated air pollution particles have decreased heart rate variability, *Eur. Respir. J. Suppl.*, 40, 76s–80s, 2003.

Donaldson, K. and Borm, P. J., The quartz hazard: a variable entity, *Ann. Occup. Hyg.*, 42, 287–294, 1998.

Donaldson, K. and Tran, C. L., Inflammation caused by particles and fibers, *Inhal. Toxicol.*, 14, 2002.

Donaldson, K. and Tran, C. L., An introduction to the short-term toxicology of respirable industrial fibers, *Mutat. Res.*, 553, 5–9, 2004.

Donaldson, K., Beswick, P. H., and Gilmour, P. S., Free radical activity associated with the surface of particles: a unifying factor in determining biological activity?, *Toxicol. Lett.*, 88, 293–298, 1996.

Donaldson, K., Jimenez, L. A., Rahman, I., Faux, S. P., MacNee, W., Gilmour, P. S., Borm, P. J. et al., Respiratory health effects of ambient air pollution particles: role of reactive species, In *Oxygen/Nitrogen Radicals: Lung Injury and Disease*, Vallyathan, V., Shi, X., and Castranova, V., eds., Marcel Dekker, New York, 2004. Volk 187 in Lung Biology in Health and Disease. Exec. ed. Lenfant C. Marcel Dekker, New York.

Donaldson, K., Tran, L., Jimenez, L., Duffin, R., Newby, D. E., Mills, N., MacNee, W., and Stone, V., Combustion-derived nanoparticles: a review of their toxicology following inhalation exposure, *Part. Fiber Toxicol.*, 2, 10, 2005.

Dreher, K. L., Jaskot, R. H., Lehmann, J. R., Richards, J. H., Mcgee, J. K., Ghio, A. J., and Costa, D. L., Soluble transition metals mediate residual oil fly ash induced acute lung injury, *J. Toxicol. Environ. Health*, 50, 285–305, 1997.

Fubini, B., Aust, A. E., Bolton, R. E., Borm, P. J. A., Bruch, J., Ciapetti, G., Donaldson, K., et al., Non-animal tests for evaluating the toxicity of solid xenobiotics, *ATLA*, 26, 579–617, 1998 Ref Type: Generic.

Ghio, A. J. and Cohen, M. D., Disruption of iron homeostasis as a mechanism of biologic effect by ambient air pollution particles, *Inhal. Toxicol.*, 17, 709–716, 2005.

Ghio, A. J., Jaskot, R. H., and Hatch, G. E., Lung injury after silica instillation is associated with an accumulation of iron in rats, *Am. J. Physiol. Lung Cell Mol. Physiol.*, 11, L686–L692, 1994.

Ghio, A. J., Stonehuerner, J., Pritchard, R. J., Piantadosi, C. A., Quigley, D. R., Dreher, K. L., and Costa, D. L., Humic-like substances in air-pollution particulates correlate with concentrations of transition metals and oxidant generation, *Inhal. Toxicol.*, 8, 479–494, 1996.

Ghio, A. J., Turi, J. L., Yang, F., Garrick, L. M., and Garrick, M. D., Iron homeostasis in the lung, *Biol. Res.*, 39, 67–77, 2006.

Gilmour, M. I., Interaction of air pollutants and pulmonary allergic responses in experimental animals, *Toxicology*, 105, 335–342, 1995.

Gilmour, P. S., Brown, D. M., Lindsay, T. G., Beswick, P. H., MacNee, W., and Donaldson, K., Adverse health-effects of PM(10) particles—involvement of iron in generation of hydroxyl radical, *Occup. Environ. Med.*, 53, 817–822, 1996.

Gilmour, P. S., Brown, D. M., Beswick, P. H., MacNee, W., Rahman, I., and Donaldson, K., Free radical activity of industrial fibers: role of iron in oxidative stress and activation of transcription factors, *Environ. Health Perspect.*, 105, 1997.

Gilmour, P. S., Rahman, I., Donaldson, K., and MacNee, W., Histone acetylation regulates epithelial IL-8 release mediated by oxidative stress from environmental particles, *Am. J. Physiol. Lung Cell. Mol. Physiol.*, 284, L533–L540, 2003.

Gilmour, P. S., Morrison, E. R., Vickers, M. A., Ford, I., Ludlam, C. A., Greaves, M., Donaldson, K., and MacNee, W., The procoagulant potential of environmental particles (PM10), *Occup. Environ. Med.*, 62, 164–171, 2005.

Ginn-Pease, M. E. and Whisler, R. L., Redox signals and NF-kappaB activation in T cells, *Free Radic. Biol. Med.*, 25, 346–361, 1998.

Grynkiewicz, G., Poenie, M., and Tsien, R. Y., A new generation of Ca2+ indicators with greatly improved fluorescence properties, *J. Biol. Chem.*, 260, 3440–3450, 1985.

Henderson, R. F., Driscoll, K. E., Harkema, J. R., Lindenschmidt, R. C., Chang, I. Y., Maples, K. R., and Barr, E. B., A comparison of the inflammatory response of the lung to inhaled versus instilled particles in f344 rats, *Fundam. Appl. Toxicol.*, 24, 183–197, 1995.

Hesterberg, T. W., Miiller, W. C., Mast, R., Mcconnell, E. E., Bernstein, D. M., and Anderson, R., Relationship between lung biopersistence and biological effects of man-made vitreous fibers after chronic inhalation in rats, *Environ. Health Perspect.*, 102, 133–137, 1994.

Hubbard, A. K., Timblin, C. R., Shukla, A., Rincon, M., and Mossman, B. T., Activation of NF-kappaB-dependent gene expression by silica in lungs of luciferase reporter mice, *Am. J. Physiol. Lung Cell. Mol. Physiol.*, 282, L968–L975, 2002.

Iyer, R. and Holian, A., Involvement of the ICE family of proteases in silica-induced apoptosis in human alveolar macrophages, *Am. J. Physiol.*, 273, L760–L767, 1997.

Jimenez, L. A., Drost, E. M., Gilmour, P. S., Rahman, I., Antonicelli, F., Ritchie, H., MacNee, W., and Donaldson, K., PM(10)-exposed macrophages stimulate a proinflammatory response in lung epithelial cells via TNF-alpha, *Am. J. Physiol. Lung Cell. Mol. Physiol.*, 282, L237–L248, 2002.

Jimenez, L. A., Thompson, J., Brown, D. A., Rahman, I., Antonicelli, F., Duffin, R., Drost, E. M., Hay, R. T., Donaldson, K., and MacNee, W., Activation of NF-kappaB by PM(10) occurs via an iron-mediated mechanism in the absence of IkappaB degradation, *Toxicol. Appl. Pharmacol.*, 166, 101–110, 2000.

Karin, M., The NF-kappa B activation pathway: its regulation and role in inflammation and cell survival, *Cancer J. Sci. Am.*, 4 (suppl. 1), S92–S99, 1998 Review.

Kimber, I., Hilton, J., Basketter, D. A., and Dearman, R. J., Predictive testing for respiratory sensitization in the mouse, *Toxicol. Lett.*, 86, 193–198, 1996.

Lund, L. G. and Aust, A. E., Mobilization of iron from crocidolite asbestos by certain chelators results in enhanced crocidolite-dependent oxygen consumption, *Arch. Biochem. Biophys.*, 287, 91–96, 1991.

Maltoni, C., Minardi, F., and Morisi, L., Pleural mesotheliomas in Sprague–Dawley rats by erionite: first experimental evidence, *Environ. Res.*, 29, 238–244, 1982.

Marwick, J. A., Kirkham, P. A., Stevenson, C. S., Danahay, H., Giddings, J., Butler, K., Donaldson, K., MacNee, W., and Rahman, I., Cigarette smoke alters chromatin remodeling and induces proinflammatory genes in rat lungs, *Am. J. Respir. Cell Mol. Biol.*, 31, 633–642, 2004.

McNeilly, J. D., Jimenez, L. A., Clay, M. F., MacNee, W., Howe, A., Heal, M. R., Beverland, I. J., and Donaldson, K., Soluble transition metals in welding fumes cause inflammation via activation of NF-kappaB and AP-1, *Toxicol. Lett.*, 158, 152–157, 2005.

Meyer, M., Schreck, R., and Baeuerle, P. A., H_2O_2 and antioxidants have opposite effects on activation of NF-kappa B and AP-1 in intact cells: AP-1 as secondary antioxidant-responsive factor, *EMBO J.*, 12, 2005–2015, 1993.

Mills, N. L., Tornqvist, H., Robinson, S. D., Gonzalez, M., Darnley, K., MacNee, W., Boon, N. A., Donaldson, K., et al. , Diesel exhaust inhalation causes vascular dysfunction and impaired endogenous fibrinolysis, *Circulation*, 112, 3930–3936, 2005.

Molinelli, A. R., Madden, M. C., Mcgee, J. K., Stonehuerner, J. G., and Ghio, A. J., Effect of metal removal on the toxicity of airborne particulate matter from the Utah Valley, *Inhal. Toxicol.*, 14, 1069–1086, 2002.

Nemmar, A., Hoylaerts, M. F., Hoet, P. H. M., Dinsdale, D., Smith, T., Xu, H., Vermylen, J., and Nemery, B., Ultrafine particles affect experimental thrombosis in an in vivo hamster model, *Am. J. Resp. Crit. Care Med.*, 166, 998–1004, 2002.

Nemmar, A., Hoylaerts, M. F., Hoet, P. H., Dinsdale, D., Smith, T., Xu, H., Vermylen, J., and Nemery, B., Ultra-fine particles affect experimental thrombosis in an in vivo hamster model, *Am. J. Respir. Crit. Care Med.*, 166, 998–1004, 2002.

Nemmar, A., Hoylaerts, M. F., Hoet, P. H., and Nemery, B., Possible mechanisms of the cardiovascular effects of inhaled particles: systemic translocation and prothrombotic effects, *Toxicol. Lett.*, 149, 243–253, 2004.

Oberdörster, G., Sharp, Z., Elder, A. P., Gelein, R., Kreyling, W., and Cox, C., Translocation of inhaled ultrafine particles to the brain, *Inhal. Toxicol.*, 16, 437–445, 2004.

Oberdörster, G., Oberdörster, E., and Oberdörster, J., Nanotoxicology: an emerging discipline evolving from studies of ultra-fine particles, *Environ. Health Perspect.*, 113, 823–839, 2005.

Parekh, A. B. and Penner, R., Store depletion and calcium influx, *Physiol. Rev.*, 77, 901–930, 1997.

Piette, J., Piret, B., Bonizzi, G., Schoonbroodt, S., Merville, M. P., Legrand, P., and Bours, V., Multiple redox regulation in NF-kappaB transcription factor activation, *Biol. Chem.*, 378, 1237–1245, 1997 Review.

Radomski, A., Jurasz, P., Onso-Escolano, D., Drews, M., Morandi, M., Malinski, T., and Radomski, M. W., Nanoparticle-induced platelet aggregation and vascular thrombosis, *Br. J. Pharmacol.*, 146, 882–893, 2005.

Rahman, I. and MacNee, W., Oxidative stress and regulation of glutathione in lung inflammation, *Eur. Respir. J.*, 16, 534–554, 2000.

Rice, T. M., Clarke, R. W., Godleski, J. J., Al Mutairi, E., Jiang, N. F., Hauser, R., and Paulauskis, J. D., Differential ability of transition metals to induce pulmonary inflammation, *Toxicol. Appl. Pharmacol.*, 177, 46–53, 2001.

Schins, R. P. F. and Donaldson, K., Nuclear factor kappa B activation by particles and fibers, *Inhal. Toxicol.*, 12 (suppl. 12), 317–326, 2000 Ref Type: Generic.

Schlesinger, R. B., Ben-Jebria, A., Dahl, A. R., Snipes, M. B., and Ultman, J., Disposition of inhaled toxicicants, In *Handbook of Human Toxicology*, Massaro, E. J., ed., CRC Press, Boca Raton, FL, 493–550, 1997.

Schwartz, J., Air pollution and blood markers of cardiovascular risk, *Environ. Health Perspect.*, 109 (suppl. 3.), 405–409, 2001.

Searl, A., Buchanan, D., Cullen, R. T., Jones, A. D., Miller, B. G., and Soutar, C. A., Biopersistence and durability of nine mineral fiber types in rat lungs over 12 months, *Ann. Occup. Hyg.*, 43, 143–153, 1999.

Seger, R. and Krebs, E. G., The MAPK signaling cascade, *FASEB J.*, 9, 726–735, 1995.

Sen, C. K., Khanna, S., Reznick, A. Z., Roy, S., and Packer, L., Glutathione regulation of tumor necrosis factor-alpha-induced NF- kappa B activation in skeletal muscle-derived L6 cells, *Biochem. Biophys. Res. Commun.*, 237, 645–649, 1997.

Shi, T., Schins, R. P., Knaapen, A. M., Kuhlbusch, T., Pitz, M., Heinrich, J., and Borm, P. J., Hydroxyl radical generation by electron paramagnetic resonance as a new method to monitor ambient particulate matter composition, *J. Environ. Monit.*, 5, 550–556, 2003.

Shukla, A., Timblin, C., BéruBé, K., Gordon, T., McKinney, W., Driscoll, K., Vacek, P., and Mossman, B. T., Inhaled particulate matter causes expression of nuclear factor (NF)-kappaB-related genes and oxidant-dependent NF-kappaB activation in vitro, *Am. J. Respir. Cell Mol. Biol.*, 23, 182–187, 2000.

Shukla, A., Flanders, T., Lounsbury, K. M., and Mossman, B. T., The gamma-glutamylcysteine synthetase and glutathione regulate asbestos-induced expression of activator protein-1 family members and activity, *Cancer Res.*, 64, 7780–7786, 2004.

Squadrito, G. L., Cueto, R., Dellinger, B., and Pryor, W. A., Quinoid redox cycling as a mechanism for sustained free radical generation by inhaled airborne particulate matter, *Free Radic. Biol. Med.*, 31, 1132–1138, 2001.

Stone, V., Tuinman, M., Vamvakopoulos, J. E., Shaw, J., Brown, D., Petterson, S., Faux, S. P. et al., Increased calcium influx in a monocytic cell line on exposure to ultra-fine carbon black, *Eur. Respir. J.*, 15, 297–303, 2000.

Sun, Q., Wang, A., Jin, X., Natanzon, A., Duquaine, D., Brook, R. D., Aguinaldo, J. G. et al., Long-term air pollution exposure and acceleration of atherosclerosis and vascular inflammation in an animal model, JAMA, 294, 3003–3010, 2005.

Suwa, T., Hogg, J. C., Quinlan, K. B., Ohgami, A., Vincent, R., and van Eeden, S. F., Particulate air pollution induces progression of atherosclerosis, *J. Am. Coll. Cardiol.*, 39, 935–942, 2002.

Tao, F., Gonzalez-Flecha, B., and Kobzik, L., Reactive oxygen species in pulmonary inflammation by ambient particulates, *Free Radic. Biol. Med.*, 35, 327–340, 2003.

Thastrup, O., Cullen, P. J., Drobak, B. K., Hanley, M. R., and Dawson, A. P., Thapsigargin, a tumor promoter, discharges intracellular $Ca2+$ stores by specific inhibition of the endoplasmic reticulum $Ca2(+)$-ATPase, *Proc. Natl. Acad. Sci. USA*, 87, 2466–2470, 1990.

Thomas, P. T. and Zelikoff, J. T., Air pollutants: modulators of pulmonary host resistance against infection, In *Air pollution and health*, Holgate, S. T., Samet, J. M., Koren, H. S., and Maynard, R. L., eds., Academic Press, London, 357–379, 1999. Ref Type: Generic.

Timblin, C. R., Janssen, Y. M., and Mossman, B. T., Transcriptional activation of the protooncogene c-jun by asbestos and H_2O_2 is directly related to increased proliferation and transformation of tracheal epithelial-cells, *Cancer Res.*, 55, 2723–2726, 1995.

Timblin, C., BéruBé, K., Churg, A., Driscoll, K., Gordon, T., Hemenway, D., Walsh, E., Cummins, A. B., Vacek, P., and Mossman, B., Ambient particulate matter causes activation of the c-jun kinase/stress-activated protein kinase cascade and DNA synthesis in lung epithelial cells, *Cancer Res.*, 58, 4543–4547, 1998.

Veronesi, B., Makwana, O., Pooler, M., and Chen, L. C., Effects of subchronic exposures to concentrated ambient particles. VII. Degeneration of dopaminergic neurons in Apo E−/− mice, *Inhal. Toxicol.*, 17, 235–241, 2005.

Wilson, M. R., Lightbody, J. H., Donaldson, K., Sales, J., and Stone, V., Interactions between ultra-fine particles and transition metals in vivo and in vitro, *Toxicol. Appl. Pharmacol.*, 184, 172–179, 2002.

Wong, B. A., Inhalation exposure systems design, methods and operation, In *Toxicology of the Lung*, Gardner, D. E., Crapo, J. D., and McClellan, R. O., eds., Taylor and Francis, Philadelphia, PA, 1–53, 1999. Chapter 1, Ref Type: Generic.

Xia, Z., Dickens, M., Raingeaud, J., Davis, R. J., and Greenberg, M. E., Opposing effects of ERK and JNK-p38 MAP kinases on apoptosis, *Science*, 270, 1326–1331, 1995.

Zhang, Q., Kusaka, Y., Sato, K., Nakakuki, K., Kohyama, N., and Donaldson, K., Differences in the extent of inflammation caused by intratracheal exposure to three ultra-fine metals: role of free radicals, *J. Toxicol. Environ. Health*, 53, 423–438, 1998.

18 Models for Testing the Pulmonary Toxicity of Particles: Lung Bioassay Screening Studies in Male Rats with a New Formulation of TiO_2 Particulates

David B. Warheit, Kenneth L. Reed, and Christie M. Sayes
DuPont Haskell Laboratory for Health and Environmental Sciences

CONTENTS

18.1 INTRODUCTION

Formulation changes in the form of altered surface treatments are known to frequently occur for a variety of commercialized particle-types. The R-100 formulation of rutile-type titanium dioxide is a well-known low-toxicity particulate. This study was designed as a pulmonary screening tool to determine whether Pigment A TiO_2 particles (TiO_2 particles that have been substantially encapsulated with pyrogenically deposited amorphous silica) may impart significant toxicity in the lungs of rats; and more importantly, how the activity of this TiO_2 formulation compares with other reference particulate materials, including standardized TiO_2 particulates. Thus, the objective of this study was to evaluate in the lungs of rats—using a well-developed, short-term pulmonary bioassay—the acute pulmonary toxicity effects of intratracheally instilled Pigment A TiO_2 particle samples and to compare the pulmonary toxicity of these samples with two other low toxicity particulate-types (reference negative controls), along with a cytotoxic particulate (reference positive control) sample. Another aim was to bridge the results obtained herein with data previously generated from inhalation studies with crystalline silica-quartz particles and with carbonyl iron particulates as the inhalation/instillation bridge materials.

Bridging studies can be useful in providing an inexpensive safety screen when assessing the hazards of new developmental compounds or when making small modifications to an existing commercial particle-type, such as surface treatments. The strength of the bridging strategy is dependent upon having good inhalation toxicity data on one of the bridging compounds. The particle-type for which inhalation data exists can then be used as a reference control material for an intratracheal instillation bridging study (see Figure 18.1). The basic idea for the bridging concept is that the effects of the instilled material serve as a control (known reference) material and then are "bridged" on the one hand to the *inhalation toxicity data* for that material, as well as to the new materials being tested. The results of bridging studies in rats are then useful as pulmonary toxicity screening (i.e., hazard) data, because consistency in the response of the inhaled and instilled control material serves to validate the responses with the newly tested particulate matter.

Numerous studies have investigated the pulmonary toxicological impacts of surfaces treatments on titanium dioxide particles. In some cases, the surface coatings had little or no impact on the lung toxicity of TiO_2 particles in rats [5,8–10]. In another study, in evaluating six different formulations of commercial TiO_2 products, we reported that surface treatments might modify the lung inflammatory response following exposures in rats to fine-sized TiO_2 particle-types [2]. The

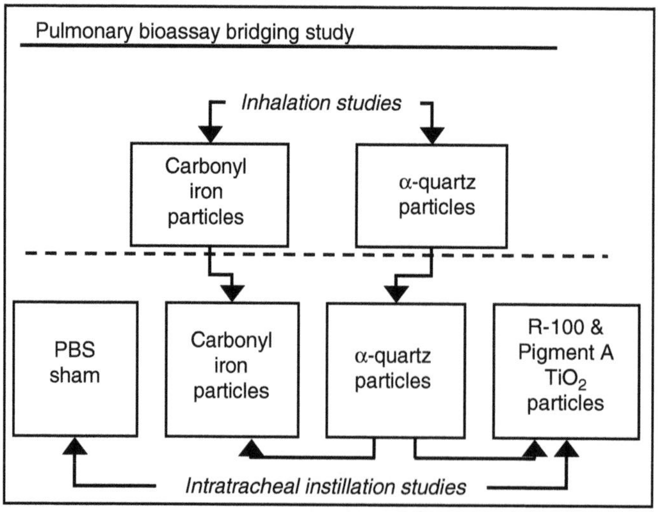

FIGURE 18.1 Schematic demonstrating the strategy for conducting pulmonary bioassay bridging studies. Bridging studies can have utility in providing an inexpensive preliminary safety screen when evaluating the hazards of new developmental compounds. The basic idea for the bridging concept is that the effects of the instilled material serve as a control (known) material and are then "bridged" to the inhalation toxicity data for that material and to the new materials being tested.

current study describes a model or pulmonary bioassay method for assessing the pulmonary hazard potential of intratracheally instilled particle-types. The chapter also discusses the relevance of this screening methodology as a possible surrogate for inhalation studies with low-solubility particle-types.

18.2 METHODS

18.2.1 General Experimental Design

The fundamental features of this pulmonary bioassay are dose-response evaluation, time-course assessments, and reference particle-types (positive and negative). The time-course studies are used to assess the sustainability of the observed affect. The major endpoints of this study were the (1) time-course and dose/response intensity of pulmonary inflammation and cytotoxicity (bronchoalveolar lavage (BAL) parameters), (2) airway and lung parenchymal cell proliferation, and (3) histopathological evaluation of lung tissue (see Figure 18.2).

The lungs of rats were exposed via intratracheal instillation with single doses of 1 or 5 mg/kg crystalline silica (α-quartz) particles, carbonyl iron particles, R-100 fine-TiO_2 particles, or to Pigment A fine-TiO_2 particles coated with amorphous SiO_2. The intratracheal instillation route of entry technique is not a substitute for the more physiologically relevant inhalation method of exposure. However, the intratracheal instillation method of exposure can be a qualitatively reliable screen for assessing the pulmonary toxicity of particles [1,2]. All particles were prepared in a volume of phosphate-buffered saline (PBS) solution and subjected to probe sonication for at least 15 minutes. Groups of PBS-instilled rats served as controls. The lungs of PBS and particle-exposed rats were evaluated by BAL fluid analyses at 24 h, 1 week, 1 month, and 3 months postexposure (pe). For lung cell proliferation and histopathology studies, additional groups of animals were instilled with the particle-types listed above as well as a PBS solution.

For the lung tissue studies, additional groups of animals (4 rats/group) were instilled with the particle-types listed above plus the vehicle control, i.e., PBS. These studies were dedicated to lung tissue analyses, but only the high-dose groups (5 mg/kg) and PBS controls were utilized in the morphology studies. These studies consisted of cell proliferation assessments and histopathological evaluations of the lower respiratory tract. Similar to the BAL fluid studies, the intratracheal instillation exposure period was followed by 24-h, 1-week, 1-month, and 3-month recovery periods.

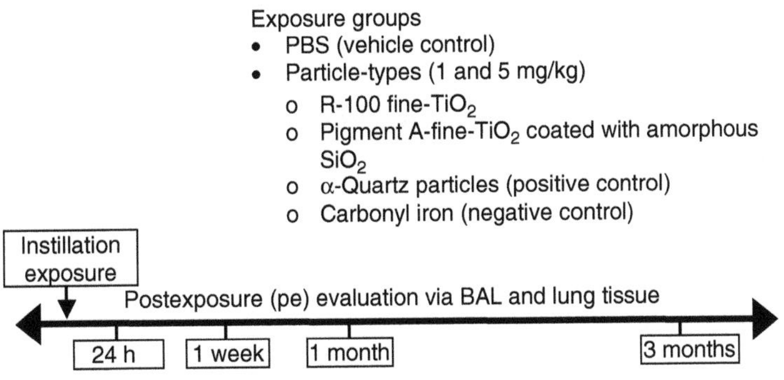

FIGURE 18.2 Experimental design for bridging case study.

18.2.2 Animals

Groups of male Crl:CD®(SD)IGS BR rats (Charles River Laboratories, Inc., Raleigh, North Carolina) were used in this study. The rats were approximately 8 weeks old at study start (mean weights in the range of 240–255 g). All procedures using animals were reviewed and approved by the Institutional Animal Care and Use Committee and the animal program is fully accredited by the Association for Assessment and Accreditation of Laboratory Animal Care (AAALAC).

18.2.3 Particle Types

The R-100 fine-titanium dioxide particles (~99 wt.% titanium dioxide, ~1 wt.% alumina) possessing an average particle diameter of ~300 nm and an average BET surface area of ~6 m^2/g were obtained from the DuPont Company. A patented chloride process produced Pigment A fine-titanium dioxide particles with an amorphous SiO_2 surface coating (~96 wt.% titanium dioxide, ~1 wt.% alumina, ~3 wt.% amorphous silica (particle encapsulating)) possessing an average particle diameter of ~290 nm and an average BET surface area of ~7.9 m^2/g were also obtained from the DuPont Company (Table 18.1). Both of the DuPont TiO_2 particle samples were in the rutile crystal phase and hydrophilic in nature.

Crystalline silica particles (α-quartz, Min-U-Sil 5) ranging in size from 0.3 to 3 μm were obtained from the US Silica Company. Carbonyl iron (CI) particles ranging in size from 0.8 to 3.0 μm were obtained from GAF Corporation.

18.2.4 Lung Cell Proliferation Studies

Groups of particulate-exposed rats and corresponding controls were pulsed 24 h after instillation, as well as 1 week, 1 month, and 3 months postexposure, with an intraperitoneal injection of 5-bromo-2′deoxyuridine (BrdU) dissolved in a 0.5 N sodium bicarbonate buffer solution at a dose of 100 mg/kg body weight. The animals were euthanized 6 h later by pentobarbital injection. Following cessation of spontaneous respiration, the lungs were infused with a neutral buffered formalin fixative at a pressure of 21 cm H_2O. After 20 min of fixation, the trachea was clamped, and the heart and lungs were carefully removed en bloc and immersion-fixed in formalin. In addition, a 1 cm piece of duodenum (which served as a positive control) was removed and stored in formaldehyde. Subsequently, parasagittal sections from the right cranial and caudal lobes and regions of the left lung lobes as well as the duodenal sections were dehydrated in 70% ethanol and sectioned for histology. The sections were embedded in paraffin, cut, and mounted on glass slides. The slides were stained with an anti-BrdU antibody, with an AEC (3-amino-9-ethyl carbazole) marker, and counter-stained with aqueous hematoxylin. A minimum of 1000 cells/animal were counted in terminal bronchiolar and alveolar regions. For each treatment group, immunostained nuclei in airways (i.e., terminal bronchiolar epithelial cells)

TABLE 18.1
Characterization of R-100 and Pigment A fine-TiO_2, as Well as Carbonyl Iron, and α-Quartz Particulates

Particle	Crystallinity	Surface Area (m^2/g)	Average Particle Diameter (nm)
R-100 fine-TiO_2	Rutile	6	~300
Pigment A coated with amorph SiO_2	Rutile	8	~290
Carbonyl iron	—	Not determined	~1500
Crystalline silica	α-quartz	5	~480

or lung parenchyma (i.e., epithelial, interstitial cells, or macrophages) were counted by light microscopy at 1000× magnification [3,4].

18.2.5 Bronchoalveolar Lavage Methods

The lungs of sham and particulate-exposed rats were lavaged with a warmed PBS solution as described previously. Methodologies for cell counts, differentials, and pulmonary biomarkers in lavaged fluids were conducted as previously described [3,4]. Briefly, the first 12 mL of lavaged fluids recovered from the lungs of PBS or particulate-exposed rats was centrifuged at 700 g, and 2 mL of the supernatant was removed for biochemical studies. All biochemical assays were performed on BAL fluids using a Roche Diagnostics (BMC)/Hitachi® 717 clinical chemistry analyzer using Roche Diagnostics (BMC)/Hitachi® reagents. Lactate dehydrogenase (LDH), alkaline phosphatase (ALP), and lavage fluid protein were measured using Roche Diagnostics (BMC)/Hitachi® reagents. LDH is a cytoplasmic enzyme and is used as an indicator of cell injury. ALP activity is a measure of Type II alveolar epithelial cell secretory activity, and increased ALP activity in BAL fluids is considered to be an indicator of Type II lung epithelial cell toxicity. Increases in BAL fluid micro protein (MTP) concentrations generally are consistent with enhanced permeability of vascular proteins into the alveolar regions, indicating a breakdown in the integrity of the alveolar-capillary barrier.

18.2.6 Lung Histopathology Studies

The lungs of rats exposed to particulates or PBS controls were prepared for light microscopy by tracheobronchial airway infusion under pressure (21 cm H_2O) at time periods of 24 hours, 1 week, 1 month, and 3 months postexposure. Sagittal sections of the left and right lungs were made using a razor blade. Tissue blocks were dissected from left, right upper, and right lower regions of the lung and were subsequently prepared for light microscopy (paraffin embedded, sectioned, and hematoxylin–eosin stained) [3,4].

18.2.7 Statistical Analyses

For analyses, each of the experimental values were compared to their corresponding sham control values for each time point. A one-way analysis of variance (ANOVA) and Bartlett's test were calculated for each sampling time. When the F test from ANOVA was determined to be significant, the Dunnett's test was utilized to compare means from the control group and each of the groups exposed to particulates. Significance was judged at the 0.05 probability level.

18.3 RESULTS

18.3.1 Lung Weights

Lung weights of rats were enhanced with increasing age on the study (i.e., increased postexposure time periods following intratracheal instillation exposures). Lung weights in high-dose quartz-exposed rats were slightly increased relative to controls at 1 week, and at 1 month and 3 months postexposure (data not shown).

18.3.2 Lung Cell Proliferation Results

Tracheobronchial cell proliferation rates (percentage of immunostained cells with BrdU) were measured only in high-dose (5 mg/kg) particulate-exposed rats and corresponding controls at 24 h, 1 week, 1 month, and 3 months postexposure (pe). Although increases in cell labeling indices were measured in R-100 fine-TiO_2 as well as α-quartz-exposed animals at 24 h postexposure, these effects were not sustained (data not shown).

Lung parenchymal cell proliferation rates (percentage of immunostained cells with BrdU) were measured only in high-dose (5 mg/kg) particulate-exposed rats and corresponding controls at 24 h, 1 week, 1 month, and 3 months postexposure (pe). Small but significant transient increases in lung cell proliferation indices were measured in carbonyl iron particle or in Pigment A fine-TiO_2 particle-exposed rats at 24 h, but these effects were not sustained at any other postexposure time points. Significantly larger increases in cell proliferation indices were measured in the lungs of α-quartz exposed rats measured from 24 h postexposure through 3 months postexposure (data not shown).

To summarize, exposures to 5 mg/kg quartz particles produced increased tracheobronchial cell proliferation compared to PBS controls, but increases were statistically significant only at 24 h postexposure. In contrast to tracheobronchial cell labeling indices, exposures to 5 mg/kg quartz particles produced substantially greater lung parenchymal cell proliferation rates at all time points postexposure, suggesting a greater likelihood to result in lung cell genotoxicity or other adverse pulmonary effects over time with continued exposures.

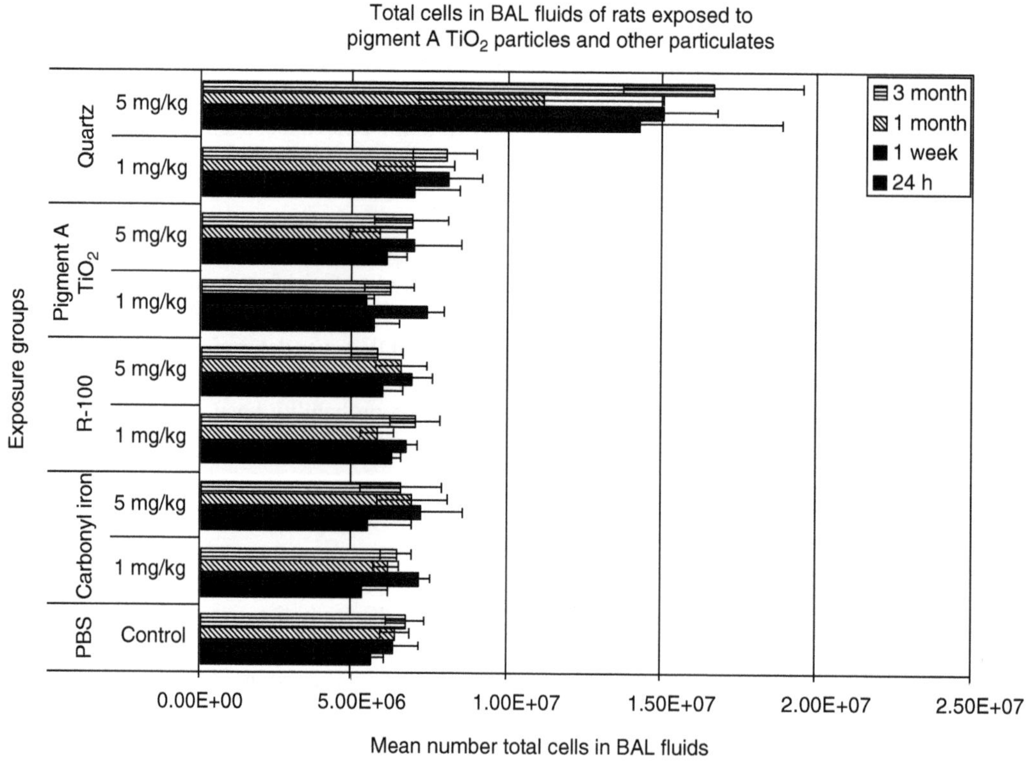

FIGURE 18.3 Total number of cells in BAL fluids recovered from particulate-exposed rats and controls as evidenced by % neutrophils (PMN) in BAL fluids at 24 h, 1 week, 1 month, and 3 months postexposure (pe). Values given are means ± S.D. The numbers of BAL cells recovered from the lungs of high-dose quartz groups were substantially higher than any other groups for all postexposure time periods.

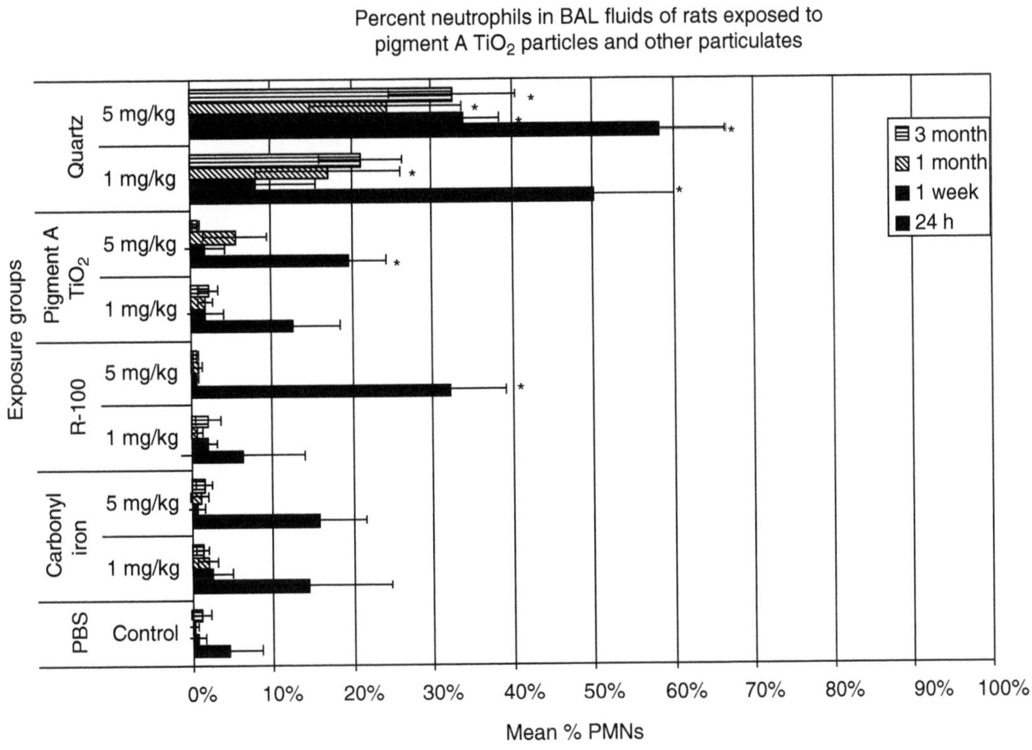

FIGURE 18.4 Pulmonary inflammation in particulate-exposed rats and controls as evidenced by % neutrophils (PMN) in BAL fluids at 24 h, 1 week, 1 month, and 3 months postexposure (pe). Values given are means ± S.D. Intratracheal instillation exposures of several particle-types produced a short-term, pulmonary inflammatory response, as evidenced by an increase in the percentages/numbers of BAL-recovered neutrophils, measured at 24 h postexposure. However, only the exposures to quartz particles (1 and 5 mg/kg) produced sustained pulmonary inflammatory responses, as measured through 3 months postexposure. $*p<0.05$.

18.3.3 Bronchoalveolar Lavage Fluid Results

18.3.3.1 Pulmonary Inflammation

The numbers of cells recovered by BAL from the lungs of high-dose α-quartz-exposed (5 mg/kg) groups were substantially higher than any of the other groups for all postexposure time periods (Figure 18.3). Intratracheal instillation exposures of virtually all particle-types produced a short-term pulmonary inflammatory response, as evidenced by an increase in the percentages/numbers of BAL-recovered neutrophils, measured at 24 h postexposure. However, only the exposures to α-quartz particles (1 and 5 mg/kg) produced sustained pulmonary inflammatory responses, as measured through 3 months postexposure (Figure 18.4).

18.3.3.2 Bronchoalveolar Lavage (BAL) Fluid Parameters

Transient and reversible increases in (BAL) fluid LDH values, as an indicator of general cytotoxicity, were measured in the lung fluids of high-dose (5 mg/kg) R-100 fine-TiO_2-exposed rats at 1 week postexposure, but were not sustained through the other postexposure time periods. In contrast, exposures to high-dose (5 mg/kg) α-quartz particles produced a persistent enhancement in BAL fluid LDH values throughout the 3-month post-instillation exposure period (Figure 18.5A).

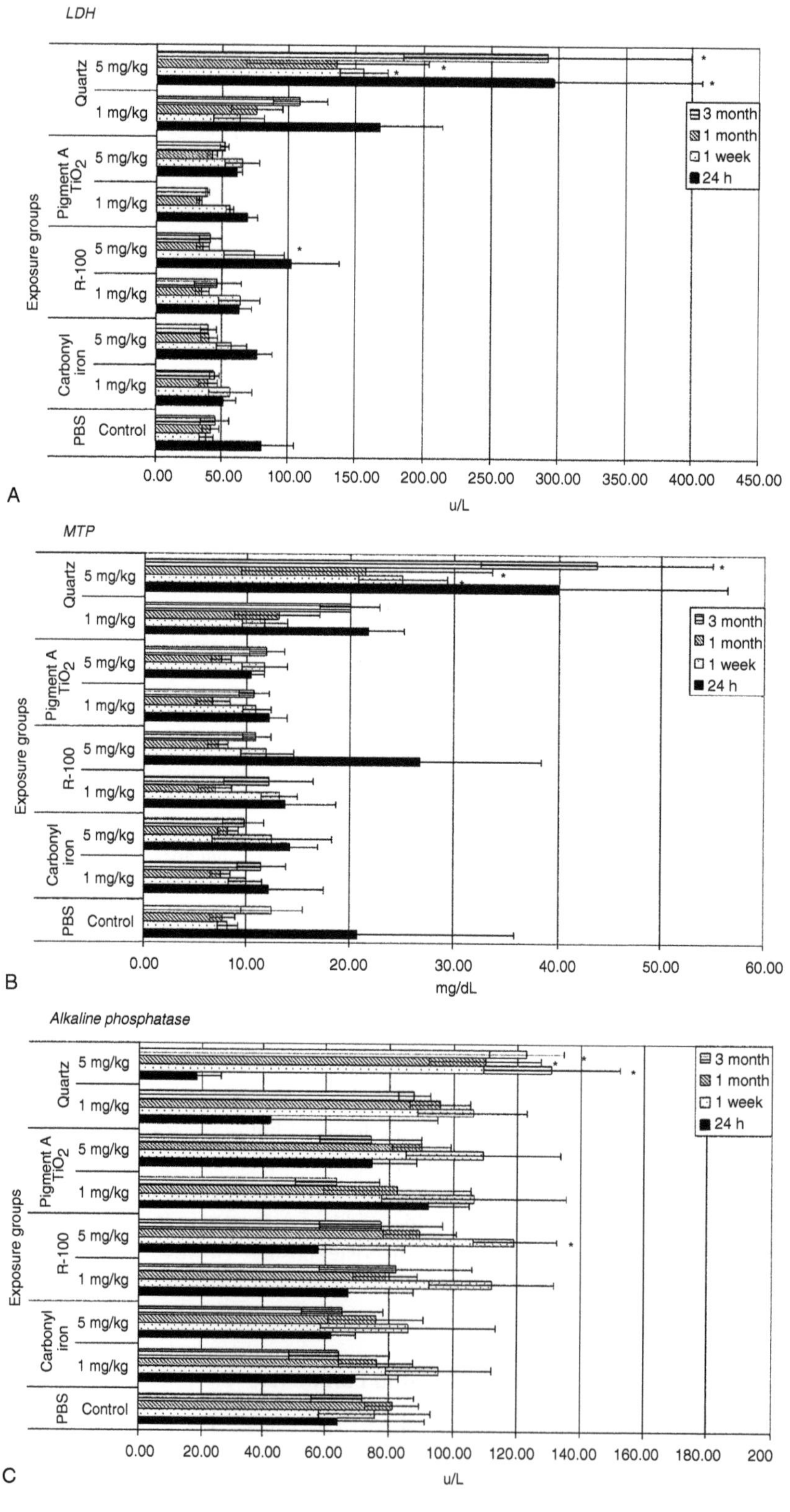

FIGURE 18.5 (See facing page.)

Similarly, transient increases in BAL fluid microprotein (MTP) values were measured in the lung fluids of high-dose (5 mg/kg) R-100 fine-TiO_2-exposed rats at 24 h postexposure, but were not different from control values at 1 week postexposure. In contrast, exposures to 5 mg/kg α-quartz particles produced persistent increases in BAL fluid microprotein values at 24 h, 1 week, 1 month, and 3 months postexposure (Figure 18.5B). Transient increases in BAL fluid ALP values were measured only in the lungs of R-100 fine-TiO_2-exposed rats at 1 week postexposure (5 mg/kg), but substantial increases in BAL fluid ALP values were measured at 1 week through 3 months postexposure in rats exposed to 5 mg/kg α-quartz particles (Figure 18.5C).

To summarize the results from BAL fluid biomarker studies, intratracheal instillation exposures to α-quartz particles resulted in sustained, dose-dependent, lung inflammatory responses, associated with cytotoxic and lung permeability effects, measured from 24 h through 3 months postexposure. Exposures to high-dose R-100 fine-TiO_2 particles (5 mg/kg) produced small but transient pulmonary inflammatory responses, and these effects were not sustained. Exposures to carbonyl iron particles or to Pigment A fine-TiO_2 particles produced a brief neutrophilic response at 24 h postexposure; however, this was in large part related to the bolus dose associated with the intratracheal instillation exposure methodology.

18.3.3.3 Pulmonary Histopathological Evaluations

Histopathological analyses of lung tissues revealed that pulmonary exposures to carbonyl iron, to R-100 fine-TiO_2 particles, or to Pigment A fine-TiO_2 particles in rats produced no significant adverse effects when compared to PBS-exposed controls, as evidenced by the normal lung architecture observed in the exposed animals at postinstillation exposure time periods ranging from 24 h to 3 months (Figure 18.6A and B). Histopathological analyses of lung tissues at several time points postexposure demonstrated no differences between the R-100 fine-TiO_2-exposed rats and those exposed to Pigment A fine-TiO_2 particles (Figure 18.6B). A light micrograph of a lung tissue section of a rat instilled with 5 mg/kg carbonyl iron particles at 1 month postexposure demonstrated appropriate alveolar macrophage phagocytic responses and normal lung architecture. Lung tissue sections from rats instilled with 5 mg/kg R-100 fine-TiO_2 particles appeared very similar histologically to the lung tissue sections recovered from the Pigment A fine-TiO_2 particle-exposed rats at each postexposure time period, and demonstrated normal pulmonary architecture.

Histopathological analyses of lung tissues in rats exposed to α-quartz particulates in rats revealed that pulmonary exposures produced dose-dependent pulmonary inflammatory responses characterized by neutrophils and accumulations of foamy (lipid-containing) alveolar macrophage. In addition, lung tissue thickening as a prerequisite to the development of fibrosis was observed and progressive over postexposure time periods (Figure 18.6C and Figure 18.6D).

18.4 DISCUSSION

This chapter presents a case study and methodology designed to investigate the hazard potential of changing the formulation (i.e., surface coating) of a TiO_2 particle-type. More importantly, the study assessed how the activity of this TiO_2 formulation, identified as Pigment A fine-TiO_2 particles,

FIGURE 18.5 BAL fluid analyses for particulate-exposed rats and corresponding controls at 24 h, 1 week, 1 month, and 3 months postexposure (pe). (A) LDH (lactate dehydrogenase), (B) MTP (microprotein values), and (C) alkaline phosphatase. Values given are means ± S.D. Transient and reversible increases in BAL fluid values were measured in the lungs of high-dose (5 mg/kg) R-100 fine-TiO_2-exposed rats at 1 week postexposure. In contrast, exposures to 5 mg/kg α-quartz particles produced sustained increases in BAL fluid LDH values through the 3-month postexposure period. $^*p < 0.05$.

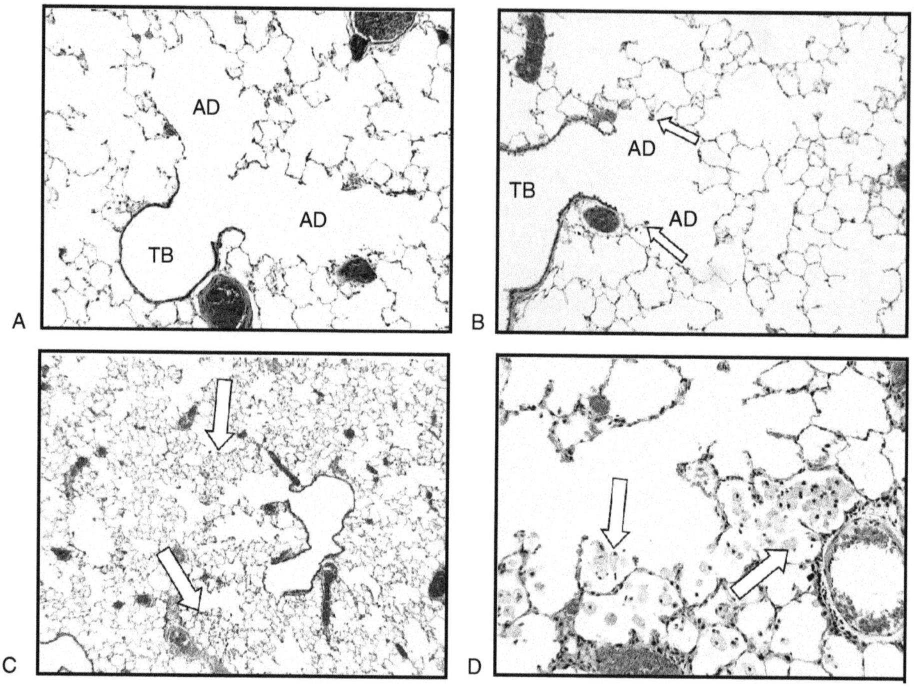

FIGURE 18.6 Light micrographs of lung tissue from a rat exposed to (A) carbonyl iron particles (5 mg/kg) at 1 month postexposure and (B) Pigment A fine-TiO_2 particles (5 mg/kg) at 3 months postinstillation exposure (magnification = 100×). Light micrographs of lung tissue from a rat exposed to α-quartz particles (5 mg/kg) at 3 months postinstillation exposure using (C) low magnification (40×) and (D) high magnification (200×). For (A) and (B), these micrographs illustrate the terminal bronchial (TB) and corresponding alveolar ducts (AD) and demonstrate the lung architecture and normal macrophage phagocytosis of each particle (arrows). For (C) and (D), the arrows demonstrate accumulation of foamy alveolar macrophages in the alveolar regions of quartz-exposed rats.

compared with other negative control, reference particle-types such as R-100 fine-TiO_2 particles or carbonyl iron particles. The combination of BAL and lung tissue studies concomitant with an experimental design consisting of dose-response, time-course evaluations, and the inclusion of reference particle-types provides a powerful tool for assessing the acute pulmonary toxicity of this new particle-type.

The bioassay described herein provides evidence that toxicological information on a particle's various surface treatments can be assessed in a routine systematic manner. A beneficial feature of the bioassay is the ability to compare, via bridging strategies, the effects of inhaled versus instilled particulate materials. In this regard, pulmonary bridging studies can generate important preliminary hazard data when assessing the safety of new developmental or commercial compounds or when making modifications to existing chemical products, such as surface coatings on particulates. The strength of the bridging strategy is dependent upon having good inhalation toxicity data for comparisons to the instillation data. The materials for which there is inhalation data can then be used as control particle-types for comparing with an intratracheal instillation bridging study (Figure 18.1). The basic idea for the bridging concept is that the effects of the instilled material serve as a control (known) material and then are "bridged" to the inhalation toxicity data for that material, as well as to the new materials being tested. The results of bridging studies in rats are then useful as preliminary pulmonary toxicity screening (i.e., hazard) data, because consistency in the response of the inhaled and instilled control material serves to validate the responses with the newly tested particle-type. It should be noted, however, that these pulmonary screening types of studies are not to be used as substitutes for longer-term, more substantive inhalation toxicity studies, such as 90-day inhalation studies or 2-year inhalation bioassay studies.

Based upon the data developed from this pulmonary bioassay, it was concluded that Pigment A fine-sized TiO_2 particles coated with amorphous silica do not cause significant pulmonary toxicity and the pulmonary effects measured were not significantly different from the effects produced by R-100 fine-TiO_2 particles.

In a recent pulmonary bioassay study in rats, we evaluated the pulmonary toxicity of inhaled as well as intratracheally instilled TiO_2 particle-types with various surface coatings, ranging from 0 to 6% alumina (Al_2O_3) and/or 0 to 11% amorphous silica (SiO_2). The effects from exposures to the coatings on TiO_2 particles were compared to animals exposed to base TiO_2 particles as well as air exposed (inhalation study) or PBS-instilled (instillation study) controls. In the inhalation toxicity study, groups of rats were exposed to aerosols of TiO_2 particle-types for 4 weeks at high-dose concentrations ranging from 1130 to 1300 mg/m^3 and pulmonary tissues were assessed by histopathological techniques at several postexposure time periods, including immediately after, as well as 2 weeks, 3, 6, and 12 months postexposure. In a second study, the lungs of rats were intratracheally instilled with nearly identical TiO_2 particle-types (relative to the inhalation study) at doses of 2 and 10 mg/kg body weight; and the lungs of saline-instilled and TiO_2-exposed rats were assessed both using pulmonary lavage biomarkers and by histopathology/cell proliferation evaluations of lung tissues at 24 h, 1 week, 1 month, and 3 months postexposure. The findings from the two studies were compared and they demonstrated that for both studies (inhalation and instillation), only the TiO_2 particles with the largest compositions of alumina and amorphous silica coatings produced pulmonary fibrosis in the inhalation study, and the most sustained and intense lung inflammatory and cytotoxicity response in the instillation study. The results indicate that surface coatings may influence the pulmonary toxicity of inhaled particle-types. Perhaps more importantly, the results demonstrated that the intratracheal instillation studies served as an accurate screening tool for the findings in the inhalation study [2].

In an earlier study, the acute pulmonary toxicity potential of fine-titanium dioxide particles made hydrophobic by surface application of octyltriethoxysilane (OTES) was assessed [5]. In similar-type pulmonary bioassay studies, at higher doses (2 and 10 mg/kg), the toxicity of OTES-coated TiO_2 particles was not significantly different from the hydrophilic R-100 fine-TiO_2 particles. R-100 fine-TiO_2 has a mean particle diameter of 300 nm and a surface area of $\sim$6 m^2/g, while the OTES-treated TiO_2 particles have a primary particle size of 230 nm and a surface area of $\sim$8 m^2/g.

Surface coatings on titanium dioxide particles have been a subject of interest over the past few years. The potential health concern for the effects of hydrophobic coatings was initially raised by Pott and colleagues [6,7]. These investigators recently reported that intratracheal instillation exposure to ultrafine hydrophobic-coated titanium dioxide particles (T-805 sample) resulted in unexpected mortality of exposed rats [7,8]. In their study, two rats treated with 6 mg hydrophobic ultrafine-TiO_2 particles (T-805) demonstrated immediate symptoms of respiratory damage when compared to animals treated with other dusts, and the rats survived less than one-half hour; 3 mg also induced a fatal effect; and 1 mg seemed to be tolerated with some limitations. Subsequent studies were carried out with weekly instillations of 0.5 mg hydrophobic TiO_2 (T-805), and still produced some mortality. Moreover, nearly all of the rats exposed to equivalent dosages of hydrophilic ultrafine-titanium dioxide particles (P-25 sample) survived the intratracheal instillation exposures. A confounding factor of the Pott study which had not been addressed was the potential toxicity of 1% Tween, which was selectively added as a detergent to the T-805 sample but not to the P-25 sample. Thus, it was conceivable that the detergent significantly contributed to the toxic effects observed in the T-805-exposed rats.

The results reported by Pott and colleagues have not been confirmed by a variety of other investigators who have compared the pulmonary hazard effects of hydrophilic and hydrophobic-coated titanium dioxide particles. As discussed above, we have previously evaluated in rats the pulmonary toxicity of instilled hydrophilic versus hydrophobic fine-TiO_2 particles, using a pulmonary bioassay methodology. The results demonstrated that only the high-dose (10 mg/kg) hydrophilic fine-TiO_2 particles and those with particle-types containing a surfactant, Tween 80, produced a transient and reversible pulmonary inflammatory response, which was resolved within 1 week postexposure. In that study, we concluded that the OTES hydrophobic coating on the fine-TiO_2 particle does not cause significant pulmonary toxicity.

It is interesting to note that studies reported by other investigators have demonstrated that the hydrophobic coatings did not produce enhanced toxicity. Hohr and colleagues [8] evaluated acute pulmonary inflammation in rats after intratracheal instillation of surface modified (hydrophilic and hydrophobic) fine-TiO_2 (180 nm) or ultrafine-TiO_2 (20–30 nm) particles at equivalent mass (1 or 6 mg) or calculated surface area doses (100, 500, 600, and 3000 cm^2). These investigators reported that BAL fluid biomarkers of lung inflammatory responses correlated with the administered surface dose delivered to the lungs. Moreover, the exposures to the hydrophobic-coated TiO_2 particles produced reduced pulmonary inflammation relative to hydrophilic-coated particulates, although these minor effects were not significantly different between the two samples. The authors concluded that the surface area, rather than the hydrophobic surface coating, influences the acute pulmonary inflammatory response produced following intratracheal instillation of either fine or ultrafine-TiO_2 particles.

In another study comparing effects of surface treatments, Oberdörster [9] exposed rats via intratracheal instillation to two different types of aggregated ultrafine-TiO_2 particle samples (particle size of both types reported to be $\sim$20 nm) at doses of 50 or 500 µg. One surface treatment was silane-coated, making the particle surface hydrophobic, while the other particle sample was uncoated and hydrophilic in nature. Exposures to hydrophobic-coated, ultrafine-TiO_2 particles produced a reduced pulmonary inflammatory response at 24 h postexposure when compared to identical doses of the uncoated, hydrophilic TiO_2 particles. Oberdörster reported that his findings appear to conflict with an earlier report by Pott and coworkers [6,7].

The results of numerous studies demonstrate that inhalation exposure to hydrophilic, fine-TiO_2 particles or carbonyl iron particles in rats produces low pulmonary toxicity and induces adverse inflammatory effects only at substantial particle overload concentrations [4,11–15]. In the study reported herein, the highest dose of instilled hydrophilic, R-100 fine-TiO_2 particles coated with amorphous SiO_2 (5 mg/kg) or carbonyl iron particles produced only a minor, transient lung inflammatory response measured at 24 hours postexposure. Accordingly, the inhalation toxicity data and the instillation toxicity results for hydrophilic, R-100 fine-TiO_2 particles and carbonyl iron particles in rats are consistent, i.e., clearly demonstrating that pulmonary exposures produce few adverse lung effects. These lung toxicity results evaluated after TiO_2 particulate exposures in rats clearly contrast with the pulmonary effects measured following inhalation [3] or intratracheal instillation exposures in rats [16] to crystalline silica (α-quartz) particles, which produces sustained pulmonary inflammatory responses leading to the development of lung fibrosis in both inhalation and instillation models.

In summary, this chapter has demonstrated the utility of a screening methodology for assessing the acute toxicity of particle exposures in the lungs of rats. We have utilized a multidisciplinary approach by conducting dose response, time course studies, and utilizing reference particle-types. In addition, we have combined BAL-based investigations concomitant with lung tissue evaluations. Accordingly, based on the findings in the case study described herein, intratracheally instilled Pigment A TiO_2 particles with amorphous silica surface treatment exhibited no significant differences in pulmonary toxicity responses when compared to R-100 TiO_2 particles or carbonyl iron particles. This was a clear indication of the low toxicity of this particulate-type. In contrast, the positive control, reference particle-type, quartz, produced persistent lung inflammatory responses along with the development of adverse lung tissue effects. Therefore, we conclude that the lung bioassay methodology described in this chapter can be utilized as a model to evaluate the pulmonary toxicity of particle-types with different surface treatments.

ACKNOWLEDGMENTS

This study was supported by DuPont Titanium Technologies.

Dr. Peter Jernakoff supplied the Pigment A TiO_2 particulates. Denise Hoban, Elizabeth Wilkinson, and William L Batton conducted the BAL fluid biomarker assessments. Carolyn Lloyd, Lisa Lewis, and John Barr prepared lung tissue sections and conducted the BrdU cell proliferation staining methods. Don Hildabrandt provided animal resource care. We thank Drs. Peter Jernakoff, Kevin Leary, and Brian Coleman for helpful comments on this chapter.

REFERENCES

1. Driscoll, K. E., Costa, D. L., Hatch, G., Henderson, R., Oberdörster, G., Salem, H., and Schlesinger, R. B., Intratracheal instillation as an exposure technique for the evaluation of respiratory tract toxicity: uses and limitations, *Toxicol. Sci.*, 55, 24–35, 2000.
2. Warheit, D. B., Brock, W. J., Lee, K. P., Webb, T. R., and Reed, K. L., Comparative pulmonary toxicity inhalation and instillation and studies with different TiO_2 particle formulations: impact of surface treatments on particle toxicity, *Toxicol. Sci.*, 88, 514–524, 2005.
3. Warheit, D. B., Carakostas, M. C., Hartsky, M. A., and Hansen, J. F., Development of a short-term inhalation bioassay to assess pulmonary toxicity of inhaled particles: comparisons of pulmonary responses to carbonyl iron and silica, *Toxicol. Appl. Pharmacol.*, 107, 350–368, 1991.
4. Warheit, D. B., Hansen, J. F., Yuen, I. S., Kelly, D. P., Snajdr, S., and Hartsky, M. A., Inhalation of high concentrations of low toxicity dusts in rats results in pulmonary and macrophage clearance impairments, *Toxicol. Appl. Pharmacol.*, 145, 10–22, 1997.
5. Warheit, D. B., Reed, K. L., and Webb, T. R., Development of pulmonary bridging studies. Pulmonary toxicity studies in rats with triethoxyoctylsilane (OTES)-coated, pigment-grade titanium dioxide particles, *Exp. Lung Res.*, 29, 593–606, 2003.
6. Pott, F., Althoff, G. H., Roller, M., Hohr, D., and Friemann, J., High acute toxicity of hydrophobic ultra-fine titanium dioxide in an intratracheal study with several dusts in rats, In *Relationships Between Respiratory Disease and Exposure to Air Pollution*, Mohr, U., Dungworth, D. L., Brain, J. D., Driscoll, K. E., Grafstrom, R. C., and Harris, C. C., eds., ILSI Press, Washington, DC, pp. 270–272, 1998.
7. Pott, F., Roller, M., Althoff, G. H., Hohr D., and Friemann, J., Acute lung toxicity of hydrophobic titanium dioxide in an intratracheal carcinogenicity study with nineteen dusts in rats, In *Health Effects of Particulate Matter in Ambient Air*, Proceedings of an International Conference sponsored by the Air and Waste Management Association and the Czech Medical Association, Vostal, J., ed., pp. 301–307, 1998b.
8. Hohr, D., Steinfartz, Y., Schins, R. P. F., Knaapen, A. M., Martra, G., Fubini, B., and Borm, P. J. A., The surface area rather than the surface coating determines the acute inflammatory response after instillation of fine and ultra-fine TiO_2 in the rat, *Int. J. Hyg. Environ. Health*, 205, 239–244, 2002.
9. Oberdörster, G., Pulmonary effects of inhaled ultra-fine particles, *Int. Arch. Occup. Environ. Health*, 74, 1–8, 2001.
10. Rehn, B., Seiler, F., Rehn, S., Bruch, J., and Maier, M., Investigations on the inflammatory and genotoxic lung effects of two types of titanium dioxide: untreated and surface treated, *Toxicol. Appl. Pharmacol.*, 189, 84–95, 2003.
11. Lee, K. P., Henry Norman, W., Trochimowicz, H. J., and Reinhardt, C. F., Pulmonary response to impaired lung clearance in rats following excessive titanium dioxide dust deposition, *Environ. Res.*, 41, 144167, 1986.

12. Muhle, H., Bellmann, B., Creutzenberg, O., Dasenbrock, C., Ernst, H., Kilpper, R., MacKenzie, J. C. et al., Pulmonary response to toner upon chronic inhalation exposure in rats, *Fundam. Appl. Toxicol.*, 17, 280–299, 1991.
13. Hext, P. M., Current perspectives on particulate induced pulmonary tumors, *Hum. Exp. Toxicol.*, 13, 700–715, 1994.
14. Vu, V. T., Use of hazard and risk information in risk management decisions: solid particles and fibers under EPA's TSCA and EPCRA, *Inhal. Toxicol.*, 8, 181–191, 1996.
15. Bermudez, E., Mangum, J. B., Asgharian, B., Wong, B. A., Reverdy, E. E., Janszen, D. B., Hext, P. M., Warheit, D. B., and Everitt, J. I., Long-term pulmonary responses of three laboratory rodent species to subchronic inhalation of pigmentary titanium dioxide particles, *Toxicol. Sci.*, 70, 86–97, 2002.
16. Zhang, D. D., Hartsky, M. A., and Warheit, D. B., Time course of quartz and TiO_2 particle induced pulmonary inflammation and neutrophil apoptotic responses in rats, *Exp. Lung Res.*, 28, 641–670, 2002.

19 Air Pollution and Human Brain Pathology: A Role for Air Pollutants in the Pathogenesis of Alzheimer's Disease

Lilian Calderón-Garcidueñas
The Center for Structural and Functional Neurosciences,
University of Montana

William Reed
Department of Pediatrics and Center for Environmental Medicine,
University of North Carolina at Chapel Hill

CONTENTS

19.1 INTRODUCTION

Adverse health effects associated with chronic exposures to air pollutants (indoor, outdoor, and occupational settings) are an important issue for millions of people around the world. As the world population becomes older, significant increases in neurodegenerative diseases such as Alzheimer's have been projected over the next decades (Brookmeyer, Gray, and Kawas 1998; Hebert et al. 2003). Alzheimer's disease (AD) is an irreversible, fatal brain disorder that presently affects 4.5 million people in the United States and it is projected that it will affect between 13 and 16 million by

2050 (Brookmeyer, Gray, and Kawas 1998; Hebert et al. 2003). Alzheimer's patients have a major medical, social, and economic impact, thus any factors that could modify these projections need to be pursued and integrated into multidisciplinary studies of AD. The role played by the environment in the pathogenesis of AD is unclear (Brown, Lockwood, and Sonawane 2005). Our findings suggest that exposures to significant levels of particulate matter and photo-oxidants may accelerate the appearance of precursors of Alzheimer's disease in sentinel animals and in humans.

In this chapter, we will review the pathophysiology of AD as it is currently understood and summarize comparative pathology, human neuropathology, and clinical studies of residents of cities with significant chronic concentrations of particulate matter, endotoxins, ozone, and a myriad of other air pollutants. We discuss how air pollutants might promote AD indirectly by causing systemic inflammation or directly by causing brain injury following their entry into the brain via known pathways.

19.2 MOLECULAR BASIS OF ALZHEIMER'S DISEASE PATHOGENESIS

Alzheimer's brains exhibit two pathological hallmarks: (1) the accumulation of β-amyloid peptides (Aβ) in the extracellular space in the form of neuritic plaques and (2) intraneuronal filamentous tangles (neurofibrillary tangles, NFTs) containing hyperphosphorylated tau protein (reviewed in Selkoe 2001). Aβ peptides are 37–43 amino acid proteolytic fragments of β-amyloid precursor protein (APP), an alternatively spliced transmembrane protein expressed by all cells. Normal neurons primarily express the 695 amino acid form of APP. Aβ peptides are generated by the proteolytic cleavage of APP by two proteases, β- and γ-secretase (Selkoe 2004). Neuritic plaques are foci of extracellular Aβ deposition that are associated with axonal and dendritic injury (Selkoe 2001). A large part of the fibrillar Aβ found in plaques is the 42 amino acid-isoform (Aβ42) that is more hydrophobic and prone to aggregation than other Aβ isoforms. Precursor lesions of neuritic plaques are referred to as "diffuse plaques" (Selkoe 2001).

Genetic studies of familial AD (FAD), an inherited, early onset form of AD, suggest that the generation of Aβ42 plays a contributory role in AD pathogenesis. Mutations in any of three genes, APP, presenillin-1, and presenillin-2 cause a specific increase in Aβ42 generation that is associated presenillin with FAD (Scheuner et al. 1996). All three genes might be expected to regulate Aβ42 generation, as APP is the precursor of Aβ peptides and the presenilins are essential components of γ-secretase (reviewed in Selkoe and Kopan 2003). An increase in APP gene dose as occurs in Down's syndrome (3 copies of chromosome 21 rather than 2) is associated with early onset AD as well (Lemere et al. 1996). Finally, a major risk factor for the development of AD is the epsilon 4 polymorphism of the apolipoprotein E gene (Corder et al. 1993), whose protein product enhances Aβ42 stability and accumulation (Strittmatter et al. 1993; reviewed in George-Hyslop 2000). These findings support the amyloid cascade hypothesis of AD pathogenesis.

The amyloid cascade hypothesis postulates that increased production of the Aβ42 results in its accumulation and oligomerization in limbic and association cortices leading to the gradual deposition of Aβ42 oligomers as diffuse plaques and to subtle effects of the oligomers on synaptic efficacy. These early changes are believed to cause microglial and astrocytic activation, proliferative responses in microglia (gliosis), altered neuronal ionic homeostasis, and widespread synaptic dysfunction and neurodegeneration (Selkoe 2001; Cleary et al. 2005). In this hypothesis, an increase in the generation of Aβ42 is the essential and self-propagating pathogenic event in AD.

An alternative pathogenic mechanism proposes that AD is a consequence of a failure of the control of neuronal differentiation (reviewed in Arendt 2003; Nagy 2005; and Webber et al. 2005). In general, brain neurons rest in a postmitotic, differentiated state. However, adaptive behaviors such as learning and memory require that synapses are continually lost or formed and strengthened or weakened in a remodeling process that relies upon cellular repair machinery. It is hypothesized that normal synaptic turnover together with genetic factors that cause instability in the repair machinery or external factors that result in neuronal injury (e.g., oxidative stress and inflammation) or both could result in excessive demand upon the repair machinery and increase the risk of dedifferentiation with subsequent cell cycle activation (Arendt 2003).

Alternatively, cell cycle activation in neurons could be mediated by inappropriate mitogenic stimuli caused by altered expression or mutation of molecular components of signal transduction pathways that regulate the cell cycle. Although the function of APP is still poorly understood, limited evidence suggests that APP may regulate neuronal survival (Koo and Kopan 2004) by signaling to the nucleus, a process that probably involves its proteolytic cleavage by β and γ secretases. Thus, overexpression of APP and mutations in APP and presenilins might disturb neuronal cell cycle arrest initiating AD pathogenesis.

Reentry into the cell cycle is believed to be a natural consequence of aging, however, in normal aging, brain neurons arrest in G1 phase. In contrast, AD neurons appear to undergo DNA replication (S phase) (Yang, Geldmacher, and Herrup 2001) and become trapped G2 as suggested by the aberrant expression of G2-specific cell cycle regulating proteins in AD brain (Nagy 2005). Although the mechanism of arrest in G2 is unclear, it presumably occurs because neurons are incapable of completing the cell cycle by undergoing mitosis. In this alternate hypothesis, the apparent failure of the G1/S transition checkpoint is believed to be the essential pathogenic event in AD.

G2 phase is characterized by the activation of specific cyclin-dependent kinases (cdk) that ensure the progression through G2 to mitosis. The activation of G2-specific cdks is associated with a gradual destabilization of the microtubule (MT) cytoskeleton. In neurons, MTs are essential for axonal transport. Thus, cell cycle activation is a plausible cause of the axonopathy and transport deficits that are observed early in AD pathogenesis. Moreover, transport deficits inhibit anterograde and retrograde transport of APP, causing the buildup of APP in the neuronal cell body and an increase in Aβ42 generation (Stokin et al. 2005).

Neurons in AD brain upregulate the expression of protein inhibitors of the cell cycle, such as glycogen synthase kinase-3 (reviewed in Bhat, Budd Haeberlein, and Avila 2004) and cyclin-dependent kinase inhibitors (Arendt et al. 1996; reviewed in Nagy 2005). It is suggested that this phenomenon is caused by the cellular stress incurred by the inability of neurons in G2 to undergo mitosis and is presumably intended to inhibit progression to mitosis. However, there are likely side effects of these changes in gene expression. For example, glycogen synthase kinase-3 could phosphorylate the soluble tau protein, which has been released from depolymerizing MTs, thus promoting development of NFTs. Indeed, a number of stress-activated kinases are capable of phosphorylating tau protein (Lovestone and Reynolds 1997) and may also promote the development NFTs.

In the alternative hypothesis, the pathogenic event is irreversible, because the neuron is left with no mechanism for progressing to mitosis or returning to G0 phase. Rather it progressively degenerates as axonopathy and transport deficits worsen and Aβ42 and hyperphosphorylation of tau protein build up eventually forming neuritic plaques and NFTs.

The alternative hypothesis of AD pathogenesis introduces a mechanism by which external events, such as exposure to air pollutants, could affect AD pathogenesis. Perturbations of the neuronal microenvironment, such as toxicant-induced oxidative stress or inflammation, could cause enough damage to elicit reentry into the cell cycle. In other words, damage to a neuronal network that retains synaptic remodeling ability could present a mitogenic stimulus and trigger the development of AD.

Alternatively, both toxicant-induced brain oxidative stress and inflammation could accelerate the consequences of cell cycle reentry by neurons. The "two hit hypothesis" of AD pathogenesis postulates that both cell cycle reentry and oxidative stress are necessary for the development of Alzheimer's disease (Zhu et al. 2004). This suggestion seems reasonable given that neurons trapped in G2 are probably more vulnerable insults. Thus external factors that cause oxidative stress or inflammation could significantly accelerate AD pathogenesis.

19.3 ALZHEIMER'S DISEASE PATHOGENESIS AND COX2

Vane (1971) showed that the anti-inflammatory action of nonsteroidal anti-inflammatory drugs (NSAIDs) depends on their ability to inhibit cyclooxygenases, which in turn results in a diminished synthesis of prostaglandins. In the early 1990s, cyclooxygenases were shown to exist as at least two distinct isoforms: COX1 and COX2. COX2 has emerged as the isoform that is primarily responsible for the synthesis of the prostanoids involved in acute and chronic inflammatory states (Hinz and Brune 2002).

Under basal conditions, COX2 has limited constitutive neuronal distribution in the CNS and contributes to synaptic activity and memory consolidation (Breder, Dewitt, and Kraig 1995; Breder and Saper 1996; Minghetti 2004). COX2 expression is induced by various proinflammatory stimuli, such as cytokines, growth factors, and tumor promoters (Hinz and Brune 2002; Minghetti 2004).

Age is the most important risk factor for development of AD. The number of people with the disease doubles every 5 years beyond age 65. Oxidative stress is one of the crucial factors participating in the aging process. Cyclooxygenase-derived reactive oxygen species (ROS) generation increases with age, and cyclooxygenase-mediated prostanoid synthesis is one of the major sources of ROS in the aging process (Kim et al. 2000). Transgenic mice overexpressing human COX2 in hippocampal neurons develop neuronal apoptosis and cognitive deficits in an age-dependent manner, and are more susceptible to β-amyloid toxicity with potentiation of redox impairment (Ho et al. 1999; Bazan 2001; Bazan and Lukiw 2002). COX2 expression in hippocampal neurons may be a predictor of early AD (Ho et al. 2001) and chronic increased COX2 production in brain may have a number of consequences, including free radical mediated cellular damage, vascular dysfunction, alterations in cellular metabolism and neuronal cell cycle, and increases in total Aβ content (Naslund et al. 2000; Strauss et al. 2000; Ho et al. 2001; Xiang et al. 2002a; Scali et al. 2003). COX2 influences processing of APP and promotes amyloid plaque deposition in a mouse model of AD (Xiang et al. 2002b). That COX2 plays a role in AD pathogenesis is also supported by epidemiological studies showing an association between long-term use of NSAIDs and a reduced risk of developing AD (Aisen 2002), by the protective effects of COX2 inhibitors in models of AD (Giovannini et al. 2003), and by gene expression profiling showing significantly upregulated stress induced proteins, including COX2, in human brain (Lukiw 2004).

19.4 THE MEXICO CITY ENVIRONMENT

Mexico City represents an extreme of urban growth and environmental pollution (Chow et al. 2004). It is a megacity that covers an area of 2000 km^2 surrounded by a series of volcanic and discontinuous mountain ranges that limit the natural ventilation of the basin. The basin has more than 30,000 industrial facilities and 3.5 million vehicles, with an estimated annual emission of 2.6 million tons of particulate and gaseous air pollutants. The critical air pollutants are ozone (O_3) and particulate matter (PM). Figure 19.1 illustrates 8 h O_3 concentrations observed during January 2005 at three monitoring stations. The higher concentrations of ozone are registered in the SW, downwind from the areas where the ozone precursors are produced. Figure 19.2 shows the annual concentrations of $PM_{2.5}$ micrometers or less in aerodynamic diameter ($PM_{2.5}$) observed in NE, SW and downtown Mexico City.

Pollutant levels in Mexico City vary within a relatively narrow range throughout the year, so its residents are exposed all year long to significant burden air pollutants. The pollution levels have been sustained or worsened in the last 20 years (Bravo and Torres 2002), so exposures of current children and teenagers are truly lifelong, having begun in utero. Moreover, there is a relatively low mobility of Mexico City residents, so individuals tend to be exposed to the same environment for a prolonged period. Thus Mexico City presents an opportunity to study chronic health effects associated with prolonged year-round exposures to severe air pollution.

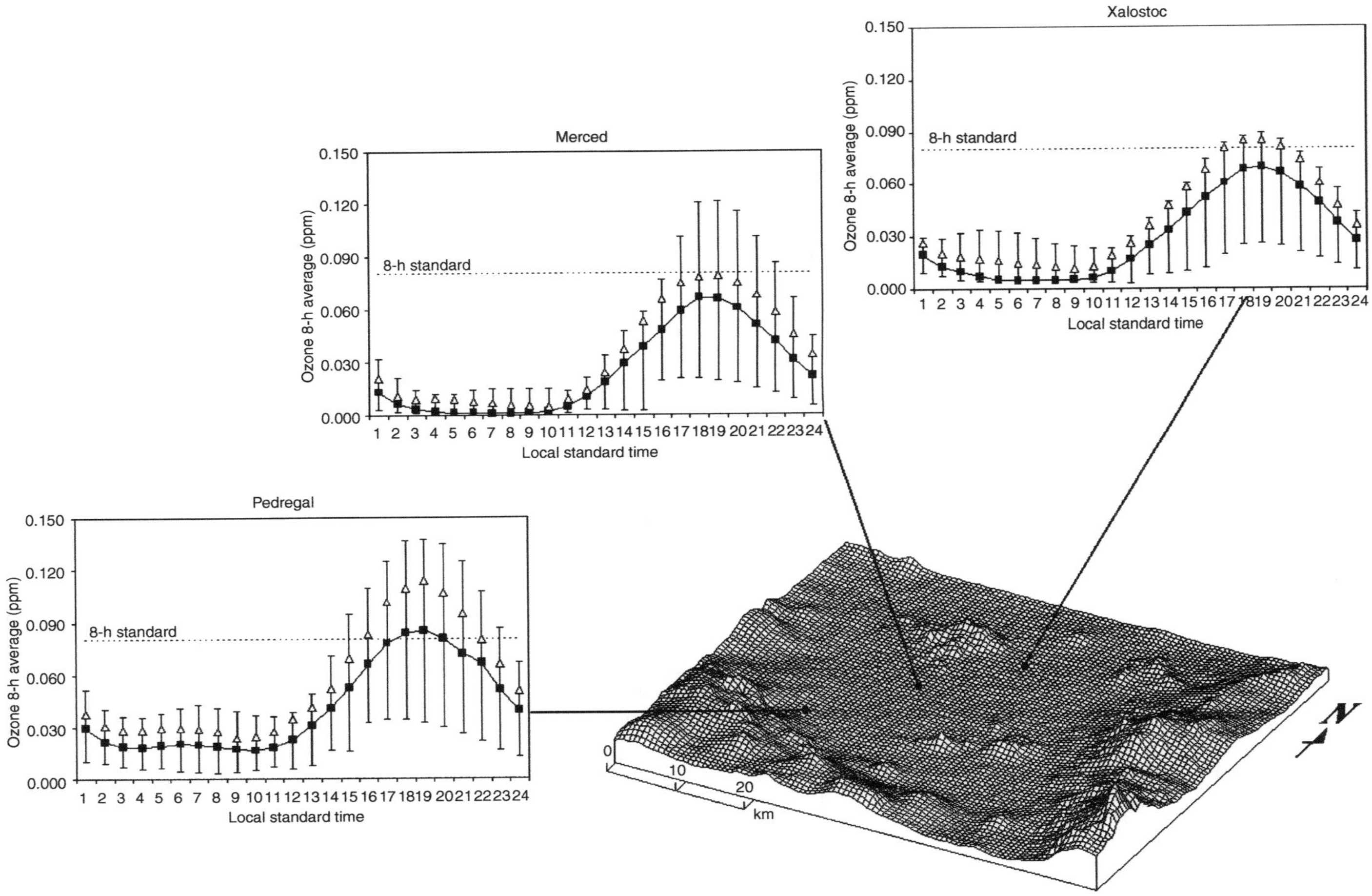

FIGURE 19.1 Spatial and temporal profile of O_3 air pollution in metropolitan Mexico City. The P_{50} (Δ), and the monthly diurnal average of 8 h O_3 concentrations (□) observed during January 2005 are illustrated at three representative monitoring stations: Xalostoc, located in a northeast industrial area; Merced, located downtown; and Pedregal, a residential area located in the southwest. The higher concentrations of ozone are registered in Pedregal, downwind of the areas where the ozone precursors are produced.

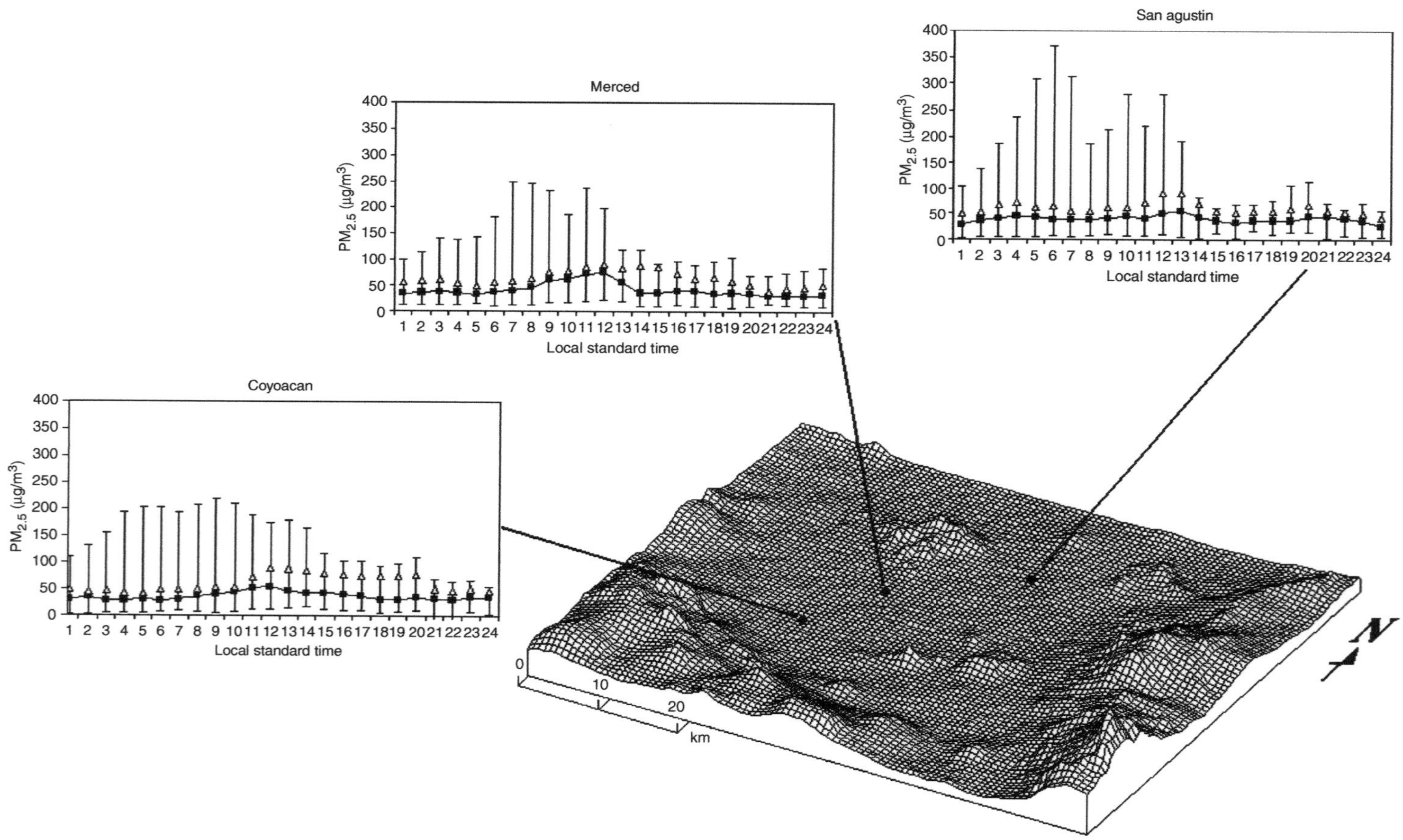

FIGURE 19.2 Spatial and temporal profile of $PM_{2.5}$ air pollution in metropolitan Mexico City. Typically, the highest $PM_{2.5}$ concentrations are observed in San Agustin in the northeast and Merced downtown, with the lowest levels in Pedregal in the southwest. $PM_{2.5}$ annual concentrations are above the annual standard for all three stations.

The potentially toxic components of PM air pollution include acids, polyaromatic hydrocarbons, metals, biological products such as lipopolysaccharide (LPS), and inorganic carbon particles. It is suspected that O_3, polyaromatic hydrocarbons, transition metals, and inorganic carbon particles have a common mechanism of toxicity—the depletion of cellular anti-oxidant defenses (cellular oxidative stress) through the generation of ROS (i.e., hydrogen peroxide, superoxide, hydroxyl radical, and singlet oxygen). However, the mechanism by which ROS are elicited varies significantly with the pollutant. The pollutant-induced oxidative stress probably activates stress-response signal transduction pathways by the activation or inactivation of oxidant-sensitive components of these pathways (e.g., transcription factors and phosphatases). Some pollutants, such as the transition metal vanadium (a potent phosphatase inhibitor itself), may activate these pathways by ROS-independent mechanisms as well. The stress-response pathways regulate the expression of genes encoding antioxidant synthesis and reducing enzymes, detoxification enzymes, and they also regulate immune, inflammatory, cell survival, and apoptotic responses. LPS activates many of the same pathways by acting through a specific receptor complex composed of the LPS binding protein CD14 and a heterodimeric plasma membrane protein complex composed of Toll-like receptor-4 (TLR4) and MD2. Chronic activation of the stress response pathways culminates in a chronic inflammatory response in targeted tissues.

We have focused on toxicity due to metals and LPS. The most abundant metals in Mexico City PM are: Ca, Fe, K, Zn, and Pb, while metals typically present in motor vehicle exhaust and fuel oil combustion products, Cr, Ni, V and S, are present in lower concentrations (Chow et al. 2002). LPS content in PM samples show a range of 15.3 to 20.6 ng/mg of PM and SE samples show the highest endotoxin concentrations 59 EU/mg PM (Alfaro-Moreno et al. 2002; Osornio-Vargas et al. 2003).

19.5 COX2 AND IL-1β EXPRESSION, Aβ42 ACCUMULATION, AND NEUROPATHOLOGY IN THE BRAINS OF DOGS AND HUMANS EXPOSED TO SEVERE URBAN AIR POLLUTION

Healthy Mexico City dogs experience chronic upper and lower respiratory tract inflammation and breakdown of both the respiratory and olfactory epithelial barriers. The brains of these dogs exhibited endothelial and astrocytic upregulation of cycloxygenase-2 (COX2) expression in olfactory bulb, frontal cortex, and hippocampus, three critical targets in Alzheimer's disease (Calderón-Garcidueñas et al. 2003a). There was also activation of neuronal NF-κB, increased induced nitric oxide synthase (iNOS) expression in cortical endothelial cells as early as 4 weeks of age, and breakdown of the blood brain barrier (BBB) (Calderón-Garcidueñas et al. 2002). iNOS derived nitric oxide (NO) contributes to the generation of peroxynitrite and BBB breakdown (Winkler 2001), leading to vasogenic edema and secondary brain damage (Thiel and Audus 2001; Chao et al. 1992). Thus the increase in iNOS expression may be related to the breakdown of the BBB. Furthermore, NO may also contribute to the maintenance, self-perpetuation, and progression of neurodegenerative processes (Grammas et al. 1997).

Olfactory bulb and hippocampal apurinic/apyrimidinic sites in genomic DNA were significantly higher in exposed Mexico City dogs versus controls (Calderón-Garcidueñas et al. 2003a), suggestive of increased oxidative stress. There was significant DNA damage in the olfactory bulb of young Mexico City dogs several months before cortical Aβ42 diffuse plaques were detected (Calderón-Garcidueñas et al. 2003a). In Mexico City dogs olfactory mucosa pathology appeared by 4 months of age and thus appears to be a precursor of the olfactory bulb pathology (Calderón-Garcidueñas et al. 2003a).

Even young dogs (<1 year) showed accumulation of Aβ42 in neurons, glial cells and blood vessels, and the presence of Aβ42-positive diffuse plaques. Dogs from Mexico City exhibited white matter perivascular gliosis as early as 3 mo of age and astrogliosis increased significantly with age (Figure 19.3).

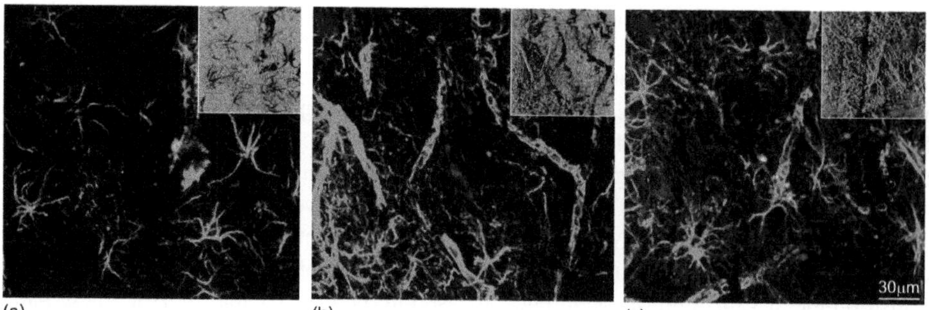

(a) (b) (c)

FIGURE 19.3 **(See color insert)** Reactive gliosis and astrocytic proliferation in the frontal cortex white matter of healthy Mexico City dogs. Glial cells and proliferating astrocytes were localized in paraffin sections of frontal cortex of Mexico City dogs by immunohistochemistry using fluorescein-labeled anti-glial fibrillary acidic protein (GFAP, green) and phycoerythrin-labeled anti-bromodeoxyuridine (BrdU, red), respectively, and examined by confocal microscopy: (a) 3-year-old male, (b) 5-year-old female, and (c) 14-year-old female. Gliosis worsens with age. Images represent maximum intensity projections, showing the maximum intensity of all layers along the viewing direction. The inserts represent 3D reconstructions from the same data sets. (Pictures were taken by Dr. Barbara Rothen-Rutishauser Ph.D. Institute of Anatomy, University of Bern, Bern, Switzerland.)

These observations were consistent with an acceleration of an Alzheimer's-like pathology in apparently healthy dogs.

The foregoing studies using dogs as sentinels suggested that chronic exposure to severe air pollution might have adverse effects on human brain. To examine this possibility, we conducted a study using autopsy brain samples from Mexican subjects, all lifelong residents of two large cities with severe air pollution, Mexico City and Monterrey, and five small cities with low levels of air pollution. Evidence of chronic respiratory tract inflammation was present in all residents of cities with severe air pollution.

COX2 mRNA abundance was measured by real-time RT-PCR analysis of total RNA isolated from brain tissues. There was a significant elevation of COX2 mRNA levels in frontal cortex and hippocampus of the high exposure group (Figure 19.4a and d), along with an elevation of COX2 immunoreactivity in frontal cortex confirmed by quantitative image analysis of COX2 immunoreactivity (IR) (Figure 19.4b). In subjects from the low exposure group, COX2 IR was confined to neuronal cell bodies, whereas subjects from the high exposure group exhibited COX2 staining in neuronal cell bodies and dendrites, as well as strong COX2 staining of the endothelium in the frontal cortex (Figure 19.4c). COX2 IR was largely confined to neurons in the hippocampus (Figure 19.4f).

There was a strong positive association between COX2 mRNA levels and oxidative DNA damage as measured by apurinic/apyrimidinic (AP) sites ($r = 0.89$, $p = 0.001$) in frontal cortex (Calderón-Garcidueñas et al. 2004). The positive correlation between COX2 mRNA and AP sites could be a consequence of COX2-mediated prostanoid synthesis, a major source of ROS that are capable of damaging DNA. The DNA damage in frontal cortex suggests that oxidative stress could be a relevant and early event. Oxidative damage is an early event in AD, and it is greatest early in the disease and decreases with disease progression (Nunomura et al. 2001; Perry et al. 2002).

In normal brain expression, the 695 amino acid form of APP (APP695) is much greater than the expression of the 751 amino acid form (APP751). AD brain is characterized by a reversal in the relative expression of APP isoforms as APP751 expression increases dramatically. We measured the APP751/APP695 mRNA ratio by real-time RT-PCR in frontal cortex and hippocampus. There

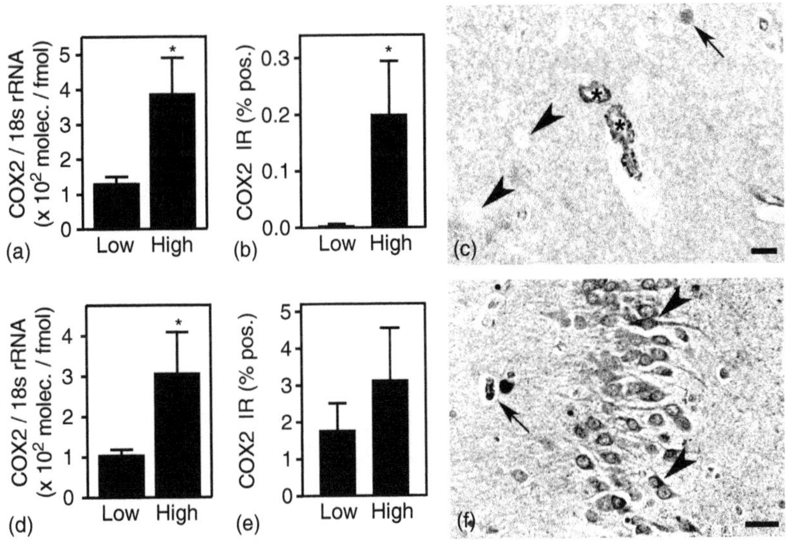

FIGURE 19.4 **(See color insert)** COX2 expression in frontal cortex (a–c) and hippocampus (d–f). COX2 mRNA abundance was measured by RT-PCR and normalized for 18s rRNA levels. COX2 protein was localized in sections of paraffin-embedded tissues by IHC and its abundance was measured by quantitative image analysis. COX2 mRNA was significantly elevated in the high exposure group in both frontal cortex (a, $p = 0.009$), and hippocampus (d, $p = 0.04$). COX2 immunoreactivity (IR) was significantly elevated in the high exposure group in frontal cortex (b, $p = 0.01$), but not in hippocampus (e). Means $\pm$ SEMs are shown in A, B, D, and F. (c) Representative COX IHC in frontal cortex from a subject in the high exposure group showing strong staining of endothelial cells in the capillaries (*), and pyramidal neurons (arrow), while other neurons were negative (arrowheads). Scale = 20 μm. (f) Representative COX IHC in dentate gyrus from a subject in the high exposure group showing COX2 positive neurons (arrowheads) and capillaries (short arrow). Scale = 15 μm. (Calderón-Garcidueñas, L. et al., Brain inflammation and Alzheimer's-like pathology in individuals exposed to severe air pollution, *Toxicol. Pathol.*, 32, 650–658, 2004. With permission.)

was no statistically significant difference between the APP751/APP695 ratios in the high and low exposure groups. However, there was a significant positive correlation between COX2 mRNA in frontal cortex and the APP751/APP695 ratio in the frontal cortex of the high exposure group only (Calderón-Garcidueñas et al. 2004).

Aβ42 accumulation in frontal cortex and hippocampus was measured by quantitative immunohistochemistry. Aβ42 was detected in the perikaryon of pyramidal frontal cortex neurons and in cortical and white matter astrocytes (Figure 19.5a) and subarachnoid and cortical blood vessels (Figure 19.5b) in subjects from the high exposure group. Aβ42 accumulation in the frontal cortex (Figure 19.5c) and hippocampus (Figure 19.5d) of the high exposure group was significantly elevated compared to the low exposure group. Three subjects in the high exposure group had rare diffuse Aβ42 plaque-like staining in the frontal cortex of (32, 38, and 43 years old). The diffuse Aβ42 plaques were associated with reactive astrocytes (e.g., Figure 19.5e) or apoptotic nuclei (not shown). None of the three subjects carried the apolipoprotein E ε4 allele (Calderón-Garcidueñas et al. 2004), a risk factor for the development of Alzheimer's disease.

In a follow-up autopsy study, the olfactory bulb was examined for evidence of COX2 expression and Aβ42 accumulation. In accordance with findings in dogs, there was a significant up-regulation of COX2 and IL-1β mRNA expression and Aβ42 accumulation in the olfactory bulb of highly exposed subjects compared to controls (Figure 19.6). Aβ42 accumulation in olfactory bulb was also seen in arterial smooth muscle cells starting as early as the second decade of life (Figure 19.7).

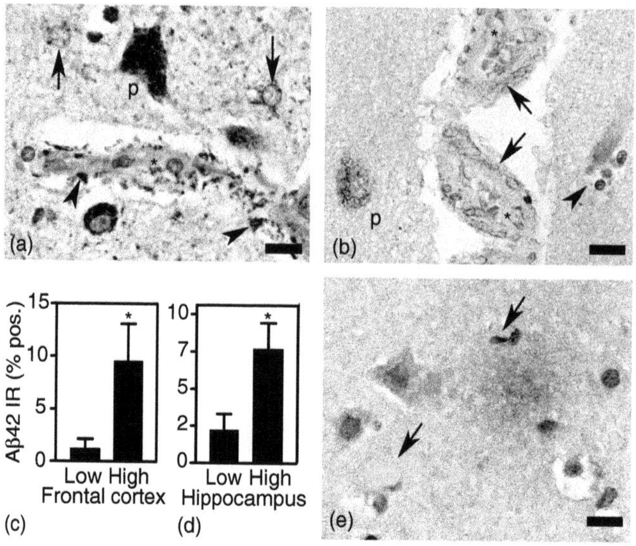

FIGURE 19.5 **(See color insert)** Aβ42 accumulation in frontal cortex and hippocampus. Aβ42 was localized in sections of paraffin-embedded tissues by IHC. (a) Anti-Aβ42 stained pyramidal neurons (p), astrocytes (arrows) and astrocytic processes (arrowheads) around blood vessels (*). (b) In addition to accumulation in pyramidal neurons, (p) Aβ42 was deposited in smooth muscle cells (arrows) in cortical arterioles (*). A dead neuron surrounded by glial cells is indicated (arrowhead). (c and d) Quantitative image analysis of Aβ42 IHC showed a significant increase in Aβ42 immunoreactivity (Aβ42 IR) in both frontal cortex (c, * $p = 0.04$) and hippocampus (d, * $p = 0.001$) in the high exposure group. Means ± SEMs are shown. (e) Aβ42 IHC of frontal cortex from a 38-year-old subject from Mexico City showing diffuse plaque-like staining with surrounding reactive astrocytes (arrows). Scale = 20 μm. (Calderón-Garcidueñas, L. et al., Brain inflammation and Alzheimer's-like pathology in individuals exposed to severe air pollution, *Toxicol. Pathol.*, 32, 650–658, 2004. With permission.)

The major neuropathological findings in exposed humans included: (1) breakdown of the BBB, as indicated by the presence extravascular red blood cells, hemosiderin-laden macrophages, reactive astrocytes, and apoptotic nuclei (Figure 19.8a and Figure 19.8c), (2) age-related progressive reactive gliosis in the supratentorial white matter (GFAP-positive cells) (Figure 19.8b), and (3)

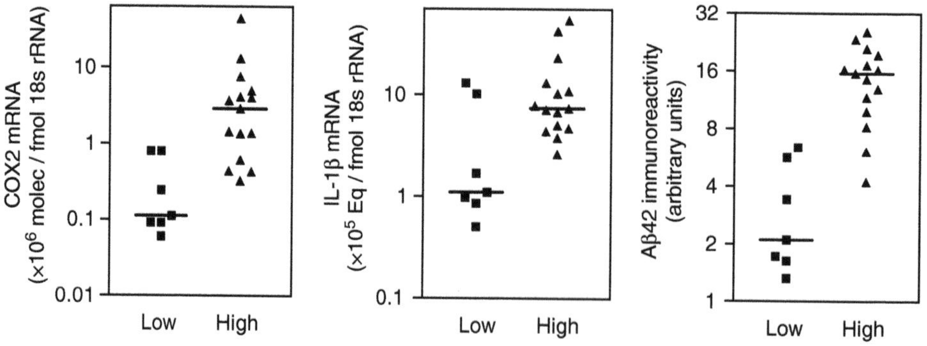

FIGURE 19.6 COX2, IL-1β and Aβ42 expression in olfactory bulb of low versus high exposed subjects. COX2 and IL1β mRNA abundances were measured by RT-PCR and normalized for 18 s rRNA. Aβ42 protein was localized in paraffin-embedded tissues by IHC and its abundance was measured by quantitative image analysis. COX2 and IL-1β mRNA were significantly elevated in the high exposure group ($p = 0.002$ and 0.024, respectively). Highly exposed subjects showed a significant increase in Aβ42 immunoreactivity as well.

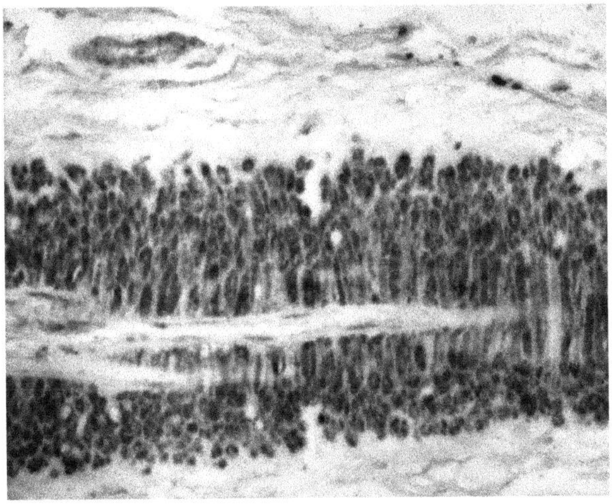

FIGURE 19.7 A light micrograph of a subarachnoid blood vessel in the olfactory bulb of a 14-year-old boy from northeast Mexico City. IHC for Aβ42 shows marked accumulation of the toxic peptide in smooth muscle cells.

senile plaques of the diffuse type in subjects in their 20 s and mature plaques in subjects in their 30 s (not shown).

Our findings in humans were largely consistent with those in dogs and supported the notion that chronic exposure to severe air pollution is associated with significant changes in the brain microenvironment, most notably inflammation and Aβ42 accumulation. These two conditions, known to precede the appearance of neuritic plaques and neurofibrillary tangles (hallmarks of Alzheimer's disease), were seen in humans with no known risk factors for AD.

19.6 CLINICAL STUDIES OF MEXICO CITY CHILDREN

Apparently healthy children living in Southwest Metropolitan Mexico City have abnormal nasal, radiologic, spirometric, and peripheral blood findings, most likely related to their lifelong exposure

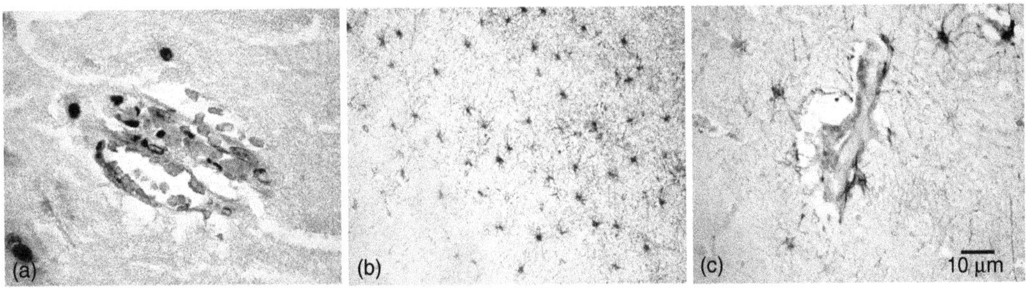

FIGURE 19.8 **(See color insert)** Reactive pathology and breakdown of the BBB in frontal cortex of an 11-year-old Mexico City male. Paraffin embedded frontal cortex was stained with hematoxylin and eosin (a) and anti-GFAP (b and c). (a) There are numerous perivascular macrophages with hemosiderin-like granules and free RBC surrounding a blood vessel. A neuron is seen in the lower left corner. (b) Same 11-year-old child showing reactive gliosis in subcortical frontal white matter (GFAP). (c) A 17-year-old Mexico City male with reactive astrocytes around cortical frontal blood vessels. The vessel is surrounded by macrophages loaded with hemosiderin-like pigment (GFAP). Leaking blood vessels are indirect evidence of a breakdown of the BBB.

to air pollutants. Nasal abnormalities indicative of a breakdown of the nasal respiratory barrier were present in 22% of children. Hyperinflation and interstitial markings seen in chest x-rays were described in 67 and 49% of children, respectively. Ten percent of children exhibited a mild restrictive pattern of lung function by spirometry. Interstitial markings, reflecting conditions whose common denominator is an inflammatory bronchiolar, peribronchiolar, and/or alveolar duct process, were associated with a decrease in predicted values of FEF_{25-75}, FEF_{75}, and the FEV_1/FVC ratio. Thus continuous exposure to air pollutants was associated with evidence of chronic upper and lower respiratory tract injury and inflammation in Mexico City children.

Blood smear findings included neutrophils with toxic granulations, and schistocytes indicative of systemic inflammation and endothelial damage. Mexico City children had more serum IL-10 and IL-6 and less IL-8 and IL-1β than age/gender/socioeconomic matched controls, indicating a shift in the circulating cytokine profile in favor of anti-inflammatory cytokines (Calderón-Garcidueñas et al. 2003b). These children also exhibited significantly higher plasma concentrations of endothelin-1 (ET-1) (Calderon-Garciduenas et al. 2005) and PGE2, and increased expression of the LPS binding protein mCD14 by blood monocytes (Calderon-Garciduenas et al. 2005). These findings established that Mexico City children have systemic disturbances in addition to respiratory tract injury and inflammation.

Cohorts of Mexico City and control children have been evaluated for neuropsychological function as well. Compared to controls, Mexico City children showed significant deficits in areas of attention, concentration, and other higher order cognitive abilities. T_1-weighted magnetic resonance images (MRI) demonstrated mild frontal atrophy and increased volume of the inferior horn of the lateral ventricles suggestive of a decreased volume of the hippocampal formation (Figure 19.9). Hyperintense white matter anterior frontal lesions were also seen in T_2-weighted MRI images in MC children but not in controls (not shown). Autopsies of healthy Mexico City children who died suddenly showed significant white matter blood vessel leakage and perivascular gliosis (Figure 19.8) that could explain the hyperintense frontal T_2-weighted images (insert Figure 19.9 the MR1 picture).

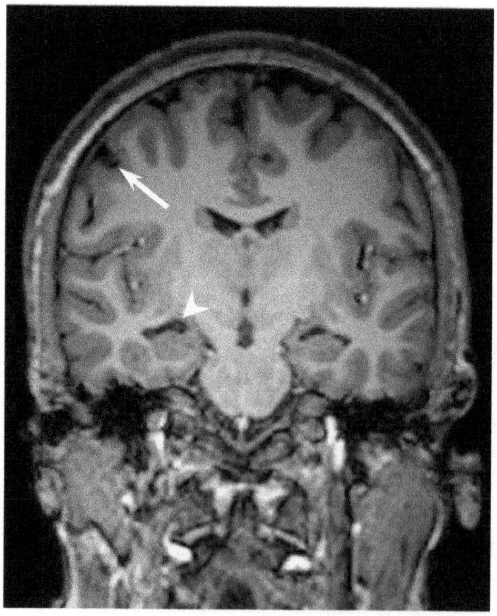

FIGURE 19.9 T_1-weighted coronal brain magnetic resonance image of a 14-year-old girl, a life long resident of Mexico City, showing mild cortical frontal atrophy (arrow) and increased volume of the inferior horn of the lateral ventricles (arrowhead).

19.7 POTENTIAL MECHANISMS OF AIR POLLUTANT-INDUCED INFLAMMATION AND NEURODEGENERATION

19.7.1 Air Pollutant-Induced Systemic Inflammation

Systemic infections and chronic inflammation of peripheral organs that elicit systemic inflammation, however mild, could have significant effects on the brain and/or contribute to the progression of chronic neurodegenerative diseases with a clear-cut intrinsic inflammatory component, such as AD. Chronic exposure to air pollutants that cause chronic respiratory tract inflammation could elicit a systemic inflammatory response. We have clinical evidence of such chronic inflammation in the respiratory tract and systemic involvement in Mexico City children (see Section 19.6). A systemic inflammatory response is characterized by altered levels of circulating cytokines, both pro- and anti-inflammatory (Perry 2004). These mediators of inflammation in turn could elicit inflammation in the central nervous system by: (1) accessing brain regions that lack a BBB, (2) chain activation of brain endothelium, perivascular macrophages, and microglia in brain regions with a BBB, (3) active transport of cytokines across the BBB, and (4) activation of sensory afferents of the vagus and presumably the trigeminal nerve which communicate with the brainstem.

Cells forming the BBB are in the best position to transfer the information from the circulation to the brain parenchyma and molecules produced by the BBB cells bind to their cognate receptors expressed at the surface of neurons responsible for triggering the hypothalamus–pituitary–adrenal (HPA) axis (Rivest 2001). Brain capillaries exhibit both constitutive and induced expression of receptors for TNF-α, IL-1β, and IL-6 (Ericsson et al. 1995; Nadeau and Rivest 1999; reviewed in Rivest 2001). Rivest's proposed cascade of events for systemic/local brain inflammation applies very well to the scenario of chronic inflammation affecting the upper and lower respiratory tracts seen in residents of highly polluted regions. In short, he proposes that circulating inflammatory mediators target cells of the BBB, IL-1β being the early cytokine effector. As a mediator of the innate immune response, IL-1β induces COX2 gene expression in the cerebral endothelium, leading to PGE2 synthesis and the activation of circuits that alter neuronal activity and activate transduction events leading to further inflammation in the brain parenchyma. Because inflammation significantly contributes to AD pathogenesis, the activation of inflammatory cascades in the brain, as a consequence of sustained air pollutant-induced inflammation of the respiratory tract, might initiate or promote AD pathogenesis.

Elmsquist (1997a, 1997b) suggests an alternate pathway by which cytokines could signal the central nervous system (CNS) via nerves such as the vagus. This is an intriguing possibility in air pollution exposed subjects because both LPS and IL-1β are recognized by chemosensory receptors located in vagal paraganglia in the vagus nerve at several levels including cervical, thoracic, and abdominal regions (Elmquist, Scammell, and Saper 1997b). Thus, in air pollution-exposed subjects, the induction of cytokine expression in the CNS could take place via peripheral activation of the vagus nerve.

Finally, there is evidence that increased concentrations of glucocorticoids have adverse effects on the brain, particularly in areas such as the hippocampus. Thus, an alteration of the hypothalamic–pituitary–adrenal (HPA) axis in response to environmental stimuli could have an impact on brain plasticity and neurodegeneration (Rivest 2001; Fassbender et al. 2004).

19.7.2 Transport of PM-Associated Metals to the Brain

It is well documented that metals instilled into the nasal cavity or inhaled ultrafine carbon particles accumulate in olfactory bulb in rodents (Sunderman 2001; Dorman et al. 2002; Oberdörster et al. 2004). Transport of metals and ultra-fine particles into cortical areas in rodents has been shown as well (Henriksson 1997; Dorman 2002; Oberdörster 2004). The initial uptake of and transport of toxicants is mediated by chemosensory neurons of the olfactory nerve that are embedded in the olfactory epithelium in the nose. Their dendrites are in direct contact with the contents of the lumen of the olfactory epithelium, while their axons project into the olfactory bulb.

In addition to the olfactory pathway, the trigeminal pathway may also be a route of entry of air pollutants into the brain. The trigeminal nerve is the most important sensorial nerve in the head and is responsible for the innervation of the nasal and oral cavities, the paranasal sinuses. and the teeth. The uptake and clearance of inhaled manganese by the trigeminal nerve has been shown in rats and mice (Lewis et al. 2005).

Metals might also reach the brain through the systemic circulation. Ultrafine PM can be transported in the systemic circulation via macrophage-like cells (Calderón-Garcidueñas et al. 2001) and the escape of inhaled particles into the blood circulation has been documented in humans (Geys et al. 2005). Nanoparticles (particles < 100 nm in diameter) of anthropogenic and engineered materials are capable of significant transcytosis across epithelial and endothelial barriers (Oberdörster et al. 2004; Geys et al. 2005). Thus, combustion-derived PM carrying metals and aromatic hydrocarbons could elicit inflammatory and oxidative stress responses after reaching the brain endothelium via the systemic circulation.

To determine whether PM components are taken up by the brain with chronic exposure to urban air pollution, we used inductively-coupled plasma mass spectroscopy (ICP-MS) to measure tissue burden of Ni and V, two metal components of Mexico City PM that are derived from fossil fuel combustion. Both Ni and V were detectable in the olfactory epithelium, olfactory bulb, and frontal cortex of Mexico City dogs (Calderón-Garcidueñas et al. 2003a). Ni and V were present in significant amounts in the lung, heart (left ventricle), frontal cortex, and hippocampus of Mexico City residents (e.g., Figure 19.10). In addition, we detected subtle increases in the concentration of V in olfactory bulb and hippocampus in a small sampling of Mexico City residents (Figure 19.11). A study involving a larger subject population is underway to confirm these preliminary observations.

Both V and Ni are capable of causing oxidative stress. V catalyzes the formation of ROS that can directly injure cells and causes indirect damage by inducing the expression of leukocyte chemoattractants that recruit inflammatory cells into tissues (Kawanishi et al. 2002). Ni, on the other hand, suppresses ROS scavenging mediated by glutathione and protein-bound sulfhydryl groups, resulting in increased concentrations of ROS, enhanced lipid peroxidation, DNA damage, and altered calcium homeostasis (Stohs and Bagchi 1995). Both metals activate stress-responses *in vitro*, including those mediated by the transcription factor NF-κB (Goebeler et al. 1995; Chen et al. 1999). Thus Ni and V are likely capable of disturbing the brain microenvironment and initiate or promote AD pathogenesis.

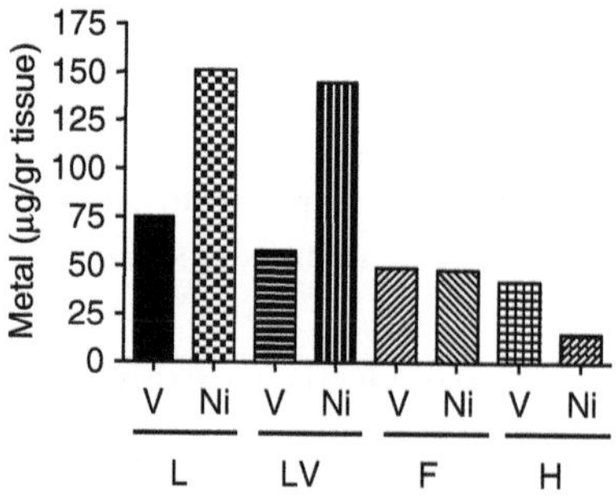

FIGURE 19.10 Vanadium (V) and nickel (Ni) were measured by ICP-MS in lung (L), heart left ventricle (LV), frontal cortex (F), and hippocampus (H) in a 43-year-old male Mexico City resident. There are significant amounts of Ni and V both in the lungs and heart, while vanadium is higher than Ni in the hippocampus.

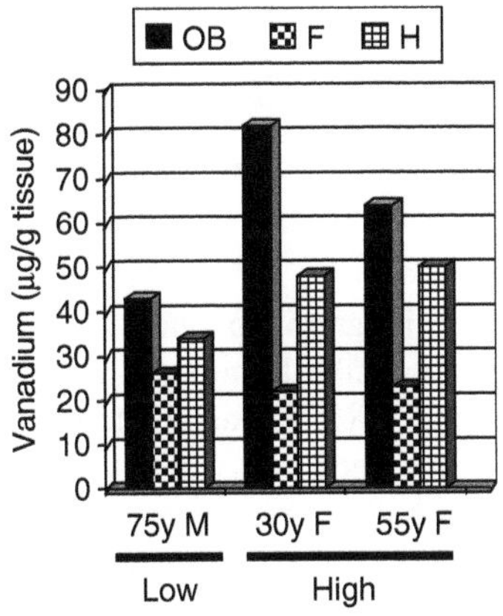

FIGURE 19.11 Metal content in olfactory bulb (OB), frontal cortex (F), and hippocampus (H) of two Mexico City residents ages 30 and 55 (high exposure) versus a 75-year-old resident of a less polluted city (low exposure) was measured by inductively coupled plasma mass spectrometry (ICP-MS). The olfactory bulb and the hippocampi exhibit the highest concentrations of vanadium in the Mexico City residents, while anterior frontal cortex has the lowest.

Other more abundant metal components of Mexico City PM that may be taken up by the brain could exacerbate AD pathogenesis by interacting with Aβ42. Aβ42 binds Cu, Fe, and Zn and metal binding induces a β-sheet-like conformational change in Aβ resulting in enhanced aggregation (Bossy-Wetzel, Schwarzenbacher, and Lipton 2004). Further, Aβ is able to potentiate the ability of metal ions, including Fe, Cu, and Al, to generate ROS (Bondy, Guo-Ross, and Truong 1998).

19.7.3 LPS Toxicity

Clinical studies of Mexico City children (Section 19.6) showed a significant increase in the expression of the LPS-binding protein CD14 on the plasma membrane surface (mCD14) of peripheral blood monocytes. A similar finding in a study of farm families suggested that the upregulation of monocyte mCD14 is a consequence of chronic exposure to LPS (Lauener et al. 2002). Taken together, these findings suggest that the children may be responding to LPS in Mexico City PM. Interestingly, LPS may be playing a role in the pathogenesis of the white matter alterations seen on magnetic resonance images (see Section 19.6), because the small number of children who carried the Asp299Gly TLR4 polymorphism that were hyporesponsive to LPS did not show white matter alterations (unpublished observations). Additional studies are underway to confirm these preliminary observations in a larger study population.

LPS could also be playing a role in evoking the robust expression of COX2 by the brain microvasculature of dogs and humans chronically exposed to severe air pollution (Calderón-Garcidueñas et al. 2003a) (Figure 19.4). Robust expression of COX2 mRNA is induced in brain perivascular cells after systemic intraperitoneal (IP) or intravenous (IV) administration of LPS in the mouse (Breder and Saper 1996; Elmquist et al. 1997a). COX2 is also induced in rat brain endothelial cells by IP LPS challenge, and because this induction does not require intact vagal afferent functions, Matsamura proposed that a humoral mediator could be participating in the brain endothelial COX2 induction (Matsumura et al. 2000).

Circulating LPS causes a rapid transcriptional activation of genes encoding its receptors as well as a variety of pro-inflammatory molecules (Rivest 2003). LPS plays a major role as a stimulus to the BBB and the HPA axis (Rivest 2001). Intravenous and intraperitoneal injections of low to moderate doses of LPS induce COX2 mRNA and protein along brain capillaries, choroid plexus, and leptomeninges (Rivest 2001). Thus, exposure to highly inflammogenic ultra-fine and fine PM and to PM components like LPS with direct or indirect access to the brain could trigger the responses that Rivest describes as the cerebral innate immune system response (Rivest 2003).

19.8 SUMMARY

Healthy dogs exhibited chronic brain inflammation and an acceleration of Alzheimer's-like pathology, with their brain findings being similar to humans. Humans exposed to severe air pollution had significantly higher expression of COX2, a potent biologically active mediator of inflammation and oxidative stress, and increased neuronal and astrocytic accumulation of intraneuronal, astrocytic, and vascular Aβ42 in olfactory bulb, frontal cortex, and hippocampus compared to residents of cities with low levels of air pollution. Inflammation and Aβ42 accumulation were seen in the brain of subjects with no history of neurological or cognitive impairment, and no known risk factors for dementia. Early diffuse plaques in teenagers and young adults and microvascular white matter pathology in children were striking findings as well. The early presence of olfactory bulb pathology in exposed subjects, including teenagers, raises the possibility that the nose is a portal of entry of pollutants to the brain. The identification of COX2 as an early marker of neuroinflammation in subjects exposed to air pollutants suggests that NSAIDs may be of use in diminishing the effects of chronic exposure to air pollutants on the brain. Defining the health effects of chronic exposure to a polluted urban atmosphere and their potential impact on the brain may improve the understanding of the pathogenesis of neurodegenerative diseases such as Alzheimer's. In addition, the identification and mitigation of environmental factors that influence AD pathogenesis may be one approach to limiting the future impact of AD. Our findings justify additional epidemiological and toxicological studies that further characterize the association between chronic exposure to air pollutants and the risk of developing Alzheimer's and other neurodegenerative diseases.

ACKNOWLEDGMENTS

This work was supported by grants 1KO1NS046410-01A1, 1R21 ES0123293-01A1, and Environmental Protection Agency cooperative agreement CR829522. Although the research described in this article has been funded wholly or in part by the United States Environmental Protection Agency through cooperative agreement CR829522 with the Center for Environmental Medicine, Asthma, and Lung Biology at the University of North Carolina at Chapel Hill, it has not been subjected to the Agency's required peer and policy review, and therefore does not necessarily reflect the views of the Agency and no official endorsement should be inferred. Mention of trade names or commercial products does not constitute endorsement or recommendation for use. We thank Dr. Ricardo Torres-Jardon from the Universidad National Autonoma de Mexico for Figures 19.1 and Figure 19.2.

REFERENCES

Aisen, P. S., Evaluation of selective COX-2 inhibitors for the treatment of Alzheimer's disease, *J. Pain Symptom Manage.*, 23, S35–S40, 2002.

Alfaro-Moreno, E., Martinez, L., Garcia-Cuellar, C., Bonner, J. C., Murray, J. C., Rosas, I., Rosales, S. P., and Osornio-Vargas, A. R., Biologic effects induced *in vitro* by PM10 from three different zones of Mexico City, *Environ. Health Perspect.*, 110, 715–720, 2002.

Arendt, T., Synaptic plasticity and cell cycle activation in neurons are alternative effector pathways: the Dr. Jekyll and Mr. Hyde concept of Alzheimer's disease or the yin and yang of neuroplasticity, *Prog. Neurobiol.*, 71, 83–248, 2003.

Arendt, T., Rodel, L., Gartner, U., and Holzer, M., Expression of the cyclin-dependent kinase inhibitor p16 in Alzheimer's disease, *Neuroreport*, 7, 3047–3049, 1996.

Bazan, N. G., COX-2 as a multifunctional neuronal modulator, *Nat. Med.*, 7, 414–415, 2001.

Bazan, N. G. and Lukiw, W. J., Cyclooxygenase-2 and presenilin-1 gene expression induced by interleukin-1β and amyloid β42 peptide is potentiated by hypoxia in primary human neural cells, *J. Biol. Chem.*, 277, 30359–30367, 2002.

Bhat, R. V., Budd Haeberlein, S. L., and Avila, J., Glycogen synthase kinase 3: a drug target for CNS therapies, *J. Neurochem.*, 89, 1313–1317, 2004.

Bondy, S. C., Guo-Ross, S. X., and Truong, A. T., Promotion of transition metal-induced reactive oxygen species formation by β-amyloid, *Brain Res.*, 799, 91–96, 1998.

Bossy-Wetzel, E., Schwarzenbacher, R., and Lipton, S. A., Molecular pathways to neurodegeneration, *Nat. Med.*, 10 (suppl.), S2–S9, 2004.

Bravo, H. A. and Torres-Jardon, R., Air pollution levels and trends in the Mexico city metropolitan area, In *Urban Air Pollution and Forests: Resources at Risk in the Mexico City Air Basin*, Fenn, M., Bauer, L., and Hernández, T., eds., Springer-Verlag, New York, 121–159, 2002.

Breder, C. D. and Saper, C. B., Expression of inducible cyclooxygenase mRNA in the mouse brain after systemic administration of bacterial lipopolysaccharide, *Brain Res.*, 713, 64–69, 1996.

Breder, C. D., Dewitt, D., and Kraig, R. P., Characterization of inducible cyclooxygenase in rat brain, *J. Comp. Neurol.*, 355, 296–315, 1995.

Brookmeyer, R., Gray, S., and Kawas, C., Projections of Alzheimer's disease in the United States and the public health impact of delaying disease onset, *Am. J. Public Health*, 88, 1337–1342, 1998.

Brown, R. C., Lockwood, A. H., and Sonawane, B. R., Neurodegenerative diseases: an overview of environmental risk factors, *Environ. Health Perspect.*, 113, 1250–1256, 2005.

Calderón-Garcidueñas, L., Gambling, T. M., Acuna, H., Garcia, R., Osnaya, N., Monroy, S., Villarreal-Calderón, A., Carson, J., Koren, H. S., and Devlin, R. B., Canines as sentinel species for assessing chronic exposures to air pollutants: Part 2. Cardiac pathology, *Toxicol. Sci.*, 61, 356–367, 2001.

Calderón-Garcidueñas, L., Azzarelli, B., Acuna, H., Garcia, R., Gambling, T. M., Osnaya, N., Monroy, S. et al., Air pollution and brain damage, *Toxicol. Pathol.*, 30, 373–389, 2002.

Calderón-Garcidueñas, L., Maronpot, R. R., Torres-Jardon, R., Henríquez-Roldán, C., Schoonhoven, R., Acuña-Ayala, H., Villarreal-Calderón, A. et al., DNA damage in nasal and brain tissues of canines exposed to air pollutants is associated with evidence of chronic brain inflammation and neurodegeneration, *Toxicol. Pathol.*, 31, 524–538, 2003a.

Calderón-Garcidueñas, L., Mora-Tiscareno, A., Fordham, L. A., Valencia-Salazar, G., Chung, C. J., Rodriguez-Alcaraz, A., Paredes, R. et al., Respiratory damage in children exposed to urban pollution, *Pediatr. Pulmonol.*, 36, 148–161, 2003b.

Calderón-Garcidueñas, L., Reed, W., Maronpot, R. R., Henriquez-Roldan, C., Delgado-Chavez, R., Calderón-Garcidueñas, A., Dragustinovis, I. et al., Brain inflammation and Alzheimer's-like pathology in individuals exposed to severe air pollution, *Toxicol. Pathol.*, 32, 650–658, 2004.

Calderon-Garciduenas, L., Romero, L., Barragan, G., and Reed, W., Upregulation of plasma endothelin-1 (ET-1) levels and CD14 expression on peripheral blood monocytes in healthy children exposed to urban air pollution, *FASEB J.*, 19 (4), A489, 2005.

Chao, C. C., Hu, S., Molitor, T. W., Shaskan, E. G., and Peterson, P. K., Activated microglia mediate neuronal cell injury via a nitric oxide mechanism, *J. Immunol.*, 149, 2736–2741, 1992.

Chen, F., Demers, L. M., Vallyathan, V., Ding, M., Lu, Y., Castranova, V., and Shi, X., Vanadate induction of NF-κB involves IκB kinase β and SAPK/ERK kinase 1 in macrophages, *J. Biol. Chem.*, 274, 20307–20312, 1999.

Chow, J. C., Watson, J. G., Edgerton, S. A., and Vega, E., Chemical composition of PM2.5 and PM10 in Mexico City during winter 1997, *Sci.Total Environ.*, 287, 177–201, 2002.

Chow, J. C., Watson, J. G., Shah, J. J., Kiang, C. S., Loh, C., Lev-On, M., Lents, J. M., Molina, M. J., and Molina, L. T., Megacities and atmospheric pollution, *J. Air Waste Manag. Assoc.*, 54, 1226–1235, 2004.

Cleary, J. P., Walsh, D. M., Hofmeister, J. J., Shankar, G. M., Kuskowski, M. A., Selkoe, D. J., and Ashe, K. H., Natural oligomers of the amyloid-β protein specifically disrupt cognitive function, *Nat. Neurosci.*, 8, 79–84, 2005.

Corder, E. H., Saunders, A. M., Strittmatter, W. J., Schmechel, D. E., Gaskell, P. C., Small, G. W., Roses, A. D., Haines, J. L., and Pericak-Vance, M. A., Gene dose of apolipoprotein E type 4 allele and the risk of Alzheimer's disease in late onset families, *Science*, 261, 921–923, 1993.

Dorman, D. C., Brenneman, K. A., McElveen, A. M., Lynch, S. E., Roberts, K. C., and Wong, B. A., Olfactory transport: a direct route of delivery of inhaled manganese phosphate to the rat brain, *J. Toxicol. Environ. Health A.*, 65, 1493–1511, 2002.

Elmquist, J. K., Breder, C. D., Sherin, J. E., Scammell, T. E., Hickey, W. F., Dewitt, D., and Saper, C. B., Intravenous lipopolysaccharide induces cyclooxygenase 2-like immunoreactivity in rat brain perivascular microglia and meningeal macrophages, *J. Comp. Neurol.*, 381, 119–129, 1997a.

Elmquist, J. K., Scammell, T. E., and Saper, C. B., Mechanisms of CNS response to systemic immune challenge: the febrile response, *Trends Neurosci.*, 20, 565–570, 1997b.

Ericsson, A., Liu, C., Hart, R. P., and Sawchenko, P. E., Type 1 interleukin-1 receptor in the rat brain: distribution, regulation, and relationship to sites of IL-1-induced cellular activation, *J. Comp. Neurol.*, 361, 681–698, 1995.

Fassbender, K., Walter, S., Kuhl, S., Landmann, R., Ishii, K., Bertsch, T., Stalder, A. K. et al., The LPS receptor (CD14) links innate immunity with Alzheimer's disease, *FASEB J.*, 18, 203–205, 2004.

George-Hyslop, P. H., Molecular genetics of Alzheimer's disease, *Biol. Psychiatry*, 47, 183–199, 2000.

Geys, J., Coenegrachts, L., Vercammen, J., Engelborghs, Y., Nemmar, A., Nemery, B., Hoet, P. H., *In vitro* study of the pulmonary translocation of nanoparticles: a preliminary study, *Toxicol. Lett.*, 2005.

Giovannini, M. G., Scali, C., Prosperi, C., Bellucci, A., Pepeu, G., and Casamenti, F., Experimental brain inflammation and neurodegeneration as model of Alzheimer's disease: protective effects of selective COX-2 inhibitors, *Int. J. Immunopathol. Pharmacol.*, 16, 31–40, 2003.

Goebeler, M., Roth, J., Brocker, E. B., Sorg, C., and Schulze-Osthoff, K., Activation of nuclear factor-κB and gene expression in human endothelial cells by the common haptens nickel and cobalt, *J. Immunol.*, 155, 2459–2467, 1995.

Grammas, P., Botchlet, T. R., Moore, P., and Weigel, P. H., Production of neurotoxic factors by brain endothelium in Alzheimer's disease, *Ann. NY Acad. Sci.*, 826, 47–55, 1997.

Hebert, L. E., Scherr, P. A., Bienias, J. L., Bennett, D. A., and Evans, D. A., Alzheimer disease in the US population: prevalence estimates using the 2000 census, *Arch. Neurol.*, 60, 1119–1122, 2003.

Henriksson, J., Tallkuist, J., and Tjalve, H., Uptake of nickel into the brain via olfactory neurons in rats, *Toxical. Lett.*, 91, 153–162, 1997.

Hinz, B. and Brune, K., Cyclooxygenase-2 — 10 years later, *J. Pharmacol. Exp. Ther.*, 300, 367–375, 2002.

Ho, L., Pieroni, C., Winger, D., Purohit, D. P., Aisen, P. S., and Pasinetti, G. M., Regional distribution of cyclooxygenase-2 in the hippocampal formation in Alzheimer's disease, *J. Neurosci. Res.*, 57, 295–303, 1999.

Ho, L., Purohit, D., Haroutunian, V., Luterman, J. D., Willis, F., Naslund, J., Buxbaum, J. D., Mohs, R. C., Aisen, P. S., and Pasinetti, G. M., Neuronal cyclooxygenase 2 expression in the hippocampal formation as a function of the clinical progression of Alzheimer disease, *Arch. Neurol.*, 58, 487–492, 2001.

Kawanishi, S., Hiraku, Y., Murata, M., and Oikawa, S., The role of metals in site-specific DNA damage with reference to carcinogenesis, *Free Radic. Biol. Med.*, 32, 822–832, 2002.

Kim, H. J., Kim, K. W., Yu, B. P., and Chung, H. Y., The effect of age on cyclooxygenase-2 gene expression: NF-κB activation and IκBα degradation, *Free Radic. Biol. Med.*, 28, 683–692, 2000.

Koo, E. H. and Kopan, R., Potential role of presenilin-regulated signaling pathways in sporadic neurodegeneration, *Nat. Med.*, 10 (suppl. 1), S26–S33, 2004.

Lauener, R. P., Birchler, T., Adamski, J., Braun-Fahrlander, C., Bufe, A., Herz, U., von Mutius, E. et al., Expression of CD14 and Toll-like receptor 2 in farmers' and non-farmers' children, *Lancet*, 360, 465–466, 2002.

Lemere, C. A., Blusztajn, J. K., Yamaguchi, H., Wisniewski, T., Saido, T. C., and Selkoe, D. J., Sequence of deposition of heterogeneous amyloid β-peptides and APOE in Down syndrome: implications for initial events in amyloid plaque formation, *Neurobiol. Dis.*, 3, 16–32, 1996.

Lewis, J., Bench, G., Myers, O., Tinner, B., Staines, W., Barr, E., Divine, K. K., Barrington, W., and Karlsson, J., Trigeminal uptake and clearance of inhaled manganese chloride in rats and mice, *Neurotoxicology*, 26, 113–123, 2005.

Lovestone, S. and Reynolds, C. H., The phosphorylation of tau: a critical stage in neurodevelopment and neurodegenerative processes, *Neuroscience*, 78, 309–324, 1997.

Lukiw, W. J., Gene expression profiling in fetal, aged, and Alzheimer hippocampus: a continuum of stress-related signaling, *Neurochem. Res.*, 29, 1287–1297, 2004.

Matsumura, K., Kaihatsu, S., Imai, H., Terao, A., Shiraki, T., and Kobayashi, S., Cyclooxygenase in the vagal afferents: is it involved in the brain prostaglandin response evoked by lipopolysaccharide?, *Auton. Neurosci.*, 85, 88–92, 2000. Dec. 20

Minghetti, L., Cyclooxygenase-2 (COX-2) in inflammatory and degenerative brain diseases, *J. Neuropathol. Exp. Neurol.*, 63, 901–910, 2004.

Nadeau, S. and Rivest, S., Effects of circulating tumor necrosis factor on the neuronal activity and expression of the genes encoding the tumor necrosis factor receptors (p55 and p75) in the rat brain: a view from the blood-brain barrier, *Neuroscience*, 93, 1449–1464, 1999.

Nagy, Z., The last neuronal division: a unifying hypothesis for the pathogenesis of Alzheimer's disease, *J. Cell Mol. Med.*, 9, 531–541, 2005.

Naslund, J., Haroutunian, V., Mohs, R., Davis, K. L., Davies, P., Greengard, P., and Buxbaum, J. D., Correlation between elevated levels of amyloid β-peptide in the brain and cognitive decline, *JAMA*, 283, 1571–1577, 2000.

Nunomura, A., Perry, G., Aliev, G., Hirai, K., Takeda, A., Balraj, E. K., Jones, P. K. et al., Oxidative damage is the earliest event in Alzheimer disease, *J. Neuropathol. Exp. Neurol.*, 60, 759–767, 2001.

Oberdörster, G., Sharp, Z., Atudorei, V., Elder, A., Gelein, R., Kreyling, W., and Cox, C., Translocation of inhaled ultrafine particles to the brain, *Inhal. Toxicol.*, 16, 437–445, 2004.

Osornio-Vargas, A. R., Bonner, J. C., Alfaro-Moreno, E., Martinez, L., Garcia-Cuellar, C., Ponce-de-Leon-Rosales, S., Miranda, J., and Rosas, I., Proinflammatory and cytotoxic effects of Mexico City air pollution particulate matter *in vitro* are dependent on particle size and composition, *Environ. Health Perspect.*, 111, 1289–1293, 2003.

Perry, V. H., The influence of systemic inflammation on inflammation in the brain: implications for chronic neurodegenerative disease, *Brain Behav. Immun.*, 18, 407–413, 2004.

Perry, G., Nunomura, A., Hirai, K., Zhu, X., Perez, M., Avila, J., and Castellani, R. J., Is oxidative damage the fundamental pathogenic mechanism of Alzheimer's and other neurodegenerative diseases?, *Free Radic. Biol. Med.*, 33, 1475–1479, 2002.

Rivest, S., How circulating cytokines trigger the neural circuits that control the hypothalamic–pituitary–adrenal axis, *Psychoneuroendocrinology*, 26, 761–788, 2001.

Rivest, S., Molecular insights on the cerebral innate immune system, *Brain Behav. Immun.*, 17, 13–19, 2003.

Scali, C., Giovannini, M. G., Prosperi, C., Bellucci, A., Pepeu, G., and Casamenti, F., The selective cyclooxygenase-2 inhibitor rofecoxib suppresses brain inflammation and protects cholinergic neurons from excitotoxic degeneration *in vivo*, *Neuroscience*, 117, 909–919, 2003.

Scheuner, D., Eckman, C., Jensen, M., Song, X., Citron, M., Suzuki, N., Bird, T. D. et al., Secreted amyloid β-protein similar to that in the senile plaques of Alzheimer's disease is increased in vivo by the presenilin 1 and 2 and APP mutations linked to familial Alzheimer's disease, *Nat. Med.*, 2, 864–870, 1996.

Selkoe, D. and Kopan, R., Notch and Presenilin: regulated intramembrane proteolysis links development and degeneration, *Annu. Rev. Neurosci.*, 26, 565–597, 2003.

Selkoe, D. J., Alzheimer's disease: genes, proteins, and therapy, Physiol, *Rev.*, 81, 741–766, 2001.

Selkoe, D. J., Alzheimer disease: mechanistic understanding predicts novel therapies, *Ann. Intern. Med.*, 140, 627–638, 2004.

Stohs, S. J. and Bagchi, D., Oxidative mechanisms in the toxicity of metal ions, *Free Radic. Biol. Med.*, 18, 321–336, 1995.

Stokin, G. B., Lillo, C., Falzone, T. L., Brusch, R. G., Rockenstein, E., Mount, S. L., Raman, R. et al., Axonopathy and transport deficits early in the pathogenesis of Alzheimer's disease, *Science*, 307, 1282–1288, 2005.

Strauss, K. I., Barbe, M. F., Marshall, R. M., Raghupathi, R., Mehta, S., and Narayan, R. K., Prolonged cyclooxygenase-2 induction in neurons and glia following traumatic brain injury in the rat, *J. Neurotrauma.*, 17, 695–711, 2000.

Strittmatter, W. J., Weisgraber, K. H., Huang, D. Y., Dong, L. M., Salvesen, G. S., Pericak-Vance, M., Schmechel, D., Saunders, A. M., Goldgaber, D., and Roses, A. D., Binding of human apolipoprotein E to synthetic amyloid β peptide: isoform-specific effects and implications for late-onset Alzheimer disease, *Proc. Natl Acad. Sci. USA*, 90, 8098–8102, 1993.

Sunderman, F. W. Jr., Nasal toxicity, carcinogenicity, and olfactory uptake of metals, *Ann. Clin. Lab. Sci.*, 31, 3–24, 2001.

Thiel, V. E. and Audus, K. L., Nitric oxide and blood–brain barrier integrity, *Antioxid. Redox. Signal.*, 3, 273–278, 2001.

Vane, J. R., Inhibition of prostaglandin synthesis as a mechanism of action for aspirin-like drugs, *Nat. New Biol.*, 231, 232–235, 1971.

Webber, K. M., Raina, A. K., Marlatt, M. W., Zhu, X., Prat, M. I., Morelli, L., Casadesus, G., Perry, G., and Smith, M. A., The cell cycle in Alzheimer disease: a unique target for neuropharmacology, *Mech. Ageing Dev.*, 126 (10), 1019–1025, 2005.

Winkler, F., Koedel, U., Kastenbauer, S., and Pfister, H. W., Differential expression of nitric oxide synthases in bacterial meningitis: role of the inducible isoform for blood-brain barrier breakdown, *J. Infect. Dis.*, 183, 1749–1759, 2001.

Xiang, Z., Ho, L., Valdellon, J., Borchelt, D., Kelley, K., Spielman, L., Aisen, P. S., and Pasinetti, G. M., Cyclooxygenase (COX)-2 and cell cycle activity in a transgenic mouse model of Alzheimer's disease neuropathology, *Neurobiol. Aging.*, 23, 327–334, 2002a.

Xiang, Z., Ho, L., Yemul, S., Zhao, Z., Qing, W., Pompl, P., Kelley, K. et al., Cyclooxygenase-2 promotes amyloid plaque deposition in a mouse model of Alzheimer's disease neuropathology, *Gene. Expr.*, 10, 271–278, 2002b.

Yang, Y., Geldmacher, D. S., and Herrup, K., DNA replication precedes neuronal cell death in Alzheimer's disease, *J. Neurosci.*, 21, 2661–2668, 2001.

Zhu, X., Raina, A. K., Perry, G., and Smith, M. A., Alzheimer's disease: the two-hit hypothesis, *Lancet Neurol.*, 3, 219–226, 2004.

20 Biologically Based Lung Dosimetry and Exposure–Dose–Response Models for Poorly Soluble Inhaled Particles*

Lang Tran
Institute of Occupational Medicine

Eileen Kuempel
Risk Evaluation Branch,
CDC National Institute for Occupational Safety and Health

CONTENTS

* Disclaimer: The findings and conclusions in this chapter are those of the authors and do not necessarily represent the view of the National Institute for Occupational Safety and Health.

20.1 INTRODUCTION

Inhaled airborne particles may be deposited in the respiratory tract with a probability that depends on the physical properties of the particles, the velocity of the air, and the structure of the airways. Once deposited, particles may be retained at the site of deposition, translocated elsewhere in the body, or cleared by the biological processes specific to each region of the respiratory tract.

The major regions of the human respiratory tract include the extrathoracic (nasopharynx or head airways), thoracic (tracheobronchial airways), and alveolar (pulmonary or gas-exchange) (ICRP 1994). These regions differ in structure and function (Miller 1999; McClellan 2000). The functions of the extrathoracic and thoracic regions include air conditioning and conducting, while the main function of the alveolar region is the gas exchange. Clearance of particles depositing in the alveolar region occurs primarily by alveolar macrophage (AV)-mediated clearance to the thoracic region, where they are cleared via the "mucociliary escalator" and then expectorated or swallowed. All regions of the respiratory tract include lymphatic tissue. The extrathoracic region drains to the extrathoracic lymph nodes, and the thoracic and alveolar regions drain to the thoracic (also called hilar) lymph nodes. Particles that are not cleared from the lungs may enter the lung interstitium and translocate to the lymph or blood circulation.

Several terms have been adopted to describe particles based on their size and probability of deposition within the respiratory tract. Inhalable particles are those capable of depositing anywhere in the respiratory tract. Thoracic particles are those capable of depositing in the lung airways. Respirable particles are those capable of depositing in the gas exchange region of the lungs (ACGIH 2005). The respirable particle size distribution includes the ultrafine or nanoparticles (primary particle diameter <0.1 μm), fine particles (<2.5 μm), and coarse particles with diameters <10 μm.

20.1.1 Comparison of Human and Rodent Lung Structure and Physiology

Humans and rodents have in common the major respiratory tract regions, but differ in the structural and physiological details of each region. For example, rats are obligate nose breathers, while humans breathe through either the mouth or nose, depending on the level of exertion and other factors. The nasal airways in rats are more extensive, and the particle deposition fractions in this region are greater than in humans. Conversely, particle deposition fractions in the tracheobronchial region are greater in humans than in rats. Deposition occurs primarily by particle-airway impaction in that region. Rats have an asymmetric (monopodial) branching system of tracheobronchial airways, while primates including humans have a symmetric (bipodial or tripodial) branching system (Crapo et al. 1990). Humans have respiratory bronchioles leading to the alveolar ducts while rats do not, instead having terminal bronchioles leading directly to the alveolar ducts. Yet, the alveolus structure, where gas exchange occurs, is similar in rodents, humans, and other mammals (Mauderly 1996). Because of the structural and size differences in the human and rat respiratory tract, the particle sizes that are inhalable differ in rats and humans (Ménache, Miller, and Raabe 1995).

Humans also differ from rats in physiological factors such as breathing and metabolic rates. Normal alveolar clearance is approximately 10 times faster in rats than in humans (Snipes 1989). Tracheobronchial clearance is relatively rapid in both rats and humans (retention half-times from hours to days), although in humans it has been shown that some particles that deposit in the airways are cleared more slowly (Stahlhofen, Scheuch, and Bailey 1995). The fraction of slowly cleared particles from the lung airways has been shown to increase with decreasing particle size from 6 to <1 μm geometric diameter (Kreyling and Scheuch 2000). This may be an important retention mechanism for nanoparticles, as well. The particle concentration in the lung airways (and particularly at airway bifurcations and centriacinar region) has been associated with both cancer and non-cancer lung diseases (Churg and Stevens 1988; Churg et al. 2003).

Particles that deposit in the alveolar region are associated with the slowest clearance phase in both rats and humans, with normal retention half-times of approximately 2 months in rats and from months to years in humans (Bailey, Fry, and James 1985). The rate of alveolar clearance can depend on the particle exposure concentration and duration in both rats and humans. For example, in coal miners, little or no clearance of particles was observed to occur after retirement from mining (Freedman and Robinson 1988; Kuempel et al. 1997). In rats (and mice and hamsters) with sufficiently high exposures, "overloading" of lung clearance has been observed at greater lung burdens and longer retention times than expected based on studies at lower exposures (Morrow 1988; Muhle et al. 1990; Elder et al. 2005).

20.1.2 Lung Dosimetry Models

The differences in human and rat lung structure and physiology that influence the kinetics of particle deposition and clearance can be described using biologically based mathematical models. Also called lung dosimetry models, these models describe the relationship between the external exposure to airborne particles and the internal dose of particles in the lungs. Biomathematical models that describe the exposure–dose relationship of a toxicant over time are called toxicokinetic models, while those describing the dose–response relationship are called toxicodynamic models. Although less common, models that describe the exposure, dose, and response relationships are called toxicokinetic/toxicodynamic models.

Lung dosimetry models have been developed for several species, but mostly in rats and humans. These models often focus separately on the processes of particle deposition or clearance/retention, although some have been integrated in software programs for humans (ICRP 1994; NCRP 1997; CIIT and RIVM 2002) and rats (CIIT and RIVM 2002). In several earlier rat models, the lungs have been described as a single compartment, with a dose-dependent clearance rate coefficient to account for overload (Yu et al. 1988; Yu and Rappaport 1997; CIIT and RIVM 2002). Other rat models described the lung clearance of insoluble particles during chronic exposure in terms of clearance to the tracheobronchial region, transfer to lymph nodes, and sequestration within the alveolar region (Vincent et al. 1987; Jones et al. 1988; Strom, Johnson, and Chan 1989; Stöber, Morrow, and Hoover 1989; Stöber, Morrow, and Morawietz 1990a, Stöber et al. 1990b). Of the human lung dosimetry models, many have focused on particle deposition. The human multiple path particle deposition (MPPD) model (CIIT and RIVM 2002 includes options for lung morphology based on data by Yeh and Schum (1980), Mortensen et al. (1988), or Koblinger and Hofmann (1990). Other deposition models include an empirical (data-based) model of ultrafine aerosol deposition in the human tracheobronchial airways (Zhang and Martonen 1997) and a stochastic model of particle deposition, with parameters described as statistical distributions based on experimental measurement, which allows for intra- and inter-individual variation in deposition due to lung structure and geometry (Koblinger and Hofmann 1985). Human lung dosimetry models have recently been reviewed by Martonen, Rosati, and Isaacs (2005).

In this chapter, two biomathematical models of the long-term clearance and retention of inhaled particles in rats or humans are described in detail. These include a biologically based model of exposure–dose–response in rats (Tran et al. 1999, 2000) and a human exposure–dose model calibrated and validated using data from two independent cohorts of coal miners in the U.S. (Kuempel 2000; Kuempel et al. 2001a, 2001b) and the U.K. (Tran and Buchanan 2000). The features of each of these models are unique compared with other existing models. The rat model is the only toxicokinetic/toxicodynamic model currently available for poorly soluble particles. The human model structure is biologically based and is the only clearance/retention model to be validated using human particle lung burden data. The structures of these human and rat models are compatible, which facilitates biologically based extrapolation from the rat to the human for those parameters that are not available for humans. Finally, examples are provided of using these models in risk assessment of occupational exposure to poorly soluble particles.

20.2 MATHEMATICAL MODEL OF THE RETENTION AND CLEARANCE OF PARTICLES FROM THE RAT LUNGS

Mathematically, the deposition and clearance process is a dynamic system, which can be described as a series of compartments. For example, in this model, X_i represents the quantity of free particles on the alveolar surface. Generally, the change in the particle burden in compartment i, dX_i/dt, is described by equations of the form

$$\frac{dX_i}{dt} = D + I_{ij} - O_{ik} \tag{20.1}$$

where

$D =$ input from outside the system to compartment i,
$I_{ji} =$ input from compartment j to compartment i,
$O_{ik} =$ output from compartment i to compartment k.

Equation 20.1 is called the "mass balance" Equation (because, over a set of compartments, mass is preserved).

If the rate of transfer of particles from compartment j to compartment i is assumed to be directly proportional to the mass of particles resident in compartment j, i.e.,

$$I_{ij} = k_{ij}X_j \tag{20.2}$$

then Equation 20.2 is called the "mass action" type and k_{ij} the "transfer rate" is the fraction per unit time.

For multiple inputs and outputs Equation 20.1 can be generalized as

$$\frac{dX_i}{dt} = D + \sum_{j=1}^{m} I_{ji} - \sum_{k=1}^{n} O_{ik} \quad i = 1, \ldots, l \tag{20.3}$$

where m is the number of compartments that output to compartment i, n is the number of compartments that receive output from compartment i and l is the total number of compartments which make up the system.

A system of equations such as Equation 20.3 can represent the dynamics of the retention and clearance of particles/fibres in the alveolar region of the lung.

20.2.1 Structure of the Rat Biomathematical Lung Model

The model is defined by a set of differential equations, which describe the rates at which the quantities of particles in the various compartments are assumed to change. Below we describe

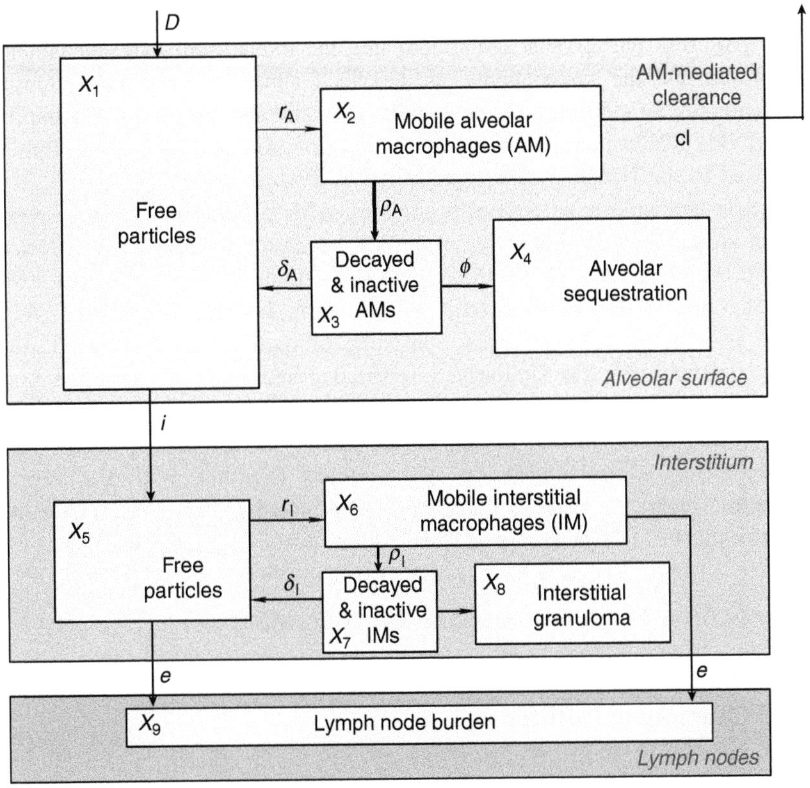

FIGURE 20.1 Schema of the compartments (X_1–X_9) and the transfer rates between compartments.

these compartments, the scientific assumptions about the translocations between them, and the rate parameters governing these processes.

20.2.1.1 Compartments of the Model

Our mathematical model describes the progress over time of the retention of particles and the alveolar macrophage (AM)-mediated clearance process in the pulmonary region, together with the particle redistribution and the overload phenomena. Figure 20.1 shows the nine conceptual compartments describing the location of inhaled particles, plus the main translocation routes between them, including AM-mediated clearance (Tran et al. 1999, 2000).

In Figure 20.1, inhaled particles in the respirable size range can reach the alveolar region of the lung, where they come into contact with epithelial cells. The mass (mg) of free particles on the alveolar surface is represented by compartment X_1. As the result of this contact, these particles are readily transferred into the interstitium (compartment X_5 represents the amount of free particles in the interstitium). This process is likely to be dependent on particle size (Ferin, Oberdörster, and Penney 1992; Oberdörster, Ferin, and Lehnert 1994; Geiser et al. 2005). However, the particle-epithelial cell contact also generates chemotactic signals that attract AMs to the site of particle deposition (Warheit et al. 1988; Reynolds 2005). The ensuing phagocytosis by AMs endeavors to clear the alveolar surface of particles (and thus prevent interstitialisation). Subsequently, the ingested particles are removed by migrating AMs to the mucociliary escalator. (Compartment X_2 represents the amount of particles inside mobile, active AMs.) However, these cells have a finite lifespan. AMs eventually decay and become inactive (compartment X_5 represents the amount of particles inside decayed AMs) and release their particle load onto the alveolar surface for re-phagocytosis by other, more effective, AMs. Free particles that cross the alveolar epithelium into the interstitium may encounter interstitial macrophages (IMs) and the same events, as described above, are repeated (compartment X_6 represents the amount of particles inside mobile IMs and X_7 represents the particle amount inside decayed IMs). However, from the interstitium, some particles (both free and inside IMs) are removed to the lymph nodes (represented by compartment X_9).

As the particle-epithelial cells contact progresses, AMs become increasingly retained in the alveolar region where they phagocytose *until* they become overloaded. As overloaded AMs decay, this load becomes increasingly difficult to redistribute to more effective AMs (i.e., the macrophages that ingest this particle load will, in turn, become overloaded). Gradually, a "sequestration" pool of particles emerges, consisting of particles in overloaded AMs. This is represented by compartment X_4. Similarly, interstitial granulomas are assumed to be derived from overloaded IMs. The amount of particles sequestered in granulomas is represented by compartment X_8 in the model. Table 20.1 gives a summary description of each of the compartments. The retention of particle-laden AMs occurs together with the recruitment of polymorphonuclear leukocyte (PMN) cells into the affected region—this is the hallmark of the inflammatory process (Donaldson and Tran 2002).

20.2.2 Mathematical Formulation of the Rat Lung Model

20.2.2.1 The Mathematical Description of the Normal (Non-Overload) Retention and Clearance of Particles

20.2.2.1.1 On the Alveolar Surface

The rate of change of the mass of free particles (in mg day^{-1}) consists *primarily* of the deposition of particles from the aerosol, the phagocytosis by AMs, and the interstitialisation of these particles, and *secondarily,* of the release of particles from macrophages which reach the end of their lifecycle

TABLE 20.1
The Compartments in the Model Representing the Location of Particles and the Level of Inflammation

Symbol	Location of Particles
	On the alveolar surface
X_1	Free on alveolar surface
X_2	Successfully phagocytosed by alveolar macrophages
X_3	In inactive alveolar macrophages, can be released for re-phagocytosis
X_4	Sequestered in overloaded, immobile alveolar macrophages
	In the interstitium
X_5	Free in interstitium
X_6	Successfully phagocytosed by interstitial macrophages
X_7	Attached to inactive interstitial macrophages, can be re-released for phagocytosis
X_8	Interstitial granuloma
	At the lymph nodes
X_9	Thoracic lymph nodes
	PMN recruitment
PMN	Number of PMN cells in the alveolar region

$$\frac{dX_1}{dt} = D - r_A X_1 - iX_1 + \delta_A X_3 \tag{20.4a}$$

where

X_1 is the mass (mg) of free particles remaining on the alveolar surface;
D is the dose rate of particles deposited on the alveolar surface (mg day^{-1}), calculated from Equation 2.4b;
r_A is the rate of phagocytosis by AMs (day^{-1});
i is the rate of interstitialisation (day^{-1});
X_3 is the mass (mg) of particles in macrophages in the inactive phase of their lifecycle; and
δ_A is the death rate for inactive macrophages.

The deposited dose rate D of deposited particles (in mg day^{-1}) is calculated as

$$D = \text{Concentration} \times \text{Ventilation rate} \times \text{Daily Exposure period} \times \text{Alveolar deposition fraction} \times (5/7) \times (6/100) \tag{20.4b}$$

where

Concentration is the aerosol concentration (mg m^{-3});
Ventilation rate is the breathing ventilation rate of the rat (l minute^{-1});
Daily Exposure period is the duration of each daily exposure (hr day^{-1});
Alveolar deposition fraction is the fraction of the inhaled particles of a given size deposited in the alveolar region;
(5/7) converts the concentration for a five-days-per-week inhalation pattern into the equivalent average concentration for the 7-days week; and
(6/100) converts the units of the breathing rate to match the time and volume units of the concentration and exposure period.

The alveolar deposition fraction, used in Equation 20.4b, was derived in two ways: (i) from the assumption that inhaled particles are of the (Mass Median Aerodynamic Diameter) MMAD size, and also (ii) from the measured particle size distribution, and using experimental data on the alveolar deposition efficiency for particle inhaled (Raabe et al. 1988).

The transfer rate coefficients (D, r_A, i, etc.) in these equations are shown in Figure 20.1 next to their translocation routes. The coefficients i, δ_A are approximately constant when the lung burden is low, but at higher lung burden the macrophage mediated clearance becomes impaired and the transfer rates become functions of the alveolar particle surface area, s_{alv}, and the form of this dependence is described later. This assumes that the dependence is on the sum of particles which are available to the AMs, i.e., dependence on $s_{alv} = s(X_1 + X_2 + X_3 + X_4)$, where s is particle-specific surface area (in unit of area *per* unit of mass). The phagocytosis rate is left constant for the range of particles to be modeled presently. However, it is envisaged that phagocytosis will become less effective as AMs are expected to clear larger epithelial areas, (covered by particles with larger surface areas). This is likely to be true for nanoparticles. However, data is currently lacking for a reasonably accurate model.

Equation 20.4a, with these coefficients written as functions of s_{alv}, becomes

$$\frac{dX_1}{dt} = D - r_A X_1 - i(s_{alv})X_1 + \delta_A(s_{alv})X_3 \tag{20.4c}$$

Particles that have been phagocytosed by macrophages will subsequently either be removed from the alveolar region by way of macrophage migration and the mucociliary escalator or be released onto the alveolar surface upon the necrosis of AMs. So the rate of change of the mass of phagocytosed particles in active AMs (i.e., X_2) is

$$\frac{dX_2}{dt} = r_A X_1 - \mathrm{cl}(S_{\mathrm{alv}})X_2 - \rho_A X_2 \qquad (20.5)$$

where cl is the AM-mediated clearance rate (day^{-1}), r_A is the phagocytosis rate (day^{-1}), and ρ_A is the transfer rate (day^{-1}) from active AMs to inactive AMs. When this clearance is unaffected by overload, cl is estimated to be 0.015 day^{-1} (Stöber, Morrow, and Hoover 1989). When clearance is affected by overload, then the dependence of cl on s_{alv} is described by Equation 20.13. The phagocytosis rate r_A is assumed to be independent of the particle surface area as AMs are assumed to be locally mobile in the alveolar region and able to phagocytose particles. Also, ρ_A is assumed to be unaffected by the particle surface area.

The mass of particles inside inactive AMs, X_3, is described by

$$\frac{dX_3}{dt} = \rho_A X_2 - \delta_A(s_{\mathrm{alv}})X_3 - \phi(s_{\mathrm{alv}})X_3 \qquad (20.6)$$

where δ_A is the release rate of particles from inactive AMs back to the alveolar surface and ϕ is the rate of transfer into the alveolar sequestration compartment. Note that for a certain choice of δ_A and ϕ, $\delta_A + \phi = \text{constant}$.

Equations 20.4 through Equation 20.6 describe the dynamics of translocation of particles on the alveolar surface when the lung defenses are not overloaded. The fourth compartment on the alveolar surface, X_4, becomes involved once the lung becomes overloaded with particles. The rate of change of the amount of particles in the alveolar sequestration compartment (X_4), representing the mass of particles trapped inside overloaded macrophages, is

$$\frac{dX_4}{dt} = \phi(s_{\mathrm{alv}})X_3 \qquad (20.7)$$

20.2.2.1.2 *In the Interstitium*

Once particles are interstitialised, they will be phagocytosed readily by IMs. Interstitialised particles that escape phagocytosis, together with the particles phagocytosed by Ims, may eventually be removed to the lymph nodes. Let X_5 be the mass of free particles that are interstitialised, then

$$\frac{dX_5}{dt} = i(s_{\mathrm{alv}})X_1 - e(s_{\mathrm{inst}})X_5 - r_I X_5 + \delta_I(s_{\mathrm{inst}})X_7 \qquad (20.8)$$

where

e is the removal rate (day^{-1}) of particles to the lymph nodes;
r_I and δ_I, respectively, the rates of phagocytosis by macrophages and release from inactive macrophages, are assumed to have the same value for IMs as for AMs;
and s_{inst} is the interstitial burden in unit of surface area, i.e. $s_{\mathrm{inst}} = s(X_5 + X_6 + X_7 + X_8)$.

Equation 20.8 for IMs is comparable to Equation 20.4c for AMs—the first term on the right hand side of Equation 20.8 is the transfer from alveolar surface (instead of deposition in Equation 20.4c), the second and third terms include X_5 instead of X_1, and the last term includes X_7 instead of

X_3. Similarly, the mass of particles phagocytosed by IMs is

$$\frac{dX_6}{dt} = r_I(s_{inst})X_5 - e(S_{inst})X_6 - \rho_I X_6 \tag{20.9}$$

where the removal rate to lymph nodes (e) is assumed to be the same for IMs as for interstitialised free particles.

The mass of particles trapped in interstitial granulomas is described by

$$\frac{dX_7}{dt} = \rho_I X_6 - \delta(s_{inst})X_7 - \upsilon(s_{inst})X_7 \tag{20.10}$$

where the transfer rate of particles from active IMs to inactive IMs (ρ_I) and the release rate from inactive IMs (δ_I) are also assumed to have the same dependence on the relevant burden (interstitial or alveolar particle surface area), and also the same non-overload values as for AMs; υ is the rate (day^{-1}) of interstitial granuloma formation which occurs when the IM defense of the interstitium becomes impaired.

The conditions relating to the transfer of particles to interstitial granuloma (X_8) are linked with overload and therefore are described in the section on overload (later). However, the mass of particles trapped in interstitial granulomas is described by

$$\frac{dX_8}{dt} = \upsilon(s_{inst})X_7 \tag{20.11}$$

20.2.2.1.3 At the Lymphatic Level

The mass of particles accumulated in the mediastinal lymph nodes is the sum of the transfer from free interstitialised particles (X_5) and particles in IMs (X_6)

$$\frac{dX_9}{dt} = e(X_5 + X_6) \tag{20.12}$$

20.2.2.2 Mathematical Description of Overload

As described earlier, the impairment of pulmonary clearance during exposure due to overload correlated with the increase in the rate of recruitment of PMNs. The PMN level, in turn, correlated with particle surface area. This impairment of clearance can be described mathematically as a function, θ, of alveolar particle burden (in terms of mass or surface area), which varies between 0 and 1. As θ is a multiplier of the rate parameters, these parameters are fully functioning when $\theta \approx 1$.

Mathematical expressions were developed to describe this progressive impairment. Similar equations were used in other models (e.g., Yu et al. 1988; Stöber, Morrow, and Hoover 1989; Tran, Jones, and Donaldson 1997). Note that all of these functional forms are essentially chosen for practical reasons (i.e., they integrate well with the models in which they form a part). For example, Tran, Jones, and Donaldson 1997 used an exponential decay form

$$\theta(m_{alv}) = e^{-\lambda.(m_{alv}-m_{crit})^{\beta}} \quad \text{for } m_{alv} > m_{crit}$$

$$\theta(m_{alv}) = 1 \quad \text{for } m_{alv} \leq m_{crit}$$

where m_{alv} is the particle mass in the alveolar region and m_{crit} is the critical mass from which impairment begins to manifest. λ and β are parameters controlling the rate and form of decay. This function has two limitations. First, the parameters of this function cannot be related to some tangible entity, such as mass or surface area. So, it is difficult to judge the plausibility of different values which (λ and β) give a good fit with data. Finally, there is a deterministic boundary at m_{crit} below which there is no impairment—i.e., the equation above provides an abrupt switch over to impairment of clearance. While there is some evidence that this might be the case (Muhle et al. 1990), it is more plausible that impairment would likely progress continuously. Thus, a new functional form for θ in terms of alveolar surface burden, s_{alv}, is introduced

$$\theta(s_{\text{alv}}) = 1 - \frac{1}{\left(1 + \left(\frac{s_{1/2}}{s_{\text{alv}}}\right)^{\beta}\right)} \tag{20.13}$$

This functional form is similar to that used by Yu and Rappaport (1997) to describe retardation of clearance of insoluble dust. The function is dependent on two parameters, namely $s_{1/2}$ and β. The former, $s_{1/2}$, represents the level of particle surface area such that the impairment is half of its original value; while the latter, β, controls the steepness of the impairment. Figure 20.2 shows the behavior of θ for two different sets of values for β and $s_{1/2}$ over a range of values of s_{alv}. One advantage this function has over the earlier functions from the literature is that one of its parameters, $s_{1/2}$, is readily interpretable and will be useful in the comparison of the effects of different dusts on their retention and clearance.

Since particle surface area affects clearance by mobile macrophages, we assume here that the clearance rate is modified as

$$cl(s_{\text{alv}}) = \theta(s_{\text{alv}})cl \tag{20.14}$$

where *cl*, on the right-hand side of Equation 20.14, is the time-independent rate for low lung burdens. Thus, as the particle burden on the alveolar surface (in terms of surface area) increases, mobile macrophages are increasingly retained on the alveolar surface, as described by Equation 20.14. During this phase, particles released by inactive AMs upon death will be less likely to be removed by mobile AMs to the mucociliary escalator (i.e., the transfer rate δ_A, back to the alveolar surface to be re-phagocytosed *and* then cleared by AMs, decreases with increasing alveolar lung burden). Instead, these particles are re-phagocytosed by retained AMs leading to transfer at a rate,

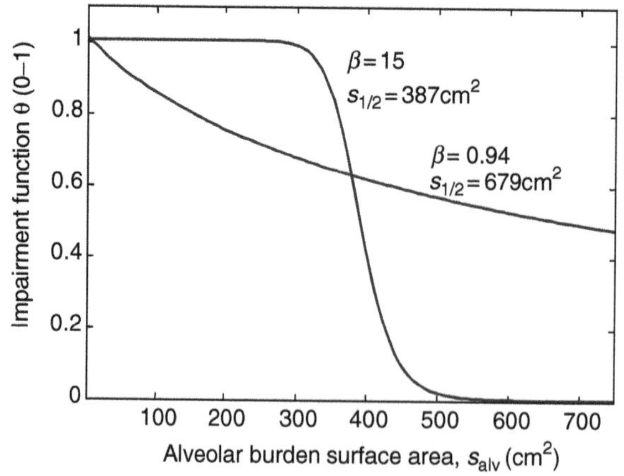

FIGURE 20.2 The impairment function for particle surface area between 0 and 750 cm^2 and two different sets of values for (β, $s_{1/2}$).

ϕ the sequestration rate (day^{-1}), into an alveolar sequestration compartment (X_4). In this case, ϕ increases as impairment develops

$$\phi(s_{\text{alv}}) = (1 - \theta(s_{\text{alv}}))\phi \tag{20.15}$$

and

$$\delta_{\text{A}}(s_{\text{alv}}) = \theta(s_{\text{alv}})\delta_{\text{A}} \tag{20.16}$$

For particles with large surface area, as inhalation progresses, there is increasing contact between these particles and epithelial cells, potentially causing damage to these cells. It is assumed that a damaged epithelium will allow greater access of particles into the interstitium (Adamson and Hedgecock 1995). Therefore, the rate of particle interstitialisation increases concurrently with the progression of impairment. Mathematically, the rate of interstitialisation can be modeled as

$$i(s_{\text{alv}}) = i_{\text{normal}}\theta(s_{\text{alv}}) + (1 - \theta(s_{\text{alv}}))i_{\text{max}} \tag{20.17}$$

where i_{normal} is the rate of interstitialisation under normal conditions and i_{max} is the maximum rate of interstitialisation under complete impairment. So, according to this equation, initially $i(s_{\text{alv}}) = i_{\text{normal}}$ ($\neq 0$), because AM defense is not absolutely effective and there is always some interstitialisation taking place. Once impairment starts, $i(s_{\text{alv}})$ increases from i_{normal} towards i_{max}.

At the interstitial level, we assume that interstitial granuloma will be formed when the defense of the interstitium becomes impaired. There is, however, an absence of data regarding the particulate burden in the interstitium. Therefore, for the present, we are restricted to constructing the framework for this part of the model. This framework is presented to show how the concepts can be included, although the choice of values for the transfer rates will be limited to being plausible (but unsupported) and they will also be chosen so as not to affect the predictions of quantities which can be tested by the existing data (for lymph node burdens).

For the current model, we assume that the impairment of clearance for IMs by dust loading has the same form of dependence on dust loading as for the AMs. We also assume that the impairment of motility follows the same dependence on the impairment function θ, thus

$$\upsilon(s_{\text{inst}}) = (1 - \theta(s_{\text{inst}}))\upsilon \tag{20.18}$$

The differential equations (Equation 20.4 through Equation 20.18), describing the kinetics of the retention and clearance of particles under normal circumstance (i.e., low exposure and non-impairment of AM defense mechanisms) and for the overload situation, constitute the current mathematical model. The model provides a quantitative, scientifically based representation of the mechanisms of removal of particles from the lung.

The above equations describing the effect of particulate overload describe the process that results in a higher proportion of the lung burden entering the interstitium. The presence of more particles in the interstitium makes more particles available for transfer to the mediastinal lymph nodes. However, there does not appear to be a reason why a higher proportion of the interstitialised particles should be transferred to lymph nodes, so the coefficient for transfer from interstitium to mediastinal lymph nodes (X_9) remains constant.

20.2.2.3 Mathematical Description of PMN Recruitment

In this section, the original model is extended to describe the inflammatory recruitment of PMN cells. There is an association between the mean number of PMNs in the bronchoalveolar lavage

(BAL) fluid and the mean lymph node burden, expressed as surface area (Tran et al. 1999). Since particles found in the lymph nodes were originally interstitialised, the net rate of PMN recruitment is assumed to be proportional to the rate of particle interstitialisation (expressed as particle surface area) and a PMN removal rate that is attributed to normal lifecycle of this type of cell. Thus,

$$\frac{\mathrm{dPMN}}{\mathrm{d}t} = \mathrm{Rec}.\, i(s_{\mathrm{alv}}).s.\, X_1 - \mathrm{Rem.\, PMN} \tag{20.19}$$

where PMN represents the number of PMNs ($\times 10^6$) in the BAL fluid. Rec is the number of PMNs recruited *per* unit of dust interstitialised (as surface area). Rem is the removal rate of PMNs (day^{-1}). The specific particle surface area is s and X_1 is the mass of free particles on the alveolar surface.

20.2.2.4 Summary of Model Parameters

The translocations between the compartments of the model are expressed by *transfer rates* (labeled in Figure 20.1 and defined in Table 20.2). These rates determine the fraction of mass of particles per unit time, which are translocated from one compartment to another (e.g., r, the phagocytosis rate of macrophages, represents the fraction of particles transferred from X_1, the compartment of free particles on the alveolar surface, to X_2 the compartment of successfully phagocytosed particles). In addition to the transfer rates, there are parameters belonging to the impairment function (e.g., β and $s_{1/2}$ in Equation 20.13) and those belonging to the deposited dose D (e.g., breathing rate, deposition fraction, etc.).

TABLE 20.2
The Parameters of the Mathematical Model

Parameters	Symbol	Unit
Deposition		
Deposited dose rate, function of breathing rate, deposition efficiency and exposure concentration	D	mg day^{-1}
Kinetics in Macrophages		
Phagocytosis rate by AMs or IMs[a]	r_A, r_I[a]	day^{-1}
AM-mediated clearance of particles	cl	day^{-1}
Transfer rate of particles from active to inactive AMs or IMs[a]	ρ_A, ρ_I	day^{-1}
Release rate of particles back to the alveolar surface or interstitium for re-phagocytosis[a]	δ_A, δ_I	day^{-1}
Kinetics of Particles		
Normal interstitialisation rate of free particles	i_{normal}	day^{-1}
Maximum interstitialisation rate of free particles	i_{max}	day^{-1}
Removal rate of particles to the lymph nodes	e	day^{-1}
Overload and Sequestration		
Alveolar sequestration rate	ϕ	day^{-1}
Rate of formation of interstitial granuloma	υ	day^{-1}
PMN Recruitment		
PMN recruitment rate	Rec	No. of cells recruited *per* unit of particle surface area burden
PMN removal rate	Rem	day^{-1}

[a] The subscripts A and I indicate that the coefficients apply, respectively, to the alveolar and interstitial macrophages.

20.2.3 Model Parameters

20.2.3.1 Parameter Values

Table 20.3 shows the values of the parameters used in the calculation of the deposited dose D (Equation 20.4b). Table 20.4 shows the values of the parameters drawn from previous studies, to be used in the model. The parameters include the phagocytosis rate, which is based on experimental evidence described by Stöber et al. (1989, 1990a, 1990b), indicating that phagocytosis usually takes place within 2–6 h and so, following Stöber, we use the 6 h estimate. This is equivalent to a phagocytosis rate of approximately $1/6 = 0.166\ h^{-1}$, or equivalently 4 day^{-1} when expressed in the same units as the other rates in Table 20.2. The macrophage mediated clearance rate has been estimated as ranging from 0.01 day^{-1} to approximately 0.02 day^{-1}, with the value of 0.015 day^{-1} being commonly applicable (e.g., Stöber, Morrow, and Hoover 1989). Estimates of the time scales for the macrophage normal life cycle also based on the evidence presented by Stöber et al. (1990a, 1990b) were used to estimate the rate of transfer from active to inactive macrophages (ρ) and for release from inactive macrophages either for re-phagocytosis (transfer rate δ) or after overload to become trapped in a succession of overloaded macrophages (transfer rate ϕ). For example, if the time scale for the active phase of the life cycle is T_a days, then a population of macrophages in kinetic equilibrium would have a fraction of $1/T_a$ of the active macrophages pass from active to inactive phase each day, so $\rho 1/T_a$. Similarly, there would be a rate of death and release of particles from inactive macrophages δ equal to $1/T_i$, where T_i is the time scale of the inactive phase. Both T_a and T_i have been originally estimated by Stöber et al. (1989, 1990a, 1990b), and values of T_a ($=28$ days) and T_i ($=7$ days) from their studies, corresponding to a reasonable estimate of AM total life cycle of 35 days (van Oud Alblas and van Forth, 1986) were used in our model.

The rate of particulate deposition into the lung was estimated from the volume inhaled, the aerosol concentration, and the alveolar deposition fraction. The values estimated for the breathing rate and alveolar deposition fraction and the plausible range are listed in Table 20.3. A wide range of values for the breathing rate is plausible, as various studies have used markedly different estimates, as shown in the first row of Table 20.3. The alveolar deposition fraction has been measured in rats as a function of the particle aerodynamic diameter by Raabe et al. (1988), giving estimates of 7% for the TiO_2 particles of MMAD$\sim$2.1 μm. Table 20.4 shows all the parameters of the model.

TABLE 20.3
Factors Affecting the Deposited Dose

Breathing rate (*l*/min)	0.1–0.3	Stöber et al. (1994) and Yu et al. (1994)
	0.154	Value used in this study
Target concentrations (mg m^{-3})	50 mg m^{-3}	TiO_2
Exposure regimen	7 h/day 5 days/week	
Correction factor (to treat exposure over 5 days as continuous over the week)	$5/7 = 0.714$	(Also used by Morrow (1988))
TiO_2 deposition fraction	0.07	Original estimates derived from *in vivo* data used in this study, and consistent with values from Raabe et al. (1977) and Raabe et al. (1988)

TABLE 20.4
The *a Priori* Fixed Model Parameters

Parameters	Symbol	Value	Unit
Deposition			
Deposited dose rate, function of breathing rate, deposition efficiency and exposure concentration	D	*see* Table 20.3	mg day^{-1}
Kinetics in Macrophages			
Phagocytosis rate by AMs or IMs[a]	r_A, r_I	4	day^{-1}
AM-mediated clearance of particles	cl	0.015	day^{-1}
Transfer rate of particles from active to inactive AMs or IMs[a]	ρ_A, ρ_I	0.036	day^{-1}
Release rate of particles back to the alveolar surface or interstitium for re-phagocytosis[a]	δ_A, δ_I[a]	0.14	day^{-1}
Transfer from overloaded IM to granuloma	ν	0.14	day^{-1}
Kinetics of Particles			
Interstitialisation of free particles, normal rate	i_{normal}	0.03	day^{-1}
Interstitialisation of free particles, maximum rate	i_{max}	1.8	day^{-1}
Removal rate of particles to the lymph nodes	e	0.1	day^{-1}
Overload and Sequestration			
Alveolar sequestration rate	ϕ	0.14	day^{-1}
Rate of formation of interstitial granuloma	ν	0.14	day^{-1}
Impairment Function			
Overload threshold	$s_{1/2}$	387	cm^2
Overload constant	β	15	
PMN Recruitment			
PMN recruitment rate	Rec	0.025	No. of cells *per* unit of s.a. burden
PMN removal rate	Rem	0.01	day^{-1}

[a] The subscripts A and I indicate that the coefficients apply, respectively, to the alveolar and interstitial macrophages.

20.3 EXPERIMENTAL DATA

Two contrasting, poorly soluble "low-toxicity" mineral dusts were used to compare the dose response relationships at exposure concentrations calculated to produce volumetrically similar alveolar deposition rates (Tran et al. 1999). Then, if the dose–response relationships were determined solely by volumetric loading (Morrow 1988), the results would show similarity between the two dusts.

The chosen dusts (Table 20.5) provided contrasting particle sizes with similar densities. Target concentrations (Table 20.6) were calculated from expected alveolar deposition fractions for the size distribution of each dust, accounting for elutriation in the aerosol sampler.

TABLE 20.5
Physical Characteristics of the Test Particles

Dust	Density (g/cm^3)	MMAD (μm)	Specific Surface Area (m^2/g)
Titanium dioxide (TiO_2)	4.25	2.1	6.67
Barium Sulphate ($BaSO_4$)	4.5	4.3	3.13

TABLE 20.6
Target Exposure Concentrations of Respirable TiO_2 and $BaSO_4$

	"Low" Concentration	"High" Concentration
Titanium dioxide (TiO_2)	25 mg m^{-3}	50 mg m^{-3}
Barium Sulphate ($BaSO_4$)	37.5 mg m^{-3}	75 mg m^{-3}
Duration of experiment	203 days	119 days

Rats were sacrificed at 6 time points during exposure for measurement of (i) lung burden, (ii) burden in mediastinal hilar lymph nodes, and (iii) numbers of AMs, lymphocytes, and neutrophils (PMN) in BAL fluid. Groups of 6 rats were used for particulate burdens, and further groups of 6 for BAL.

The lung burdens at the early time points showed that similar mass deposition rates were achieved. However, the $BaSO_4$ lung burdens appeared to latterly approach a steady state level, indicating effective clearance whereas TiO_2 lung burdens continued to increase, (Figure 20.3) consistent with overload, lymph node burdens were higher for TiO_2 (Figure 20.4). Mean numbers of PMNs (inflammation) also increased more rapidly for TiO_2 (Figure 20.5). However,

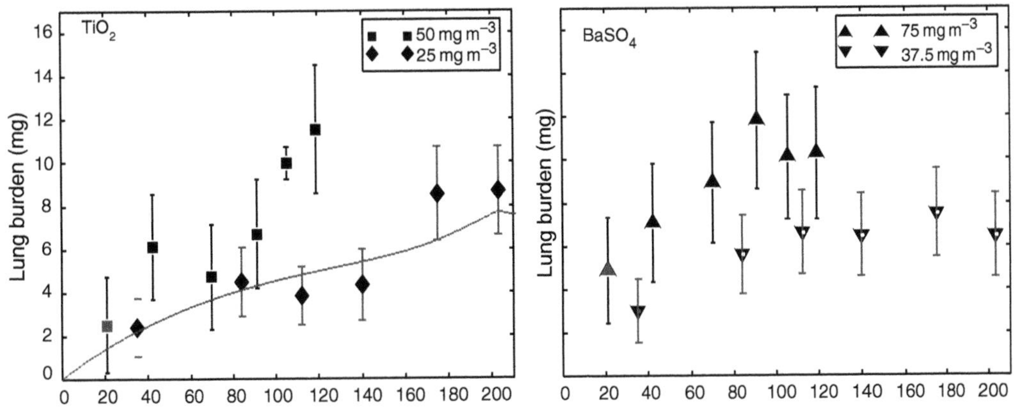

FIGURE 20.3 Mean lung burdens during exposure and model predictions (lines).

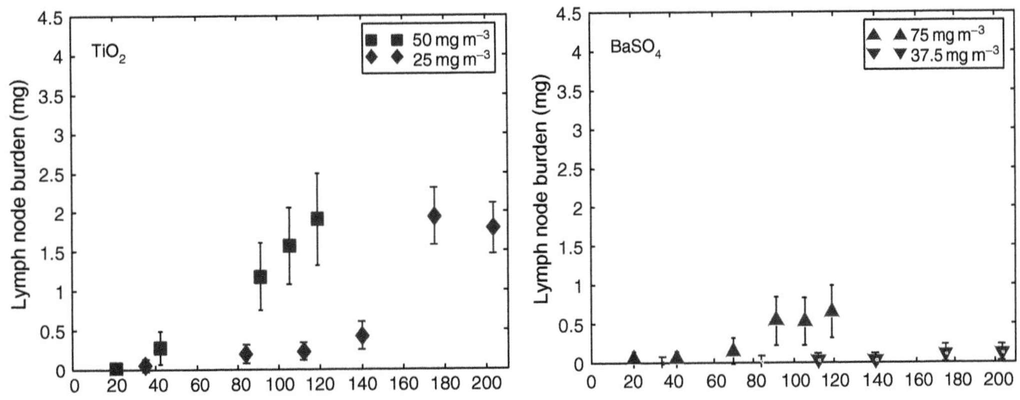

FIGURE 20.4 Mean lymph node burden during exposure and model predictions (lines).

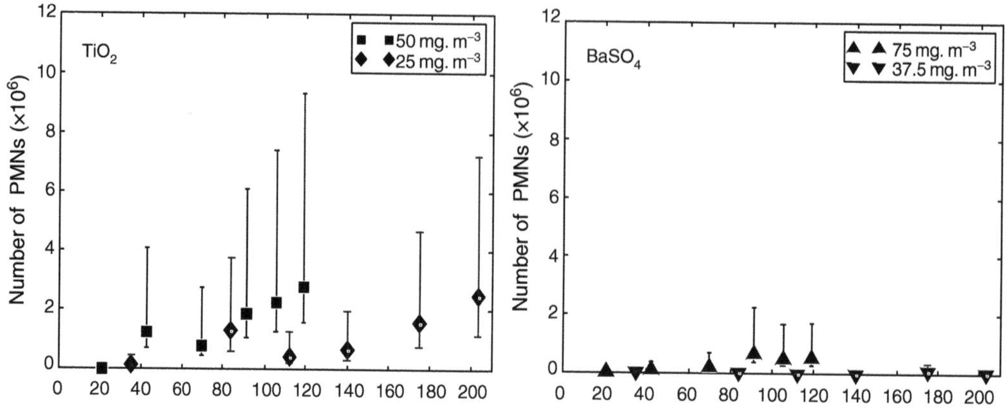

FIGURE 20.5 Mean number of PMN during exposure and model predictions (lines).

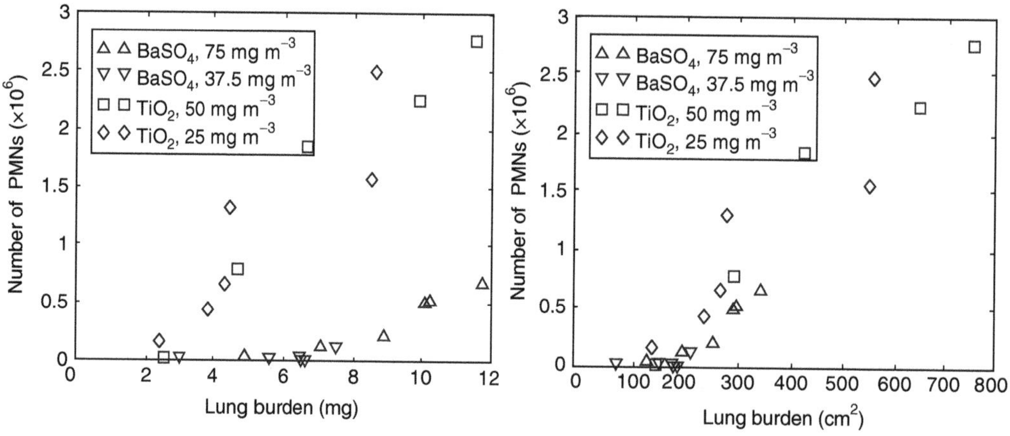

FIGURE 20.6 Mean number of PMN compared to mean lung burdens expressed as mass surface area.

if lung burdens were expressed as surface area, the PMN data for both dusts could be described by a common trend (Figure 20.6).

20.4 STRATEGY FOR MODEL CALIBRATION AND VALIDATION

The lung, lymph node burden, and number of PMNs from the "high" exposure TiO_2 50 mg m^{-3} experiment were used to estimate the non-fixed parameters. Specifically,

1. The lung burden data were used to estimate β and $s_{1/2}$ and the factor for reducing the breathing rate parameter specific to the "high" exposure TiO_2.
2. The lymph node burden *data* were *used* to estimate the translocation rate e.
3. The PMN data to estimate Rec and Rem.

Once these parameters were estimated, the model was fully identified and calibrated. The next step was to validate the model by predicting the outcomes of the "low" exposure TiO_2 experiment and both the $BaSO_4$ experiments and comparing (i.e., checking visually for consistency) with the data from these experiments. (A re-calibration of the model would have been carried out if necessary.)

The model calibration led to the predictions for the "high" exposure TiO_2 shown as the higher lines in Figure 20.3 through Figure 20.5. Model simulations for the "low" exposure TiO_2 experiment and both the $BaSO_4$ experiments (lower lines in Figure 20.3 through Figure 20.5) agreed well visually with the data, thus, we considered the model validated.

20.5 MODEL EXTRAPOLATION TO HUMANS

The four main areas in the exposure–dose–response relationships that are expected to differ between rats and humans are: (i) Exposure Concentration, (ii) Deposited Dose, (iii) Retention and Clearance, and (iv) Cell Recruitment. Occupational human exposure is usually at lower airborne concentration and a longer duration than the exposure in animal studies. The deposited dose is influenced by the ventilation rate and the deposition fraction. Both of these parameters are dependent on the morphology of the lung and are expected to differ between species. Once deposited, particles are either retained or cleared; the retained particles are either interstitialised (parameters i_{normal}, i_{max}) and removed to the lymph nodes (e) or cleared by macrophages (cl). The retention and clearance of particles is known to vary between species (Bailey, Fry, and James 1985). The impairment of particle clearance following overload ($s_{1/2}$) is also known to be species dependent (Bermudez et al. 2002, 2004; Elder et al. 2005).

Table 20.7 lists the model parameters that were either estimated or scaled to humans. The remaining parameters were kept fixed at the values estimated from animal data.

20.5.1 Method for Extrapolation

The following method was used in extrapolating (animal based) model parameters to their human equivalents.

1. For Exposure (Concentration, t_{start}, t_{final})
   We replaced the parameters for concentration and duration with the relevant human occupational equivalents (e.g., 4 mg m^{-3} and working life time of 45 years).
2. For Deposited Dose (Ventilation rate, Deposition fraction)
   We used data available from Hattis et al. (2001) (see Table 20.8).
3. For Retention and Clearance (cl, e, i_{normal}, i_{max})
   We scaled parameters inversely with the ratio of pulmonary surface area to the power 0.25, in accord with the method of Ings (1990). The derivation of this extrapolation factor for kinetic parameters is provided in O'Flaherty (1989).
   e.g., $cl_{human} = cl_{rat}$ (rat pulmonary surface area /human pulmonary surface area)$^{0.25}$. This produces an estimate of the lung clearance rate for humans that is consistent with other estimates of the rate for humans (Bailey, Fry, and James 1985).

TABLE 20.7
Model Parameters to Be Converted to Human Equivalents (For the Definition of the Parameters for Retention, Clearance, and Cell Recruitment, see Table 20.2)

Exposure	Deposited Dose	Retention and Clearance	Cell Recruitment
Concentration	Ventilation rate	cl	Rec
t_{start}	Deposition fraction	e	
t_{final}		$s_{1/2}$	
Daily exposure period		i_{normal}	
		i_{max}	

TABLE 20.8
Central Value for Ventilation Rate, Deposition Fraction and Clearance Rate and Their Distribution

Parameters	Distribution	$\log_{10}$(GSD)	Mean
Ventilation rate	Log-normal	0.12	1.7 (m^3/h)
Deposition fraction[a]	Log-normal	0.30	0.092
Clearance (cl)	Log-normal	0.21	0.0036 (day^{-1})

[a] This depends on the particle MMAD. Abbreviations: MMAD, mass median aerodynamic diameter; GSD, geometric standard deviation.

Source: From Hattis, D., Goble, R., Russ, A., Banati, P., Chu, M., *Risk Anal.*, 21(4), 585–599, 2001.

4. For Threshold Burden ($s_{1/2}$)
   Following Morrow (1988), we expressed the critical lung burden in units of mg/g lung of rat then multiplied by human lung weight to get an absolute value for this parameter for humans. We then converted into particle surface area units using the specific surface area of the TiO_2.
5. For Cell Recruitment (Rec)
   The recruitment of PMN and their removal are events that take place in relation to the particle dose interstitialised from the rat alveolar epithelial surface area. As humans have a much larger surface area, the recruitment rate for PMNs is scaled down with the ratio of pulmonary surface areas (rat/human).
6. For Parameter Distribution
   This approach was applied to all model parameters (Table 20.9) except for the ventilation rate, deposition fraction, and clearance rate, for which we have independent information from Hattis et al. (2001). These parameters' distribution characteristics are given in Table 20.8. One thousand randomly generated parameter sets for humans were generated.

20.5.2 Results

20.5.2.1 Results from Parameter Extrapolation

Table 20.9 shows the mean values for each of the parameters, for rats and for humans. The data available from Hattis et al. (2001) were used to construct the distribution of values for the ventilation rate, deposition fraction, and clearance rate in humans lung surface area is from parent (1992). The results are shown in Figure 20.7.

20.5.2.2 Simulation Results

The human-scaled model was run for the 1000 human parameter sets, for a human population with a 45-year exposure at 4 mg m^{-3} of respirable dust (TiO_2), working on an 8-h shift per day and 250 days per year (Tran et al. 2003). The results for the three main assays: lung burden, lymph node burden, and number of PMN cells, are shown in Figure 20.8.

For each assay, the upper curves represent the 95th, 90th, 85th, and 70th percentiles of the variation. The two lowest curves for each assay are generated using the central values in Table 20.9 and the 5th percentile. At a level of 4 mg m^{-3}, the extrapolated model predicted the occurrence of overload in approximately 30% of the human population. This is indicated

TABLE 20.9
The Rat Based Model Parameters and Their Extrapolated Counterparts

Parameter	Rat	Human
Exposure		
Concentration (mg m^{-3})	4	4
t_{start} (yr)	0	0
t_{end} (yr)	2	45
Deposited Dose		
Deposition fraction	0.06	0.32
Ventilation rate (m^3/h)	0.18	13.5
Hours exposed (h/day)	7	8
Retention and Clearance		
cl (day^{-1})	0.015	0.0036
i_{normal} (day^{-1})	0.03	0.0072
i_{max} (day^{-1})	1.8	0.4347
e (day^{-1})	0.1	0.0242
$m_{1/2}$ (mg)	5.8	4.05×10^3
Cell Recruitment		
Rec	0.025	8.67×10^{-5}
Lung Parameters		
Lung weight (gm)	1.43	1000
Lung surface area (cm^2)	4865	1430000

in the retardation of clearance in the build up of lung burden for the 95th, 90th, and 85th percentile curves. The pattern of overload diminished at lower percentiles, and from the 70th percentile, pulmonary clearance of dust is unimpaired. To find the level of airborne concentration such that 95 percent of the population can avoid overload, further extrapolations at lower exposure concentrations were needed. Simulations with stepwise decreases in concentration were made until concentration reached a level such that the 95 percent of the population does not develop overload within a lifetime working exposure. This critical exposure level was found to be 1.3 mg m^{-3}. Also, at this concentration, the predicted PMN level was low compared to the population of AM (7×10^9 cells) (Crapo et al. 1983; Dethloff and Lehnert 1988). Therefore, 1.3 mg m^{-3} could be interpreted as a Non-Observed-Adverse-Effect-Level (NOAEL) for humans (Trans et al. 1999).

Figure 20.8 and Figure 20.9 show the results of the further extrapolations: the upper curve represents the 95th percentile level for each assay. The central curve is obtained from the central values in Table 20.9 and the lower curve represents the fifth percentile.

20.6 HUMAN LUNG DOSIMETRY MODEL

The work above has illustrated an approach to obtain a human lung dosimetry model by extrapolating the animal-based results. Now, we introduce a biologically based human dosimetric lung model, developed to describe the long-term clearance and retention of respirable particles in the lungs of humans. This model has a unique advantage of being validated with human data.

The data used in developing this model included working lifetime exposure histories of lung burden data from an autopsy study in U.S. coal miners (Vallyathan et al. 1996). Several models describing the plausible mechanisms of particle retention and clearance in humans were initially investigated, including: (1) overloading as observed in rat lungs, (2) particle sequestration in the lung interstitium, and (3) a combination of both processes. The fits of these models to the coal miner

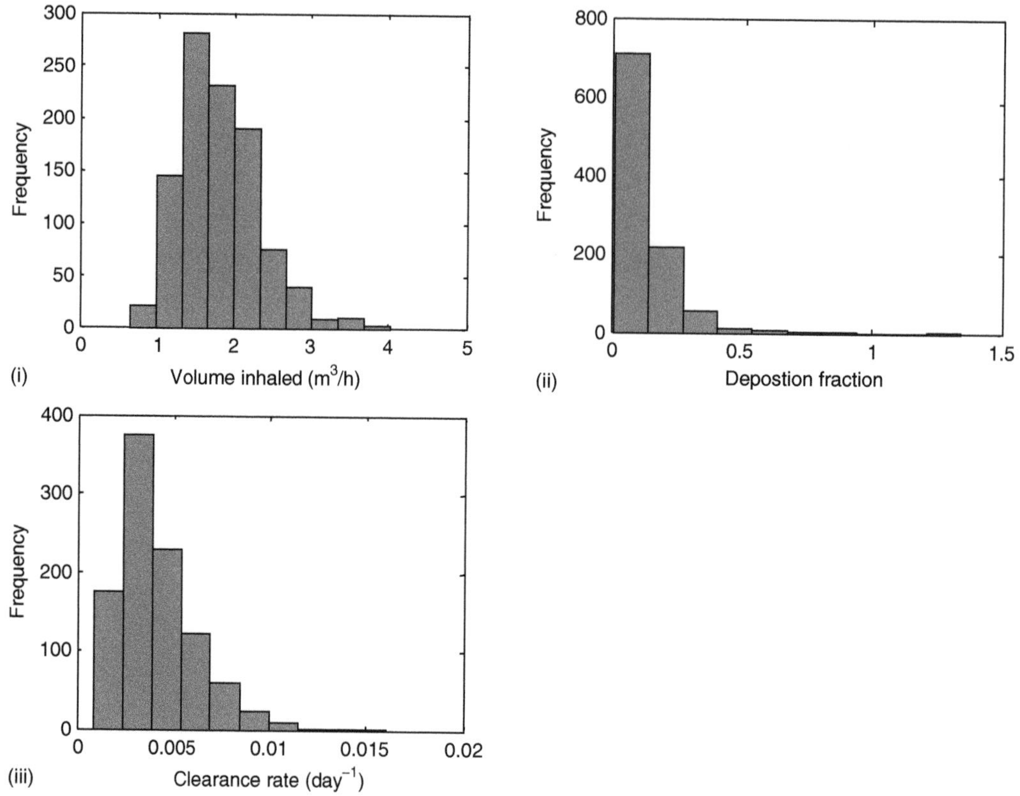

FIGURE 20.7 The distribution for (i) ventilation rate, (ii) deposition fraction, (iii) clearance rate (From Hattis, D., Goble, R., Russ, A., Banati, P., and Chu, M., *Risk Anal.*, 21(4), 585–599, 2001.)

data were evaluated using statistical methods to determine which mechanism may best describe the particle clearance kinetics in the lungs of these humans (Kuempel 2000).

The results showed that a one-compartment rat lung overload model under predicted the lung burdens in the miners with relatively low lifetime exposures and overpredicted the lung burdens in the miners with the higher exposures. This is because at low exposures, the rat model is a simple, first-order kinetic model that predicts effective clearance and very little particle retention in retired miners' lungs. However, at high exposures, the rat model predicts impaired clearance and much higher retained burdens than those actually observed in coal miners.

The model structure that provided the best fit to the coal miner data was a three-compartment, higher-order model with an interstitial or sequestration compartment (Kuempel et al. 2001a). This model includes alveolar, interstitial, and hilar lymph node compartments (Figure 20.10). The form of the model that provides the best fit to the lung dust burden data in these coal miners includes a first-order interstitialisation process and either no dose-dependent decline in alveolar clearance or much less decline than expected from rodent studies.

Key: D is the dose rate of deposited particles; first order rate coefficients include alveolar-macrophage mediated clearance of particles to the tracheobronchial region (K_T), transfer of particles into the pulmonary interstitium (K_I), and translocation of particles to the hilar lymph nodes (K_{LN}); F is an exponential decay function that describes overloading as a dose-dependent decline in K_T; and M is the particle mass in a given lung region, including that cleared to the tracheobronchial region (M_T), or retained in the alveolar (M_A), interstitial (M_I), gas-exchange (M_{LU}), or hilar lymph node (M_{LN}) regions.

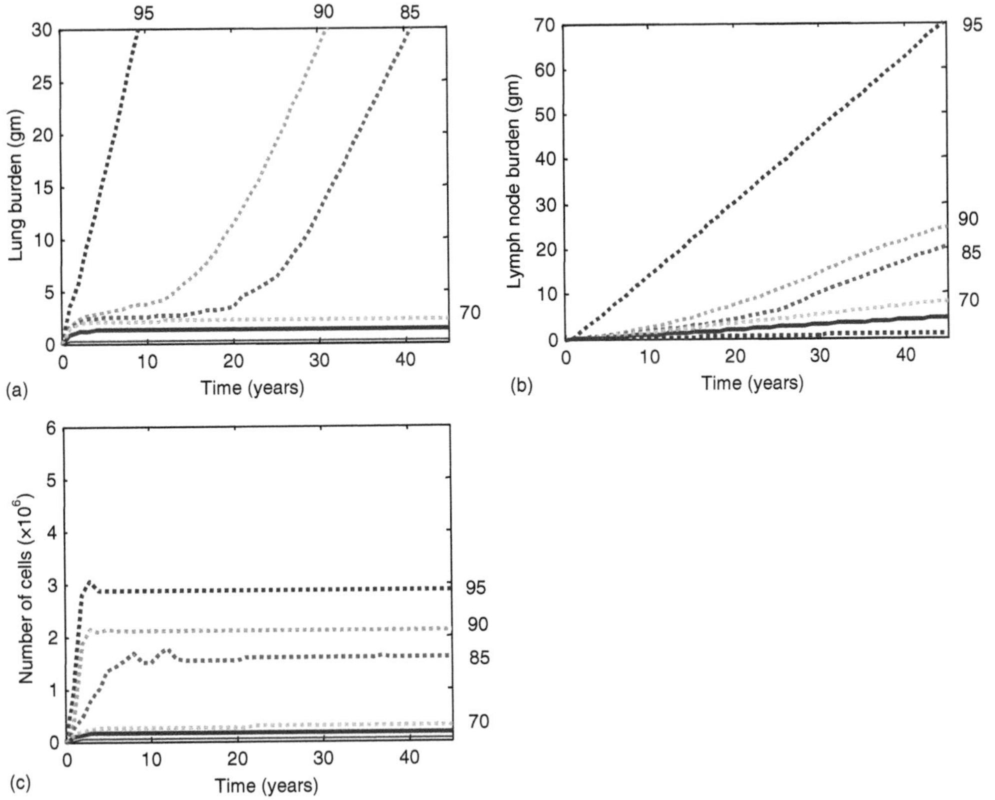

FIGURE 20.8 Simulation of the dose–response to TiO_2 in humans. (a) Lung burden, (b) Lymph node burden, and (c) Number of PMN cells. Each curve represents the 95th, 90th, 85th, and 70th percentile of the variation in each assay. The two lowest curves for each assay are generated using respectively the central value in Table 20.9 and the 5th percentiles.

20.6.1 Model Equations and Description

This three-compartment human lung model describes the kinetics of particle mass transfer in the lungs. The model consists of a series of nonlinear differential equations that are integrated over time to predict individuals' lung and lymph node particle burdens.

The rate of change of particle mass in the alveoli (M_A) at any time (t) is defined as

$$dM_A/dt = R_D - R_T - R_I \tag{20.20}$$

where R_D is the deposition rate (mg/yr) of inhaled, respirable particles into the alveoli (described in Equation 20.21). R_T is the clearance rate (mg/yr) of particles from the alveoli to the tracheobronchi (Equation 20.22). R_I is the transfer rate (mg/yr) of particles from the alveoli into the interstitium (Equation 20.24).

$$R_D = F_D \times C_I \times V_I \times d \tag{20.21}$$

where F_D is the fractional deposition (fraction of the inhaled particle mass that is deposited in the alveolar region of the lungs; C_I is the airborne concentration of dust inhaled (mg/m^3); V_I is the

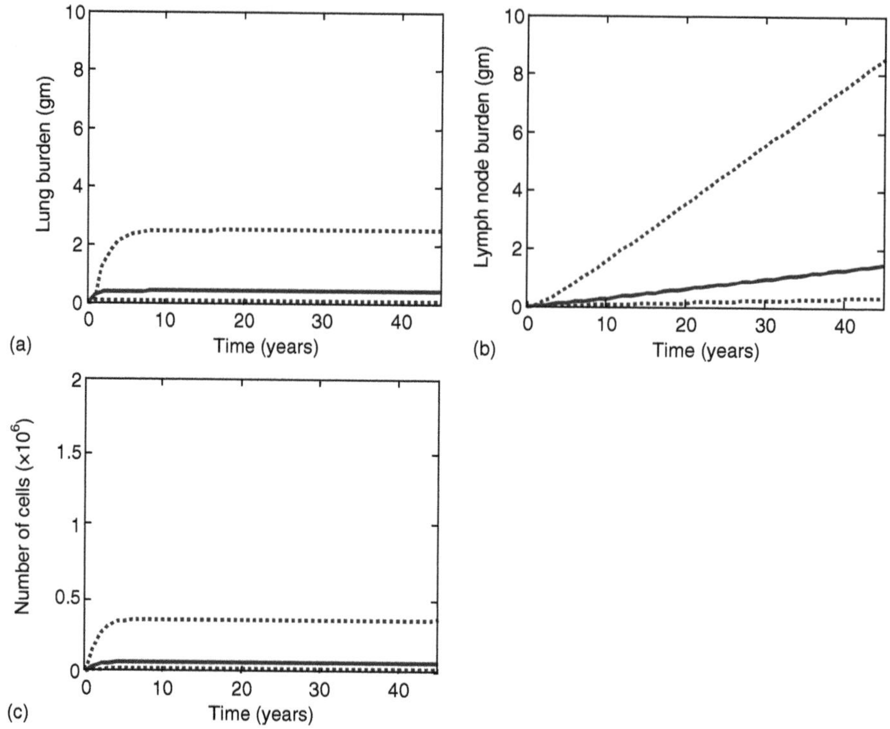

FIGURE 20.9 The dose–response at 1.3 mg m^{-3}. (a) Lung burden, (b) lymph node burden, and (c) number of PMN cells. For each assay, the upper curve represents the 95th percentile of the variation; the two lowest curves are generated using respectively the central values in Table 20.9 and the 5th percentiles.

volume of air inhaled in an 8-h work day (m^3/d); and d is the days worked per year (d/yr) (estimated as 5 days/week×50 weeks/year).

$$R_T = K_T \times 365 \times F \times M_A \tag{20.22}$$

where K_T is the first-order rate coefficient (d^{-1}) for particle clearance from the alveoli to the tracheobronchi. M_A is defined in Equation 20.20; 365 is the days per year, to convert R_T to units of yr^{-1}; and F is defined in Equation 20.23a and Equation 20.23b.

$$F = 1 \quad \text{when } M_A < M_{min} \tag{20.23a}$$

$$F = \exp\{-B[(M_A - M_{min})/(M_{max} - M_{min})]^C\} \quad \text{when } M_A > M_{min} \tag{20.23b}$$

where F is a dose-dependent modifying factor of K_T. M_{min} and M_{max} are constants representing the human-equivalent minimum and maximum critical lung dust burdens at which the dose-dependent decline in the alveolar clearance rate coefficient begins and reaches a maximum, respectively, as predicted from rodent studies (see next section); M_A is defined in Equation 20.20. When $M_A < M_{min}$, F is set equal to 1; when $M_A > M_{min}$, F equals a value (between 0 and 1) that is determined by B. C is a shape parameter (set to 1 in this model).

$$R_I = K_I \times 365 \times M_A \tag{20.24}$$

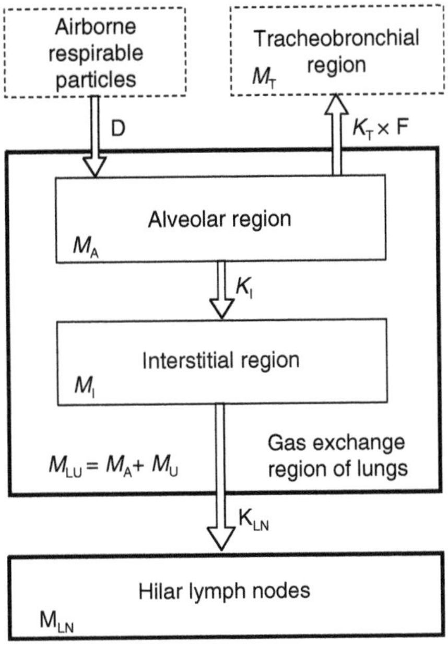

FIGURE 20.10 Three-compartment human lung dosimetry model. (From Kuempel, E. D., O'Flaherty, E. J., Stayner, L. T., Smith, R. J., Green, F. H. Y., and Vallyathan, V., *Reg. Toxicol. Pharmacol.*, 34, 69–87, 2001a. With permission.)

where K_I is the first-order rate coefficient (d^{-1}) for transfer of particles from the interstitium to the lung-associated (hilar) lymph nodes. M_A is defined in Equation 20.20; and 365 converts R_I to units of $year^{-1}$.

The rate of change of particle mass in the interstitium (M_I) at any time (t) is defined as

$$dM_I/dt = R_I - R_{LN} \tag{20.25}$$

where R_I is defined in Equation 20.20 and Equation 20.25. R_{LN} is the translocation rate (mg/yr) of particles from the interstitium to the lung-associated (hilar) lymph nodes, as follows

$$R_{LN} = K_{LN} \times 365 \times M_I \tag{20.26}$$

where K_{LN} is the first-order rate coefficient (d^{-1}) for translocation of particles from the interstitium to the hilar lymph nodes. M_I is defined in Equation 20.25; and 365 converts R_{LN} to units of yr^{-1}.

The rate of change of particle mass in the hilar lymph nodes (M_{LN}) at any time (t) is defined as

$$dM_{LN}/dt = R_{LN} \tag{20.27}$$

where R_{LN} is defined in Equation 20.26.

The mathematical equations in the three-compartment model reflect the biological structures and kinetic processes in human lungs that influence respirable particle clearance and retention. The build-up of particles in the alveolar compartment is determined by the rates of particle deposition, AM-mediated clearance, and particle transfer into the interstitium (Equation 20.20). The deposition of particles in the alveolar compartment is assumed to occur at a rate proportional to the concentration in inhaled air, i.e., a first-order process (Equation 20.21). AM-mediated clearance is described as a first-order process at lung burdens below those potentially causing impairment of clearance (Equation 20.22). At higher lung dust burdens, AM-mediated clearance can be described as dose-dependent, declining exponentially (Equation 20.23). The model also assumes that some particles will escape phagocytosis by AMs and enter the interstitium at a constant rate (Equation 20.24); this will occur even at low lung dust burdens, below the estimated human-equivalent dose associated with overloading of alveolar clearance. The buildup of particles in the interstitium is described by the difference in the rates of particles entering from the alveoli and particles leaving to the lung-associated lymph nodes (Equation 20.25). The rate of particle translocation to the lymph nodes is also assumed to be first-order (Equation 20.26), and the accumulation of particles in the lymph nodes occurs only by particles passing through the interstitium (Equation 20.27). The rate coefficients for these processes are estimated by the model fitting and parameter estimation (Kuempel et al. 2001a). By suitable integration, the model equations also enable prediction of the amount of dust in the lung and lymph node compartments at any point in time.

20.6.2 Model Parameter Description and Estimation

The initial model parameter values were based on data available in the literature. Some of these parameters were fixed while others were allowed to vary to optimize the model fit to the data. The fixed values include: fractional deposition (F_D) in the alveolar region of the lungs for particles with mass median aerodynamic diameter of 5 μm, assuming mouth breathing at inhalation rate of 1.7 m^3/h (ICRP 1994); the volume of air inhaled (V_I) in an 8-h work day (m^3/d), for heavy work, defined as 7 h of light exercise and 1 h of heavy exercise, for a reference worker (Caucasian, age 30 years, height 176 cm, weight 73 kg) (ICRP 1994); and d is the days worked per year (d/yr) (estimated as 5 days/week×50 weeks/year). The fixed parameters in the expression describing the overloading of alveolar clearance include the estimated human-equivalent lung dust burdens associated with the beginning of decline in the alveolar clearance rate coefficient and the leveling-off of that decline (M_{min} and M_{max}, respectively, in Equation 20.23a and Equation 20.23b above).

The parameter values that were allowed to vary in optimizing the fit of the model to the data were K_T, K_{LN}, K_I, and B. Because a primary objective was to evaluate clearance kinetics of particles in humans, including the possibility of overloading of alveolar clearance as observed in rodents, the deposition parameters were fixed at the average human values in the literature (ICRP 1994), and the clearance parameters were iteratively varied to determine the best fit of the model to the data. A systematic grid search approach was used to determine the parameter values that provided the best fit of the model to the data (Kuempel et al. 2001a). A sensitivity analysis of the model parameter values was performed (Kuempel et al. 2001b). The distribution of clearance rate coefficient was estimated, which provided quantitative information about the plausible particle clearance rates in this cohort (Kuempel et al. 2001b). These values were consistent with values previously reported for long-term particle clearance in humans (Bailey, Fry, and James 1985). The model parameters are described in Table 20.10.

The optimized parameter values and statistical model fit are shown in Table 20.11. The model with the best fit (lowest mean squared error) was the three-compartment model with interstitial-sequestration compartment but with no dose-dependent decline in AM-mediated clearance.

From Table 20.11, the optimal choice, i.e., the three-compartment model with first-order interstitialisation and effective (no overload) alveolar clearance, was chosen and simulated, assuming the mean exposure to respirable coal mine dust among all miners in this study (3 mg/m^3 for 36 years) (Kuempel et al. 2001a). The results are presented graphically in Figure 20.11.

A sensitivity analysis was performed to determine which model parameters are most influential in the model prediction of two key variables: the lung and lymph nodes burden

TABLE 20.10
Description of Variables and Constants in Three-Compartment Human Lung Dosimetry Model

Abbreviation	Units	Description
F_D	None	Fractional deposition of airborne respirable dust in alveolar region
V_I	m^3/day	Volume of air inhaled in 8-h day, heavy work
D	Days/year	Days exposed/year
C_I	mg/m^3	Mean concentration of respirable coal mine dust inhaled, by job[a]
D	Years	Duration of exposure, by job[a]
K_T	day^{-1}	Rate coefficient, alveolar macrophage-mediated clearance to tracheobronchi
K_I	day^{-1}	Rate coefficient, transfer from alveoli to interstitium
K_{LN}	day^{-1}	Rate coefficient, translocation from interstitium to hilar lymph nodes
F	None	Exponential decay function, dose-dependent reduction in K_T
B	None	Slope modifier of F
C	None	Shape modifier of F
M_{min}	mg	Minimum lung dust burden associated with beginning of dose-dependent decline in K_T
M_{max}	mg	Maximum lung dust burden associated with leveling off of dose-dependent decline in K_T

[a] Individual work history data for each miner (input data).
Source: From Kuempel, E. D., O'Flaherty, E. J., Stayner, L. T., Smith, R. J., Green, F. H. Y., and Vallyathan, V., *Reg. Toxicol. Pharmacol.*, 34, 69–87, 2001a. With permission.

(Kuempel et al. 2001b). Each model parameter of the optimal parameter set was allowed to change by ± 10 percent of the central value given in Table 20.11. The summary statistics, mean squared error and mean bias, were calculated for the two variables. The results of this exercise are shown in Table 20.12.

It is clear that any change which results in a higher retention of lung burden, such as a higher deposition, lower clearance, or higher interstitialization, will lead to a higher mean squared error for lung burden. In particular, the predicted lung burden is most sensitive to the deposition rate. For the predicted lymph nodes burden, as expected, the deposition rate and the translocation rate to the lymph nodes are most sensitive.

20.6.3 Application of Human Lung Dosimetry Modeling in Risk Assessment

In this section, we describe an alternative biomathematical modelling approach to that described in Section 20.5. In Section 20.5, a rat-based exposure-dose-response model is extrapolated to humans to predict the working lifetime exposure concentration of TiO_2 that is not likely to result in pulmonary inflammation, based on the rat model. Here, a human-based lung dosimetry model (Sections 20.7.1 and 20.7.2) describing the exposure-dose relationship of respirable particles is used in conjunction with a statistical model of the rat dose-response relationship for particle surface area dose in the lungs and initiation of pulmonary inflammation following inhalation exposure to finesized TiO_2 or $BaSO_4$. This illustrates two different biornathematical modeling approaches to the same rat data (Tran et al. 1999). First, the relationship between particle surface area dose and pulmonary inflammation in rats was investigated using a statistical approach. Statistical models do not explicitly model the biological mechanisms; rather, they involve fitting a mathematical expression to the data. The data used here are from the subchronic inhalation study in rats exposed to fine TiO_2 or $BaSO_4$ (Tran et al. 1999) described earlier. Specifically, individual rat data were obtained for PMN count in the lungs. Different groups of rats were used to estimate retained particle lung burden.

TABLE 20.11
Optimized Parameter Values and Statistical Fit of the Three-Compartment Human Lung Dosimetry Models, by Degree of Overloading of Alveolar Macrophage-Mediated Clearance in Human Interstitial/Sequestration Model

Model Parameter[a]	Model Structure and Optimized Parameter Value		
	No Overload	50% Overload	90% Overload
F_D	0.12	—[b]	—[b]
V_I (m^3/day)	13.5	—[b]	—[b]
d (days/year)	250	—[b]	—[b]
K_T (years)	1×10^{-3}	1.5×10^{-3}	1.4×10^{-3}
K_I (day^{-1})	4.7×10^{-4}	7.0×10^{-4}	3×10^{-4}
K_{LN} (day^{-1})	1×10^{-5}	1×10^{-5}	1×10^{-5}
B (day^{-1})	0.0001	0.69	2.3
C	1	—[b]	—[b]
M_{min}	1.05×10^2	—[b]	—[b]
M_{max}	1.05×10^5	—[b]	—[b]

Dataset	Statistical Fit of Model to Lung Burden and Lymph Node Burden Data: Mean Squared Error (MSE)		
	No Overload	50% Overload	90% Overload
All Miners ($n = 131$)	79.3	85.8	231
Miners with hilar lymph node burden data ($n = 57$)	94.7	106	354
	1.31[c]	1.39[c]	2.15[c]
Miners with no post-exposure duration ($n = 11$)	70.0	68.9	148

[a] Parameter description provided in Table 20.10.
[b] Fixed at values in no overload model.
[c] MSE for hilar lymph dust node burden (g).
Source: From Kuempel, E. D., O'Flaherty, E. J., Stayner, L. T., Smith, R. J., Green, F. H. Y., and Vallyathan, V., *Reg. Toxicol. Pharmacol.*, 34, 69–87, 2001a. With permission.

The following piecewise linear model fit was fit to the PMN data (Figure 20.12) (Kuempel et al. 2005, NIOSH 2005).

$$\mu_{PMN}(d_i) = \begin{cases} \beta_0 & d_i < \gamma \\ \beta_0 + \beta_1(d_i - \gamma) & d_i \geq \gamma \end{cases} \qquad (20.28)$$

where d_i is the dose for the ith rat; γ is the threshold parameter; B_0 is the mean PMN count for dose of zero or up to γ; and B_1 describes the effect of dose above γ on the mean PMN count.

This nonlinear equation was fit to the data using maximum likelihood estimation of the parameter values. Approximate confidence limits were found using profile likelihood and validated using parametric bootstrap methods. The maximum likelihood estimate (MLE) of the threshold dose was 139 cm^2 (95% confidence interval estimates: 130–144 cm^2), based on the model fit to the TiO_2 and $BaSO_4$ data from Tran et al. (1999). The threshold dose (MLE and CI) was considered the "critical" dose for initiation of pulmonary inflammation in rats, and this critical dose per g of lung tissue was assumed to be the same in rats and humans (i.e., equal sensitivity to equivalent dose assumed) (Jarabek et al. 2005) (Table 20.13). The estimated critical lung dose as particle surface

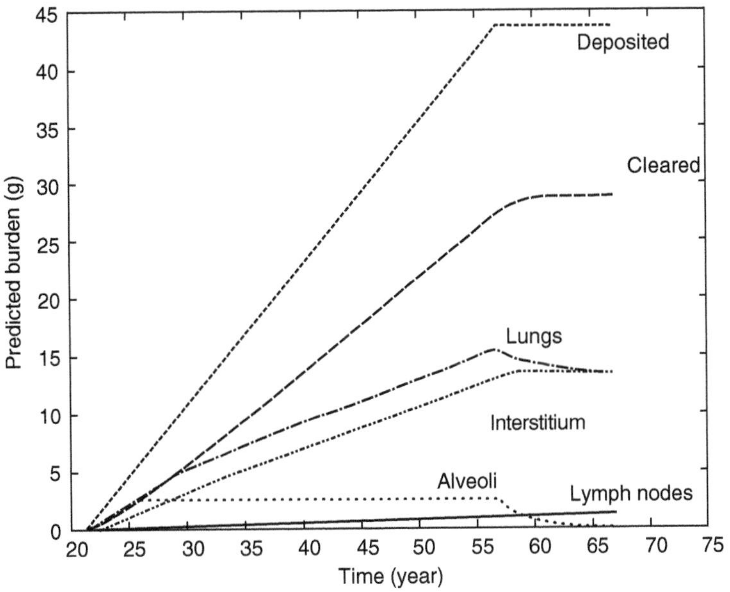

FIGURE 20.11 Predicted mass of particles retained in various lung compartments over time "Deposited" refers to the alveolar region of the lungs, and "cleared" means from the alveoli to the mucociliary clearance path in the tracheobronchial region. (From Kuempel, E. D., O'Flaherty, E. J., Stayner, L. T., Smith, R. J., Green, F. H. Y., and Vallyathan, V., *Reg. Toxicol. Pharmacol.*, 34, 69–87, 2001a. With permission.)

area is the same for fine and ultrafine particles (Table 20.13) due to the consistent relationship between particle surface area dose and PMN response (whether particle size is fine or ultrafine) (Tran et al. 1999; Oberdörster et al. 1994). To calculate the critical dose as particle mass per whole lung in humans, the critical dose as particle surface area (m^2 TiO_2/g lung) was divided by the particle specific surface area (m^2 TiO_2/g TiO_2) and multiplied by the human lung mass (assumed to be 1000 g vs. 1 g in rats) (Table 20.13) (Kuempel et al. 2005; NIOSH 2005).

The critical dose estimates as particle surface area were converted to mass dose because the lung dosimetry models are mass-based, as are workplace exposure limits. In addition to the interstitial/sequestration model described here (Section 20.6), the MPPD human lung dosimetry model (CIIT & RIUM 2002) was used for comparision. These models were used to estimate the airborne concentrations of either fine or ultrafine TiO_2 over a 45-year working lifetime that would be associated with an increase in pulmonary inflammation, derived from the rat data (Table 20.13) (Kuempel et al. 2005; NIOSH 2005). The arithmetic mean parameter estimates for K_T and K_{LN} from Tran and Buchanan (2000) were used in the interstitial/sequestration model since those values were based on the larger U.K. miner cohort with more detailed exposure data than the U.S. study. The two lung dosimetry models (MPPD and interstitial/sequestration) provided similar estimates, although the MPPD model predicted higher airborne concentrations by a factor of approximately two, due to prediction of lower lung burdens by the same factor. In other words, the interstitial/sequestration model predicted higher average lung burdens for a given external exposure than did the MPPD model. The MPPD model uses the ICRP (1994) clearance model, which includes three first-order clearance rate coefficients for the alveolar region and has been shown previously to predict lower lung burdens than the interstitial/sequestration model (Kuempel and Tran 2002). Deposition was similar in both models, as the estimated fractional deposition from the MPPD model (CIIT & RIUM 2002) was also used as the deposition fraction in the interstitial/sequestration clearance model.

TABLE 20.12
Sensitivity of Best Group-Fit Model Parameters for Deposition and Clearance among Miners with Lymph Node Data ($n = 57$)

Parameter and Specified Change from Initial Value[a]	Mean Squared Error	Mean Bias (g)	Mean Predicted Lung Dust Burden (g)	Percent Change in Output with 10% Change in Input[b]	Mean Predicted Lymph Node Dust Burden (g)	Percent Change in Output with 10% Change in Input[c]
Default values[d]	95.2	+0.99	14.2	—[e]	1.41	—[e]
F_D +10%	101.0	−0.59	15.6	+9.8	1.55	+9.9
F_D −10%	95.5	+2.3	12.9	−9.2	1.30	−7.8
K_T +10%	94.0	+2.0	13.3	−6.3	1.33	−5.7
K_T −10%	98.6	−0.15	15.2	+7.0	1.50	+6.4
K_I +10%	98.1	+0.03	15.0	+5.6	1.50	+6.4
K_I −10%	93.8	+2.0	13.2	−7.0	1.31	−7.1
K_{LN} +10%	95.0	+1.0	14.0	−1.4	1.54	+9.2
K_{LN} −10%	95.5	+1.0	14.3	+0.7	1.28	−9.2

[a] F_D, fractional deposition; first-order rate coefficients; K_T, alveolar macrophage-mediated clearance of particles to the tracheobronchi; K_I, transfer of particles to the interstitium; K_{LN}, translocation of particles to the hilar lymph nodes.
[b] Output is mean predicted lung burden.
[c] Output is mean predicted lymph node burden.
[d] $F_D = 0.12$; $K_T = 8.8 \times 10^{-4}\ d^{-1}$; $K_I = 4.5 \times 10^{-4}\ d^{-1}$; $K_{LN} = 1.0 \times 10^{-5}\ d^{-1}$.
[e] Not applicable because percent change is relative to the default values.
Source: From Kuempel, E.D., Tran, C.L., Smith, R. J., and Bailer, A. J., *Reg. Toxicol. Pharmacol.*, 34, 88–101, 2001b. With permission.

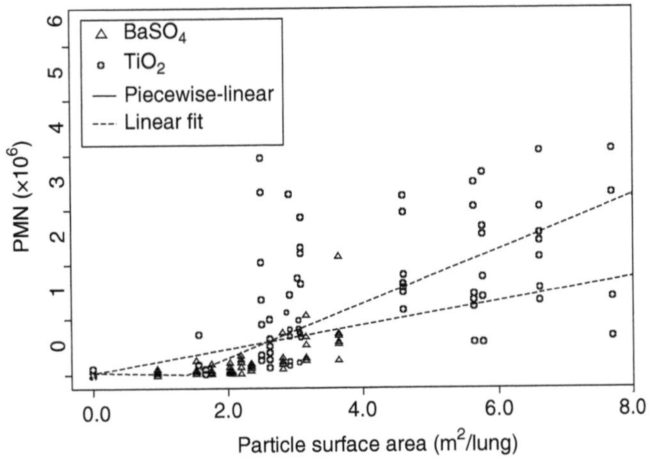

FIGURE 20.12 Piecewise-linear and linear model fits to rat data on pulmonary inflammation (PMN count) and particle surface area dose of titanium dioxide. (Data from Tran, C. L., Cullen, R. T., Buchanan, D., Jones, A. D., Miller, B. G., Searl, A., Davis, J. M. G., and Donaldson, K. Investigation and Prediction of pulmonary responses to dust. Part II. In *Investigations into the Pulmonary Effects of Low Toxicity Dusts*. Parts I and II., Health and Safety Executive, Suffolk, UK, Contract Research Report 216/1999, 1999.)

TABLE 20.13
Airborne Mass Concentrations of Fine and Ultrafine TiO_2 (Over a 45-Year Working Lifetime) and Human-Equivalent Lung Burdens Associated with Pulmonary Inflammation in Rats

	Human-Equivalent Critical Lung Dose: MLE (95% LCL)		Mean Airborne Exposure Concentration: MLE (95% LCL) (mg/m^3)	
Particle size	Particle Surface Area (cm^2/g lung)	Particle Mass (g/lung)	Multiple-Path Particle Deposition (MPPD) Lung Model (CIIT & RIVM 2002) (mg/m^3)	Interstitial Lung Model (Kuempel et al. (2001a, 2001b) and; Tran and Buchanan 2000) (mg/m^3)
Fine (2.1 μm MMAD, 2.2 GSD; 6.68 m^2/g)	139 (130)	2.1 (1.9)	2.0 (1.8)	1.1 (0.9)
Ultrafine (0.8 μm MMAD, 1.8 GSD; 48 m^2/g)	139 (130)	0.29 (0.22)	0.22 (0.17)	0.10 (0.08)

Abbreviations: MMAD, mass median aerodynamic diameter; GSD, geometric standard deviation; MLE, Maximum likelihood estimate; 95% LCL, 95% lower confidence limit.

Source: From Kuempel, E. D., Wheeler, M., Smith, R., and Bailer, A. J., *Nanomaterials: A Risk to Health at Work? First International Symposium on Occupational Health Implications of Nanomaterials*, 12–14 October 2004, Buxton, Derbyshire, UK., Health and Safety Executive, U.K., Buxton, 111, 2005; NIOSH, 2005. NIOSH Current Intelligence Bulletin: Evaluation of Health Hazard and Recommendations for Occupational Exposure to Titanium Dioxide. Unpublished Public Review Draft, November 22, 2005. Cincinnati, OH: U.S. Department of Health and Human Services, Public Health Service Centers for Disease Control and Prevention, National Institute for Occupational Safety and Health, 158, Available at: http://www.cdc.gov/niosh/docs/preprint/tio$_2$/pdfs/TIO$_2$Draft.pdf.

20.7 DISCUSSION

20.7.1 The Contribution of Dosimetric Modeling to Particle Toxicology

In modeling complex biological processes, it is often possible to describe broad trends using simple regression or other statistical models. However, when it is desired to model the evolution of a process over time, and particularly when the status of the process at any time point influences its subsequent course, it is necessary to develop dynamic models based around differential equations. In either case, the model will be more plausible if based upon an understanding of the underlying biological mechanisms.

A model is a mathematical equation or system of equations that, given quantitative inputs, predicts certain outputs; and, given the same inputs, the output prediction will always be the same. In considering the relevance of model predictions to real-life situations involving populations of animals or of humans, there is an additional need to allow for the variation to be expected in any such population.

The present work has extended our earlier deterministic rat lung dosimetry model by allowing stochastic variation in the parameters, which in turn induces variation in the outputs. Since the input variation is under our control and known, we can investigate the relationships between input and output variation. Such investigations are commonly labeled "sensitivity analysis" when the focus is on the effect of small variations in the inputs, and "uncertainty analysis" when we consider the entire range of variation (Saltelli, Chan, and Scott 2000). In the rat model presented here, we have used Monte Carlo simulation, with plausible assumptions for parameter variation, to generate a pseudo-sample of 1000 instances from an idealized population of rats. By extrapolating the results from this exposure–dose–response model in rats to humans, we have predicted exposure concentrations of poorly soluble particles such as TiO_2 that are not expected to result in lung clearance overload and the onset of inflammation in most workers. We believe that our approach to this problem is novel, by using biomathematical lung models in the risk assessment of poorly soluble particles including nanoparticles.

It is clear that these extensions to the basic process of dynamic modeling introduce a number of new assumptions, and that the plausibility of the results rests in large part on the plausibility of the assumptions, in particular on the implied ranges of variability. At the same time, biomathematical lung models can play an integral role in research on the toxicokinetics of inhaled particles. These models are useful in the design of experimental studies, such as by identifying the research and data needs to validate key parameters and by facilitating the generation and testing hypotheses on biological mechanisms.

The rat lung dosimetry model described here was extrapolated to humans by using human parameter values where available (e.g., the deposition fraction) and by using appropriate allometric scaling methods when human parameter values were not available and needed to be estimated. Data available from literature (Hattis et al. 2001) regarding the variation in some key parameters in humans, such as the deposition fraction and clearance rate, have shown a much wider variation than the observed variation in animal data. This was no surprise as the rats are all from the same strain and the same supply.

A Monte Carlo simulation of the rat biomathematical model extrapolated to humans,predicted that for a working life-time exposure to fine TiO_2 at 4 $mg \cdot m^{-3}$, only 70 percent of the population would avoid overload (Section 20.5.2.2). A concentration level such that 95 percent of the population would avoid overload was found to be 1.3 $mg \cdot m^{-3}$. At this level, the PMN number was not elevated (Figure 20.9). Therefore, 1.3 $mg \cdot m^{-3}$ is an estimate of a working lifetime exposure concentration to fine-sized TiO_2 that is not expected to result in lung disease associated with pulmonary inflammation, based on the rat biomathematical lung model extrapolated to humans.

We have also taken a direct approach at modeling the human exposure-dose relationship for respirable particles. For humans, the overloading of lung clearance of particles as observed in rats has not been demonstrated exposure-dose relationship for respirable particles, although a study of lung dust burdens in retired coal miners showed that particle clearance rates were reduced or even undetectable in some miners (Freedman and Robinson 1988). This finding is consistent with particle overloading in the lungs of rodents, in which clearance becomes impaired at high lung dust burdens and the impairment continues after exposure ceases. However, the findings are also consistent with a sequestration process, in which particles are transferred (as a first-order process) to the interstitial region of the lungs and retained, with very slow clearance to the hilar lymph nodes, as shown in the lung dosimetry modeling in coal miners. We have constructed various models representing different degrees of overload and found that the model structure that provided the best fit to the coal miner data was a higher-order model with an interstitial or sequestration compartment (Kuempel 2000; Kuempel et al. 2001a). The model representing the rat-based overload kinetics did not improve the model fit to the data, although a lesser degree of overloading could not be ruled out. This model structure was validated in an independent study of U.K. coal miners (Tran and Buchanan 2000). In addition, in that study where individual miners' working lifetime quartz exposure data were available, it was shown that quartz translocated to the hilar lymph nodes at a faster rate than coal dust. The optimized parameter values were consistent with but not identical in the two cohorts.

The human model structure with an interstitial or sequestration compartment is consistent with the observations in retired coal miners (Freedman and Robinson 1988) and with observation of particle retention in the interstitium of human lungs (Nikula et al. 1997). This human model structure is compatible with some of the animal models, which include both sequestration compartments and dose-dependent overloading (Strom, Johnson, and Chan 1989; Stöber et al. 1990a, 1990b; Stöber, 1999; Tran et al. 1999, 2000), although the additional overloading pathway was not needed to fit the coal miner data. A principal difference in this human model compared to the ICRP and NCRP models is that it treats the alveolar and interstitial regions of the lung as separate compartments, which reflects both the biological structure of the lungs and the disposition of particles in these regions.

Further research is needed on whether interspecies differences in particle retention sites within the lungs may influence the sensitivity to a given lung dust burden, and influence the disease response. The findings from this study also suggest that a human lung dosimetry model without an interstitialization or sequestration compartment would not be adequate for predicting the end-of-life lung dust burdens in humans—at least not among those with high dust exposures such as coal miners (Kuempel and Tran 2002). Other data sets, particularly those including individuals with low exposures, are needed for further validation of this model and are ongoing.

Finally, we have also used a more empirical approach, based on a statistical model, to estimate a threshold lung burden of fine and ultrafine TiO_2 at which pulmonary inflammation begins in rats (Section 20.6.3). This estimate of a critical lung dose at which pulmonary inflammation begins is assumed to be equivalent in rats and humans. That is, in the absence of human data, an equivalent particle surface area dose in the lungs in either species is assumed to elicit an equal inflammation response. Next, the human lung dosimetry models were used to estimate the working lifetime exposure concentration that would lead to an equivalent critical lung burden (Table 20.13). These results show that, as expected, the mass airborne concentrations associated with a given particle surface area dose in the lungs is lower for ultrafine TiO_2 than for fine TiO_2. This is due to the dose-response relationship between particle surface area dose and pulmonary inflammation, and to the higher surface area per unit mass of ultrafine TiO_2 compared to fine TiO_2 . Despite the different biomathematical and statistical models used to fit the rat data, the risk estimates are similar. The lower bound estimates for fine-sized TiO_2 (95% LCL of 0.9 or 1.8 mg/m^3) (Table 20.13), based on the statistical modelling of the rat doseresponse data and the human biomathematical exposure-dose model, are similar to the 95% LCL (1.3 mg/m^3) estimated earlier using the rat biornathematical exposuredose-response model extrapolated to humans (Section 20.5.2.2). This consistency across models, rat and human, with species-appropriate parameter values, provides some validation of the models for use in risk assessment. These estimates are provided for illustration only of the use of biornathematical lung models in risk assessment, and it is beyond the scope of this chapter to discuss the many issues that are considered in a full risk assessment. For example, it should be noted that none of these exposure estimates include adjustment by "safety" or "uncertainty" factors, which are often used in risk assessment to account for uncertainty in factors including extrapolation from animals to humans, interindividual variability, and subchronic to chronic effects (Jarabek et al. 2005). The use of biologically-based models may eliminate the need for some of these uncertainty factors, for example, by accounting for the kinetic differences that influence the exposure-dose relationship in animals and humans.

20.7.2 Issues in the Dosimetry of Nanoparticles

An important area for future lung dosimetry model development is extension of current validated models to include the translocation of particles. Short-term rodent studies have shown that nanoparticles are retained in the lungs to a greater extent than larger respirable particles (Ferin, Oberdörster, and Penney 1992; Oberdörster, Ferin, and Lehnert 1994), which may be due to less effective phagocytosis by AMs (Renwick, Donaldson, and Clouter 2001; Renwick et al. 2004) and increased entry of the nanoparticles into the intersititialization. Nanoparticles may also enter the blood circulation and translocate to other organs, although the rate of translocation may depend on the chemical composition of the particles (Kreyling et al. 2002; Oberdörster et al. 2002; Geiser et al. 2005). In contrast, in a study of the long-term (up to 6 months) clearance of iridium nanoparticles, the lung retention was found to be similar to that reported previously for other poorly soluble, micrometer-sized particles (Semmler et al. 2004). Comparison of observed vs. model-predicted lung burdens in rats show that some rat lung dosimetry models (Tran et al. 2000; Tran, Graham, and Buchanan 2001; Tran et al. 2002; CIIT and RIVM 2002) predict reasonably well the retained mass lung burdens in rats exposed by chronic inhalation to ultrafine or fine poorly soluble particles (Kuempel et al. 2006). More study is needed on the role of particle characteristics, such as chemical composition, surface charge, and size, on the translocation of inhaled particles. In humans, the translocation rate of quartz particles to the lung-associated lymph nodes was estimated to be greater than that for coal particles (Tran and Buchanan 2000), a finding consistent with pathology data (Seaton and Cherrie 1998).

Because of the limited and somewhat contradictory data, it is not certain to what extent the current mass-based lung dosimetry models may predict the clearance and translocation of respirable particles of various size and composition. These models describe the mass transfer of *all* particles that deposit in a given region of the respiratory tract. Since lung deposition models describe particle size-specific deposition in each major region of the respiratory tract, these models inherently allow for particle size-specific clearance to the extent that the particles are subject to clearance by the biological mechanisms of a given respiratory tract region. However, if the efficiency of the particle clearance mechanisms within a lung region vary with particle size, as in the lung airways (Kreyling and Scheuch 2000), then current clearance/retention models may need to be revised to account for this difference. In addition, current models may need to be extended to include pathways for the direct translocation of nanoparticles to the blood circulation and to other organs beyond the lungs. Some models include pathways for dissolution of soluble particles (e.g., Yu, Yoon, and Chen 1991; ICRP 1994). In addition, current models may need to be extended to include other routes of exposure besides inhalation, including dermal, ingestion, or even translocation of nanoparticles along the olfactory nerve into the brain, as reported by Oberdörster et al. (2004) in rats. Before lung dosimetry models in rats or humans are extended to describe the disposition of inhaled nanoparticles, the model structure needs to be validated using existing data of fine and ultrafine particles. The rat and human models described in this chapter are biologically based, validated models of the long-term clearance and retention in the alveolar region of the lungs. Given the many existing lung dosimetry models for particle deposition and clearance, for practical purpose, it would be worthwhile to harmonize these various model structures to the extent possible, including integrating validated models in one format for use and additional development.

ACKNOWLEDGMENT

The authors would like to acknowledge Matt Wheeler, NIOSH, for fitting the piecewise linear model to the rat data and for preparing Figure 20.12.

REFERENCES

ACGIH, TLVs® and BEIs® Based on the Documentation of the Threshold Limit Values for Chemical Substances and Physical Agents & Biological Exposure Indices. American Conference of Governmental Industrial Hygienists, Cincinnati, Ohio, U.S.A., Appendix C, 74–77, 2005.

Adamson, I. Y. and Hedgecock, C., Patterns of particle deposition and retention after instillation to mouse lung during acute injury and fibrotic repair, *Exp. Lung Res.*, 21 (5), 695–709, Sep–Oct. 1995.

Bailey, M. R., Fry, F. A., and James, A. C., Long-term retention of particles in the human respiratory tract, *J. Aerosol. Sci.*, 16(4), 295–305, 1985.

Bermudez, E., Mangum, J. B., Asgharian, B., Wong, B. A., Reverdy, E. E., Janszen, D. B., Hext, P. M., Warheit, D. B., and Everitt, J. I., Long-term pulmonary responses of three laboratory rodent species to subchronic inhalation of pigmentary titanium dioxide particles, *Toxicol. Sci.*, 70 (1), 86–97, 2002.

Bermudez, E., Mangum, J. B., Wong, B. A., Asgharian, B., Hext, P. M., Warheit, D. B., and Everitt, J. I., Pulmonary responses of mice, rats, and hamsters to subchronic inhalation of ultrafine titanium dioxide particles, *Toxicol. Sci.*, 77, 347–357, 2004.

Churg, A. M. and Stevens, B., Association of lung cancer and airway particle concentration, *Environ. Res.*, 45(1), 58–63, 1988.

Churg, A., Brauer, M., del Carmen Avila-Casado, M., Fortoul, T. I., and Wright, J. L., Chronic exposure to high levels of particulate air pollution and small airway remodeling, *Environ. Health Perspect.*, 111 (5), 714–718, 2003.

CIIT and RIVM, Multiple-path particle deposition: a model for human and rat airway particle dosimetry, v.1.0. Research Triangle Park, NC: CIIT Centers for Health Research (CIIT) and The Netherlands: National Institute for Public Health and the Environment (RIVM), 2002.

Crapo, J. D., Young, S. L., Fram, E. K., Pinkerton, K. E., Barry, B. E., and Crapo, R. O., Morphometric characteristics of cells in the alveolar region of mammalian lungs, *Am. Rev. Respir. Dis.*, 128, S42–S46, 1983.

Crapo, J. D., Chang, Y. L., Miller, F. J., and Mercer, R. R., Aspects of respiratory tract structure and function important for dosimetry modeling: Interspecies comparisons, In *Principles of Route-to-Route Extrapolation for Risk Assessment*, Gerrity, J. R. and Henry, C. J., eds., Elsevier, New York, 15–32, 1990.

Dethloff, L. A. and Lehnert, B. E., Pulmonary interstitial macrophages: isolation and flow cytometric comparisons with alveolar macrophages and blood monocytes, *J. Leukocyte Biol.*, 43, 80–90, 1988.

Donaldson, K. and Tran, C. L., Inflammation caused by particles and fibers, *Inhal Toxicol*, 14 (1), 5–27, 2002.

Elder, A, Gelein, R., Finkelstein, J. N., Driscoll, K. E, Harkema, J., Oberdörster, G., Effects of subchronically inhaled carbon black in three species. I. retention kinetics, lung inflammation, and histopathology, *Toxicol Sci.*, [Epub ahead of print], 2005.

Ferin, J., Oberdörster, G., and Penney, D. P., Pulmonary retention of ultrafine and fine particles in rats, *Am. J. Respir. Cell Mol. Biol.*, 6, 535–542, 1992.

Freedman, A. P. and Robinson, S. E., Noninvasive magnetopneumographic studies of lung dust retention and clearance in coal miners, In *Respirable Dust in the Mineral Industries: Health Effects, Characterization and Control*, Frantz, R. L. and Ramani, R. V., eds., The Pennsylvania State University, University Park, PA, 181–186, 1988.

Geiser, M., Rothen-Rutishauser, B., Kapp, N., Schurch, S., Kreyling, W., Schulz, H., Semmler, M., Im Hof, V., Heyder J, J., and Gehr, P., Ultra-fine particles cross cellular membranes by nonphagocytic mechanisms in lungs and in cultured cells, *Environ. Health Perspect.*, 113 (11), 1555–1560, 2005.

Hattis, D., Goble, R., Russ, A., Banati, P., and Chu, M., Human interindividual variability in susceptibility to airborne particles, *Risk Anal.*, 21(4), 585–599, 2001.

ICRP, Human respiratory tract model for radiological protection. A report of a task group of the International Commission on Radiological Protection, Elsevier Science Inc., Tarrytown, New York, ICRP Publication No. 66, 1994.

Ings, R.M., Interspecies scaling and comparisons in drug development and toxicokinetics, *Xenobiotica*, 20 (11), 1201–1231, 1990.

Jarabek, A. M., Asgharian, B., and Miller, F. J., Dosimetric adjustments for interspecies extrapolation of inhaled poorly soluble particles (PSP), *Inhal. Toxicol.*, 17 (7–8), 317–334, 2005.

Jones, A. D., Vincent, J. H., McMillan, C. H., Johnston, A. M., Addison, J., McIntosh, C., Whittington, M. S., Cowie, H., Parker, I., Donaldson, K., and Bolton, R. E., Animal studies to investigate the deposition and clearance of inhaled mineral dusts, Final report on CEC Contract 7248/33/026. Institute of Occupational Medicine, Edinburgh, (IOM Report TM/88/05), 1988.

Koblinger, L. and Hofmann, W., Analysis of human lung morphometric data for stochastic aerosol deposition calculations, *Phys. Med. Biol.*, 30 (6), 541–556, 1985.

Koblinger, L. and Hofmann, W., Monte Carlo modeling of aerosol deposition in human lungs. Part I: simulation of particle transport in stochastic lung structure, *J. Aerosol Sci.*, 21, 661–674, 1990.

Kreyling, W. G. and Scheuch, G., Clearance of particles deposited in the lungs, In *Particle-Lung Interactions*, Gehr, P. and Heyder, J., eds., Marcel Dekker, Inc., New York, 323–376, 2000.

Kreyling, W. G., Semmler, M., Erbe, F., Mayer, P., Takenaka, S., Schulz, H., Oberdörster, G., and Ziesenis, A., Translocation of ultrafine insoluble iridium particles fiom lung epithelium to extrapulmonary organs is size dependent but very low. *J. Toxicol. Env. Health Pt A*, 65 (20), 1513–1530, 2002.

Kuempel, E. D., Comparison of human and rodent lung dosimetry models for particle retention, *Drug Chem. Toxicol.*, 23 (1), 203–222, 2000.

Kuempel, E. D. and Tran, C. L., Comparison of human lung dosimetry models: implications for risk assessment, *Ann. Occup. Hyg.*, 46 (suppl. 1), 337–341, 2002.

Kuempel, E. D., O'Flaherty, E. J., Stayner, L. T., Attfield, M. D., Green, F. H. Y., and Vallyathan, V., Relationships between lung dust burden, pathology, and lifetime exposure in an autopsy study of U.S. coal miners, *Ann. Occup. Hyg.*, 41 (suppl. 1), 384–389, 1997.

Kuempel, E. D., O'Flaherty, E. J., Stayner, L. T., Smith, R. J., Green, F. H. Y., and Vallyathan, V., A biomathematical model of particle clearance and retention in the lungs of coal miners. Part I. Model development, *Reg. Toxicol. Pharmacol.*, 34, 69–87, 2001a.

Kuempel, E. D., Tran, C. L., Smith, R. J., and Bailer, A. J., A biomathematical model of particle clearance and retention in the lungs of coal miners. Part II. Evaluation of variability and uncertainty, *Reg. Toxicol. Pharmacol.*, 34, 88–101, 2001b.

Kuempel, E. D., Wheeler, M., Smith, R., and Bailer, A. J., Quantitative risk assessment in workers using rodent dose–response data of fine and ultrafine titanium dioxide. [Abstract], In *Nanomaterials: A Risk to Health at Work? First International Symposium on Occupational Health Implications of Nanomaterials*, 12–14 October 2004, Buxton, Derbyshire, UK., Health and Safety Executive, U.K., Buxton, 111, 2005.

Kuempel, E. D., Tran, C. L., Bailer, A. J., Castranova, V., Lung dosimetry models in rats and humans: use and evaluation for risk assessment of nanoparticles, *Inhal. Toxicol.*, 2006 *in press*.

Martonen, T. B., Rosati, J. A., and Isaacs, K. K., Modeling deposition of inhaled particles, In *Aerosols Handbook: Measurement, Dosimetry, and Health Effects*, Ruzer, L. S. and Harley, N. H., eds., CRC Press, Boca Raton, Florida, 113–155, 2005.

Mauderly, J. L., Lung overload: the dilemma and opportunities for resolution, In *Particle Overload in the Rat Lung and Lung Cancer, Implications for Human Risk Assessment*, Proceedings of a conference held at the Massachusetts Institute of Technology, Mauderly, J. L, and McCunney, R. J., eds., March 29–30, 1995, Taylor and Francis, Washington, DC, 1–28, 1996.

McClellan, R. O., Particle interactions with the respiratory tract, In *Particle-Lung Interactions*, Gehr, P. and Heyder, J., eds., Marcel Dekker, Inc., New York, 1–63, 2000. Chapter 1.

Ménache, M. G., Miller, F. J., and Raabe, O. G., Particle inhalability curves for humans and small laboratory animals, *Ann. Occup. Hyg.*, 39, 317–328, 1995.

Miller, F. J., Dosimetry of particles in laboratory animals and humans, In *Toxicology of the Lung*, Gardner, D. E., Crapo, J. D., and McClellan, R. O., eds. 3rd ed., Taylor and Francis, Philadelphia, PA, 513–556, 1999. Chapter 18.

Morrow, P. E., Possible mechanisms to explain dust overloading of the lungs, *Fund. Appl. Toxicol.*, 10, 369–384, 1988.

Mortensen, J. D. et al., Age related morphometric analysis of human lung casts, *Extrapolation of Dosimetric Relationships for Inhaled Particles and Gases*, Academic Press, San Diego, CA, 59–68, 1988.

Muhle, H., Creutzenberg, O., Bellmann, B., Heinrich, U., Ketkar, M., and Mermelstein, R., Dust overloading of lungs: investigations of various materials, species differences, and irreversibility of effects, *J. Aerosol Med.*, 3 (suppl. 3), S111–S128, 1990.

NCRP, Deposition, retention, and dosimetry of inhaled radioactive substances, *National Council on Radiation Protection and Measurements,* Bethesda, MD. Report No. 125, 253, 1997.

Nikula, K. J., Avila, K. J., Griffith, W. C., and Mauderly, J. L., Lung tissue responses and sites of particle retention differ between rats and cynomolgus monkeys exposed chronically to diesel exhaust and coal dust, *Fundam. Appl. Toxicol.*, 37, 37–53, 1997.

NIOSH, 2005. NIOSH Current Intelligence Bulletin: Evaluation of Health Hazard and Recommendations for Occupational Exposure to Titanium Dioxide. Unpublished Public Review Draft, November 22, 2005. Cincinnati, OH: U.S. Department of Health and Human Services, Public Health Service Centers for Disease Control and Prevention, National Institute for Occupational Safety and Health, 158, Available at: http://www.cdc.gov/niosh/docs/preprint/tio$_2$/pdfs/TIO$_2$Draft.pdf.

Oberdörster, G., Ferin, J., and Lehnert, B. E., Correlation between particle size in vivo particle persistence, and lung injury, *Environ. Health Perspect.*, 102 (suppl. 5), 173–179, 1994.

Oberdörster, G., Sharp, Z., Atudorei, V., Elder, A., Gelein, R., Lunts, A., Kreyling, W., and Cox, C., Extra-pulmonary translocation of ultrafine carbon particles following whole-body inhalation exposure of rats, *J. Toxicol. Environ. Health Pt. A*, 65 (20), 1531–1543, 2002.

Oberdörster, G., Sharp, Z., Atudorei, V., Elder, A., Gelein, R., Kreyling, W., and Cox, C., Translocation of inhaled ultrafine particles to the brain, *Inhal. Toxicol.*, 16 (6–7), 437–445, 2004.

O'Flaherty, E. J., Interspecies conversion of kinetically equivalent doses, *Risk Analysis*, 9 (4), 587–598, 1989.

Parent, R. A., *Treatise on Pulmonary Toxicology: Comparative Biology of the Normal Lung*, Vol. 1, CRC Press, Boca Raton, 1992.

Raabe, O.G., Yeh, H.C., Newton, G.J., Phalen, R.F., and Velasquez, D.J., Deposition of inhaled monodisperse aerosols in small rodents. In Inhaled Particles IV. Proceedings of an' international symposium organised by the British Occupational Hygiene Society, Edinburgh, 22–26, September 1975. Vol. 1. Oxford: Pergamon Press: 3–21, 1977.

Raabe, O.G., Al-Bayati, M.A., Teague, S.V., Rasolt, A., Regional deposition of monodisperse coarse and fine aerosol particles in small laboratory animals, In *Inhaled Particles VI. Proceedings of an international symposium and workshop on Lung Dosimetry organized by the British Occupational Hygiene Society in co-operation with the Commission of the European Communities*, Cambridge, 2–6, September 1985. Pergamon Press, Oxford, 53–64. (Ann. Occup. Hyg.; 32 (suppl.1)), 1988.

Renwick, L. C., Donaldson, K., and Clouter, A., Impairment of alveolar macrophage phagocytosis by ultrafine particles, *Toxicol. Appl. Pharmacol.*, 172 (2), 119–127, 2001.

Renwick, L. C., Brown, D., Clouter, A., and Donaldson, K., Increased inflammation and altered macrophage chemotactic responses caused by two ultrafine particles, *Occup. Environ. Med.*, 61, 442–447, 2004.

Reynolds, H. Y., Lung inflammation and fibrosis: an alveolar macrophage-centered perspective from the 1970s to 1980s, *Am. J. Respir. Crit. Care Med.*, 171 (2), 98–102, 2005.

Saltelli, A., Chan, K., and Scott, M., eds., Sensitivity Analysis, John Wiley & Sons publishers, Probability and Statistics series, 2000.

Seaton, A. and Cherrie, J. W., Quartz exposures and severe silicosis: a role for the hilar nodes, *Occup. Environ. Med.*, 55 (6), 383–386, 1998.

Semmler, M., Seitz, J., Erbe, F., Mayer, P., Heyder, J., Oberdörster, G., and Kreyling, W. G., Long-term clearance kinetics of inhaled ultrafine insoluble iridium particles from the rat lung, including transient translocation into secondary organs, *Inhal. Toxicol.*, 16 (6–7), 453–459, 2004.

Snipes, M. B., Long-term retention and clearance of particles inhaled by mammalian species, *CRC Crit. Rev. Toxicol.*, 20 (3), 175–211, 1989.

Stahlhofen, W., Scheuch, G., and Bailey, M. R., Investigations of retention of inhaled particles in the human bronchial tree, *Radiat. Prot. Dosim.*, 60, 311–319, 1995.

Stöber, W., Morrow, P. E., and Hoover, M. D., Compartmental modeling of the long-term retention of insoluble particles deposited in the alveolar region of the lung, *Fund. Appl. Toxicol.*, 13, 823–842, 1989.

Stöber, W., Morrow, P. E., and Morawietz, G., Alveolar retention and clearance of insoluble particles in rats simulated by a new physiology-oriented compartmental kinetics model, *Fund Appl. Toxicol.*, 15, 329–349, 1990a.

Stöber, W., Morrow, P. E., Morawietz, G., Koch, W., and Hoover, M., Developments in modeling alveolar retention of inhaled insoluble particles in rats, *J. Aerosol Med.*, 3 (suppl. 1), 129–154, 1990b.

Stöber, W., Morrow, P.E., Koch, W., Morawietz, G., Alveolar clearance and retention of inhaled soluble particles in rats simulated by a model inferring macrophage particle load distributions. *J. Aerosol Sci.* 25, 975–1002, 1994.

Strom, K. A., Johnson, J. T., and Chan, T. L., Retention and clearance of inhaled submicron carbon black particles, *J. Toxicol. Environ. Health*, 26, 183–202, 1989.

Tran, C. L,, Buchanan, D., Development of a biomathematical lung model to describe the exposure-dose relationship for inhaled dust among UK coal miners. Institute of Occupational Medicine, Edinburgh, UK, IOM Research Report TM/00/02, 2000.

Tran, C. L., Jones, A. D., and Donaldson, K., Overloading of particles and fibres, *Ann. Occup. Hyg.*, 41(suppl. 1), 237–243, 1997.

Tran, C. L., Cullen, R. T., Buchanan, D., Jones, A. D., Miller, B. G., Searl, A., Davis, J. M. G., and Donaldson, K., Investigation and prediction of pulmonary responses to dust. Part II, In *Investigations Into the Pulmonary Effects of Low Toxicity Dusts. Parts I and II.*, Health and Safety Executive, Suffolk, UK, Contract Research Report 216/1999, 1999.

Tran, C. L., Buchanan, D., Cullen, R. T., Searl, A., Jones, A. D., and Donaldson, K., Inhalation of poorly soluble particles II. Influence of particle surface area on inflammation and clearance, *Inhal. Toxicol.*, 12, 1113–1126, 2000.

Tran, C.L,, Graham, M,, and Buchanan, D., A biomathematical model for rodent and human lungs describing exposure, dose, and response to inhaled silica. Institute of Occupational Medicine, Edinburgh, U.K., IOM Research Report TM/01/04, 35, 2001.

Tran, C. L., Kuempel, E. D., and Castranova, V., A model of exposure, dose and response to inhaled silica, *Ann. Occup. Hyg.*, 46 (suppl. 1), 14–17, 2002.

Tran, C. L., Miller, B. G., and Jones, A. D., Risk assessment of inhaled particles using a physiologically based mechanistic model. Institute of Occupational Medicine Research Report 141, prepared for the Health and Safety Executive, 47, 2003.

Vallyathan, V., Brower, P. S., Green, F. H. Y., and Attfield, M. D., Radiographic and pathologic correlation of coal workers' pneumoconiosis, *Am. J. Respir. Crit. Care. Med.*, 154, 741–748, 1996.

van Oud Alblas, A. B. and van Furth, R., Origin, kinetics, and characterization of pulmonary macrophages in the normal steady state, *J. Exp. Med.*, 136, 186–192, 1986.

Vincent, J. H., Jones, A. D., Johnston, A. R., Bolton, R. E., and Cowie, H., Accumulation of inhaled mineral dust in the lung and associated lymph nodes: implications to exposure and dose in occupational lung disease, *Ann. Occup. Hyg.*, 31, 375–393, 1987.

Warheit, D. B., Overby, L. H., Beroge, G., and Brody, A. R., Pulmonary macrophages are attracted to inhaled particles through complement activation, *Exp. Lung Res.*, 14, 51–66, 1988.

Yeh, H. C. and Schum, G. M., Models of human lung airways and their application to inhaled particle deposition, *Bull. Math. Biol.*, 42, 461–480, 1980.

Yu, R. C. and Rappaport, S. M., A lung retention model based on Michaelis-Menten-like kinetics, *Environ. Health Perspect.*, 105 (5), 496–503, 1997.

Yu, C. P., Morrow, P. E., Chan, T. L., and Yoon, K. J., A non-linear model of alveolar clearance of insoluble particles from the lung, *Inhal Toxicol*, 1, 97–107, 1988.

Yu, C. P., Yoon, K. J., and Chen, Y. K., Retention modeling of diesel exhaust particles in rats and humans, *J. Aerosol Med.*, 4 (2), 79–115, 1991.

Yu, C. P., Zhang, L., Oberdörster, G., Mast, R. W., Glass, L. R., and Utell, M. J., Clearance of rehctory ceramic fibres (RCF) fiom the rat lung: development of a model. *Environ. Res.* 65, 243–253, 1994.

Zhang, Z. and Martonen, T. B., Deposition of ultrafine aerosols in human tracheobronchial airways, *Inhal Toxicol.*, 9 (2), 99–110, 1997.

21 Nanoparticles in Medicine

Paul J. A. Borm
Centre of Expertise in Life Sciences, Hogeschool Zuyd

Detlef Müller-Schulte
Magnamedics GmbH

CONTENTS

21.1 INTRODUCTION

Recent years have witnessed unprecedented growth of research and applications in the area of nanoscience and nanotechnology. There is increasing optimism that nanotechnology, as applied to medicine, will bring significant advances in the diagnosis and treatment of disease. Anticipated applications in medicine include drug delivery, diagnostics, nutraceuticals, and production of biocompatible materials (ESF 2005; Ferrari 2005; Vision Paper 2005). Engineered nanoparticles (<100 nm) or nanostructured materials (NSM) are important tools to realize these applications. The reason why these nanoparticles (NP) are attractive for such purposes is based on their important and unique features, such as their surface to mass ratio that is much larger than that of other particles, their quantum properties, and their ability to adsorb and carry other compounds. NP have a large (functional) surface that is able to bind, adsorb, and carry other compounds such as drugs, probes, and proteins. However, many challenges must be overcome if the application of nanotechnology is to realize the anticipated improved understanding of the patho-physiological basis of disease, bring more sophisticated diagnostic opportunities, and yield improved therapies. One of the most challenging problems that nanotechnology is facing is posed by research data with combustion-derived nanoparticles (CDNP), such as diesel exhaust particles (DEP). Research has demonstrated that exposure to CDNP is associated with a wide variety of effects (review: Donaldson et al. 2005) including pulmonary inflammation, immune adjuvant effects (Granum and Lovik 2002), and systemic effects including blood coagulation and cardiovascular effects (review: Borm and Kreyling 2004; Oberdörster, Oberdörster, and Oberdörster 2005a; Oberdörster et al. 2005b). Since cut-off size for their definition (100 nm) is the same, now both terms are used as equivalent. The meeting of the worlds of nanoscience and engineered nanoparticles along with Particle Toxicology and its know-how of ultrafines, has led to an impressive series of workshops over the past years. However, little exchange of methods and concepts has taken place and therefore the aim of this chapter is to discuss applications of nanoscale materials in nanomedicine along with their toxicological properties.

21.2 PARTICLES IN NANOMEDICINE: HEALTH BENEFITS AND PERSPECTIVES

Nanomedicine uses nano-sized tools for the diagnosis, prevention, and treatment of disease to increase understanding of the underlying complex pathophysiological mechanisms. The ultimate goal is to improve quality of life. The aim of nanomedicine may be broadly defined as the comprehensive monitoring, control, construction, repair, defense, and improvement of all human biological systems, working from the molecular level using engineered devices and nanostructures to ultimately achieve medical benefit. In this context, nanoscale should be taken to include active components or objects in the size range from one nanometre to hundreds of nanometres. They may be included in a micro-device (that might have a macro-interface) or a biological environment. The focus, however, is always on nano interactions within a framework of a larger device or biologically, within a sub-cellular (or cellular) system. These definitions originate from a working group initiated in early 2003 by the European Science Foundation (ESF 2005).

21.2.1 Imaging and Diagnostic Tools

In the course of extending our knowledge in the field of nanoparticles, the medical area has witnessed in the last few years increasing attempts to exploit the intriguing perspectives nanotechnology offers in medical diagnostics and analytics as well as look for a substitution or assistant modalities for conventional x-ray analysis. It is realistic to say that nanotechnological tools will be routinely used in diagnosis long before being approved for the treatment of diseases. Some tools are already on the market or available on short-term, but many others still need considerable development (Table 21.1).

21.2.1.1 Magnetic Particles

This in particular applies to *magnetic nanoparticles* and *luminescent species* that are applied for the detection of pathogenic cells or tissue (Table 21.2). The principle is based on the specific targeting

TABLE 21.1
Anticipated Time Lines for Nano Based Diagnostic Tools

Time Interval	Anticipated Development	Particles Used
2005–2010	Cell sorting, on site detection in packaging, cell separation	Magnetic or fluorescent beads with specific antibodies
2010–2015	Encapsulation, coating of contrast enhancement agents	Magnetic beads, quantum dots, polymers with fluorescent markers
2010–2015	Biomimetic sensors	Gels encapsulating antibodies and drugs
2015–2020	Transfection nanodevice, implantable devices, multifunctional cameras	Complex multifunctional nanomaterials

Source: Modified from ESF, *European Science Foundation Policy Briefing: ESF Scientific Forward Look on Nanomedicine 2005*. IREG Strasbourg, France, ISBN, 2-912049-520, 2005.

of these nanovectors onto the target tissue or cell which can afterwards be monitored either by magnetic resonance imaging (MRI), ultrasound, or optical screening procedures such fluorescence molecular tomography (Graves et al. 2005). Due to the enormous know how accumulated in the science area in the last decades, a broad spectrum of various nanocarriers have been developed which are able to meet the medical needs and prerequisites. This applies to (i) the magnetic properties, (ii) size adaptation, and (iii) specific tissue targeting.

One of the key aims in modern diagnostics is undoubtedly the improvement and increase of detection sensitivity, leading ultimately to a detection of neoplastic chances in cancer in the earliest possible stage. This aspect certainly constitutes the focus of present research and development. The magnetic properties of nanoparticles are primarily exploited in MRI—where the contrast of the target tissue results from the diverse signal intensities this tissue delivers in response to a specifically applied radio frequency pulse. This response is a function of the proton density and magnetic relaxation time, which on its part is determined by the biochemical structure, and properties of the

TABLE 21.2
Overview of Particles and Their Medical or Pharmaceutical Applications

Particle Class	Materials	Size (nm)	Payload	Application
Liposomes	Lipid-mixtures	50–100	10% volume	Drug delivery
Magneto-liposomes	Lipid mixtures	50–100	Ferrofluids	Imaging, diagnostics, separation
Dendrimers	Branched polymers	5–50	Drugs	Drug delivery
Fullerenes	Carbon based carriers			Photodynamics
Polymer carriers	Dextrane Polylactic acid Polysaccharides Poly(cyano)acrylates Polyethyleinemine	50–50,000	Drugs	Drug delivery
Polymer magnetic carriers	All above containing ferrofluids	50–50,000	Various	Diagnostics, separation
Quantum dots	CdSe, Zns, Si	2–10	None	Diagnostics
Latex beads	Polystyrene	20–2000	Markers (FITC)	Diagnostics
Silica-beads	Amorphous silica	20–10,000	Ferrofluids	Diagnostics

tissue. The effect of the magnetic nanoparticles on the relaxation times is to create local field inhomogeneity, which consequently shortens the relaxation times resulting in an enhancement of the image contrast. In practice and especially for *in vivo* applications, the magnetic nanoparticles are coated with a functional polymer that fulfils two major prerequisites:

- The polymer serves as matrix for attaching disease relevant moieties such as peptides, antibodies, or other small molecules that will bind to the pathological target.
- The coating should assist the overall blood compatibility and in particular the increase of the blood half-lives, which is a crucial point as such particles are cleared by the liver and spleen within a few minutes. A review by Bulte and Brooks (1997) summarizes the diverse approaches with special focus on variety of coatings and type of nanocarrier on the contrast behavior.

The potential of superparamagnetic iron oxide particles (SPIO) in *MR lymphography* using different administration routes and the usage of differently modified magnetic liposomes for the specific imaging of an adenocarcinoma in a rat model have been addressed by Kresse, Wagner, and Taupitz (1997) and Päuser et al. (1997). Likewise, Guimaraes et al. (1994) demonstrated the imaging of hyperplastic and tumorous lymph nodes in rodents using commercial SPIOs. Parallel to the preparation and optimization of the different nanocarriers and iron oxides, respectively, which was subject of diverse research (Tiefenauer, Kuhne, and Andres 1993; Grimm et al. 2000; Lawaczeck and Menzel 2004), the mode of applicability of these particles is primarily determined by the targeting possibility. This topic is basically addressed by the coupling of such targeting mediating species that show a high specific affinity towards the according target tissue or cell. Zhao et al. (2002) investigated the influence of the HIV-1 Tat peptide derivatized iron oxide nanoparticles for the uptake in cells. This peptide, which translocates exogenous molecules into cells, facilitates the cellular uptake of the nanoparticles in an exponential fashion and results in a 100-fold increase in cell labeling efficiency. Further approaches addressing this topic pertain to the coating with galactosides (Weissleder et al. 1990) for the targeting of hepatocytes via the asialoglycoprotein receptor, anti-carcinoembryonic antigen CEA coupled magnetic nanoparticles as models for tumor targeting (Tiefenauer, Kuhne, Andres 1993), and antibody coated nanoparticles directed against the HT-29 surface antigen showing their basic applicability as relaxation and targeting agents (Cerdan et al. 1989). Lanza et al. (2004) concisely discuss the diverse aspects of molecular imaging including the physical magnetic parameters, the possibilities of active and passive targeting into the diverse organ tissues, ligand coupling strategies with the emphasize on antibodies, avidin and aptamers, and the spectrum of diverse nanovectors ranging from gadolinium loaded perfluorocarbon-lipid particles targeted towards fibrin, liposomes, and αβ-integrin-targeted nanoparticles specially designed for the detection of angiogenesis.

Jaffer and Weissleder (2005) reviewed the different molecular imaging systems, including MRI, nuclear, and optical imaging. They also conducted a survey about the diverse imaging agents, their clinical applications with particular emphasis on cancer, atherosclerosis, apoptosis, and inflammatory enzyme activity imaging. Apart from nanocarriers, Mikawa et al. (2001) describe a new approach in MRI contrast enhancement by using water-soluble gadolinium metallofullerenes used for both *in vivo* and *in vitro* tests. They could show a 20-fold higher relaxivity than that of the commercial MRI contrast agent Magnevist®. *In vivo* MRI at lung, liver, spleen, and kidney of mice corroborated these findings.

21.2.1.2 Fluorescent Nanoparticles

Other types of nanoparticles that have attracted much interest in the last few years in the area of diagnostics are the optical markers in the form of *fluorescent nanoparticles*. Imaging of cell-surface receptors, antigens, and gene expressions using conventional fluorescent molecular probes or fluorescent-tagged antibodies to trace tumor cells, apoptosis, and metastases have been described (Graves et al. 2005). Despite their broad usage and successful application, which has resulted in the development of fluorescence molecular tomography, these probes have generally some disadvantages with regard to photo stability and show low *in vivo* biodegradability of tagged biomolecules, or they show inappropriate bio distribution. Hence, there is an ongoing tendency to use the potential of these fluorescent markers in combination with a polymer carrier to exploit their full potential for molecular imaging in medical diagnostics.

The number of fluorescent molecules potentially used as markers in bioscience is underlined by the fact that there are a number of companies specializing in the commercialization of such compounds. In this review, we will focus solely on two issues, namely, dye-doped silica carriers and quantum dots, as these appear to provide the most promising perspectives in this area. Based on the well-known Stöber suspension process for the preparation of monodisperse silica nanoparticles, several researchers have used this intriguing technology to create fluorescent markers. Because of the pronounced functionality of the silica matrix, isothiocyanate derivatized fluorophores in the form of rhodamine and fluorescein were directly coupled to the matrix. This approach was intensively studied by van Blaaderen (2006), aiming to develop the coupling chemistry on derivatized nanoparticles, their fluorescence properties in different solvents, and the diverse physical characterization using, e.g., light scattering and transmission electron microscopy. An alternative route (Santra and Biomoleküle et al. 2005) uses a W/O micro emulsion of a fluorescein-isothiocyanate derivatized silane precursors to obtain dye-doped nanocarriers. To demonstrate the bio imaging potential of these markers, the nanocarriers were modified with the Tat peptide and folic acid. Incubating these carriers with A-549 and lung adenocarcinoma cells revealed an extensive cell labeling. A further basic method of preparing dye labeled particles was recently described (Graf et al. 1999) using a multi-step core-shell approach whose aim is to protect the encapsulated fluorophores such as rhodamine B, coumarin, and pyrene with a silane layer.

21.2.1.3 Quantum Dots

Compared to conventional organic fluorescent dyes, *quantum dots*—nanocrystalline semiconductors for bioimaging purposes mainly composed of III–V materials, e.g., GaP, GaAs, and INAs or II–VI materials, e.g., CdSe, CdS, ZnSe—represent a novel class of marker systems with unique optical properties (Figure 21.1). This is first of all reflected by the size- and composition-tunable fluorescence emission from visible to infrared light. By reducing the size of the nanocrystal, we observe a blue shift and vice versa in the emission (Murphy 2002; Parak et al. 2003). Together with the large absorption coefficient across a wide spectral range, the high quantum yield, longer fluorescent decay times in comparison to conventional dyes, and pronounced photo stability, these nanocrystals are an ideal marker for bioimaging. However, to apply these nanocrystals for *in vivo* imaging, two pre-requisites need to be fulfilled: the quantum dots must be water dispersible, and, due to the high toxicity of the basic chemical constituents, they have to be coated with a biocompatible matrix. This prevents direct contact with the biological tissue thus preventing toxicity (Chen and Yao 2004). This is achieved by coating the nanocarriers with a range of diverse polymers and surfactants. Most promising approaches fulfilling the basic purposes are siloxanes, polyethylene glycol, phospholipids, carboxymethyldextran, mercaptoacetic acid, dithiothreitol, glutathione, or synthetic polymers, e.g., in form of a block copolymers (Hirai, Okubo, and Komasawa 2001; Winter et al. 2001; Dubertret et al. 2002; Chen, Ji, and Rosensweig 2003; Parak et al. 2003; Gao et al. 2004). Apart from the stabilization, the chemical functionality of the coatings also provides a basis to attach target-finding bioligands to the surface that paves a way in *in vivo* imaging. Promising approaches in this field include the attachment of specific antibodies or peptides for neuron targeting (Winter), DNA coupling for the detection of nucleotide polymorphism (Parak et al. 2003), and antibody and streptavidin for labeling breast cancer cells (Wu et al. 2003; Gao et al. 2004). Apart from the bioimaging application, quantum dots are also very

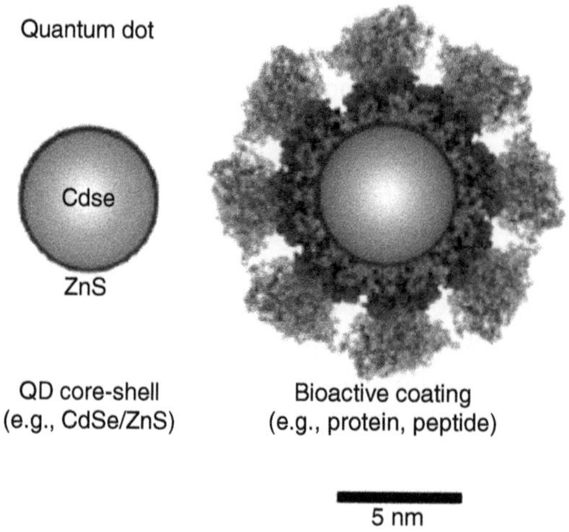

FIGURE 21.1 QDs consist of a metalloid core and a cap/shell that shields the core and renders the QD bioavailable. The further addition of biocompatible coatings or functional groups can give the QD a desired bioactivity. (Reproduced from Hardman, R., *Environ. Health. Perspect.*, 114, 165–172, 2006. With permission.)

suitable for optical coding in the scope of bioassays. Han et al. (2001) recently described multicolor polystyrene CdSe/ZnS nanocrystals for multiplexed optical coding of biomolecules. The extremely high number of possible codes when using multiple wavelengths and multiple intensities are exemplarily demonstrated in a DNA hybridization assay. We recently described a new approach for highly fluorescent marker systems. Using a novel inverse sol–gel suspension technique, quantum dots and fluorescence dyes can be easily encapsulated in a one-step procedure into silica nano- and microbeads (Müller-Schulte 2004, Müller-Schulte et al. 2005). The intriguing aspect of this novel technique is that it allows a synthesis of fluorescent carriers within a few minutes thus providing a significant time saving in comparison to established methods, as well as a simultaneous combinatorial encapsulation of diverse fluorescent compounds and magnetic colloids. This opens up novel perspectives for cell screening and bioassays.

21.2.2 Nanomaterials and Nanodevices

Nanomaterials and nanodevices are critical in nanomedicine. On the one hand, the principles of materials science may be employed to identify biological mechanisms and develop medical therapeutics. On the other hand, the opportunities of nanomedicine depend on the appropriate nanomaterials and nanodevices to realize their potentials. Nanomaterials and nanodevices for nanomedicine are produced largely based on nanoscale assemblies for targeting and ligand display.

The field of *medical implants* is fast growing since new, light, durable, and biocompatible implants can be constructed on a tailor-made basis. Medical implants are being used in every organ of the human body. Ideally, medical implants must have biomechanical properties comparable to those of autogenous tissues without any adverse effects. In each anatomic site, studies of the long-term effects of medical implants must be undertaken to determine accurately the safety and performance of the implants. Today, implant surgery has become an interdisciplinary undertaking involving a number of skilled and gifted specialists. Applications can be identified for each implant site, from orthopaedics, dentistry, to cardiovascular surgery. *Artificial hips* on one hand need to be fixed steadily in the bone while on the other hand the knob needs to be flexible, biocompatible, and not subject to wearing. The success of total hip replacement depends in part on the materials, design, and processing of the materials used in the implant. During surgery, the painful parts of the damaged hip are replaced with artificial hip parts, which make up the prosthesis—a device that substitutes or supplements a joint. To duplicate the action of a ball-and-socket hip joint, the prosthesis has three parts (Figure 21.2):

- The stem, usually made from metal which has to be fixed in the bone
- The ball or head, made of ceramic or metal
- The shell and accompanying liner, with the shell made of metal and the liner made of cross-linked polyethylene. The liner may also be made of ceramic or metal.

21.2.3 Drug Delivery Tools

Drug delivery and related pharmaceutical development in the context of nanomedicine should be viewed as science and technology of nanometre size scale complex systems (10–1000 nm), consisting of at least two components, one of which is an active ingredient (Duncan 2003; Ferrari 2005). The whole system leads to a special function related to treating, preventing, or diagnosing diseases, sometimes called smart-drugs or theragnostics (LaVan, McGuire, and Langer 2003). Depending on the origin, the materials employed include synthetic or semi-synthetic polymers and natural materials such as lipids, polymers, and proteins (Aston 2005). The primary goals for research of nanobiotechnologies in drug delivery include:

- Faster development of new safe medicines,
- More specific drug delivery and targeting, and
- Greater safety and biocompatibility.

Currently, the development of a new drug is estimated at over 700 billion euro, and the number of FDA approvals has been gradually decreasing over the past decade. It is anticipated that with the

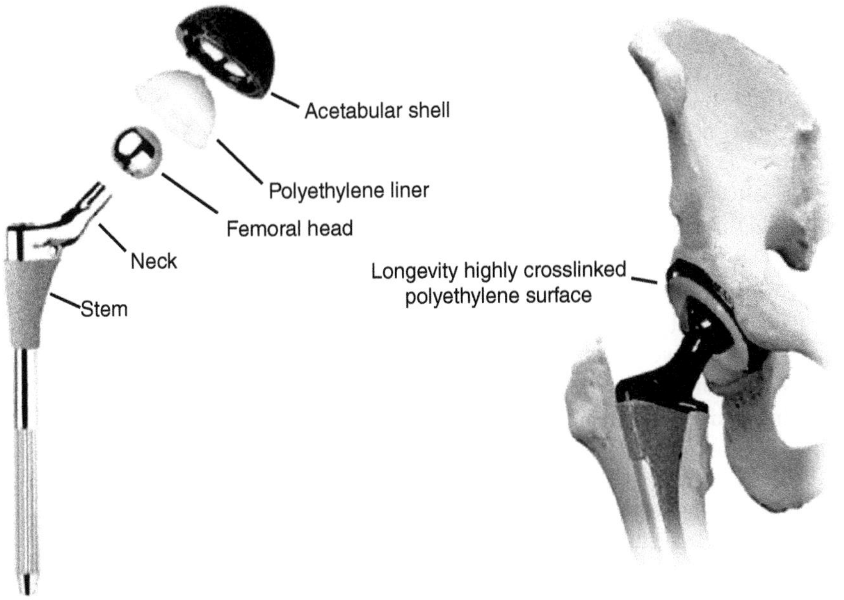

FIGURE 21.2 Schematical depiction of the building parts of an artificial hip, which is probably the most successful medical implant over the past 20 years. Dependent on the material and the wear forces, every step causes the release of a significant amount of particles (wear debris) into the surrounding tissue.

help of nanoscience, so-called orphan drugs may be further developed and drugs that were eliminated in the development of the process may be reactivated. However, the pharmaceutical industry is currently showing little interest and major developments occur in academia. It is therefore quite obvious that the nanomaterials segment, which includes several long-established markets such as carbon black rubber filler, catalytic converter materials, and silver nanoparticles used in photographic films and papers, presently accounts for over 97.5% of global nanotechnology sales. By 2008, the nanomaterials share of the market will have shrunk to 74.7% of total sales (BCC 2004). Nanotools will have increased their share to 4.3% ($1.2 billion), and nanodevices will have established a major presence in the market with a 21% share ($6.0 billion). The Life Science applications are thought to induce the latter increase.

The main issues in the search for appropriate carriers as drug delivery systems pertain to the following topics that are basic prerequisites for design of new materials. They comprise (i) biocompatibility, (ii) biodistribution, (iii) functionality, (vi) targeting and (v) drug incorporation and release ability. Certainly none of the so far developed carriers fulfill all of these parameters to the full extent; the progress made in nanotechnology, inter alia, emerging from the progress in the polymer-chemistry, however, can provide an intriguing basis to tackle this issue in a promising way. The following paragraphs briefly discuss different particle types currently used in nanomedicine for drug delivery.

21.2.3.1 Liposomes

One of the earliest approaches in the field of drug delivery concerns liposomes that are nanovesicles composed of phospholipids. Starting with the pioneering work by Gregoriadis (1973) and later by Fendler (1977), the cell-like structure of these vectors and their enormous chemical and physical versatility in terms of size, lipid composition, surface charge, fluidity, surface functionality, and drug incorporation ability make these carriers the material of choice when approaching these subjects. Due to the potential variety of the liposomes, researchers have pursued diverse routes in the development in order to fulfill the above-defined parameters. Bio- and blood compatibility represent the starting prerequisites for a carrier design because they directly determine the blood half-life and hence the pathway and biodistribution of a particle. Several attempts have been described to design liposomes avoiding the uptake by the immune system—particularly the reticuloendothelial system (RES), to clear foreign bodies—and to enhance blood half-life from generally less than one hour to several hours. This objective has been addressed by modifying the liposome surface. Iga et al. (1994) tested various lipid compositions and found that the incorporation of negatively charged polyoxyethylene-stearyl-derivatives revealed the greatest reduction in RES uptake. This is caused by the negative charge and the specific chain length of the polyoxyethylenes. The most promising approach in enhancing liposome bloodhalf-life comes from preparations, which incorporated either monosialogangliosides or polyethyleneglycols into the lipid membrane. Several authors (Gabizon et al. 1990; Allen and Hansen 1991; Torchilin et al. 1994) have comprehensively addressed this issue leading to the concept of so-called *stealth liposomes*. Sphingomyelin–egg phosphatidylcholine–cholesterol-composed liposomes with a half-life of over 20 h show a dosage-independence of blood clearance (Allen and Hansen 1991). The reason for this effect is mainly attributed to the sterical hindrance of opsonins. Apart from directly influencing the blood-half lives, the composition and derivatization of the lipids also exert an impact on the tissue distribution of the vesicles. This and the aspects concerning, amongst others, the stability, RES uptake, pharmacokinetics, lipid composition, particles size, surface modification, and surface charge have been concisely reviewed by Woodle and Lasic (1992). Following the development of providing appropriate routes for fundamentally improving the biocompatibility and biodistribution, the application of these nanovectors notably in cancer therapies also require functionality, i.e., the possibility to attach targeting finding moieties to the liposome surface thus enabling tissue targeting. Several research groups, whereby mainly tumor antibodies were attached via the functional lipid phosphatidylethanolamine, have addressed this objective. Jones and Hudson describe anti-placental alkaline phosphatase antibodies linked to immunoliposomes that could be effectively targeted to tumor cells. Similarly, specific antibody-linked polyethyleneglycol liposomes containing entrapped doxorubicin, targeted to KLN-205 lung carcinoma cells, were capable of reducing the tumor in a mice model significantly (Ahmad et al. 1993). The almost unlimited application potential of the liposome technology is furthermore underlined by research done by Flasher, Konopka, and Chamow (1994), who targeted CD4 coated liposomes to immunodeficiency virus type 1-infected cells opening up novel perspectives in another highly explosive medical area.

The final parameter to be addressed and which determines the efficacy and applicability of a drug carrier is the incorporation capacity for a specific drug and its release once the target tissue has been reached (Figure 21.3). Several authors (Kim 1993; Allen 2002; Park 2002) have reviewed this issue and describe a multitude of well-known anti-cancer drugs in liposome formulations, the targeting aspect, and the medical therapeutic background. Based on the amount of publications, liposome based papers represent by far the biggest portion in this list followed by polymer nanoparticles. Despite some drawbacks of the liposome concept regarding mainly lack of chemical, biological, and physical stability, the commercialization of liposome tumor therapeutics amounting to more than $200 million US in the year 2003 underlines the promising basis of this technology (Wagner and Wechsler 2004).

21.2.3.2 Magnetoliposomes

In the last two decades, there has been an increasing trend to render particle carriers magnetic. While magnetic nano- and microparticles and -beads are already fully established in the area

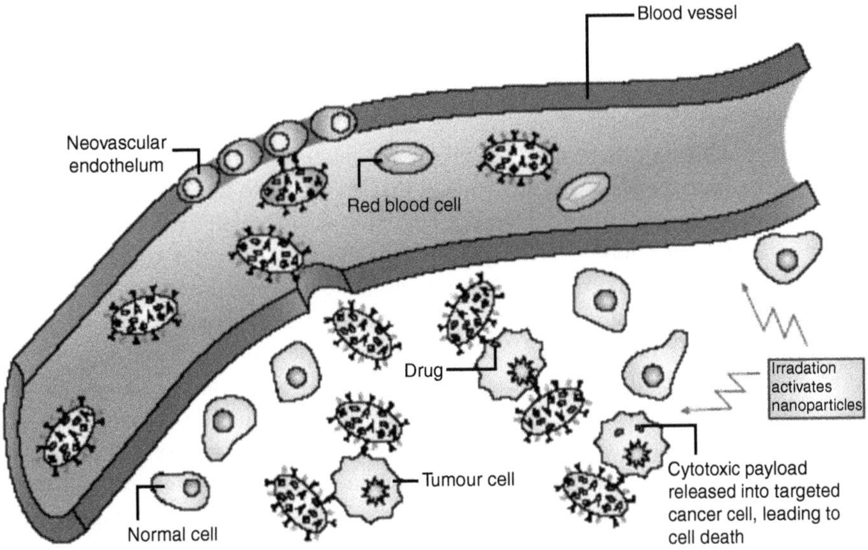

FIGURE 21.3 (**See color insert**) Multi-component targeting strategies using nanoparticles in treating cancer. Nanoparticles extravasate into the tumor stroma through the fenestrations of the angiogenic vasculature, demonstrating targeting by enhanced permeation and retention. The particles carry multiple antibodies, which further target them to epitopes on cancer cells, and direct antitumour action. Nanoparticles are activated and release their cytotoxic action when irradiated by external energy. Not shown: nanoparticles might preferentially adhere to cancer neovasculature and cause it to collapse, providing anti-angiogenic therapy. The red blood cells are not shown to scale; the volume occupied by a red blood cell would suffice to host 1–10 million nanoparticles of 10 nm diameter. (Reproduced from Ferrari, M., *Nat. Rev.*, 5, 161–171, 2005. With permission.)

of molecule and cell separation, this concept is just beginning to penetrate the medical area. The intriguing perspectives are: (i) to maneuver or target these carriers into special locations or tissues using a hand or a electromagnet and (ii) to inductively heat up these magnetic particles in an high frequency magnetic field (induction coil) within the scope of tumor hyperthermia.

De Cuyper and his group conducted pioneering work for the synthesis and modification of magnetoliposomes. This technique comprises a dialysis procedure of preformed lipid vesicles in the presence of lauric acid-coated magnetite nanoparticles (De Cuyper and Joniau 1988; De Cuyper 1996). Hodenius et al. (2002) described the modification of the magnetoliposomes with biotinylated lipids for streptavidin targeting. Viroonchatapan et al. (1995) presented thermosensitive dextran magnetite-incorporated liposomes with varying magnetite contents for application in hyperthermia and for potential cancer treatments. Ito et al. (2004) described real-life applications of magnetic immunoliposomes for tumor treatment. Anti-HER2 antibodies were attached to the vesicles and showed 60% incorporation into SKBr3 cells. Subsequent exposure of the cells to 42.5°C using an alternating magnetic field resulted in a strong cytotoxic effect. Shinkai et al. (1994) prepared magnetoliposomes for hyperthermia treatment of cancer by coating a special lipid composition onto magnetite particles. The nanovectors carrying antibodies directed to the surface antigens of human colonic cancer cells BM314 and glioma cells U251-SP showed a 12 times more efficient cell uptake than control cells. Zhang et al. (2005) investigated the potential of negatively charged paclitaxel magnetic liposomes as carriers for breast carcinoma via parenteral administration. Their studies demonstrated that these magnetoliposomes could be delivered to tumors more effectively than conventional vesicles, resulting in a higher potency on the therapy of breast cancer than other formulations.

A new potential approach using surface modified magnetoliposomes (MP) in HIV-infection was described by Müller-Schulte et al. (1997), as shown in Figure 21.4. Due to the dramatic development of this infection worldwide, the lack of effective drugs, as well as a disappointing perspective for vaccination treatments, new therapeutic measures are highly desirable. The approach described exploits the heat sensitivity of HIV-viruses that are known to be irreversibly inactivated at 50°C–60°C. Thus heat treatment is applied in this method: magnetoliposomes with a size of 50–100 nm are injected into the patient. To direct the MP to the site of the infection, one can exploit the same biochemical mechanism that HIV uses to infect the target cells (mainly T4 helper/inducer cells and macrophages). This initial infection is brought about by the interaction of the HIV envelope protein gp120 with the CD4 receptor of the T4 helper cell. This leads to the incorporation of the virus in the target cell, which then triggers all the required steps for virus proliferation. To target the liposomes, these CD4 receptors are chemically bound to the vesicles. With this receptor, the MP imitate the target cells and can hence bind to the gp120 envelope protein of HIV, as well as the HIV infected cells which also contain the gp120 ligand. The concomitant presence of the gp120 envelope protein on the infected cells is a result of the infection and proliferation process respectively (budding process). This opens up the exiting perspective to simultaneously target the MP to the HIV infected cells. After MP are administered and attached to the target organs (HIV and infected cells), the magnetic liposomes are inductively heated up to the appropriate temperatures of 50°C–60°C using an external high frequent alternating magnetic field (induction heating). Such induction devices are state-of-the-art. The special technique makes it possible for only the viruses and infected cells to be heated up, not the residual tissue. This is achieved by selecting specific frequencies that abrogate coupling of the tissue so that all the energy is adsorbed by the liposomes. This selective heating also allows the application of higher temperatures (>50°C), thus shortening the overall duration of treatment to a few minutes. The course of the disease can be monitored after each treatment by using, e.g., the T4 count. Depending on infection status, an additional treatment can be applied any time.

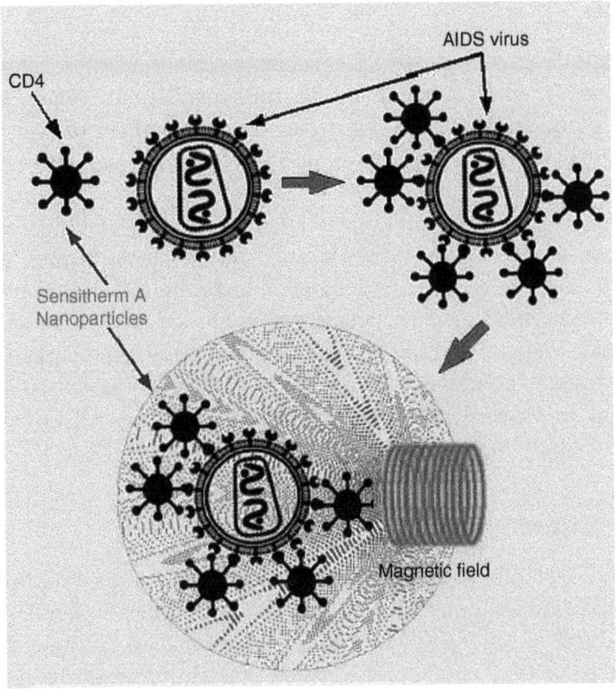

FIGURE 21.4 Graph demonstrating a strategy to bind and destroy HIV infected cells by magnetoliposomes (MP). To direct the MP to the site of the infection, one can exploit the same biochemical mechanism that HIV uses to infect the target cells (mainly T4 helper/inducer cells and macrophages). This initial infection is brought about by the interaction of the HIV envelope protein gp120 with the CD4 receptor of the T4 helper cell. This leads to the incorporation of the virus in the target cell, which then triggers all the required steps for virus proliferation. To target the liposomes, these CD4 receptors are chemically bound to the vesicles. With this receptor, the MP imitate the target cells and can hence bind to the gp120 envelope protein of HIV as well as the HIV infected cells which also contain the gp120 ligand.

21.2.3.3 Dendrimers

The class of dendrimers constitutes another type of promising carriers for drug delivery purposes. Although the preparation of these highly branched polymer particles has been investigated for the last two decades and their intriguing chemical and physical properties concerning shape, size, and functionality reviewed (Tomalioa et al. 1990), their potential application in biosciences has just begun to conquer this area. The benefits emerging from these nanodevices are (i) their exact adaptable size in the range of 5–50 nm, (ii) their chemical variability and (iii) the high loading capacity with active agents which is 25% higher than with conventional polymer–drug-conjugates which lie in the range of not more than 10% (Malik, Evagorou, and Duncan 1999). Notably, the adaptable small size of the dendrimers in combination with a biocompatible polymer is a precondition for an appropriate tissue distribution and for targeting and lowering the toxicity. The class of dendrimers best examined so far is poly(amidoamines), which exhibit pronounced biocompatibility, water solubility, non-immunogenicity, and high chemical functionality. All of these properties distinguish the vectors and make them an ideal platform for the development of drug carrier systems (Patri, Majoros, and Baker 2002). Methotrexate and folic acid conjugated dendrimers have been designed as a carrier for the targeted delivery of chemotherapeutic and imaging agents to specific cancer cells (Majoros et al. 2005), underlining the potential of this kind of nanodevices in medical therapy.

21.2.3.4 Fullerenes

Fullerenes are a class of compounds that are in a comparable stage of development as the dendrimers. The exact spherical carbon structure of fullerenes consists of regularly arranged five and six carbon rings that provide an interesting basis for the development of well defined devices for medical therapy. However, although discovered about 20 years ago and despite the enormous quantity of papers dealing with the issues such as synthesis, derivatization, investigation of the physical structure and various properties, applications in the medical therapeutic area are rather scarce. Karam, Mitch, and Coursey (1997) describe fullerenes containing gamma ray emitting Tc-99m, the first direct encapsulation of a radionuclide during fullerene formation. These appropriately derivatized "radioendofullerenes" could potentially be used as a radiopharmaceutical for medical imaging. Later Wang et al. (2004) reviewed the structure and properties of fullerenes and derivatives, their ability to function as carriers for singlet oxygen photosensitizers and their potential applications in photodynamic therapy. The perspectives of fullerenes in the field of pharmaceutics were recently reviewed by Gorman (2002).

21.2.3.5 Polymer Carriers

Polymers as basic material for the design of drug carriers certainly surpass the liposomes with regards to variability. This has resulted in an enormous amount of diverse synthetic, semi-synthetic, and natural polymers. Above all, a copolymerization or combination of these three basic polymers leads to an almost unlimited number of various materials, bringing science a step closer to fulfilling the ultimate goal of achieving optimally designed carrier systems. In view of this wide field, we will focus on those approaches that provide the most promising therapeutic perspectives. As basic synthetic material, polylactic acid, polyacrylates, polycyanoacrylates, poly(ethyleneimine), and silica gels were in the focus of recent research. Semisynthetic materials include modified polysaccharides such as agarose, or poly(amino acids) such as poly(L-lysine) and poly(glutamic acid), whereas natural polymers comprise hyaluronic acid, polysaccharides, dextrin, chitosan, gelatin, peptides, and proteins. The main criteria for selecting theses polymers are basic biocompatibility and functionality. Brannon-Peppas and Blanchette (2004) have reviewed the aspect of biocompatibility by resuming that small sizes and hydrophilicity of the nanocarriers are one of the major parameters to circumvent elimination by the RES system. Targeting of the nanodevices to tumor tissue, addressing important physiological aspects such as tissue permeability, angiogenesis, and tumor vasculature, particularly specifying a number of concrete anti-angiogenic drugs, are also discussed.

The topic of biocompatibility is furthermore addressed by using either natural hydrophilic polymers such as polysaccharides, serum albumin, gelatin, or chitosan, or applying biodegradable polymers whose most prominent representative are poly(lactid) or poly(lactid-*co*-glycolic) acid. Nanocarriers made of these polymers loaded with paclitaxel showed an *in vitro* payload release of over 50% within the first 24 h (Brannon-Peppas and Blanchette 2004). Protein encapsulated polylactid acid nanocarriers were also used to demonstrate the positive influence of PEGylation on biocompatibility (Quellec et al. 1998) thus constituting the basis for a novel stealth concept among synthetic polymers. Two hundred nanometers nanoparticles made from chitosan and carboxymethyl cellulose might be further candidates to fulfill the basic requirements of biocompatibility. Enzymatic degradation of these carriers revealed an over 80% release of DNA fragment within 6 h (Watanabe and Iwamoto 2005) thus supplying a basis for appropriate drug delivery systems. Chitosan nanoparticles were the subject of recent research testing diverse formulations that demonstrated their potential for nasal insulin delivery devices (Dyer et al. 2002). Completing the spectrum of diverse carrier systems, Lemarchand and Gref (2004) highlight the background and aspects of synthesis, characterization, pharmacokinetics, targeting, and application of polysaccharide derivatives in cancer therapy and diagnostics. It was shown that polysaccharides nanoparticles are able to modulate the biodistribution of the loaded drug. Other nanoparticles prepared from polyacrylates and poly(cyanoacrylates) using these carriers as immobilization or encapsulated basis for antisense oligonucleotides and diverse anti cancer drugs, underline the broad palette of the synthetic polymer approach (Kriwet and Walter 1998; Zobel et al. 2000; Kreuter 2004). In particular, the cyanoacrylate derivatives loaded with doxorubicin, loperamide, and tubocurarine in combination with parallel administration of specific surfactants such as Tween 20 are discussed as potential transport systems to the brain (Kreuter 2004).

Recently, Balthasar et al. (2005) and Bharali et al. (2005) described carriers on the basis of gelatine and silica gels. The latter focuses on modified silica nanoparticles which function after transplantation into a mouse brain as an efficient non-viral transfection vectors expressing EGF. The approach benefits from the outstanding properties of these carriers in form of monodispersity, aqueous suspension stability, and surface functionalization, thus providing a platform for an effective manipulation of stem cells and *in vivo* targeted brain therapy.

21.2.3.6 Magnetic Nanocarriers

Intensive research has been carried out in the field of polymers to extend the applicability of the conventional polymer carriers by introducing magnetic properties. Basically, two routes are pursued for the preparation of such particles:

- The first method uses a suspension technology whereby magnetic nanoparticles or magnetic colloids are simultaneously encapsulated into the polymer matrix (Müller-Schulte 2003).
- The second approach uses coating of preformed magnetic nanoparticles (e.g., magnetite or iron oxides) with functional polymers or surfactants. The coating thereby fulfils analogous requirements as the above described polymers—biocompatibility and functionality with regards to ligand attachment, targeting, and drug loading (Bergemann et al. 1999).

The integral aspects of magnetic nanoparticles (usually ferrofluid) regarding synthesis, surface modification using different polymeric and non-polymeric stabilizers, ligand targeting, magnetic properties, and the application in tissue repair, drug delivery, and hyperthermia were recently comprehensively reviewed by Mornet et al. (2004); and Gupta and Gupta (2005). Apart from the more fundamental research in this field, there are two approaches highlighting a concrete and promising perspective for practical medical applications. This pertains to a clinical phase I study conducted by Lübbe et al. (1996) on 14 tumor patients using 100 nm polysaccharide-coated magnetic nanoparticles to which epidoxorubicin was physically adsorbed. The particles were well tolerated and it was demonstrated that the ferrofluid could be successfully directed to the tumors in about one half of the patients. The same group also conducted tests on rats and immunosuppressed mice showing no intolerance with the epirubicin-bound ferrofluids. The treatment resulted in complete tumor suppression in an experimental human kidney as well as in a xenotransplanted colon carcinoma model Lübbe et al. (1996).

Another important application of ferrofluids lies in hyperthermia. The initial concept, developed by Gordon (1987) and extended by the introduction of the Curie-temperature concept (Müller-Schulte 1995), imply the application of ferro- or ferrimagnetic particles directed to the tumor site which are then heated up in external high frequency magnetic fields. This induction heating results in a temperature increase in the tumor cells, desirably 42°C–43°C, which kills tumor cells as they are more sensitive towards heat than normal cells. Maier-Hauff et al. (2005) have entered a phase I clinical trial for the treatment of malignant gliomas using magnetic fluid hyperthermia.

We refer to other review articles that pertain to the various aspects of magnetic carriers in biomedicine, concerning both the therapeutic and diagnostic aspect as well as basic magnetic issues (Pankhurst et al. 2003; Tartaj et al. 2003; Kalambur et al. 2005; Neuberger et al. 2005). Here, we describe a novel approach by one of the authors (D.M.S.) best described as a contactless controllable drug carrying system based on thermosensitive magnetic nano- and microparticles Figure 21.5. This concept can be regarded as a key contribution in the area of smart or stimuli-responsive drug carriers whose general aim is to exploit an internal or external stimulus to control or manipulate the drug release, whereby the stimulus might be a thermal, magnetic, or electrical one or the pH, ultrasound, or ionic strength. Stimulus-responsive hydrogels in form of poly-*N*-isopropylacrylamide derivatives and poly(acrylamide-*co*-maleic acid) were recently investigated as novel drug carriers (Bajpai, Bajpai, and Kalla 2002; Soppimath et al. 2002) representing examples for thermo- and pH sensitive carriers.

Hirsch et al. (2003) developed another intriguing novel approach aiming at the stimuli responsive concept. Their basic principle was to use gold coated silica nanoparticles which, when exposed to near-infrared light, are heated up because of the plasmon resonance of the gold layer. Human breast carcinoma cells incubated with these composites *in vitro* appeared to have undergone irreversible damage on exposure to the infrared light, whereas cells not treated with these nanoparticles displayed no loss of viability. The restriction of this approach, however, is that the penetration depth of the infrared is feasible only for a few centimeters below the skin.

In the novel stimulus responsive concept, spherically shaped polymer particles (beads) consisting of hydroxypropyl cellulose or poly-*N*-isopropylacrylamide serve as thermosensitive matrix. The unique feature of these polymers is that a phase transition accompanied by a drastic shrink process occurs at temperatures above 30°C (*lower critical solution temperature*), a process that is triggered by a phase separation. By encapsulating ferro(i)magnetic colloids into the polymer matrix and exposing these magnetic beads to an external high frequency magnetic field (induction coil, magnetic amplitude 10–50 kA/m, 0.3–0.5 MHz), the subsequent shrink process can be

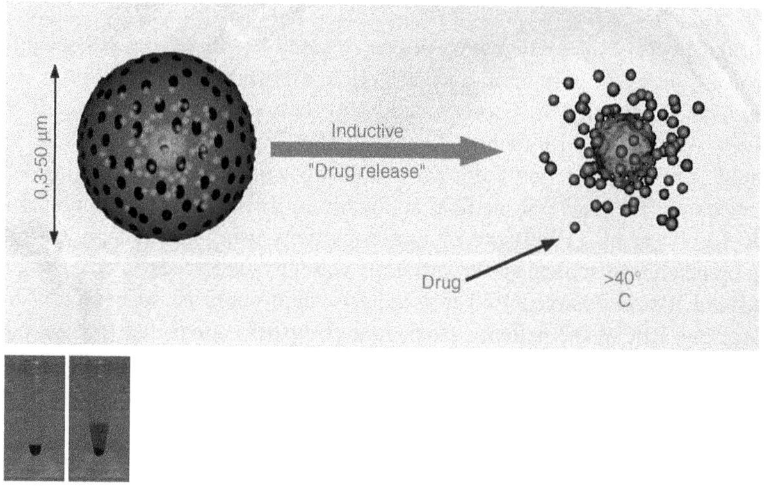

FIGURE 21.5 (See color insert) Graph illustrating a contactless controllable drug carrying system based on thermosensitive magnetic nano- and micro particles. The insert shows the application of the system with Rhodamine B encapsulated beads that is released after heating up to 45°C. By encapsulating ferro (i)magnetic colloids into the polymer matrix and exposing these magnetic beads to an external high frequency magnetic field (induction coil, magnetic amplitude 10–50 kA/m, 0.3–0.5 MHz), the subsequent shrink process can be remotely induced.

remotely induced. This process is reversible, i.e., the shape of the polymer goes back to its original form once the initial temperature is reached again. This principle can be exploited for the construction of controllable stimulus-response drug carriers. For this purpose, biologically active substances (e.g., anticancer drugs) are encapsulated together with the magnetic colloid into the polymer matrix. The subsequent inductive heating above the critical phase transition temperature, which is simultaneously accompanied by the de-swelling of the polymer, leads to an immediate release of the encapsulated drugs. Depending on the size of the polymer carrier, this drug release occurs over a period of 1–4 min, thus making these carriers an ideal controllable drug depot. Release measurements were carried out using model substances such as methylene blue and rhodamine B.

By applying the thermosensitive carrier-based therapy (Figure 21.5), drug release can be carefully tailored to the specific therapeutic requirements through the selection and formulation of drug–polymer combinations. Thus the total dose of medication and the kinetics of release are variables that can be widely controlled over minutes and hours. Moreover, polymer matrix based drug systems may increase the life span of active compounds, hence helping gain optimal conditions for therapeutic strategies. The release matrix regulates the plasma concentration peaks and reduces toxicity and other side effects.

In parallel to the drug release, the described induction heat mechanism can be used to heat tumor tissues or cells within the scope of hyperthermia. This basic technology uses the novel inductively heatable thermosensitive magnetic carriers in combination with the drug release and the hyperthermia. This opens up new perspectives for a highly intensified, targeted treatment of tumors.

21.3 HAZARDS AND RISKS OF NANOPARTICLES

21.3.1 General Concepts

To use the potential of nanotechnology in nanomedicine, full attention to safety and toxicological issues is needed. This is particularly true for applications such as drug delivery and imaging, as discussed in the previous paragraphs. In these applications, these particles are brought intentionally into the human body and environment, and some of these new applications are envisaged an important improvement of health care (Buxton et al. 2003; Ferrari 2005). Opinions started to divert when toxicologists claimed that new science, methods, and protocols are needed (Borm 2002; Donaldson et al. 2004, 2005; Nel et al. 2006). However, the need for this is now underlined by several expert reports (Oberdörster, Oberdörster, and Oberdörster 2005a; Oberdörster et al. 2005b; SCENIHR 2005) and more importantly by the following concepts:

1. Nanomaterials are developed for their unique (surface) properties in comparison to bulk materials. Since surface is the contact layer with the body tissue and a crucial determinant of particle response, these unique properties need to be investigated from a toxicological standpoint. The argument that current tests and procedures in drug and device approval are sufficient is therefore not valid.
2. Nanoparticles are attributed qualitatively different effects from micron-sized particles, such as changed body distribution, passage of the blood brain barrier, and triggering of blood coagulation pathways. Such tests are not routine in protocol toxicology of medical devices and need to be discussed.
3. Effects of combustion derived nanoparticles in environmentally exposed populations mainly occur in diseased individuals (see also Chapter 14 and Chapter 15). Typical pre-clinical screening is almost always done in healthy animals and volunteers and risks of particles may therefore be detected at a very late stage.

21.3.2 Toxicological Effects of Nanoparticles

As already discussed in other chapters of this book, NP exert some very special properties that are very relevant in the further design of toxicity testing of engineered nanomaterials. The first issue is that of cardiovascular effects and the second is of penetration and effects in the central nervous system.

21.3.2.1 Effects on Blood and Cardiovascular System

As we discussed earlier, ligand coated engineered nanoparticles are being explored and used as agents for molecular imaging or drug delivery tools. This has led to a considerable understanding of particle properties that can affect penetration in tissue without affecting tissue function. Cationic NP, including gold and polystyrene have been shown to cause hemolysis and blood clotting, while usually anionic particles are quite non-toxic. This conceptual understanding may be used to prevent potential effects of unintended NP exposure. Similarly, drug loaded nanoparticles have been used to prolong half-life or reduce side-effects and have shown which particle properties need to be modified to allow delivery, while being biocompatible (review: Gupta and Gupta 2005).

On the other hand, one is trying to find explanations for the increased risk of patients with CV diseases upon exposure to PM and/or traffic. Several toxicological studies have demonstrated that combustion and model NPs can gain access to the blood following inhalation or instillation and can enhance experimental thrombosis, but it is not clear whether this was an effect of pulmonary inflammation or particles translocated to the blood (Nemmar et al. 2002, 2003a, 2003b). High exposures to DEP by inhalation caused altered heart rate in hypertensive rats (Campen et al. 2003), interpreted as a direct effect of DEP on the pacemaker activity of the heart. Inflammation in distal sites has long been associated with destabilization of atheromatous plaques, and both instillation and inhalation of PM cause morphological evidence of atheromatous plaque increase and destabilization in rabbits (Suwa 2002) and mice (Chen 2005). Ultrafine carbon black instilled into the blood has been reported to induce platelet accumulation in the hepatic microvasculature of healthy mice in association with pro-thrombotic changes on the endothelial surface of the hepatic microvessels (Khandoga et al. 2004). Recent studies with carbon derived nanomaterials showed that platelet aggregation was induced by both single and multi-wall carbon nanotubes, but not by the C60-fullerenes that are used as building blocks for these CNT (Radomski et al. 2005). This data shows that not all nanomaterials act similar in this test, and that surface area is not the only factor playing a role here. The data also corroborate the earlier concept developed in medicine that only cationic species have an effect on blood clotting. Interestingly, this is the first study that allows bridging of data, since also a real life PM10 sample (SRM1648) was included in the test series. Actually, the PM sample showed a lower effect compared to nanoparticles, but the effect was abolished after filtering the particles out on a 0.1 μm filter (Radomski et al. 2005).

21.3.2.2 Uptake and Effects of Nanoparticles in the Brain

Nanoparticles can get access to the brain by two different mechanisms, i.e., (1) transsynaptic transport after inhalation through the olfactory epithelium and (2) uptake through the blood–brain barrier. The first pathway has been studied primarily with model particles such as carbon, Au, and MnO_2 in experimental inhalation models (review: Oberdörster, Oberdörster, and Oberdörster 2005a; Oberdörster et al. 2005b). The second pathway has been the result of extensive research and particle surface manipulation in drug delivery, as an approach to try and get drugs to the brain (review: Kreuter 2004). The latter studies suggest that the physiological barrier may limit the distribution of some proteins and viral particles after transvascular delivery to the brain, suggesting that the healthy BBB contains defense mechanisms protecting it from blood borne nanoparticle exposure. A number of pathologies, including hypertension and allergic encephalomyelitis, however, have been associated with increased permeability of the BBB to nanoparticles in experimental set ups. Reversely, the nanoparticle surface charges have been shown to alter blood–brain integrity (Lockman et al. 2004) and need consideration for brain toxicity and brain distribution profiles.

Since nanoparticles have been shown to induce the production of reactive oxygen species and oxidative stress, and oxidative stress has been implicated in the pathogenesis of neurodegenerative diseases such as Parkinson's and Alzheimer's (Kedar 2003), it is conceivable that the long term effects might include a decrease in cognitive function. Evidence for such effects is presented by studies in biopsies from city dwellers and Alzheimer's-like pathology have demonstrated increased markers of inflammation and AB42-accumulation in frontal cortex and hippocampus in association with the presence of nanoparticles (Calderon-Garciduenas et al. 2004). Recently, inhalation exposure of BALB/c mice to particulate matter showed the activation of pro-inflammatory cytokines in the brain of exposed mice (Campbell et al. 2005). Whether this is due to the fraction of combustion nanoparticles remains to be investigated.

21.3.3 Current Data on the Toxicology of Engineered Nanoparticles

In the past few years, a number of papers have described the toxicology of newly engineered nanomaterials, including fullerenes carbon nanotubes (review: Donaldson et al. 2006), and quantum dots (Hardman 2006). These papers have illustrated that apart from size and surface area, many more parameters describing the material (surface) properties have to be included.

21.3.3.1 Quantum Dots

Fluorescent nanocrystals such as quantum dots (QDs)—semiconductor nanocrystals (2–100 nm) with unique optical and electrical properties—are often made often toxic constituents and the QD core often consists of a variety of metal complexes such as semi-conductors, noble metals, and magnetic transition metals (Hardman 2006). For instance, QDs can be composed of indium phosphate (InP), indium arsenate (InAs), or zinc sulfide (ZnS), Zinc–Selenium (ZnSe), cadmium–selenium (CdSe), or cadmium–tellurium (CdTe) cores. To render these particles biologically compatible or active, modern QDs are functionalized or given secondary coatings, which improve water solubility, core durability and prevent leaching of metals from the core, and suspensions characteristics. For example QD cores can be coated with hydrophilic polyethylene glycol (PEG). Nevertheless cadmium, as one of the most used constituent metals in QD, has a biological half-life of 15–20 years in humans and upon accumulation able to cause liver and kidney response upon repetitive use.

21.3.3.2 Carbon Nanotubes

Carbon nanotubes are long carbon-based tubes that can be either single- or multi-walled and have the potential to act as biopersistent fibers. Nanotubes have aspect ratios >100, with lengths of several mm and diameters of 0.7–1.5 nm for single-walled nanotubes (SWNT) and 2–50 nm for multi-walled nanotubes (MWNT). Different geometric structures of SWNT and MWNT exhibit different cytotoxicity to macrophages (Jia et al. 2005) and the toxicity of SWNT in fibroblast was dependent on the degree of functionalization. The mechanisms involved in cell toxicity and activation as shown in different target cells include ROS generation, lipid peroxidation, oxidative stress, mitochondrial dysfunction, and changes in cell morphology (Shevdova et al. 2006). However, MWCNT have been reported to act as radical scavengers in a radical generating system (Fenoglio et al. 2006). Both SWNT and MWNT nanotubes also induced platelet aggregation, but the C60-fullerenes that are used as building blocks for these CNT (Radomski et al. 2005) were inactive. MWNT elicit pro-inflammatory effects in keratinocytes (Monteiro-Riviere et al. 2005). Several studies using intratracheal instillation of high doses of nanotubes in rodents demonstrated chronic lung inflammation, including foreign-body granuloma formation and interstitial fibrosis (Lam et al. 2004; Warheit et al. 2004; Muller et al. 2005). These studies also reveal the tendency of the nonphysiologic administration route and the unrealistic high doses to lead to asphyxiation through nanotube clumping in the airways (Warheit et al. 2004). Although it has been suggested that the granulomatous inflammation could be a biopersistent fiber affect, the high dose of the aggregated nanotubes and the presence of metal impurities (e.g., Fe) could account for artificial toxicity.

21.3.3.3 Fullerenes

Fullerenes are being explored as potential new antimicrobial agents (Yamakoshi et al. 2003) and thus may shift microbial communities if they are released into the environment via effluents. Therefore, various studies with fullerenes have been published with regard to the exotoxicity of these important building blocks in nanomaterials. Tests with un-coated, water soluble, colloidal fullerenes (nC60) showed that the 48-h LC_{50} in Daphnia magna is 800 ppb (Oberdörster 2004). In largemouth bass, although no mortality was seen, lipid peroxidation was seen in the brain and glutathione depletion in the gill after exposure to 0.5 ppm nC60 for 48 h (Oberdörster 2005). There are several hypotheses as to how lipid damage may have occurred in the brain, including direct redox activity by fullerenes reaching the brain via circulation or axonal translocation and dissolving into the lipid-rich brain tissue; oxyradical production by microglia; or reactive fullerene metabolites may be produced by cytochrome P450 metabolism.

21.3.3.4 Dendrimers

Apart from application in drug-delivery, dendrimers are being investigated for many other uses including bacterial cell killing, as gene transfer agents, and for transmembrane transport. Little published data is available on the toxicity of this class of particles. A recent review on this topic (Duncan and Izzo 2005) concluded that it would only ever be possible to designate a dendrimers as "safe" when related to a specific application. The so far limited clinical experience with dendrimers makes it impossible to designate any particular chemistry intrinsically "safe" or "toxic."

21.3.3.5 Wear Particles from Implants

Wear is a key limiter to the service life of implants and devices, but also the source of particles and nanoparticles. These particles consist of micro- and nanoparticles and may invoke immune and other responses. Particulate debris can be generated following total joint replacement as a result of either wear or corrosion. Wear of the polyethylene (PE) acetabular cup articulating against the hard metal or ceramic femoral head (see Figure 21.1) leads to the generation of PE particles. Wear of the nonbearing surfaces rubbing together, such as back-side wear of an ace tabular liner, fretting of the morse taper in modular stems, stem/cement or stem/bone fretting wear in cemented and non-cemented hip prostheses, respectively, may lead to the generation of metal and PMMA wear particles. The UHMWPE particles isolated from tissues retrieved from failed total hip replacements vary greatly insize and morphology, from large platelet-like particles up to 250 μm in length, fibrils, shreds, and sub-micrometer globule-shaped spheroids between 0.1 and 0.5 μm in diameter. The vast majority of the numbers of particles are the globular spheroids and the mode of the frequency distribution reported in a range of studies is 0.1–0.5 μm (Bell et al. 2002). Wear particles produced by this type of wear can lead to an inflammatory reaction and osteolysis (review: Ingham and Fisher 2005), however, this type of wear is not intentional whereas the wear of ultra-high molecular weight polyethylene is an inevitable consequence of the normal function of the prosthesis. *In vivo* and *in vitro* studies with wear debris particles of materials including titanium alloys, PMMA, and cobalt chrome have clearly demonstrated that particle-stimulated macrophages will elaborate a range of potentially osteolytic mediators (IL-1, IL-6, TNF-α, GM-CSF, PGE2) and bone resorbing activity (review: Ingham and Fisher 2005). Induction of bone resorbing activity in particle-stimulated macrophage supernatants has been shown to be dependent upon particle size and particle concentration. Particles with a mean size of 0.24 μm stimulated bone resorption at a ratio of 10 μm^3/cell. Larger particles failed to stimulate bone-resorbing activity in macrophage culture supernatants (Green et al. 2000). With respect to cytokine production, particles in the 0.1–1.0 μm-sized range at a volumetric concentration of 10–100 μm^3 of particles per cell are the most biologically reactive. *In vitro* studies have indicated that, of the numerous cytokines, TNF-a is a key osteolytic cytokine generated by particle-stimulated macrophages. Algan, Purdon, and Horowitz (1996) showed that addition of anti-TNF-a antibody was able to significantly inhibit bone resorption by supernatants from particle-stimulated macrophages.

21.3.4 Nanomaterials in Medicine: Future Toxicological Needs

Although there is a considerable amount of data on the toxicity of NP, this data is mainly based on a small panel of NP (combustion derived NP, TiO_2, CB) and the assumption that a lot of effects by PM are driven by the ultrafine particle fraction in it (Donaldson et al. 2002). Toxicity studies with nanomaterials that have been used for drug delivery or imaging have primarily concern the reduction of drug toxicity as in cancer treatment. With the more widespread application of nanoparticles in drug delivery systems for other diseases, a number of recommendations need to be considered:

1. Testing protocols need to include the effects that have been described by CDNP. These include effects on blood coagulation, platelet aggregation, atherosclerosis, and possibly also brain translocation and activation.
2. Simple primary testing on hematotoxicity, complement activation, and cytokine release as compatibility tests are not enough and first need to be complemented with assays looking at the biodegradation of the delivery devices, including its metabolic and distribution profiles.
3. Current protocols (OECD, FDA) need to be screened for their technical feasibility to detect the conventional toxicity of nanoparticles and new protocols need to be designed and validated for typical nanoparticles effects.
4. Current legislation discriminating testing procedures between the active constituent and the carrier as a medical device needs to be revisited to ensure full testing of nanostructured materials that are included in a drug preparation.

To do this, communication and bridging between disciplines and different sectors of nanoscience and toxicology is needed to delete the uncertainty and gap between intentional and unintentional produced NP. Nanotechnology itself is enabling only through integration from concepts in physics, biology, and chemistry. Industry and academia are now facing the challenge to develop a conceptual understanding of biological responses to nanomaterials, and use this know-how to develop safe nanomaterials. Only combining know-how on material properties, size, and quantum effects of anticipated products with expertise on biological reactions to materials will lead to safe, sustainable applications of nanomaterials. Both academia and industry may therefore reconsider their current tiered testing approaches (industry) and academic, mechanism-driven use of nanoparticles into a network with exchange of know-how and information on hazardous materials.

REFERENCES

Ahmad, I., Longenecker, M., Samuel, J., and Allen, T. M., Antibody-targeted delivery of doxorubicin entrapped in sterically stabilized liposomes can eradicate lung cancer in mice, *Cancer Res.*, 53, 1484–1488, 1993.

Algan, S. M., Purdon, M., and Horowitz, S. M., Role of tumor necrosis factor in particulate induced bone resorption, *J. Orthop. Res.*, 14, 30–35, 1996.

Allen, T. M., Ligand-targeted therapeutics in anticancer therapy, *Nat. Rev. Drug Discov.*, 2, 750–763, 2002.

Allen, T. M. and Hansen, C., Pharmacokinetics of stealth versus conventional liposomes: effect of dose, *Biochim. Biophys. Acta*, 1068, 133–141, 1991.

Aston, R., Saffie-Siebert Canham, L., and Ogden, J., Nanotechnology applications for drug delivery, *Pharm. Techn. Europe*, April, 21–28, 2005.

Bajpai, S. K., Bajpai, M., and Kalla, K. G., Colon-specific oral delivery of vitamin B-2 from poly(acrylamide-*co*-maleic acid) hydrogels: an in vitro study, *J. Appl. Polym. Sci.*, 84 (6), 1133–1145, 2002.

Balthasar, S., Michaelis, K., Dinauer, N., von Briesen, H., Kreuter, J., and Langer, K., Preparation and characterization of antibody modified gelatin nanoparticles as drug carrier system for uptake in lymphocytes, *Biomaterials*, 26, 2723–2732, 2005.

Business Communications Company, Inc. (BCC), Global nanotechnology market to reach $29 billion by 2008, 2004. http://www.bccreserach.com/editors/RGB-290.html.

Bell, J., Tipper, J. L., Ingham, E., Stone, M. H., Wroblewski, B. M., and Fisher, J., Quantitative analysis of UHMWPE wear debris isolated from the periprosthetic femoral tissues from a series of Charnley total hip arthroplasties, *Bio-med. Mater. Eng.*, 12, 189–201, 2002.

Bergemann, C., Müller-Schulte, D., Oster, J., Brassard, L., and Lübbe, A. S., Magnetic ion-exchange nano- and microparticles for medical, biochemical, and molecular biological application, *J. Magn. Magn. Mater.*, 194, 45–52, 1999.

Bharali, D. J., Klejbor, I., Stachowiak, E. K., Dutta, P., Roy, I., Kaur, N., Bergey, E. J., Prasad, P. N., and Stachowiak, M. K., Organically modified silica nanoparticles: a nonviral vector for in vivo gene delivery and expression in the brain, *Proc. Natl. Acad. Sci.*, 32, 11539–11544, 2005.

Borm, P. J. A., Particle toxicology: from coal mining to nanotechnology, *Inhal. Toxicol.*, 14, 311–324, 2002.

Borm, P. J. A. and Kreyling, W., Toxicological hazards of inhaled nanoparticles-potential implications for drug delivery, *J. Nanosci. Nanotechnol.*, 4, 521–531, 2004.

Brannon-Peppas, L. and Blanchette, J. O., Nanoparticle and targeted systems for cancer therapy, *Adv. Drug Deliv. Rev.*, 56, 1649–1659, 2004.

Bulte, J. W. M. and Brooks, R. A., Magnetic Nanoparticles as Contrast Agents for MR Imaging, In *Scientific and Clinical Applications of Magnetic Carriers*, Häfeli et al., eds., Plenum Press, New York, 527–543, 1997.

Buxton, D. B., Lee, S. C., Wickline, S. A., and Ferrari, M., National heart, lung, and blood institute nanotechnology working group. Recommendations of the National Heart, Lung, and Blood Institute Nanotechnology Working Group, *Circulation*, 108 (22), 2737–2742, Dec 2 2003.

Calderon-Garciduenas, L., Reed, W., Maronpot, R. R., Henriquez-Roldan, C., Delgado-Chavez, R., Calderon-Garciduenas, A., Dragustinovis, I. et al., Brain inflammation and Alzheimer's-like pathology in individuals exposed to severe air pollution, *Toxicol. Pathol.*, 32 650–658, 2004.

Campbell, A., Oldham, M., Becaria, A., Bondy, S. C., Meacher, D., Sioutas, C., Misra, C., Mendez, L. B., and Klinmath, A., Particulate matter in polluted air may increase biomarkers of inflammation in mouse brain, *Neurotoxicology*, 26 (1), 133–140, 2005.

Campen, M. J., McDonald, J. D., Gigliotti, A. P., Seilkop, S. K., Reed, M. D., and Benson, J. M., Cardiovascular effects of inhaled diesel exhaust in spontaneously hypertensive rats, *Cardiovasc. Toxicol.*, 3 (4), 353–361, 2003.

Cerdan, S., Lötscher, H. R., Künnecke, B., and Seelig, J., Monoclonal antibody-coated magnetite particles as contrast agents in magnetic resonance imaging of tumors, *Magn. Res. Med.*, 12, 151–163, 1989.

Chen, L. C. and Nadziejko, C., Effects of subchronic exposures to concentrated ambient particles (CAPs) in mice, *Inhal. Toxicol.*, 17 (4–5), 21724, Apr 2005.

Chen, G. Y. J. and Yao, S. Q., Lighting up cancer cells with "dots,", *Lancet*, 364, 2001–2003, 2004.

Chen, Y., Ji, T., and Rosensweig, Z., Synthesis of glyconanospheres containing luminescent CdSe–ZnS quantum dots, *Nano Lett.*, 3, 581–584, 2003.

Colvin, V. L., The potential environmental impact of engineered nanomaterials, *Nat. Biotechnol.*, 21 (10), 1166–1170, 2003.

De Cuyper, M., In *Applications of magnetoproteoliposomes in bioreactors operating in high-gradient magnetic field*, Barenholz, Y. and Lasic, D. D., eds. *Handbook of Nonmedical Applications of Liposomes*, Vol. III, CRC Press Inc., Boca Raton, FL, 323–340, 1996.

De Cuyper, M. and Joniau, M., Magnetoliposomes formation and structural charaterization, *Eur. Biophys. J.*, 15, 311–319, 1988.

Dockery, D. W., Pope, C. A. 3rd, and Xu, X., An association between air pollution and mortality in six U.S. cities, *N. Engl. J. Med.*, 329 (24), 1753–1759, 1993.

Donaldson, K., Brown, D., Clouter, A., Duffin, R., MacNee, W., Renwick, L., and Stone, V., The pulmonary toxicology of ultrafine particles, *J. Aerosol Med.*, 15, 213–220, 2002.

Donaldson, K., Stone, V., Tran, C. L., Kreyling, W., and Borm, P. J., *Nanotoxicol. Occup. Environ. Med.*, 61(9), 727–728, 2004.

Donaldson, K., Tran, L., Jimenez, L. A., Duffin, R., Newby, D. E., Mills, N., MacNee, W., and Stone, V., Combustion-derived nanoparticles: a review of their toxicology following inhalation exposure, *Part Fibre Toxicol.*, 2, 10, 2005.

Donaldson, K., Aitken, R., Tran, L., Stone, V., Duffin, R., Forrest, G., and Alexander, A., Carbon nanotubes: a review of their properties in relation to pulmonary toxicology and workplace safety, *Toxicol. Sci.*, Epub ahead of print.

Dubertret, B., Skourides, P., Norris, D. J., Noireaux, V., Brivanlou, A. H., and Libchaber, A., In vivo imaging of quantum dots encapsulated in phospholipid micelles, *Science*, 298 (5599), 1759–1762, 2002.

Duncan, R., The dawning era of polymer therapeutics, *Nat. Rev.*, 2, 347–360, 2003.

Duncan, R. and Izzo, L., Dendrimer biocompatibility and toxicity, *Adv. Drug Deliv. Rev.*, 57 (15), 2215–2237, 2005. Epub 2005 Nov 16, Review

Dyer, A. M., Hinchcliffe, M., Watts, P., Castile, J., Jabbal-Gill, I., Nankervis, R., Smith, A., and Illum, L., Nasal delivery of insulin using novel chitosan based formulations: a comparative study in two animal models between simple chitosan formulations and chitosan nanoparticles, *Pharm. Res.*, 19, 998–1008, 2002.

European Science Foundation Policy Briefing (ESF), *ESF Scientific Forward Look on Nanomedicine 2005*, IREG Strasbourg, France, ISBN 2-912049-520, 2005.

Fendler, J. H. and Romero, A., Liposomes as drug carriers, *Life Sci.*, 20, 1109–1120, 1977.

Fenoglio, I., Tomatis, M., Lison, D., Muller, J., Fonseca, A., Nagy, J. B., and Fubini, B., Reactivity of carbon nanotubes: free radical generation or scavenging activity? *Free Radic. Biol. Med.*, 40 (7), 1227–1233, 2006.

Ferrari, M., Cancer nanotechnology: opportunities and challenges, *Nat. Rev.*, 5, 161–171.

Flasher, D., Konopka, K., Chamow, S. M., Dazin, P., Ashkenazi, A., Pretzer, E., and Duzgunes, N., Liposome targeting to human-immunodeficiency-virus type 1-infected cells via recombinant soluble CD4 and CD4 immunoadhesin (CD4-IgG), *Biochim. Biophys. Acta*, 1194 (1), 185–196, 1994.

Gabizon, A., Price, D. C., Huberty, J., Bresalier, R. S., and Papahadjopoulos, D., Effect of liposome composition and other factors on the targeting of liposomes to experimental tumors: biodistribution and imaging studies, *Cancer Res.*, 50, 6371–6378, 1990.

Gao, X., Cui, Y., Levenson, R. M., Chung, L. W. K., and Nie, S., In vivo cancer targeting and imaging with semiconductor quantum dots, *Nat. Biotechnol.*, 22, 969–976, 2004.

Gordon, R. T., Use of magnetic susceptibility probes in the treatment of cancer, US Patent 4,662,359, 1987.

Gorman, J., Buckymedicine—coming soon to a pharmacy near you, *Sci. News*, 162, 26, 2002.

Graf, C., Schärtl, W., Fischer, K., Hugenberg, N., and Schmidt, M., Dye-labeled poly(organosiloxane) microgels with core-shell architecture, *Langmuir*, 15 (19), 6170–6180, 1999.

Granum, B. and Lovik, M., The effect of particles on allergic immune responses, *Toxicol. Sci.*, 65, 7–17, 2002.

Graves, E. E., Yessayan, D., Turner, G., Weissleder, R., and Ntziachristos, T., Validation on in vivo fluorochrome concentrations measured using fluorescence molecular tomography, *J. Biomed. Opt.*, 10 (4), 044019, 2005.

Green, T. R., Fisher, J., Matthews, J. B., Stone, M. H., and Ingham, E., Effect of size and dose on bone resorption activity of macrophages in vitro by clinically relevant ultra-high molecular weight polyethylene particles, *J. Biomed. Mater. Res. Appl. Biomater.*, 53, 490–497, 2000.

Gregoriadis, G., Drug entrapment in liposomes, *FEBS Lett.*, 36, 292–296, 1973.

Grimm, J., Karger, N., Lusse, S., Winoto-Morbach, S., Krisch, B., Muller-Hulsbeck, S., and Heller, M., Characterization of ultrasmall paramagnetic magnetite particles as superparamagnetic contrast agents in MRI, *Invest. Radiol.*, 35 (9), 553–556, 2000.

Guimaraes, R., Clement, O., Bittoun, J., Carnot, F., and Frija, G. M. R., Lymphography with superparamagnetic iron nanoparticles in rats, *Am. J. Radiol.*, 162, 201–207, 1994.

Gupta, A. K. and Gupta, M., Synthesis and surface engineering of iron oxide nanoparticles for biomedical applications, *Biomaterials*, 26 (18), 3995–4021, 2005.

Han, M., Gao, X., Su, J. Z., and Nie, S., Quantum dots tagged microbeads for multiplexed optical coding of biomolecules, *Nat. Biotechnol.*, 19, 631–635, 2001.

Hardman, R., A toxicological review of quantum dots: toxicity depends on physicochemical and environmental factors, *Environ. Health Perspect.*, 114, 165–172, 2006.

Hirai, T., Okubo, H., and Komasawa, I., Incorporation of CdS nanoparticles formed in reverse micelles into mesoporous silica, *J. Colloid Interface Sci.*, 235 (2), 358–364, 2001.

Hirsch, L. R., Stafford, R. J., Bankson, J. A., Sershen, S. R., Rivera, B., Price, R. E., Hazle, J. D., Halas, N. J., and West, J. L., Nanoshell-mediated near-infrared thermal therapy of tumors under magnetic resonance guidance, *Proc. Natl. Acad. Sci.*, 100 (23), 13549–13554, 2003.

Hodenius, M., De Cuyper, M., Desender, L., Müller-Schulte, D., Steigel, A., Lueken, H., *Chem. Phys. Lipid.*, 120 (1-2), 75–85, 2002.

Iga, K., Ohkouchi, K., Ogawa, Y., and Toguchi, H., Membrane modification by negatively charged stearyl-polyoxyethylene derivatives for thermosensitive liposomes, *J. Drug Target.*, 2, 259–267, 1994.

Ingham, E. and Fisher, J., The role of macrophages in osteolysis of total joint replacement, *Biomaterials*, 26, 1271–1286, 2005.

Ito, A., Juga, Y., Honda, H., Kikkawa, H., Horiuchi, A., Watanabe, Y., and Kobayashi, T., Magnetite nanoparticle-loaded anti-HER2 immunoliposomes for combination of antibody therapy with hyperthermia, *Cancer Lett.*, 212, 167–175, 2004.

Jaffer, F. A. and Weissleder, R., Molecular imaging in the clinical arena, *JAMA*, 293, 855–862, 2005.

Jia, G., Wang, H. F., and Yan, L., Cytotoxicity of carbon nanomaterials: single-wall nanotube, multi-wall nanotube, and fullerene, *Environ. Sci. Technol.*, 39 (5), 1378–1383, 2005.

Kalambur, V. S., Han, B., Hammer, B. E., Shield, T. W., and Bischof, J. C., In vitro characterization of movement, heating and visualization of magnetic nanoparticles for biomedical applications, *Nanotechnology*, 16 (8), 1221–1233, 2005.

Karam, L. R., Mitch, M. G., and Coursey, B. M., Encapsulation of Tc-99m within fullerenes: a novel radionuclidic carrier, *Appl. Radiat. Isot.*, 48, 771–776, 1997.

Kedar, N. P., Can we prevent Parkinson's and Alzheimer's disease? *J. Postgrad. Med.*, 49 (3), 236–245, 2003.

Khandoga, A., Stampfl, A., Takenaka, S., Schulz, H., Radykewicz, R., Kreyling, W., and Krombach, F., Ultrafine particles exert prothrombotic but not inflammatory effects on the hepatic microcirculation in healthy mice in vivo, *Circulation*, 109(10), 1320–1325, 2004.

Kim, S., Liposomes as carriers of cancer chemotherapy—current status and future prospects, *Drugs*, 46, 618–638, 1993.

Kresse, M., Wagner, S., and Taupitz, M., SPIO-enhanced MR lymphography, In *Scientific and Clinical Applications of Magnetic Carriers*, Häfeli, U. et al., eds., Plenum Press, New York, 545–559, 1997.

Kreuter, J., Influence of the surface properties on nanoparticle-mediated transport of drugs to the brain, *J. Nanosci. Nanotechnol.*, 4 (5), 484–488, 2004.

Kriwet, B., Walter, E., and Kissel, T., Snythesis of bioadhesive poly(acrylic acid) nano-and microparticles using an inverse emulsion polymerization method for the entrapment of hydrophilic drug candidates, *J. Control. Release*, 56, 149–158, 1998.

Lam, C. W., James, J. T., McCluskey, R., and Hunter, R. L., Pulmonary toxicity of single-wall carbon nanotubes in mice 7 and 90 days after intratracheal instillation, *Toxicol. Sci.*, 77 (1), 126–134, 2004.

Lanza, G. M., Winter, P., Caruthers, S., Schmeider, A., Crowder, K., Morawski, A., Zhang, H. Y., Scott, M. J., and Wickline, S. A., Novel paramagnetic contrast agents for molecular imaging and targeted drug delivery, *Curr. Pharm. Biotechnol.*, 5 (6), 495–507, 2004.

LaVan, D. A., McGuire, T., and Langer, R., Small scale systems for in vivo drug delivery, *Nat. Biotechnol.*, 21, 1184–1191, 2003.

Lawaczeck, R., Menzel, M., and Pietsch, H., Superparamagnetic iron oxide particles: contrast media for magnetic resonance imaging, *Appl. Organomet. Chem.*, 18, 506–513, 2004.
Lemarchand, C., Gref, R., and Couvreur, P., Polysaccharide-decorated nanoparticles, *Eur. J. Pharm. Biopharm.*, 58, 327–341, 2004.
Lockman, P. R., Koziara, J. M., Mumper, R. J., and Allen, D. D., Nanoparticle surface charges alter blood-brain barrier integrity and permeability, *J. Drug Target.*, 12, 635–641, 2004.
Lübbe, A., Bergemann, C., Riess, H., Schriever, F., Reichardt, P., Possinger, K., Matthias, M. et al., Clinical experiences with magnetic drug targeting: a phase I study with 4′-epidoxorubicin in 14 patients with advanced solid tumors, *Cancer Res.*, 56, 4686–4693, 1996.
Lübbe, A. S., Bergemann, C., Huhnt, W., Fricke, T., and Riess, H., Preclinical experiences with magnetic drug targeting: tolerance and efficacy, *Cancer Res.*, 56, 4694–4701, 1996.
Maier-Hauff, K., Jordan, A., Nestler, D., Scholz, R., Rothe, R., Feussner, A., Gneveckow, U., Wust, P., and Felix, R., Magnetic fluid hyperthermia (MFH) as an alternative treatment of malignant gliomas, *Strahlenther. Onkol.*, 44–44 (suppl 1), 2005.
Majoros, I. J., Thomas, T. P., Mehta, C. B., and Baker, J. R., Poly(amidoamine) dendrimer-based multi-functional engineered nanodevice for cancer therapy, *J. Med. Chem.*, 48, 5892–5899, 2005.
Malik, N., Evagorou, E. G., and Duncan, R., Dendrimer-platinate: a novel approach to cancer chemotherapy, *Anticancer Drugs*, 10 (8), 767–776, 1999.
Mikawa, M., Kato, H., Okumura, M., Narazaki, M., Kanazawa, Y., Miwa, N., and Shinohara, H., 2001.
Monteiro-Riviere, N. A., Nemanich, R. J., Inman, A. O., Wang, Y. Y., and Riviere, J. E., Multi-walled carbon nanotube interactions with human epidermal keratinocytes, *Toxicol. Lett.*, 155 (3), 377–384, 2005.
Mornet, S., Vasseur, S., Grasset, F., and Duguet, E., Magnetic nanoparticle design for medical diagnosis and therapy, *J. Mater. Chem.*, 14, 2161–2175, 2004.
Muller, J., Huaux, F., Moreau, N., Misson, P., Heilier, J. F., Delos, M., Arras, M., Fonseca, A., Nagy, J. B., and Lison, D., Respiratory toxicity of multi-wall carbon nanotubes, *Toxicol. Appl. Pharmacol.*, 207 (3), 221–231, 2005.
Müller-Schulte, D., Mittel zur selektiven AIDS Therapy sowie Verfahren zur Herstellung und Verwendung derselben, German Pat. Application DE 4412651, 1995.
Müller-Schulte, D., Separating, detecting or quantifying biological materials using magnetic cross-linked polyvinyl alcohol, US-Patent 6,514,688, 2003.
Müller-Schulte, D., Luminescent, spherical, non-autofluorescent silica gel particles with changeable emision intensities and emission frequencies, Eur Patent Application PCT/EP 03/03163, 2004.
Müller-Schulte, D., Füssl, F., Lueken, H., and De Cuyper, M., A new AIDS therapy approach using magnetoliposomes, In *Scientific and Clinical Applications of Magnetic Carriers*, Häfeli, U. et al., eds., Plenum Press, New York, 517–526, 1997.
Müller-Schulte, D., Schmitz-Rode, T., and Borm, P., Ultra-fast synthesis of magnetic and luminescent silica beads for versatile bioanalytical applications, *J. Magn. Magn. Mater.*, 293, 135–143, 2005.
Murphy, C. J., Optical sensing with quantum dots, *Anal. Chem.*, Oct 1 520A–526A, 2002.
Nel, A., Xia, T., Madler, L., and Li, N., Toxic potential of materials at the nanolevel, *Science*. 3, 11(5761), 622–627, 2006.
Nemmar, A., Hoylaerts, M. F., Hoet, P. H., Vermylen, J., and Nemery, J., Size effect of intratracheally instilled particles on pulmonary inflammation and vascular thrombosis, *Toxicol. Appl. Pharmacol.*, 186(1), 38–45, 2003.
Nemmar, A., Hoet, P. H., Vanquickenborne, B., Dinsdale, D., Thomeer, M., Hoylaerts, M. F., Vanbilloen, H., Mortelmans, L., and Nemery, B., Passage of inhaled particles into the blood circulation in humans, *Circulation*, 105(4), 411–414, 2002.
Nemmar, A., Nemery, B., Hoet, P. H., Vermylen, J., and Hoylaerts, M. F., Pulmonary inflammation and thrombogenicity caused by diesel particles in hamsters: role of histamine, *Am. J. Respir. Crit. Care Med.*, 168 (11), 1366–1372, 2003.
Neuberger, T., Schopf, B., Hofmann, H., Hofmann, M., and von Rechenberg, B., Superparamagnetic nanoparticles for biomedical applications: possibilities and limitations of a new drug delivery system, *J. Magn. Magn. Mater.*, 293 (1), 483–496, 2005.
Oberdörster, E., Manufactured nanomaterials (fullerenes C60) induce oxidative stress in the brain of juvenile largemouth bass, *Environ. Health Perspect.*, 112 (10), 1058–1062, 2005.
Oberdörster, G., Pulmonary effects of ultrafine particles, *Int. Arch. Occup. Environ. Health*, 74, 1–8, 2001.

Oberdörster, G., Sharp, Z., Atudorei, V., Elder, A., Gelein, R., Kreyling, W., and Cox, C., Translocation of inhaled ultrafine particles to the brain, *Inhal. Toxicol.*, 16 (6–7), 437–445, 2004.

Oberdörster, G., Oberdörster, E., and Oberdörster, J., Nanotoxicology: an emerging discipline evolving from studies of ultrafine particles, *Environ. Health Perspect.*, 113 (7), 823–839, 2005.

Oberdörster, G., Maynard, A., Donaldson, K., Castranova, V., Fitzpatrick, J., Ausman, K., Carter, J. et al., ILSI Research Foundation/Risk Science Institute Nanomaterial Toxicity Screening Working Group. Principles for characterizing the potential human health effects from exposure to nanomaterials: elements of a screening strategy, *Part Fibre Toxicol.*, 2, 8, 2005.

Pankhurst, Q. A., Connolly, J., Jones, S. K., and Dobson, J., Applications of magnetic nanoparticles in biomedicine, *J. Phys. D: Appl. Phys.*, 36 (13), R167–R181, 2003.

Parak, W. J., Gerion, D., Pellegrino, T., Zanchet, D., Micheel, C., Williams, S. C., Boudreau, R., Le Gros, M. A., Larabell, C. A., and Alivisatos, A. P., Biological applications of colloidal nanocrystals, *Nanotechnology*,14 (7), R15–R27, 2003.

Park, J. W., Liposome-based drug delivery in breast cancer treatment, *Breast Cancer Res.*, 4, 95–99, 2002.

Patri, A. K., Majoros, I. J., and Baker, R. J., Dendritic polymer macromolecular carriers for drug delivery, *Curr. Opin. Chem. Biol.*, 6, 466–471, 2002.

Päuser, S., Reszka, R., Wagner, S., Wolf, K. J., Buhr, H. J., and Berger, G., Superparamagnetic iron oxide particles as marker substances for searching tumor specific liposomes with magnetic resonance imaging, In *Scientific and Clinical Applications of Magnetic Carriers*, Häfeli, U. and Schätt, W., eds., Plenum Press, New York, 561–568, 1997.

Quellec, P., Gref, R., Perrin, L., Dellacherie, E., Sommer, F., Verbavatz, J. M., and Alonso, M. J., Protein encapsulation within polyethylene glycol-coated nanospheres. I. Physicochemical characterization, *J. Biomed. Mater. Res.*, 42, 45–54, 1998.

Radomski, A., Jurasz, P., Alonso-Escolano, D., Drews, M., Morandi, M., Malinski, T., and Radomski, M. W., Nanoparticle-induced platelet aggregation and vascular thrombosis, *Br. J. Pharmacol.*, 146 (6), 882–893, 2005.

Roy, I., Ohulchanskyy, T. Y., Bharali, D. J., Pudavar, H. E., Mistretta, R. A., Kaur, N., and Prasad, P. N., Optical tracking of organically modified silica nanoparticles as DNA carriers: A nonviral, nanomedicine approach for gene delivery, *Proc. Nat. Acad. Sci. U.S.A.*, 102(2), 279–284, 2005.

Santra, S., Dutta, D., and Biomoleküle, M., Functional dye-doped silica nanoparticles for bioimaging, diagnostics and therapeutics, *Food Bioproducts Process.*, 83, 1–5, 2005.

Sayes, C. M., Liang, F., and Hudson, J. L., Functionalization density dependence of single-walled carbon nanotubes cytotoxicity in vitro, *Toxicol. Lett.*, 161 (2), 135–142, 2006.

Scientific Committee on Emerging And Newly Identified Health Risks (SCENIHR), Opinion on The appropriateness of existing methodologies to assess the potential risks associated with engineered and adventitious products of nanotechnologies. European Commission Health and Consumer Protection Directorate-General Directorate C—Public Health and Risk Assessment.

Seaton, A., MacNee, W., Donaldson, K., and Godden, D., Particulate air pollution and acute health effects, *Lancet*, 345, 176–178, 1995.

Shinkai, M., Suzuki, M., Iijima, S., and Kobayashi, T., Antibody-conjugated magnetoliposomes for targeting cancer cells and their application in hyperthermia, *Biotechnol. Appl. Biochem.*, 21, 125–137, 1994.

Shvedova, A. A., Kisin, E. R., Mercer, R. et al., Unusual inflammatory and fibrogenic pulmonary responses to single-walled carbon nanotubes in mice, *Am. J. Physiol. Lung Cell Mol. Physiol.*, 289 (5), L698–708, 2005.

Soppimath, K. S., Aminabhavi, T. M., Dave, A. M., Kumbar, S. G., and Rudzinski, W. E., Stimulus-responsive "smart" hydrogels as novel drug delivery systems, *Drug Develop. Ind. Pharm.*, 28, 957–974, 2002.

Suwa, T., Hogg, J. C., Quinlan, K. B., Ohgami, A., Vincent, R., and van Eeden, S. F., Particulate air pollution induces progression of atherosclerosis, *Am. Coll. Cardiol.*, 39, 943–945, 2002.

Tartaj, P., Morales, M. D., Veintemillas-Verdaguer, S., Gonzalez-Carreno, T., and Serna, C. J., The preparation of magnetic nanoparticles for applications in biomedicine, *J. Phys. D: Appl. Phys.*, 36 (13), R182–R197, 2003.

Tiefenauer, L. X., Kuhne, G., Andres, and R. Y., Antibody magnetite nanoparticles—in-vitro characterization of a potential tumor-specific contrast agent for magnetic-resonance-imaging, *Bioconj. Chem.*, 4(5), 347–352, 1993.

Tomalioa, D. A., Naylor, A. M., and Goddard, W. A., Starburst-Dendrimere: Kontrolle von Grösse Gestalt, Oberflächenchemie, Topologie und Flexibilität beim Übergang von Atomen zu makroskopischer Materie, *Angew. Chem.*, 102, 119–238, 1990.

Torchilin, V. P., Omelyanenko, V. G., Papisov, M. I., Bogdanov, A. A., Trubetskoy, V. S., Herron, J. N., and Gentray, C. A., Poly(ethylene glycol) on the liposome surface: on the mechanism of polymer coated liposome longevity, *Biochim. Biophys. Acta*, 1195, 11–20, 1994.

Van Blaaderen, A., Materials science—colloids get complex, *Nature*, 439(7076), 545–546, 2006.

Van Blaaderen, A. and Vrij, A., Synthesis and characterization of colloidal dispersions of fluorescent monodisperse silica spheres, *Langmuir*, 8, 2921–2931, 1992.

Viroonchatapan, E., Ueno, M., Sato, H., Adachi, I., Nagae, H., Tazawa, K., and Horikoshi, I., Preparation and characterization of dextran magnetite-incorporated thermosensitive liposomes: an on-line flow system for quantifying magnetic responsiveness, *Pharm. Res.* 12(8), 1176–1183, 1995.

Velikov, K. P., Moroz, A., and van Blaaderen, A., Photonic crystals of core-shell colloidal particles, *Appl. Phys. Lett.*, 80(1), 49–51, 2002.

Wagner, V. and Wechsler D., Nanobiotechnologie II: Anwendungen in der Medizin und Pharmazie, Publisher: Zukünftige Technologien Consulting der VDI Technologiezentrum GmbH, Düsseldorf, Germany, 2004.

Wang, S. Z., Gao, R. M., Zhou, F. M., and Selke, M., Nanomaterials and singlet oxygen photosensitizers: potential applications in photodynamic therapy, *J. Mater. Chem.*, 14 (4), 487–493, 2004.

Warheit, D. B., Laurence, B. R., Reed, K. L., Roach, D. H., Reynolds, G. A., and Webb, T. R., Comparative pulmonary toxicity assessment of single-wall carbon nanotubes in rats, *Toxicol. Sci.*, 77 (1), 117–125, 2004. Jan

Watanabe, J., Iwamoto, S., and Ichikawa, S., Entrapment of some compounds into biocompatible nano-sized particles and their releasing properties, *Colloids Surf. B Biointerfaces*, 42, 141–146, 2005.

Weissleder, R., Reimer, P., Lee, A. S., Wittenberg, J., and Brady, T., MR receptor imaging: ultrasmall iron oxide particles targeted to asialoglycoprotein receptor, *Am. J. Radiol.*, 155, 1161–1167, 1990.

Winter, J. O., Liu, T. Y., Korgel, B. A., and Schmidt, C. E., Recognition molecule directed interfacing between semiconductor quantum dots and nerve cells, *Adv. Mater.*, 13, 1673–1677, 2001.

Woodle, M. C. and Lasic, D. D., Sterically stabilized liposomes, *Biochim. Biophys. Acta*, 1113, 171–199, 1992.

Wu, X., Liu, H., Liu, J., Haley, K. N., Treadway, J. A., Larson, J. P., Ge, N., Peale, F., and Bruchez, M. P., Immunofluorescent labeling of cancer marker her2 and other cellular targets with semiconductor quantum dots, *Nat. Biotechnol.*, 21, 41–46, 2003.

Yamakoshi, Y., Umezawa, N., Ryu, A., Arakane, K., Miyata, N., Goda, Y., Masumizu, T., and Nagano, T., Active oxygen species generated from photoexcited fullerene (c60) as potential mediciens:02-*versus 102, *J. Am. Chem. Soc.*, 125(42), 12803–12809, 2003.

Zhang, J. Q., Zhang, Z. R., Yang, H., Tan, Q. Y., Qin, S. R., and Qiu, X. L., Lyophilized paclitaxel magnetoliposomes as a potential drug delivery system for breast carcinoma via parenteral administration: in vitro and in vivo studies, *Pharm. Res.*, 22, 573–583, 2005.

Zhao, M., Kircher, M. F., Josephson, L., and Weissleder, R., Differential conjugation of tat peptide to superparamagnetic nanoparticles and its effect on cellular uptake, *Bioconjug. Chem.*, 13, 840–844, 2002.

Zobel, H. P., Junghans, M., Maienschein, V., Werner, D., Gilbert, M., Zimmermann, H., Noe, C., Kreuter, J., and Zimmer, A., Enhanced antisense efficacy of oligonucleotides adsorbed to monomethylaminoethylmethacrylate methylmethacrylate copolymer nanoparticles, *Eur. J. Pharm. Biopharm.*, 49(3), 203–210, 2000.

22 The Toxicology of Inhaled Particles: Summing Up an Emerging Conceptual Framework

Ken Donaldson
MRC/University of Edinburgh Centre for Inflammation Research, University of Edinburgh

Lang Tran
Institute of Occupational Medicine

Paul J. A. Borm
Centre of Expertise in Life Sciences, Hogeschool Zuyd

CONTENTS

22.1 OVERVIEW

The chapters in this book set out the state-of-the-science for particle toxicology as it pertains in the early twenty-first century. It points out the maturity of this area of applied science, and toxicology is, above all, an applied science. Toxicology's primary aim is to provide hazard data for risk assessment towards safe ways of working and living with the chemicals we encounter on a daily basis, which all have inherent toxicity for biological systems. In particle toxicology, we seek to provide the data that will allow us to manage the risk associated with living in atmospheres that are often complex mixtures of particles of varying toxicity. However, mechanistic particle toxicology crosses over into mainstream molecular medicine and can provide important clues to the basis of other types of disease. Our studies of the cardiovascular system, oxidative stress and molecular signaling in relation to particles offer an understanding that is generally applicable. This is eminently clear from the pages of this book where high quality and innovative research approaches are demonstrated in addressing the issues relating to particle effects. This chapter aims to draw together the various threads in the fabric of particle toxicology and present a new unifying concept, albeit simplified and incomplete, for this discipline.

It is clear that the rise in nanotechnological applications and products has and will have a huge impact on particle toxicology. Nanoparticle research has become the key area for study by particle toxicologists. It represents a considerable challenge with new portals of entry, such as the skin and the gut and new target organs such as the blood and brain. Additionally, pharmacological uses of nanoparticles have caused a realignment and whole new areas of research are opening up that build on the kind of expertise owned by particle toxicologists.

22.2 DEFINING THE PARTICLE TOXICOLOGY ENDEAVOR

Particle toxicologists study particles in two main ways:

1. Studies aimed at refining the dose-metric; or put simply, "*What* is it about particles that makes them harmful or not?"
2. Mechanistic studies; or put simply, "*How* do harmful particles cause harm?"

Studies aimed at refining the dose metric Knowledge is incomplete and we are not fully aware of the nature of the harmful dose (the quantity of the particle's physicochemistry that drives adverse effects; see below) for many particle types. If we fully understood what made any particle type harmful, we could focus in on that parameter for measurement and so improve risk management. Historically, there have been frequent mismatches between the current dose-metrics and our existing knowledge of the toxicology of some types of particles. For example, asbestos and other fibers are measured as mass of all airborne fibers visible by light microscopy longer than 5 μm (with diameter <3 μm and aspect ratio $>3{:}1$). However, toxicological research has shown that fibers that are both biopersistent and longer than about 20 μm are the pathogenic ones (Donaldson 2004). Despite this, the "old" fiber standard remains, and it takes no account of the issues of length or solubility. In the case of PM_{10}, the mass of particles around 10 μm are measured but much of this mass is harmless, e.g. salt. However, smaller particles ($PM_{2.5}$) or transition metals (Donaldson et al. 2004a) or oxidant generation (Schaumann et al. 2004) might be better predictors for health effects.

Mechanistic studies These studies aim to understand the cellular and molecular basis of the toxic effects of particles and the sum of their interactions with biological systems. Such studies contribute to risk assessment by providing a more complete framework for our understanding of how particles behave in the body, the effects they have on cells and how being in the body changes them. In combination with toxicokinetics, which describe how fast and to which extent particles get distributed to different organs or tissues, such studies allow the entire "life history" of particles in the body to be traced. These mechanistic studies offer the possibility of therapeutic intervention in the process of disease. An example is the current exploration of soluble TNF-receptors in the treatment of IPF, as a result of initial work on TNF-α in fibrotic disorders caused by quartz-containing dust (Piguet et al. 1990).

22.3 CLASSICAL TOXICOLOGY

The classical toxicology paradigm of exposure–dose–response can be used for particles as for any other toxin. Exposure, dose and response together with ADME/toxicokinetics allow us to describe the detailed history of a toxin in the body. Complete toxicokinetics analysis is not available for any pathogenic particle. Most pathogenic particles have not been considered to be metabolized and excreted in any conventional way. However, nanoparticles, because of their smallness, may undergo such changes (Oberdörster et al. 2005b). The fuller our understanding of these processes, the better we will be able to interpret toxicology studies and understand the nature of the toxic effect of any particle. Of exposure, dose and response, we are especially focused on "dose" as a key to understanding molecular and cellular toxicity as well as contributing to understanding the best metric. For nanoparticles, the dose-metric is not yet elucidated and is likely to vary with different particle types since they can be composed of a range of materials and can be different sizes and shapes. The concept of a "biologically effective dose" (BED) is an important one for particles since all particle exposures are mixed. We can hypothesize that there is one or more actual component of this total dose that actually drives the adverse effect, and this is the BED.

22.4 EXPOSURE

22.4.1 Exposure at Portal of Entry

In the past, particle toxicology was concerned almost exclusively the lungs. With the advent of nanoparticle toxicology there has been a sea change in how inhalation particle toxicology is viewed (Donaldson et al. 2004b). Following inhalation of nanoparticles, the blood and brain are seen as secondary targets for particle effects (Oberdörster et al. 2005a, 2005b).

The aerodynamic diameter is the key particle parameter that predicts whether any particle gains access to the lungs and it also determines the site of particle deposition in the lungs (Gehr et al. 2000). Aerodynamic diameter is important for deposition of bigger particles with impaction and sedimentation being the main processes, whilst these become less important as size diminishes and diffusion comes to dominate the deposition process. For fibers, interception is an important deposition process whereby the extremities of the fiber make contact with airspace walls while the center of gravity of the fiber is following the airstream at a bifurcations (Morgan and Seaton 1995). Deposition of compact particles occurs as particles fall out in accord with their weight, i.e., sedimentation. They also deposit by impaction as particles fail to negotiate bifurcations and collide with the bronchial wall; deposition at bifurcations is increased also by the normal turbulence that results from the disruption to the even flow of air at these points. Finally, the smallest particles reach the distal lungs to the point where the net flow of air is zero, where they move by molecular (Brownian) motion and they deposit efficiently by diffusion. For these reasons, deposition in the lung is highly focal as a result of the dose being applied to certain anatomical areas/hot-spots, such as the bifurcations of airways and the centriacinar region. At these hot-spots, deposition can be 100-fold higher than in adjacent areas (Balashazy et al. 2003).

22.4.2 Toxicokinetics and Translocation from the Portal of Entry

Up until recently, the translocation of particles from their site of entry to other target organs was not considered a major issue. However, because of data showing nanoparticle translocation from the lungs (Hoet et al. 2004; Nemmar et al. 2004), there is increasing concern in this regard. To date, animal studies show some translocation of radioactive nanoparticles from the lungs to the blood following instillation exposure (Nemmar et al. 2001) and to the brain following inhalation exposure (Oberdörster et al. 2004) (Figure 22.1). There is no evidence for this type of translocation following inhalation of any nanoparticle type in man at the time of writing. A flow diagram of the hypothetical

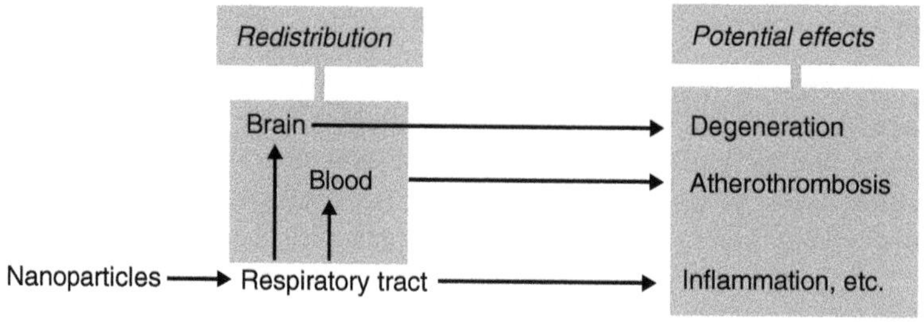

FIGURE 22.1 Outline of the potential toxicokinetic pathways for inhaled nanoparticles to translocate and have effects in the body following inhalation exposure.

fate and effects of nanoparticles is shown in Figure 22.2, based on limited animal studies with a few selected nanoparticle types.

22.4.3 Brain

Limited studies indicate that nanoparticles can gain access to the brain via the nose and the nerves that run from the olfactory epithelium into the olfactory lobes of the brain (Oberdörster et al. 2004). Nothing is known of the dosimetry in relation to exposure, nor whether this is generic property of nanoparticles. However, given the ubiquitousness of combustion-derived nanoparticles in our environment, if this is a general property of NP, then it is likely that we all have a burden of nanoparticles in our brain. Indirect evidence that this is indeed true and that there might be adverse effects on the brain comes from studies in Mexico City describing unusual brain pathology in the young (Calderon-Garciduenas et al. 2004). It is not known if nanoparticles can generally cross the blood/brain barrier, but medical nanoparticles have been designed to efficiently translocate to the brain from the blood (Kreuter et al. 2002).

22.4.4 Blood

Limited data indicates that nanoparticles can pass from the lungs to the blood (Kreyling et al. 2002a; Kreyling et al. 2002b; Nemmar et al. 2004). In the blood, particles can have a range of effects on blood components and associated cells, such as endothelial cells, monocytes and platelets (Hoet, Nemmar, and Nemery 2004). Exposure to combustion-derived nanoparticles and PM have shown effects at the level of the endothelium to impair vasomotion in a human model (Mills et al. 2005), which is known to be a risk factor for myocardial infarction. Concomitant pro-thrombotic effects that would favor thrombus propagation in the event of atherothrombosis were also reported (Mills et al. 2005). Few studies have reported the effect of engineered nanoparticles on the cardiovascular system (Radomski et al. 2005), but several studies have identified that exposure to PM or CAPS enhances severity of atherogenesis in Watanabe rabbits and ApoE mice (Suwa et al. 2002; Chen and Nadziejko 2005; Sun et al. 2005). Since particle based systems are being explored for molecular imaging in atherosclerotic disease, more research in this area is needed.

22.5 DOSE AND THE CONCEPT OF BIOLOGICALLY EFFECTIVE DOSE (BED)

The internal dose is the quantity of a toxin that gains access to the body. For inhalation exposure, because of particle clearance, of course, the fraction of the deposited dose that remains in a long-lived compartment like the interstitium or in long-lived macrophage accumulations can take part in toxic reactions, and it is less than the "total dose" that is breathed in and that deposits. This is then acted on by the milieu in the interstitium or in cells to cause dissolution of non-biopersistent particles or components. This may release harmful soluble components as well as harmless soluble components, which are cleared or metabolized. The BED is a useful concept, being that fraction of the total dose that is sufficiently biopersistent and also sufficiently active to cause an adverse effect such as oxidative stress, adduct formation, etc. This can be understood in that all realistic particle exposures are particles that are multi-component and poly-dispersed and within this total dose, we can identify sub-fractions that are likely to be more harmful or effective than others. For example, PM_{10} is measured by the mass metric, yet a variable and often substantial quantity of the mass of PM_{10} is sea salt, which is likely to be completely invisible to the lung following exposure at ambient levels, i.e., a harmless dose. Conversely, the transition metals can be seen to be driving the oxidative stress and inflammatory effects of PM in human (Schaumann et al. 2004) and animal models (Jimenez et al. 2000; Campen et al. 2001; Campen et al. 2002; Molinelli et al. 2002), yet this component would contribute very little to mass. Thus the transition metals can be seen as the BED, and their ability to cause oxidative stress is an early biological effect. Other examples of BED are the amount of "free" or "clean" quartz surfaces that are available to interact with cells that drive quartz's inflammatory effects (Donaldson and Borm 1998) and the proportion of biopersistent, long (>20 μm) fibers, which drive the pathogenicity of a fiber sample (Donaldson and Tran 2004c). The concept of the BED is shown in Figure 22.2.

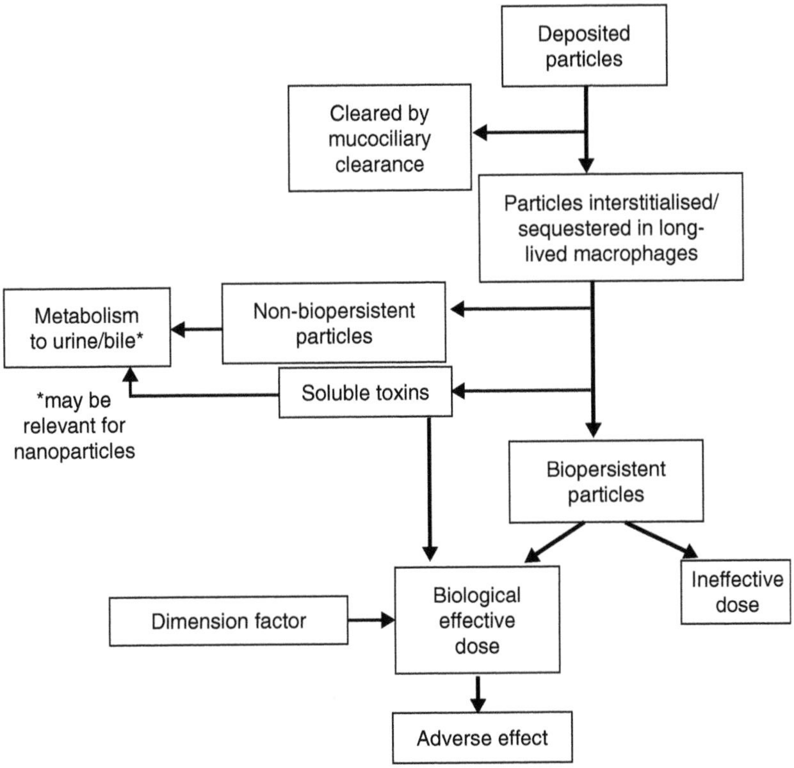

FIGURE 22.2 The relationship between deposited particles, biopersistence and the BED.

22.5.1 A Generalized Paradigm of Particle Toxicity Based on BED

These factors can be assembled into a small number of sources of BED that allow us to present a generalized paradigm for the total BED in the lungs for any specific particle type. The paradigm may be similar or different for pathogenicity of particles in other sites, such as blood or brain. For insoluble particles, it is only the surface layer that makes contact with the biological system and so can be considered to mediate the harm. For this reason, the surface area has come to the fore as a dose-metric that effectively describes the potential toxicity of particles, rather than total mass or particle number. The particle's surface is likely to vary in intrinsic reactivity/toxicity, depending on the material and (nanoscale) the pattern of which the particle is made. Therefore, intrinsic reactivity is a multiplier of surface area to obtain the total reactive surface produced by any mass of particles. In addition, we know from the asbestos and SVF experience that, for fibers, length can be important and so we can add "shape" as a factor. In addition, studies of PM_{10}, welding fume, etc. tell us that many particles are complex and contain soluble components that can have considerable toxic potential. Additionally, the length of time that the particle is likely to persist and not be cleared determines the length of time that the BED is applied to the system; thus a biopersistence factor is required. Taking these factors together, we have a paradigm that could predict the toxicity of a particle in the lungs.

The three main attributes for the biologically effective dose of a particle to a cell are:

1. Surface attribute = surface area × specific surface reactivity (i.e., reactivity per unit SA) × surface availability
2. Dimension attribute = length + diameter (mainly length if greater than a critical length)
3. Composition attribute = Volume × specific volumetric reactivity (i.e., the toxic material per unit volume) × availability (= release rate i.e., amount per unit time)

For acute effects the BED is related to the Potency, which can be best described as the sum of (1) + (2) + (3). For chronic effects the biopersistence plays a dominant role and the BED is best described by the product of biopersistence and potency:

$$\text{Biologically effective dose} = \text{Bio-persistence} \times \text{Potency}$$

This equation reflects the issue of translocation but deals specifically with the toxic outcome of the interaction between a particle and a cell. We do not think we are yet in a position where we can even approach the development of a paradigm for translocation that is based on structure. Table 22.2 shows attributes contributing to Bio for several particle types:

22.5.2 Oxidative Stress as an Early Biological Effect Marker of BED for Different Pathogenic Particles

Inspection of Table 22.1 raises questions as to how so many different chemical and physical entities (length, soluble toxins of different types, different surfaces) could all constitute some common form of "harmful dose" to the machinery of the cell. However, this may be understood in the findings that the ability to deliver oxidative stress is a common property of harmful particles. This oxidative stress can emanate from the particle itself and it can also be a consequence of the cellular and/or the inflammatory response induced by the particle (Donaldson et al. 2003; Knaapen et al. 2004).

There is a highly plausible link between oxidative stress and inflammation (Barrett et al. 1999; Tao et al. 2003; Donaldson et al. 2005b), with many oxidative stress-responsive pathways signaling for pro-inflammatory gene transcription (Piette et al. 1997; Rahman and MacNee 2000). There is also a clear link between oxidative stress and adducts such as 8-hydroxy-deoxyguanosine, the hydroxyl radical-induced adduct of guanine, which is involved in carcinogenesis (Lloyd et al. 1998; Tsurudome et al. 1999; Maeng et al. 2003). Many pathogenic particle types have been shown to activate NF-κB (Schins and Donaldson 2000) and cause inflammation (Donaldson and Tran 2002). The role of the surface is emphasized by studies showing that the ability of quartz to deliver oxidative stress is dramatically lowered by surface coating (Knaapen et al. 2002).

Combustion-derived NP have their effects by oxidative stress and inflammation (Donaldson et al. 2005a), and several engineered NP types have been described as having oxidative stressing

TABLE 22.1
Table Shows the BED for a Number of Particle Types That Are Well Studied and the Mismatch with Their Exposure Metric

Particle	Biologically Effective Dose	Current Metric
Quartz	Area of reactive (unblocked or unpassivated) surface	Respirable mass
Asbestos	Biopersistent fibers longer than ~20 μm	Fibers longer then 5 μm, >3 μm diameter and aspect ratio>3
PM_{10}	Organics/metals/surfaces	Mass by PM_{10} convention
Welding fume (NP)	Soluble transition metals	Respirable mass
Diesel soot (NP)	Organics/metals/surfaces	Contained in PM_{10}
Carbon black (NP)	Surface area	Nuisance dust standard of respirable mass

TABLE 22.2
Relative Importance of Properties Contributing to the BED of Different Particle Types

	Surface Attribute		Dimension Attribute	Composition Attribute	
Particle type	Surface Area	Surface Reactivity	Length[a]	Soluble Toxins[b]	Biopersistence[c]
Quartz	+	+++++	No	No	++++
Amphibole asbestos	+++	++	++++	+	+++
Welding fume	+++	No	No	+++++	+
ROFA	++	+	No	++++	+
PM10	++	+	No	+++	+
NP carbon black	++++	+	No	No	++++

[a] Longer than 20 μm.
[b] E.g., metals, organics.
[c] More plusses equals more SA, reactivity, soluble toxins or biopersistence.

TABLE 22.3
Oxidative Stress Mechanism for Different Particle Types

Source of oxidative stress	Exemplar Particle	Mechanism of Generation Oxidative Stress	Reference
Surface reactivity	Quartz	Chemical groups on fracture surfaces	(Vallyathan et al. 1994)
Soluble metals	ROFA, welding fume	Fenton chemistry	(Shi 2003 3196/id)
Organics	DEP, PM_{10}	Redox cycling of quinones etc.	(Squadrito et al. 2001)
Shape	Amphibole asbestos	Transition metals	(Lund and Aust 1991)

TABLE 22.4
Characteristics of Inhalation Exposure to Particles

	Typical Particle Types	Exposed Population	Exposure	Typical Responses
Occupational	Silica, asbestos, welding fume, manufactured nanoparticles, organic particles (grain, cotton)	Predominantly healthy males <65 years old	High	Pneumoconiosis, COPD, cancer, asthma
Environmental	PM_{10} containing combustion-derived nanoparticles	Everyone including susceptible and >65 years old, ill populations with pre-existing inflammation and oxidative stress	Low	Exacerbations of COPD/asthma, cardiovascular disease, diabetes, cancer

effects (Hussain et al. 2005; Manna et al. 2005; Sayes et al. 2005). Attributes contributing to oxidative stressing potential of different particle types is shown in Table 22.3.

22.6 RESPONSE

Particle-related lung diseases of various types can be seen in both occupational and environmental settings, and both the pattern and intensity of the exposure can differ quite distinctly between these two (Table 22.4).

22.6.1 The Occupational Setting

Traditional particle-associated lung diseases are those seen in occupational settings and the classical particles are quartz, asbestos, coalmine dust, etc. High airborne mass exposures, characteristic of historic workplaces, leads to the responses of pneumoconiosis and COPD (Table 22.4). In addition, there are cancers and asthma arising in workplaces due to particle exposures. The worker population can generally be seen as a healthy, predominantly male population that can in general tolerate such exposures well, at least at the commencement of their exposure, because of the "healthy worker" status. The "healthy" status within the workforce is conserved by the simple process that those who are adversely affected by exposure to the dusty atmosphere leave to take up alternative employment leaving only a healthy workforce.

22.6.2 The Environmental Setting

There is a well-documented link between exposure to environmental particles (PM_{10}) and mortality/morbidity in airways and cardiovascular disease and cancer (Pope and Dockery 1999; Pope et al. 2003). These low mass exposures commonly affect susceptible populations of patients with existing lung disease (asthma and COPD) or cardiovascular disease to produce quite a different exposure pattern and response (Table 22.4). Aged populations and those with airways disease and cardiovascular disease have pre-existing oxidative stress as part of the inflammatory components of their conditions, and this could be a factor in making them susceptible to particle effects driven by oxidative stress.

Common features of the two paradigms are encapsulated in Figure 22.3, where the common roles of oxidative stress and inflammation are emphasized. With NP, translocation and effects distal to the site of deposition come into play and render the whole equation more complex. However, the

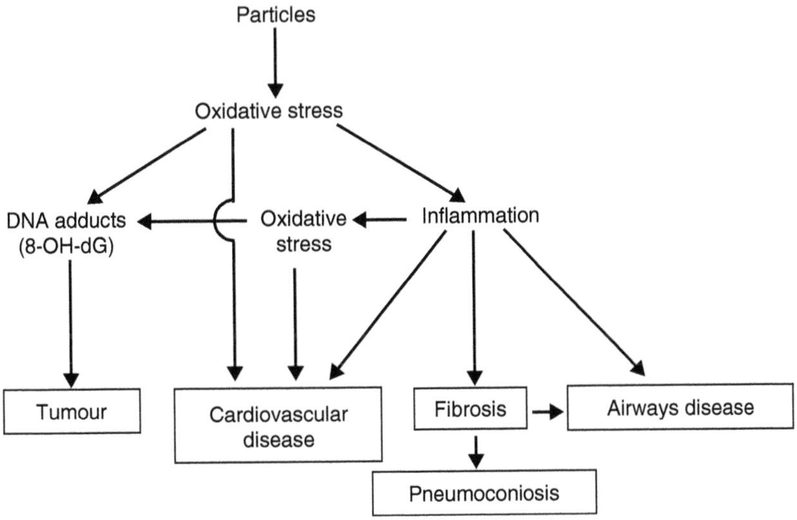

FIGURE 22.3 A paradigm describing the events that could occur following the deposition of pathogenic particles.

basic principle of particle toxicology will be seen to apply and the responses should be interpretable in the light of the foregoing discussion.

22.7 CONCLUSIONS

The shape of this chapter emerged while we were putting together and reviewing all the chapters of the book. We believe that it represents the first effort to try and develop a single conceptual framework for the adverse effects of pathogenic particles, and it is undoubtedly simplified. However, it offers hypotheses for testing and a framework to unify research and thinking in this important and fertile area of applied research. As our understanding increases, it will undoubtedly be refined and improved and we look forward to the contributions of our colleagues in this regard.

REFERENCES

Balashazy, I., Hofmann, W., and Heistracher, T., Local particle deposition patterns may play a key role in the development of lung cancer, *J. Appl. Physiol.*, 94, 1719–1725, 2003.

Barrett, E. G., Johnston, C., Oberdörster, G., and Finkelstein, J. N., Antioxidant treatment attenuates cytokine and chemokine levels in murine macrophages following silica exposure, *Toxicol. Appl. Pharmacol.*, 158, 211–220, 1999.

Calderon-Garciduenas, L., Reed, W., Maronpot, R. R., Henriquez-Roldan, C., gado-Chavez, R., Calderon-Garciduenas, A., and Dragustinovis, I. et al., *Brain inflammation and Alzheimer's-like pathology in individuals exposed to severe air pollution Toxicol. Pathol.*, 32 2004.

Campen, M. J., Nolan, J. P., Schladweiler, M. C., Kodavanti, U. P., Evansky, P. A., Costa, D. L., and Watkinson, W. P., Cardiovascular and thermoregulatory effects of inhaled PM-associated transition metals: a potential interaction between nickel and vanadium sulfate, *Toxicol. Sci.*, 64, 243–252, 2001.

Campen, M. J., Nolan, J. P., Schladweiler, M. C., Kodavanti, U. P., Costa, D. L., and Watkinson, W. P., Cardiac and thermoregulatory effects of instilled particulate matter-associated transition metals in healthy and cardiopulmonary-compromised rats, *J. Toxicol. Environ. Health A*, 65, 1615–1631, 2002.

Chen, L. C. and Nadziejko, C., Effects of subchronic exposures to concentrated ambient particles (CAPs) in mice. V. CAPs exacerbate aortic plaque development in hyperlipidemic mice, *Inhal. Toxicol.*, 17, 217–224, 2005.

Donaldson, K. and Borm, P. J., The quartz hazard: a variable entity 7, *Ann. Occup. Hyg.*, 42, 287–294, 1998.

Donaldson, K. and Tran, C. L., Inflammation caused by particles and fibers, *Inhal. Toxicol.*, 14, 2002.

Donaldson, K. and Tran, C. L., An introduction to the short-term toxicology of respirable industrial fibers, *Mutat. Res.*, 553, 5–9, 2004.

Donaldson, K., Stone, V., Borm, P. J., Jimenez, L. A., Gilmour, P. S., Schins, R. P., and Knaapen, A. M. et al., Oxidative stress and calcium signaling in the adverse effects of environmental particles (PM(10)), *Free Radic. Biol. Med.*, 34, 1369–1382, 2003.

Donaldson, K., Jimenez, L. A., Rahman, I., Faux, S. P., MacNee, W., Gilmour, P. S., Borm, P. J., Schins, R. P. F., Shi, T., and Stone, V., Respiratory health effects of ambient air pollution particles: role of reactive species, In *Oxygen/Nitrogen Radicals: Lung Injury and Disease*, Vallyathan, V., Shi, X., and Castranova, V., eds., Marcel Dekker, New York, 2004a. Volk 187 in Lung Biology in Health and Disease, Exec. Ed., Lenfant, C.

Donaldson, K., Stone, V., Tran, C. L., Kreyling, W., and Borm, P. J., Nanotoxicology, *Occup. Environ. Med.*, 61, 727–728, 2004b.

Donaldson, K., Tran, L., Jimenez, L., Duffin, R., Newby, D. E., Mills, N., MacNee, W., and Stone, V., Combustion-derived nanoparticles: a review of their toxicology following inhalation exposure, *Part. Fiber Toxicol.*, 2, 10, 2005a.

Donaldson, K., Tran, L., Jimenez, L. A., Duffin, R., Newby, D. E., Mills, N., MacNee, W., and Stone, V., Combustion-derived nanoparticles: a review of their toxicology following inhalation exposure, *Part. Fiber Toxicol.*, 2, 10, 2005b.

Gehr, P., Brand, P., and Heyder, J., Particle deposition in the respiratory tract, In *Particle Lung Cell Interactions*, Gehr, P. and Heyder, J., eds., Vol. 143, Marcel Dekker, New York, 229–322, 2000. Lung Biology in Health and Disease, Executive Editor, Lenfant, C., Ref Type: Generic

Hoet, P. H., Nemmar, A., and Nemery, B., Health impact of nanomaterials? 12, *Nat. Biotechnol.*, 22, 19, 2004.

Hussain, S. M., Hess, K. L., Gearhart, J. M., Geiss, K. T., and Schlager, J. J., In vitro toxicity of nanoparticles in BRL 3A rat liver cells 1, *Toxicol. In Vitro*, 19, 975–983, 2005.

Jimenez, L. A., Thompson, J., Brown, D. A., Rahman, I., Antonicelli, F., Duffin, R., Drost, E. M., Hay, R. T., Donaldson, K., and MacNee, W., Activation of NF-kappaB by PM(10) occurs via an iron-mediated mechanism in the absence of IkappaB degradation, *Toxicol. Appl. Pharmacol.*, 166, 101–110, 2000.

Knaapen, A. M., Albrecht, C., Becker, A., Hohr, D., Winzer, A., Haenen, G. R., Borm, P. J., and Schins, R. P., DNA damage in lung epithelial cells isolated from rats exposed to quartz: role of surface reactivity and neutrophilic inflammation, *Carcinogenesis*, 23, 1111–1120, 2002.

Knaapen, A. M., Borm, P. J., Albrecht, C., and Schins, R. P., Inhaled particles and lung cancer. Part A: mechanisms, *Int. J. Cancer*, 109, 799–809, 2004.

Kreuter, J., Shamenkov, D., Petrov, V., Ramge, P., Cychutek, K., Koch-Brandt, C., and Alyautdin, R., Apolipoprotein-mediated transport of nanoparticle-bound drugs across the blood–brain barrier, *J. Drug Target.*, 10, 317–325, 2002.

Kreyling, W., Semmler, M., Erbe, F., Mayer, P., Takenaka, S., Oberdörster, G., and Ziesenis, A., Minute translocation of inhaled ultrafine insoluble iridium particles from lung epithelium to extrapulmonary tissues, *Ann. Occup. Hyg.*, 46 (suppl.1), 223–226, 2002a.

Kreyling W. G., Semmler M., Erbe F., Mayer P., Takenaka, S., Shculz H., Oberdörster, G., and Ziesenis A., Ultrafine insoluble iridium particles are negligibly translocated from the lung epithelium to extra-pulmonary organs, 2002b.

Lloyd, D. R., Carmichael, P. L., and Phillips, D. H., Comparison of the formation of 8-hydroxy-2′-deoxyguanosine and single- and double-strand breaks in DNA mediated by fenton reactions, *Chem. Res. Toxicol.*, 11, 420–427, 1998.

Lund, L. G. and Aust, A. E., Iron-catalyzed reactions may be responsible for the biochemical and biological effects of asbestos 7, *Biofactors*, 3, 83–89, 1991.

Maeng, S. H., Chung, H. W., Yu, I. J., Kim, H. Y., Lim, C. H., Kim, K. J., Kim, S. J., Ootsuyama, Y., and Kasai, H., Changes of 8-OH-dG levels in DNA and its base excision repair activity in rat lungs after inhalation exposure to hexavalent chromium, *Mutat. Res.*, 539, 109–116, 2003.

Manna, S. K., Sarkar, S., Barr, J., Wise, K., Barrera, E. V., Jejelowo, O., Rice-Ficht, A. C., and Ramesh, G. T., Single-walled carbon nanotube induces oxidative stress and activates nuclear transcription factor-kappaB in human keratinocytes 1, *Nano. Lett.*, 5, 1676–1684, 2005.

Mills, N. L., Tornqvist, H., Robinson, S. D., Gonzalez, M., Darnley, K., MacNee, W., and Boon, N. A. et al., Diesel exhaust inhalation causes vascular dysfunction and impaired endogenous fibrinolysis 1 *Circulation*, 12, 3930–3936, 2005.

Molinelli, A. R., Madden, M. C., Mcgee, J. K., Stonehuerner, J. G., and Ghio, A. J., Effect of metal removal on the toxicity of airborne particulate matter from the Utah Valley, *Inhal. Toxicol.*, 14, 1069–1086, 2002.

Morgan, W. K. C. and Seaton, A., *Occupational Lung Diseases*, Saunders, Philadelphia PA, 1995. Ref Type: Generic.

Nemmar, A., Vanbilloen, H., Hoylaerts, M. F., Hoet, P. H., Verbruggen, A., and Nemery, B., Passage of intratracheally instilled ultrafine particles from the lung into the systemic circulation in hamster, *Am. J. Respir. Crit. Care Med.*, 164, 1665–1668, 2001.

Nemmar, A., Hoylaerts, M. F., Hoet, P. H., and Nemery, B., Possible mechanisms of the cardiovascular effects of inhaled particles: systemic translocation and prothrombotic effects, *Toxicol. Lett.*, 149, 243–253, 2004.

Oberdörster, G., Sharp, Z., Elder, A. P., Gelein, R., Kreyling, W., and Cox, C., Translocation of inhaled ultrafine particles to the brain, *Inhal. Toxicol.*, 16, 437–445, 2004.

Oberdörster, G., Maynard, A., Donaldson, K., Castranova, V., Fitzpatrick, J., Ausman, K., and Carter, J. et al., Principles for characterizing the potential human health effects from exposure to nanomaterials: elements of a screening strategy 1 *Part. Fiber Toxicol.*, 2 2005a.

Oberdörster, G., Oberdörster, E., and Oberdörster, J., Nanotoxicology: an emerging discipline evolving from studies of ultrafine particles, *Environ. Health Perspect.*, 113, 823–839, 2005b.

Piette, J., Piret, B., Bonizzi, G., Schoonbroodt, S., Merville, M. P., Legrand, P., and Bours, V., Multiple redox regulation in NF-kappaB transcription factor activation, *Biol. Chem.*, 378, 1237–1245, 1997. [Review] [86 refs].

Piguet, P. F., Collart, M. A., Grau, G. E., Sappino, A. P., and Vassalli, P., Requirement of tumor-necrosis-factor for development of silica- induced pulmonary fibrosis, *Nature*, 344, 245–247, 1990.

Pope, C. A., Burnett, R. T., Thurston, G. D., Thun, M. J., Calle, E. E., Krewski, D., and Godleski, J., Cardiovascular mortality and long-term exposure to particulate air pollution: epidemiological evidence of general pathophysiological pathways of disease, *Circulation*, 109, 71–77, 2003.

Pope, C. A. and Dockery, D. W., Epidemiology of particle effects, In *Air Pollution and Health*, Holgate, S. T., Samet, J. M., Koren, H. S., and Maynard, R. L., eds., Academic Press, San Diego CA, 673–705, 1999.

Rahman, I. and MacNee, W., Regulation of redox glutathione levels and gene transcription in lung inflammation: therapeutic approaches, *Free Radic. Biol. Med.*, 28, 1405–1420, 2000.

Sayes, C. M., Gobin, A. M., Ausman, K. D., Mendez, J., West, J. L., Colvin, V. L., Nano-C(60) cytotoxicity is due to lipid peroxidation, *Biomaterials*, 2005.

Schaumann, F., Borm, P. J., Herbrich, A., Knoch, J., Pitz, M., Schins, R. P., Luettig, B., Hohlfeld, J. M., Heinrich, J., and Krug, N., Metal-rich ambient particles (particulate matter 2.5) cause airway inflammation in healthy subjects 1, *Am. J. Respir. Crit. Care Med.*, 170, 898–903, 2004.

Schins, R. P. F. and Donaldson, K., Nuclear factor kappa B activation by particles and fibers, *Inhal. Toxicol.*, 12 (suppl. 3), 317–326, 2000. Ref Type: Generic.

Squadrito, G. L., Cueto, R., Dellinger, B., and Pryor, W. A., Quinoid redox cycling as a mechanism for sustained free radical generation by inhaled airborne particulate matter, *Free Radic. Biol. Med.*, 31, 1132–1138, 2001.

Sun, Q., Wang, A., Jin, X., Natanzon, A., Duquaine, D., Brook, R. D., and Aguinaldo, J. G. et al., Long-term air pollution exposure and acceleration of atherosclerosis and vascular inflammation in an animal model 1 *JAMA*, 294 2005.

Suwa, T., Hogg, J. C., Quinlan, K. B., Ohgami, A., Vincent, R., and van Eeden, S. F., Particulate air pollution induces progression of atherosclerosis, *J. Am. Coll. Cardiol.*, 39, 935–942, 2002.

Tao, F., Gonzalez-Flecha, B., and Kobzik, L., Reactive oxygen species in pulmonary inflammation by ambient particulates, *Free Radic. Biol. Med.*, 35, 327–340, 2003.

Tsurudome, Y., Hirano, T., Yamato, H., Tanaka, I., Sagai, M., Hirano, H., Nagata, N., Itoh, H., and Kasai, H., Changes in levels of 8-hydroxyguanine in DNA, its repair and OGG1 mRNA in rat lungs after intratracheal administration of diesel exhaust particles, *Carcinogenesis*, 20, 1573–1576, 1999.

Vallyathan, V., Castranova, V., Pack, D., Leonard, S., Hubbs, A., Shumaker, J., Ducatman, B., et al., Potential role of free-radicals in acute silicosis, *Faseb J.*, 8, A, 1994.

Index

B

C

D

E

F

G

H

I

L

M

N

O

P

Q

R

S

U

V

W

Milton Keynes UK
Ingram Content Group UK Ltd.
UKHW050456071024
449327UK00015B/402

9 780367 389611